P9-AQI-327

A Primer of Multivariate Statistics,

Third Edition

A Primer of Multivariate Statistics,
Third Edition

Richard J. Harris
University of New Mexico

2001 LAWRENCE ERLBAUM ASSOCIATES, PUBLISHERS
Mahwah, New Jersey London

The final camera copy for this work was prepared by the author, and therefore the publisher takes no responsibility for consistency or correctness of typographical style. However, this arrangement helps to make publication of this kind of scholarship possible.

Lawrence Erlbaum Associates, Inc., Publishers
10 Industrial Avenue
Mahwah, NJ 07430

Cover design by Kathryn Houghtaling Lacey

Library of Congress Cataloging-in-Publication Data

Harris, Richard J.
A primer of multivariate statistics / Richard J. Harris.—3rd ed.
p. cm.
Includes bibliographical references and index.
ISBN 0-8058-3210-6 (alk. paper)
1. Multivariate analysis. I. Title.

QA278 .H35 2001
519.5'35—dc21

2001033357
CIP

Books published by Lawrence Erlbaum Associates are printed on acid-free paper, and their bindings are chosen for strength and durability.

Printed in the United States of America
10 9 8 7 6 5 4 3 2 1

Dedicated to classmates and colleagues at Lamar (Houston) High School, Class of '58; at Caltech '58–'61; in the Stanford Psychology doctoral program '63–'68; and at the University of New Mexico '68–'01. Thanks for setting the bar so high by your many achievements and contributions.

A Primer of Multivariate Statistics
Third Edition

Richard J. Harris
University of New Mexico

Preface to the Third Edition

Well, as I looked over the syllabus and the powerpoint presentations for my multivariate course I realized that the material I was offering UNM students had progressed beyond what's available in the first two editions of the *Primer of Multivariate Statistics*. (You did ask "Why did you bother revising the *Primer?*", didn't you?) That of course begs the question of why I didn't just adopt one of the current generation of multivariate textbooks, consign the *Primer of Multivariate Statistics* to the Museum of Out-of-Print Texts, and leave the writing of statistics texts to UNM colleagues and relatives. (See M. B. Harris, 1998, and Maxwell & Delaney, 2000.)

First, because somebody out there continues to use the *Primer*, which at the turn of the millenium continues to garner about fifty *SCI* and *SSCI* citations annually. Sheer gratitude for what this has done for my citation counts in my annual biographical reports suggests that I ought to make the *Primer* a more up-to-date tool for these kind colleagues.

Second, because I feel that the current generation of multivariate textbooks have moved too far in the direction of pandering to math avoidance, and I vainly (probably in both senses) hope that renewed availability of the *Multivariate Primer* will provide a model of balance between how-to and why in multivariate statistics.

Third, because of the many new examples I have developed since the second edition. I have found that striking, "fun" data sets are also more effective. As you work your way through this third edition, look for the Presumptuous Data Set, the N = 6 Blood Doping dataset, the Faculty Salary data showing an overall gender bias opposite in direction to that which holds in every individual college, and the tragic tale of the Beefy Breasted Bowery Birds and their would-be rescuer, among other dramatic datasets both real and hypothetical.

Fourth, because a good deal of new material (ammunition?) has become available in the effort to convince multivariate researchers and authors that they really should pay attention to the emergent variables (linear combinations of the measures that were "fed into" your MRA or Manova or Canona) that actually produced that impressive measure of overall relationship – and that these emergent variables should be interpreted on the basis of the linear combination of the original variables that generates or estimates scores on a given emergent variable, *not* on structure coefficients (zero-order correlations). Several of the new datasets mentioned in the previous paragraph reinforce this point. I am especially pleased to have found a colleague (James Grice) with the intellectual commitment and the skill actually to carry out the Monte Carlo analyses necessary to show how much more closely regression-based estimates of (orthogonal) factors mimic the properties of those factors than do loadings-based estimates. (Cf. the discussion in Chapter 7 of Dr. Grice's dissertation and of Harris & Grice, 1998.)

Fifth, because structural equation modeling has become so ubiquitous in the literature of so many areas, so user friendly, so readily available, and so *un*constrained by cpu time that it simply must be represented in any attempt at a comprehensive treatment of multivariate statistics. The focus of this third edition is still on the "classic" multivariate techniques that derive emergent variables as deterministic linear combinations of the original measures, but I believe that it provides enough of a taste of latent-variable approaches to give the reader a feel for why she should consider diving into more detailed treatments of confirmatory factor analysis, SEM, etc.

Some of the above reasons for my feeling compelled to produce a third edition can also be seen as reasons that you and all the colleagues you can contact should consider reading the result and prescribing it for your students:

- New coverage of structural equations modeling (Chapter 8), its manifest-variable special case, path analysis (Chapter 2), and its measurement-model-only special case, confirmatory factor analysis (Chapter 7).
- New, more interesting and/or compelling demonstrations of the properties of the various techniques.

Additionally, the ease with which multivariate analyses can now be launched (and speedily completed) from anyone's desktop led me to integrate computer applications into each chapter, rather than segregating them in a separate appendix.

Unsurprisingly, I feel that the strengths of the first two editions have been retained in the *Multivariate Primer* you're about to read:

One of the reviewers of the second edition declared that he could never assign it to his students because I took too strong a stand against one-tailed tests, thereby threatening to curb his right to inculcate unscientific decision processes in his students. (Well, he may have put it a bit differently than that.) As a glance through Harris (1997a, b) will demonstrate, I feel at least as strongly about that issue and others as I did fifteen years ago, and although I try to present both sides of the issues, I make no apology for making my positions clear and proselytizing for them. Nor do I apologize for leaving a lot of references from the "good old" (19)70's, 60's, and 50's and even earlier (look for the 1925 citation) in this edition. Many of the techniques developed "back then" and the analyses of their properties and the pros and cons of their use are still valid, and a post-Y2K date is no guarantee of ... well, of anything.

Naturally there have been accomplices in putting together this third edition. First and foremost I wish to thank Publisher Larry Erlbaum, both on my behalf and on behalf of the community of multivariate researchers. Larry is a fixture at meetings of the Society for Multivariate Experimental Psychology, and the home Lawrence Erlbaum Associates has provided for *Multivariate Behavioral Research* and for the newly launched series of multivariate books edited by SMEP member Lisa Harlow has contributed greatly to the financial stability of the society and, more importantly, to making multivariate techniques broadly available. Personally I am grateful for Larry's quick agreement to consider my proposal for LEA to publish a third edition of the *Multivariate Primer* and for his extraordinary forbearance in not abandoning the project despite my long delay in getting a formal prospectus to him and my even longer delay getting a complete draft to Editors Debra Riegert and Jason Planer (Editor Lane Akers having by then moved on to other projects). Elizabeth L. Dugger also made a major contribution to the book (and to helping me maintain an illusion of literacy) through her extremely detailed copyediting, which went beyond stylistic and grammatical considerations to pointing out typos (and displays of sheer illogicality) that could only have been detected by someone who was following the substance of the text in both its verbal and its mathematical expression.

I would also like to thank three reviewers of the prospectus for the third edition, Kevin Bird (University of New South Wales), Thomas D. Wickens (UCLA) and Albert F. Smith (Cleveland State University) for their support and for their many constructive criticisms and helpful suggestions. Dr. Bird was a major factor in getting the second edition of the *Primer* launched ("feeding" chapters to his multivariate seminar as fast as I could get them typed and copied), and he was on hand (via the wonder of intercontinental email) to be sure the third edition didn't stray too far from the core messages and organization of the first two editions. In addition to reviewing the prospectus, Dr. Wickens has also helped maintain my motivation to prepare a third edition by being one of the intrepid few instructors who have continued to use the second edition in their courses despite its having gone out of print. He has also made my twice-a-decade teaching of a seminar on mathematical modeling in psychology easier (thus freeing up more of my time to work on the *Primer* revision) by making available to me and my students copies of his out-of-print text on Markov modeling (*Models for Behavior*, 1982).

They also serve who only have to read the darned thing. Thanks to the UNM graduate students whose attentive reading, listening, and questioning (beloved are those who speak up in class) have helped me see where better examples, clearer explanations, etc. were needed. I'm especially grateful to the most recent class (Kevin Bennett, Heather Borg-Chupp, Alita Cousins, Winston Crandall, Christopher Radi, David Trumpower,and Paula Wilbourne), who had to forego a complete hardcopy, instead pulling the most recent revision of each chapter off the web as we progressed through the course. Paula was especially brave in permitting her dissertation data to be subjected to multivariate scrutiny in class.

I've thanked Mary, Jennifer, and Christopher for their patience with my work on the first two editions, but a family member (Alexander) who arrived after publication of the second edition surely deserves to get *his* name mentioned in a *Primer* preface. Thanks, Alex, for being willing to rescue a member of the hopeless generation (with respect to *real* computers, anyway) on so many occasions, and for putting up with many months of "once the book's shipped off" excuses.

So, why are you lingering over the witty prose of a preface when you could be getting on with learning or reacquainting yourself with multivariate statistics?

Dick Harris November 2000

Contents

1 The Forest before the Trees

2 Multiple Regression: Predicting One Variable from Many

3 Hotelling's T^2: Tests on One or Two Mean Vectors

4 Multivariate Analysis of Variance: Differences Among Several Groups on Several Measures

5 Canonical Correlation: Relationships Between Two Sets of Variables

6 Principal Component Analysis: Relationships Within a Single Set of Variables

7 Factor Analysis: The Search for Structure

8 The Forest Revisited

Digression 1
Finding Maxima and Minima of Polynomials

Digression 2
Matrix Algebra

A Primer of Multivariate Statistics,
Third Edition

1
The Forest before the Trees

1.0 WHY STATISTICS?

This text and its author subscribe to the importance of sensitivity to data and of the wedding of humanitarian impulse to scientific rigor. Therefore, it seems appropriate to discuss my conception of the role of statistics in the overall research process. This section assumes familiarity with the general principles of research methodology. It also assumes some acquaintance with the use of statistics, especially significance tests, in research. If this latter is a poor assumption, the reader is urged to delay reading this section until after reading Section 1.2.

1.0.1 Statistics as a Form of Social Control

Statistics is a form of social control over the professional behavior of researchers. The ultimate justification for any statistical procedure lies in the kinds of research behavior it encourages or discourages. In their descriptive applications, statistical procedures provide a set of tools for efficiently summarizing the researcher's empirical findings in a form that is more readily assimilated by the intended audience than would be a simple listing of the raw data. The availability and apparent utility of these procedures generate pressure on researchers to employ them in reporting their results, rather than relying on a more discursive approach. On the other hand, most statistics summarize only certain aspects of the data; consequently, automatic (e.g., computerized) computation of standard (cookbook?) statistics without the intermediate step of "living with" the data in all of its concrete detail may lead to overlooking important features of these data. A number of authors (see especially Anscombe, 1973, and Tukey, 1977) offered suggestions for preliminary screening of the data so as to ensure that the summary statistics finally selected are truly relevant to the data at hand.

The inferential applications of statistics provide protection against the universal tendency to confuse aspects of the data that are unique to the particular sample of subjects, stimuli, and conditions involved in a study with the general properties of the population from which these subjects, stimuli, and conditions were sampled. For instance, it often proves difficult to convince a subject who has just been through a binary prediction experiment involving, say, predicting which of two lights will be turned on in each of several trials that the experimenter used a random-number table in selecting the sequence of events. Among researchers, this tendency expresses itself as a proneness to generate complex post hoc explanations of their results that must be constantly revised because they are based in part on aspects of the data that are highly unstable from one replication of the study to the next. Social control is obtained over this tendency, and the"garbage rate" for published studies is reduced, by requiring that experimenters first demonstrate that their results cannot be plausibly explained by the null hypothesis of no true relationship in the population between their independent and dependent variables. Only after this has been established are experimenters permitted to foist on their colleagues more complex explanations. The scientific community generally accepts this control over their behavior because

1. Bitter experience with reliance on investigators' informal assessment of the generalizability of their results has shown that some formal system of "screening" data is needed.

2. The particular procedure just (crudely) described, which we may label the ***null hypothesis significance testing (NHST) procedure,*** has the backing of a highly developed mathematical model. If certain plausible assumptions are met, this model provides rather good quantitative estimates of the relative frequency with which we will falsely reject (Type I error) or mistakenly fail to reject (Type II error) the null hypothesis. Assuming again that the assumptions have been met, this model also provides clear rules concerning how to adjust both our criteria for rejection and the conditions of our experiment (such as number of subjects) so as to set these two "error rates" at prespecified levels.

3. The null hypothesis significance testing procedure is usually not a particularly irksome one, thanks to the ready availability of formulae, tables, and computer programs to aid in carrying out the testing procedure for a broad class of research situations.

1.0.2 Objections to Null Hypothesis Significance Testing

However, acceptance is not uniform. Bayesian statisticians, for instance, point out that the mathematical model underlying the null hypothesis significance testing procedure fits the behavior and beliefs of researchers quite poorly. No one, for example, seriously entertains the null hypothesis, because almost any treatment or background variable will have ***some*** systematic (although possibly miniscule) effect. Similarly, no scientist accepts or rejects a conceptual hypothesis on the basis of a single study. Instead, the scientist withholds final judgment until a given phenomenon has been replicated on a variety of studies. Bayesian approaches to statistics thus picture the researcher as beginning each study with some degree of confidence in a particular hypothesis and then revising this confidence in (the subjective probability of) the hypothesis up or down, depending on the outcome of the study. This is almost certainly a more realistic description of research behavior than that provided by the null hypothesis testing model. However, the superiority of the Bayesian approach as a descriptive theory of research behavior does not necessarily make it a better prescriptive (normative) theory than the null hypothesis testing model. Bayesian approaches are not nearly as well developed as are null hypothesis testing procedures, and they demand more from the user in terms of mathematical sophistication. They also demand more in terms of ability to specify the nature of the researcher's subjective beliefs concerning the hypotheses about which the study is designed to provide evidence. Further, this dependence of the result of Bayesian analyses on the investigator's subjective beliefs means that Bayesian "conclusions" may vary among different investigators examining precisely the same data. Consequently, the mathematical and computational effort expended by the researcher in performing a Bayesian analysis may be relatively useless to those of his or her readers who hold different prior subjective beliefs about the phenomenon. (The *may*s in the preceding sentence derive from the fact that many Bayesian procedures are robust across a wide range of prior beliefs.) For these reasons, Bayesian approaches are not employed in the Primer. Press (1972) has incorporated Bayesian approaches wherever possible.

An increasingly "popular" objection to null hypothesis testing centers around the contention that these procedures have become ***too*** readily available, thereby seducing researchers and journal editors into allowing the tail (the inferential aspect of statistics) to wag the dog (the research process considered as a whole). Many statisticians have appealed for one or more of the following reforms in the null hypothesis testing procedure:

1. Heavier emphasis should be placed on the ***descriptive*** aspects of statistics, including, as a minimum, the careful examination of the individual data points before, after, during, or possibly instead of "cookbook" statistical procedures to them.

2. The research question should dictate the appropriate statistical analysis, rather than letting the ready availability of a statistical technique generate a search for research paradigms that fit the assumptions of that technique.

3. Statistical procedures that are less dependent on distributional and sampling assumptions, such as randomization tests (which compute the probability that a completely random reassignment of observations to groups would produce as large an apparent discrepancy from the null hypothesis as would sorting scores on the basis of the treatment or classification actually received by the subject) or jackknifing tests (which are based on the stability of the results under random deletion of portions of the data), should be developed. These procedures have only recently become viable as high-speed computers have become readily available.

4. Our training of behavioral scientists (and our own practice) should place more emphasis on the hypothesis-generating phase of research, including the use of post hoc examination of the data gathered while testing one hypothesis as a stimulus to theory revision or origination. Kendall (1968), Mosteller and Tukey (1968), Anscombe (1973), and McGuire (1973) can serve to introduce the reader to this "protest literature."

1.0.3 Should Significance Tests be Banned?

Concern about abuses of null hypothesis significance testing reached a peak in the late 1990s with a proposal to the American Psychological Association (APA) that null hypothesis significance tests (NHSTs) be banned from APA journals. A committee was in fact appointed to address this issue, but its deliberations and subsequent report were quickly broadened to a set of general recommendations for data analysis, framed as specific suggestions for revisions of the data-analysis sections of the APA publication manual—*not* including a ban on NHSTs (Wilkinson and APA Task Force on Statistical

Inference, 1999).

Most of the objections to NHST that emerged in this debate were actually objections to researchers' misuse and misinterpretation of the results of NHSTs—most notably, treating a nonsignificant result as establishing that the population effect size is exactly zero and treating rejection of H_0 as establishing the substantive importance of the effect. These are matters of education, not of flawed logic. Both of these mistakes are much less likely (or at least are made obvious to the researcher's readers, if not to the researcher) if the significance test is accompanied by a confidence interval (CI) around the observed estimate of the population effect—and indeed a number of authors have pointed out that the absence or presence of the null-hypothesized value in the confidence interval matches perfectly (at least when a two-tailed significance test at level α is paired with a traditional, symmetric $(1-\alpha)$-level CI) the statistical significance or nonsignificance of the NHST. This has led to the suggestion that NHSTs simply be replaced by CIs. My recommendation is that CIs be used to *supplement*, rather than to replace, NHSTs, because

1. The *p*-value provides two pieces of information not provided by the corresponding CI, namely an upper bound on the probability of declaring statistical significance in the wrong direction (which is at most half of our *p* value; Harris, 1997a, 1997b) and an indication of the likelihood of a successful exact replication (Greenwald, Gonzalez, Harris, & Guthrie, 1996).

2. Multiple-*df* overall tests, such as the traditional *F* for the between-groups effect in one-way analysis of variance (Anova), are a much more efficient way of determining whether there are any statistically significant patterns of differences or among the means or (in multiple regression) statistically reliable combinations of predictors than is examining the confidence interval around each of the infinite number of possible contrasts among the means or linear combinations of predictor variables.

The aspect of NHST that Les Leventhal and I (Harris, 1997a, 1997b; Leventhal, 1999a, 1999b; Leventhal & Huynh, 1996) feel *should* be banned (or, more accurately, modified) is the way its underlying logic is stated in almost all textbooks—namely as a decision between just two alternatives: the population parameter θ is exactly zero or is nonzero in the case of two-tailed tests and $\theta > 0$ or $\theta \leq 0$ in the case of one-tailed tests. As Kaiser (1960b) pointed out more than forty years ago, under this logic the only way to come to a decision about the sign (direction) of the population effect is to employ a one-tailed test, and under neither procedure is it possible to conclude that your initial hypothesis about this direction was wrong. That is, researchers who take this decision-making logic seriously are "thus faced with the unpalatable choice between being unable to come to any conclusion about the sign of the effect … and violating the most basic tenet of scientific method" – namely, that scientific hypotheses must be falsifiable (Harris, 1997a, p. 8).

This book thus adopts what Harris (1994, 1997a, 1997b) referred to as three-valued hypothesis-testing logic and what Leventhal and Huynh (1996) labeled the directional two-tailed test. Specifically, every single-*df* significance test will have one of *three* possible outcomes: a conclusion that $\theta > 0$ if and only if (iff) $\hat{\theta}$ (the sample estimate of θ) falls in a right-hand rejection region occupying a proportion α_+ of $\hat{\theta}$'s null distribution; a conclusion that $\theta < 0$ iff $\hat{\theta}$ falls at or below the $100(\alpha_-)^{th}$ percentile of the null distribution; and a conclusion that we have insufficient evidence to be confident of whether $\theta > 0$ or $\theta < 0$ iff $\hat{\theta}$ falls within the nonrejection region. (I of course agree with the Wilkinson and APA Task Force 1999 admonition that one should "never use the unfortunate expression, 'accept the null hypothesis.'") When $\alpha_+ \neq \alpha_-$ we have Braver's (1975) *split-tailed* test, which preserves the one-tailed test's greater power to detect population effects in the predicted direction without its drawbacks of zero power to detect population effects in the nonpredicted direction and violation of scientific method.

The reader wishing a more thorough immersion in the arguments for and against NHST would do well to begin with Harlow, Mulaik, and Steiger (1997)'s engagingly titled volume, *What If There Were No Significance Tests?*

1.0.4 Math Modeling's the Ultimate Answer

The ultimate answer to all of these problems with traditional statistics is probably what Skellum (1969) referred to as the "mathematization of science," as opposed to the (cosmetic?) application of mathematics to science in the form of very broad statistical models. Mathematization of the behavioral sciences involves the development of mathematically stated theories leading to quantitative predictions of behavior and to derivation from the axioms of the theory of a multitude of empirically testable predictions. An excellent discussion of the advantages of this approach to theory construction vis-à-vis the more typical verbal-intuitive approach was provided by Estes (1957), and illustrations of its fruitfulness were provided by Atkinson, Bower, and Crothers (1965), Cohen (1963), and Rosenberg (1968). The impact on statistics of the adoption of mathematical approaches to theory construction is at least twofold:

1. Because an adequate mathematical model must account for variability as well as regularities in behavior, the appropriate statistical model is often implied by the axioms of the model itself, rather than being an ad hoc addition to what in a verbal model are usually overtly deterministic predictions.

2. Because of the necessity of amassing large amounts of data in order to test the quantitative details of the model's predictions, ability to reject the overall null hypothesis is almost never in doubt. Thus, attention turns instead to measures of goodness of fit relative to other models and to less formal criteria such as the range of phenomena handled by the model and its ability to generate counterintuitive (but subsequently confirmed) predictions.

As an example of the way in which testing global null hypotheses becomes an exercise in belaboring the obvious when a math model is used, Harris' (1969) study of the relationship between rating scale responses and pairwise preference probabilities in personality impression formation found, for the most adequate model, a correlation of .9994 between predicted and observed preference frequency for the pooled "psychometric function" (which was a plot of the probability of stating a preference for the higher rated of two stimuli as a function of the absolute value of the difference in their mean ratings). On the other hand, the null hypothesis of a perfect relationship between predicted and observed choice frequencies could be rejected at beyond the 10^{-30} level of significance, thanks primarily to the fact that the pooled psychometric function was based on 3,500 judgments from each of 19 subjects.

Nevertheless, the behavioral sciences are too young and researchers in these sciences are as yet too unschooled in mathematical approaches to hold out much hope of mathematizing all research; nor would such complete conversion to mathematically stated theories be desirable. Applying math models to massive amounts of data can be uneconomical if done prematurely. In the case of some phenomena, a number of at least partly exploratory studies need to be conducted first in order to narrow somewhat the range of plausible theories and point to the most profitable research paradigms in the area. Skellum (1969), who was cited earlier as favoring mathematization of science, also argued for a very broad view of what constitute acceptable models in the early stages of theory construction and testing. (See also Bentler and Bonnett's 1980 discussion of goodness-of-fit criteria and of the general logic of model testing.) Null hypothesis testing can be expected to continue for some decades as a quite serviceable and necessary method of social control for most research efforts in the behavioral sciences.

Null hypotheses may well be merely convenient fictions, but no more disgrace need be attached to their fictional status than to the ancient logical technique of reductio ad absurdum, which null hypothesis testing extends to probabilistic, inductive reasoning.

As becomes obvious in the remaining sections of this chapter, this book attempts in part to plug a "loophole" in the current social control exercised over researchers' tendencies to read too much into their data. (It also attempts to add a collection of rather powerful techniques to the descriptive tools available to behavioral researchers. Van de Geer [1971 in fact wrote a textbook on multivariate statistics that deliberately omits any mention of their inferential applications.) It is hoped that the preceding discussion has convinced the reader—including those who, like the author, favor the eventual mathematization of the behavioral sciences—that this will lead to an increment in the quality of research, rather than merely prolonging unnecessarily the professional lives of researchers who, also like the author, find it necessary to carry out exploratory research on verbally stated theories with quantities of data small enough to make the null hypothesis an embarrassingly plausible explanation of our results.

1.0.5 Some Recent Developments in Univariate Statistics

Although the focus of this text is on multivariate statistics, there are a few recent developments in the analysis of single variables that I feel you should be aware of and that, given the glacial pace of change in the content of introductory statistics textbooks, you are unlikely to encounter elsewhere. One of these (the proposed switch from two-alternative to three-alternative hypothesis testing) was discussed in section 1.0.3. The others to be discussed here are the MIDS and FEDS criteria as alternatives to power computations and Prior Information Confidence Intervals (PICIs). Discussion of another univariate development, the subset-contrast critical value, is deferred to Chapter 4's discussion of univariate Anova as a prelude to discussing its multivariate extension.

1.0.5.1 The MIDS and FEDs criteria as alternatives to power calculation[1]. There are two principal reasons that a researcher should be concerned with power, one a priori and one post hoc.

1. To insure that one's research design (usually boiling down to the issue of how many subjects to run, but possibly involving comparison of alternative designs) will have an acceptably high chance of producing a statistically significant result (or results) and thereby permit discussion of hypotheses that are considerably more interesting than the null hypothesis.

2. To determine, for an already-completed study, whether those effects that did *not* achieve statistical significance did so because the population effects are very small or because your design, your execution, and/or unforeseen consequences (e.g., individual differences on your response measure being much larger than anticipated or an unusually high rate of no-shows) have led to a very low-power study (i.e., to imprecise estimation of the population parameter).

These are certainly important considerations. However, computation of power for any but the simplest designs can be a rather complicated process, requiring both a rather split personality (so as to switch back and forth between what you do when you're acting as if you believe the null hypothesis and the consequences of your behavior while wearing that hat, given that you're *really* sampling from a population or populations where the null hypothesis is wrong) and consideration of *noncentral* versions of the usual χ^2 and F distributions.

A lot of the work involved in these types of power calculations has been done for you by Cohen (1977) and presented in the form of tables that can be entered with a measure of effect size (e.g., $(\mu - \mu_0)/\sigma$) to yield the sample size needed to achieve a given power, for a number of common tests. Nonetheless, complexity is one of the factors responsible for the very low frequency with which researchers actually report the power of their tests or use power computations in deciding upon sample size for their studies.

The other factor is the intrusion of traditional practice (why should I have to run 40 subjects per cell when there are a lot of studies out there reporting significant results with only 10 subjects per cell?) and practical considerations (I can't ask for 200% of the department's subject pool for the semester, or I can't get enough hours of subject-running time out of my research assistants to collect so much data), with the end result that the effort involved in making power calculations is apt to go for nought.

What is needed is a rough-and-ready approximation to needed sample size that can be used in a wide variety of situations, that requires no special tables, and that, therefore, involves a degree of effort commensurate with the benefits derived therefrom.

---

[1] Adapted from a more detailed treatment in section 2.5 of Harris (1994).

Such a method ought to increase the number of researchers who do pause to take power into consideration. Such a method is provided, Harris and Quade (1992) argue, by the Minimally Important Difference Significant (MIDS) criterion.

The MIDS criterion for sample size is defined as follows:

> **MIDS (Minimally Important Difference Significant) Criterion for Sample Size**
> Simply set up a hypothetical set of results (means, proportions, correlations, etc.) from the situation where your hypothetical results represent the smallest departure from the null hypothesis you would consider as having any theoretical or practical significance. (This is the minimally important difference, or MID.) Then adjust sample size until these hypothetical results barely achieve statistical significance. Use that sample size in your study.

Harris and Quade (1992) showed that use of the MIDS criterion sets your power for the case where the population discrepancy equals the MID to very nearly 50% for any single-*df* test statistic. This may sound like unduly low power. However, recall that it refers to the probability of getting a statistically significant result *if the population effect (difference, correlation, etc.) is just barely worth talking about*, for which 50% power seems about right. (You wouldn't want to have more than a 50/50 chance of detecting a result you consider of no practical or theoretical significance, but you certainly would want more than a 50/50 chance of detecting a result that *is* worth talking about.)

Moreover, unless you're in the habit of conducting studies of effects you're convinced have no practical or theoretical significance, the power of your test for the effect size you *expect* to find will be considerably greater than .5. Harris and Quade (1992) showed that, for both z ratio and *t* tests, the actual power when using the MIDS criterion and an alpha level of .05 is .688, .836, or .975 if the ratio of the actual population effect size to the minimally important effect size is 1.2, 1.5, or 2, respectively. (Corresponding figures for tests employing an alpha of .01 are powers of .740, .901, or .995.) These figures are high enough so that we're almost always safe in setting our power for the MID at 50% and letting our power for nontrivial population effect sizes "take care of itself."

If you find it impossible to specify the MID (because, e.g., you would consider any nonzero population effect important), or if you simply prefer to focus on the likelihood of being able to publish your results, then you can employ the FEDS (Fraction of Expected Difference Significant) criterion, which, although not explicitly stated by Harris and Quade, is implied by the finding cited in the preceding paragraph that the power of a significance test depends almost entirely on α, N, and the ratio of population effect size to the sample effect size used in computing your significance test on hypothetical data. Thus, for instance, all we have to do to insure 80% power for a .05-level test is to select an N such that a sample effect size .7 as large as the true population effect size would just barely reach statistical significance—assuming, of course, that the estimate of the population standard deviation you employed is accurate and that the population effect size really is as large as you anticipated.

Using the MIDS criterion requires no computations or critical values beyond those you would use for your significance test, anyway, and using the FEDS criterion to achieve 80% (97%) power for a given true population effect size requires in addition only that you remember (or look up in Table 1.1) the "magic fraction" .7 (.5).

Table 1.1 Fraction of Population Effect Size That Must Be Statistically Significant in Order to Achieve a Given Level of Power for your Significance Test

Expected Power	FED Required to Be Significant for Alpha = .05	.01	.001
.50000	1.00000	1.00000	1.00000
.60000	.88553	.91045	.92851
.66667	.81983	.85674	.88425
.75000	.74397	.79249	.82989
.80000	.69959	.75373	.79633
.85000	.65411	.71308	.76047
.90000	.60464	.66777	.71970
.95000	.54371	.61029	.66672
.97000	.51031	.57798	.63631
.97500	.50000	.56789	.62671
.98000	.48832	.55639	.61571
.99000	.45726	.52545	.58583
.99900	.38809	.45460	.51570

Note: Power figures are for "correct power," that is, for the probability that the test yields statistical significance in the same direction as the true population difference.

1.0.5.2 Prior-Information Confidence Intervals (PICIs). Combining the emphasis in section 1.0.3 on three-valued logic, including split-tailed tests, with NHST critics' emphasis on confidence intervals, it seems natural to ask how to construct a confidence interval that shows the same consistency with a split-tailed test that the classic symmetric CI shows with the two-tailed test.

The traditional CI-generating procedure yields the shortest confidence interval of any procedure that takes the general form

$$\bar{Y} - z_{\alpha_+}\sigma_{\bar{Y}} \le \mu \le \bar{Y} + z_{\alpha_-}\sigma_{\bar{Y}}$$

where z_P is the $100(1\text{-}P)^{th}$ percentile of the unit-normal distribution and $\alpha_+ + \alpha_- = \alpha$, the complement of the confidence level of the CI (and also the total Type I error rate allocated to the corresponding significance test). That shortest CI is yielded by choosing $\alpha_+ = \alpha_- = \alpha/2$ in the preceding equation, yielding a CI that is symmetric about $\bar{Y}$.

However, if we wish a CI of this form to be consistent with a split-tailed significance test, in that our H_0 is rejected if and only if the CI doesn't include μ_0, then α_+ and α_- must match the proportions of α assigned to the right-hand and left-hand rejection regions, respectively, for our NHST, thus sacrificing precision of estimation for consistency between CI and NHST. The resulting CI, which was labeled by Harris and Vigil (1998) the CAST (Constant-Alpha Split-Tailed) CI, can be thought of as the end result of split-tailed NHSTs carried out on all possible values of μ_0, retaining in the CAST all values of μ_0 that are not rejected by our significance test—provided that we employ the same (thus, constant) α_+ for all of these significance tests.

But the rationale for carrying out a split-tailed significance test in the first place implies that α_+ (and thus α_+/α_-) should *not* be constant, but should reflect the weight of the evidence and logic suggesting that μ is greater than the particular value of μ_0 being tested and should thus be a monotonic decreasing function of μ_0. (It cannot logically be the case, for instance, that the likelihood that $\mu > 110$ is greater than the likelihood that $\mu > 120$.) Harris and Vigil (1998) therefore suggest that the researcher who finds a split-tailed significance test appropriate should use as her CI the PICI (Prior Information Confidence Interval) given by the recursive formula,

$$\bar{Y} - z_{\alpha_+(LL)}\,\sigma_{\bar{Y}} \leq \mu \leq \bar{Y} + z_{\alpha_-(UL)}\,\sigma_{\bar{Y}}$$

where LL and UL are the upper and lower limits of the CI in the equation, and where

$\alpha_+(x) = \alpha\{1/[1 + \exp(d(x - \mu_{exp}))]\}$; $\alpha_-(x) = \alpha - \alpha_+(x)$; and

$\exp(w) = e^w$;

Further, d is a constant greater than zero that determines how rapidly $\alpha_+(x)$ decreases as x increases; and μ_{exp} can be any number but yields optimal properties (including a subtantially narrower CI—Harris & Vigil give examples of 20% reduction in width as compared to the traditional, symmetric CI) when the value chosen is close to the actual value of the population mean.

1.1 WHY MULTIVARIATE STATISTICS?

As the name implies, ***multivariate statistics*** refers to an assortment of descriptive and inferential techniques that have been developed to handle situations in which sets of variables are involved either as predictors or as measures of performance. If researchers were sufficiently narrowminded or theories and measurement techniques so well developed or nature so simple as to dictate a single independent variable and a single outcome measure as appropriate in each study, there would be no need for multivariate statistical techniques. In the classic scientific experiment involving a single outcome measure and a single manipulated variable (all other variables being eliminated as possible causal factors through either explicit experimental control or the statistical control provided by randomization), questions of *patterns* or *optimal combinations* of variables scarcely arise. Similarly, the problems of *multiple comparisons* do not becloud the interpretations of any *t* test or correlation coefficient used to assess the relation between the independent (or predictor) variable and the dependent (or outcome) variable.

However, for very excellent reasons, researchers in all of the sciences—behavioral, biological, or physical—have long since abandoned sole reliance on the classic univariate design. It has become abundantly clear that a given experimental manipulation (e.g., positively reinforcing a class of responses on each of *N* trials) will affect many somewhat different but partially correlated aspects (e.g., speed, strength, consistency, and "correctness") of the organism's behavior. Similarly, many different pieces of information about an applicant (for example, high school grades in math, English, and journalism; attitude toward authority; and the socioeconomic status of his or her parents) may be of value in predicting his or her grade point average in college, and it is necessary to consider how to combine all of these pieces of information into a single "best" prediction of college performance. (It is widely known – and will be demonstrated in our discussion of multiple regression – that the predictors having the highest correlations with the criterion variable when considered singly might contribute very little to that combination of the predictor variables that correlates most highly with the criterion.)

As is implicit in the discussion of the preceding paragraph, multivariate statistical techniques accomplish two general kinds of things for us, with these two functions corresponding roughly to the distinction between ***descriptive*** and ***inferential*** statistics.

On the descriptive side, they provide rules for combining the variables in an optimal way. What is meant by "optimal" varies from one technique to the next, as is made explicit in the next section. On the inferential side, they provide a solution to the ***multiple comparison*** problem. Almost any situation in which multivariate techniques are applied could be analyzed through a series of univariate significance tests (e.g., *t* tests), using one such univariate test for each possible combination of one of the predictor variables with one of the outcome variables. However, because ***each*** of the univariate tests is designed to produce a significant result $\alpha \times 100\%$ of the time (where α is the "significance level" of the test) when the null hypothesis is correct, the probability of having at least one of the tests produce a significant result when in fact nothing but chance variation is going on increases rapidly as the number of tests increases. It is thus highly desirable to have a means of explicitly controlling the experimentwise error rate. Multivariate statistical techniques provide this control.[2]

---

[2] A possible counterexample is provided by the application of analyses of variance (whether univariate or multivariate) to studies employing factorial designs. In the univariate case, for instance, a *k*-way design (*k* experimental manipulation combined factorially) produces a summary table involving 2^{k-1} terms—*k* main effects and $k(k-1)/2$ two-way interactions, and so on—each of which typically yields a test having a Type I error rate of .05. The experimentwise error rate is thus $1 - (.95)^{2^{k-1}}$. The usual multivariate extension of this analysis suffers from exactly the same degree of compounding of Type I error rate, because a separate test of the null hypothesis of

They also, in many cases, provide for post hoc comparisons that explore the statistical significance of various possible explanations of the overall statistical significance of the relationship between the predictor and outcome variables. The descriptive and inferential functions of mutivariate statistics are by no means independent. Indeed, the multivariate techniques we emphasize in this primer (known as ***union–intersection*** procedures) base their tests of significance on the sampling distribution of the "combined variable" that results when the original variables are combined according to the criterion of optimality employed by that particular multivariate technique.[3] This approach provides a high degree of interpretability. When we achieve statistical significance, we also automatically know which combination of the original set of variables provided the strongest evidence against the overall null hypothesis, and we can compare its efficacy with that of various a priori or post hoc combining rules-including, of course, that combination implied by our substantive interpretation of the optimal combination. (For instance, $.01\ Y_1 - .8931\ Y_2 + 1.2135\ Y_3$ would probably be interpreted as essentially the difference between subjects' performance on tasks 2 and 3, that is, as $Y_3 - Y_2$. Such a simplified version of the optimal combination will almost always come very close to satisfying the criterion of optimality

---

no difference among the groups on any of the outcome measures (dependent variables) is conducted for each of the components of the among-group variation corresponding to a term in the univariate summary table. In the author's view, this simply reflects the fact that the usual univariate analysis is appropriate only if each of the terms in the summary table represents a truly a priori (and, perhaps more important, theoretically relevant) comparison among the groups, so that treating each comparison independently makes sense. Otherwise, the analysis should be treated (statistically at least) as a one-way Anova followed by specific contrasts employing Scheffé's post hoc significance criterion (Winer, 1971, p. 198), which holds the experimentwise error rate (i.e., the probability that any one or more of the potentially infinite number of comparisons among the means yields rejection of H_0 when it is in fact true) to at most α. Scheffé's procedure is in fact a multivariate technique designed to take into account a multiplicity of *independent* variables. Further, Anova can be shown to be a special case of multiple regression, with the overall F testing the hypothesis of no true differences among any of the population means corresponding to the test of the significance of the multiple regression coefficient. This overall test is not customarily included in Anova applied to factorial designs, which simply illustrates that the existence of a situation for which multivariate techniques are appropriate does not guarantee that the researcher will apply them.

[3] The union-intersection approach is so named because, when following this approach, (a) Our overall null hypothesis, $H_{0,ov}$, is rejected whenever one or more of the univariate hypotheses with respect to particular linear combinations of the original variables is rejected; the rejection region for $H_{0,ov}$ is thus the *union* of the rejection regions for all univariate hypotheses. (b) We fail to reject $H_{0,ov}$ only if each of our tests of univariate hypotheses also fails to reach significance; the nonrejection region for $H_{0,ov}$ is thus the *intersection* of the nonrejection regions or the various univariate hypotheses.

as effectively as does the absolutely best combination, but it will be much more interpretable than the multiple-decimal weights employed by that combination.) On the other hand, if you are uninterested in or incapable of interpreting linear combinations of the original measures but instead plan to interpret each variable separately, you will find that a series of univariate tests with ***Bonferroni-adjusted*** critical values satisfies the need for control of experimentwise error rate while providing less stringent critical values (and thus more powerful tests) for the comparisons you can handle. In this case, Section 1.1.1 below will satisfy all of your needs for handling multiple measures, and you may skip the remainder of this book.

1.1.1 Bonferroni Adjustment: An Alternative to Multivariate Statistics

The Bonferroni inequality (actually only the best known of a *family* of inequalities proved by Bonferroni, 1936a, 1936b) simply says that if n_t significance tests are carried out, with the ith test having a probability of α_i of yielding a decision to reject its *Ho,* the overall probability of rejecting one ,or more of the null hypotheses associated with these tests is less than or equal to the sum of the individual alphas, that is,

$$\alpha_{ov} \leq \sum \alpha_i .$$

Thus, if we wish to carry out a total of n_t univariate tests while holding our experimentwise error rate (the probability of falsely declaring one or more of these comparisons statistically significant) to at most α_{exp}, we need only set the α_i for each test to α_{exp}/n_t or to any other set of values that sum to α_{exp} .

This procedure is valid for any finite number of tests, whether they are chi square, *F*, *t*, sign, or any other statistic, and whether the tests are independent of each other or highly correlated. However, the inequality departs farther from equality and the overall procedure therefore becomes increasingly conservative, as we choose more highly intercorrelated significance tests. (Interestingly, as Derivation 1.1 at the back of this book demonstrates, the Bonferroni adjustment procedure provides exact, rather than conservative, control of the ***per experiment*** error rate, which is the average *number* of Type I errors made per repetition of this set of tests with independent samples of data.)

Thus, for instance, you might wish to examine the efficacy of each of 13 predictors of obesity, while keeping your experimentwise alpha level to .05 or below. To do this, you need only compare each of the 13 correlations between a predictor and your ponderosity index (your operational definition of obesity) to the critical value of Pearson *r* for α_i = .05/13 = .00385. If you have a priori reasons for expecting, say, caloric intake and time spent in exercising at 50% or further above the resting pulse rate to be especially important predictors, it would be perfectly legitimate to test each of these two predictors at the α_i = .02 level and each of the remaining 11 at the .01/11 = .000909 level. Similarly, if you have measured the results of an anxiety manipulation in terms of four different dependent measures, and simply wish to know whether your subjects' mean response to the three levels of anxiety differs significantly in terms of any one or more of the four dependent variables, you need only carry out four univariate F tests, each at the .05/4 = .0125 level (assuming your desired experimentwise alpha = .05) or perhaps with the dependent variable you are especially interested in tested at the .02 level (for greater power) and each of the remaining three tested at the .01 level.

1.1.2 Why Isn't Bonferroni Adjustment Enough?

Bonferroni adjustment is a very general technique for handling multiple variables—as long as you are satisfied with or limited to explanations that treat each variable by itself, one at a time. However, if the descriptive benefits of a truly mutivariate approach are desired—that is, if linear combinations that yield stronger relationships than are attainable from any single variable are to be sought—Bonferroni-adjusted univariate tests are useless, because the number of combinations implicitly explored by a union-intersection test is infinite. Fortunately, for any one sample of data, the degree of relationship attainable from the best linear combination of the measures has an upper bound, and the sampling distribution of this upper bound can be determined. It is the possibility that the optimal combination will have a meaningful interpretation in its own right that justifies adoption of a multivariate procedure rather than simply performing Bonferroni-adjusted univariate tests.

Both of these possibilities are quite often realized, as indeed there are sound empirical and logical reasons to expect them to be. For instance, the sum or average of a number of related but imperfect measures of a single theoretical construct can be expected to be a much more reliable measure of that construct than any single measure. Similarly, although caloric intake and amount of exercise may each be predictive of body weight, the difference between these two measures is apt to be an especially significant prognosticator of obesity. Multivariate statistics can be of considerable value in suggesting new, emergent variables of this sort that may not have been anticipated—but the researcher must be prepared to think in terms of such combinations if this descriptive aspect of mutivariate statistics is not to lead to merely a waste of statistical power.

What, then, are the univariate techniques for which multivariate optimization procedures have been worked out, and what are the optimization criteria actually employe

1.2 A HEURISTIC SURVEY OF STATISTICAL TECHNIQUES

Most statistical formulae come in two versions: a ***heuristic*** version, which is suggestive of the rationale underlying that procedure, and an algebraically equivalent ***computational*** version, which is more useful for rapid and accurate numerical calculations. For instance, the formula for the variance of a sample of observations can be written either as

$$\sum\left(Y_i - \bar{Y}\right)^2 / (N-1),$$

which makes clear the status of the variance as approximately the mean of the squared deviations of the observations about the sample mean, or as

$$N\sum Y^2 - \left(\sum Y\right)^2 / [N(N-1)],$$

which produces the same numerical value as the heuristic formula but avoids having to deal with negative numbers and decimal fractions. The present survey of statistical techniques concentrates—in the same way as heuristic versions of statistical formulae—on what it is that each technique is designed to accomplish, rather than on the computational tools used to reach that goal. In general, in any situation we will have m predictor variables, which may be either discrete (e.g., treatment method or gender of subject) or continuous (e.g., age or height), and p outcome variables. The distinctions among various statistical techniques are based primarily on the sizes of these two sets of variables and are summarized in Table 1.2

The techniques listed in the right-hand column of Table 1.2 have not been explicitly labeled as univariate or multivariate. Here, as is so often the case, common usage and logic clash. If we define multivariate techniques (as the present author prefers) as those applied when two or more variables are employed either as independent (predictor) or dependent (outcome) variables, then higher order Anova must certainly be included and possibly one-way Anova as well. However, Anova and Ancova involving only one control variable, and thus only bivariate regression, have traditionally not been included in treatments of multivariate statistics. (Note that bivariate regression is a univariate statistical technique.)

If we define multivariate techniques as those applied to situations involving multiple ***dependent*** measures, then multiple regression is excluded, which again runs counter to common practice. To some extent the usual practice of including multiple regression but excluding Anova (despite the fact that the latter can be viewed as a special case of the former) from multivariate statistics reflects the historical accident that multivariate techniques have come to be associated with correlational research, whereas Anova is used primarily by "hard-nosed" researchers who conduct laboratory experiments.[4] It is hoped that the present text will weaken this association. However, the issues involved in higher order Anova have become so involved and specialized, and there are so many excellent texts available in this area of statistics (e.g., Winer, 1971), that detailed treatment of Anova is omitted from this book on practical, rather than logical, grounds.

#[5] At a practical level, what distinguishes techniques considered univariate from those considered multivariate is whether matrix manipulations (see sections 1.3 and 2.2.3 and, at the end of the book, Digression 2) are required. As we shall see in subsequent chapters, any of the traditional approaches to statistics can be defined in terms of matrix

[4] For a review of the history of multivariate statistics, see Harris (1985).

[5] Paragraphs or whole sections preceded by the symbol # may be skipped without loss of continuity.

operations. Univariate statistical techniques are simply those in which the matrix operations "degenerate" into scalar operations—that is, matrix operations on 1 x 1 matrices. (If you have had no prior experience with matrix operations, you will probably

Table 1.2 Statistical Techniques

Predictor (Independent) Variables	Outcome (Dependent) Variables	Criterion for Combining Variables	Name of technique(s)
1 discrete, 2 levels	1		t test
1 discrete, > 2 levels	1		One-way analysis of variance (Anova)
> 2 discrete	1		Higher-order Anova
1 continuous	1		Pearson r, bivariate Regression
≥ 2 continuous	1	Maximize correlation of combined variable with outcome variable	Multiple correlation, Multiple regression Analysis (MRA)
Mixture of discrete, continuous	1	Maximize correlation of continuous predictors with outcome variable within levels of the discrete variable	Analysis of covariance (Ancova)
1 discrete, 2 levels	≥ 2	t ratio on combined variable as large as possible	Hotelling's T^2, discriminant analysis
1 discrete, > 2 levels	≥ 2	Maximize one-way F on combined variable	One-way multivariate analysis of variance (Manova)
≥ 2 discrete	≥ 2	F on combined variable maximized for each effect	Higher-order Manova
Mixture of discrete, continuous	≥ 2	Maximize correlation of combined predictor variable within levels of the discrete variable(s)	Multivariate Ancova (Mancova)
≥ 2 continuous	≥ 2	Maximize correlation of combined predictor variable with combined outcome variable	Canonical correlation, canonical analysis (Canona)
≥ 2 continuous		Maximize variance of combined variable	Principal component Analysis (PCA)
≥ 2 continuous		Reproduce correlations among original variables as accurately as possible	Factor analysis (FA)

Note: In every technique except factor analysis, the combined variable is simply a weighted sum of the original variables. The predictor-outcome distinction is irrelevant in principal component analysis and in factor analysis, where only one set of variables is involved. In Manova, Ancova, Canona, PCA, and FA, several combined variables are obtained. Each successive combined variable is selected in accordance with the same criterion used in selecting the first combined variable, but only after the preceding variables have been selected, and it is subject to the additional constraints that it be uncorrelated with the preceding combined variables.

find that this definition makes more sense to you after reading Chapter 2.) This operational definition of multivariate statistics resolves the paradox mentioned in the preceding paragraph that Anova is generally considered a univariate technique, despite its being a special case of multiple regression, which is commonly viewed as a multivariate technique. When the matrix formulae for a multiple regression analysis (MRA) are applied to a situation in which the predictor variables consist of k - 1 dichotomous group-membership variables (where, for i = 1, 2, ... k - 1, X_i = 1 if the subject is a member of group i, 0 if he or she is not), and when these group-membership variables are uncorrelated (as they are when the same number of observations is obtained for each combination of levels of the independent variables), the matrix operations lead to relatively simple single-symbol expressions that are the familiar Anova computational formulae. Consequently, matrix operations and the connection with MRA need not arise at all in carrying out the Anova. However, unequal cell sizes in a higher order Anova lead to correlated group-membership variables and force us to employ the matrix manipulations of standard MRA in order to obtain uncorrelated least-squares estimates of the population parameters cor-responding to the effects of our treatments. Thus higher order Anova with equal cell sizes is a univariate technique, while higher order Anova with unequal cell sizes is a multivariate technique.

If the union-intersection approach to multivariate significance tests is adopted, the operational definition just offered translates into the statement that multivariate techniques are those in which one or more combined variables—linear combinations of the original variables—are derived on the basis of some criterion of optimality .

1.2.1 Student's *t* Test

The most widely known of the experimental researcher's statistical tools is Student's t test. This test is used in two situations. The first situation exists when we have a single sample of observations whose mean is to be compared with an a priori value. Two kinds of evidence bear on the null hypothesis that the observed value of $\overline{X}$ arose through random sampling from a population having a mean of μ_0:

1. The difference between $\overline{X}$ and μ_0 .

2. The amount of fluctuation we observe among the different numbers in the sample.

The latter provides an estimate of σ, the standard deviation of the population from which the sample was presumably drawn, and thus—via the well-known relationship between the variance of a population and the variance of the sampling distribution of means of samples drawn from that population—provides an estimate of $\sigma_{\overline{X}}$. The ratio between these two figures is known to follow Student's t distribution with N - 1 degrees of freedom when the parent population has a normal distribution or when the size of the sample is sufficiently large for the central limit theorem to apply. (The central limit theorem says, among other things, that means of large samples are normally distributed irrespective of the shape of the parent population—provided only that the parent population has finite variance.) The hypothesis that $\mu = \mu_0$ is therefore rejected at the α significance level if and only if the absolute value of

$$t = \frac{\overline{X} - \mu}{\sqrt{\sum (X - \overline{X})^2 / [N(N-1)]}}$$

is greater than the 100(1 - α/2) percentile of Student's t distribution with N - 1 degrees of freedom.

The second, and more common, situation in which the t test is used arises when the significance of the difference between two independent sample means is to be evaluated. The basic rationale is the same, except that now a pooled estimate of σ^2, namely,

$$s_c^2 = \frac{\sum (X_1 - \overline{X}_1)^2 + \sum (X_2 - \overline{X}_2)^2}{N_1 + N_2 - 2},$$

is obtained. This pooled estimate takes into consideration only variability within each group and is therefore independent of any differences among the population means. Thus, large absolute values of

$$t = \frac{\overline{X}_1 - \overline{X}_2}{\sqrt{s_c^2 (1/N_1 + 1/N_2)}}$$

lead us to reject the null hypothesis that $\mu_1 = \mu_2$ and conclude instead that the population means differ in the same direction as do the sample means.

1.2.2 One-Way Analysis of Variance

When more than two levels of the independent variable (and thus more than two groups of subjects) are involved, the null hypothesis that $\mu_1 = \mu_2 = ... = \mu_k$ is tested by comparing a *direct measure* of the variance of the k sample means, namely,

$$\frac{\sum (X_j - \overline{X})^2}{k - 1}$$

with an *indirect estimate* of how much these k means would be expected to vary if in fact they simply represented k random samples from the same population: s_w^2 / n, where n is the size of each group and s_w^2 is computed in the same way s_c^2 was for the t test, except that k instead of only 2 sums of squared deviations must be combined. Large values of F, the ratio between the direct estimate and the indirect estimate (or, in Anova terminology, between the between-group mean square and the within-group mean square), provide evidence against the null hypothesis. A table of the F distribution provides the precise value of F needed to reject the overall null hypothesis at, say, the .05 level.

More formally, for those readers who benefit from formality, the between-group (better, among-group) mean square,

$$MS_a = \frac{n\sum(X_j - \overline{X})^2}{k-1}$$

has the expected value

$$\sigma^2 + \frac{n\sum(\mu_j - \mu)^2}{k-1} = \sigma^2 + n\sigma_\tau^2 ,$$

where n is the number of observations in each of the k groups, and σ_τ^2 represents the variance of the k population means. (A more complicated constant replaces n in the preceding expressions if sample sizes are unequal.) The within-group mean square, $MS_W = s_w^2$, has the expected value σ^2, the variance common to the k populations from which the samples were drawn. Under the null hypothesis, $\sigma_\tau^2 = 0$, so that both the numerator and denominator of $F = MS_a/MS_W$, have the same expected value. (The expected value of their ratio E(F) is nevertheless not equal to unity but to $df_w/(df_w - 2)$, where df_w is the number of degrees of freedom associated with MS_W, namely, $N - k$, where N is the total number of observations in all k groups.)

If the assumptions of independently and randomly sampled observations, of normality of the distributions from which the observations were drawn, and of homogeneity of the variances of these underlying distributions are met, then MS_a / MS_W is distributed as F. Numerous empirical sampling studies (for example, D. W. Norton's 1953 study) have shown this test to be very robust against violations of the normality and homogeneity of variance assumptions, with true Type I error rates differing very little from those obtained from the F distribution. This is true even when population variances differ by as much as a ratio of 9 or when the populations are as grossly nonnormal as the rectangular distribution. Much more detailed discussion of these and other matters involved in one-way Anova is available in a book by Winer (1971) or in his somewhat more readable 1962 edition.

Because nothing in this discussion hinges on the precise value of k, we might expect that the t test for the difference between two means would prove to be a special case of one-way Anova. This expectation is correct, for when $k = 2$, the F ratio obtained from Anova computations precisely equals the square of the t ratio computed on the same data, and the critical values listed in the $df = 1$ column of a table of the F distribution are identically equal to the squares of the corresponding critical values listed in a t table. However, we *cannot* generalize in the other direction by, say, testing every pair of sample means by an α-level t test, becausee the probability that at least one of these $k(k - 1)/2$ pairwise tests would produce a significant result by chance is considerably greater than α when k is greater than 2. The overall F ratio thus provides protection against the inflation of the overall ("experimentwise") probability of a Type I error that running all possible t tests (or simply conducting a t test on the largest difference between any pair of sample means) would produce.

Once it has been concluded that the variation among the various group means represents something other than just chance fluctuation, there still remains the problem of specifying the way in which the various treatment effects contribute to the significant overall F ratio: Is the difference between group 2 and all other groups the only difference of any magnitude, or are the differences between "adjacent" treatment means all about the same?

How best to conduct such specific comparisons among the various means and especially how to assess the statistical significance of such comparisons are at the center of a lively debate among statisticians. The approach that I favor is Scheffé's contrast method. Scheffé's approach permits testing the significance of the difference between any two linear combinations of the sample means, for example,

$$(X_1 + X_2)/2 - (X_3 + X_4 + X_5)/3$$

or

$$(5X_1 + 3X_2 + X_3) - (X_4 + 3X_5 + 5X_6).$$

The only restriction is that the weights employed within each of the two linear combinations sum to the same number or, equivalently, that the sum of the weights (some of which may be zero) assigned to all variables sum to zero.

Moreover, the significance tests for these contrasts can be adjusted, simply through multiplication by an appropriate constant, to fix at α either the probability that a *particular* preplanned contrast will lead to a false rejection of the null hypothesis or the probability that *any* one or more of the infinite number of comparisons that might be performed among the means will lead to a Type I error. This latter criterion simply takes as the critical value for post hoc comparisons the maximum F ratio attainable from any contrast among the means, namely, $(k - 1)$ times the F ratio used to test the overall null hypothesis: $(k - 1)F_\alpha(k - 1, N - k)$. Thus, the Anova F ratio is an example of the union-intersection approach, because the overall test is based on the sampling distribution of the maximized specific comparison. Therefore, the Scheffé procedure has the property that if and only if the overall F ratio is statistically significant can we find at least one statistically significant Scheffé-based specific comparison. This is not true of any other specific comparison procedure.

Finally, associated with each contrast is a "sum of squares for contrast." This sum is compared with the sum of the squared deviations of the group means about the grand mean of all observations to yield a precise statement of the percentage of the variation among the group means that is attributable to that particular contrast.

This "percentage of variance" interpretation is made possible by the fact that the sums of squares associated with any $k - 1$ *independent* contrasts among the means add to a sum identical to SS_a, the sum of squares associated with the total variation among the k means. Independence (which simply requires that the two contrasts be uncorrelated and thus each be providing unique information) is easily checked, because two contrasts are independent if and only if the sum of the products of the weights they assign to corresponding variables $\Sigma c_1 c_2$ is zero. (A slightly more complex check is necessary when sample sizes are unequal.)

The Scheffé contrast procedure may, however, be much too broad for the researcher who is only interested in certain kinds of comparisons among the means, such as a prespecified set of contrasts dictated by the hypotheses of the study, or all possible pairwise comparisons, or all possible comparisons of the unweighted means of any two distinct subsets of the means. For situations like these, where the researcher can specify in advance of examining the data the size of the set of comparisons that are candidates for examination, the use of Bonferroni critical values (see section 1.1.1) is appropriate and will often yield more powerful tests than the Scheffé approach. An additional advantage of the Bonferroni procedure is that it permits the researcher to use less stringent αs for (and thus provide relatively more powerful tests of) the contrasts of greatest interest or importance. Of course, if the Bonferroni approach is adopted, there is no reason to perform the overall F test, because that test implicitly examines all possible contrasts, most of which the researcher has declared to be of no interest. Instead, the overall null hypothesis is rejected if and only if one or more contrasts is statistically significant when compared to its Bonferroni-adjusted critical value.

The astute reader will have noticed that the Bonferroni-adjustment approach shares many of the properties of the union-intersection approach—most importantly, the consistency each provides between the results of tests of specific comparisons and the test of $H_{0,ov}$ The primary difference between the two approaches is that the family of specific comparisons considered by the union-intersection approach is infinitely large, whereas the Bonferroni approach can be applied only when a finite number of specific comparisons is selected as the family of comparisons of interest prior to examining any data. The Bonferroni approach is thus sometimes said to involve ***finite-intersection*** tests.

1.2.3 Hotelling's T^2

It is somewhat ironic that hard-nosed experimentalists, who are quick to see the need for one-way analysis of variance when the number of experimental manipulations (and thus of groups) is large, are often blithely unaware of the inflated error rates that result if t-test procedures are applied to more than one outcome measure. Any background variable (age, sex, socioeconomic status, and so on) that we employ as a basis for sorting subjects into two different groups or for distinguishing two subcategories of some population almost inevitably has many other variables associated with it. Consequently, the two groups (subpopulations) so defined differ on a wide variety of dimensions. Similarly, only very rarely are the effects of an experimental manipulation (high versus low anxiety, low versus high dosage level, and so on) confined to a single behavioral manifestation. Even when our theoretical orientation tells us that the effect of the manipulation should be on a single conceptual dependent variable, there are usually several imperfectly correlated operational definitions of (that is, means of measuring) that conceptual variable. For instance, learning can be measured by such things as percentage of correct responses, latency of response, and trials to criterion, and hunger can be measured by subjective report, strength of stomach contractions, hours because last consumption of food, and so forth. Thus all but the most trivial studies of the difference between two groups do in fact produce several different measures on each of the two groups unless the researcher has deliberately chosen to ignore all but one measure in order to simplify the data analysis task. Hotelling's T^2 provides a means of testing the overall null hypothesis that the two populations from which the two groups were sampled do not differ in their means on any of the p measures.

Heuristically, the method by ,which this is done is quite simple. The p outcome measures from each subject are combined into a single number by the simple process of multiplying the subject's score on each of the original variables by a weight associated with that variable and then adding these products together. (The same weights are, of course, used for every subject.) Somewhat more formally, the combined variable for each subject is defined by

$$W_i = w_1 X_{i,1} + w_2 X_{i,2} + \Lambda + w_p X_{i,p}$$

A univariate t ratio based on the difference between the two groups in their mean values of W is computed. A second set of weights that may produce a larger t ratio is tried out, and this process is continued until that set of weights that makes the univariate t ratio on the combined variable as large as it can possibly be has been found. (Computationally, of course, the mathematical tools of calculus and matrix algebra are used to provide an analytic solution for the weights, but this simply represents a shortcut to the search procedure described here.) The new variable defined by this set of optimal weights — optimal in the sense that they provide maximal discrimination between the two groups — is called the ***discriminant function,*** and it has considerable interpretative value in its own right. The square of the t ratio that results from the maximizing procedure is known as T^2, and its sampling distribution has been found to be identical in shape to a member of the family of F distributions. Thus, it provides a test of the overall null hypothesis of identical "profiles" in the populations from which the two samples were drawn.

There is also a single-sample version of Hotelling's T^2 that tests the overall H_0 that the population means of the p measures are each identical to a specific a priori value. Its major application is to situations in which each subject receives more than one experimental treatment or is measured in the same way on several different occasions. Such ***within-subject*** or ***repeated-measure*** designs are handled by transforming each subject's responses to the p treatments or his or her scores on the p different occasions to

p- 1 contrast scores (e.g., the differences between adjacent pairs of means), and then testing the overall null hypothesis that each of these contrast scores has a population mean of zero. (In the case in which $p = 2$, this reduces to the familiar t test for correlated means.) In section 1.2.6 and in chapter 4, this approach is compared to the unfortunately more common univariate Anova approach to repeated-measures designs.

Example 1.1. Anglo Versus Chicano Early Memories. In an as yet unpublished study, Ricardo Gonzales, Jose Carabajal, and Samuel Roll (1980) collected accounts of the earliest memories of 26 traditional Chicanos (so classified on the basis of continued use of Spanish and endorsement of prototypically Hispanic values), 11 acculturated Chicanos, and 34 Anglos—all students at the University of New Mexico. Judges, who were uninformed as to ethnicity of a participant, read each account and rated it on the following nine dependent variables:

Y_1 = Number of people mentioned
- 3 = three or more individuals mentioned
- 2 = one or two individuals besides subject mentioned
- 1 = no one but subject mentioned

Y_2 = Enclosure
- 3 = subject was enclosed within definite boundaries, such as a house or car, in memory
- 2 = subject was partially enclosed, such as inside an alleyway or fence
- 1 = subject was in a completely open space
- 0 = none of the above conditions were mentioned

Y_3 = Weather
- 3 = prominent mention of weather
- 0 = no mention

Y_4 = Readiness to judge others' feelings (rated on a continuum)
- 3 = very descriptive account of others' feelings
- 1 = short description of environment and behavior of others, but no mention of their feelings

Y_5 = Competition
- 3 = subject exhibited extreme aggressiveness
- 2 = someone else exhibited overt aggression
- 1 = subject remembers self or other having strong desire for something someone else had

Y_6 = Labor-typed activities (rated on a continuum)
- 3 = prominent mention
- 0 = no mention

Y_7 = Interaction with father (rated on a continuum)
- 3 = prominent mention
- 0 = no mention

Y_8 = Repressed emotions (rated on continuum)
3 = high evidence
0 = no evidence

Y_9 = Generosity
8 = subject displays voluntary generosity to a human
7 = subject displays voluntary generosity to an inanimate object or an animal
6 = subject observes someone else's generosity to a human
et cetera

Concentrating on the difference between the traditional Chicanos and the Anglos, Y_9 (generosity) was the only dependent variable whose two-sample t test could be considered statistically significant by a Bonferroni-adjusted criterion at a .05 level of experimentwise alpha: $t(58) = 4.015$ as compared to the Bonferroni-adjusted critical value of $t_{.05/9}(58) = 2.881$. However, rather than being content simply to examine each dependent variable by itself, the researchers conducted a two-sample T^2 on the traditional Chicano versus Anglo difference with respect to the full 9-element vector of dependent measures, obtaining a T^2 of 48.45, a value over 3 times as large as the squared t ratio for *generosity* alone. The linear combination of the measures that yielded that value of T^2 was $1.443Y_1 - .875Y_2 - .604Y_3 + .158Y_4 - 1.059Y_5 - 1.923Y_6 - .172Y_7 + .965Y_8 + .811Y_9$. Generosity and number of people are rather social themes, whereas enclosure, competition, and labor-related activities are more work-oriented themes. Thus, it seems reasonable to interpret these results as suggesting that the really big difference between traditional Chicanos and Anglos lies in the greater prevalence in the Chicanos' early memories of social themes over more competitive, work-oriented themes. This substantive interpretation was "pinned down" by computing the squared t ratio for the difference between these two groups on a simplified discriminant function, namely, $(3Y_9/8 + Y_1)/2 - (Y_2 + Y_5 + Y_6)/3$. The resulting value of t^2 was 43.05, also statistically significant when compared to the .01-level critical value demanded of T^2 (namely, 29.1). Thus, the T^2 analysis yielded a previously unsuspected but readily interpretable variable that provided considerably greater differentiation between these two groups than did any of the original dependent variables considered by themselves.

1.2.4 One-Way Multivariate Analysis of Variance

Just as Hotelling's T^2 provides a generalization of the univariate t test, so does one-way multivariate analysis of variance (Manova) provide a generalization of one-way (univariate) analysis of variance. One-way Manova is applicable whenever there are several groups of subjects, with more than one measure being obtained on each subject. Just as in the case of T^2, the test of the overall null hypothesis is accomplished by reducing the set of p measures on each subject (the *response profile* or *outcome vector*) to a single number by applying a linear combining rule

$$W_i = \sum_j w_j X_{i,j}$$

to the scores on the original outcome variables. A univariate F ratio is then computed on the combined variable, and new sets of weights are selected until that set of weights that makes the F ratio as large as it possibly can be has been found. This set of weights is the (multiple) discriminant function, and the largest possible F value is the basis for the significance test of the overall null hypothesis.

The distribution of such *maximum F statistics* (known, because of the mathematical tools used to find the maximum, as ***greatest characteristic root*** statistics) is complex, and deciding whether to accept or reject the null hypothesis requires the use of a computer subroutine or a series of tables based on the work of Heck (1960), Pillai (1965, 1967), and Venables (1973). The tables are found in Appendix A.

Actually, if both $(k - 1)$ and p are greater than unity, the Manova optimization procedure doesn't stop with computation of the (first) discriminant function. Instead, a second linear combination is sought—one that produces as large an F ratio as possible, subject to the condition that it be uncorrelated with the first (unconditionally optimal) discriminant function. Then a third discriminant function—uncorrelated with the first two (and therefore necessarily yielding a smaller F ratio than either of the first two) – is found; and so on, until a total of $s = \min(k - 1, p) = p$ or $k - 1$ (whichever is smaller) discriminant functions have been identified.

The discriminant functions beyond the first one identify additional, nonredundant dimensions along which the groups differ from each other. Their existence, however, leads to a number of alternative procedures for testing the overall $H_{o,\text{ov}}$. All of these alternatives involve combining the maximized F ratios corresponding to the s discriminant functions into a single number. The extensive literature comparing the power and robustness of these various multiple-root tests with each other and with the greatest characteristic root (gcr) statistic is discussed in section 4.5.

Example 1.2. Inferring Social Motives From Behavior. Maki, Thorngate, and McClintock (1979, Experiment 1) generated sequences of behavior that were consistent with seven different levels of concern for a partner's outcome. These levels ranged from highly competitive (negative weight given to a partner's outcome) to altruistic (concern only that the partner received a good outcome, irrespective of what the chooser got). In each of these seven chooser's-motive conditions, 10 subjects observed the chooser's behavior and then "evaluated the personality attributes of the chooser on 30 9-point bipolar adjective scales" (p. 208).

Rather than performing 30 separate one-way Anovas on the differences among the 7 conditions with respect to each of the 30 scales, the authors conducted a one-way multivariate analysis of variance (labeled as a *discriminant function analysis*) on the 30-variable outcome vector. They report that only the first discriminant function (the one yielding the largest univariate F for the differences among the seven conditions) was statistically significant and that "the larger discriminant coefficients obtained from the analysis indicated that higher ratings of agitation, stability, selfishness, unfriendliness, badness, masculinity, and—oddly enough—politeness tended to be associated with self-centered classification by this function." Moreover, means on this discriminant function for the seven levels of concern for partner's outcome were highest for the negative-concern (competitive) motives, intermediate for individualistic motives (indifference towards, i.e., zero weight for, partner's outcome), and very low for conditions in which the chooser showed positive concern for his or her partner (altruistic or cooperative motives).

Only an overall test based on all six discriminant functions was reported. However, from the reported value, it is clear that the F ratio for the first discriminant function by itself must have been at least 86.7, a highly significant value by the gcr criterion.

1.2.5 Higher Order Analysis of Variance

The procedures of one-way analysis of variance can be applied no matter how large the number of groups. However, most designs involving a large number of groups of subjects arise because the researcher wishes to assess the effects of two or more independent (manipulated) variables by administering all possible combinations of each level of one variable with each level of the other variables.

An example of such a factorial design would be a study in which various groups of subjects perform tasks of high, low, or medium difficulty after having been led to be slightly, moderately, or highly anxious, thus leading to a total of nine groups (lo-slight, lo-mod, lo-hi, med-slight, med-mod, med-hi, hi-slight, hi-mod, and hi-hi). The investigator will almost always wish to assess the ***main effect*** of each of his or her independent variables (i.e., the amount of variation among the means for the various levels of that variable, where the mean for each level involves averaging across all groups receiving that level) and will in addition wish to assess the ***interactions*** among the different independent variables (the extent to which the relative spacing of the means for the levels of one variable differs, depending on which level of a second variable the subjects received). For this reason, inclusion of contrasts corresponding to these main effects and interactions as preplanned comparisons has become almost automatic, and computational procedures have been formalized under the headings of two-way (two independent variables), three-way (three independent variables), and so on, analysis of variance.

In addition, researchers and statisticians have become aware of the need to distinguish between two kinds of independent variables: ***fixed*** variables (for example, sex of subject), identified as those variables whose levels are selected on a priori grounds and for which an experimenter would therefore choose precisely the same levels in any subsequent replication of the study, and ***random*** variables (e.g., litter from which subject is sampled), whose levels in any particular study are a random sample from a large population of possible levels and for which an experimenter would probably select a new set of levels were he or she to replicate the study. Our estimate of the amount of variation among treatment means that we would expect if the null hypothesis of no true variation among the population means were correct will depend on whether we are dealing with a fixed or a random independent variable. This selection of an appropriate "error term" for assessing the statistical significance of each main effect and each interaction in a study that involves both fixed and random independent variables can be a matter of some complexity. Excellent treatments of this and other problems involved in higher order analysis of variance are available in several texts (cf. especially Winer, 1971; Hays & Winkler, 1971; and Harris, 1994) and is not a focus of this text.

1.2.6 Higher Order Manova

It is important to realize that a multivariate counterpart exists for every univariate analysis of variance design, with the multivariate analysis always involving (heuristically) a search for that linear combination of the various outcome variables that makes the univariate F ratio (computed on that single, combined variable) for a particular main effect or interaction as large as possible. Note that the linear combination of outcome variables used in evaluating the effects of one independent variable may not assign the same weights to the different outcome variables as the linear combination used in assessing the effects of some other independent variable in the same analysis. This may or may not be desirable from the viewpoint of interpretation of results. Methods, however, are available for ensuring that the same linear combination is used for all tests in a given Manova.

A multivariate extension of Scheffé's contrast procedure is available. This extension permits the researcher to make as many comparisons among linear combinations of independent and/or outcome variables as desired with the assurance that the probability of falsely identifying anyone or more of these comparisons as representing true underlying population differences is lower than some prespecified value.

Finally, it should be pointed out that there exists a class of situations in which univariate analysis of variance and Manova techniques are in "competition" with each other. These are situations in which each subject receives more than one experimental treatment or is subjected to the same experimental treatment on a number of different occasions. Clearly the most straightforward approach to such *within-subject* and *repeated-measures* designs is to consider the subject's set of responses to the different treatments or his or her responses on the successive trials of the experiment as a single outcome vector for that subject, and then apply Manova techniques to the N outcome vectors produced by the N subjects. More specifically, within-subjects effects are tested by single-sample T^2 analyses on the vector of grand means of contrast scores (grand means because each is an average across all N subjects, regardless of experimental condition). Each of these N subjects yields a score on each of the p - 1 contrasts among his or her p responses. Interactions between the within-subjects and between-subjects effects are tested via a Manova on this contrast-score vector. (Between-subjects effects are tested, as in the standard univariate approach, by a univariate Anova on the average or sum of each subject's responses.)

However, if certain rather stringent conditions are met (conditions involving primarily the uniformity of the correlations between the subjects' responses to the various possible pair of treatments or trials), it is possible to use the computational formulae of univariate Anova in conducting the significance tests on the within-subjects variables. This has the advantage of unifying somewhat the terminology employed in describing the between- and within-subjects effects, although it is probable that the popularity of the univariate approach to these designs rests more on its avoidance of matrix algebra.

Example 1.3. Fat, Four-eyed, and Female. In a study that we'll examine in more detail in chapter 6, Harris, Harris, and Bochner (1982) described "Chris Martin" to 159 Australian psychology students as either a man or a woman in his or her late twenties who, among other things, was either overweight or of average weight and who either did or did not wear glasses. Each subject rated this stimulus person on 12 adjective pairs, from which 11 dependent variables were derived.

A 2 x 2 x 2 multivariate analysis of variance was conducted to assess the effects of the 3 independent variables and their interactions on the means of the 11 dependent variables and the linear combinations thereof. Statistically significant main effects were obtained for wearing glasses and for being overweight. The results for the obesity main effect were especially interesting. On every single dependent variable the mean rating of "Chris Martin" was closer to the favorable end of the scale when he or she was described as being of average weight. Thus we appear to have a simple, straightforward, global devaluation of the obese, and we might anticipate that the discriminant function (which yielded a maximized F ratio of 184.76) would be close to the simple sum or average of the 11 dependent variables. Not so. Instead, it was quite close to

$$Y_5 \text{ (Outgoing)} + Y_9 \text{ (Popular)} - Y_2 \text{ (Active)} - Y_8 \text{ (Attractive)} - Y_{11} \text{ (Athletic)}.$$

In other words, the stereotype of the overweight stimulus person that best accounts for the significant main effect of obesity is of someone who is more outgoing and popular than one would expect on the basis of his or her low levels of physical attractiveness, activity, and athleticism. This is a very different picture than would be obtained by restricting attention to univariate Anovas on each dependent variable.

1.2.7 Pearson *r* and Bivariate Regression

There are many situations in which we wish to assess the relationship between some outcome variable (e.g., attitude toward authority) and a predictor variable (e.g., age) that the researcher either cannot or does not choose to manipulate. The researcher's data will consist of a number of pairs of measurements (one score on the predictor variable and one on the outcome variable), each pair having been obtained from one of a (hopefully random) sample of subjects. Note that bivariate regression is a univariate statistical technique. The predictor variable (traditionally labeled as X) will be of value in predicting scores on the outcome variable (traditionally labeled as Y) if, in the population, it is true either that high values of X are consistently paired with high values of Y, and low X values with low Y values, or that subjects having high scores on X are consistently found to have low scores on Y, and vice versa. Because the unit and origin of our scales of measurement are usually quite arbitrary, "high" and "low" are defined in terms of the number of standard deviations above or below the mean of all observations that a given observation falls. In other words, the question of the relationship between X and Y is converted to a question of how closely $z_x = (X - \mu_x)/\sigma_x$ matches $z_y = (Y - \mu_y)/\sigma_y$, on the average. The classic measure of the degree of relationship between the two variables in the population is

$$\rho_{xy} = \frac{\sum z_x z_y}{N},$$

where the summation includes all N members of the population. Subjects for whom X and Y lie on the same side of the mean contribute positive cross-product ($z_x z_y$) terms to the summation, while subjects whose X and Y scores lie on opposite sides of their respective means contribute negative terms, with ρ_{xy} (the Pearson product-moment coefficient of correlation) taking on its maximum possible value of + 1 when $z_x = z_y$ for all subjects and its minimum possible value of - 1 when $z_x = -z_y$ for all subjects. The researcher obtains an estimate of ρ through the simple expedient of replacing μ and σ terms with sample means and standard deviations—whence,

$$r_{xy} = \frac{\sum z_x z_y}{N-1} = \frac{\sum (X-\overline{X})(Y-\overline{Y})}{\sqrt{\sum (X-\overline{X})^2 \sum (Y-\overline{Y})^2}}.$$

The Pearson r (or ρ) is very closely related to the problem of selecting the best linear equation for predicting Y scores from X scores. If we try out different values of b_0 and b_1 in the equation $\hat{z}_y = b_0 + b_1 z_x$ (where $\hat{z}_y$ is a predicted value of z_y), we will eventually discover (as we can do somewhat more directly by employing a bit of calculus) that the best possible choices for b_0 and b_1 (in the sense that they make the sum of the squared differences between $\hat{z}_y$ and z_y as small as possible) are $b_0 = 0$ and $b_1 = r_{xy}$. [The formula for predicting raw scores on Y from raw scores on X follows directly from the z-score version of the ***regression equation*** by substituting $(X - \overline{X})/s_x$ and $(Y - \overline{Y})/s_y$ for z_x and z_y, respectively, and then isolating $\hat{Y}$, the predicted raw score on Y, on the left-hand side of the equation.] Furthermore, the total variation among the Y scores s_y^2 can be partitioned into a component attributable to the relationship of Y to X, $s_{\hat{y}}^2$, and a second component representing the mean squared error of prediction, $s_{y \bullet x}^2$. (The square root of this latter term is called the ***standard error of estimate***.) The ratio between $s_{\hat{y}}^2$ and s_y^2— that is, the percentage of the variance in Y that is "accounted for" by knowledge of the subject's score on X—is identically equal to r^2_{xy}, the square of the Pearson product-moment correlation coefficient computed on this sample of data. (This suggests, incidentally, that r^2_{xy} is a more directly meaningful measure of the relationship between the two variables than is r_{xy}. Tradition and the fact that r^2 is always smaller and less impressive than r have, however, ensconced r as by far the more commonly reported measure.) This close relationship between the correlation coefficient and the linear regression equation (linear because no squares or higher powers of z_X are employed) should alert us to the fact that r is a measure of the degree of *linear* relationship between the two variables and may be extremely misleading if we forget that the relationship may involve large nonlinear components. The problem of possible curvilinearity is a persistent (although often unrecognized) one in all applications of correlation and regression.

Quite large sample values of r can of course arise as a consequence of random fluctuation even if ρ in the population from which. the subjects were sampled truly equals zero. The null hypothesis that $\rho = 0$ is tested (and hopefully rejected) via a statistic having a t distribution, namely,

$$t = \frac{r}{\sqrt{(1-r^2)/(N-2)}} .$$

The researcher seldom has to employ this formula, however, as tables of critical values of r are readily available.

The Pearson r was developed specifically for normally distributed variables. A number of alternative measures of correlation have been developed for situations in which one or both measures are dichotomous or rank-order variables. However, with few exceptions (such as those provided by Kendall's tau and the Goodman–Kruskal measures of the relationship between nominal variables), the resulting values of these alternative measures differ little from Pearson r. In one set, the measures (typified by Spearman's coefficient of rank-order correlation, the phi coefficient, and the point biserial coefficient) are numerically identical to Pearson r when it is blindly applied to the decidedly nonnormal ranks or 0–1 (dichotomous) measures. The measures in the other set (typified by the tetrachoric coefficient and biserial r) represent an exercise in wishful thinking—the numerical values of these measures are equal to the values of Pearson r that "would have been" obtained if the normally distributed measures that presumably underlie the imperfect data at hand had been available for analysis.[6]

---

[6] The nature of the situations in which the Pearson r, as compared with alternative measures of correlation such as Spearman's rho, is appropriate is sometimes stated in terms of the strength of measurement required for the operations involved in computing r (e.g., addition, subtraction, or multiplication of deviation scores) to be meaningful. Stevens (1951, 1968) and others (e.g., Senders, 1958; Siegel, 1956) would restrict "parametric" statistics such as the Pearson r and the t test to situations in which the data have been shown to have been obtained through a measurement process having at least interval scale properties (versus "weaker" scales such as ordinal measurement).

A number of other authors (e.g., Anderson, 1981, and an especially entertaining article by Lord, 1953) pointed out that statistical conclusions are valid whenever the distributions of

None of the descriptive aspects of Pearson r are altered by applying it to "imperfect" data, and the critical values for testing the hypothesis of no relationship that have been derived for the alternative measures are often nearly identical to the corresponding critical values for the Pearson r applied to impeccable data. Therefore, this book does not use the more specialized name (phi coefficient, for example) when Pearson r values are computed on dichotomous or merely ordinal data.

One particularly revealing application of the Pearson r to dichotomous data arises (or shoud arise) whenever a t test for the difference between two independent means is appropriate. If the dependent measure is labeled as Y and each subject is additionally assigned an X score of 1 or 0, depending on whether he or she was a member of the first or the second group, then the t test of the significance of the difference between the r computed on these pairs of scores and zero will be numerically identical to the usual t test for the significance of the difference between $\overline{X}_1$ and $\overline{X}_2$. This is intuitively reasonable, because in testing the hypothesis that $\mu_1 = \mu_2$ we are essentially asking whether there is any relationship between our independent variable (whatever empirical operation differentiates group 1 from group 2) and our outcome measure. More importantly, $r^2 = t^2/[t^2 + (N - 2)]$ provides a measure of the percentage of the total variation among the subjects in their scores on the dependent measure that is attributable to (predictable from a knowledge of) the group to which they were assigned. Many experimentalists who sneer at the low percentage of variance accounted for by some paper-and-pencil measure would be appalled at the very low values of r^2 underlying the highly significant (statistically) values of t produced by their experimental manipulations.

1.2.8 Multiple Correlation and Regression

We often have available several measures from which to predict scores on some criterion variable. In order to determine the best way of using these measures, we collect a sample of subjects for each of whom we have scores on each of the predictor variables, as well as a score on the criterion variable. We wish to have a measure of the overall degree of relationship between the set of predictor variables and the criterion (outcome) measure. Mystical, complex, and Gestalt-like as such an overall measure sounds, the coefficient of

---

numbers from which the data are sampled meet the assumptions (typically, normality and homogeneity of variance) used to derive the particular techniques being applied, irrespective of

the measurement process that generated those numbers. Moreover, the validity of parametric statistics (e.g., the correspondence between actual and theoretically derived Type I error rates) is often affected very little by even relatively gross departures from these assumptions. I agree with this latter position, although I (along with most of Stevens's critics) agree that level of measurement may be very relevant to the researcher's efforts to fit his or her statistical conclusions into a theoretical framework. These points are discussed further in chapter 8.

multiple correlation (multiple R) is really nothing more than the old, familiar Pearson r between Y_i (our outcome measure on each subject i) and $W_i = \Sigma\, w_j X_{i,j}$, a linear combination the scores of subject i on the predictor variables. The particular weights employed are simply those we discover [by trial and error or, somewhat more efficiently, by calculus and matrix algebra] produce the largest possible value of R. These weights turn out to be identical to those values of the b_j in the ***multiple regression equation***

$$\hat{Y} = b_1 X_1 + b_2 X_2 + \cdots + b_m X_m$$

that make $\sum_i (\hat{Y}_i - Y_i)^2$ as small as possible. The null hypothesis that the population value of multiple R is truly zero is tested by comparing the amount of variance in Y accounted for by knowlege of scores on the Xs with the amount left unexplained.

When Y and the Xs have a multivariate normal distribution (or the Xs are fixed and the Ys are normally distributed for each combination of values of the Xs), the ratio

$$F = \frac{(N - m - 1)R^2}{m(1 - R^2)}$$

has an F distribution with m and N - m - 1 degrees of freedom. Specific comparisons among particular b coefficients are provided by procedures analogous to Scheffé's contrast methods for Anova. Indeed, as the last few sentences suggest, Anova is just a special case of multiple regression in which the predictor variables are all group membership variables with X_j equal to 1 if subject i is at level j of the independent variable, to -1 if he or she is at the last level, and to zero in all other cases.[7]

The interaction terms of higher order Anova are derived from multiple regression either through contrasts computed on the b_j terms or by including crossproducts ($x_i x_j$) of the group-membership variables as separate predictor variables. This approach to taking into account possible nonadditive relationships among the predictor variables is of course available (although seldom used) when the predictor variables are continuous, rather than discrete. (One should be careful to use cross products of *deviation* scores, rather than *raw-score* cross products, when assessing interactions involving continuous variables. See section 2.7 for details.) Similarly, curvilinear relationships between any of the predictor variables and Y can be provided for (and tested for significance) by including as additional predictors the squares, cubes, and so on of the original predictors. Unfortunately, adding these variables cuts into the degrees of freedom available for

---

[7] Dichotomous (for example, one/zero) coding of group membership is adequate for one-way designs, but this trichotomous coding must be used when interaction terms are to be examined. Also, the group-membership scores must be weighted by cell sizes in the unequal-n factorial designs in which main effects (also called "factors" of the design despite the potential confusion with the latent variables of *factor* analysis discussed in section 1.2.13 and in chapter 8) are *not* to be corrected for the confounding with other effects produced by having, for example, a higher proportion of highly educated respondents in the high than in the low or moderate fear-arousing conditions.

significance tests. Even if the population value of multiple R is truly zero, the expected value of the *sample* R is equal to $\sqrt{m/(N-1)}$, where m is the number of predictor variables and N is the number of subjects. Thus, as the number of predictor variables approaches the number of subjects, it becomes increasingly difficult to discriminate between chance fluctuation and true relationships.

Example 1.4. Chicano Role Models, GPA, and MRA. Melgoza, Harris, Baker, and Roll (1980) examined background information on and the grades earned by each of the 19,280 students who took one or more courses at the University of New Mexico (UNM) during the spring semester of 1977. (All names were replaced with code numbers before the data were released to the researchers.) In addition to examining differences between the 5,148 Chicano students (so identified by the Records Office on the basis of self-report or surname) and the 14,132 Anglos (in New Mexican parlance, almost anyone who does not identify himself or herself as being of Hispanic heritage) in their preferences for and academic performance in various subject areas, the authors were interested in testing the hypothesis that "Chicano students do better academically if they have had Chicano professors" (p. 147). Indeed, *t* tests showed such an effect:

> We next examine the "modeling effect," that is, the difference in performance in Anglo-taught courses (thus controlling for any effect of ethnicity of instructor) by Chicanos who have taken versus those who have not taken courses taught by a role model, that is a same-sex Chicano or Chicana instructor. We find that Chicanos who have taken one or more courses from Chicano instructors have a higher mean GPA (2.45) than those who have not had a Chicano instructor (2.21), yielding a $t(2245) = 4.20$, $p < .001$. Similarly, Chicanas who have taken one or more courses from a Chicana have a higher mean GPA (2.77) than those who have not (2.39, yielding a $t(2022) = 2.27$, $p < .05$. (p. 156)

However, previous analyses had revealed large, statistically significant effects on GPA of such factors as class (freshman, sophomore, and so on), age, and size of the town where the subject attended high school. It seemed possible that the observed differences in mean GPAs might owe more to differences in these background variables between students who take courses taught by Chicano instructors and those who do not than to the postive effects of having had a role model In order to test for this possibility, two multiple regression analyses were carried out (one for male students and one for females) employing as predictors these background variables, together with a group-membership variable indicating whether or not the student had taken one or more courses from a Chicano instructor. Mean GPA in Anglo-taught courses served as the dependent variable. The regression coefficient for the group-membership variable (i.e., for the "modeling effect") was quite small and statistically nonsignificant in each of these analyses, thus indicating that whether or not a student had taken one or more courses from a Chicano instructor added nothing to the ability to predict his or her GPA beyond that provided by knowing his or her scores on the background variables. In fact, regression analyses including only the group-membership variable and class in school yielded statistically nonsignificant regression coefficients for the modeling effect for both male and female students. As the authors concluded:

> Apparently, the overall tendency for Chicanos who have had a Chicano instructor to achieve higher GPAs was due to the fact that students who are upperclassmen have higher GPAs and are also more likely to have had a course taught by a Chicano instructor, simply by virtue of having taken more courses. Of course, this does not rule out the possibility of an indirect effect [of having had a Chicano instructor] as a role model; that is, Chicanos who have had such a role model may be more likely to remain in school than those who have not. (p. 156)

1.2.9 Path Analysis

Path analysis is a technique that is designed to test the viability of linear, additive models of the causal relationships among directly observable variables by comparing the correlations implied by each model against those actually observed.

In the case where the "causal flow" is unidirectional (called ***recursive models*** in path analysis)—that is, where no variable X that is a cause of some variable Y is itself caused (even indirectly) by—testing the model is accomplished by carrying out a series of multiple regression analyses (MRAs), so that developing skill in carrying out MRAs brings with it the additional benefit of enabling one to test alternative causal models.

For instance, one possible explanation of our results on the impact of a role model on Chicanos' and Chicanas' GPAs is that the number of years one has spent at UNM (YAUNM) has a positive causal effect on the likelihood of having been exposed to a role model (RM), which in turn has a positive causal effect on one's GPA. This relationship can be represented by the path diagram,

YAUNM → RM → GPA,

and by the statement that the impact of YAUNM on GPA is entirely mediated by RM. This model implies that there should be a positive correlation between RM and GPA (which prediction is supported by the results of our t tests) and that there should be a positive correlation between YAUNM and RM (not reported earlier, but true and statistically significant). So far, so good. However, the model also implies that an MRA of GPA predicted from YAUNM and RM should yield a positive, statistically significant regression coefficient for RM and a near-zero, statistically nonsignificant regression coefficient for YAUNM—just the opposite of what we found when we carried out that MRA. (The basic principle, by the way, is that path coefficients in path analysis of recursive models are estimated by regression coefficients obtained in a series of MRAs, each of which takes one of the variables that has an arrow pointing to it as Y and all other variables that precede Y in the "causal chain" as predictors.) An alternative explanation is the one offered in the paper, which can be symbolized as

RM → YAUNM → GPA,

and can be expressed verbally via the statement that the impact of RM on GPA is entirely mediated by ("channeled through") YAUNM. This model is entirely consistent with the obtained correlations and regression analyses.

Unfortunately, so is

RM ← YAUNM ← GPA,

which says that the correlation between RM and GPA is a spurious one, due entirely to both RM and GPA being positively affected by years at UNM. (The longer you "stick around," the more likely it is that you'll encounter a role model of your ethnicity, and the higher your GPA is; instructors tend to grade more easily in upper-division classes and are very reluctant to give anything as low as a C in a graduate class.) As shown in section 2.9, two path models that differ only in the direction of an arrow (such as the last two considered) have identical implications for the observable correlations among the variables and therefore cannot be distinguished on the basis of those correlations.

1.2.10 Canonical Correlation

More often than the statistical techniques used in the literature would suggest, we have several outcome measures as well as several predictor variables. An overall measure of the relationship between the two sets of variables is provided by canonical R, which is simply a Pearson r calculated on two numbers for each subject:

$$W_i = \Sigma\, w_j X_{i,j} \text{ and } V_i = \Sigma\, v_j X_{i,j}$$

where the Xs are predictor variables and the Ys are outcome measures. Heuristically, the w_j and the v_j (which are the ***canonical coefficients*** for the predictor and outcome measures, respectively) are obtained by trying out different sets of weights until the pair of sets of weights that produces the maximum possible value of canonical R (R_c) has been obtained.

As in Manova, when each set of variables contains two or more variables, the analysis of their interrelationships need not stop with the computation of canonical R. A second pair of sets of weights can be sought that will produce the maximum possible Pearson r between the two combined variables, subject to the constraint that these two new combined variables be uncorrelated with the first two combined variables. A total of $\min(p, m) = p$ or m (whichever is smaller) pairs of sets of weights (each of which has a corresponding coefficient of canonical correlation, is uncorrelated with any of the preceding sets of weights, and accounts for successively less of the variation shared by the two sets of variables) can be derived.

A canonical R of 1.0 could involve only one variable from each set, because all other variables are uncorrelated. Consequently, considerable effort has been expended recently (e.g., Cramer & Nicewander, 1979; DeSarbo, 1981; and Van den Wollenberg, 1977) on developing measures of ***redundancy*** (the percentage of the variance in one set predictable from its relationships to the variables in the other set) and ***association*** (some function of the canonical Rs). We discuss these efforts in section 5.4.4.

Just as univariate analysis of variance can be considered a special case of multiple regression, multivariate analysis of variance can be considered a special case of canonical correlation—the special case that arises when one of the two sets of variables consists entirely of group-membership variables. All of the various statistical techniques for examining relationships between sets of measures can thus be seen as various special cases of canonical analysis, as Fig. 1.1 illustrates.

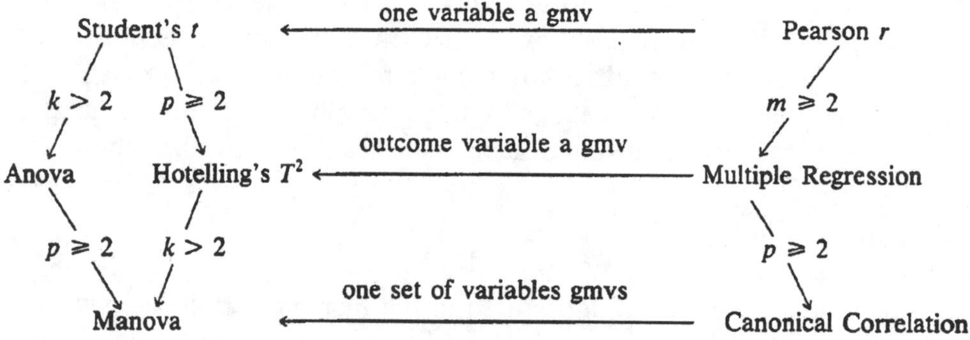

Figure 1.1 **Multivariate Analyses of Between-Set Relationships**

Note. **A southwesterly arrow represents an increase in the number of independent variables, while a southeasterly arrow represents an increase in the number of dependent variables. A westerly arrow indicates that one set of variables consists of group-membership variables (gmvs).**

p = **the number of dependent (outcome) measures.**
k = **the number of levels of the discrete (usually manipulated) independent variables.**
m = **the number of measures and (usually continuous) independent variables.**

It is possible to generalize canonical correlation—at least in its significance-testing aspects—to situations in which more than two sets of variables are involved. For instance, we might have three paper-and-pencil measures of authoritarianism, ratings by four different psychiatrists of the degree of authoritarianism shown by our subjects in an interview, and observations based on a formal recording system, such as Bales' interaction process analysis, of the behavior of each subject in a newly formed committee. We could then test the overall null hypothesis that these three sets of measures of authoritarianism have nothing to do with each other. Unfortunately the generalization is based on the likelihood-ratio approach to significance testing and therefore tells us nothing about the magnitude of any relationship that does exist or about ways to combine the variables within each set so as to maximize some overall measure of the degree of interrelationship among the sets.

Example 1.5. Television Viewing and Fear of Victimization. Doob and MacDonald (1979) suspected that previous reports that people who maintain a heavy diet of television viewing are more fearful of their environment (e.g., estimate a higher likelihood of being assaulted) than those who devote fewer hours to watching television may be the result of a confound with actual incidence of crime. People who live in high-crime areas might have a constellation of background characteristics that would also predispose them toward a high level of exposure to television, thus accounting for the reported correlation. From 34 questions dealing with fear of crime, the 9 questions that loaded most highly on the first principal factor (see section 1.2.13) were selected as dependent variables. Respondents were also asked to report their level of exposure to various media during the preceding week, leading to measures of exposure to television in general, television violence in particular, radio news, and newspapers. These four exposure measures were combined with the subject's age and sex, whether he or she lived in the city or the suburbs (of Toronto), whether the area of residence was a high- or a low-crime area on the basis of police statistics, and with the interaction between these last two variables to yield a set of nine predictor variables. A canonical analysis of the relationships between these nine predictors and the nine fear-of-crime measures was conducted, yielding two pairs of canonical variates that were statistically significant by the gcr criterion (canonical *R*s of .608 and .468, respectively). On the basis of their standardized (z-score) canonical variate weights, these two R_cs were interpreted as follows:

> The first pair of canonical variates suggests that those who do not see crimes of violence as a problem in their neighborhood (Question 1), who do not think that a child playing alone in a park is in danger (Question 3); and who do not think that they themselves are likely to be the victims of an assault (Question 6), but who are afraid that their houses might be broken into (Question 7) and who do not walk alone at night (Question 8) tend to be females living in low-crime (city) areas. The second pair of canonical variates appears to indicate that people who have areas near them that they will not walk in at night (Question 9) and who fear walking alone at night (Question 11), but who do not think that they will be victims of a violent crime (Question 6) tend to be females living in high-crime (city) areas who listen to a lot of radio news. (p. 176)

Notice that neither total television viewing nor television violence plays a role in either of these canonical variates. Indeed, their standardized weights ranked no higher than fifth in absolute value among the nine predictors in the first canonical variate and eighth in the second canonical variate for predictors. This relative unimportance of television viewing when the other predictors are taken into account was further corroborated by a multiple regression analysis (see Section 1.2.8) in which the score on the first principal factor of the 34 fear-of-crime questions was the dependent variable. This MRA produced regression coefficients for total television viewing and for television violence that were far from statistically significant. As the authors summarized their analyses: "In summary, then, it appears that the amount of television watched did not relate to the amount of fear a person felt about being a victim of crime when other, more basic variables were taken into account" (p. 177).

1.2.11 Analysis of Covariance

The analyses we have considered so far have included only two kinds of variables: predictor variables and outcome variables. (This predictor-outcome distinction is blurred in many situations in which we are only interested in a measure of the overall relationship between the two sets, such as the Pearson r or canonical R.) All of the variables in either set are of interest for the contribution they make to strengthening the interrelations between the two sets. In many situations, however, a third kind of variable is included in the analysis. We might, rather uncharitably, call these variables "nuisance" or "distractor" variables. A more neutral term, and the one we adopt, is ***covariates***. A covariate is a variable that is related to (covaries with) the predictor and/or the outcome variables, and whose effects we wish to control for statistically as a substitute for experimental control. For instance, suppose that we were interested in studying the relative efficacy of various diet supplements on muscular development. Therefore, we would administer each supplement to a different randomly selected group of adult males. Then, after each group had used the supplement for, say, 6 months, we would measure, for example, the number of pounds of pressure each could generate on an ergograph (a hand-grip type of measuring device). We could then run a one-way analysis of variance on the ergograph scores. However, we know that there are large individual differences in muscular strength, and although the procedure of randomly assigning groups will prevent these individual differences from exerting any systematic bias on our comparisons among the different supplements, the high variability among our subjects in how much pressure they can exert *with or without* the diet supplement will provide a very noisy "background" (high error variability) against which only extremely large differences in the effectiveness of the supplements can be detected. If this variability in postsupplement scores due to individual differences in the genetic heritage or past history of our subjects could somehow be removed from our estimate of error variance, we would have a more precise experiment that would be capable of reliably detecting relatively small differences among the effects of the various diet supplements. This is what analysis of covariance (Ancova) does for us. Ancova makes use of an infrequently cited property of regression analysis, namely, that the expected value of any particular b coefficient derived in the analysis is a function only of the population value of the regression coefficient for that particular variable and not of the population parameters representing the effects of any other variable. As a consequence of this property, when a group-membership variable and a covariate are included in the same regression analysis, the resulting estimate of the effect of a membership in that group is independent (in the sense described earlier) of the effect of the covariate on the subject's performance. Our overall test of the differences among the group means is, then, a test of the statistical significance of the increase in R^2 that results when the group membership variables are added to the covariates as predictors of the dependent variable. (For a single-*df* contrast, represented by a single group-membership variable, this is identical to a test of the null hypothesis that the population regression coefficient is zero.)

Including the covariates in our regression analysis uses up some of the degrees of freedom in our data. To see this, we need only consider the case in which the number of covariates is only one less than the number of subjects, so that a multiple *R* of 1.0 is obtained; that is, each subject's score on the outcome variable is "predicted" perfectly from knowledge of his or her scores on the covariates, regardless of whether there is any true relationship in the population between the covariates and the outcome measure. Thus, there would be no "room" left to estimate the true contribution of between-group differences to scores on our dependent variable. Multivariate analysis of covariance (Mancova) is an extension of (univariate) Ancova and consists of an analysis of canonical correlation in which some of the variables included in the predictor set are covariates. We have seen this process and the sorts of conclusions it leads to illustrated in Examples 1.4 and 1.5. (But see Delaney & Maxwell, 1981, Maxwell, Delaney, & Dill, 1983, and Maxwell, Delaney, & Manheimer, 1985, for discussions of some of the subtleties of Ancova, including some of the ways in which Ancova differs from Anova performed on MRA residual scores, and comparison of the power of Ancova to that of an expanded Anova in which each covariate is added to the Anova design as an explicit blocking factor.)

This independence property is also of potentially great value in factorial-design experiments in which the various groups are of unequal size. One consequence of the unequal group sizes is that the means for the various levels of one independent variable are no longer independent of the means for the various levels of other manipulated variables. Because the estimates of the main effects of the various independent variables are, in analysis of variance, based solely on these means, the estimates of the main effects are interdependent. Variations in means and drug dosage levels in a treatment setting provide a good illustration of this point. A high mean for subjects receiving the highest dosage level—as compared to the mean for those at various other levels—may be solely attributable to the fact that more of the subjects who have been made highly anxious (as opposed to slightly or moderately anxious subjects) received the high dosage. Thus, *anxiety* is the important variable. This interdependence among estimates of treatment effects is completely absent when a multiple regression including both covariates and group-membership variables is performed on the data.

One common situation arises in which Ancova, Anova, and Manova are all in "competition." This occurs when, as in the ergograph example used to introduce Ancova, each subject is measured both before and after administration of some set of treatments (different program of instruction, various drugs, assorted persuasive communications, and so on). Probably the most common approach is to conduct an Anova on change scores, that is, on the difference between the pretreatment and posttreatment score for each subject. Alternately, the two scores for each subject could be considered as an outcome vector, and a Manova could be run on these vectors. Finally, Ancova could be employed, treating the baseline score as the covariate and the posttest score as the single outcome measure. The major disadvantage of Manova in this situation is that it is sensitive to all differences among the groups on either measure, whereas subjects will usually have been assigned to groups at random so that we know that any between-group differences in baseline scores are the result of random fluctuation. When choosing between the change score approach and Ancova, Ancova is usually the superior choice on purely statistical grounds. It makes use of the correction formula that removes as much as possible of the error variance, whereas the correction formula used in the change score analysis is an a priori one that by definition will effect less of a reduction in error variance. Furthermore, change scores have a built-in tendency, regardless of the nature of the manipulated variable or the outcome measure, to be negatively correlated with baseline scores, with subjects (or groups) scoring high on the premeasure tending to produce low change scores. The corrected scores derived from Ancova are uncorrelated with baseline performance. The change score is, however, a very "natural" measure and has the big advantage of ready interpretability. (However, see the work of C.W. Harris [1963], Cronbach & Furby [1970], Nicewander & Price [1978, 1983], Messick [1981], and Collins & Horn (1991) for discussions of the often paradoxical problems of measuring change .)

1.2.12 Principal Component Analysis

The statistical techniques we have discussed so far all involve relationships *between* sets of variables. However, in ***principal component analysis*** (PCA) and in ***factor analysis*** (FA) we concentrate on relationships *within* a single set of variables. Both of these techniques—although authors differ in whether they consider principal component analysis to be a type of factor analysis or a distinct technique—can be used to reduce the ***dimensionality*** of the set of variables, that is, to describe the subjects in terms of their scores on a much smaller number of variables with as little loss of information as possible. If this effort is successful, then the new variables (components or factors) can be considered as providing a description of the "structure" of the original set of variables.

The "new variables" derived from the original ones by principal component analysis are simply linear combinations of the original variables. The ***first principal component*** is that linear combination of the original variables that maximally discriminates among the subjects in our sample, that is, whose sample variance is as large as possible. Heuristically, we find this first principal component by trying out different values of the w_js in the formula $W_i = \Sigma\, w_j X_{i,j}$ and computing the variance of the subjects' scores until we have uncovered that set of weights that makes s_W^2 as large as it can possibly be. Actually, we have to put some restriction on the sets of weights we include in our search (the usual one being that the sum of the squares of the weights in any set must equal unity), because without restrictions we could make the variance of Wi arbitrarily large by the simple and uninformative expedient of using infinitely large w_j terms . At any rate, once we have found the first principal component, we begin a search for the ***second principal component***: that linear combination of the p original variables that has the largest possible sample variance, subject to the two constraints that (a) the sum of the squares of the weights employed equals unity and (b) scores on the second PC are uncorrelated with scores on the first PC. This process is continued until a total of p principal components have been "extracted" from the data, with each successive PC accounting for as much of the variance in the original data as possible subject to the condition that scores on that PC be uncorrelated with scores on any of the preceding PCs. The sum of the variances of subjects' scores on the p different PCs will exactly equal the sum of the variances of the original variables. Moreover, as implied by the increasing number of restrictions put on the permissible sets of weights included in the search for the PCs, each successive PC will have a lower associated sample variance than its predecessor. If the original variables are highly interrelated, it will turn out that the first few PCs will account for a very high percentage of the variation on the original variables, so that each subject's scores on the remaining PCs can be ignored with very little loss of information. This condensed set of variables can then be used in all subsequent statistical analyses, thus greatly simplifying the computational labor involved in, for example, multiple regression or Manova. Because the weights assigned the original variables in each principal component are derived in accordance with purely internal criteria, rather than on the basis of their relationship with variables outside this set, the value of multiple R or the significance level obtained in a Manova based on subjects' scores on the reduced set of PCs will inevitably be less impressive than if scores on the original variables had been employed. In practice, the loss of power is usually rather minimal—although we show in chapter 6 an example in which dropping the last of 11 PCs, accounting for only, 3.5% of the interindividual variance, leads to a 31.8% drop in the magnitude of our maximized F ratio. A more important justification than simplified math for preceding statistical analyses on a set of variables by a PCA of that set is the fact that the PCs are uncorrelated, thus eliminating duplication in our interpretations of the way subjects' responses on this set of variables are affected by or themselves affect other sets of variables.

Unlike the results of any of the other techniques we have discussed, the results of a PCA are affected by linear transformations of the original variables. The investigator must therefore decide before beginning a PCA whether to retain the original units of the variables or to standardize the scores on the different variables by converting to z scores (or by using some other standardization procedure). This decision will naturally hinge on the meaningfulness of the original units of measurement.

By examining the pattern of the weights b_i assigned to each variable i in the linear combination that constitutes the jth principal component, labels suggestive of the meaning of each PC can be developed. The PCs, when so interpreted, can be seen as providing a description of the original variables, with their complex intercorrelations, in terms of a set of uncorrelated latent variables (factors) that might have generated them. The principal-component weights, together with the variance of each PC, would in turn be sufficient to compute the correlations between any two original variables, even if we did not have access to any subject's actual score on any variable or any PC. (This follows from the well-known relationship between the variance of a linear combination of variables and the variances of the individual variables entering into that linear combination.)

1.2.13 Factor Analysis

Notice that the "explanation" of the relationships among the original variables provided by PCA is not a particularly parsimonious one, because p PCs are needed to reproduce the intercorrelations among the original p variables. (The fact that the PCs are uncorrelated and ranked in order of percentage of variance accounted for may, nevertheless, make it a very useful explanation.) If we eliminate all but the first few PCs, we obtain a more parsimonious description of the original data. This is done, however, at the expense of possible systematic error in our reproduction of the intercorrelations, because there may be one or two variables that are so much more highly related to the "missing" PCs than to those that are included as to make our estimates of the intercorrelations of other variables with these one or two variables highly dependent on the omitted data. Note also that PCA uses all of the information about every variable, although it is almost certain that some of the variation in subjects' scores on a given variable is unique variance, attributable to influences that have nothing to do with the other variables in the set. We might suspect that we could do a better job of explaining the relationships among the variables if this unique variance could somehow be excluded from the analysis. Note also that the criterion used to find the PCs ensures that each successive PC accounts for less of the variance among the original variables than its predecessors. The investigator may, however, have strong grounds for suspecting that the "true" factors underyling the data are all of about equal importance—or for that matter, that these latent variables are not uncorrelated with each other, as are the PCs.

Factor analysis (FA) refers to a wide variety of techniques that correct for one or more of these shortcomings of PCA. All factor analysis models have in common the explicit separation of unique variance from common variance, and the assumption that the intercorrelations among the p original variables are generated by some smaller number of latent variables. Depending on how explicit the researcher's preconceptions about the nature of these underlying variables are, each original variable's ***communality*** (defined by most authors as the percentage of that variable's variance that is held in common with other variables) may either be produced as an offshoot of the analysis or have to be specified in advance in order to arrive at a factor-analytic solution. A factor-analytic solution always includes a table indicating the correlation (loading) of each original variable with (on) each latent variable (factor), with this table referred to as the ***factor structure.*** The usual choice is between advance specification of the number of factors, with the analysis yielding communalities, and advance specification of communalities, with the analysis yielding the number of factors.

What is gained by employing FA versus PCA is the ability to reproduce the original pattern of intercorrelations from a relatively small number of factors without the systematic errors produced when components are simply omitted from a PCA. What is lost is the straightforward relationship between subjects' scores on the various factors and their scores on the original variables. In fact, estimating subjects' scores on a given factor requires conducting a multiple regression analysis of the relationship between that factor and the original variables, with the multiple R of these estimates being a decreasing function of the amount of unique variance in the system. (The indeterminacy in the relation between the factors and the original measure is, however, eliminated if Guttman's [1953, 1955, 1956] image analysis procedure—which involves assuming the communality of each variable to be equal to its squared multiple correlation with the remaining variables—is employed.)

A second loss in almost all methods of factor analysis is the uniqueness of the solution. A given factor structure simply represents a description of the original intercorrelations in terms of a particular frame of reference. That pattern of intercorrelations can be equally well described by any other frame of reference employing the same number of dimensions (factors) and the same set of communalities. Unless additional constraints besides ability to reproduce the intercorrelations are put on the analysis, any one of an infinite number of -interrelated factor structures will be an acceptable solution.

Principal factor analysis (PFA), which is essentially identical to PCA except for the exclusion of unique variance, assures the uniqueness of the factor structure it produces by requiring that each successive factor account for the maximum possible percentage of the common variance while still remaining uncorrelated with the preceding factors. The ***triangular-decomposition method*** requires that the first factor be a general factor on which each variable has a nonzero loading, that the second factor involve all but one of the original variables, that the third factor involve $p - 2$ of the original p variables, and so on. All other commonly used factor methods use some arbitrary mathematical constraint (usually one that simplifies subsequent mathematical steps) to obtain a preliminary factor structure; then they rotate the frame of reference until a factor structure that comes close to some a priori set of criteria for ***simple structure*** is found. Probably the most common of all approaches to factor analysis is rotation of a solution provided by PFA, despite the fact that such rotation destroys the variance-maximizing properties of PFA. This is probably primarily due to the mathematical simplicity of PFA compared with other methods of obtaining an initial solution.

The ***maximum-likelihood method*** generates factors in such a way as to maximize the probability that the observed pattern of correlations could have arisen through random sampling from a population in which the correlations are perfectly reproducible from the number of factors specified by the researcher. The ***minimum-residual (minres) method*** searches for factor loadings that produce the smallest possible sum of squared discrepancies between the observed and reproduced correlations.

The ***multiple-group method*** requires that the researcher specify in advance various subsets of the original variables, with the variables within each subset being treated as essentially identical. This is the only commonly used method that yields an initial solution involving correlated factors. The researcher's willingness to employ correlated factors in order to obtain a simpler factor structure is usually expressed in his or her choice of criteria for rotation of the frame of reference employed in the initial solution.

The ***centroid method*** is a technique for obtaining an approximation to PFA when all computations must be done by hand or on a desk calculator. It has very nearly disappeared from use as computers have become nearly universally available.

Finally, there is a growing and promising tendency to specify the characteristics of the factor structure as precisely as possible on purely a priori grounds, to use optimization techniques to "fill in" remaining details, and then to use the final value of the optimization criterion as an indication of the adequacy of the researcher's initial assumptions. Given the emphasis of the minres and maximum-likelihood methods on the goodness of fit between observed and reproduced correlations, the reader is correct in guessing that one of these two methods is the most likely choice for conducting such a ***confirmatory factor analysis.***

As can be seen from the preceding discussion, factor analysis is a very complex area of multivariate statistics that shows rather low internal organization and minimal relationship to other multivariate statistical techniques. Adequate discussion of this important area requires a textbook to itself, and Harman (1976), Mulaik (1972), and Comrey (1992), among other authors, have provided such texts. This book is confined to a fairly full discussion of PFA and a cursory survey of the other techniques.

Example 1.6. Measuring Perceived Deindividuation. Prentice-Dunn and Rogers (1982) wished to test (among other things) the hypothesis that an internal state of deindividuation (lessened self-awareness) has a causal effect on aggression. The internal state of each of the 48 male subjects was assessed via a retrospective questionnaire (administered after the subject had finished a teaching task that appeared to require him to deliver 20 electric shocks to a learner) containing 21 items derived from an existing scale and from the authors' own previous research. These 21 items were subjected to a principal factor analysis, followed by varimax rotation of the two factors that **a *scree analysis*** (looking for the point beyond which the percentage of variance accounted for by successive factors drops sharply to a plateau) had suggested should be retained. We are not told what method was used to estimate communalities. The two rotated factors were interpreted on the basis of items having "loadings of .4 or above on only one factor and on interpretability of the data" (p. 509), resulting in an 11-item factor (positive loadings of .83 to .50 for such items as *Feelings of togetherness*, *Felt active and energetic*, *Time seemed to go quickly*, and *Thinking was somewhat altered*), labeled as the *Altered Experience* factor, and a 4-item factor (positive loadings of .84 to .63 for *Aware of the way my mind was working*, *Alert to changes in my mood*, *Aware of myself*, *and Thought of myself a lot*), labeled as *Private Self-Awareness*. Though factor score coefficients were not reported (thus we cannot check the implication of the preceding interpretations that each of the included items would have a positive and nearly equal coefficient in the formula for estimating scores on its factor), the subjects' factor scores were estimated, and these two sets of scores were used in subsequent path (regression) analyses both as dependent variables influenced by experimental manipulations of accountability and attention and as (mediating) independent variables influencing aggression.

1.2.14 Structural Equation Modeling

Structural equation modeling (SEM) is an increasingly popular technique that combines factor analysis (the "measurement model"), which relates sets of directly observable variables to underlying conceptual (latent) variables, with path analysis of the (causal?) relationships among those conceptual variables. It was designed for and primarily intended for confirmatory use—that is, to test alternative models of the causal relationships among the conceptual variables.

1.3 LEARNING TO USE MULTIVARIATE STATISTICS

1.3.1 A Taxonomy of Linear Combinations

The astute reader will have noticed a theme pervading the sections of the preceding survey of statistical techniques that dealt with multivariate techniques: each technique's derivation of an emergent variable defined as a linear combination of the measures submitted to the technique for analysis. As pointed out in section 1.1.1, if the researcher is uninterested in or incapable of interpreting linear combinations of her or his measures, then Bonferroni-adjusted univariate tests provide a more powerful way of examining those measures one at a time (while controlling familywise alpha at an acceptable level) than would a fully multivariate analysis of the set of measures. There is ample anecdotal evidence available demonstrating that many researchers (and their statistical advisers) do indeed fall under the second proviso in that, while they may verbally describe their hypotheses and their results in terms that imply linear combinations of their independent and dependent variables, their explicit significance tests are restricted to single dependent variables and to pairwise differences among the levels of their independent variables. It is my hope that by devoting a subsection of this first chapter to the general issue of what linear combinations of variables represent and how they can be interpreted, I may increase the number of readers who can make full use of multivariate statistical techniques.

Readily interpretable linear combinations of variables fall into three broad classes: simple *averages* of subsets of the measures, *profiles* consisting of being high on some measures and low on others, and *contrasts* among the original measures.

1.3.1.1 Averages of subsets of the measures. Researchers seldom attempt to assess a personality trait or an ability by asking a single question or measuring performance on a single trial of a single task. Rather, we subject our research participants and our students to hours-long batteries of items, ask them to memorize dozens of pairings of stimuli, or record their electrophysiological responses to hundreds of trials of a signal-detection task, and then average subsets of these responses together to come up with a much smaller number (sometimes just one) of derived measures in justified confidence that whatever we are attempting to measure will as a result be more reliably (and probably more validly) assessed than if we had relied on a single response.

The general form of such averages is

$$Y = a_1X_1 + a_2X_2 + \cdots + a_pX_p,$$

where the X_is are the p original measures; the a_is are the weights we assign to these measures in coming up with a new, combined (composite) score; Y is the new, emergent variable (or scale score) on which we focus our subsequent interpretations and statistical analyses; and the a_is are all either zero or positive (or become so after "reflecting" scores on the X_is so that a high score on each X_i is reasonably interpreted as reflecting a high score on the conceptual variable thought to underly the various measures—it is generally good practice, e.g., to have some of the items on a putative Machismo scale worded so that a highly "macho" participant would strongly agree with it while other items would elicit strong disagreement from such a high-machismo person).

When first beginning the process of developing a new scale it is not unusual to find that some of the many questions we ask and some of the many kinds of tasks we ask the participants to perform simply don't "carry their weight" in measuring what we want to measure and/or don't "hang together with" the other measures (e.g., have a low part–whole correlation) and are thus dropped from the scale (i.e., are each assigned an a_i of zero). However, it is generally found that, for X_is that are measured in a common metric, differences among the nonzero weights assigned to the variables make very little difference—besides which, it is almost always much easier to come up with a substantive interpretation of an unweighted average of the X_is than it is to explain why one variable receives a weight that's 23.6193% larger or smaller in absolute value than some other variable.

But, given a set of positive a_is generated by a multivariate technique in full, multi-decimal glory, how are we to decide which to include in our interpretation of this new, emergent variable and which to "zero out"? Where to draw the line between high and negligible weights is somewhat subjective, akin to applying the familiar scree test to eigenvalues in a principal components or factor analysis (cf. chapters 6 and 7.) Starting with the highest value of any weight a_i, one keeps assigning +1s to variables having lower and lower a_is until there is a sharp drop in magnitude and/or the ratio between this coefficient and the largest one gets to be larger than, say, 2 or 3. All original variables whose a_is are smaller than this are ignored in one' interpretation (i.e., are assigned weights of zero in the simplified linear combination implied by the interpretation).

Of course, before engaging in this kind of comparison of magnitudes of a_i you should be sure that the X_is are expressed in a common metric. For instance, if the optimal combination of your measures is .813(weight in kilograms) + .795(height in meters), the relative magnitude of the weights for weight and height cannot meaningfully ascertained, because this combination is identical both to .000813(weight in grams) + .795(height in meters) and to .813(weight in kilograms) + .000795(height in mm).

Where the original units of measurement are not commensurable from measure to measure, a kind of common metric can be imposed by converting to the equivalent linear combination of *z*-scores on the X_is, thus leading to an interpretation of your emergent variable as that which leads to having a high score, on average, on those X_is that are assigned nonzero weights, relative to one's peers (i.e., to the mean of the population distribution for each included original variable). Indeed, many variables can be expected to have their effects via comparison with others in your group. For example, a height of 5 feet 6 inches would make you a giant if you're a member of a pigmy tribe but a shrimp among the Watusi and would thus be likely to have opposite effects on self-esteem in the two populations. But where the original units of measurement are commensurable, an average of (or differences among) raw scores is apt to be more straightforwardly interpretable than an average of (or differences among) z-scores. To borrow an example from chapter 3, it's much more straightforward to say that the difference between distance from water and distance from roads (in meters) is greater for one species of hawk than it is for another species than it is to say that the R ed-Shouldered hawks display a greater difference between their percentile within the distribution of distances from water and their percentile within the distribution of distances from roads than do Red-Tailed hawks.

If we find that, even after reflecting original variables, our composite variable performs best (correlates more highly with a criterion measure, yields a higher overall *F* for differences among our experimental conditions, etc.) when some of the a_is are positive while other X_is are assigned negative weights, then this linear combination is best interpreted as either a profile or a contrast.

1.3.1.2. Profiles. Another very natural way of interpreting that aspect of a set of variables that relates most strongly to another variable or set of variables is in terms of that pattern of high scores on p_1 of the variables, accompanied by low scores on p_2 of the variables, that a person with a high score on the underlying conceptual variable would be expected to display, while a person with a very low score on the conceptual variable would be expected to display the mirror-image pattern. Such a pattern of highs and lows can be referred to as a ***profile*** of scores, and the rank ordering of scores on the underlying conceptual variable can be matched perfectly via a linear combination of the observed variables in which the p_1 a_is associated with the first set of X_is mentioned above are all +1, while the p_2 "negative indicators" are all assigned a_is of −1—and of course the variables that are in neither set receive weights of zero.

A straightforward way of deciding which X_is to assign +1s to, which to assign −1s to, and which to "zero out" when attempting to provide a profile interpretation of the linear combination of the measures yielded by some multivariate procedure is to apply the "scree test" described in section 1.3.1.1 separately to those variables whose multi-decimal combining weights are positive and to the absolute values of the negative a_is. Of course this procedure should only be applied to variables that have (or have been transformed – perhaps to *z*-scores—to have) a common metric, and only after having decided whether the raw scores, *z*-scores, or some other transformation provide the most meaningful such common metric.

1.3.1.3 Contrasts—among original variables as among means. The interpretation of an emergent variable as a profile (cf. the preceding section)—that is, as a

(+1,-1,0) combination of the original variables—especially makes sense where the variables are essentially bipolar in nature (e.g., happy/sad or inactive/active), so it is simply a matter of which end of the scale is being referenced. When the variables are essentially unipolar it is often better to take the additional step of dividing each +1 weight by the number of such weights and each –1 weight by their number—that is, to interpret the combination as the difference between the simple averages of two subsets of the dependent variables. Those readers familiar with analysis of variance (and who isn't, after having read sections 1.2.2 and 1.2.5?) will recognize this as a ***contrast*** among the original variables—specifically, a subset contrast [also referred to as a (k_1, k_2) contrast], directly analogous to contrasts among means in Anova. The analogy is especially close to within-subjects Anova, where a test of the significance of a contrast among the means for different levels of our within-subjects factor can be carried out by computing a score for each subject on the specified contrast among that subject's scores on the original measures and then conducting a single-sample t or F test of the H_0 that individual scores on the contrast have a population mean of zero.

There are of course other kinds of contrasts among means or variables than subset contrasts, with the most common alternative interpretation of a set of contrast coefficients being as a proposed pattern of the population means. I and many other statistics instructors have found that far and away the most difficult concept for students (and, apparently, statisticians, given the nearly exclusive focus on pairwise comparisons among developers of multiple-comparison procedures) to grasp when studying Anova is that of the relationship between contrast coefficients and patterns of differences among means. Following is a (perhaps overly verbose) attempt on my part to clarify this relationship, put together for use by students in my Anova course.

Contrast coefficients provide a convenient common language with which to describe the comparisons among means you wish to examine. Without such a common language, we would have to maintain a huge encyclopedia that listed one set of formulae for testing the difference between a single pair of means; another set of formulae for testing the difference between the average of three of the means versus the average of two of the means; another set for testing the difference between the average of two of the means versus a single mean; another set for testing whether the relationship between the population mean on your dependent variable and the value of your independent variable is linear, given equal spacing of those values in your study; another set for testing for a linear relationship to unequally spaced values of your independent variable where the distance (on the independent-variable axis) between levels 2 and 3 is twice that between levels 1 and 2; another set for testing linearity when the spacing of the levels of your independent variable is proportional to 1, 2, 4, 7; and so on, ad infinitum. With the help of contrast coefficients, however, we can learn a single set of formulae for testing any of these, provided only that we can "crowbar" our null and/or alternative hypotheses into the generic form, $\sum c_j \mu_j = \text{or} < \text{or} > 0$.

Well, then, how do we come up with ideas (hypotheses) about differences among the k population means, and how do we translate these hypotheses into the generic $\sum c_j \mu_j$ form?

There are three sources of ideas for hypotheses to test:

1. Logical, theory-driven analyses of how subjects' responses should be affected by the differences in what we did to the participants in the *k* different groups or by their membership in the categories represented by those groups.
2. Results of prior studies. We generally expect that the results of our study will be similar to the results of previous studies (including our own pilot studies) that have used the same or highly similar experimental manipulations, naturally occurring categories, and/or dependent variables.
3. Post hoc examination of the data—that is, looking at the sample means and trying to summarize how they appear to differ. (Computing $\bar{Y}_j - \bar{\bar{Y}}$ for each group and then picking simpler, rounded-off numbers that closely approximate those deviation scores is a good way to pick contrast coefficients that will yield a large F_{contr} that accounts for a high proportion of the variation among the means—that is, a high proportion of SS_b . You can then try to interpret the simplified coefficients as a comparison of two subsets of the means or as a simple pattern of population means, reversing the process, described in the next section, of converting a prediction into contrast coefficients.)

Once we have a prediction (or, in the case of post hoc examination of the sample means, postdiction) about a particular pattern of differences among the means, we can translate that predicted pattern into a set of contrast coefficients to use in testing our hypothesis in one of two ways:

1. If our hypothesis is that one subset of the groups has a higher mean on the dependent variable (on average) than does another subset of the groups, we can test this hypothesis via a subset contrast, i.e. by choosing contrasts coefficients of zero for any group that isn't mentioned in our hypothesis, $1/k_a$ for each of the k_a groups in the subset we think will have the higher average mean, and $-1/k_b$ for each of the k_b group in the subset we think will have the lower average mean. (Once we've computed these contrast coefficients, we could convert to a set of integer-valued contrast coefficients that are easier to work with numerically by multiplying all of the original contrast coefficients by $k_a\, k_b$.) For instance,

If your hypothesis is that groups 1, 3, and 5 have higher population means (on average) than do groups 2 and 4, you would use contrast coefficients of

(1/3, -1/2, 1/3, -1/2, 1/3) or (2, -3, 2, -3, 2).

If you wanted to test the difference between groups 2 and 4 versus groups 1 and 5, you would use contrast coefficients of

(1/2, -1/2, 0, -1/2, 1/2) or (1, -1, 0, -1, 1).

If you wanted to test the hypothesis that group 3 has a lower population mean than group 5, you would use contrast coefficients of

(0, 0, -1, 0, 1) or any multiple thereof.

If you wanted to test the hypothesis that group 4 has a higher population mean than do the other 4 groups, averaged together, you would use

(-1/4, -1/4,-1/4, 1, -1/4) or (-1, -1, -1, 4, -1).

And so forth, ad infinitum.

2. If our hypothesis is that the population means, when plotted along the vertical axis of a graph whose horizontal axis is the dimension along which the values of our independent variable are situated, form a pattern that is described by a specific set of numbers, then we can test that hypothesized pattern by using as our contrast coefficients (a) zero for any groups not mentioned in our hypothesis, and, for groups that are mentioned, (b) the numbers describing the hypothesized pattern, minus the mean of those numbers (so that the resulting deviation scores sum to zero, as good little contrast coefficients ought). (Remember that the sum of the deviations of any set of numbers about their mean equals zero.) Thus, for instance:

If we hypothesize that the population means for our 5 groups will be proportional to 3, 7, 15, 18, and 22, then we should use contrast coefficients of (-10, -6, 2, 5, 9) —which we got by subtracting 13 (the mean of the 5 original numbers) from each of the original numbers.

If our predicted pattern is 2, 4, 8, 16, 32, then we should use contrast coefficients of (-10.4, -8.4, -4.4, 3.6, 19.6) —that is, each of the original numbers, minus 12.4 (the mean of the 5 original numbers).

If our hypothesis is that groups 3 through 5 will have population means proportional to 7,9,14, then we should use contrast coefficients of (0, 0, -3, -1, 4)—that is, zero for the two groups our hypothesis doesn't mention and the original numbers minus 10 (their mean) for the mentioned groups.

If we are using an independent variable set at 2, 4, 8, and 16 (e.g., number of marijuana cigarettes smoked before driving through an obstacle course), and some theory or half-crazed major adviser predicts that population mean scores on our dependent variable (e.g., number of orange cones knocked down while driving the course) will be proportional to the square root of the value of the independent variable for that group, then we can test that hypothesis by (a) taking out our pocket calculator and computing the square roots of 2, 4, 8, and 16 to derive the prediction that the group means should be proportional to 1.414, 2, 2.828, and 4; (b) converting these numbers to contrast coefficients by subtracting 2.561 (the mean of the 4 numbers) from each, yielding contrast coefficients of (-1.147, -.561, .267, 1.439); and (c) plugging into the generic formula for a t_{contr} or an F_{contr}.

The somewhat abstract discussion of these last three subsections will of course be “fleshed out” by specific examples as you work your way through the chapters of this book. At a bare minimum, however, I hope that this section has given you an appreciation of the enormous number of choices of ways of interpreting any given linear combination of variables that emerges from one of your multivariate analyses, so that you won’t be tempted to agree with the most common complaint about canonical analysis (chapter 5)—namely, that it’s impossible to interpret canonical variates, nor to adopt the most common approach to T^2 (chapter 3) and to multivariate analysis of variance (chapter 4)—namely, carrying out the overall, multivariate test but then ignoring the linear combination of the dependent variables that "got you there" and instead carrying out only fully a priori, univariate t or F tests on single dependent variables.

1.3.2 Why the Rest of the Book?

It must be conceded that a full understanding of the heuristic descriptions of the preceding section would put the student at about the 90th percentile of, say, doctoral candidates in psychology in terms of ability to interpret multivariate statistics and in terms of lowered susceptibility to misleading, grandiose claims about multivariate techniques. This heuristic level of understanding would, further, provide a constant frame of reference to guide the student in actually carrying out multivariate analyses on his or her own data (which usually means submitting these data to a computer program)-suggesting, for instance, several checks on the results—and would establish a firm base from which to speculate about general properties that a particular technique might be expected to have and the con-sequences of changing some aspect of that technique. This first chapter is an important part of the book, and the student is urged to return to it often for review of the general goals and heuristic properties of multivariate statistical procedures. The points covered in this chapter are all reiterated in later portions of the book, but they are nowhere else found in such compact form, so free of distracting formulae, derivations, and so on, as in this first chapter.

Why, then, the rest of the book? In part because of the "catch" provided in the second sentence of this section by the phrase, *full understanding*. As you no doubt noticed in your initial introduction to statistics, whether it came through a formal course or through self-instruction with the aid of a good text, the warm, self-satisfied feeling of understanding that sometimes follows the reading of the verbal explanation of some technique—perhaps supplemented by a brief scanning of the associated formulae — evaporates quickly (often producing a decided chill in the process) when faced with a batch of actual data to which the technique must be applied. True understanding of any statistical technique resides at least as much in the fingertips (be they caressing a pencil or poised over a desk calculator or a PC keyboard) as in the cortex. The author has yet to meet anyone who could develop an understanding of any area of statistics without performing many analyses of actual sets of data—preferably data that were truly important to him or her. In the area of multivariate statistics, however, computational procedures are so lengthy as to make hand computations impractical for any but very small problems. Handling problems of the sort you will probably encounter with real data demands the use of computer programs to handle the actual computations. One purpose of the rest of the book is therefore ***to introduce you to the use of "canned" computer programs and to discuss some of the more readily available programs for each technique***. One approach to gaining an understanding of multivariate statistics would be a "black box" approach in which the heuristic descriptions of the preceding section are followed by a set of descriptions of computer programs, with the student being urged to run as many batches of data through these programs as possible, with never a thought for what is going on between the input and output stages of the computer run. Such a procedure has two important shortcomings:

1. It puts the researcher completely at the mercy of the computer programmer(s) who wrote the program or adapted it to the local computer system.
2. It renders impossible an analytic approach to exploring the properties of a particular technique.

The first of these shortcomings subsumes at least four limitations. First, no computer program is ever completely "debugged," and a knowledge of the steps that would have been followed had the researcher conducted the analysis "by hand" can be of great value in detecting problems in the computer's analysis of your data—which will of course contain that "one in a million" combination of conditions that makes this program "blow up" despite a long past history of trouble-free performance. It is also of use in generating small test problems that can be computed both by hand and by the computer program as a check on the latter. Second, the person who programmed the particular analysis may not have provided explicitly for all of the subsidiary analyses in which you are interested, but a few supplementary hand calculations (possible only if you are familiar with the computational formulae) can often be performed on the output that is provided by the program to give you the desired additional information. This is related to a third limitation—it is difficult to "pick apart" a canned program and examine the relationships among various intermediate steps of the analysis, because such intermediate steps are generally not displayed. Finally, input formats, choices between alternative computational procedures, the terminology used in expressing results, and so on, differ from one package of canned programs to another, and even from one version to the next release of the same program in the same package. Knowledge that is tied too closely, to a specific program is knowledge that can be far too easily outdated. A corollary of this last limitation is that the "nuts and bolts" of getting, say, a Manova program to analyze data on your local computer cannot be set out in this text; it must be provided by your instructor and/or your local computing center staff. We can, however, use programs current at the time of the writing of this book to illustrate the kinds of facilities available and the kinds of interpretative problems involved in deciphering computer output.

With respect to the second shortcoming, computer programs can only tell us what happens when a particular set of data, with all of its unique features, is subjected to a particular analysis. This is a very important function, and when the analysis is applied to a wide assortment of data, it can begin to suggest some general properties of that technique. However, we can never be certain that the property just deduced will hold for the next set of data except by examination of the formulae—in algebraic form—used by the computer. (Actually, the heuristic formulae or the stated goals of the technique or the mathematical derivation that gets us from the latter to the former may be of greater use than the computational formulae in deriving general properties. The main point is that black-box use of canned programs is never a sufficient base for such generalizations.) A second "service" provided by the present text is thus ***the display of the formulae that could be used by the researcher to conduct an analysis by hand.***

This is generally feasible only if no more than two or three variables are involved or if the variables are uncorrelated. Beyond this low level of complexity, use of ***scalar*** formulae (the form with which the reader is probably most familiar, in which each datum is represented by a separate algebraic symbol) becomes impractical, and use instead of ***matrix algebra*** (in which each *set* of variables is represented by a separate symbol) becomes essential. Thus, for instance, the variance of the linear combination, $a_1X_1 + a_2X_2$, of two variables is written as $a_1 s_1^2 + a_2 s_2^2 + 2a_1a_2s_{12}$ in scalar form. However, the scalar expression for the variance of a linear combination of 20 variables would involve 210 terms, whereas the variance of any linear combination of variables—whether there are 2 or 20 or 397 of them—can be written in matrix algebraic form as **a'Sa**. The matrix algebraic form is usually easier to remember and intuitively more comprehensible than the basic algebraic form. A subsidiary task of the rest of the text will be to ***develop a familiarity with the conventions governing the use of these highly compact matrix formulae and equations.*** Finally, both the basic and matrix algebra expressions will usually be presented in both heuristic and computational form.

The matrix algebraic expressions for multivariate statistics are much easier to remember, much easier to manipulate, and intuitively more comprehensible than the scalar formulae. The facility with which you can explore the properties of multivariate techniques, handle analyses that aren't provided for by the computer programs available to you, derive interesting new versions of formulae, and squeeze more information out of your computer output than the programmer had explicitly provide for will thus be severely limited if you do not become familiar with at least elementary matrix algebra. However, these are activities that are more likely to appeal to "quantitative" majors than to the rather larger subpopulation of students and researchers who are primarily (if not only) interested in how multivariate statistics can help them interpret (and perhaps sanctify for publication) their data. Matrix notation can be quite confusing and distracting for the uninitiated, and getting initiated into the joys of matrix algebra can easily take up 2–3 weeks of course time that could otherwise have been spent on learning computer syntax for an additional technique or two. This third edition of the *Primer* therefore deemphasizes matrix algebra, in keeping with the growing number of multivariate courses (such as the one I teach) that consciously sacrifice the full facility with multivariate statistics that acquaintance with matrix algebra makes possible in favor of fitting more techniques (e.g., path analysis, confirmatory factor analysis, and structural equations) into a single semester.

If you should, after reading this edition of the Primer, decide that you wish to develop a deeper understanding of multivariate statistics, you should read through Digression 2 and attend carefully to the matrix equations presented throughout this Primer. In the meantime you can go a long way towards avoiding the "black box syndrome" so often associated with reliance on canned computer programs by:

1. Using the two- and three-variable scalar formulae provided in this book to work out a few simple analyses against which to check the computer package you are using for possible bugs, and to be sure that you're reading its output correctly.
2. Always supplementing the initial run of your data through a multivariate program with additional univariate analyses (whether carried out by hand or via a second or third

computer run) on simplified, substantively interpretable versions of the optimal (but many-decimaled) linear combinations of your original variables that that first computer run has identified for you. (As I'll emphasize throughout this text, if you didn't ask the program to print out the optimal combining weights, you should be doing Bonferroni-adjusted univariate analyses, rather than a truly multivariate analysis.) Thus, for instance, an initial one-way Manova (cf. section 1.2.4) on the differences among four groups in their profile of means on 12 dependent variables should be followed up by a run in which scores on at least one simplified version of the discriminant function are computed, along with the four group means on this new variable, the univariate F for differences among these four means (to be compared, of course, to the maximized F on the multi-decimaled discriminant function itself), and Fs for one or more specific contrasts among the four means.

For many readers, the three kinds of information mentioned so far—the goals of each technique, the formulae used in accomplishing these goals, and the computer programs that actually carry out the required computations—will provide a sufficient working understanding of multivariate statistics. However, an intrepid few—those who had had prior exposure to math or to statistics or who dislike the hand-waving, black magic approach to teaching—will want to know something about the method (primarily mathematical) used to get from the goals of a technique to the specific formulae used. The process of filling in this gap between goals and formulae, of justifying the latter step-by-step in terms of the former, is known as a proof. There are several such proofs throughout this book. It is important to keep in mind that the proofs are *not* essential to (though they may be of help in deepening) a working understanding of multivariate statistics. The reader who is willing to take the author's statements on faith should have no compunctions about skipping these proofs entirely (especially on a first reading).

To make this easier, the proofs are, as much as possible, put in a separate Derivations appendix toward the end of the book. Further, paragraphs or entire sections that may be skipped without loss of continuity are indicated by the symbol # in front of the section number or before the first word of the paragraph. On the other hand, an attempt has been made to make even the proofs understandable to the mathematically naive reader. Where a high-powered mathematical tool such as differential calculus must be used, a "digression" is provided at the back of the book in which an intuitive explanation and a discussion of the reasonableness of that tool are presented. Considerable emphasis is also put on analyzing the properties of each statistical technique through examination of the "behavior" of the computational formulae as various aspects of the data change.

Each chapter introduces a new technique or set of multivariate techniques. The general plan of each chapter is roughly as follows: First, the heuristic description of the technique is reiterated, with a set of deliberately simplified data serving to illustrate the kind of questions for which the technique is useful. Next the formulae used in conducting an analysis by hand are displayed and are applied to the sample data, the relationship between formulae and heuristic properties is explored, and some questions (both obvious and subtle) about the technique are posed and answered. Then some available computer programs are described. The student is then asked to work a demonstration problem, that is, to go through the hand computations required for an analysis of some deliberately simplified data, performing a large number of different manipulations on various aspects of the data to demonstrate various properties of the technique. (Yes, Virginia, a *t* test computed on the discriminant function really does give the same value as all those matrix manipulations used to find T^2.) These hand calculations are then compared with the results of feeding the data to a canned computer program. It has been the author's experience that at least 75% of the real learning accomplished by students comes in the course of working these demonstration problems. Do not skip them.

There has been enough talk about how the text is going to go about teaching you multivariate statistics. To borrow one of Roger Brown's (1965) favorite Shakespearean quotes:

> Shall we clap into? roundly, without hawking
> or spitting or saying we are hoarse . . . ?
>
> *As You Like It* V. iii

QUIZ 1: SEE HOW MUCH YOU KNOW AFTER READING JUST ONE CHAPTER!

1. A researcher conducts a number of MRAs (multiple regression analyses) to predict each of 10 measures of academic achievement from various sociometric measures. She finds that adding birth month to age and high school GPA leads in every case to an increase in the R^2 of the predictors with the outcome measure. How convincing a demonstration of the value of astrology is this?

2. Suppose the researcher now decides to do an analysis of canonical correlation (a Canona) on the 3 background variables versus the 10 outcome measures. Having the results of her MRAs in hand, can you set a lower bound on the possible value of R^2 that will result from this analysis?

3. Upon doing the Canona, she discovers that the computer has reported not just one but *three* pairs of canonical variates. Could she improve on R^2 by taking as her predictor variable and her outcome variable some linear combination of the three different canonical variates?

4. How is a Manova affected by adding two more dependent measures? (Consider both the effect on the resulting maximized *F* ratio and the effect on the critical value for your test statistic.)

5. Consider some set of data (e.g., your most recent journal article) that involves more than one dependent measure. How might (or did) multivariate statistical techniques

contribute to the analysis? Which techniques? Would you mind sharing your data with the class for practice in conducting analyses?

SAMPLE ANSWERS TO QUIZ I

1. Not very impressive. Adding any predictor—even a column of N random numbers—cannot lower R^2 and, unless the new predictor is very carefully constructed, it will in fact increase R^2. To show that this is so, consider how you might go about choosing the regression weights for the new equation. Clearly, you can always keep the old weights for the predictors that were already in the equation and assign a weight of zero to the new predictors, thereby guaranteeing that the correlation between that linear combination and Y will be identical to the old value of R^2, and you almost inevitably will be able to capitalize on chance to find a set of weights yielding a higher correlation with Y. Before we get excited about an increase in multiple R^2 produced by adding a predictor, we need to ask (a) how large is the increase and (b) is it statistically significant?

2. Yes. The squared canonical correlation cannot be lower in value than the largest of the previously obtained squared multiple Rs. The optimization procedure, which looks for the best pair of linear combinations of the two sets of measures, could always use the regression weights from the MRA that yielded the highest value of R^2 as its canonical weights for the three background variables and zero for all measures except for the one that served as Y in the MRA yielding this highest value of R^2 as its canonical weights for the 10 outcome measures. The Canona optimization procedure thus has this pair of sets of weights to fall back on; therefore, it cannot do worse than this in maximizing the squared correlation between its two choices of linear combinations.

3. This is the "trickiest" of the questions, in that your intuitions urge you to answer "Yes." However, any linear combination of the three canonical variates is a linear combination of three linear combinations of the original variables and is therefore itself a linear combination of the three original variables. For instance, if $Q_1 = X_1 + X_2 + X_3$; $Q_2 = 2X_1 - X_2 + 3X_3$; and $Q_3 = 2X_2 - X_3$; then $2Q_1 - Q_2 - Q_3 = 0 \cdot X_1 + 1 \cdot X_2 + 0 \cdot X_3 = X_2$. But, by definition, the highest possible squared correlation between any linear combination of the background variables and any linear combination of the achievement measures is R_c^2; therefore, even though the particular linear combination we're looking at was derived by combining canonical variates, it still cannot yield a squared correlation with any linear combination of the measures in the other set higher than the (first) squared canonical correlation. This does not, however, rule out the possibility of finding some *non*linear combining rule that would yield a higher squared correlation. Note, by the way, that it is not enough to answer "No" to this question on the grounds that the canonical variates for either set are uncorrelated with each other, because we know that three measures, each of which correlates highly with some outcome measure, will yield a higher R^2 if they are

uncorrelated with each other than if they are highly intercorrelated. Admittedly, a proof is (as my freshman calculus professor, Victor Bohnenblust, used to remind us) whatever will convince your audience, but in this case anyone who has read Chapter 2 will be unlikely to be convinced, because applying the same argument to MRA leads to an incorrect conclusion.

3. By the same reasoning we applied in question 1, the maximized F ratio must go up (because the maximization routine has the fallback option of assigning both new dependent variables weights of zero while retaining the discriminant function weights from the previous Manova). However, because we know that this is true even if we add two columns of random numbers to the data, it must also be true that the expected value of the maximized F ratio (and its percentiles) under the null hypothesis must be greater for $p + 2$ than for p dependent variables, and its critical value must then also be greater. (We could make this argument a bit tighter by considering how the sample space for the test statistic derived from $p + 2$ measures could be constructed from the corresponding points in the p-measure sample space.)[8]

5. It is my fond hope that reading Chapter 1 will provide more than enough understanding of multivariate techniques to be able to tell when they could or should be used. Although the use of multivariate statistics is becoming increasingly common (especially in the areas of social, personality, and developmental psychology), these techniques are still greatly underused—particularly by authors who publish in the more hard-nosed experimental journals.

The suggestion of using data of real interest to students in the class is one that must be evaluated by the instructor and the class. On the pro side is the greater realism and the potentially greater motivating value of such data. On the con side is the fact that such examples are likely to be "messier," involve more data, and involve more work for the instructor than would contrived or thoroughly familiar data sets. Furthermore, what one student finds interesting may not always inspire the enthusiasm of his or her classmates.

---

[8] The answers to questions 1–4 are designed to show you how many of the properties of multivariate techniques are implicit in the definitions of their goals (what each seeks to maximize), without having to get into the mathematics of how the maximization is actually accomplished. Nor are the applications trivial. Question 1 was inspired by a journal article that made almost exactly that argument, and knowing (Question 4) that critical values for multivariate test statistics must increase as the number of variables involved goes up can be very useful as a check on the critical values one pulls out of a complicated set of tables such as the gcr tables.

2
Multiple Regression: Predicting One Variable From Many

As pointed out in Chapter 1, ***multiple regression*** is a technique used to predict scores on a single outcome variable Y on the basis of scores on several predictor variables, the X_is. To keep things as concrete as possible for a while (there will be plenty of time for abstract math later), let us consider the hypothetical data in Data Set 1.

Data Set 1

Subject	X_1	X_2	X_3	Y		Intercorrelation matrix X_1	X_2	X_3	Y	Variance
1	102	17	4	96	X_1	1.0	.134	.748	.546	17.5
2	104	19	5	87	X_2		1.0	.650	−.437	5.0
3	101	21	7	62	X_3			1.0	−.065	5.0
4	93	18	1	68	Y				1.0	190.5
5	100	15	3	77						
$\bar{X}$	100	18	4	78						

Say that X_1, X_2, and X_3 represent IQ as measured by the Stanford–Binet, age in years, and number of speeding citations issued the subject in the past year, respectively. From these pieces of information, we wish to predict Y—proficiency rating as an assembly-line worker in an auto factory. In fact, these are purely hypothetical data. Hopefully you will be able to think of numerous other examples of situations in which we might be interested in predicting scores on one variable from scores on 3 (or 4 or 2 or 25) other variables. The basic principles used in making such predictions are not at all dependent on the names given the variables, so we might as well call them $X_1, X_2, \ldots, X_n$, and Y. The Xs may be called independent variables, predictor variables, or just predictors, and Y may be referred to as the *dependent variable,* the *predicted variable,* the *outcome measure*, or the *criterion.* Because in many applications of multiple regression all measures (Xs and Y) are obtained at the same time, thus blurring the usual independent-dependent distinction, *predictor–predicted* would seem to be the most generally appropriate terminology. In Data Set 1 we have three predictor variables and one predicted variable.

At this point you might question why anyone would be interested in predicting Y at all. Surely it is easier (and more accurate) to look up a person's score on Y than it is to look up his or her score on three other measures (X_1, X_2, and X_3) and then plug these numbers into some sort of formula to generate a predicted score on Y. There are at least three answers to this question. First, we may be more interested in the prediction formula itself than in the predictions it generates. The sine qua non of scientific research has always been the successive refinement of mathematical formulae relating one variable to one or more other variables, for example, $P = VT/C$, $E = mc^2$, Stevens's power law versus Thurstonian scaling procedures, and so on. Second, and probably the most common reason for performing multiple regression, is that we may wish to develop an equation that can be used to predict values on Y for subjects for whom we do not already have this information. Thus, for instance, we might wish to use the IQ, age, and number of speeding citations of a prospective employee to predict his probable performance on the job as an aid in deciding whether to hire him. It seems reasonable to select as our prediction equation for this purpose that formula that does the best job of predicting the performance of our present and past employees from these same measures. The classic way of approaching this problem is to seek from our available data the best possible estimates of the parameters (free constants not specified on a priori grounds) of the population prediction equation. Not unexpectedly, this "best" approximation to the population prediction equation is precisely the same as the equation that does the best job of predicting Y scores in a random sample from the population to which we wish to generalize. Finally, we may wish to obtain a measure of the overall degree of relationship between Y, on the one hand, and the Xs, on the other. An obviously relevant piece of information on which to base such a measure is just how good (or poor) a job we can do of predicting Y from the Xs. Indeed, one of the outputs from a multiple regression analysis (MRA) is a measure called the coefficient of multiple correlation, which is simply the correlation between Y and our predicted scores on Y and which has properties and interpretations that very closely patallel those of Pearson's correlation coefficient for the bivariate case.

(If the mention of Pearson r does not bring an instant feeling of familiarity, you should reread section 1.2.7. If the feeling of familiarity is still not forthcoming, a return to your introductory statistics text is recommended. This prescription is basically the same for any univariate statistical technique mentioned from this point on with which you are not thoroughly familiar. Try rereading the relevant section of chap. 1 and, if further jogging of your memory is necessary, reread the relevant chapter(s) of the book you used in your introductory statistics course. If this is no longer available to you or has acquired too many traumatic associations, other good texts such as those by M. B. Harris [1998], Kirk [1998], Glass and Stanley [1970], or Hays and Winkler [1971] can be consulted. Consulting an unfamiliar text, however, requires additional time for adjusting to that author's notational system.)

2.1 THE MODEL

We now return to the problem of developing a prediction equation for Data Set 1. What kind of prediction technique would you recommend? Presumably you are thoroughly familiar with the pros and cons of clinical judgment versus statistical prediction and are at least willing to concede the usefulness of confiningg our attention to numerical formulae that can be readily used by anyone and that do not require Gestalt judgments or clinical experience. Given these restrictions, you will probably come up with one of the following suggestions, or something closely related:

1. Take the mean of the subject's scores on X_1, X_2, and X_3 as his or her predicted score on Y. (Why is this an especially poor choice in the present situation?)

2. Take the mean of the subject's z scores on the three predictor variables as his or her predicted z score on Y.

3. Use only that one of X_1, X_2, and X_3 that has the highest correlation with Y, predicting the same z score on Y as on this most highly correlated predictor variable.

4. Use the averaging procedure suggested in (2), but base the prediction on a *weighted* average in which each z score receives a weight proportionate to the absolute value of the correlation of that variable with the predicted variable. (A weighted average is a weighted sum divided by the sum of the weights, or equivalently, a weighted sum in which the weights sum to 1.0. Thus, for instance, the suggested weighting procedure would involve taking

$$\frac{.546z_1 + .437z_2 + .065z_3}{1.048} = .520z_1 + .418z_2 + .062z_3$$

as our predicted score.)

You may wish to try out each of these suggestions, plus your own, on Data Set 1 and compare the resulting fit with that provided by the multiple regression equation that we will eventually develop. The important point for the present purposes is that you are unlikely to have suggested a formula involving cubes, tangents, hyperbolic sines, and so on. Instead, in the absence of compelling reasons for assuming otherwise, there is a strong tendency to suggest some sort of *linear* combination of the Xs, which is equivalent to assuming that the true relationship between Y and the Xs is described by the equation

$$Y_i = \beta_0 + \beta_1 X_{i,1} + \beta_2 X_{i,2} + \cdots + \beta_m X_{i,m} + \varepsilon_i \tag{2.1}$$

or

$$Y_i = \hat{Y}_i + \varepsilon_i, \tag{2.2}$$

where Y_i is subject i's score on the outcome variable; $X_{i,j}$ is subject i's score on the predictor variable j; β_j is the regression coefficient, that is, the weight applied to predictor variable j in the population regression equation for predicting Y from the X_js; and ε_i is subject i's ***residual score***, that is, the discrepancy between his or her actual score on Y and the score predicted for him or her, $\hat{Y}$, on the basis of the presumed linear relationship between Y and the Xs. Equation (2.1) is referred to as *linear* because none of the Xs are squared, cubed, or in general raised to any power other than unity or zero. (β_0 can be thought of as $\beta_0 X_0^0$.) Thus a plot of Y as a function of any one of the Xjs ($j = 1, 2, \ldots m$) would produce a straight line, as would a plot of Y as a function of the composite variable $\hat{Y}$. This is not a serious limitation on the kinds of relationships that can be explored between Y and the X_js, however, because a wide variety of other relationships can be reduced to an equation of the same form as (2.1) by including as predictor variables various transformations of the original X_js. Thus, for instance, if an experimenter feels that the true relationship between Y, X_1, and X_2 is of the form $Y = cX_1^u X_2^v$ he or she can, by taking logarithms of both sides of the equation, transform the relationship to

$$\log Y = \log(c) + u \log X_1 + v \log X_2,$$

that is, to

$$Y^* = \beta_0 + \beta_1 X_1^* + \beta_2 X_2^*,$$

where Y^* and X_i^* are transformed values of Y and X_i, namely,

$$Y^* = \log Y, \qquad \beta_0 = \log(c), \qquad \beta_1 = u, \qquad \beta_2 = v,$$
$$X_1^* = \log X_1, \quad \text{and } X_2^* = \log X_2.$$

In other words, by taking as our basic variables the logarithms of the scores on the original, untransformed variables, we have produced a *linear* prediction equation to which the techniques discussed in this chapter are fully applicable. Note that there is a single set of βs that is used for all individuals (more generally, sampling units) in the population, whereas ε_i is generally a different number for each individual. Note, too, that inclusion of the ε_i term makes our assumption that Equation (2.1) accurately describes the population a tautology, because ε_i will be chosen for each subject so as to force the equality in that equation. However, it will naturally be hoped that the ε_i terms will be independent of each other, will have a mean of zero, and will have a normal distribution—in fact, we will make these assumptions in all significance tests arising in multiple regression analysis. Any evidence of nonrandomness in the ε_is casts serious doubt on the adequacy of a linear regression equation based on these m predictor variables as a predictor of Y. It is therefore always wise to examine subjects' residual scores for such things as a tendency to be positive for moderate values of $\hat{Y}$ and negative for extreme values of $\hat{Y}$ (indicating curvilinearity in the relationship between Y and the X_js, something that is possibly correctable by including squared terms) or a systematic trend related to the number of observations sampled prior to this one (indicating lack of independence in the sampling process). Finally, note that the βs are population parameters that must be estimated from the data at hand, unless these data include the entire population to which we wish to generalize.

We could discuss Equation (2.1) in terms of m-dimensional geometry, with the hypothesis that m independent variables are sufficient to produce perfect prediction of scores on Y (that is, that all the ε_is are truly zero) being equivalent to the hypothesis that in an $(m + 1)$-dimensional plot of Y as a function of the m X_js, all the data points would lie on an m-dimensional "hyperplane." However, most nonmathematicians find it difficult to visualize plots having more than three dimensions. The author therefore prefers to think of Equation (2.1) as expressing simple, two-dimensional relationship between the single variable Y and the single ***composite variable*** $\hat{Y}$ (which does in fact yield a single number for each subject). Thus, multiple regression accomplishes what sounds like a complex, Gestalt-like exploration of the relationship between Y and the m predictor variables by the simple expedient of reducing the problem right back to a univariate one by constructing a single composite variable that is simply a weighted sum of the original m X_js.

This technique for reducing a multivariate problem to essentially a univariate one is used over and over again as we learn about other multivariate statistical techniques. The trick, of course, comes in picking the weights used in obtaining the weighted sum of the original variables in an optimal way. We now turn to this problem.

2.2 CHOOSING WEIGHTS

Even if we had the entire population of interest available to us, it would still be necessary to decide what values of $\beta_0, \beta_1, \beta_2, \ldots$ provide the best fit between the Xs and Y. As it turns out, two different and highly desirable criteria lead to precisely the same numerical values for $b_1, b_2, \ldots$--our *estimates* of the βs. These criteria are:

1. That the sum of the squared errors of prediction, $E = \sum e_i^2 = \sum (Y_i - \hat{Y}_i)^2$, be as small as possible.
2. That the Pearson product-moment correlation between Y and $\hat{Y}$ be as high as possible.

We examine each of these criteria and then show that the values of $b_1, b_2, \ldots$ that these two criteria produce also satisfy a third desirable criterion, namely, that each estimate of a given b_j be a function only of the parameter it is designed to estimate.

2.2.1 Least–Squares Criterion

Let us return to the hypothetical data in Data Set 1. Eventually, for each subject we will compute a predicted score Y_i that is simply a linear combination of that subject's scores on the various predictor variables. We want to select the weights used in obtaining that linear combination such that

$$E = \sum (Y_i - \hat{Y}_i)^2$$

is as small as possible. Let us work up to the general procedure gradually.

If we are interested in predicting Y from a single predictor variable, say X_1, we know from elementary statistics that our best choice for b_1 would be the value $r_{1y}(s_y/s_1)$, and our best choice for b_0 would be $b_0 = \overline{Y} - b_1\overline{X}_1$. It may be instructive to consider how these choices arise. The expression for E can be rewritten as

$$E = (Y_i - b_0 - b_1 X_{i,1})^2.$$

We could simply try out different values of b_0 and b_1 until we saw no room for further improvement. For instance, taking $b_0 = 0$ and $b_1 = 1$ produces an E of 62 + 172 + 392 + 252 + 232 = 3000, and trying out various values of b_1 while keeping $b_0 = 0$ reveals that a b_1 of about .78 produces the smallest E, namely, about 610. (The optimum values of b_0 and b_1 produce an E of about 535.) We could then try several other values of b_0, finding for each that value of b_1 that minimizes E. Then, drawing a curve connecting these points would probably allow us to guess the best choice of b_1 fairly accurately. However, even with this aid, the search will be a laborious one, and it should be clear that as soon as we add a second predictor variable, thus giving us three numbers to find, this kind of trial–and–error search will become prohibitively time–consuming.

The alternative, of course, is to try calculus. If you have not been introduced to calculus already, you have three options:

1. You may accept the results of any application this text makes of calculus on faith.
2. You may consult Digression 1, which presents (at a simplified level) all the calculus you need for finding maxima and minima of quadratic functions such as the expression for E.
3. You may test any result arrived at via calculus either by trying out slightly different criterion or by the slightly more elegant process of adding a constant term to each numerical values and observing their effect on the optimization of the algebraic expressions produced by calculus and then showing that the optimization criterion "deteriorates" unless each of these added values equals zero. (This, however, does not entirely rule out the possibility that you have identified a *local* minimum or maximum – a problem that plagues computerized optimization routines that, like many programs for *structural equation modeling*, discussed in chap. 8, rely on trial-and-error search for the optimal combination of parameters.)

At any rate, all that calculus does is to allow us to obtain an algebraic expression for the *rate* at which E changes as a function of each of the variables-in this case, b_0 and b_1. Recognizing that at the point at which E is at a minimum, its rate of change in all directions (with respect to any variable) must be temporarily zero, we then need only set the algebraic expressions for rate of change equal to zero and solve the resulting two simultaneous equations for b_0 and b_1. The details of this process are given in Derivation 2.1. (See the separate Derivations section, Appendix C of this book.) The end result is

$$b_0 = \frac{\sum Y - b_1 \sum X}{N} = \overline{Y} - b_1 \overline{X};$$

and

$$b_1 = \frac{\sum x_1 y}{\sum x_1^2} = \frac{\sum x_1 y \sqrt{\sum y^2 / \sum x_1^2}}{\sqrt{\sum x_1^2 \sum y^2}} = r_{1y}(s_y / s_1).$$

Note that the present text follows the convention of letting capital letters stand for "raw" observations and lowercase letters for deviation scores. Thus,

$$\sum x_1 x_2 = \sum (X_1 - \overline{X}_1)(X_2 - \overline{X}_2).$$

If we now use both X_1 and X_2 as predictors, we obtain

$$b_0 = \overline{Y} - b_1 \overline{X}_1 - b_2 \overline{X}_2,$$

$$b_1 = \frac{\sum x_1 y \sum x_2^2 - \sum x_2 y \sum x_1 x_2}{\sum x_1^2 \sum x_2^2 - (\sum x_1 x_2)^2}$$

$$= \frac{s_2^2 - s_{12} s_{2y}}{s_1^2 s_2^2 - s_{12}^2},$$

and

$$b_2 = \frac{s_1^2 - s_{12} s_{1y}}{s_1^2 s_2^2 - s_{12}^2},$$

where

$$s_{ij} = \frac{\sum (X_i - \overline{X}_i)(X_j - \overline{X}_j)}{N - 1} = r_{ij} s_i s_j$$

and

$$s_{iy} = \frac{\sum (X_i - \overline{X}_i)(Y - \overline{Y})}{N - 1} = r_{iy} s_i s_y$$

for i and j = 1 or 2. Again, details are provided in Derivation 2.1.

Note that $s_{ij} = \Sigma x_i x_j/(N - 1)$ is the sample ***covariance*** of the variables X_i and X_j, as $s_{iy} = \Sigma x_i y/(N - 1)$ is the sample covariance of X_i with Y. Further, the covariance of a variable with itself is its variance, that is, $s_{ii} = \Sigma x_i x_i/(N - 1) = s_i^2$; and s_j (written with a single subscript) = $\sqrt{s_i^2}$ = the standard deviation of variable ***i.*** Thus in the preceding expressions, s_1^2 refers to the variance of X_1, whereas s_{12}^2 is the square of the covariance of X_1 with X_2.

It is assumed that the reader is thoroughly familiar with summation notation and its various simplifying conventions—such as omitting the index of summation and the range of the summation when these are clear from the context. If just what is being summed in the preceding equations is not perfectly clear, an introductory text such as that of M. B. Harris (1997) or Kirk (1998) or the one used in your introductory statistics course should be consulted for a review of summation notation.

If we now add X_3 to the set of predictors, we find (as the reader may wish to verify) that

$$b_0 = \overline{Y} - b_1\overline{X}_1 - b_2\overline{X}_2 - b_3\overline{X}_3,$$

and

$$\begin{aligned}
\left(\sum x_1^2\right)b_1 + \left(\sum x_1x_2\right)b_2 + \left(\sum x_1x_3\right)b_3 &= \sum x_1y;\\
\left(\sum x_1x_2\right)b_1 + \left(\sum x_2^2\right)b_2 + \left(\sum x_2x_3\right)b_3 &= \sum x_2y;\\
\left(\sum x_1x_3\right)b_1 + \left(\sum x_2x_3\right)b_2 + \left(\sum x_3^2\right)b_3 &= \sum x_3y;
\end{aligned}$$

whence

$$b_1 = \frac{s_{1y}(s_2^2s_3^2 - s_{23}^2) + s_{2y}(s_{13}s_{23} - s_{12}s_3^2) + s_{3y}(s_{12}s_{23} - s_{13}s_2^2)}{D};$$

$$b_2 = \frac{s_{1y}(s_{13}s_{23} - s_{12}s_3^2) + s_{2y}(s_1^2s_3^2 - s_{13}^2) + s_{3y}(s_{12}s_{13} - s_{23}s_1^2)}{D};$$

$$b_3 = \frac{s_{1y}(s_{12}s_{23} - s_{13}s_2^2) + s_{2y}(s_{12}s_{13} - s_{23}s_1^2) + s_{3y}(s_1^2s_2^2 - s_{12}^2)}{D};$$

where

$$\begin{aligned}
D &= s_1^2(s_2^2s_3^2 - s_{23}^2) + s_{12}(s_{13}s_{23} - s_{12}s_3^2) + s_{13}(s_{12}s_{23} - s_{13}s_2^2)\\
&= (s_1s_2s_3)^2 - (s_1s_{23})^2 - (s_2s_{13})^2 - (s_3s_{12})^2 + 2s_{12}s_{23}s_{13}.
\end{aligned}$$

Before we consider what the general formula for 111 predictors would look like, let us apply the formulae we have developed so far to Data Set 1. First, if we are solely interested in X_1 as a predictor, then our best-fitting regression equation is

$$\begin{aligned}
\hat{Y} &= r_{1y}(s_y/s_1)X_1 + (\overline{Y} - b_1\overline{X}_1)\\
&= .546\sqrt{190.5/17.5}X_1 + (78 - b_1\cdot 100)\\
&= 1.80X_1 - 102.
\end{aligned}$$

Application of this equation to each of the subjects' scores on X_1 yields predicted values of 81.6, 85.2, 79.8, 65.4, and 78.0 for subjects 1-5, respectively, and a total sum of squared deviations (errors of prediction) of approximately 535:

$$(81.6 - 96)^2 + (85.2 - 87)^2 + (79.8 - 62)^2 + (78.0 - 77)^2$$

$$= 14.4^2 + 1.8^2 + 17.8^2 + 2.6^2 + 1.0^2$$
$$= 535.2\,.$$

(It shouldn't be necessary to write down the separate squares before summing; if your five-dollar pocket calculator can't cumulate a sum of squared terms in one pass, shoot six dollars on one that can.)

In order to apply the formulae for the two-predictor case (and, subsequently, for the three-predictor case), we need to know the covariances of scores on our predictor and outcome variables. We could, of course, compute these directly, and the reader will probably wish to do so as a check on the alternative method, which is to "reconstruct" them from the intercorrelations and variances provided for you in the data table. We need only multiply a given correlation coefficient by the product of the standard deviations involved in that correlation to obtain the covariance of the two variables. (Be sure you understand why this works.) This gives values of 1.25, 31.5, and -13.5 for s_{12}, s_{1y}, and s_{2y}, respectively. Plugging into the formulae for b_0, b_1, and b_2 gives a final regression equation of

$$\hat{Y} = 2.029X_1 - 3.202X_2 - 67.3,$$

which yields predictions of 85.3, 82.9, 70.4, 63.8, and 87.6 for the five subjects, and a total sum of squared deviations of 333.1. Moving now to the three-predictor case, we find, after extensive calculations, that our regression equation is

$$\hat{Y} = 5.18\,X_1 + 1.70\,X_2 - 8.76\,X_3 - 435.7,$$

which yields predictions of 86.7, 91.7, 62.0, 68.0, and 81.7 for the five subjects and a sum of squared errors of prediction of 130.7. Table 2.1 summarizes the results of our regression analysis of Data Set 1.

Table 2.1 Multiple Regression analyses of Data Set 1

Predictors Included	b_0	b_1	b_2	b_3	$\hat{Y}_1$	$\hat{Y}_2$	$\hat{Y}_3$	$\hat{Y}_4$	$\hat{Y}_5$	$\sum e^2$	R	F^a
1	-102.0	1.80	0	0	81.6	85.2	79.8	65.4	78.0	535.2	.546	1.271
1, 2	-67.3	2.03	−3.21	0	85.3	82.9	70.4	63.8	87.6	333.1	.740	1.287
1, 2, 3	-435.7	5.18	1.70	−8.76	86.7	91.7	62.0	68.0	81.7	130.7	.910	1.611

[a] The basis for this column is explained in section 2.3

Note that as we added predictors, we obtained increasingly better fit to the *Y*s, as measured by our criterion, the sum of the squared errors. Note also that the weight assigned X_1 in the regression equation was positive and moderately large for all three equations, but the weight assigned X_2 shifted drastically from -3.20 (larger in absolute value than b_1) when only X_1 and X_2 were being used to 1.70 (about 1/3 as large as b_1)

when all three predictor variables were employed.

However, we should be careful not to base judgments of the relative importance of the different predictors on ***raw-score regression coefficients*** when (as in the present case) (a) the standard deviations of the various predictors are quite different and (b) there is no metric (unit of measurement) common to all predictors. For instance, if we choose to measure IQ (X_1 in our original scenario) in "SAT points"[1], rather than in IQ points and age (X_2) in centuries, rather than years, while retaining the original scores on number of speeding citations (X_3), we will get identical predicted scores on Y, using all three predictors from

$$\hat{Y} = 5.18(\text{IQ in IQ points}) + 1.70(\text{age in years}) - 8.76(\text{\# of citations}) - 435.7$$

(our original equation) and from

$$\begin{aligned}\hat{Y} &= (5.18/.16)(\text{IQ in SAT points}) + (1.70\bullet 100)(\text{age in decades})\\ &\quad - 8.76(\text{\# of citations}) - 435.7\\ &= 32.375(\text{IQ in SAT points}) + (170)(\text{age in decades})\\ &\quad - 8.76(\text{\# of citations}) - 435.7,\end{aligned}$$

which changes the rank order of the magnitudes of the regression coefficients from $b_3 > b_1 > b_2$ to $b_2 > b_1 > b_3$. A comparison that can be reversed just by changing unit of measurement is clearly not a meaningful one. A useful (and common) method of restoring meaningfulness to comparisons of regression coefficients is to convert all variables to z scores (which enforces a common metric via the unit-independent relationship between z scores and percentiles of a normal distribution and the similarly unit-independent relationship between z score and minimum possible percentile represented by Chebyshev's inequality) and then derive the regression equation for predicting z scores on Y (z_Y) from z scores on the Xs (z_1, z_2, z_3, etc.). Applied to Data Set 1, this yields

$$\hat{z}_y = \frac{\hat{y}}{\sqrt{190.5}} = \frac{1.80x_1}{\sqrt{190.5}} = 1.80\left(\frac{\sqrt{17.5}}{\sqrt{190.5}}\right)\frac{x_1}{\sqrt{17.5}} = .546z_1 \text{ when using } X_1 \text{ as the sole predictor;}$$

$$= 2.029\left(\frac{s_1}{s_y}\right)z_1 - 3.202\left(\frac{\sqrt{5}}{s_y}\right)z_2 \;= .615z_1 - .519z_2 \text{ when using } X_1 \text{ and } X_2;$$

$$= 5.18\left(\frac{s_1}{s_y}\right)z_1 + 1.70\left(\frac{s_2}{s_y}\right)z_2 - 8.76\left(\frac{s_3}{s_y}\right)z_3 = 1.570z_1 + .275z_2 - 1.419z_3 \quad \text{for } X_1, X_2, \& X_3.$$

Finally, note that X_3 was an important contributor to the regression equation employing all three predictors, even though its correlation with the outcome variable was

[1] IQ in SAT points = (Stanford-Binet IQ - 100)(.16)

only -.065; that is, used by itself it would account for only about 0.4% of the variance in Y scores. These shifts in sign and magnitude of a given regression coefficient, which result as the context in which that variable is imbedded is changed (that is, as the other predictors employed are changed), are quite common in applications of multiple regression analysis.

They should not be surprising, because the contribution variable j makes to predicting Y will of course depend on how much of the information it provides about Y (as reflected in r^2_{jy}) is already provided by the other predictors (as reflected partially in its correlations with the other Xs). Social psychologists among readers of this text will recognize a very close parallel between this general phenomenon of multiple regression analysis and Wishner's (1960) analysis of changes in the "centrality" of a given personality trait (i.e, the magnitude of the effect on a subject's impression of a person when that person's reported position on one trait is changed) as the nature of the other traits mentioned in the description is varied. Messick and Van de Geer (1981) provided a number of interesting examples of this same sort of context dependence for relationships among discrete variables. (For example, it is possible for men and women to be equally likely to be admitted to each department of a university, although for the university as a whole a much smaller percentage of female than of male applicants are admitted.) Now let us consider our second criterion for deriving the bs.

To make the point even more clearly and to provide a data set that can be used to accentuate the differences among competing approaches to measuring variable importance and interpreting the results of regression analyses, consider Data Set 1b (henceforth referred to as the "Presumptuous Data Set", because it crams so much didactic value into so few data points), as shown in Table 2.2.

Table 2.2 Data Set 1b: A Presumptuous Data Set

Participant	X_1	X_2	X_3	Y		Correlation Matrix X_1	X_2	X_3	Y
1	-2	-1	-2	1		1.0	.6	.9	-.2
2	-1	-2	-1	-2	→		1.0	.8	.6
3	0	0	0	0				1.0	.0
4	2	2	1	2					1.0
5	1	1	2	-1					

The reader will of course wish to verify that using X_2 (clearly the best single predictor) alone yields $\hat{z}_y = .6z_2$ and an R^2 of .36; using X_1 and X_2 (the two predictors with the highest zero-order correlations with Y) yields $\hat{z}_y = (-7/8)z_1 + (9/8)z_2$ and R^2 = .85; but using X_2 and X_3 yields $\hat{z}_y = (5/3)z_2 + (-4/3)z_3$. That is, adding to our first regression equation a predictor that by itself has absolutely no relationship raises the proportion of variation in scores on Y that is accounted for by knowing scores on our predictors from 36% to 100%! Of course we can't improve on perfect prediction, so our prediction equation using all three predictors is just $\hat{z}_y = (0)z_1 + (5/3)z_2 + (-4/3)z_3$. We show later (section 2.6) that one of the most common measures of importance of contribution to the regression equation would declare that X_3's contribution to the three-predictor regression equation is precisely zero, and one of the most common recommendations for interpreting regression equations would lead us to interpret this one as indicating that participants with high scores on X_2 and, to a lesser extent, low scores on X_1 (no mention of X_3 as being at all relevant) are those most likely to have high scores on Y.

2.2.2 Maximum Correlation Criterion

It would be nice to have an overall measure of how good a job the regression equation derived from a multiple regression analysis does of fitting the observed *Y*s. A fairly obvious measure is the Pearson product-moment correlation between predicted and observed values $r_{y\hat{y}}$. This has been computed for each of the three sets of predictions generated for Data Set 1 and is listed in the column headed *R* in Table 2.1. (The reader will, of course, wish to verify these computations.) This also provides an overall measure of the degree of relationship between the set of predictor variables and the outcome variable, because it is the correlation between *Y* and a particular linear combination of the *X*s. We naturally wish to be as optimistic as possible in our statement about the strength of this relationship, so the question arises as to whether the values of b_0, b_1, b_2, and so on, can be improved on in terms of the value of $r_{y\hat{y}}$ to which they lead. Intuitively, the answer would seem to be no, because we know that there is an extremely close relationship between the magnitude of *r* and the adequacy of prediction in the univariate case. However, to put this hunch on firmer ground, let us ask what would have happened had we used the "optimism criterion" (had chosen *b*s that make r_{yw} as large as possible, where $W = b_0 + b_1 X_1 + \cdots$) to derive the *b*s . The maximum-*R* criterion is of no help to us in the univariate case, because we know that any linear transformation of the *X* scores will leave the correlation between *X* and *Y* unaffected. There may be a deep logical (or philosophical) lesson underlying this fact, but it is more likely simply an indication that prediction optimization was used as a criterion in developing the formula for the Pearson *r*. It should also be clear that the maximum-*R* criterion will be of no value in picking a value of b_0, because the correlation between $W = b_0 + b_1X_1 + \cdots + b_mX_m$ and *Y* is unchanged by addition or subtraction of any constant whatever. However, one of the properties of univariate regression equations is that they always predict that a subject who scores at the mean of one variable will also score at the mean of the other variable. Thus, if we can use the maximum-R criterion to pick xalues of b_1,b_2, and so on, $b_0 + b_0$ can be computed from the requirement that $\hat{Y} = \overline{Y}$ when $W = \overline{W}$, that is,

$$\overline{Y} = \overline{W} = b_1\overline{X}_1 + b_2\overline{X}_2 + \ldots + b_m\overline{X}_m + b_0.$$

Because the value of r is unaffected by linear transformations of Y, we could apply a transformation to Y [namely, converting to $z_y/(N - 1)$] that would guarantee that $\Sigma\, y^2 = 1$, thereby simplifying our arithmetic somewhat. Also, we know that requiring that $R_{Y \bullet X_1, X_2} = r_{wy}$ be as large as possible is not enough to uniquely specify b_1 and b_2, because the correlation of $b_1X_1 + b_2X_2$ with Y is identical to the correlation of $c(b_1X_1 + b_2X_2) = (cb_1)X_1 + (cb_2)X_2$ with Y. We therefore need to put some additional restriction on the permissible values of b_1 and b_2. The end result of applying a bit of calculus (cf. Derivation 2.2) is that it is indeed true that the same weights that minimize the sum of the squared errors of prediction also yield a linear combination of the original variables that correlates more highly with Y than any other linear combination of the predictor variables.

The three-variable case is about the limit of usefulness of scalar (i.e., nonmatrix) algebraic expressions for the regression coefficients. Beyond $m = 3$, we either have to rely on the computer to find optimal weights or learn to use matrix-algebraic expressions. In this text I'll concentrate on computer use. However, I can't resist trying to dazzle you with all that a bit of matrix algebra can do for you in terms of understanding multivariate statistics, in hopes of motivating you to bolster your skills in matrix algebra (and perhaps to read Digressions 1–3) some day.

2.2.3 The Utility of Matrix Algebra

Looking back over the formulae we have developed so far, we can see some patterns that are suggestive of the form the multiple regression problem and its solution might take in the general case of m predictors. For instance, it is fairly clear (and we gave some reasons earlier for believing) that b_0 will always be computed from the values of the other coefficients as

$$b_0 = \overline{Y} - b_1\overline{X}_1 - b_2\overline{X}_2 - \ldots - b_m\overline{X}_m.$$

Similarly, as a quick glance through Derivation 2.1 reveals, the pattern of coefficients involved in the simultaneous equations that must be solved for the values of $b_1, b_2, \ldots, b_m$ seems to be very closely related to the sample covariances, s_{ij} and s_{iy}, as follows:

$$\begin{array}{cccccccccccc}
s_1^2b_1 & + & s_{12}b_2 & + & s_{13}b_3 & + & \cdots & + & s_{1m}b_m & = & s_{1y} \\
s_{12}b_1 & + & s_2^2b_2 & + & s_{23}b_3 & + & \cdots & + & s_{2m}b_m & = & s_{2y} \\
s_{13}b_1 & + & s_{23}b_2 & + & s_3^2b_3 & + & \cdots & + & s_{3m}b_m & = & s_{3y} \\
\vdots & & \vdots & & \vdots & & \vdots & & \vdots & \vdots & \vdots \\
s_{1m}b_1 & + & s_{2m}b_2 & + & s_{3m}b_3 & + & \cdots & + & s_m^2b_m & = & s_{my}
\end{array}$$

What is not clear, however, is what will be the general form of the expressions for b_1, b_2, and so on, that result from solving these equations. It seems strange that the apparent regularity of the equations would not be paralleled by predictability of their solutions. This is precisely the sort of situation that leads a mathematician to suspect that a new level of operating is needed—an algebra designed to handle arrays of numbers or symbols as a whole, rather than having to deal with each symbol or number separately. Such a "higher order" algebra has been developed and is known as ***matrix algebra.*** Basically, there are three advantages to matrix algebra:

1. It *summarizes* expressions and equations very compactly.
2. It facilitates our *memorizing* these expressions.
3. It greatly *simplifies* the procedures for deriving solutions to mutivariate problems.

The first function is illustrated by the following summary of the solution for the general *m*-predictor case:

$$b_0 = \overline{Y} - \overline{\mathbf{X}}\mathbf{b}, \;\; \mathbf{b} = \mathbf{S}_x^{-1}\mathbf{s}_{xy}; \;\; R^2 = \frac{(\mathbf{b}'\mathbf{s}_{xy})^2}{(\mathbf{b}'\mathbf{S}_x\mathbf{b})s_y^2} = \frac{\mathbf{b}'\mathbf{s}_{xy}}{s_y^2}.$$

This is a bit more compact than the expressions derived earlier for, say, the 3-predictor case, and it is many orders of magnitude less complex than would be, say, the expressions for the 30-variable case written in "single-symbol" form. The second, mnemonic function of matrix algebra is in part a direct consequence of the first, because it is always easier to remember a compact expression than one requiring dozens (or hundreds) of symbols. This advantage is greatly enhanced, however, by the fact that the matrix expressions usually turn out to be such direct and obvious generalizations of the formulae for the univariate (in the present context, single-predictor) case. For example, in multiple regression, the set of *m* *b*s corresponding to the *m* predictors is computed from the matrix equation $\mathbf{b} = \mathbf{S}_x^{-1}\mathbf{s}_{xy}$, where $\mathbf{S}_x$ is the ***covariance matrix*** of the *m* predictors (the obvious generalization of s_x^2); $\mathbf{s}_{xy}$ is a column vector listing the covariance of each predictor variable with the outcome variable (the obvious generalization of s_{xy}; and the -1 exponent represents the operation of ***inverting*** a matrix, which is analogous to the process of inverting (taking the reciprocal of) a single number. To see how matrix algebra accomplishes the third function (that of simplifying derivations), let us review the reasoning that produced the matrix-form solution for the multiple regression problem. First, we want to minimize

$$E = \sum(Y - \hat{Y})^2 = \sum(y - \hat{y})^2 = \sum y^2 - 2\sum y\hat{y} + \sum \hat{y}^2$$
$$= \mathbf{Y}'\mathbf{Y} - 2\mathbf{b}'\mathbf{s}_{xy} + \mathbf{b}'\mathbf{S}_x\mathbf{b},$$

whence

$$\frac{d\mathbf{E}}{d(\mathbf{b})} = \mathbf{0} - 2\mathbf{s}_{xy} + 2\mathbf{S}_x\mathbf{b} = \mathbf{0} \;\Leftrightarrow\; \mathbf{S}_x\mathbf{b} = \mathbf{s}_{xy}.$$

Because the *b*s are what we are after, it would be convenient if we could isolate the vector **b** on the left-hand side by "dividing" both sides by the matrix $\mathbf{S}_x$. The closest matrix analog of division is multiplication by the inverse of a matrix (just as division by a scalar is equivalent to muliplication by its inverse). It can be shown that there exists a matrix $\mathbf{S}_x^{-1}$, the inverse of $\mathbf{S}_x$ that has the property that $\mathbf{S}_x^{-1}\mathbf{S}_x = \mathbf{S}_x\mathbf{S}_x^{-1} = \mathbf{I}$, the identity matrix (which has ones on the main diagonal and zeros elsewhere), whence

$$\mathbf{S}_x^{-1}\mathbf{S}_x\,\mathbf{b} = \mathbf{S}_x^{-1}\mathbf{s}_{xy},$$

whence

$$\mathbf{b} = \mathbf{S}_x^{-1}\mathbf{s}_{xy}.$$

[Actually, as pointed out in Digression 2, $\mathbf{S}_x^{-1}$ exists only if $\mathbf{S}_x$ has what is called full rank, that is, if and only if $|\mathbf{S}_x| \neq 0$. However, whenever $\mathbf{S}_x$ is singular, that is, has a zero determinant, one or more of the predictors can be expressed as an exact linear combination of the other predictors, so that this redundant predictor(s) can be eliminated from the regression procedures by deleting the corresponding row and column of $\mathbf{S}_x$ and the corresponding member of $\mathbf{s}_{xy}$, thus eliminating the singularity without losing any predictive ability. (Ease of theoretical interpretation may, however, influence your decision as to which of the redundant predictors to eliminate.)]

The preceding derivation is a great deal simpler than attempting to solve explicitly for the relationship between each b_j and particular elements of $\mathbf{s}_{xy}$ and $\mathbf{S}_x$. It is also much more general than the single-symbol approach, because it holds for all sizes of the set of predictor variables. Of course, if you wish an explicit algebraic expression for the relationship between each b_j and particular elements of $\mathbf{s}_{xy}$ and $\mathbf{S}_x$, you will find the derivational labor involved in developing explicit expressions for $\mathbf{S}_x^{-1}$ of the same order of magnitude (although always somewhat less, because of the straightforwardness of the matrix inversion process) as solving the single-symbol equations. However, such expressions are seldom needed for any but the simplest cases, such as $m \leq 3$, in which the full formulae may be useful in testing computer programs. It is, of course, to readily available matrix inversion computer programs that we turn when numerical solutions for the *b*s are desired in a problem of practical size. Before we turn to a discussion of the use of "canned" computer programs (section 2.4), you may wish to examine Derivation 2.3, which shows how much simpler it is to use matrix algebra than "ordinary" (scalar) algebra in deriving *b*s that satisfy the maximum-*r* criterion.

2.2.4 Independence of Irrelevant Parameters

We now wish to confirm the statement that the procedures for estimating the bs satisfy the condition of "independence of irrelevant parameters." We need a more precise way of expressing this condition. Let us assume that we have a set of expressions for computing the *b*s from the observed values on *Y* or *y*. These will be of the form

$$b_j = k_{1j}y_1 + k_{2j}y_2 + \cdots + k_{Nj}y_N,$$

or, in matrix notation, **b** = **Ky**, where each row of **K** contains the coefficients for computing the corresponding *b* from the *y*s. Now, we also have a theoretical expression for each *y*, namely

$$y_i = \beta_1 x_{i1} + \beta_2 x_2 + \cdots + \beta_m x_{im} + \varepsilon_i,$$

that is, $\mathbf{y} = \mathbf{x}\boldsymbol{\beta} + \boldsymbol{\varepsilon}$ in matrix form. Each of our sample regression coefficients, b_j, can therefore be expressed as a function of the population parameters, β_1, β_2, ... , β_m and $\sum_i k_{ij}\varepsilon_{ij}$ by substituting for each value of y_i its theoretical expression in terms of the βs. This gives us an expression for each b_j as a linear combination of the βs and the error term, where the combining weights for the linear combination are a function of the values of β_1 in the equation cited at the beginning of this section. But because the ε_is are assumed to be randomly distributed around zero (and thus to have an expected value of zero), the expected value of each b_j (its mean value in the population of all possible samples of data) will be only the linear combination of the βs, where once again the combining weights are a function of the k_{ij}s.

For example, in the two-predictor case applied to Data Set 1, we have

$$b_1 = (45y_1 + 75\,y_2 + 5\,y_3 - 140\,y_4 + 15\,y_5)/1375$$

and

$$b_2 = (-80\,y_1 + 50\,y_2 + 205\,y_3 + 35\,y_4 - 210\,y_5)/1375.$$

(You'll naturally want to check these expressions—computed from a simple matrix–algebraic expression—by plugging the values of y_1 through y_5 into the preceding two expressions and showing that they equal our previously computed values of b_1 and b_2.)

Substituting into the expressions for b_1 and b_2 our theoretical expressions for the *y*s gives

$$\begin{aligned} E(b_1) &= [45(2\,\beta_1 - \beta_2) + 75(4\beta_1 + \beta_2) + 5(\beta_1 + 3\,\beta_2) - 140(-7\,\beta_1) + 15(-3\beta_2)]/1375 \\ &= \beta_1; \end{aligned}$$

and

$$\begin{aligned} E(b_2) &= [-80(2\beta_1 - \beta_2) + 50(4\beta_1 + \beta_2) + 205(\beta_1 + 3\beta_2) + 35(-7\,\beta_1) - 210(-3\,\beta_2)]/1375 \\ &= \beta_2; \end{aligned}$$

where, for present purposes, E(*b*) stands for the expected value of *b*.

The fact that the expected value of each of our b_js is a function only of the particular β_j of which it is intended to be an estimate is what we mean by ***independence of irrelevant parameters***, and it is very easy to show (using matrix algebra—it would be very difficult to do using only scalar algebra) that the estimates provided by multiple regression analysis always have this property.

One potentially useful offshoot of this demonstration arises from the recognition that the k_{ij} terms are completely independent of the y_is. Thus if the same set of predictor variables were to be related to each of several outcome measures, it might prove efficient to compute the k_{ij}s and then apply them in turn to each of the Y variables to compute the several different regression equations.

Further, when our predictors are actually group-membership variables being used to represent an analysis of variance design (section 2.3.2), examination of the k_{ij}s for each b_j allows us to write expressions for the contrast among the population means for the various subgroups that is being estimated by each regression coefficient.

2.3 RELATING THE SAMPLE EQUATION TO THE POPULATION EQUATION

Let us consider for a moment the four-predictor case applied to Data Set 1. Let us assume that our five subjects had scores of 1, 2, 3, 4, and 5 on. X_4. Thus X_4 has correlations of -.567, -.354, -.424, and -.653 with X_1, X_2, X_3, and Y, respectively. It has the highest correlation with the outcome variable of any of the predictor variables; consequently, we might anticipate that it would contribute greatly to the regression equation and lower $\sum e_i^2$ considerably from its three-predictor value of 130.7. However, even though X_4 would account for about 43% of the variation in Y even if it were used by itself, it also shares a great deal of variance with the other three predictors. Thus, the information it provides about Y is somewhat redundant and might contribute relatively little to our ability to predict Y over what the first three predictors already provide. To find out which factor is more important, we carry out the regression analysis, finding that $\hat{Y}$ = $2.38182X_1 - 2.03636X_2 - 4.09091X_3 - 5.60000X_4 - 90.36364$ = 96, 87, 62, 68, and 77 (to 4 decimal places) for the 5 subjects, whence $R^2 = 1$ to within rounding error.

In other words, inclusion of this very arbitrary predictor variable has produced a near-perfect fit between predicted and observed scores on Y. It is important to ask whether this result was a lucky accident due to the particular values chosen for X_4 or whether instead the fact that $\sum e_i^2$ is not *exactly* zero is due solely to inaccuracy in numerical computations.

The latter turns out to be the case, as the reader may verify by trying out another set of values for X_4. No matter what numbers are selected, you will find that, except for round-off errors, the regression analysis procedures for predicting the five subjects' Y scores on the basis of their scores on four predictor variables, regardless of what numbers are used for X_4 (or, for that matter, X_1, X_2, and X_3), will lead to a multiple R of 1.0. This should lead us to suspect that our R of 1 is telling us more about the efficiency of the optimization process than about any underlying relationship between these variables and

Y in the population from which the data were "sampled." It has been shown (e.g., Wishart, 1931) that the expected value of a squared sample coefficient of multiple correlation based on N subjects and m predictors, ***none of which have any true relationship to the outcome variable*** in terms of their population correlations with Y, is equal to $m/(N - 1)$. This also clearly establishes that the sample value of R^2 is a biased estimate of the corresponding population coefficient. An unbiased estimate of R^2 is available, namely,

$$\hat{R}^2_{\text{pop}} = 1 - \frac{N-1}{N-m-1}(1-R^2). \tag{2.3}$$

Drasgow and Dorans (1982) found that this simple estimate was just as unbiased (showed as low an average departure from the true population value) as a more complicated expression involving $(1 - R^2)^4$ that was developed by Herzberg (1969) from an even more complicated, infinite-series formula proposed by Olkin and Pratt (1958).

Cattin (1980) and Drasgow and Dorans (1982) discussed formulae for estimating the value of R^2 to be expected on cross-validation, that is, if the coefficients derived from this sample are applied to a new sample. The problem, of course, is that our optimization procedure has no way of knowing what portion of the variation among the variables is truly repesentative of the population and what portion represents "wild" behavior of the particular sample. It therefore utilizes ***all*** variation in picking values for b_1, b_2, and so on. This is very reminiscent of two similar situations in elementary statistics:

1. The need to use $N - 1$, rather than N, in the denominator of s^2 in order to compensate for the fact that in taking deviations about $\overline{X}$, rather than about the (unfortunately unknown) true population value of the mean, we guarantee that Σx^2 will be smaller than $N\sigma_x^2$.
2. The fact that the Pearson r between two variables when only two observations on their relationship are available must be either + 1 or - 1 (or undefined if the two scores for either variate are identical).

The geometric interpretation of this latter fact is the statement that a straight line (the line representing the regression equation) can always be passed through any two points (the two points on the scatter diagram that summarize the two subjects' scores on X and Y). A similar interpretation holds for the two-predictor case in multiple regression. We now have a three-dimensional scattergram, and perfect fit consists of passing a plane through all points on this three-dimensional scattergram—which, of course, is always possible when there are only three points. In the general case, we will have an $(m + 1)$-dimensional scatter plot for which we seek the best fitting m-dimensional figure. Perfect fit will always be possible when the number of points on the scatter diagram (the number of subjects) is less than or equal to $m + 1$.

The case in which $N \leq m + 1$ represents the most extreme capitalization on chance.

In all cases, however, there will be a need to assess how much information the regression equation that has been computed for our sample of data provides about the *population* regression equation. If we have selected an appropriate model for the data, so that the residuals show no systematic relationship to, for example, the magnitude of the predicted value of Y or the order in which the observations were drawn, and if Y is normally distributed while the X_{ij}s are fixed constants for each subject, then it can be shown (as friendly mathematical statisticians have done) that our estimates of the true population variance of residual scores and of the true population variance of predicted scores are independent of each other and each follow a chi-square distribution, whereas the b_js are normally distributed. We can thus, for instance, take $F = R^2(N - m - 1)/ [(1 - R^2)m]$ as a test statistic to be compared with the critical value of the F distribution with m and $N - m - 1$ degrees of freedom in deciding whether the sample provides strong enough evidence to reject the null hypothesis of no true relationship between the predictor variables and the outcome variable. If the $X_{i,j}$s are *not* fixed constants, but are themselves random variables that together with Y are distributed in accordance with the multivariate normal distribution, the distributions of our sample estimates of the entries of the covariance matrix follow a Wishart distribution (a multivariate generalization of the chi-square distribution) and the b_js follow a complex distribution that is *not* exactly multivariate normal. However, the statistical tests that apply in this case (known as the *variance-component* model, in contrast to the *fixed* model discussed in the preceding paragraph) are identical to those derived from the fixed model, as is demonstrated by the fact that Winer (1971) and Morrison (1976) arrived at the same formulae working from the assumptions of the fixed and variance-component model, respectively. What is not as clear are the consequences for inferential uses of multiple regression of having some of the X_js fixed constants and others random variables (this case is known as the *mixed* model). Inferential procedures for the mixed model in analysis of variance—which, as shown in chapter 5, can be viewed as a special case of multiple regression in which the predictor variables are all dichotomous, group-membership variables—are well known and generally involve using different "error terms" for tests involving only the fixed bs than for those involving regression coefficients associated with random variables. The more general case (the mixed model with other than dichotomous variables) does not appear to have been extensively studied—perhaps because multiple regression has typically been applied in correlational research involving measured, rather than manipulated, variables. Fixed variables (implying zero error of measurement in assigning subjects scores on that predictor variable) are rarely encountered in this context.

Table 2.3 summarizes the various significance tests that are available to us if our initial assumption of random residuals is tenable. In any study, one should plot the residuals and examine them for gross departures from randomness. (This will, however, be a truly meaningful procedure only if a large number of subjects, say at least 30 more than the number of predictor variables employed, is used.) Draper and Smith (1981), Daniel and Wood (1971), Anscombe (1973), Mosteller and Tukey (1968, 1977), and

Table 2.3 Summary of Significance Tests for Multiple Regression

Source	df	Sum of Squares
Total variance of z_y	N - 1	1.0
Variance attributable to all m predictors	m	R^2
Variance due to addition or Subtraction of k predictors	k	Change in R^2 on adding the k predictors to or deleting them from the regression equation
H_0 that β_j has a true population regression weight of θ	1	$(b_z - \theta)^2/V$, where $V = r_{ij}$, the jth main diagonal entry of $\mathbf{R}_x^{-1}$
H_0 that predictor j has a true population value of zero	1	Above with $\theta = 0$, or change in R^2 when X_j is added last
Contrast among β_js, $\sum a_j b_{z_j} = 0$	1	$\left(\sum a_j b_{z_j}\right)^2 / U$, where $U = \mathbf{a}'\mathbf{R}_x^{-1}\mathbf{a}$, which equals a'a when the predictors are uncorrelated
Deviation from prediction = error = residual	N - m - 1	$(1 - R^2)$

Note. Each test consists of dividing *MS* (= *SS*/*df*) for that source by the *MS* for deviation from prediction and comparing the resulting value to the critical value for your chosen significance level of an *F* distribution having the specified degrees of freedom. As presented in a computer program's output, the *SS* entries may all be multiplied by the sum of squared deviations of *Y* about its mean (in keeping with Anova tradition) or by the variance of *Y* (for the sake of idiosyncrasy). In either of these cases, the regression weights referred to will be raw score (unstandardized) weights.

Tukey (1977) all argued convincingly for the usefulness of graphical methods in detecting departures of the data from the assumptions of the linear regression model.

In particular, Draper and Smith (1966) demonstrate that the usual sum of squares for residual from regression is actually composed of two components: error variance and systematic departures from linearity. Where there are several "repeat" observations (sets of subjects having exactly the same scores on all predictor variables), the sum of squared deviations about a mean *Y* score within each of these sets of repeat observations can be used to compute a mean square for error that is independent of systematic departures from the model being employed, which in turn can be used to test the statistical significance of the departures from the model, using

$$F = \frac{MS_{\text{dep}}}{MS_{\text{err}}},$$

where MS_{err} is the mean square for error just described and

$$MS_{\text{dep}} = \frac{SS_{\text{res}} - SS_{\text{err}}}{df_{\text{res}} - df_{\text{err}}}.$$

The major drawback to the use of this test is that it is often difficult to find *any* repeat observations in a study in which the predictors are organismic variables not subject to experimental control.

Daniel and Wood (1971) suggested a "factorial" approach to developing the most adequate model before considering a multiple regression analysis complete. Once possible sources of improvement in the fit of the'model have been identified, primarily by graphical means, a series of 2^s MRAs is conducted, where s is the number of such possible sources of improvement and the 2^s MRA "runs" through the computer consist of all possible combinations of the presence or absence of each improvement in a particular run. These sources of improvement might include deleting one or more outliers (observations that do not seem to belong on the same scatter diagram as the others), including the square or the cube of one of the predictor variables or a cross-product of two of the predictors as additional predictor variables, performing a square-root or logarithmic transformation on the predicted variable Y, and so on. This would seem to be an excellent procedure for an exploratory study. However, three cautions need to be considered before (or while) adopting this factorial approach:

1. The usual significance levels for tests conducted on the overall R^2, on individual regression coefficients, and so on, do not apply to the final regression equation that results from this factorial exploration, because those probability levels were derived under the assumption that this is a single, preplanned analysis. This is, of course, no problem if the researcher is not interested in hypothesis testing but only in model development.
2. The number of runs can become prohibitively large if the researcher is not rather selective in generating possible sources of improvement.

(c) This approach to model development should not be confused with the development of mathematical models of behavior from a set of axioms about the processes that generate the observed behavior (cf. Atkinson, et al., 1965, and Estes, 1957, for discussion of this latter approach). Daniel and Wood's factorial approach is essentially a systematization of the "curve-fitting" approach to the development of theories of learning, which proved to be such a dead end. There are for any single relationship, such as that between a specific set of predictors and a single outcome variable, an almost limitless number of quite different looking functions relating the two that nevertheless provide very nearly equally good fits to the data. Moreover, a function developed on purely empirical grounds without an underlying axiomatic justification cannot be generalized to situations involving slightly different sets of predictors or different experimental conditions. Where Y or the X_j are not normally distributed, the multivariate equivalent of the central limit theorem so crucial in univariate statistics assures us that the statistical tests outlined in Table 2.3 remain valid for large N (cf. Ito, 1969).

Unfortunately, there have been few empirical sampling studies of how large an N is needed for this robustness of multiple regression analysis under violations of normality assumptions to evidence itself. We can expect the overall test of the hypothesis that the population value of R is zero will display considerably more robustness than tests on individual regression coefficients. I recommend interpreting these latter tests with great caution if the Xs and the Ys have grossly nonnormal distributions and $N - m$ is less than about 50. Other authors have been even more pessimistic. Cooley and Lohnes (1971) cited Marks's (1966) recommendation, on the basis of a series of computer-generated cross-validation studies (in which a prediction equation is derived from one sample and then tested for predictive ability on a second random sample), that the sample regression coefficients be ignored altogether in developing a prediction equation, replacing them with the simple correlation between each predictor i and Y, r_{iy}, whenever N is less than 200. Cooley and Lohnes considered this recommendation overly pessimistic, though they do not provide an alternative "critical value" of N.

Actually, the sample sizes used by Marks were 20, 80, and 200; thus, his data would really only support a recommendation that the regression weights not be taken seriously unless $N > 80$—not too different from this book's earlier suggestion that $N - m$ be > 50. More common is a recommendation that the ratio of N to m be some number, for example, 10. However, ratio rules clearly break down for small values of m (would you really feel comfortable with a single-predictor regression coefficient based on 10 cases?). Green (1991), on the basis of extensive Monte Carlo runs, concluded that *neither* a difference rule nor a ratio rule is adequate. He provided a more complicated formula that requires that the magnitude of the population multiple ρ^2 you wish to detect be taken into account.

More importantly, Marks's findings—which include higher predictor-criterion correlations using unity (1.0) as the weight for each predictor than from the regression coefficients computed on a previous sample in 65% of the cases—seem to be at odds with the well-known theorem (cf. Draper & Smith, 1966, p. 59) that for data that fit the assumptions of normality and linearity the sample regression coefficient b_i provides an unbiased estimate of the corresponding population parameter β_i and has a sampling distribution with lower variance than any other unbiased estimator that is linearly related to the Ys.

These findings have nevertheless been supported by further Monte Carlo and empirical studies summarized by Dawes and Corrigan (1974). In particular, Schmidt (1971) found that for population correlation matrices that involved no "suppressor"variables (variables whose correlations with the criterion are negligible, but that receive sizeable weights in the population regression equation), unit weights out-performed least-squares regression weights unless the ratio of N to m was 25 or more, whereas for his matrices that did include such suppressor variables, a ratio of 15 to 1 was necessary. This suggests that the crucial factor determining whether regression weights will outperform unity weights (or simple predictor–criterion correlations) is how different the population regression weights are from the a priori weights of unity or r_{iy}. For instance, Dawes and Corrigan (1974, p. 105) pointed out that Marks's (1966) simulations were subject to the condition that "the partial correlation between any two predictors, partialing out the criterion variable, was zero." It can readily be shown that this implies that the true population regression weight for predictor variable X_i will be equal to

$$r_{xy} / [\sum_j r_{iy}^2 + (1 - r_{iy}^2)]$$

that is, that the population regression weights are very nearly directly proportional to the simple predictor-criterion correlations. Marks thus inadvertently "loaded" his Monte Carlo runs in favor of r_{iy} as the weighting factor—and probably unity weights as well, depending on how variable his population values of r_{iy} were. (Marks's paper is unavailable except in abstract form.) For more recent "rounds" in the equal-weights controversy, see Wainer (1976, 1978) and Laughlin (1978, 1979).

Perhaps the best remedy for a tendency to read more from a set of regression coefficients than their reliability warrants would be to construct confidence intervals about them. For a single coefficient chosen on a priori grounds, the 95% confidence interval (CI) about the z-score weights is given by

$$b_{z_j} \pm \sqrt{\frac{r_{jj}(1 - R^2)}{N - m - 1}},$$

where $t_{.05}$ is the 97.5th percentile of the t distribution having N - m - 1 degrees of freedom, and r_{jj} is the jth main diagonal entry of the inverse of the correlation matrix. Few computer programs provide the inverse of the correlation matrix, but almost all provide the option of printing out the 95% confidence interval around each regression coefficient. (Be sure that you know, either from the labeling of the output or from careful reading of the manual for the program you're using, whether the confidence interval is for the raw-score regression coefficient or for the z-score regression coefficient.) For those readers who like to know what their computer program is doing to generate those confidence intervals:

For a set of simultaneous confidence intervals about the individual b_js as well as about any linear combination of the z-score weights, use

$$\mathbf{a'b}_z \pm \sqrt{\frac{(\mathbf{a'R}_x^{-1}\mathbf{a})(m)[F_\alpha(m, N - m - 1)](1 - R^2)}{N - m - 1}}$$

as the (1 - α)-level CI about $\mathbf{a'b}$, (e.g., the sum of or a contrast among the various b_js), with the confidence intervals about individual coefficients being given (as a special case) as

$$b_{z_j} \pm \sqrt{\frac{r_{jj}(m)(1 - R^2)[F_\alpha(m, N - m - 1)]}{N - m - 1}}$$

Use of these simultaneous CIs yields a probability of at least (1 - α) that, for any given sample of observations, all of the (potentially infinite number of) CIs we construct enclose their corresponding population values.

If you should be interested in using a confidence level other than 95% or in examining all m confidence intervals with Bonferroni adjustment for the multiplicity of CI's constructed, you need only substitute the appropriate value of α in the preceding equations, or multiply the width of the confidence interval as reported by your program by $t_{(\text{chosen } \alpha)/m} / t_{.05}$, where (chosen α) = 1 - (your desired confidence level expressed as a proportion) (e.g., .01 for a 99% CI).

The rationale underlying the significance tests of regression coefficients may seem a bit obscure. However, as we pointed out in discussing the independence-of-irrelevant-parameters property in section 2.2.4, the matrix algebraic expression for each regression coefficient can be written as a linear combination (weighted sum) of the observed scores on Y. Because each Y score is an independent sample from a normal distribution with a variance estimated by the MS for residual, the variance of each b_j across successive samples is just MS_{res} multiplied by the sum of the squares of the weights by which the Y_i are 'multiplied in computing b_j. For z-score regression weights, this turns out (Derivation 2.5) to be the main diagonal entry of $\mathbf{R}_x^{-1}$. Similarly, the covariance of b_{z_i} and b_{z_j} depends on the cross products of the weights used to compute these two coefficients. Thus our best estimate of the variance of $\sum c_j b_{z_j}$ is given by $(\mathbf{c}'\mathbf{R}_x^{-1}\mathbf{c})\cdot MS_{\text{res}}$ and the F tests are, as usual, of the form

$$\frac{(\text{An observed statistic - Its assumed population value})^2}{\text{An estimate of the variance of the statistic}}$$

Finally, note that the entries for degrees of freedom *(df)* in Table 2.2 assume that the m predictors are not linearly related, that is, cannot be related perfectly by a linear function. More generally, we should replace m by the rank of the matrix ($\mathbf{x}'\mathbf{x}$), that is, by the number of predictors we can specify without any of them being perfectly computable as a linear combination of the others. Where $\mathbf{R}_x$ or $\mathbf{x}'\mathbf{x}$ is of less than full rank we cannot obtain its inverse; thus we must resort to one of the techniques discussed in section 3.2, namely dropping one of the redundant predictors or working instead with the subjects' scores on the principal components (cf. Chapter 6) of $\mathbf{x}'\mathbf{x}$, $\mathbf{S}_x$ or $\mathbf{R}_x$.

2.3.1 Raw-Score Variance Versus Σy^2 versus z-Score Variance as the Basis for MRA

A glance back through the chapter shows that many of the formulae have alternative forms. Thus, for instance, $R^2 = \Sigma\, r_{iy}\, b_{zi} = \Sigma\, s_{iy}\, b_i / s_y^2 = \Sigma\, b_i\, \Sigma x_i y / \Sigma y^2$, or, in matrix terminology,

$$R^2 = \mathbf{r}'_{xy}\mathbf{R}_x^{-1}\mathbf{r}_{xy} = \frac{\mathbf{s}'_{xy}\mathbf{S}_x^{-1}\mathbf{s}_{xy}}{s_y^2}$$

$$= \frac{(\mathbf{x}'\mathbf{y})(\mathbf{x}'\mathbf{x})^{-1}\mathbf{x}'\mathbf{y}}{\sum y^2}.$$

The equivalence between the last two forms should be obvious, because

$$s_{iy} = \Sigma\, x_i\, y_i\,/(N - 1).$$

Consequently, whether we base our calculations on sums of squares and cross-products of deviation scores or on variances and covariances should have no effect, with the constant factor obligingly canceling out.

That starting from the correlations among the Xs and Y should also yield the same value of R^2 may seem a bit less obvious until we recognize that the correlation between two variables is the same as the covariance of their z-score forms, so that each r_{ij} is simply the covariance we obtain if we transform our original variables to standardized (z-score) form before beginning analysis.

However, the regression coefficients obtained via correlation-based formulae will refer to the prediction of scores on z_y from knowledge of scores on the z_j, rather than to the prediction of Y from the X_i. Similarly, the values of the various sums of squares in Table 2.3 will represent portions of Σy^2 (the traditional Anova approach), of s_y^2 (the traditional way of talking about r_{xy}^2 as a portion of shared variance), or of $\sum z_y^2/(N - 1) = 1.0$, depending on which of the three matrices we use as the basis for our analysis. This may seem confusing, but you will encounter no difficulty as long as you remember that the results obtained from alternative formulae, while perhaps numerically different, are nevertheless effectively equivalent and translatable back and forth from z scores to raw scores, and vice versa. Thus, for instance, if application of b_{z_i}s to z scores yields a predicted z_y of -1.2, it will turn out that application of the raw-score weights b to x scores yields a $\hat{Y}$ that corresponds to a $\hat{z}_y$ of -1.2. More generally, $\hat{z}_y = (\hat{Y} - \overline{Y})/s_y$ and $\hat{Y} = \overline{Y} + \hat{z}_y s_y$. A bit of algebra will permit one to translate results stated in terms of raw scores or variances into completely equivalent results stated in terms of z scores or correlations. However, by far the simplest way to avoid mistakes is to stay within a single system. If your regression coefficients were computed with the aid of $\mathbf{R}_x$ (and are therefore z-score weights), be sure that you apply them to z scores, rather than to raw scores, in generating predicted scores. If your MS for the effect of adding two predictors involved $\Sigma\, y^2$, be sure that your MS_{res} does, too. Only confident algebraists should mix correlations, covariances, and sums of cross-products of deviation scores in their calculations. This leaves, however, two problems:

1. How can you tell which system a computer program has used, so that your subsequent use of its output stays within that system?
2. Which system should you use when you have a choice?

Answering the first question is made more difficult by the fact that computer programmers do not always follow the sage advice you've been given. In particular, it is not uncommon for a program to report *z*-score regression weights but to base its summary table on an additive partitioning of Σy^2; however, there is always consistency within the summary table. Many programs report both *z*-score and raw-score regression weights. If so, the former will almost always be labeled either as "standardized" weights or as "betas." The second usage does not, alas, mean that the authors have found a magic way to compute the population regression weights, β, but is simply a way of distinguishing *b* from b_z on a printer that doesn't allow subscripts.

In trying to determine whether the summary table is based on deviation scores, covariances, or correlations (i.e., on a partitioning of Σy^2, s_y^2, or $s_{z_y}^2 = 1$), the simplest clue (if it is available) is the entry for SS_{total} , which should equal $\Sigma y^2 = (N - 1)\, s_y^2$, s_y^2 itself, or 1.0, respectively. [A fourth possibility is that SS_{total} will equal $N - 1$, indicating that the table provides a partitioning of $\sum z_y^2$ and that $(N - 1)\mathbf{R}$ was used to generate regression coefficients, and so on.] If "Total" is not listed as a source in the summary table, simply add the *SS* entries for regression due to all *m* predictors and for error or residual. Table 2.4 provides a summary of alternative formulae in the three systems.

The answer to the second question is "it depends"—primarily on the uses to which you wish to put your results. In deciding which predictors to retain in the equation, the *z*-score weights (or their corresponding "usefulness" measures; see section 2.6) are of more use than the raw-score weights. However, in choosing the simplified weights (usually 1, 0, and -1; see section 2.3.3), the raw-score coefficients *b* should be employed if the units of measurement are mutually compatible; otherwise, the simplified regression variate should be based on the *z*-score regression weights b_z .

Insofar as significance tests are concerned, all tests of a given hypothesis yield exactly the same *F* ratio, whether based on deviation scores, covariances, or correlations. The only apparent exceptions to this are the test of the H_0 that b_1 has a specific, nonzero value and the test of a contrast among the β_js. They appear to be exceptions in that the test of, say, H_0 that $\beta_{z_1} = \beta_{z_2}$ yields a different *F* ratio than H_0 that $\beta_1 = \beta_2$. This is in fact not an exception, because the two tests are of different hypotheses; $\beta_{z_1} = \beta_{z_2}$ if and only if $\beta_1(s_1/ s_y) = \beta_2(s_2/ s_y)$, so that the hypotheses are equivalent if and only if $s_1 = s_2$. If instead, for instance, $s_1 = 10$ while $s_2 = 2$, the null hypothesis that $\beta_{z_1} = \beta_{z_2}$ (i.e., that the *z*-score regression weights for variables 1 and 2 are identical in the population) is equivalent to H_0 that $\beta_1 = \beta_2/5$, whereas H_0 that $\beta_1 = \beta_2$ is equivalent to H_0 that

Table 2.4 Alternative MRA Formulae

Statistic	$\mathbf{R}_x$ - based	$\mathbf{S}_x$ - based	$\mathbf{x'x}$ – based
Regression coefficients	$\mathbf{b}_z = \mathbf{R}_x^{-1}\mathbf{r}_{xy}$ $b_{z_j} = b_j(s_j / s_y)$	$\mathbf{b} = \mathbf{S}_x^{-1}\mathbf{s}_{xy}$ $b_j = b_{z_j}(s_y / s_j)$	$= (\mathbf{x'x})^{-1}\mathbf{x'y}$ $b_j = b_{z_j}(s_y / s_j)$
R^2	$\mathbf{b'}_z\mathbf{r}_{xy} = \sum_j b_{z_j} r_{jy}$ $= \mathbf{r'}_{xy}\mathbf{R}_x^{-1}\mathbf{r}_{xy}$ $= \mathbf{b'}_z\mathbf{R}_x\mathbf{b}_z$	$= \frac{\mathbf{b's}_{xy}}{s_y^2} = \frac{\sum s_{jy} b_j}{s_y^2}$ $= \frac{\mathbf{r'}_{xy}\mathbf{R}_x^{-1}\mathbf{r}_{xy}}{s_y^2}$ $\mathbf{b'}S_x\mathbf{b}/s_y^2$	$= \frac{\mathbf{b'x'y}}{\sum y^2} = \frac{\sum b_j \sum x_j y}{\sum y^2}$ $= \frac{\mathbf{y'x(x'x)^{-1}x'y}}{\sum y^2}$ $= \mathbf{b'(x'x)b}/\sum y^2$
SS_{total}	1.0	s_y^2	$\sum y^2$
SS due to regression	R^2	$\mathbf{b's}_{xy} = R^2 s_y^2$	$\mathbf{b'x'y} = R^2 \sum y^2$
SS due to adding predictors	ΔR^2	$\Delta R^2 s_y^2$	$\Delta R^2 \sum y^2$
SS for contrast among βs	$\frac{(\mathbf{c'}_z\mathbf{b}_z)^2}{\mathbf{c'}_z\mathbf{R}_x^{-1}\mathbf{c}_z}$	$\frac{\left(\sum c_j b_j\right)^2}{\mathbf{c'S}_x^{-1}\mathbf{c}}$	$\frac{(\mathbf{c'b})^2}{\mathbf{c'(x'x)^{-1}c}}$
SS_{res}	$1 - R^2$	$(1 - R^2)s_y^2$	$(1 - R^2)\sum y^2$

Note: Computations involving bold-face terms (matrices and vectors) require use of matrix algebra or of a computer program.

$\beta_{z_1} = 5\beta_{z_2}$. (The test of the hypothesis that $\beta_j = 0$, on the other hand, is identical to the test of H_0 that $\beta_{z_j} = 0$, because a constant times zero still equals zero.)

2.3.3 Specific Comparisons

Rejection of the overall H_o should almost never be the final step of a multiple regression analysis. Instead, the researcher will almost always be concerned with the related problems of (a) interpreting the regression variate and (b) considering the possibility that a subset of only p of the m predictors may perform almost as well as the full set in predicting scores on Y. The first problem involves providing an intuitively or theoretically meaningful interpretation of $\hat{Y}$, that linear combination of the X_js that best predicts scores on Y and that we may refer to as the ***regression variate,*** thereby emphasizing that it is best interpreted as a new, emergent variable in its own right rather than as a listing of separate univariate predictors. Whatever interpretation we put on the regression variate, it is unlikely to justify, say, use of .832 rather than .834 for b_{z_3}. Instead, most interpretations will imply either raw-score or z-score weights of zero, 1.0, or -1.0. (We must keep in mind that simple processes can sometimes generate complex-appearing coefficients. The point remains, however, that sampling variation precludes finding precisely these values.)

The adequacy of one's interpretation of the regression variate should be assessed by testing for statistical significance the ***simplified regression variate*** ($\hat{Y}_{\text{simp}}$) your interpretation implies and by comparing its squared correlation with Y to R^2 (which is, of course, the squared correlation between Y and the regression variate expressed in its full, multidecimal glory). If your substantive interpretation is an apt one, this latter comparison will show very little decrease in the proportion of variance in Y explained by the simplified (and interpretable) variate compared to that provided by $\hat{Y}$ itself.

The critical value for the test of $r^2_{Y\hat{Y}_{\text{simp}}}$ is simply that value of R^2 (i.e., of $r^2_{Y\hat{Y}}$) that would have been required to consider R^2 statistically significant, namely,

$$R^2_{\text{crit}} = \frac{F_\alpha(m, N-m-1)}{[(N-m-1)/m] + F_\alpha(m, N-m-1)}.$$

We can in fact use R^2_{crit} as a post hoc critical value for tests not only of $\hat{Y}_{\text{simp}}$, but also of individual predictors and linear combinations thereof. After all, $\hat{Y}_{\text{simp}}$ represents the maximum possible capitalization on chance (use of the data to guide our selection of tests), so any other linear combination of the X_js we might propose, whether or not we first examine the data, involves less than maximal capitalization on chance and must be considered significant on a post hoc basis if its squared correlation with Y exceeds that (R^2_{crit}) set for the deliberately maximized combination of measures.

Comparing $r^2_{Y\hat{Y}_{\text{simp}}}$ to R^2_{crit} is equivalent to (and indeed was derived from) simply substituting $r^2_{Y\hat{Y}_{\text{simp}}}$ for R^2 in our F test of the statistical significance of R^2, that is, to comparing

$$F = \frac{r^2_{Y\hat{Y}_{\text{simp}}} / m}{(1 - r^2_{Y\hat{Y}_{\text{simp}}})/(N-m-1)}$$

to $F_\alpha(m, N - m - 1)$. However, we know that $(1 - R^2)/(N - m - 1)$ is an unbiased estimate of error variance, whereas $(1 - r^2_{Y\hat{Y}_{\text{simp}}})/(N - m - 1)$ overestimates error variance—by a considerable amount if we consider very nonoptimal choices of $\hat{Y}$, as we might very well do in using R^2_{crit} for post hoc exploration of a variety of different linear combinations. It is thus legitimate (and, as it turns out, preserves the basic union–intersection logic of this post hoc exploration procedure) to compute our F instead as

$$F = \frac{r^2_{Y\hat{Y}_{\text{simp}}} / m}{(1 - R^2)/(N - m - 1)}.$$

This is easily shown to be equivalent to using

$$r^2_{\text{crit}} = [m/(N - m - 1)](1 - R^2)F_\alpha(m, N - m - 1)$$

as our critical value for squared correlations between Y and post hoc choices of $\hat{Y}_s$.

This post hoc exploration of the data can continue ad infinitum with at most a probability of α_{ov} (the alpha level chosen for the test of R^2) of falsely rejecting one or more null hypotheses. You should therefore feel at liberty to test all linear combinations of predictors that seem theoretically or empirically interesting. If, on the other hand, you are solely interested in individual predictors (i.e., in choosing the best single predictor of Y from among the X_js), you should not have performed the multiple regression analysis in the first place but should instead have tested each value of r^2_{jy} against the Bonferroni-adjusted critical value,

$$\frac{t^2_{\alpha/m}}{t^2_{\alpha/m} + (N - 2)},$$

where α is the desired experimentwise error rate.

2.3.3 Illustrating Significance Tests

Our discussion of significance tests in MRA has thus far been couched in (matrix) algebraic terms. To begin the process of consolidating this newly acquired statistical knowledge in the fingertips, we should apply some of these techniques to real data. In an effort to give you an idea of the process, Example 2.1 describes an analysis of real data conducted in a multivariate statistics class at the University of New Mexico.

Example 2.1 *Locus* of *Control, the CPQ, and Hyperactivity.* The data for this exercise were provided by Linn and Hodge's (1982) followup to Dick Linn's (1980) master's thesis. In his thesis, Linn investigated the problem of identifying hyperactive children on the basis of the Connors Parent Questionnaire (CPQ), the Locus of Control scale (LOC), and two sustained-attention measures obtained in the laboratory: CPT–Correct (CPT–C) and CPT–Error (CPT–E). It was already known that the CPQ is moderately correlated with hyperactivity: The zero-order correlation with a dichotomous clinical judgment was .8654, r^2 = .749, for Linn's sample of 16 children diagnosed as hyperactive and 16 "normals"; $F(1,30) = (.749/.241)(30) = 89.5$. However, the CPQ is based on retrospective reports by the parents of their child's behavior at various ages, and many researchers would feel more comfortable supplementing or replacing the CPQ with data based on the child's own responses (the LOC) and on his or her observable behavior (the two CPT measures). We'll consider first the alternative of simply substituting some linear combination of LOC, CPT–C, and CPT–E for the CPQ. A multiple regression analysis with CPQ scores serving as the outcome measure (Y) and the other three scores as predictors (X_1 through X_3) is obviously relevant here; only if R^2 is quite high would we wish to consider this substitution strategy (as opposed to the strategy of *supplementing* the CPQ with the other three measures).

The data were well within the limits (~12 variables and ~100 subjects) of STAT/BASIC, the IBM package of on–line, interactive statistical programs available on the University of New Mexico's computer, so the scores of the 32 children on five variables (the four we've mentioned plus a fifth variable that incorporated the clinical judgment of hyperactivity or not as a 0–1 dichotomy) were typed into a STAT/BASIC file via the keyboard of a CRT (cathode ray tube) terminal for editing and for easy access by students in the multivariate statistics class. The on-line correlation program was then run to obtain the 5 *x* 5 correlation and covariance matrices (**R** and **S**). The appropriate subsets of these matrices were then typed into MATLAB (Moler, 1982), a program that allows a user sitting at a CRT or computer terminal to enter matrix commands and have the computer carry out these matrix operations interactively.

Actually, the STAT/BASIC correlation program reports correlation coefficients to only three decimal places. Because some of the detailed comparisons the multivariate students were expected to consider might require greater accuracy than three significant digits, it was decided to enter the covariance matrix into MATLAB and then let MATLAB (which includes an option to have results of calculations reported to 12 decimal places) compute a more accurate correlation matrix from the variances and covariances, using the fundamental relationship

$$r_{ij} = \frac{s_{ij}}{s_i s_j}.$$

This process yielded the following variances and covariances (arranged conveniently in a rectangular array known to *cognoscenti* as the ***variance-covariance matrix***):

$$\mathbf{S} = \begin{array}{l} \\ \text{LOC} \\ \text{CPT-C} \\ \text{CPT-E} \\ \text{CPQ} \\ \text{Hyper} \end{array} \begin{array}{c} \begin{array}{ccccc} \text{LOC} & \text{CPT-C} & \text{CPT-E} & \text{CPQ} & \text{Hyper} \end{array} \\ \begin{bmatrix} 17.007 & -42.277 & 30.414 & 85.897 & 0.952 \\ -42.277 & 498.750 & -315.055 & -509.000 & -5.468 \\ 30.414 & -315.055 & 365.870 & 355.333 & 4.500 \\ 85.897 & -509.000 & 355.333 & 1658.254 & 17.903 \\ 0.952 & -5.468 & 4.500 & 17.903 & .258 \end{bmatrix} \end{array}$$

Note: The entry in the CPQ row of the CPQ column (1658.254) gives the variance of CPQ scores, s^2_{CPQ}. The entry in the CPT–C row of the LOC column gives the covariance between CPT–C scores and LOC scores, $s_{\text{CPT–C,LOC}}$. From the variances and covariances we can compute the correlations among these five variables (and arrange them in a rectangular form known as the ***correlation matrix***):

$$\mathbf{R} = \begin{array}{l} \\ \text{LOC} \\ \text{CPT-C} \\ \text{CPT-E} \\ \text{CPQ} \\ \text{Hyper} \end{array} \begin{array}{c} \begin{array}{ccccc} \text{LOC} & \text{CPT–C} & \text{CPT–E} & \text{CPQ} & \text{Hyper} \end{array} \\ \begin{bmatrix} 1 & -.4590 & .3856 & .5115 & .4545 \\ -.4590 & 1 & -.7375 & -.5597 & -.4820 \\ .3856 & -.7375 & 1 & .4562 & .4632 \\ .5115 & -.5597 & .4562 & 1 & .8655 \\ .4545 & -.4820 & .4632 & .8655 & 1 \end{bmatrix} \end{array}$$

The first three rows and columns of the preceding two matrices give us $\mathbf{S}_x$ and $\mathbf{R}_x$, respectively, for the present analysis, while the first three and fourth elements of the fourth column provide $\mathbf{s}_{xy}$ and $\mathbf{r}_{xy}$, respectively. From these, we calculate raw-score regression coefficients as $\mathbf{b} = \mathbf{S}_x^{-1}\,\mathbf{s}_{xy}$

$$\begin{array}{c} \qquad\qquad\qquad\qquad\qquad \begin{bmatrix} 85.90 \\ -509.00 \\ 355.33 \end{bmatrix} \\ = \begin{bmatrix} .07496 & .00530 & -.00167 \\ & .00477 & .00367 \\ & & .00630 \end{bmatrix} \begin{bmatrix} 3.148 \\ -.670 \\ .133 \end{bmatrix}. \end{array}$$

MATLAB was used to compute the raw-score regression coefficients, the z-score regression coefficients, and R^2 for the prediction of CPQ scores from LOC, CPT–C, and CPT–E, but we can also compute these statistics from the scalar formulae of section

2.2.3. Our arithmetic will be considerably simpler (though still not simple) if we compute the b_zs, because $s^2_{z_i}$ always equals 1.0, $s_{z_i,z_y} = r_{iy}$, and $s_{z_i,z_j} = r_{ij}$. Thus we have

$$D = 1 - r_{12}^2 - r_{13}^2 - r_{23}^2 + 2r_{12}\,r_{23}\,r_{13}$$
$$= 1 - (-.4590)^2 - (-.7375)^2 - .3856^2 + 2(-.459)(-.7375)(.3856) = .35779$$
$$b_{z_1} = b_{z_{LOC}} = [r_{1y}(1 - r_{23}^2) + r_{2y}(r_{13}\,r_{23} - r_{12}) + r_{3y}(r_{12}\,r_{23} - r_{13})]/D$$
$$= [.5115(.45609) - .5597(.17462) + .4562(-.04709)]/.35779$$
$$= .3188$$
$$b_{z_2} = b_{z_{CPT-C}} = r_{1y}(r_{13}\,r_{23} - r_{12}) + r_{2y}(1 - r_{13}^2) + r_{3y}(r_{12}\,r_{13} - r_{23})$$
$$= [.5115(.17462) - .5597(.85131) + .4562(.56051)]/.35779$$
$$= -.3674$$
$$b_{z_3} = b_{z_{CPT-E}} = r_{1y}(r_{12}\,r_{23} - r_{13}) + r_{2y}(r_{12}\,r_{13} - r_{23}) + r_{3y}(1 - r_{12}^2)$$
$$= [.5115(-.04709) - .4820(.56051) + .4632(.78932)]$$
$$= .0623$$

With the z-score regression coefficients in hand, we can then compute R^2 as the sum of the products of each coefficient times the corresponding r_{iy} to get $R^2 = .3188(.5115) - .3674(-.5597) + .0623(.4562) = .3971$. We can also get the raw-score regression coefficients from the relationship, $b_i = b_{zi}\,(s_y/s_i)$. Thus, for instance,

$$b_2 = -.3674\sqrt{1658.254/498.756} = -.6699.$$

As you will wish to verify, the other two are $b_{LOC} = 3.1482$ and $b_{CPT\text{-}E} = .1326$.
As a check on the consistency of our calculations so far, we can also compute R^2 as $\Sigma\, b_i\, s_{iy}/s_y^2 = [3.1482(85.9) - .6699(-509.0) + .1326(355.33)]/1658.254 = .3971$.

Computer Break 2.1: CPQ vs. LOC, CPT–C, CPT–E. You should now get familiar with your local computer package's procedures for reading in a covariance or correlation matrix from which to begin its MRA calculations, and use this procedure to verify our hand calculations thus far. In SPSS this is accomplished via the Matrix Data command and the matrix-input subcommand of REGRESSION, as follows:

```
MATRIX DATA VARIABLES = LOC, CPT-C, CPT-E, CPQ, HYPER /
  CONTENTS = N CORR / FORMAT = FREE FULL
BEGIN DATA .
 32 32 32 32
 1       -.4590  .3856  .5115  .4545
 -.4590   1     -.7375 -.5597 -.4820
 .3856  -.7375   1      .4562  .4632
 .5115  -.5597  .4562   1      .8655
 .4545  -.4820  .4632  .8655   1
END DATA .
REGRESSION  MATRIX = IN (COR = *) /
```

```
    DESCRIPTIVES/VARIABLES = LOC TO HYPER/
     STATISTICS = Defaults CHA  CI HISTORY/
    DEPENDENT = CPQ/
    ENTER CPTC CPTE/ENTER LOC/
     DEPENDENT = HYPER/ENTER CPQ/ ENTER LOC CPTC CPTE /
    Dep = Hyper/ Enter LOC CPTC CPTE /Enter CPQ/
    Scatterplot (Hyper, *Pred) / Save Pred (PredHypr) .
Subtitle  Followups to MRAs  .
Compute cpqsimp = LOC/4.124 - CPTC/22.333  .
COMPUTE ATTENTN = CPTC - CPTE  .
CORRELATIONS LOC to ATTENTN / Missing = listwise / Print = sig twotail  .
```

Before we get any further enmeshed in computational details, we should pause to reiterate the basic simplicity and concreteness of what we have been doing in this MRA. We have determined—via mathematical shortcuts, but with results perfectly equivalent to conducting a years-long trial-and-error search—that the very best job we can do of predicting scores on the CPQ from some linear combination of the scores on CPT–C, CPT–E, and LOC is to combine the raw scores on these predictors with the weights given in **b** or to combine subjects' *z* scores on the predictors in accordance with the weights given in $\mathbf{b}_z$. In other words, if we compute each of the 32 childrens' scores on the new variable $\hat{y}$ = 3.148(LOC) - .670(CPT–C) + .133(CPT–E), or instead compute the 32 scores on the new variable, $\hat{z}_y = .319z_{LOC} - .367z_{CPT\text{-}C} + .062z_{CPT\text{-}E}$, we will find that the plain old ordinary Pearson *r* between either set of 32 numbers and the corresponding 32 scores on the CP will be (within the limits of roundoff error) exactly equal to our multiple *R* of $\sqrt{.397} = .630$. No other linear combination of these predictors can produce a higher correlation with the CPQ for this sample of data. This, then, brings us back to the question of how representative of the population regression equation our results are. It's clear that R^2 is too low to justify simply substituting our regression variate for the CPQ in subsequent analyses, but we would still like to explore just what the relationship is between the CPQ and the three predictors. First, can we be confident that the population multiple correlation coefficient isn't really zero? (If not, then we can't be confident that we know the sign of the population correlation between our regression variate and *Y*, or of the correlation between *Y* and any other linear combination of the predictors.) The test of this overall null hypothesis is given by

$$F_{ov} = \frac{R^2/m}{(1-R^2)/(N-m-1)} = \frac{.397/3}{(1-.397)/(32-3-1)} = .1323/.02153 = 6.145$$

with 3 and 28 degrees of freedom. Looking up $F_\alpha(3,28)$ in the DFH = 3 column and the DFE = 28 row of Table A.4 and noting that this is very different from $F_\alpha(28, 3)$, we find that an *F* of 2.94 is needed for significance at the .05 level and 4.57 for significance at the .01 level. Because our obtained *F* is larger than either of these two values, we can safely reject the null hypothesis and conclude that our sample R^2 is significantly larger than zero at the .01 level. Because scores on the LOC could be multiplied by an arbitrary constant without losing any essential meaning—that is, since the choice of unit of measurement for scores on the LOC is arbitrary—we should base our interpretation of the regression variate on *z*-score weights. It indeed looks very close to $b_{z_1} - b_{z_2}$. Taking as our simplified regression variate $\hat{z}_{Y_{simp}} = \hat{z}_{simp} = z_{LOC} - z_{CPT\text{-}C}$, we can compute the squared correlation between $\hat{z}_{simp}$ and Y (or y or z_y, because any linear transformation of Y leaves its correlation with any other variable unchanged) very readily.

To do this we could (if we had the raw data available) compute a score for each subject on $\hat{z}_{simp}$ and then compute the correlation between that column of numbers and the column of scores on Y by hand, or via a second computer run. If the statistical package you're using has (as does SPSS) a subprogram that computes *z*-scores and saves them for use by other subprograms in the same run, you can use that subprogram to generate the *z*-scores and then use the data-transformation facilities (e.g., formulae for calculating combinations of spread-sheet columns, or SPSS COMPUTE statements) to calculate $z_1 - z_2$ before using it as input to your correlation subprogram. If there's no easy way to generate *z*-scores for internal use, or if you're doing your calculations by hand, you'll find it to be much easier to first convert $\hat{z}_{simp}$ to the corresponding linear combination of raw scores on X_1 and X_2, namely,

$$\hat{z}_{simp} = z_1 - z_2 = (X_1 - \overline{X_1})/s_1 - (X_2 - \overline{X_2})/s_2$$
$$= (1/s_1)X_1 - (1/s_2)X_2 + \text{a constant term that doesn't affect the correlation}$$
$$= .2425\,X_1 - .04478\,X_2 + \text{constant}$$

or any multiple thereof.

Please note carefully that even though we are now calculating a linear combination of the raw scores on the predictor, we are actually testing our interpretation (simplification) of the *z*-score regression variate.

However, in the present case we don't have the raw data, so we need some way to calculate the correlation between our simplified regression variate and *Y*, knowing only the variances of and covariances (or correlations) among the *X*s and between the *X*s and *Y*. This is one place where I'll have to fudge a bit on my pledge to deemphasize matrix algebra and sneak a bit in the back door to enable you to carry out this all-important step of testing your simplified regression variate. (Otherwise you would be faced with the task of coming up with an interpretation of your multi-decimaled regression variate that explains why the weight for z_3 should be .062 instead of .061 or .059 or .064 or . . . .) I thus offer to you Equation (2.4) for the covariance between two linear combinations of variables.

Covariance between two linear combinations of variables

Let $W = a_1X_1 + a_2X_2 + \cdots + a_mX_m$ and $Q = b_1X_1 + b_2X_2 + \cdots + b_mX_m$.
Then s_{WQ}, the covariance between these two linear combinations of the *X*s, is given by

$$s_{WQ} = a_1 b_1 s_1^2 + a_2 b_2 s_2^2 + \ldots + a_m b_m s_m^2 + [(a_1 b_2 + a_2 b_1) s_{12} + (a_1 b_3 + a_3 b_1) s_{13} + \cdots + (a_{m-1} b_m + a_m b_{m-1}) s_{m-1,m}] \quad (2.4)$$
$$= \sum_{i=1}^{m} a_i b_i s_i^2 + \sum_{i<j} (a_i b_j + a_j b_i).$$

The covariance of a variable with itself is, of course, its variance, and the covariance of two variables expressed in *z*-score form is their correlation, so Equation (2.4) provides all the tools you need to compute the correlation between two linear combinations of either raw or *z* scores. It is, however, an awkward and error-prone equation to employ if you don't organize your arithmetic carefully. The following procedure yields the same answers as Equation (2.4) but greatly reduces the chance of leaving out some crucial substitution:

1. Write each a_i to the left of row *i* of the covariance matrix.
2. Write each b_j at the top of column *j* of the covariance matrix.
3. In cell (*i,j*) of the covariance matrix (i.e., at the intersection of row *i* with column *j*) write the product, $a_i b_j$ (the product of the coefficient at the left of that row with the coefficient at the top of that column).
4. Now compute the sum of cross-products of the two numbers in each cell—that is, add up the $a_i b_j s_{ij}$ products across all m^2 cells. This is s_{wq}, the covariance between the two linear combinations. (It is also the result of pre-multiplying **S** by the row vector **a**' and postmultiplying it by the column vector **b**—but you don't need to know that to compute the needed covariances.)

After a few times using this procedure you'll begin to develop further shortcuts—such as getting the sum of cross-products of covariances with column-heading coefficients for a given row and then multiplying that result by the row coefficient to "polish off" that row—and you'll soon find that you don't really need to write down intermediate products, but can carry out the entire calculation in one pass through your pocket calculator. It's probably best, though, to follow the full procedure at first and let these shortcuts arise naturally from your increasing confidence and skill.

Alternatively, of course, we cold resort to the matrix equation,

$$r^2_{\hat{z}_{\text{simp}},Y} = \frac{(\mathbf{a}'_z \mathbf{r}_{xy})^2}{\mathbf{a}'_z \mathbf{R}_x \mathbf{a}_z},$$

where $\mathbf{a}_z' = (1, -1, 0)$.

Returning now to our specific problem of computing the correlation between $\hat{z}_{y,\text{simp}}$ and *Y*, we use the first four rows and columns of the correlation matrix with a_is of 1, -1, and 0 and a single b_1 of 1.

Thus $r^2_{\hat{z}_{\text{simp}},Y}$ which we can label as r^2_{simp} on behalf of the typesetter, equals

$$[.5115 - (-.5597)]^2/[1 + 1 - 2(-.4590)] = 1.0712^2/2.9180 = .393,$$

which compares quite favorably to our R^2 of .397. Moreover, we can test r^2_{simp} for statistical significance on a fully post hoc basis by substituting it for R^2 in the numerator of F_{ov}, yielding $F = (.393/3)/.02167 = 6.044$ with 3 and 28 *df*, $p < .01$. [Note that we used $(1 - R^2)/28$, rather than $(1 - r^2_{\text{simp}})/28$, in the denominator of this F ratio.]

These two significance tests—of R^2 and of r^2_{simp} — are quite sufficient for the purpose of describing the optimal prediction of the CPQ from scores on the three predictors. However, for didactic purposes, we'll examine a few other tests. First, the statistical significance of $b_{z_{LOC}}$ can be tested via F for the test of H_0 that $\beta_{z_1} = 0$, which equals

$$\frac{b^2_{z_1}}{r_{11}(MSE)},$$

where r_{11} is the (1, 1) entry in $\mathbf{R}_x^{-1}$ and where $MSE = (1 - R^2)/28$, because the rest of the terms in this F ratio are obviously coming from the **R**-based "system." Thus we have $F = .319/[1.2748(.02153)] = 3.708$ with 1 and 28 *df*, nonsignificant, even by a purely a priori criterion. This method requires knowledge of the (1,1) entry of the inverse of the correlation matrix. However, every MRA computer program I've ever encountered provides this F test—or the equivalent t test—for each of the regression coefficients. You'll find, then, that your output includes an F for the statistical significance of b_{LOC} of 3.708 with 1 and 28 *df*, as computed earlier.

On the other hand, a test of the difference between b_{z_1} and b_{z_2} is not routinely produced, and in fact is not, to my knowledge, included as an option in any current MRA program. However, many MRA programs (e.g., SPSS's REGRESSION subprogram) include as an option the printing of the covariance matrix of the regression coefficients, and/or the inverse of the correlation matrix.

If the output includes the "covariance matrix for the regression coefficients," you'll have to read your manual carefully to determine whether this is the covariance matrix for raw-score ("unstandardized") or *z*-score ("standardized") regression coefficients. If this is not clear, then compute an F for the statistical significance of a single regression coefficient under the assumption that the main diagonal entry corresponding to that coefficient is the variance of that *z*-score coefficient and again under the assumption that it is the variance of the raw-score coefficient, and compare to the F reported by your program for the statistical significance of that coefficient.

If the program outputs the inverse of the correlation matrix, then multiplication of those numbers by $(1 - R^2)/(N - m - 1)$ gives (as it turns out) the covariance matrix of the *z*-score coefficients. You can then compute the statistical significance of any linear combination of the regression coefficients, $c_1b_1 + c_2b_2 + \ldots + c_mb_m$, from

$$F_{\sum c_i b_i} = \frac{(\sum c_i b_i)^2}{\text{variance of} \sum c_i b_i}$$

where the variance in the denominator is computed via Equation (2.4), with the b_is taking on the role of *X*s. In the present case, this yields an *F* of 8.162 .

The corresponding matrix-algebraic approach computes our *F* as

$$\frac{(\mathbf{c}'_i\, \mathbf{b}_i)^2}{(\mathbf{c}'_i\, \mathbf{R}_x^{-1}\mathbf{c}_i)MSE},$$

where $\mathbf{c}_i' = (1, -1, 0)$.

Thus we have

$$F = \frac{[.319-(-.367)]^2}{[1.2748+2.3797-2(.4883)]MSE} = \frac{.686^2}{2.6779(.02153)} = 8.162$$

with 1 and 28 *df*, $p < .01$ as an a priori test, Treated as a fully post hoc test, we would divide our numerator by 3 (because we are using all three *df* when we look at the results before choosing contrast coefficients) and compare the resulting *F* to $F_\alpha(3, 28)$) whence we have $F = 2.721$ with 3 and 28 *df*, nonsignificant at the .05 level.

Finally, we could compute a fully post hoc .95-level confidence interval around β_{z_1} - β_{z_2} via

$$\mathbf{c}'\mathbf{b}_z \pm \sqrt{(\mathbf{c}'\mathbf{R}_x^{-1}\mathbf{c})(m)[F_\alpha(m, N-m-1)](1-R^2)/(N-m-1)}$$
$$= .686 \pm \sqrt{2.6779(3)(2.94)MSE} = .686 \pm \sqrt{.50852}$$
$$= .686 \pm .713,$$

or, in scalar terms,

$$\sum c_i b_{z_i} \pm \sqrt{mF_\alpha(m, N-m-1)(\text{variance of} \sum c_i b_{z_i})}$$
$$= .686 \pm \sqrt{.05766(3)(2.94)}$$
$$= -.027 \text{ to } 1.499,$$

which is an interval that includes zero, consistent with the result of our fully post hoc *F* test of this contrast. Thus, even though the difference between *z* score on LOC and *z* score on CPT–C (our simplified regression variate) is significantly related to score on the CPQ by a fully post hoc test, we can't be confident at the .95 level (again, on a fully post hoc basis) that the difference between these two predictors' *z*-score regression coefficients is not zero.

2.3.4 Stepwise Multiple Regression Analysis

The process described at the end of Section 2.3.3 (in which a simplified version of the regression variate is interpreted and tested for statistical significance) effectively reduces the number of predictors to those receiving nonzero weights in the simplified regression variate. An alternative procedure (more accurately, collection of procedures) for selecting a subset of the predictors while giving up relatively little in ability to predict Y is ***stepwise multiple regression analysis.***

Stepwise MRA takes two basic forms: step-down analysis and step-up MRA. The former begins with the full equation involving all m predictors and eliminates variables one at a time. This approach relies on the fact that the test of significance of b_j is equivalent to the test of the H_0 that adding X_j to the equation last (or deleting it first) has no effect whatever on the population value of R^2. At each stage, then, the variable whose F test for significance of contribution (b_j) to the equation is smallest can be dropped with the least impact on R^2. The step-up procedure, on the other hand, begins with that single predictor whose squared correlation with Y is highest. It then adds to the equation that X_j whose addition will yield the largest increment to R^2. This is actually a rather efficient computational technique for constructing the final regression equation. Available computer programs differ in whether they employ step-up (the most common approach), step-down, or some combination of the two approaches and in the flexibility they provide the user in specifying criteria for adding or deleting predictors (including forcing some variables to be included regardless of their performance statistics) and for stopping the whole procedure. If you wish to employ a stepwise analysis, rather than simply going directly to a solution for the 111 predictors of interest to you (which for many programs simply means that the results of the successive steps are computed, but not printed), you'll need to read quite carefully the manual for the program you're using. A few general points about stepwise MRA can, however, be offered:

1. R^2 and the values of the regression coefficients for a particular set of predictors are independent of the order in which the X_js were added or deleted to arrive at this set.
2. The magnitude or statistical significance of the increment to R^2 provided by adding X_j to the equation at any particular stage of a stepwise MRA need bear no relationship to its zero-order correlation with Y (r_{iy}) nor to its contribution at any other stage of the analysis (i.e., in the presence of a different set of predictors.)
3. No combination of stepup and step-down procedures can guarantee that the set of $k < m$ predictors you wind up with will be the "best" set of that size, in the sense of yielding a higher R^2 than any other subset of size k. (There is, however, at least one computer program—SAS's PROC RSQUARE—that does exactly that by examining the regression model for each of the $m!/[k!(m - k)!]$ subsets of predictors, although it is enormously time-consuming for $m > 10$.

4. In general,. stepwise MRA is probably best reserved for situations in which an a priori ordering of predictors in terms of theoretical or practical (e.g., cost of data collection) considerations dictates the sequence in which subsets of predictors are added.

Example 2.1 *Revisited.* An example of such a situation is provided by Linn and Hodge's (1981) study, introduced in section 2.3.4. You will recall that our MRA of CPQ predicted from LOC, CPT–C, and CPT–E, although revealing a statistically significant ability to predict CPQ from the three other variables, yielded too low a value of R^2 to warrant simply substituting our regression variate for CPQ. However, we still need to consider the possibility that these same three measures provide a useful supplement to the CPQ in the prediction of hyperactivity. To determine whether LOC, CPT–C, and CPT–E provide a statistically significant increment to the prediction of hyperactivity over that provided by the CPQ alone, the R^2 based on all four measures was computed, yielding a value of .760, clearly a nonsignificant increment. (You may wish to verify that $F = .412$.)

Note, however, that the computer output for the step in which the three additional predictors were forced into the equation reported only an F value of 21.35 with 4 and 27 df, $p < .001$. As is typical of stepwise programs, SPSS's REGRESSION summary table gives only the test for significance of R^2 based on all predictors currently in the equation; it is up to the user to construct tests of the increments to R^2 from the output provided. On the other hand, SPSS's REGRESSION includes an option (STATISTICS = CHA) to print the change in R^2 and its associated F at each step. Note, too, that this tiny, nonsignificant increment to R^2 does not necessarily mean that LOC, CPT–C, and CPT–E are unrelated to hyperactivity. An MRA employing only these three predictors yields an R^2 of .318, with a corresponding $F(3,28) = 4.34$, $p < .05$.

2.4 COMPUTER PROGRAMS FOR MULTIPLE REGRESSION

I suspect that you are by now impressed with the difficulty of carrying out an MRA "by hand" in even the 3-predictor case, and that you can imagine how very difficult hand analysis of four-or-more-predictor cases would be. Fortunately, computers are now so generally available, together with "canned" programs designed to handle most commonly desired statistical analyses, that there is no longer a need for the many involved discussions of methods to reduce the labor of hand computations and provide intermediate checks on accuracy that filled early textbooks on multivariate statistics. I do, however, have some precautions and words of advice with respect to the use of computer programs to share with you, following which we'll examine together some useful programs for regression analysis.

2.4.1 Computer Logic and Organization

Computers are fast, reliable, stupid, and rigid. The speed of computers needs no further comment, except to note that one's conception of which computational procedures are "practical" must be revised with each jump in the speed of the computer, to the point that iterative procedures involving repeating a complicated analysis with slightly different parameter values (e.g., guesses as to factor loadings) until the successive guesses coverage on an optimal set of values are quite feasible on Y2K laptop computers.

An aspect of computers' "behavior" that deserves more emphasis than it is popularly accorded is their high reliability, that is, their ability to (indeed, their inability not to) perform any set of instructions programmed into the computer hundreds of thousands of times without once departing from those specifications in the slightest detail. This is especially important in numerical computations, where a human operator employing a desk calculator will almost inevitably make at least one error in the course of entering numbers and recording results in any but the shortest string of computations. Elaborate precautions involving devising various internal checks on the calculations and having each analysis performed at least twice must therefore be taken against these errors in any "hand" analysis of more than the most trivial size. These precautions add to the time differential between hand and computer calculations.

A less desirable aspect of the computer's reliability is its rigidity, a rigidity which in a human calculator would warrant the label "stupidity." Although the computer will not "misplace" or "forget" the location of the decimal point in a number once it has stored that number in one of its memory cells, neither will it recognize that the comma in an entry of "1,000" in a correlation matrix being input to it must "really" be a decimal point, nor will it note that the data on one line have been entered consistently one space too far to the left, requiring a simple departure from the prespecified format for reading that line. (Actually, checks for these kinds of goofs in data preparation, together with instructions as to how to handle them before proceeding with the analysis, can be built into the computer program, but it is virtually impossible to anticipate in advance all of the contingencies that a reasonably intelligent human calculator operator can recognize and handle "reasonably" as they arise.) The point is that a computer performs what it is specifically instructed to do by some human programmer, and nothing else. The computer is in many respects like a human calculator operator who is extremely dependable and extremely fast, but who has no "initiative" or "reasonableness" or "common sense" or "intelligence" that is not specifically incorporated into his or her instructions.

A ***canned program*** is a program written to perform a specific function or set of functions (e.g., conduct multiple regression analyses) and then saved on a direct access storage device (such as a diskette or your PC's hard drive) in such a way that it can be called into the computer's working memory (RAM) within a matter of seconds. A canned program sacrifices flexibility (e.g., the almost complete flexibility afforded by programming your analyses in a general-purpose language such as C++ or the old standby, FORTRAN) for ease of learning and ease of use.

Most canned programs nowadays are organized into ***statistical packages*** of programs. These are a collection of subprograms, each performing a particular statistical procedure, and all sharing a common executive program for handling data input, transforming the data in various ways before analysis, and so on. One advantage of the package over simply collecting separate programs is the high similarity in the conventions concerning preparation of the commands ("syntax") for the use of the various programs in the package. Another is that by becoming familiar with just two or three of the most widely disseminated statistical packages an individual can be confident of finding familiar programs for most of his or her needs wherever he or she goes. Yet a third advantage is that the widespread use of these packages makes their subprograms the most thoroughly tested, most nearly (but never completely!) "bug-free" of all available programs. For these reasons, the present text concentrates its discussion of computer programs available for various procedures on those programs found within two nationally disseminated packages, SPSS and SAS. (The third most ubiquitous package back in 1985, BMDP, has been absorbed by SPSS.)

Before getting into a procedure-oriented discussion of particular programs, it may prove helpful to provide a few general "tips."

2.4.2 Sage Advice on Use of Computer Programs

Perhaps the most difficult step in becoming familiar with computer operations is the first one—that of learning the details of access to and use of the computing facilities available to you. For students and faculty members, one of the best ways to learn these things is to call the receptionist (or, better, the Help Desk staff) at your college's computing center and ask. (This may also be all that it takes at some computing centers to obtain an account good for the duration of your stay at that college. At other installations, a formal application with many signatures may be required.) Most computing centers will have a brief handout in which such matters as how to obtain a user number and what remote facilities are available at what hours are discussed. They also usually provide short courses early each semester on various software packages and other services. As desktop and laptop computers have acquired many times the processing speed and storage capacity of top-of-the-line mainframe computers just a few years older, computing (and instruction on and help with same) have become more decentralized—but your local computing center is still an excellent source of information, help, and deep discounts on statistical software from distributors eager to get students and their advisers "hooked" on their systems.

Throughout the process of acquainting yourself with computer operations, the following points should be kept in mind:

1. ***No program is ever completely debugged.*** Assuming otherwise is to court disaster. As a corollary to this principle:

2. ***Check the output*** of ***any program frequently, in as many ways as possible.*** Some of the kinds of checking procedures available to the user include: (a) Compare this program's answers to the output of one or more other programs. (b) Work some simple problems involving only two or three variables by hand, and then submit these same

problems to the program in question. Especially easy checks are provided by contrived data in which all variables are uncorrelated, or in which all intercorrelations between any pair of variables are identical. However, just because a series of sample problems with known answers are solved correctly does not guarantee that the program is bug-free. For instance, when the old SPSS CANCORR program first became available at the University of New Mexico, it worked perfectly on the sample problem provided in the Update Manual, but it either failed to run at all or gave incorrect answers when analyses were attempted for which the two sets of variables were of different size. It took several months for the problem to be noticed, and subsequently corrected in the next revision of the package. And in a bit of dèjá vu, I discovered (Harris, 1999b) that Versions 8.0 – 10.0 of SPSS for Windows were mishandling Canona, giving different answers depending on which set of measures was named first. This one turned out to be a matter of miscommunication between data entry and subprogram, in that specifying matrix input led (perhaps still leads) the MANOVA subprogram to select the variables in each set, not on the basis of the variable names specified in the subprogram, but on the basis of the order in which they appeared in the original MATRIX DATA statement.

3. Each month brings at least one change in local or nationwide operating procedures, ***so keep in touch*** *(and maintain friendly relations)* ***with your computing center and/or vendor support staff***.

4. ***No one ever has a program—even a canned one—run correctly the first time he or she attempts to use it,*** so do not get discouraged. On the other hand,

5. ***It is never safe to assume that an error you locate*** *in the control statements you prepare for that program* ***is the only error you have made.*** It is very tempting to make a single change and resubmit the commands in the hope that the problem has been corrected, especially because most failures in running a canned program come from errors that are detected very early in program execution and therefore waste only a few milliseconds of computer time. This may in fact be a very reasonable strategy if (a) the system you're using is not yet running into capacity problems, (b) you are submitting this run interactively so you can scan your output on your monitor before (or instead of, if an error is detected) routing it to a hardcopy printer, and (c) it is a simple enough program not to require mounting tapes or other peripheral devices nor to require more than a few minutes to be completed. If these conditions are not true (especially if the output is to go directly to a printer and thus require cutting down more trees to immortalize your errors), you should check over the entire file after a failure.

6. ***The most common sources of program failure, the ones to check for first, are*** (a) exceeding the limits you set in some systems control statement on the amount of time or number of lines of output your program would require; (b) inadvertently typing a special character such as a comma in place of a decimal point; (c) moving a data field over one or more spaces so that, for example, the line positions that are supposed to contain the subject's score on variable 5 have two decimal points, one of them "belonging" to the subject's score on variable 6; (d) omitting command terminators, such as SAS's semicolons or SPSS PC's periods at the end of each complete command; (e) putting a command terminator at the end of a subcommand, so that the program tries to

read the next subcommand as an entirely new program; (f) if using freefield input, omitting one or more scores for one of the subjects (missing data codes being especially likely candidates for omission), thus leading the program to read your next control statement as data while attempting to find the missing score(s) for the last subject; (g) getting your control statements out of order, especially if you have had to add statements to your syntax; and (h) failing to provide on the first systems control line(s) of a batch–processed job (which some of us troglodytes still use) all of the information needed by the operator as to what disk packs or other special storage devices will be required by the program.

(7) ***If you think you may need help with a program, opt for syntax over menus.*** Increasingly, the old system of preparing a set of commands that are then submitted to a program for analysis is being replaced by selection of choices from a menu (or, most often menus). As of this writing menus haven't completely replaced command language files and syntax windows, and asking for online help with a program is still very likely to reveal several "available only through syntax" notations, but the replacement process has gone far enough that recent versions of the Student Pack for SPSS PC don't even provide a syntax window. In addition to the loss of flexibility this entails, it also represents a loss of what used to be a major advantage of computer programs over pocket-calculator (even programmable-pocket-calculator) computations, namely, the record of the data you had entered and the operations you had requested that was preserved in your input file. If you point-and-click your way to an analysis whose results puzzle you, there's no way for a Help Desk consultant or your 10-year-old son (nowadays likely to be a good source of help with computing) to know, and you may not be able to remember, the sequence of point-and-click operations that led to this mess.

Enough grand generalities. On to programs useful for regression analysis.

2.4.3. Computerized Multiple Regression Analysis

2.4.3.1 MATLAB. MATLAB is a matrix-manipulation program initially developed at the University of New Mexico (UNM) by Cleve Moler (1982) and now nationally distributed. Commands are entered in essentially the same way they would be written in a matrix-algebraic formula. Details of input conventions (e.g., enclosing each matrix in square brackets and ending commands with semicolons if you *don't* want to print the result of that command immediately) are given in the online beginner's manual (Mathworks, Inc., 1998).

The MATLAB commands to carry out the initial phase of multiple regression analysis "from scratch" (i.e., beginning with the raw data) are illustrated below:

```
X = [Entries of X, the data matrix for the predictors, with
     spaces between entries; each row (case)  begun on a new line
     or separated from previous row by a semicolon]
Sx = (X'*X - X*X'/N)/(N-1)
Yp = [Subjects' scores on the outcome variable,Y, in row form];
Y = Yp'
```

```
Sxy = X'*Y - X*Yp/N
braw = inv(Sx)*Sxy
Xbar = (X'*ones(N,1))/N
Ybar = (Y'*ones(N,1))/N
b0 = Ybar - Xbar'*braw
Stdx = sqrt(diag(Sx));
Stdy = sqrt((Y'*Y - Y*Yp/N)/(N-1))
bsubz = braw*stdx/stdy
```

2.4.3.2 SPSS REGRESSION, Syntax Window. SPSS is the most widely distributed of all statistical packages. There are (mostly minor) variations among the versions of SPSS written for different operating systems (e.g., Unix vs. Windows 98) and among successive editions ("versions") of SPSS for the same operating system. You'll therefore need to make liberal use of the online tutorials and online help as well as keeping on good terms with your university's or your company's help desk personnel to puzzle out why the sample commands provided in this book give you error messages instead of results the first time you attempt to run them because, for example, the version of SPSS you're using no longer permits multiple "Dep =" subcommands within a single REGRESSION command. I'm nonetheless hopeful that this *Primer*'s examples of syntax that worked way back in 2000 will give you a feel for the basic structure of the SPSS commands being illustrated, as well as providing advice as to choices you must make among available options and as to supplementary "hand" (pocket-calculator) analyses you should carry out.

The basic setup for a "from-scratch" analysis of Data Set 1 is as follows:

```
SET LENGTH = NONE WIDTH=80 .
DATA LIST FREE /X1 to X3, Y .
BEGIN DATA .
102 17 4 96
104 19 5 87
101 21 7 62
 93 18 1 68
100 15 3 77
END DATA .
REGRESSION WIDTH=72/ DESCRIPTIVES /
   VARIABLES = X1 TO X3, Y/
   STATISTICS = DEFAULTS CHA/DEPENDENT = Y /
     Enter X1 / Enter X2 / Enter X3 .
```

Note that each command (but not data lines and not subcommands) ends with a period. This is required in SPSS for Windows, but not for Unix-based SPSS. Note, too, that only commands (not continuation lines or subcommands) begin in the first space on a line.

The "CHA" specification talks SPSS into printing the F test for the increment to R^2 provided by adding to our equation the variables mentioned in each successive "Enter" subcommand. (SPSS does not, however, tell you what the *df* for this F_{incr} are.)

If the overall test of R^2 yields statistical significance, follow-up analyses will include post hoc tests of a simplified version of the regression variate and of any other linear combinations of the predictors that seem interesting after you've examined the data. For example, having determined that the regression variate for Data Set 1 is $1.571z_1 + .275z_2 - 1.419z_3$, we could test the interpretation that it's being at a higher percentile on X_1 than on X_3 that is especially predictive of having a high score on Y via

```
Compute zysimp = X1/4.1833 - X3/2.2361 .
Correlations zysimp Y / Print = Sig Twotail .
```

Note that 4.1833 and 2.2361 are the standard deviations of X_1 and X_3, respectively, and that we don't bother "centering" X_1 and X_3 (expressing them as deviations from their respective sample means) because that just changes the additive constant (the intercept), which is irrelevant to the correlation between them. Note, too, that you will have to compute the post hoc critical value for the followup tests "by hand."

If working from an already-computed covariance or correlation matrix, you use the MATRIX DATA command in place of the DATA LIST command; put a correlation matrix, rather than the raw data, between BEGIN DATA and END DATA; and put the MATRIX IN (*) subcommand immediately after REGRESSION. See the example of a path analysis via MRA in section 2.9.

2.4.3.3 SPSS REGRESSION , Point-and-Click. An excellent source of detailed, step-by-step instructions and examples for statistical analysis via menu selection within SPSS for PCs is Green, Salkind, and Akey's (2000) *Using SPSS for Windows: Analyzing and Understanding Data.* The basic procedure for a regression analysis is to enter your raw data into the Data Editor window in the obvious way (but see your instructor or your Help Desk personnel or Green et al, 2000, if it doesn't seem obvious); define and label your variables by highlighting the column header for each one and then clicking on "Data" and "Define variable"; then make the following menu selections:

Analyze (or "Statistics" in the Student Pack and in some other versions of SPSS)
  Regression
    Linear

When the "Linear Regression" window comes up, highlight your dependent variable's name and click on the right arrow to move it into the "Dependent" slot; then highlight the predictor variables you want to enter into the equation first (holding down the control key until all have been selected) and move them into the "Independent(s) " slot by clicking the appropriate variable; unless you have a darned good reason otherwise, leave "Enter" as the selected "Method"; click on "Next" and repeat predictor-variable selection for the next block of predictors; etc. Then click on "Statistics", "Plot", and "Save" (not necessarily in that order) to make menu selections for, e.g., "R squared change" under "Statistics". Come back to the main window each time by clicking on "Continue". Then either click on "OK" to launch the analysis or click on "Paste" to have SPSS paste the commands your menu selections generate into the syntax window, go to the syntax window (an option under the "Window" button on the taskbar), and click on "Run /All" to launch your analysis.

2.4.3.4 SAS PROC REG and PROC RSQUARE. At my university SAS is available only on Unix; neither I nor the computer pods have the PC version of SAS, so I can't be of much help with the point-and-click version of SAS. On Unix you set up a file with your commands and data in it and remember to give it a name ending in ".sas"; then, at the system prompt ("%" at UNM) you enter

```
% sas filename.sas
```

and, when the command finishes executing (signaled by the appearance of a new system prompt), you discover that your commands and a brief synopsis of what SAS did with them appear in a file named "filename.log", and the output appears in a file named "filename.lst". (That's a letter "el" in "lst", not a numeric "one".) The syntax for an initial analysis of Data Set 1 is:

```
TITLE 'MRA on Data Set 1 Using Version 6 Proc REG';
OPTIONS NOCENTER LS=72;
DATA MRAHW;
     INPUT X1 X2 X3 Y;
     CARDS;
102 17 4 96
104 19 5 87
101 21 7 62
 93 18 1 68
100 15 3 77
;
PROC REG CORR SIMPLE USSCP;
  Step1: MODEL Y = X1 / stb ;
  Step2: MODEL Y = X1 X2 /  STB  ;
  Step3: MODEL Y = X1 X2 X3 / STB ;
```

Note that the "stb" option is necessary to get the *z*-score regression coefficients; that each command *and* each subcommand must end with a semicolon; and that as of this writing the semicolon following the last data "card" must go on a separate line, for no good reason I've been able to figure out.

As pointed out earlier, no combination of step-up and step-down regression analyses can guarantee that you'll find that subset of predictors that yields the largest possible R^2 for a given number of predictors. PROC RSQUARE does, as in:

```
TITLE 'MRAs on All Possible Subsets of Predictors, Data Set 1';
OPTIONS NOCENTER LS=72;
```

```
DATA DataSet1;
     INPUT X1 X2 X3 Y;
     CARDS;
102 17 4 96
104 19 5 87
101 21 7 62
 93 18 1 68
100 15 3 77
;
PROC RSQUARE  Simple Corr;
  MODEL Y = X1 X2 X3 / Select = 3;
```

This run yields, among other output, the following:

```
N = 5     Regression Models for Dependent Variable: Y

Number in      R-square   Variables in Model
  Model
      1       0.29763780   X1
      1       0.19133858   X2
      1       0.00419948   X3
 ---------------------------
      2       0.80625149   X1 X3
      2       0.56280601   X1 X2
      2       0.27478383   X2 X3
 ------------------------------
      3       0.82852143   X1 X2 X3
----------------------------------
```

The "Select =" option was unnecesary in this run; with more predictors it can save a lot of paper. It specifies the maximum number of subsets of predictors (ordered from highest to lowest resulting R^2) that is to be printed for each subset size. Thus, for example, if we had had 10 predictors, RSQUARE would have had to examine and rank order 45 pairs of predictors, 120 triples of predictors, and so on; specifying "Select = 4" would instruct the program to print only the best 4 pairs, the best 4 triples, and so on.

Each of these programs has a few thousand additional "bells and whistles" that could be added to our requested analysis: computation and plotting of various residuals from the models, options for various stepwise approaches, alternative methods of handling missing data, printing of influence statistics that tell us how different the R^2 would be if each case were the only predictor-and-outcome vector omitted, etc. Feel free to explore the options available in the particular program (and the particular version) you are using and try out those that seem useful. But you should also feel free to ignore options that you don't understand, because new methods are constantly being developed and no one—not even academics whose research centers on developing such methods—can keep up with them all.

2.5 SOME GENERAL PROPERTIES OF COVARIANCE MATRICES

There are some general properties of covariance matrices and their inverses of which you should be aware, even though you aren't expected to be conversant with the matrix algebra that renders them obvious. First, it turns out that the main diagonal entries of the inverse of a covariance or correlation matrix are closely related to the squared multiple correlations of each variable with the remaining variables. In particular (cf. Section D2.11 in Digression 2 for proofs),

$$c_{ii} = [s_i^2(1 - R_{i\bullet oth}^2)]^{-1} \quad \text{and} \quad r_{ii} = \frac{1}{1 - R_{i\bullet oth}^2},$$

whence

$$R_{i\bullet oth}^2 = 1 - \frac{1}{s_i^2 c_{ii}} = 1 - \frac{1}{r_{ii}}$$

and

$$b_{ij\bullet oth} = -\frac{c_{ij}}{c_{ii}},$$

where c_{ij} is the entry in the *i*th row and *j*th column of $\mathbf{S}^{-1}$, the inverse of the covariance matrix of a set of variables, and. is also equal to $(N - 1)d_{ij}$ where d_{ij} is the (i, j)th entry of $(\mathbf{x'x})^{-1}$; $R_{i\bullet oth}^2$ is the squared coefficient of multiple correlation between variable *i* and all variables other than *i* that are represented in **S**; and $b_{i,j\bullet oth}$ is the regression coefficient corresponding to variable *j*'s contribution to the regression equation that best predicts scores on variable *i* from scores on ***oth,*** the set of variables other than *i* involved in the regression equation.

There are two important consequences of these relationships. The first is that it is quite easy for a computer program to generate the squared multiple correlation (SMC)of each variable in a set with all other variables in that set. These SMCs are indeed often used as initial estimates of the *communalities* of the variables in a set about to be factor analyzed (cf. chap. 7).

Second, the expression for r_{ii} in terms of $R_{i\bullet oth}^2$ provides yet another insight into why r_{ii} appears in the denominator of the test of significance for the single sample regression coefficient b_i in Table 2.3. We can now rewrite that test as

$$F = \frac{b_{z_i}^2}{r_{ii} MS_{\text{res}}} = b_{z_i}^2 \frac{1 - R_{i \bullet oth}^2}{MS_{\text{res}}}$$

Thus the F ratio increases as the magnitude of X_i's z-score regression coefficient increases (differences in raw-score coefficients due to differential standard deviations being ignored), but it is "discounted" to the extent that scores on X_i are predictable from scores on the other predictors, so that some of X_i 's contribution could be "replaced" by other predictors' contributions, were X_i to be dropped.

This relationship provides an opportunity for milking some information from canned regression analysis computer programs that is not ordinarily provided explicitly by such programs. Any such program will provide a significance test of the difference between b_i and a hypothesized value of zero. BMDO2R employs the F test outlined in Table 2.3, whereas other programs (e.g., SAS PROC REG and SPSS's REGRESSION) report a t test that is simply the square root of the F ratio. Rearranging the expression for the F ratio so as to isolate $R_{i \bullet oth}^2$ yields $R_{i \bullet oth}^2 = 1 - F_i MS_{\text{res}} / b_{z_i}$. Thus, although the multiple correlation between each predictor and the other predictor variables is not usually reported by canned programs, it can be readily obtained from values that are reported. Further, because $b_{z_i}^2 \; (1 - R_{i \bullet oth}^2) = \Delta R_i^2$ = the drop in R^2 that 'results from dropping X_i from the equation (or the increase from adding it last), the "usefulness" measures of the various individual predictors' contributions to the equation (see section 2.6) are simply proportional to the F ratios for the significance of their corresponding regression coefficients.

The ready availability of all of this additional information, although very convenient for such things as item analysis in which the predictors are scores on individual items of some attitude scale or battery of tests, also raises the specter of the multiple-comparison problem in yet another guise. It might be tempting to obtain the inverse of the covariance matrix of some large set of variables, use the main diagonal entries of this inverse to compute R^2 for each variable predicted from the other variables, and test each R^2 for significance. It would therefore be useful to have available, before we start this fishing expedition, a test of the overall null hypothesis that none of the off-diagonal entries in the *population* covariance matrix differ from zero. This is equivalent to testing the hypothesis that the true population correlation between each pair of variables is zero. Bartlett (1954) provided such a test (called the ***sphericity*** test), namely, $\chi^2 = -[N - 1 - (2m + 5)/6] \ln |\mathbf{R}|$ with $m(m - 1)/2$ *df*, where m is the number of variables involved in the correlation matrix **R**, and ln |**R**| represents the natural logarithm of the determinant of **R**.

That |**R**| should be involved in this test is intuitively reasonable, because the smaller |**R**| is, the closer the system is to a perfect linear relationship among the variables (the direct opposite of no relationship, scores on any one being predictable from a linear combination of the other) and the more closely -ln |**R**| approaches positive infinity. Lawley (1940) showed that if N is large or the sample coefficients small, a good approximation to this chi-square test is provided by replacing ln |**R**| with $\sum_i \sum_{<j} r_{ij}^2$, the sum of the squared above-diagonal entries of **R**. This test is not included in most MRA programs, so you'll probably be able to carry it out only if you have access to a matrix-manipulation program such as MATLAB or SPEAKEASY.

2.6 MEASURING THE IMPORTANCE OF THE CONTRIBUTION OF A SINGLE VARIABLE

One of the issues that often arises in applications of multiple regression analysis is that of the relative importance of the contribution of a particular variable to the predictive ability represented by R^2. This turns out to be closely related to the issue of interpreting the regression coefficients in an application of multiple regression analysis. Both of these issues are discussed by Darlington (1968), on whose paper a number of the comments in this section are based.

The crucial question to ask in assessing the contribution of a given variable is "Why?" Is this measure of importance to be used in deciding whether to retain the variable in the regression equation? Or is it to be used instead in assessing which among a number of variables is the most important determinant of scores on the outcome measure?

In the first case, the obvious measure to use is the drop in R^2 that results when that variable is eliminated from the regression equation. This measure was referred to by Darlington as the "usefulness" of the variable. Looking back to Table 2.3, we can see that this can be written as

$$U_i = R^2 = R^{*2} = r'_{xy}(b_z - b^*_z),$$

where R^{*2} is the squared multiple R obtained when variable i is omitted and $\mathbf{b}^*_z$ is the vector of regression coefficients in which the ith position is set to 0 and the m - 1 other b_js are computed on the basis of the first m - 1 predictors only. This would appear to require two separate multiple regression analyses.

However, it is possible to show (Derivation 2.6) that U_i equals $b^2_{z_i}(1 - R^2_{i \bullet oth})$ where $b^2_{z_i}$ is the square of the normalized regression coefficient for predictor i. That is, it is the regression coefficient used in the equation for predicting z scores on the predictors and also = $b_i(s_i/s_y)$, and $R^2_{i \bullet oth}$ is the squared multiple correlation between variable i and the remaining m - 1 predictor variables. In Section 2.5, we pointed out that $R^2_{i \bullet oth}$

could be computed from the ith main diagonal entry of R_x^{-1} or indirectly from the t or F ratio obtained in testing the hypothesis that $\beta_i = 0$. Thus the output of any of the common computer programs for multiple regression analysis provides all of the information needed to compute U_i for each predictor variable.

The second goal is much more ambiguous, and often inappropriate. Many researchers who ask this question are really interested in assessing the value of the variable in a very different context than that in which the prediction equation was derived. For instance, Steele and Tedeschi (1967) attempted to develop an index based on the payoffs used in a two-person, two-choice experimental game that would be predictive of the overall percentage of cooperative choices (C) made in that game. Their approach was to generate 210 different indices, select from among these the 15 that correlated most highly with C, compute the corresponding regression equation, and then compare b_z^2 for the various predictors. If, as their discussion implies, they were interested in generating a single best predictor of C, this procedure was clearly inappropriate. Such a goal would seem to dictate selection of that variable the absolute value of whose simple correlation with C was highest. Putting the various variables into a regression analysis was irrelevant to the goal of selecting a single best predictor, because a regression coefficient provides information about that variable's contribution only within the highly specific context of the other m - 1 predictors. As we have seen in Data Set 1, a variable having a very low correlation with Y considered by itself may receive quite high weight in a regression analysis involving two other predictors, and vice versa. Also a variable that is very important in the three-predictor case may receive near-zero weight when a fourth predictor is added.

Given that we have computed the regression equation for the set of predictors we actually plan to employ, there are still several possible measures of the relative importance of the various predictors in that equation. The most straightforward is $b_{z_i}^2$, because it indicates how variable scores on $b_{z_i} z_i$ will be, and thus how much influence scores on X_i will have in determining the ranking of subjects on $\hat{z}_y$. However, the squares of the z-score regression coefficients do not provide an additive "partition" of R^2, because

$$R^2 = \mathbf{b}_z'\mathbf{r}_{xy} = \mathbf{b}_z'\mathbf{R}_x\mathbf{R}_x^{-1}\mathbf{r}_{xy} = \mathbf{b}_z'\mathbf{R}_x\mathbf{b}_z = \sum b_{zi}^2 + 2\sum_{i<j}\sum b_{zi} b_{zj} r_{ij},$$

In other words, $b_{z_i}^2$ ignores the contributions variable X_i makes through its correlations with the other predictors. If we arbitrarily assign one-half of a pair of predictors' "joint contribution " $2b_{z_i}b_{z_j}$ to each of the two predictors, we obtain

$$b_{z_i}^2 + (\sum_{j \neq i} b_{z_i} r_{ij}) b_{z_i} = b_{z_i} (\sum_{\text{all } j} b_{z_j} r_{ij})$$

which, as it turns out, also equals $b_{z_i} r_{iy}$—as a measure that *does* provide an additive partition of R^2. However, it lacks any operational interpretation like that given earlier for U_i (the drop in R^2 that would result if X_i were dropped from the regression equation), an interpretation that makes U_i a serious candidate for a measure of importance. Moreover, $b_{z_i} r_{iy}$ can be negative, suggesting that predictor *i hurts* our predictability, that is, we could achieve a higher R^2 if X_i were dropped from the equation. This is of course nonsense, because dropping *any* predictor can only lower (or, at best, leave unchanged) R^2. For instance, X_2 in Data Set 1's three-predictor case has an r_{2y} of -.437 and a b_{z_2} of $1.70\sqrt{5/190.5}$ = .275, so that X_2's "contribution" to R^2 by the additive partitioning equals .275(- .437) = -.120. However, rather than X_2's addition to X_1 and X_2 leading to a drop in R^2 of .12, it *increases* R^2 from .806 to .828. In other words, beguiling as the arithmetic property of adding up to R^2 is, the $b_{z_i} r_{iy}$ terms (and therefore their further breakdown into direct and indirect contributions) are essentially meaningless and uninterpretable as measures of the contributions of the different predictors to R^2.

Cooley and Lohnes (1971) made perhaps the most radical suggestion in their recommendation that r_{iy} be used as the measure of the importance of X_i to the endeavor of predicting scores on *Y*. They base this recommendation primarily on the fact that this simple correlation with *Y* is proportional to that variable's "loading on," that is, its correlation with, $\hat{Y}$, is the linear composite of the *X*s that the regression analysis proposes as indicative of the true relationship between *Y* and the *X*s and is thus similar in that respect to a factor in factor analysis. Traditionally the emphasis in interpreting factors has been on examining those original variables that correlate most highly with (have the highest loadings on) a given factor (cf. chaps. 6 and 7). Probably, however, Cooley and Lohnes' argument should be reversed, and the often rather striking discrepancy in magnitude or sign between b_{z_i} and r_{iy} should serve as evidence that the emphasis in interpreting factors would more appropriately be put on factor score coefficients (corresponding to b_{z_i}) than on the indirect criteria of factor loadings (corresponding to r_{iy}).

Table 2.5 illustrates the various importance measures as they apply to a couple of data sets we've worked with and one (the salary bias example) we'll encounter in section 2.9. Had we included a column of loadings of original variables on the regression variate it would have been directly proportional to the r_{iy} column—that is, loadings on the regression variate provide no information we didn't already have once we had computed the zero-order correlations with *Y* and thus have nothing to do with the regression equation. Note, too, how silly the parsing of "importance of contribution" gets

Table 2.5. Measures of Importance in MRA

Data Set	Variable	r_{iy}	b_{z_i}	$b_{z_i}^2(1 - R_{i\bullet oth}^2) = \Delta R^2$	$b_{z_i} r_{iy}$	=	$b_{z_i}^2$	+	$2\sum_{i>j} b_{z_j} Y_{ij}$
Presump-	X1	-.2	0	0	0	=	0	+	0
tuous	X2	.6	5/3	.922	1	=	25/9	-	16/9
Data Set	X3	0	-4/3	.150	0	=	16/9	-	16/9
					---		----		----
					1	=	41/9	-	32/9
(2,-1,-1)	X1	-.246	-.2025	.014	.050	=	.041	+	.009
(0, 1,-1)	X2	.944	1.0960	.769	1.035	=	1.201	-	.166
Gender	X3	.260	-.2515	.0265	-.065	=	.063	-	.128
Reversal	X4	-.446	.0398	.001	-.018	=	.002	-	.020
Paradox	X5	.020	-.0717	.005	.001	=	.005	-	.006
					-----		-----		-----
					1.003	=	1.312	-	.311
Linn	LOC	.5115	.319	.080	.1632	=	.1018	+	.0614
(Y =	CPT-C	-.5597	-.367	.057	.2054	=	.1347	+	.0707
CPQ)	CPT-E	.4562	.062	.061	.0283	=	.0038	+	.0245
					-----		-----		-----
					.3969	=	.2403	+	.1566

when using the $b_{z_i} r_{iy}$ measure or its further decomposition into direct and indirect effects. For instance, this measure indicates that, among the Presumptuous Data Set variables, X_2 accounts for 100% of the variance in Y and is the only predictor to make any contribution whatever to the regression equation—yet if we were to use only X_2 we would account for 15% less of the variation in Y than we do if we also include the "totally unimportant" X_3. Even more startlingly, the further decomposition of X_2's contribution would have us believe that X_2 would have accounted for 278% of Y's variance, were it not for a negative 178% indirect effect that pulled X_2's contribution down to only 100%.

Finally, perhaps the most appropriate resolution to the issue of the importance of contributions is to recognize that the relationship summarized by R^2 is a relationship between Y and the set of predictor variables as a whole. Rather than attempting to decompose this relationship, and to assign "credit" to the original variables, it might be, more rewarding theoretically to try to interpret this new composite variable Y substantively, as discussed in section 2.3.3. Similarly the (0, 1, -1) contrast (Engineering School vs. College of Education) makes a 103.5% overall contribution to prediction of faculty salaries (according to the correlation x regression measure), whereas a couple of the other predictors supposedly pull overall predictability down (even though we know that adding a predictor can never reduce the squared population multiple correlation).

2.7 ANOVA VIA MRA

We won't look at univariate analysis of variance (Anova) in detail until chapter 4. However, most readers will have had some exposure to Anova in a previous course; if not, return to this section after reading section 4.1. The most important point to recognize is that the Anova model can be expressed either as a means model, namely,

$$Y_{ij} = \mu_j + \varepsilon_j,$$

or as an effects model, namely,

$$Y_{ij} = \mu + \alpha_j + \varepsilon_i$$

where $\mu = \Sigma\mu_j/k$, $\alpha_j = \mu_j - \mu$, and k = the number of groups or levels in the study. The translation of the effects model into MRA is actually simpler than for the means model, so we focus on the effects model.

We begin our translation by defining $m = k - 1$ group-membership variables such that $X_j = 1$ if the subject is in group j and zero if he or she is not (for $1 < j < k - 1$). The regression coefficient β_j for the jth predictor variable equals α_j, and β_0 (the intercept) represents μ. This would appear to omit β_k (the deviation of the mean of the kth group from the mean of the means), but from the way in which the α_js are defined we know that $\Sigma\alpha_j = 0$ and thus

$$\alpha_k = 0 - \sum_{j=1}^{k-1} \alpha_j = -\beta_1 - \beta_2 - \cdots - \beta_{k-1},$$

so that each subject in group k receives a score of -1 on each of the predictors. It is necessary to take this indirect approach because, had we included X_k in our MRA, its perfect linear dependence on the first k - 1 predictors would have led to an **x'x** matrix with a determinant of zero ($|\mathbf{x}'\mathbf{x}| = 0$). Thus, our complete definition of X_j is that it equals 1 if the subject is at level j of our factor, -1 if he or she is at level k, and 0 otherwise. To make this a bit more explicit, consider Table 2.6.

Because the models are equivalent, the results of our MRA should be equivalent to the results we would have obtained had we performed a one-way Anova. This is indeed the case. The overall F for the significance of R^2 will have a numerical value identical to the overall F for H_0: $\mu_1 = \mu_2 = \mu_3 = \cdots = \mu_k$; the sample regression coefficients **b** will be unbiased estimates of the corresponding α_j ; and the test for the significance of a contrast among the b_js will be equivalent to a test of the corresponding contrast among the α_j terms and thus among the μ_j terms. [For instance, $2\alpha_1 - \alpha_2 - \alpha_3 = 2(\mu_1 - \mu) - (\mu_2 - \mu) - (\mu_3 - \mu) = 2\mu_1 - \mu_2 - \mu_3 + (1 + 1 - 2)\mu$.]

Table 2.6. Relationship Between MRA and Anova Effects Model

	Scores on predictors for all subjects in this group						
Group	X_1	X_2	X_3 ...	X_{k-1}	MRA expression		Anova expression
1	1	0	0	0	$\beta_0 + \beta_1 + \varepsilon_i$	=	$\mu + \alpha_1 + \varepsilon_i$
2	0	1	0	0	$\beta_0 + \beta_2 + \varepsilon_i$	=	$\mu + \alpha_2 + \varepsilon_i$
3	0	0	1	0	$\beta_0 + \beta_3 + \varepsilon_i$	=	$\mu + \alpha_3 + \varepsilon_i$
.	.	.	.	.	.	.	.
$k-1$	0	0	0	1	$\beta_0 + \beta_{k-1} + \varepsilon_i$	=	$\mu + \alpha_{k-1} + \varepsilon_i$
k	-1	-1	-1	-1	$\beta_0 - \beta_1 - \beta_2 - \ldots$	=	$\mu - \alpha_1 - \alpha_2 - \cdots$
						=	$\mu + \alpha_k + \varepsilon_i$

If we use the post hoc critical value of section 2.3.3—namely, $mF_\alpha(m, N - m - 1) = (k - 1)F_\alpha(k - 1, N - k)$—for this test (rather than the a priori critical value of Table 2.3), we thereby employ Scheffé's procedure for post hoc comparisons. (Note that this is clearly a case in which the X_j are measured in comparable units and thus in which our interpretations of Y should be based on raw-score regression weights.)

As Derivation 2.7 points out, we would have arrived at the same overall F had we performed a MRA on the first $(k - 1)$ dichotomous group-membership variables of the form $X_j = 1$ if the subject is in group j, 0 if he or she is not, simply omitting X_k and letting membership in group k be represented by a score of 0 on X_1 through X_{k-1}. (If we can't find the subject in the first $k - 1$ groups, he or she must be in the kth.) However, conversion of the resultant **X** matrix to deviation scores (**x**) implicitly redefines our model, such that each β_j is a direct estimate of α_j. Any linear combination of the β_j (Σc_j not necessarily = 0) is thus a contrast among the μ_j; for example,

$$2(\mu_1 - \mu_k) + 3(\mu_2 - \mu_k) = 2\mu_1 + 3\mu_2 - 5\mu_k .$$

Before considering an example, we'll examine the two-factor Anova model, namely,

$$Y_{ijk} = \mu + \alpha_{j\bullet} + \beta_{\bullet k} + \alpha\beta_{jk} + \varepsilon_i$$

where

$$\mu = \frac{\sum_j \sum_k \mu_{jk}}{JK}; \quad \mu_{j\bullet} = \frac{\sum_k \mu_{jk}}{K}; \quad \mu_{\bullet k} = \frac{\sum_j \mu_{jk}}{J};$$

$$\alpha\beta_{jk} = (\mu_{jk} - \mu) - \alpha_{j\bullet} - \beta_{\bullet k} = \mu_{jk} - \mu_{j\bullet} - \mu_{\bullet k} + \mu ;$$

and Y_{ijk} equals the score of the *i*th subject to receive the combination of level *j* of the *J*-level factor A with level *k* of the *K*-level factor B. The generalization to three or more factors is straightforward but tedious.

Our translation of the factorial Anova model into MRA is very similar to our translation of the one-way effect model, except that now each main effect of a k-level factor is represented by k - 1 level-membership variables (i.e., by a number of X_js equal to the degrees of freedom for that factor), each two-way interaction is represented by a set of predictor variables each of which is a product of one of the level-membership variables (lmvs) representing factor A with one of the lmvs representing factor B (the number of such predictors therefore being equal to the product of the degrees of freedom for the two factors), and so on. The overall test of the significance of any particular main effect or interaction is the MRA test for the increment to R^2 yielded when its corresponding set of predictor variables is added last, which for any single-*df* effect is equivalent to testing the statistical significance of its regression coefficient. Specific contrasts among the levels of a given factor are equivalent to the corresponding contrasts among that subset of μ_js.

An alternative approach to coding each main effect is to choose k - 1 contrasts among the levels of that factor. The contrasts must not be perfectly linearly dependent, but they need not necessarily be orthogonal. (Indeed, the usual lmvs are a set of contrasts of each level against the last level—clearly a nonorthogonal set.) A subject's score on the predictor variable representing a particular contrast is simply the contrast coefficient for the level of that factor present in the treatment he or she received. All this may be facilitated somewhat by considering a couple of examples.

Example 2.2 *In-Group/Out-Group Stereotypes* Bochner and Harris (1984) examined the differences between male and female social psychology students in their ratings on 10 traits (5 desirable, 5 undesirable) of their own group (native Australians) or of Vietnamese settlers in Australia. Taking overall favorability of the ratings on this RI-item Likert scale as the dependent variable, the resulting 2 x 2 Anova led to the coding of MRA lmvs shwon in Table 2.7. Note that the scores on X_1, X_2, and X_3 are identical to the main contrast coefficients for the main effect of sex, the main effect of group rated, and the sex-by-group interaction, respectively. The entries into an MRA program to analyze these data would consist of 12 cards (rows) with X_1, X_2, and X_3 scores of 1, 1, and 1 and with the fourth entry being that subject's score on Y; followed by 21 rows of 1, -1, - 1, Y_i; 40 rows of -1, 1, - 1, Y_i ; and 46 rows of -1, -1, 1, Y_i .

The resulting Fs for β_1, β_2, and β_3 are 1.90, 12.88, and 10.63 (each with 1 and 115 degrees of freedom), indicating highly significant effects of group and of the sex-by-group interaction. In fact, examination of the means makes it clear that the significant groups effect is almost entirely due to the male subjects, with females showing almost no tendency to rate their own group higher than Vietnamese. It should be noted that these results are identical to those yielded by an unweighted-means analysis of the data. As Derivation 2.8 demonstrates, the MRA approach (also commonly referred to as the

Table 2.7. Coding of MRA Level-Membership Variables for Study of Stereotypes

Group	n_j	lmv Scores X_1	X_2	X_3	$\bar{Y}_j$	s_j^2
Males rating Australians	12	1	1	1	4.934	.319
Males rating Vietnamese	21	1	-1	-1	3.900	.504
Females rating Australians	40	-1	1	-1	4.650	.396
Females rating Vietnamese	46	-1	-1	1	4.600	.665

least-squares approach) to nonorthogonal designs and the computationally much simpler unweighted-means analysis (Winer, 1971) yield identical estimates and significance tests for any between-groups completely factorial design in which each factor has two levels (see also Horst & Edwards; 1981, 1982).

Example 2.3 *Negative Shares and Equity Judgments* We use Experiment 1 of Harris, Tuttle, Bochner, and Van Zyl (1982) to illustrate coding of a factor with more than two levels. Each subject in this study responded to two different hypothetical situations (an anagrams task in which a performance bonus was awarded the group and a partnership devoted to selling plants at a flea market). The subject was asked to allocate final outcomes directly for one of the tasks and to specify for each participant the difference between final outcome and that participant's individual contribution to the group effort for the other task. The difference between the outcomes-allocation condition and this latter, expenses-allocation condition in the amount of money allocated to the lowest-input partner was the dependent variable. For about half the subjects (those in the some-negative condition), achieving equal final outcomes required that at least one group member receive a negative share of the expenses (i.e., be allowed to take money from the "pot" being collected to meet general expenses); for the remaining subjects, no negative numbers needed to be used.

In addition to the two-level factor, the need to balance which allocation condition was assigned to the anagrams situation and which of the two situations the subject responded to first yielded a four-level "nuisance" factor whose nonsignificance was devoutly desired. This led to the alternative codings of MRA predictor variables (one set of seven effect-coded predictors vs. a set of seven contrast-coded predictors) shown in Table 2.8. (None of the effects in this analysis came close to statistical significance, except for the test of the intercept—the *grand mean* in Anova terminology—which yielded a highly significant overall effect of allocation condition, thereby replicating earlier findings .)

Lengthy as it may have seemed, the preceding discussion of MRA as an approach to Anova was very compact, relative to the extant literature on alternative coding schemes and analytic procedures for unequal-*n*s Anova (e.g., Keren & Lewis, 1977; Overall &

Table 2.8 Alternative Codings of MRA Predictor Variables, Equity Study

Group		lmvs							Contrast coding						
Pos	Nuis	X_1	X_2	X_3	X_4	X_5	X_6	X_7	X_1	X_2	X_3	X_4	X_5	X_6	X_7
AP	1st, An	1	1	0	0	1	0	0	1	3	0	0	3	0	0
AP	1st, FM	1	0	1	0	0	1	0	1	−1	2	0	−1	2	0
AP	2nd, An	1	0	0	1	0	0	1	1	−1	−1	1	−1	−1	1
AP	2nd, FM	1	−1	−1	−1	−1	−1	−1	1	−1	−1	−1	−1	−1	−1
SN	1st, An	−1	1	0	0	−1	0	0	−1	3	0	0	−3	0	0
SN	1st, FM	−1	0	1	0	0	−1	0	−1	−1	2	0	1	−2	0
SN	2nd, An	−1	0	0	1	0	0	−1	−1	−1	−1	−1	1	1	−1
SN	2nd, FM	−1	−1	−1	−1	1	1	1	−1	−1	−1	−1	1	1	1

Spiegel, 1969, 1973a, 1973b; Rawlings, 1972, 1973; Timm & Carlson, 1975; Wolf & Cartwright, 1974). The bulk of this debate centers around two issues:

1. Whether the most appropriate test of a given Anova effect is the statistical significance of its increment to R^2 when added last (as advocated 'in this section) or, instead, the increment it provides when it is first added within a specifically ordered stepwise regression analysis.
2. Whether the effect or contrast coding recommended here should be employed or, instead, whether each of our group-membership or contrast scores should be divided by the sample size for that group.

My own attempts to grapple with these issues have lead to the following observations and recommendations: First, it is very seldom appropriate to apply a stepwise MRA procedure to an Anova design, because we are usually interested in examining each effect in the context of (and therefore unconfounded with) the other effects in our model. Moreover, the parameters actually being estimated when testing "early" contributions in a stepwise analysis are generally not those specified in the initial construction of the MRA model. For instance, in Example 2.2 (ratings of Australians vs. Vietnamese settlers), the computation of the appropriate $\mathbf{K} = (\mathbf{x'x})^{-1}\,\mathbf{x'}$ matrices shows that our estimate of β for the main effect of subject gender is

$$.5\left(\frac{12\overline{Y}_{\text{MA}} + 21\overline{Y}_{\text{MV}}}{33} - \frac{46\overline{Y}_{\text{FA}} + 40\overline{Y}_{\text{FA}}}{86}\right)$$

if tested by itself (or when added first), $(\overline{Y}_{\text{MA}} + \overline{Y}_{\text{MV}} - \overline{Y}_{\text{FA}} - \overline{Y}_{\text{FA}})/4$ if tested in the context of all three predictors (i.e., when added last), and $.194(\overline{Y}_{\text{MA}} - \overline{Y}_{\text{FA}}) +$

$.303(\overline{Y}_{MV} - \overline{Y}_{FV})$ when tested in terms of its increment to R^2 when added to the ethnicity main effect but not corrected for the interaction term.

The first of these estimates is reasonable if your interest is in predicting the difference between a randomly selected male and a randomly selected female, ignoring (and thus taking advantage of any confounds with) which kind of stimulus person is being rated. The second (testing the regression coefficient for the effect in the full three-predictor equation) estimates the gender effect we would have obtained had we had equal numbers of subjects in each cell and eliminates any confounding of effects with each other. This would appear to be the most appropriate kind of test when we're interested in assessing each effect unconfounded with other effects, as is usually the case in experimental designs and is often the case in correlational studies, as well. The last estimate (obtained when increment to R^2 is tested after some, but not all, other effects have been entered into the equation) is a complex weighting of the male–female difference for the two stimuli being rated. (The weights are *not* proportional to the number of Australians vs. Vietnamese settlers rated, which would have yielded weights of .487 and .513.) The actual hypotheses being tested in the intermediate steps of this sort of sequential approach become much more complex as the number of effects tested increases. (Even less appropriate is the not uncommon practice of deciding whether, for example, the C main effect should be considered statistically significant by taking a "majority vote" of the results of the significance tests of its contribution when added at various points in various sequences of stepwise tests.)

Further, the alternative coding (weighting contrast scores by reciprocals of sample sizes) estimates very different parameters than the ones on which we have focused. In particular (assuming that our tests are based on regression coefficients in the full model), the means involved in estimating main effects are simple averages of all the responses at a particular level of the factor, irrespective of the confounds introduced by the disproportionate distribution of levels of the other factor. These are almost never appropriate comparisons for an experimental design, and they are appropriate for predictive purposes only if you want to know how best to predict the effect of a given factor in complete ignorance of your subjects' positions on the other factors.

Before leaving this issue, let me bedevil you with one more example—one I've found useful in illustrating some of the above points to students in my class on univariate Anova.

Example 2.4 *Gender Bias in Faculty Salaries?* The Faculty Senate's Pay Equity Committee at Hypothetical University has examined the records of the three largest colleges at Hypo. U. in order to determine whether, as the Women Faculty's Caucus has charged, there is bias against females in the awarding of salaries. The results of their examination are reported in Table 2.9.

Table 2.9 Mean Faculty Salary[a] at Hypo. U. as f(College, Gender)

	Gender							
	Males			Females				
College	Mean	*SD*	*n*	Mean	*SD*	*n*	Unweighted Mean	Weighted Mean
Engineering	30	(1.491)	[55]	35	(1.414)	[5]	32.5	30.416
Medicine	50	(1.423)	[80]	60	(1.451)	[20]	55	52
Education	20	(1.451)	[20]	25	(1.423)	[80]	22.5	24
Unweighted Mean	33.333		[160]	40		[100]	36.667	
Weighted Mean	39.032			32.142			36.026	

[a]Salaries expressed in thousands of dollars.

Examination of the means suggests that there is no bias against females in determining salaries within any one of the three colleges. If anything, the bias appears to be going in the opposite direction, with females receiving higher salaries than males, on the average, within each of the three colleges. It might therefore seem reasonable to conclude that there is no bias against females for the three colleges as a whole (ie., no "main effect" for Gender such that females receive lower salaries). However, whether we conclude that there is a statistically significant tendency for females to receive *higher* salaries than men or just the opposite depends heavily on which of several possible MRA-based analyses we adopt. Setting up this unequal-*ns* Anova for analysis via MRA yields the complete data table for the Gender-Bias data set displayed in Table 2.10.

In MRA-based Anovas, the regression coefficient corresponding to a given contrast variable provides an estimate of the magnitude of that contrast; the test of the statistical significance of the regression coefficient (i.e., the test of the null hypothesis that that coefficient is zero in the population regression equation) is related to a test of the hypothesis that that contrast among the population means is zero-, and the test of the increment to R^2 provided by adding all contrast variables representing a given effect is an overall test of that effect. Thus, for example, b_3, the regression coefficient for X_3, the sole contrast variable representing gender provides an estimate of the direction and magnitude of gender bias. Because we assigned a coefficient of +1 to males and - 1 to females, a positive b_3 indicates a bias in favor of males, whereas a negative b_3 indicates a tendency for females to be paid more than males.

It would seem natural to base our test of the statistical significance of gender bias on the statistical significance of its regression coefficient in the full model (i.e., in the

Table 2.10 Data for MRA-Based Anova of Gender-Bias Data Set

	Score on Contrasts for					
Group	College		M/F	CxG		Income
Eng - M	2	0	1	2	0	Y001
	2	0	1	2	0	Y002
	.	.	.	.	.	.
	2	0	1	2	0	Y054
	2	0	1	2	0	Y055
Eng - F	2	0	-1	-2	0	Y056
	2	0	-1	-2	0	Y057
	2	0	-1	-2	0	Y058
	2	0	-1	-2	0	Y059
	2	0	-1	-2	0	Y060
Med - M	-1	1	1	-1	1	Y061
	-1	1	1	-1	1	Y062
	-1	1	1	-1	1	Y063
	.	.	.	.	.	.
	-1	1	1	-1	1	Y140
Med - F	-1	1	-1	1	-1	Y141
	-1	1	-1	1	-1	Y142
	.	.	.	.	.	.
	-1	1	-1	1	-1	Y159
	-1	1	-1	1	-1	Y160
Educ - M	-1	-1	1	-1	-1	Y161
	.	.	.	.	.	.
	-1	-1	1	-1	-1	Y178
	-1	-1	1	-1	-1	Y179
	-1	-1	1	−1	-1	Y180
Educ - F	-1	-1	-1	1	1	Y181
	-1	-1	-1	1	1	Y182
	.	.	.	.	.	.
	-1	-1	-1	1	1	Y259
	-1	-1	-1	1	1	Y260

regression equation derived by using all five predictor variables)—which turns out to be logically and (matrix-) algebraically equivalent to testing the statistical significance of the increment to R^2 that results from adding X_3 to the equation last. If we do that, we obtain a b_3 of -3.333, which turns out to be exactly one-half of the difference between the mean of the three male means and the mean of the three female means—i.e., one-half of the Gender contrast as we have been computing it. Because the associated *F* of 560.70 with 1 and 254 *df* is highly significant, it would appear that we can safely conclude that there is a statistically significant tendency for females to receive higher salaries than males.

However, many authors prefer to test the various main effects and interactions sequentially, entering the sets of contrast variables in a particular order and testing each main effect or interaction on the basis of the increment to R^2 it provides *at the particular point at which its set of predictors was added to the equation.* Given the context dependence of MRA, we know that the test of each new set of predictor (contrast) variables takes into account their relationships to (correlations with) other effects already in the equation, but is "uncorrected for" any effects that have not yet been added to the equation. We also know that the sign and magnitude of the regression coefficient for each contrast (i.e., our estimate of the sign and magnitude of the magnitude of the corresponding contrast among the population means) may differ, depending on the point at which it is added to the equation.

Let's see what happens to our estimates of the gender bias as we try out various orderings of our tests.

1. If we make gender the first effect entered into our equation, we get a b_3 at that point of 3.445, indicating a tendency for *males* to receive higher salaries than females. (The associated *F* is 1428.78 by the formula provided earlier in this section. Most MRA programs would test the R^2 for gender against $1 - R^2_{gender}$ rather than against $1 - R^2$ for the full model, and would thus yield an F of 18.459, which is valid only if all effects not yet entered into the equation are zero—which the college main effect certainly is not.)
2. If we add gender after the two college contrast variables, but before the C x G variables, we get a b_3 of -3.593 and an *F* of 908.6, indicating this time that *females* tend to receive higher salaries.
3. If we add gender after the two C x G contrast variables, but before the college main effect, we get a b_3 of 1.648 and an *F* of 284.0, indicating this time a statistically significant bias in favor of males and/or against female faculty.
4. As we already know from the full-model analysis, if we make Gender the last effect added to the equation we get a b_3 of -3.333 and an F of 560.7, indicating once again a bias *in favor of females*.

What is our committee to make of all of this? One still sees occasionally in the literature a “majority vote” approach to the question of what to do when the results of various sequential analyses disagree. Usually the vote” is whether a given effect is statistically significant, with the researcher being unaware that *what* effect [i.e., what contrast(s) among the means] is being tested also varies across the various orders. What should be done instead is to decide which order of entry of the gender effect is most relevant to what the committee *means* by “gender bias” and adopt that answer. It turns out to be a relatively straightforward matter (involving a bit of matrix inversion and multiplication) to get, for each contrast variable in a regression equation, that linear combination of the population means that is being estimated by its regression coefficient (cf. section 2.2.4.) This linear combination has a very simple interpretation in two special cases:

1. When a contrast variable is the first one added to the regression equation, what is actually being tested is a pattern of differences among the *weighted* means of the various levels of that factor, ignoring all other factors. Thus, for instance, if we simply compute a mean score for all male faculty, regardless of their college affiliation,_we get [55(30) + 80(50) + 20(20)1/155] = 39.032 thousands of dollars, whereas a similar computation gives us a weighted mean of $32,142 for female faculty. Thus the average male faculty member, ignoring which college he's in, has a salary $6,890 higher than the average female faculty member sampled from these data without regard to college affiliation. This difference of 6.890 thousands of dollars is exactly twice the b_3 obtained by entering gender into the equation first. Clearly, testing a main effect first (uncorrected for any other effects) is equivalent to simply running a one-way Anova on differences among the levels of that factor, ignoring other factors involved—that is, leaving in (taking advantage of?) any and all confounds between this factor and other main effects and interactions. Provided that our differential sample sizes are indeed reflective of differential population proportions of the various combinations of conditions, this might be a more ecologically valid test of the overall effect of this factor in the population. However, while it's clear that the 155 male faculty members receive higher average salaries than the 105 female faculty members in our sample, this is also clearly due to the underrepresentation of females in the two colleges with higher average salaries, rather than to any bias against female faculty within colleges. If there is bias (rather than, e.g., self selection) operating against females here, it is bias against hiring them into high-paying colleges, rather than bias in assignment of salaries once they get in.

2. When a contrast variable is added last to the equation, or (equivalently) is tested on the basis of its regression coefficient in the full model, the resulting test is indeed of the specified contrast among the population means. In the present case, for instance, the *unweighted mean* of the three male means is (30 + 50 + 20)/3 = 33.333, whereas the unweighted mean of the three female means is 40.000; the difference of -6.667 is exactly twice the b_3 we. obtained from the full-model analysis and from any sequential analysis where X_3 is "last in". This accurately reflects the fact that the average salary differential within particular colleges is $6,667 *in favor of females*—although it of course "misses" the overall bias caused by underrepresentation of females in the College of Engineering and the College of Medicine.

Finally, if an effect is added "in the middle" somewhere—corrected for some but not all of the other effects in this factorial design—the contrasts among the k means that are actually being tested are generally very complicated and very difficult to describe. (Kirk, 1981, provides some general formulae for these cases.) For instance, when gender is corrected for College but not for the College x Gender interaction, the actual contrast among the 6 means being tested is a

$$\begin{matrix} .758 & -.758 \\ 2.645 & -2.645 \\ 2.645 & -2.645 \end{matrix}$$

contrast, whereas testing Gender corrected for the interaction but not for the College main effect is essentially testing a

$$\begin{matrix} 4.248 & -.666 \\ 3.036 & -1.878 \\ .087 & -4.827 \end{matrix}$$

contrast. The relevance of these two escapes me.

Of course, most designs will show less dramatic differences among the results of various sequential tests. Nonetheless, these hypothetical data are useful in demonstrating that what is being tested (rather than just the statistical significance of a given effect) depends very much on the order of testing. I glean from this example as well the "moral" that only the two "extreme" cases—testing each effect uncorrected for any others or testing each effect corrected for all others—are likely to be of interest. Moreover, in any research setting where you are interested in "teasing out" possible causal effects, the full-model (rather than the completely uncorrected) approach is likely to be most relevant. Finally, of course, if the differential n_js are due to essentially random variation around a "failed" equal-*ns* design, you should be doing an unweighted-means analysis, rather than an MRA-based (least-squares) analysis.

2.8 ALTERNATIVES TO THE LEAST-SQUARES CRITERION

Throughout this chapter, we have assumed that the least-squares (LSQ) criterion—minimizing the average squared error of prediction—is an appropriate basis for selecting regression coefficients. There are indeed strong justifications for such a choice. For one thing, the LSQ criterion incorporates the entirely reasonable assumption that the seriousness (cost) of an error of prediction goes up much faster than linearly with the magnitude of the error. Big errors generally *do* have much direr consequences than do small errors. If we're predicting the strength of an aircraft wing, for instance, small errors will fall well within the margin of safety engineered into the construction process, but a large error will produce some very irate test pilots among the survivors.

However, this built-in emphasis on keeping really large errors to a minimum also makes LSQ regression highly sensitive to large deviations from the general trend that are not representative of the basic process being studied: recording errors, a sneeze that causes the subject to miss the ready signal, the one-in-ten-thousand deviation that makes it into a sample of 20 observations, and so on. Detection and elimination of such ***outliers*** is what makes computer enhancement of pictures transmitted from spacecraft so successful and what motivates much of the literature on graphical techniques cited in section 2.3. A number of authors have suggested, however, that this sensitivity to outliers should be corrected by using procedures that consider all observations but weight large errors less extremely than does the LSQ criterion. Huynh (1982) demonstrated that such approaches do indeed improve the robustness of sample regression coefficients when outliers are present but produce estimates nearly identical to LSQ weights when outliers are not present.

Statistically, the principal advantage of the LSQ criterion is that it produces BLUE (Best Linear Unbiased Estimators). That is, among all possible unbiased estimators of the βs that are expressible as linear combinations of the scores on Y [remember our $\mathbf{K} = (\mathbf{x'x})^{-1}\mathbf{x}$ matrix], LSQ estimators have the lowest possible variance across successive samples. The technique of *ridge regression* (Darlington, 1978; Morris, 1982; Price, 1977;) gets around this strong mathematical endorsement of LSQ estimators by allowing a little bit of bias (systematic over- or underestimation of the corresponding β_js) to creep into the estimators in exchange for a potentially large decrease in variability in the presence of "wild" observations. However, both Rozeboom (1979) and Morris (1982) caution that ridge regression estimates may *not* show improved robustness relative to LSQ estimators, and specifying the conditions under which the trade-off will prove beneficial is not a simple matter.

Finally, Herzberg (1969) suggested getting around the oversensitivity of LSQ estimators to random fluctuations (error variance) in the data by performing a preliminary principal component analysis (see chap. 6) on the data and then carrying out MRA on only the PCs accounting for above-average variance. Because the later (lower-variance) components are likely to represent primarily error variance, this should produce regression estimates based more on the "enduring" aspects of the data than an MRA of the original variables. Pruzek and Frederick (1978) discussed some advantages of carrying out the MRA on factor scores estimated from a factor analysis of the variables (see chap. 7) rather than on principal components.

2.9 PATH ANALYSIS

Path analysis is a technique that is designed to test the viability of linear, additive models of the causal relationships among directly observable variables by comparing the correlations implied by each model against those actually observed. It is, on the other hand, a technique that is most frequently used by researchers to provide a "catalog" of which variables cause which other variables, with no commitment to any particular functional relationship (or even to whether the causal impact is positive or negative), and with minimal attention to whether or not the conditions known to be necessary to path analysis's validity are met. Despite this history of misuse, however, a researcher employing path analysis is always forced, at a bare minimum, to organize his or her thoughts about relationships among the variables.

In the case where the "causal flow" is unidirectional (called ***recursive models*** in path analysis)—that is, where no variable X that is a cause of some variable Y is itself caused (even indirectly) by Y—testing the model is accomplished by carrying out a series of multiple regression analyses, so that your recently acquired skill in carrying out MRAs becomes highly relevant.

In this section we review path analytic terminology; the assumptions required to consider a path analysis a valid test of a set of causal assumptions; MRA-based estimation of the path coefficients; decomposition of the correlation between any two variables into direct, indirect, spurious, and unanalyzable components (with quite a few words of caution); and an overall test of one's model.

2.9.1 Path-Analytic Terminology.

If X is hypothesized to be a cause of Y ($X \rightarrow Y$ on a path diagram), X is said to be ***exogenous*** and Y, ***endogenous*** with respect to each other. With respect to the model as a whole, those variables for whom no causes are specified (those having no straight arrows pointing at them in the path diagram) are called the exogenous variables; those variables not specified as causes of any other variables (having no straight arrows pointing from them to other variables) are called the endogenous variables; and those that are both caused by one or more other variables and causes of one or more other variables are both exogenous and endogenous, depending on which causal link is being focused on.

A variable X_1 that is directly connected (by a single arrow) to another variable Y is said to be a ***direct cause*** of Y. The magnitude of this causal relationship (equivalent to a z-score regression weight and indeed estimated by it in a recursive model) is described by the ***path coefficient*** p_{Y1}, where the first subscript refers to the variable directly caused and the second subscript refers to the presumed causal variable.

A variable X_1 that is indirectly connected to Y via a unidirectional causal chain (e.g., X_1 is a cause of X_2 which is a cause of X_3 which is a cause of Y) is said to be an ***indirect cause*** of Y. The variables appearing between X_1 and Y in this unidirectional causal chain (X_2 and X_3 in the example we've been using) are said to ***mediate*** (or be ***mediators of***) the relationship between X_1 and Y. Somewhat controversially, the product of the path coefficients involved in this unidirectional causal chain (e.g., $p_{Y3} \cdot p_{32} \cdot p_{21}$) is traditionally considered a measure of an indirect component of the total correlation between X_1 and Y. Of course X_1 may have both a direct and several indirect relationships to Y.

If two variables Y_1 and Y_2 are both hypothesized to be caused (directly or indirectly) by the same variable X (both have arrows pointing to them from X or are the endpoints of unidirectional causal chains beginning with X in the path diagram), they are said to be ***spuriously correlated***. The product of the path coefficients linking Y_1 and Y_2 to X (in the simplest case, $p_{1X} \cdot p_{2X}$) is traditionally (but controversially) said to measure a spurious component of the total correlation between Y_1 and Y_2.

If two variables X_1 and X_2 are hypothesized to be correlated with each other, but the model does not specify the direction of causation (which could be $X_1 \rightarrow X_2$ or $X_2 \rightarrow X_1$ or both), these two variables are connected by a curved, double-headed arrow in the path diagram and the relationship between them is described as an ***unanalyzed correlation***. The magnitude of this relationship is indicated by r_{12}. Any product of the coefficients corresponding to paths linking two variables is said to represent an unanalyzable component of their relationship if (a) it involves one or more r_{ij}s and (b) the component's designation as direct, indirect, or spurious would be changed by whether each r_{ij} were replaced by p_{ij} or by p_{ji}. (More about this in section 2.9.4.)

As indicated earlier, a path-analytic model in which no unidirectional causal path leads from any variable back to that same variable is said to be *recursive*. (This terminology will seem strange to those familiar with recursive computer programs, which are those that can call themselves, but we're stuck with it.) Nonrecursive path models require more complex analytic techniques than do recursive models (see, e.g., James & Singh, 1978). We'll thus confine our attention in the remainder of this section to recursive models.

2.9.2 Preconditions for Path Analysis.

James, Mulaik, and Brett (1982) provide an especially clear, forceful presentation of the conditions (they identified eight) that must be met if statistical support for or against a given path model is to be taken as support for or against the assumptions about causal relationships expressed by that model. Several of these represent, to researchers with any acquaintance with research methodology, "obvious" caveats to the use of any set of empirical data to test any set of theoretical propositions. For instance, JMB's Condition 5 ("self-contained functional equations") points out that if we omit an important variable from our analysis, our conclusions about the variables we do include may be wrong. Their Condition 6 ("specification of boundaries") says that we should specify the subjects and conditions (e.g., adults of normal intelligence motivated to perform well and provided with essential items of information) to which a given theory is expected to hold. And their Condition 8 ("operationalization of variables") says that relationships among observable variables have implications for relationships among conceptual variables only if the former are reliable and valid measures of the latter.

Nevertheless, path analysis can be useful in refining such "intuitively obvious" conditions. For instance, consideration of the algebra involved in estimating path coefficients makes it clear that an omitted variable poses a problem only when the variable(s) omitted is ***relevant***, in that it is correlated at a nontrivial level with both a cause and an effect (endogenous variable) in our model, but not so highly correlated with already-included causes that it has no unique effects on the included endogenous variables.

Among the less obvious conditions stipulated by James et al. is their Condition 7 ("stability of the structural model"), which says that where causal links occur via a number of (unmeasured) intervening processes, the two variables involved in any such causal relationship must be measured over time intervals that permit those processes to have taken place.

Most crucial for our understanding of path analysis, however, are three preconditions that will come as a surprise to those users and/or observers of path analysis who think of it as a technique for proving causality from correlations, namely, their Conditions 2 ("theoretical rationale for causal hypotheses"), 3 ("specification of causal order"), and 4 ("specification of causal direction").

Condition 2 makes it clear that whereas path analysis can provide evidence *against* a postulated causal relationship (e.g., if that relationship implies a correlation that proves nonsignificant), the consistency of the empirical data with the causal assumptions provides only weak evidence *for* those causal assumptions in the absence of a clear theoretical basis for the assumptions in terms of plausible processes that could generate such a causal link.

Conditions 3 and 4 stipulate that path analysis can never provide evidence about the *direction* of causal relationships. The general principle, which for recursive models follows directly from the equivalence between path and regression coefficients (next section), is that:

Any two path models that differ only in the direction of a causal path (e.g., $X \rightarrow Y$ vs. $X \leftarrow Y$ or $X_1 \rightarrow X_2 \rightarrow Y$ vs. $X_1 \leftarrow X_2 \rightarrow Y$) imply exactly the same pattern of observed correlations and thus cannot be distinguished.

Thus support for any particular causal direction must be based on logic (e.g., knowing that one of the two variables preceded the other in time) or empirical evidence external to the path analysis.

Note that the second example given demonstrates that path analysis *cannot* be used to distinguish between indirect causation and spurious correlation—though it can be used to distinguish between the hypotheses that X_2 mediates the relationship between X_1 and Y ($X_1 \rightarrow X_2 \rightarrow Y$) and that X_1 mediates the relationship between X_2 and Y ($Y \leftarrow X_1 \leftarrow X_2$), because this distinction is based on whether p_{Y1} or p_{Y2} is zero.

James et al. also specified two ways in which the empirically observed correlations (more precisely, the path coefficients estimated therefrom) must be consistent with our path model if we are to consider it supported:

Condition 9 requires the estimate of each path stipulated by the model as non-zero to be statistically significant.

Condition 10 requires that the estimate of each path stipulated by the model as zero be statistically nonsignificant or, if statistically significant, of negligible magnitude.

Rather amazing—but consistent with my earlier claim that path analysts tend to treat the technique as providing simply a catalog of what causes what—is the omission from this otherwise extensive list of any mention of the *signs* of the causal paths, that is, of whether higher values of X tend to cause (generate?) higher or lower values of Y. We shall thus include in our list of preconditions, Condition 4' (corresponding to James et al.'s Condition 4): specification of causal direction *and sign*. And in testing the empirical fit of any path model we shall replace their Condition 9 with *Condition 9'* that each path stipulated as nonzero yield an estimate that is statistically significant *and* of the sign stipulated by the model.

Carrying out these tests requires of course that we be able to estimate path coefficients and test them for statistical significance. It is to these estimates and tests that we turn next.

2.9.3 Estimating and Testing Path Coefficients

Kenny (1979), JMB (1982), and others have shown that the path coefficients relating a set of exogenous variables $X_1, X_2, ..., X_m$ to a single endogenous variable Y are equivalent to the z-score regression coefficients yielded by an ordinary least-squares (OLS) regression analysis of Y predicted from X_1 through X_m if and only if two conditions are met:

1. The model being tested is recursive.
2. The ***disturbance term*** (ε in Equations 2.1 and 2.2) for this endogenous variable is uncorrelated with the disturbance term for any variable that precedes this Y in the causal order.

Condition (2) implies that the disturbance (error) terms in the equations for the various endogenous variables are mutually uncorrelated. Because violation of condition (2) implies that we have omitted a relevant variable from our model, it seems reasonable to proceed as if this condition is met, because otherwise we have failed to meet one of the preconditions for valid interpretation of our path analysis as reflecting on the causal relationships among our observed variables. (Put another way, we are putting the burden of satisfying condition (2) on our preanalysis identification of relevant variables.)

At any rate, for recursive models with uncorrelated disturbances we estimate the path coefficients via a series of MRAs, taking as the dependent variable for each MRA one of the endogenous variables (one variable having a straight arrow leading into it) and as the predictor variables for that MRA all other variables (perforce only those that come earlier in the causal ordering) that are presumed by the model to have non-zero direct paths into Y.

Condition 9' tests of the validity of the model are then yielded by the tests of the statistical significance of individual regression coefficients—with of course a statistically significant sample coefficient that has the wrong sign providing especially strong evidence against the model. To guard against a large number of chance results, Bonferroni-adjusted α_is (with n_t = the number of nonzero paths in the model) should be employed.

The Condition 10 test of any given path coefficient presumed zero in the model could be carried out by conducting an additional MRA in which the corresponding omitted variable is added to the set of predictors and the statistical significance of its regression coefficient is examined. It should of course be statistically nonsignificant at, say, the .10 a priori level. (Because we're hoping for *non*-significance, we should adjust our α upward, rather than downward.) It is usually, however, more efficient to run a single additional, full-model MRA in which *all* variables that precede Y in the causal ordering are used as predictors. The statistical (non)significance of each of the variables exogenous with respect to Y that is presumed by the model to have a zero direct path into Y can then be tested in one MRA.

If one or more path coefficients fails its Condition 9' or Condition 10 test, the model is considered rejected. It is usually considered legitimate to proceed to test various revised models that eliminate paths that yielded non-significant regression coefficients in the initial analysis and/or add paths that had been presumed zero but yielded significant regression coefficients—provided that the researcher is open with his or her readers about the exploratory nature of these subsequent analyses.

The researcher will probably also be interested in how well the path model can account for the observed correlations. This is accomplished by comparing the correlations among all the variables implied by the model to those actually observed. The primary tool in generating the ***reproduced correlations*** for comparison with the observed ones (and the first step as well in decomposing correlations into direct, indirect, spurious, and unanalyzed components) is the set of ***normal equations*** relating regression coefficients to observed correlations. It was pointed out in section 2.2.3 that the normal equations for a given outcome (endogenous) variable take on a very simple form in which each correlation between a given predictor (exogenous) variable and Y is equated to a linear combination of the regression (path) coefficients, with the weight assigned predictor i in the equation for r_{jy} being equal to the correlation between predictors i and j.

The end result is that the coefficients of the to-be-estimated regression coefficients in this set of simultaneous equations follow the pattern of the matrix of correlations among the predictors, and the constants on the other side of the equals signs are simply the zero-order correlations between the predictors and Y. Put a bit more concretely, if the causal ordering among the variables is $X_1, X_2, X_3, X_4, \cdots$ (no higher-numbered variable having a direct or indirect path into a lower-numbered variable) then the normal equations take on the form:

$$r_{12} = (1)p_{21}$$

$$r_{13} = (1)p_{31} + r_{12}\,p_{32}$$
$$r_{23} = r_{12}\,p_{31} + (1)p_{32}$$

$$r_{14} = (1)p_{41} + r_{12}p_{42} + r_{13}p_{43}$$
$$r_{24} = r_{12}p_{41} + (1)p_{42} + r_{23}p_{43}$$
$$r_{34} = r_{13}p_{41} + r_{23}p_{42} + (1)p_{43}$$

----

etc.

Once we have written out the normal equations, we cross out each path that our model specifies to be zero, yielding the ***reduced-model equations*** (or just "reduced equations"). Thus, for instance, if our model specifies that only X_2 and X_3 have direct effects on X_4 (i.e., that $p_{41} = 0$), the reduced-model equations involving X_4 as the predicted variable become

$$r_{14} = r_{12}p_{42} + r_{13}p_{43}$$
$$r_{24} = (1)p_{42} + r_{23}p_{43}$$
$$r_{34} = r_{23}p_{42} + (1)p_{43}$$

(If no paths are assumed zero, we have a ***fully recursive*** model, and our reproduced correlations will exactly equal the observed correlations. Comparison of observed and reproduced correlations is thus a meaningless exercise for such a ***just-identified*** path model.)

These reduced equations can then be used to generate the reproduced correlations, *provided* that we carry out our calculations in order of causal priority (in the present case, calculating r_{12} first, then r_{13} and r_{23} next, etc.), *and* that at each stage we use the reproduced correlations computed in previous stages, in place of the actually observed correlations. Alternatively, we can use the *decomposition equations*, which express each observed correlation as a function only of the path coefficients, for this purpose as well as that of decomposing correlations into direct, indirect, spurious, and unanalyzable components.

2.9.4 Decomposition of Correlations into Components

The ***decomposition equations*** provide the basis for analyzing observed correlations into direct, indirect, spurious, and unanalyzable components. (These components, must, however, be taken with several grains of salt. Like the similar decomposition of R^2 discussed in section 2.6, individual components can be larger than unity or negative in contexts where that makes no sense, so the validity of their usual interpretation is dubious. Nevertheless such decompositions are reported often enough that you need to be familiar with whence they come.)

The decomposition equations are obtained from the reduced equations through successive algebraic substitutions. We express the predicted correlation between exogenous variables and our first endogenous variable as a function solely of non-zero path coefficients and any unanalyzed correlations among the exogenous variables.

We then substitute these expressions for the corresponding correlations wherever they occur in the reduced equations for the endogenous variable that comes next in the causal ordering. After collecting terms and simplifying as much as possible, the resulting theoretical expressions for correlations between this endogenous variable and its predictors are substituted into the reduced equations for the next endogenous variable, and so on.

The result will be an algebraic expression for each observed correlation between any two variables in our model that will be the sum of various products of path coefficients and unanalyzed correlations. As pointed out earlier, we could now substitute the regression estimates of the path coefficients into these decomposition equations to generate our reproduced correlations.

For purposes of decomposition into components, however, we must first classify each product of terms as representing either a direct, an indirect, a spurious, or an unanalyzable component. This is accomplished via the following rules:

1. A ***direct component*** is one that consists solely of a single path coefficient. That is, the direct component of the correlation between X and Y is simply p_{YX}.

2. If the product involves only two path coefficients and no unanalyzed correlations, (a) it is an ***indirect component*** if the path coefficients can be rearranged such that the second index of one matches the first index of the other (e.g., $p_{43}p_{31}$ or $p_{53}p_{32}$); (b) It is a ***spurious component*** if the path coefficients have the same second index (e.g., $p_{43}p_{13}$ or $p_{53}p_{23}$).

3. If the product involves three or more path coefficients and no unanalyzed correlations, reduce it to a product of two coefficients by replacing pairs of coefficients of the form $p_{ik}\,p_{kj}$ with the single coefficient p_{ij}. Then apply the rules in (2) to determine whether it is an indirect or a spurious component.

4. If the product involves an unanalyzed correlation of the form r_{ij}, it could always be a spurious component, because r_{ij} could have been generated by X_i and X_j's sharing a common cause. To check whether it could also be an indirect component, thus leaving us uncertain as to whether it is an indirect or a spurious component, generate two new products: one involving replacing r_{ij} with p_{ij} and one replacing it with p_{ji}. If both new products would receive a common classification of spurious according to rules (2) and (3), classify this component as spurious. But if one of the new products would be classified as an indirect component and the other, as a spurious component, classify this component as unanalyzable.

Once each component has been classified, collect together all the products with a common classification, substitute in the numerical estimates of each path coefficient or unanalyzed correlation involved, and compute the sum of the components within each classification.

2.9.5 Overall Test of Goodness of Fit.

Many path analysts like to have an overall measure of goodness of fit of their path model to the observed correlations. There are many competitors for such a measure. The present section presents one such measure, based on the squared multiple correlations yielded by the endogenous variables predicted to be nonzero, as compared to the multiple correlations that would have been yielded by the corresponding fully recursive model (or just *full model*).

The test is carried out by computing

$$Q = \frac{(1-R_{1,f}^2)(1-R_{2,f}^2)\Lambda\ (1-R_{s,f}^2)}{(1-R_{1,r}^2)(1-R_{2,r}^2)\Lambda\ (1-R_{s,r}^2)}$$

as your overall measure of goodness of fit, where $R^2_{i,f}$ is the multiple R^2 between the ith endogenous variable and all variables that precede it in the causal ordering, whereas $R^2_{i,r}$ is the multiple R^2 obtained when predicting that same i^{th} endogenous variable from only those preceding variables the path model presumed to be nonzero. The statistical significance of this measure (i.e., a test of the H_0 that the reduced model fits the data just as well as the purely recursive model) is determined by comparing

$$W = -(N - d) \ln Q$$

to the chi-square distribution with d degrees of freedom, where d = the number of restrictions put on the full model to generate the path model you tested, and usually = the number of paths your model declares to be zero; and ln Q is the natural logarithm of Q.

2.9.6 Examples

Example 2.5 *Mother's Effects on Child's IQ.* Scarr and Weinberg (1986) used data from N = 105 mother–child pairs to examine the relationships among the following variables:

Exogenous variables:

1. WAIS: Mother's WAIS Vocablulary score.
2. YRSED: Mother's number of years of education.

Endogenous variables:

3. DISCIPL: Mother's use of discipline.
4. POSCTRL: Mother's use of positive control techniques.
5. KIDSIQ: Child's IQ score.

The causal ordering for the full model is as given by the preceding numbering: WAIS and YRSED are correlated, exogenous causes of all other variables; DISCIPL is presumed to precede POSCTRL in the causal order but has no direct path to POSCTRL; and DISCIPL and POSCTRL both directly affect KIDSIQ.

It seems reasonable to hypothesize that

(a) p31 = 0 (DISCIPL is unaffected by WAIS vocabulary);
(b) p41 = 0 (POSCTRL is unaffected by WAIS vocabulary);
(c) p43 = 0 (DISCIPL, POSCTRL don't directly affect each other);
(d) p51 > 0 (mom's IQ positively influences kid's IQ)
(e) p52 = 0 (there's no Lamarckean effect of YRSED on KIDSIQ);
(f) p42 > 0 (more educated moms will rely more on positive reinforcement);
(g) p32 < 0 (educated moms will rely less on punishment).

These seven assumptions, together with the nonzero but directionally unspecified p53 and p54 paths, constitute the model we are testing. (Note, however, that they are *not* the assumptions Scarr made.) The model can be summarized in the following path diagram:

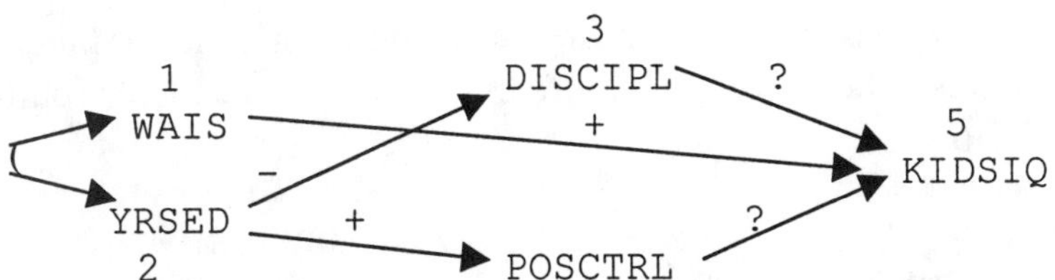

Making the usual assumptions that disturbance terms are uncorrelated with variables earlier in the causal sequence and thus with each other, *that we have the causal order right,* and that our system is self-contained, we can use MRA to estimate parameters for both the full and the reduced model. The following SPSS syntax does the "dirty work" of estimating path coefficients and testing their statistical significance for us:

```
SET LENGTH=NONE Width = 80 .
TITLE SCARR (1985) PATH ANALYSES
MATRIX DATA VARIABLES = WAIS YRSED DISCIPL POSCTRL KIDSIQ /
   CONTENTS = N CORR / FORMAT = FREE FULL
BEGIN DATA
105 105 105  105 105
1.0 .560 .397 .526 .535
.560 1.0 .335 .301 .399
.397 .335 1.0 .178 .349
.526 .301 .178 1.0 .357
.535 .399 .349 .357 1.0
END DATA
VAR LABELS WAIS MOM'S VOCAB SCORE ON WAIS /
   YRSED MOM'S YEARS OF EDUCATION/
   DISCIPL USE OF DISCIPLINE /
   POSCTRL USE OF POSITIVE CONTROL TECHNIQUES /
   KIDSIQ CHILD'S IQ SCORE /
LIST
REGRESSION MATRIX IN (*)/ VARIABLES = WAIS TO KIDSIQ/
   DEPENDENT = DISCIPL/ ENTER YRSED / ENTER WAIS /
   DEPENDENT = POSCTRL/ ENTER YRSED / ENTER WAIS DISCIPL /
   DEPENDENT = KIDSIQ/ ENTER WAIS DISCIPL POSCTRL / ENTER YRSED .
```

Note that each endogenous variable is the dependent variable in a regression equation whose predictor variables are, in the first step, all of the variables that precede this endogenous variable in the causal order and that are predicted to have a nonzero causal impact on it. The regression coefficients from this first step provide the path estimates and the significance tests of these regression coefficients provide the Condition 9' tests. In the second step, the remaining variables that precede this endogenous variable—the ones predicted to have no direct influence on it—are added. The regression coefficients for the added variables and the significance tests thereof provide the Condition 10 tests.

The SPSS output was "mined" for the following results:

Condition 9' tests: These are based on the reduced-model MRAs. There are five paths predicted by this model to be nonzero. The relevant significance tests (pulled from the SPSS output) are as follows:

Path	Pred'd Sign	Estimate	t-ratio	p value	Outcome of Cond. 9' Test
p32	-	+.335	3.608	.0005	Failed
p42	+	+.301	3.203	.0018	Passed
p54	?	+.112	1.153	.251	Failed
p53	?	+.166	1.855	.066	Failed
p51	+	+.410	3.956	.0001	Passed

An appropriate control for the fact that we carried out five tests would be to require a p value of .05/5 = .01 (or maybe .10/5) before considering the Condition 9' test passed. (We also require, of course, that the sign of the path match our a priori notions—see the test of p32.)

Condition 10 tests: These require examining the full model to see whether paths assumed by the reduced model to be zero are indeed small and statistically nonisignificant. Because we're hoping for nonsignificance, it seems appropriate to require that each path predicted to be zero yield a p value (for the test of the H_0 that it's zero) greater than, say, .10 on a fully a priori basis. There are four such paths, yielding the following results:

Path	Estimate	t-ratio	p value	Outcome of Cond. 10 Test
p31	.305	2.812	.006	Failed
p41	.533	5.027	<.001	Failed
p43	-.039	-.415	.679	Passed
p52	.119	1.189	.237	Passed

Any one of the three failed Condition 9' tests or the two failed Condition 10 tests is enough to consider the model disconfirmed. Let's nonetheless go on to the global test of reduced model. There are three endogenous variables in the full model, with associated multiple R^2 values of .1761 (DISCIPL), .2780 (POSCTRL), and .3269 (KIDSIQ). Thus the numerator of Q is (1 - .1761)(1 - .2780)(1 - .3269) = (.8239)(.7220)(.6731) = .4004. For the reduced model the corresponding R^2's are .1122, .0906, and .3174, so the denominator of Q = (.8878)(.9094)(.6827) = .5512. The reduced model employed four overidentifying restrictions (d = 4). Thus Q = .4004/.5512 = .7264 and W = -(105 - 4) ln$Q$ = -101·ln(.7264) = 32.285 can be compared to the critical value for a chi-square with 4 df, yielding $p < .001$.

Reproduced-correlations "test": This is best done after the decomposition equations have been generated.

Decomposition equations for the full model:
Step 1: Normal equations.

$$\begin{array}{lclclcl} r13 & = & p31 & + & r12\cdot p32 & & \\ r23 & = & r12\cdot p31 & + & p32 & & \end{array}$$

$$\begin{array}{lclclcl} r14 & = & p41 & + & r12\cdot p42 & + & r13\cdot p43 \\ r24 & = & r12\cdot p41 & + & p42 & + & r23\cdot p43 \\ r34 & = & r13\cdot p41 & + & r23\cdot p42 & + & p43 \end{array}$$

$$\begin{array}{lclclclcl} r15 & = & p51 & + & r12\cdot p52 & + & r13\cdot p53 & + & r14\cdot p54 \\ r25 & = & r12\cdot p51 & + & p52 & + & r23\cdot p53 & + & r24\cdot p54 \\ r35 & = & r13\cdot p51 & + & r23\cdot p52 & + & p53 & + & r34\cdot p54 \\ r45 & = & r14\cdot p51 & + & r24\cdot p52 & + & r34\cdot p53 & + & p54 \end{array}$$

Step 2: Substitute the definitions of lower-order correlations (those involved in predicting variables earlier in the causal sequence) into the normal equations, yielding the following decomposition equations:

$$\begin{array}{lccccccccccc} r13 & = & p31 & + & r12\cdot p32 & = & .3051 & + & .56(.1642) & = & .3051 & + \; .0920 \\ & & D & & U & & D & & U & & D & \;\; U \\ r23 & = & r12\cdot p31 & + & p32 & = & .56(.3051) & + & .1642 & = & .1709 & + \; .1642 \\ & & U & & D & & U & & D & & U & \;\; D \end{array}$$

$$\begin{array}{lclclclcl} r14 & = & p41 & + & r12\cdot p42 & + & r13\cdot p43 & & \\ & = & p41 & + & r12\cdot p42 & + & p43(p31 & + & r12\cdot p32) \\ & = & p41 & + & r12\cdot p42 & + & p43\cdot p31 & + & p43\cdot p32\cdot r12 \\ & & D & & U & & I & & U \\ & = & .5326 & + & .56(.0157) & + & (-.0387)(.3051) & + & (-.0387)(.1642)(.56) \\ & & D & & U & & I & & U \end{array}$$

$= .5326$ Direct $+ .0052$ Unanalyzed $- .0113$ Indirect

$$\begin{array}{lclclclcl} r24 & = & r12\cdot p41 & + & p42 & + & r23\cdot p43 & & \\ & = & r12\cdot p41 & + & p42 & + & p43(r12\cdot p31 & + & p32) \\ & = & r12\cdot p41 & + & p42 & + & p43\cdot p31\cdot r12 & + & p43\cdot p32 \\ & & U & & D & & U & & I \\ & = & .56(.5326) & + & .0157 & + & (-.0387)(.3051)(.56) & + & (-.0387)(.1642) \\ & & U & & D & & U & & I \end{array}$$

$= .0157$ Direct $+ .2916$ Unanalyzed $- .0064$ Indirect

$$
\begin{array}{lllllllllll}
r34 & = & r13\cdot p41 & & & + & r23\cdot p42 & & & + & p43 \\
 & = & p41(p31 & + & r12\cdot p32) & + & p42(r12\cdot p31 & + & p32) & + & p43 \\
 & = & p41\cdot p31 & + & p41\cdot r12\cdot p32 & + & p31\cdot r12\cdot p42 & + & p42\cdot p32 & + & p43 \\
 & & S & & S & & S & & S & & D \\
 & = & .5326(.3051) & + & .5326(.56)(.1642) & + & .3051(.56)(.0157) & + & .0157(.1642) & - & .0387 \\
 & & S & & S & & S & & S & & D
\end{array}
$$

= -.0387 Direct + 0 Indirect + .2168 Spurious

[Note that the two effects involving r12 are spurious, rather than unanalyzed, because they would be spurious whether r12 were treated as p12 or as p21.]

$$
\begin{array}{lllllllll}
r15 & = & p51 & + & r12\cdot p52 & + & r13\cdot p53 & + & r14\cdot p54 \\
r25 & = & r12\cdot p51 & + & p52 & + & r23\cdot p53 & + & r24\cdot p54 \\
r35 & = & r13\cdot p51 & + & r23\cdot p52 & + & p53 & + & r34\cdot p54 \\
r45 & = & r14\cdot p51 & + & r24\cdot p52 & + & r34\cdot p53 & + & p54
\end{array}
$$

(Left for the reader to complete.)

Of course, one would usually not be interested in decomposing the correlations unless the model provided reasonably good fit to the observed correlations.

Example 2.6 *Gender Bias Revisited: More Light on "Suppression".* Path analysis provides an interesting perspective(s) on our hypothetical gender-bias data (Example 2.4). We can think of the differential cell sizes in our Anova model as representing the effects of gender (specifically, being female) on a college, whose *i*th level represents being hired into college *i*. We can then examine the effects of femaleness and of college into which you are hired on your salary.

Let's start with a two-college version of these data, examining only the College of Medicine and the College of Education at Hypothetical University. The (fully recursive) path model (all path coefficients being statistically significant, given the large *N*) is

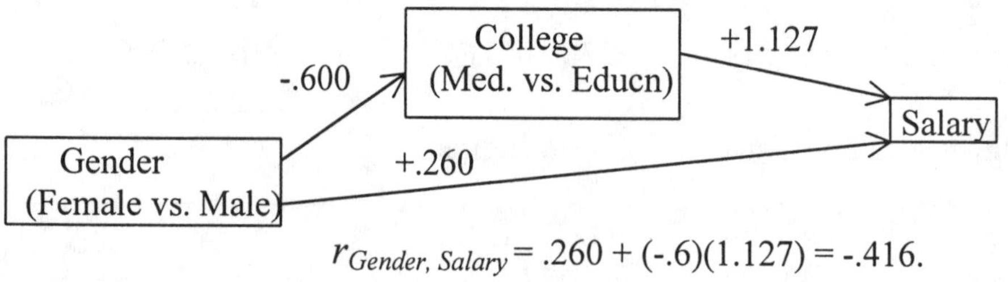

$$r_{Gender,\,Salary} = .260 + (-.6)(1.127) = -.416.$$

This makes a good deal of sense, in that it indicates that the direct effect of being female on salary is a positive one (i.e., once you have been hired into a particular college, being female is predictive of getting a somewhat higher salary), but that this positive direct effect is "overwhelmed" by the strongly negative indirect effect consisting of females' being much less likely to be hired into the College of Medicine, which pays its faculty members much more than does the College of Education, with the end result being that the overall correlation between femaleness and salary (ignoring which college you've been hired into) is moderately negative.

A number of other interesting points can be gleaned from this analysis:

1. Standardized path coefficients can be greater than 1.0 in absolute value (cf. the path coefficient of +1.127 from College to Salary).

2. Components can also be greater than 1.0 in absolute value; it's therefore clear that they are *not* in a correlation metric and do *not* (some authors' claims notwithstanding) represent the proportion of the overall correlation that's accounted for by that component. For instance, it is *not* the case that the direct effect of college on salary accounts for 113% of the overall correlation between those two variables.

3. "Suppressor effect" demonology—interpreting the positive regression coefficient for gender despite its negative zero-order correlation with salary as due to college's having suppressed the "irrelevant variance" in gender—makes far less sense (if indeed the reader can find any sense in it at all; I can't) than does the interpretation offered earlier that the negative correlation is the end result of two competing processes: a positive direct effect (measured by gender's regression coefficient) and a strongly negative indirect effect.

4. Measuring the contribution of gender to the regression equation predicting salary from gender and college on the basis of gender's loading on the regression variate (which equals $-.416/R = -.419$) is also pretty silly, involving as it does ignoring gender's path coefficient (and direct effect on salary) in favor of an overall correlation that corresponds to none of the path coefficients and none of the direct or indirect effects of gender uncovered by the path analysis.

Enough on that analysis. The obvious next step is to include all three colleges in our analysis. However, to do so requires that the college variable be represented by two group-membership variables (e.g., College of Medicine vs. Education, and Engineering vs. Medicine and Education). It's pretty clear what the predicted paths from Gender to each of these two variables, and from each to Salary, should be—but what should be the direction and sign of the path between the two group-membership variables? The obvious approach would be to leave the relationship between them an unanalyzed correlation (a curved, two-headed arrow in the diagram)—but because gender clearly precedes both of them chronologically and causally that would leave us with an unanalyzed correlation (or perhaps two single-headed arrows, one in each direction, indicating reciprocal causation) in the middle of our path diagram, and the MRA approach we've been employing would no longer work. We thus defer consideration of the three-college path analysis to chapter 8's section on structural equation modeling (SEM).

2.9.7 Some Path Analysis References

Our coverage of path analysis has been rather cursory. If you're interested in learning more about PA, you would do well to consider the following books and articles:

Kenny, D. A. (1979). *Correlation and Causality.* NY: Wiley.

James, L. R., Mulaik, S. A., & Brett, J. M. (1982). *Causal Analysis: Assumptions, Models, and Data.* Beverly Hills: Sage. [Source of Conditions 1–10.]

James, L. R., & Singh, K. (1978). An intro to the logic, assumptions, and basic analytic procedures of 2-stage least-squares. *Psychological Bulletin, 85*, 1104–1122. [Two-stage LSQ is needed when your model involves reciprocal causation.]

Lance, C. E., Cornwall, J. M., & Mulaik, S. A. (1988). Limited info parameter estimates for latent or mixed manifest and latent variable models. *Multivariate Behavioral Research, 23*, 171–187. [Includes arguments for local, rather than global, tests of fit.]

McFatter, R. M. (1979). The use of structural equation models in interpreting regression equations including suppressor and enhancer variables. *Applied Psychological Measurement, 3,* 123–135.

Smith, E. R. (1978). Specification and estimation of causal models in social psych: Comment on Tesser & Paulhus. *Journal of Personality and Social Psychology, 36,* 34–38.

Steiger, J. H. (1980). Tests for comparing elements of a correlation matrix. *Psychological Bulletin, 87*, 245–251. [Ever wonder why you so seldom see significance tests of, e.g., differences in reliabilities for two tests computed from the same sample?]

Tesser, A., & Paulhus, D. (1978). On models and assumptions: A reply to Smith. *Journal of Personality and Social Psychology, 36* , 40–42. [These two papers provide a good example of the level of technical detail that can be involved in tests of time series/reciprocal-causation models.]

Ward, J. H. (1969). Partitioning of variance and contribution or importance of a variable: A visit to a graduate seminar. *American Eductional Research Journal, 6*, 467–474. [An amusing presentation of my gripes about the $b_{zj} \cdot r_{ij}$ partitioning.]

Demonstration Problem

Each of the remaining chapters of this book provides two kinds of data for you to use in practicing the techniques presented in that chapter: (a) real data about which some researcher really cared, and (b) a highly simplified set of data to be used as the basis of a demonstration problem that asks you to do a large number of analyses by hand that serve to illustrate that the heuristic explanation of what these techniques "really" do actually hold up when you plug into the various fomulas. The demonstration problem is very important to a full understanding of, in this case, multiple regression.

Subject	X_1	X_2	X_3	Y
A	6	3	2	4.2
B	5	6	0	4.0
C	4	2	9	5.0
D	7	1	5	4.1
E	3	8	9	6.4

A. Do each of the following by hand (and/or with the help of a desk calculator),

1. Construct the cross-product matrix for the predictor variables and the vector of cross products of the predictor variables with Y.

2. Use the information from problem 1 to find the regression coefficients for predicting: (a) Y from X_1; (b) Y from X_1 and X_2; (c) Y from X_1, X_2, and X_3.

3. Calculate R^2 and R for problems 2(a)–2(c),

4. Construct summary tables for problems 2(a)–2(c) and use the tables to decide whether ρ (the true population multiple correlation coefficient) can reasonably (say, with $\alpha = .05$) be assumed to be zero.

5. For problem 2(c), calculate $\hat{Y}$ for each subject. In addition, calculate $\Sigma\hat{Y}/N$, $\sum(\hat{Y}-\bar{Y})^2$, and the Pearson product–moment correlation coefficient for the correlation between $\hat{Y}$ and Y. To what already computed statistic does each of these correspond?

6. For problem 2(c), calculate $e_i = Y_i - \hat{Y}_i$ for each subject. Also calculate $\sum e_i$, $\sum(e_i - \bar{e})^2$, and the correlation between Y_i and e_i. To what previously calculated statistic does each of these correspond?

7. Calculate the inverse of the 3 x 3 covariance matrix of the predictor variables.

8. Use the results from problem 7 to calculate the multiple correlation between (a) X_3 vs. X_1 and X_2; (b) X_2 vs. X_1 and X_3; (c) X_1 vs. X_2 and X_3.

9. Write the regression equations for Problems 8(a)–8(c).

10. For problems 2(a)–2(c) test the hypothesis that the population value of β_1 (the "weight" or "loading" of Variable 1 in the regression equation) is zero.

11. Test the significance of the increase in R^2 (and thus in predictive ability) that arises

from adding Variable 2.

12. Test the significance of the increase in R^2 that results from adding Variable 3 to the first two prediction variables.

13. Test the hypothesis that β_1 and β_3 are identical in the three-predictor equation.

14. Plot Y against $\hat{Y}$, and examine for departures from linearity.

B. Do each of the following problems by hand or desk calculator.

1. Calculate the intercorrelations among X_1, X_2, X_3, and Y.

2. Test the hypothesis that the true population value of each of the six intercorrelations is zero.

3. On the basis of the results of problem A2 (that is, the raw-score coefficients), predict what the corresponding regression coefficients will be in each case for predicting z scores on the X variables.

4. Now carry out the multiple regression analyses of problem A2, beginning in each case with the intercorrelation matrix rather than with the covariance matrix.

5. Repeat problems A4–A14, starting in each case with the intercorrelations rather than the covariances .

C. Use the computer (for example, SAS's or SPSS's REGRESSION subprogram) for each of the following problems.

1. Conduct as many of the analyses of parts A and B as possible. Do the analyses of Problems A8 and A9 in two ways:

(a) Using the computer for each multiple regression analysis, specifying a new dependent variable and a new pair of predictor variables for each "run."

(b) Using only the information provided by the computer's analysis of Y predicted from X_1, X_2, and X_3.

2. Perform the additional analyses needed to test the significance of the increase in R^2 provided by adding a quadratic component to the regression analysis in problem A2(b), that is, by adding X_1^2 as a predictor variable. Conduct the significance test.

3. Why were you not asked to compare the R^2 obtained when X_1^2, X_2^2, and X_3^2 are added as predictor variables?

Answers

A.

1.

	x_1	x_2	x_3	y
A	1	−1	−3	−.54
B	0	2	−5	−.74
C	-1	−2	4	.26
D	2	−3	0	−.64
E	-1	4	4	1.66
Mean	5	4	5	4.74

$$\rightarrow \begin{bmatrix} 10 & -13 & -15 \\ & 34 & 1 \\ & & 66 \end{bmatrix} \text{ and } \begin{bmatrix} -5.4 \\ 7.1 \\ 13.0 \end{bmatrix} \leftarrow \text{Cross-product vector}$$

Cross-product matrix

$[4.072] \leftarrow \sum y^2$

2. (a) $b_1 = -5.4/10 = -.54$; $b_0 = \overline{Y} - b_1 \overline{X} = 4.74 - (-.54)(5) = 7.44$.

(b)

$$\begin{bmatrix} b_1 \\ b_2 \end{bmatrix} = \begin{bmatrix} 10 & -13 \\ -13 & 34 \end{bmatrix}^{-1} \begin{bmatrix} -5.4 \\ 7.1 \end{bmatrix} = \frac{1}{171}\begin{bmatrix} 34 & 13 \\ 13 & 10 \end{bmatrix} \begin{bmatrix} -5.4 \\ 7.1 \end{bmatrix} = \frac{1}{171}\begin{bmatrix} -91.3 \\ 0.8 \end{bmatrix} = \begin{bmatrix} -.534 \\ .005 \end{bmatrix}$$

$b_0 = 4.74 - 5(-.534) - 4(.005) = 7.390$.

(c) $$(\mathbf{x'x})^{-1} = \frac{1}{4016}\begin{bmatrix} 2243 & 843 & 497 \\ 843 & 435 & 185 \\ 497 & 185 & 171 \end{bmatrix}; \begin{bmatrix} b_1 \\ b_2 \\ b_3 \end{bmatrix} = \frac{1}{4016}\begin{bmatrix} 334.1 \\ 941.3 \\ 852.7 \end{bmatrix} = \begin{bmatrix} .083 \\ .234 \\ .212 \end{bmatrix}.$$

3. (a) $R^2 = r_{1y} = .846^2 = -.54(-5.4)/\Sigma y^2 = 2.916/4.072 = .716$.

(b) $R^2 = [-.54(-.534) + 7.1(.005)]/4.072 = 2.916/4.072 = .716$; $R = 346$.

(The computer program's greater accuracy reveals that R in the first case is .84623, whereas it is .84629 in the two-predictor case.)

(c) $R^2 = [(-5.4)(.083) + 7.1(.234) + .212(13.0)]/4.072 = .975$; $R = .988$.

4. Summary tables. (None of the *F*s are statistically significant.)

Source	*df*	*SS*	*MS*	*F*	*df*	*SS*	*MS*	*F*	*df*	*SS*	*MS*	*F*
Regression	1	2.916	2.916	7.57	2	2.916	1.458	2.52	3	3.976	1.325	13.68
Residual	3	1.156	.386		2	1.156	.577		1	.097	.097	
Total	4	4.072			4	4.072			4	4.073		

5. Predicted scores:

Subject	$.083x_1$	$.243x_2$	$.212x_3$	$\hat{y}$	$\hat{Y}$	ε
A	.083	−.234	−.636	−.787	3.953	.247
B	.000	.468	−1.060	−.592	4.148	−.148
C	−.083	−.468	.848	.297	5.037	−.037
D	.166	−.702	.000	−.536	4.204	−.104
E	−.166	.936	.848	1.618	6.358	.042

$$\bar{\hat{Y}} = 23.7/5 = 4.74 = \bar{Y};\ \sum(\hat{Y}-\bar{Y})^2 = 3.965 = SS_{\text{regr}};$$

$$r_{Y\hat{Y}} = [(-.787)(-.54) + (-.592)(-.74) \cdots]/\sqrt{(3.965)(4.072)}$$
$$= .986 = R.$$

6. See problem A5 for error scores. $\Sigma e_i = .289$ (sum of the positive deviations) - .289 (sum of the negative deviations) = .000; $\sum(e_i - \bar{e})^2 = \sum e_i^{\ 2} = .0969 = \text{SS}_{\text{res}}$; $r_{ye} = .160 = \sqrt{1-R^2} \neq 0$. Thus our estimates of the residuals are not completely independent of the subjects' scores on *Y*.

7. $\mathbf{S}_x^{-1} = 4(\mathbf{x'x})^{-1}$ Note that $(\mathbf{x'x})^{-1}$ was given in the answers to problem 2.

8.

i	$s_i^2 c_{ii}$		$1/(s_i^2 c_{ii})$	$R^2_{i\bullet oth}$	$R_{i\bullet oth}$
3	18.5(435/1004) =	3.68	.272	.644	.802
2	8.5(171/1004) =	1.448	.691	.309	.556
1	2.5(2243/1004) =	5.585	.179	.821	.908

9. (a) $x_3 = -(497/171)x_1 - (185/171)x_2 = -2.91x_1 - 1.082x_2$;
(b) $x_2 = -(843/435)x_1 - (185/435)x_3 = -1.94x_1 - .425x_3$;

(c) $x_1 = -(843/2243)x_2 - (497/2243)x_3 = -.375x_2 - .222x_3$.

10. (a) Same as overall F, ns.
(b) $F = b_1^2(d_{ii}MS_{\text{res}}) = (.534)^2/[(34/171)(.57781)] = 2.48 = 1.575^2$, ns.
(c) $F = (.083)^2/[(2243/4016)(.097)] = 27.7/220.9 = .1259 = .355^2$, ns.

11. Increase is very close to zero. Not worth testing.

12. $F = (3.976 - 2.916)/.097 = 10.91$ with 1 and 1 *df*, ns.

13. H_0: $\beta_1 - \beta_2 = 0$. The sum of squares for contrast is
$SS_{\text{contr}} = (.212 - .083)^2/.3536 = .0471$, whence
$F = SS_{\text{contr}}/MS_{\text{res}} = .0471/.097 < 1$, ns.

14. No nonlinearity apparent.

B.

1. **R** = **DSD,** where **D** is a diagonal matrix whose entries are the reciprocals of the standard deviations of the variables, whence

$$\mathbf{R} = \begin{bmatrix} 1 & -.7050 & -.5839 & -.8463 \\ & 1 & .0111 & .6034 \\ & & 1 & .7930 \\ & & & 1 \end{bmatrix}$$

2. $\chi^2 = -(4 - 13/6) \ln|\mathbf{R}| = -(11/6) \ln|\mathbf{R}|$. However, $|\mathbf{AB}| = |\mathbf{A}| \bullet |\mathbf{B}|$ so

$$|\mathbf{R}| = |\mathbf{D}|^2 \bullet |\mathbf{S}| = 388.96/(10 \bullet 34 \bullet 66 \bullet 4.072) = .004246;$$

whence

$$\chi^2 = (11/6)(5.459) = 10.01 \text{ with 6 } df\text{, ns.}$$

The approximate test gives

$$\chi^2 = (11/6)\,\Sigma r^2 = 4.67 \text{ with 6 } df\text{, ns.}$$

3. $\hat{Y} = b_0 + b_1X_1 + b_2X_2 + \cdots + b_mX_m$;

$\hat{Y} - \overline{Y} = b_1x_1 + b_2x_2 + \cdots + b_mx_m$;

$\hat{z}_y = (\hat{Y} - \overline{Y})/s_y = (b_1/s_y)x_1 + (b_2/s_y)x_2 + \cdots + (b_m/s_y)x_m$.

However, $z_i = x_i/s_i$ whence, $x_i = s_i z_i$; whence,

$$\hat{z}_y = b_1(s_1/s_y)z_1 + b_2(s_2/s_y)z_2 + \ldots + b_m(s_m/s_y)z_m$$

$$= b_1^* z_1 + b_2^* z_2 + \ldots + b_m^* z_m ,$$

where $b_i^* = b_i (s_i / s_y)$.

Predictors	b_1^*	b_2^*	b_3^*
z_1	$-.846$	$-$	$-$
z_1, z_2	$-.836$	.014	$-$
z_1, z_2, z_3	.130	.676	.853

4. Same results as in problem A2.

5. Problem A4: $SS_{\text{regr}} = (N - 1)R^2$, $SS_{\text{res}} = (N - 1)(1 - R^2)$, $SS_{\text{tot}} = N - 1 = 4$ in each case; F unchanged.

Problem A5: Mean $\hat{z}_y = 0$; $\sum \hat{z}_y^2 = (N - 1) R^2$; $r_{z\hat{z}} = R$.

Problem A6: Mean error = 0; sum of squared errors = $(N - 1)(1 - R^2)$.

Problem A7: $R^{-1} = (\mathbf{DS}_x\mathbf{D})^{-1} = \mathbf{D}^{-1}\mathbf{S}_x^{-1}\mathbf{D}^{-1} = \begin{bmatrix} 5.5853 & 3.8705 & 3.1796 \\ & 3.6827 & 2.1823 \\ & & 2.8105 \end{bmatrix}$

Problem A8: Same multiple Rs, but obtained much more simply as $1 - 1/c_{ii}$.
Problem A9:

$$\hat{z}_1 = -.693 z_2 - .569 z_3$$
$$\hat{z}_2 = -1.051 z_1 - .593 z_3$$
$$\hat{z}_3 = -1.091 z_1 - .772 z_2$$

(Note that the coefficients in each case are simply s_i/s_y times previous coefficients.)
Problem Al0: Same Fs.
Problems All-A12: Same Fs, simpler arithmetic.
Problem A13: $F = [(.130 - .853)^2/2.26] /.0239 = 9.68$ with 3 and 1 df ns.
Not the same as for covariance-based bs.
Problem A14: Same shape, although different units, origin.

C.

1. Comparable results. Only "new" analysis is using program output from the three-predictor case to do problem A8, as follows:

$$R^2_{i\cdot\text{oth}} = 1 - [s^2 c_{ii} (N - 1)]; \quad t_i = b_i/(MS_{\text{res}}\, c_{ii});$$

whence

$$R^2_{i\cdot\text{oth}} = 1 - t_i{}^2 MS_{\text{res}}/[b_i^2 s_i^2 (N - 1)]$$
$$= 1 - 7.4(.0969)/4 = 1 - .177 = .823 \quad \text{for } i = 1;$$
$$= 1 - [2.288/(.2344)(2.915)]^2(.09685/4) = 1 - .272 = .728 \quad \text{for } i = 2;$$
$$= 1 - .355 = 645 \quad \text{for } i = 3.$$

(Substitute F for t^2 if the program you used reports F tests.) There is no simple way, however, to obtain the regression coefficients (answer to problem A9) from the given information.

2. Adding only X_1^2 gives R = .985; adding only X_2^2 gives R = 1.00. Neither F significant.

3. This would have produced negative degrees of freedom and R = 1.00, because you would be using five predictors on only five subjects.

Some Real Data and a Quiz Thereon

As part of a demonstration of Thurstone and Likert scaling techniques (cf. Selltiz, Jahoda, Deutsch, & Cook, 1959), students in the author's social psychology lab course served both as judges and as pilot subjects in the development of a scale to measure the position of students on a liberal–conservative continuum. The primary issue in comparing the lab students' ratings of how conservative various items were with the student's own conservatism scores was whether (as Thurstone contended when he first developed Thurstone scaling techniques) the scale value of an item was unaffected by the positions of the judges used. The data are given as Data Set 2 in Table 2.11.

If you are reading through this text with others—for example, as part of a class—you may wish to pool your efforts in entering these data into a file accessible to all of you.

I. Conduct each of the following analyses of the Conservatism Data:

A. Using a computer program that employs a step-down multiple regression procedure, test the null hypothesis that the judges' responses to (judgments of) statements 3, 5, 6, 12, 18, and 27 were unrelated to their own scores on the Conservatism scale. Which individual variables contribute significantly to the regression equation? Test the hypothesis that the simple sum of the judges' judgments of statements 3, 5, 6, 12, 18, and 27 is significantly related to their Conservatism–scale score. Using a t value of 1.96 for each successive step, conduct a step-down stepwise multiple regression analysis of these data.

B. Using the computer program of your choice and treating all six predictors as "free variables," conduct a step-up regression in such a way that you are certain that all six predictors are included. Now force a step-up to include variables in the order 3, 5, 6, 12,

Table 2.11 Data Set 2: Ratings of Conservatism of Statement, Multiplied by 10[a]

Subject	3	5	6	12	18	27	Conservatism score[b]
A	15	15	65	70	70	100	61
B	10	10	100	50	50	65	72
C	20	30	100	70	30	70	61
D	23	19	83	46	53	84	56
E	10	15	80	15	25	80	67
F	10	40	70	100	40	100	87
G	25	15	85	60	45	70	48
H	15	12	10	30	50	100	65
I	20	20	70	50	50	80	65
J	10	20	80	60	40	90	68
K	37	13	76	66	45	50	70
L	30	50	10	10	50	40	63
M	57	32	78	76	42	87	69
N	40	30	80	50	40	70	54
O	10	20	90	70	30	82	96
P	30	42	82	34	82	44	67
Q	30	15	75	50	75	94	76
R	15	30	80	20	50	50	71
S	20	30	95	40	35	100	86
T	15	35	100	65	95	85	73
U	5	27	85	72	40	87	53
V	80	95	100	80	50	27	59
W	1	9	84	36	49	69	57
X	20	11	100	49	30	60	70
Y	20	10	70	40	70	80	54
Z	90	15	90	69	35	80	52
1	20	40	95	80	40	35	53

[a]Ratings were made on a 10-point scale, 1.0–10.0. [Not all subjects stayed within this range.) 1.0 was labeled "Extremely liberal," and 10.0 was labeled "Extremely conservative."

[b]The conservatism scale was derived from subject's agreement or disagreement with the same statements he or she had previously rated for degree of conservatism. The most conservative possible score was 138; the most liberal, 23. Statements corresponding to the numbered columns:

3. "UNM should hire a full-time counselor to counsel legal avoidance of the draft."
5. "Birth control information and equipment should be available at the university health center to any interested coed."
6. "Police should have unlimited authority to apply force when necessary."
12. "Union shops are opposed to the American principle of individual freedom."
18. "Most people who are on welfare are not able to support themselves."
27. "The campus humor magazine .. was obscene and not suited for a .. college population."

18, and 27 and then in the order 27, 18, 12, 6, 5, and 3. Any interesting differences? Now do a step-up in which all predictors are free variables and the *F*s for inclusion and deletion are both 1.0, but the tolerance level is 50%. Any comments?

II. Answer the following "thought questions":

A . (Answerable on the basis of intuition about the nature of multiple regression.)

1. Can we, simply by examining the intercorrelation matrix, determine a minimum value of multiple *R*?

2. In conducting an item analysis of a questionnaire, each item is typically correlated with "total scale score," which equals the sum of the scores on the separate items. Could we instead do a multiple regression of the items on total scale score and use the overall magnitude of multiple *R* as an overall indication of internal consistency, with the *b*s indicating which items should be retained?

3. What other suggestion for analysis of internal consistency do you have?

4. What differences between internal consistency and validity does all this suggest?

B. (The following require some analysis or at least doodling with the data.)

1. Can we use the simple zero-order correlations with *Y* to put an upper bound on *R*?

2. Can we use a variable's correlation (Pearson *r*) with the outcome variable as a guide to how high a weight it will receive in the regression equation employing all *m* predictors to predict *Y*?

3. How adequately can we check for curvilinearity by examining separately each of the scattergrams of *y* plotted against one of the predictor variables? (In other words, how useful is the Index-Plot option of BMD02R?)

4. What is the most adequate visual check for curvilinearity? How can this be verified statistically?

5. Does curvilinearity of the relationship beyween one or more predictors and *Y* imply curvilinearity of the relationship between *Y* and $\hat{Y}$?

6. Does curvilinearity of the relationship between $\hat{Y}$ and Y imply that the relationship between *Y* and at least one of the *X*s is curvilinear?

7. Can you develop some simple formulae for the β weights and for the overall value of multiple *R* in the case in which the predictor variables are uncorrelated with each other? Would you share these formulae with the author, please?

Answers

I.

A . Own score = 47.62 - .1701 I_3 + .2394 I_5 + .0672 I_6 - .0373 I_{12} - .0655 I_{18} + .2313 I_{27}. F = .916 with 6 and 20 *df*, ns.

Clearly I_5, I_{27}, and I_3 contribute most. The quickest way to check the simple sum (or any other linear combination of the original variables) for significance of relationship to Y is to compute the correlation between this combined variable and compare this figure to the multiple R that would have been required to achieve significance. Because the multiple R represents the ultimate in capitalizing on chance (it is the correlation you obtain for the very best linear combination of the original variable), and because the overall F test for significance takes this capitalization on chance into account, therefore, whether a priori or post hoc, any linear combination of the original variables you can think up that exceeds this critical value is statistically significant. This is essentially the "union-intersection principle" that we use in Manova. In the present case, an F of 2.71 is needed for significance at the .05 level, whence solving

$$R^2(N - m - 1)/[(1 - R^2)(m)] = 2.71$$

for R gives us the critical value of R at the .05 level for these degrees of freedom, namely, R = .670. We could compute the correlation of the sum with Y and compare this to .670, but because multiple R for our sample did not exceed .670, we know that this particular linear combination will not either.

B. The first three requested analyses produce the same end result as part A. The differences arise in the speed with which R "builds up." Table 2.12 depicts the buildup for each of the requested analyses.

Table 2.12 Buildup of R^2 for Different Orders of Addition of Predictors

Order of Addition of Variables								
	"Free"		1, 2, ..., 6		27, 18, .., 3		Free with *Fs* = 1; tolerance = .50	
Step	Order	*R*	Order	*R*	Order	*R*	Order	*R*
1	6	.293	1	.275	6	.293	6	.293
2	1	.352	2	.306	5	.299	1	.3517
3	2	.439	3	.315	4	.300	2	.439
4	3	.452	4	.330	3	.311	No variable has *F* to enter greater than 1, so analysis stops here.	
5	5	.461	5	.337	2	.378		
6	4	.464	6	.464	1	.464		

II.

A.

1. Clearly R must be at least as large as the absolute value of the correlation between Y and that single X variable having the highest (in absolute magnitude) Pearson r with Y, because the optimization routine could always use regression coefficients that were zero except for that particular X variable.

2. No—or at least, not to any useful purpose. Because total scale score is just the sum of the individual item scores, the regression equation will of course simply have coefficients of 1.0 for each variable, resulting in a multiple R of 1.0, regardless of what items were used.

3. Take the multiple R between each item and the rest of the items (excluding itself) as indicative of how closely related to the other items it is. This is easily computed from the main diagonal entries of the intercorrelation matrix, and it gives us the most optimistic possible picture of the degree of internal consistency.

4. The suggestion in answer A3 employs only the intercorrelation matrix of the scores on the items of the scale and uses no information at all about how the items correlate with some external criterion. Thus internal consistency and validity are logically (and often empirically) independent.

B.

1. Consideration of a set-theoretic interpretation of correlation in terms of "common variance" suggests that R^2 must be less than or equal to $\sum r_{iy}^2$. We can draw the various variables as overlapping circles, with the overlapping area representing the amount of variance variables i and j "share," namely, $s_{ij}^2 = s_i^2 s_j^2 r_{ij}^2$ (see Fig. 2.1). It seems fairly clear that the portion of its variance that Y shares with the entire set of variables $X_1, X_2, \ldots, X_m$ (which equals R^2) must be less than or equal to the sum of the portions shared separately with $X_1, X_2, \ldots, X_m$, because any overlap between X_i and X_j in the variance they share with Y would be counted twice in the simple sum $\sum r_{iy}^2$. The overlap interpretation

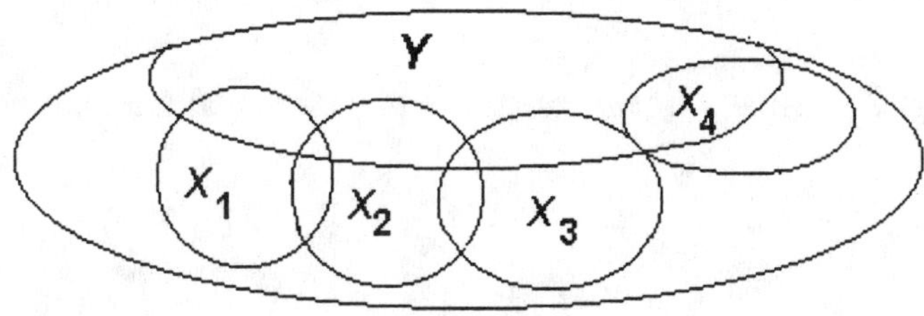

Fig. 2.1 Venn diagram of correlations among Y and four predictors.

suggested above strongly suggests that we should be able to prove that

$$s^2_{y\bullet x_1 . x_2,\cdots,x_m} = s_{1y} + s_{2y\bullet 1} + \ldots + s_{my\bullet 1,2,\cdots,m-1},$$

where $s_{iy.\mathbf{c}}$ is the covariance of X_i with y, with the effects of the variables in the set **c** "partialed out," in which case, of course, the covariance so corrected will be less than the covariance between X_i and y before the predicted scores (predicted from linear regression equations involving the variables in **c**) have been subtracted. However, clear or not, the conjecture is wrong. For a disproof we need look no farther than Data Set 1. Considering the three-predictor case, for instance, $\sum r_{iy}^2 = .493$, whereas $R^2 = .828$. So much for Venn diagrams. Creager (1969, in response to Pugh, 1968) specifically described the dangers of inferring properties of MRA from "overlapping rings." For instance, both he and Cohen and Cohen (1975, p. 87) pointed out that some of the areas representing overlap among Y and two or more predictors may be negative. For the moment, I know of no upper bound on R^2 other than 1.0.

2. We certainly cannot take r_{iy} as a minimum value. For instance, in the two-predictor equation for the z-score form of the Demonstration Problem, $r_{2y} = .604$, but $b_{z_2} = .014$. Similarly, it cannot be taken as a maximum value, because, for example, in the three predictor case $r_{2y} = .604$ and $r_{3y} = .793$, but b_{z_2} (the regression coefficient for X_2 in the z-score equation) = .676 and $b_{z_3} = .853$. However, if all the predictor variables are uncorrelated, the regression coefficients in the z-score regression equation are $b_{z_i} = r_{iy}$ (see quiz problem B7).

3. Not very adequately. It is, for instance, possible for the relationships of X_1 and X_2 to Y to be curvilinear, while the relationship between $\hat{Y}$ and Y is perfectly linear. For example, we can have $Y = X_1 + X_1^2 = X_1 + X_2$ if $X_2 = X_1^2$. In this case $Y = (X_2)^{.5} + X_2$, so that the separate relationship of X_1 to Y and X_2 to Y are both curvilinear, whereas the relationship of Y to X_1 and X_2 considered together is perfectly linear.

4. A plot of $\hat{Y}$ versus Y. If this is linear, then $Y = a + c\hat{Y}$ is an adequate description of the data, whence

$$\begin{aligned}\hat{Y} &= a + c(b_0 + b_1X_1 + b_2X_2 + \ldots + b_mX_m) \\ &= b_0' + b_1'X_1 + b_2'X_2 + \ldots + b_m'X_m\end{aligned}$$

is an adequate description of the data (no X^2 terms are needed), where $b_0' = a + cb_0$ and $b_i' = cb_i$.

5. No. See quiz Problem B3.

6. Yes. If $Y = a + c\hat{Y} + d\hat{Y}^2$, then $Y = a + c\hat{Y} + d(b_0^2 + b_1^2X_1^2 + \cdots + b_m^2X_m^2 + \cdots 2\Sigma\Sigma b_ib_jr_{ij})$. Thus if no X^2 terms or X_iX_j terms were required in the regression equation for predicting Y, d in the univariate regression of Y on $\hat{Y}$ would vanish.

7. Yes. In that case, $\mathbf{S}_x$ is a diagonal matrix whose entries are the variances of the predictor variables, whence

$$\mathbf{S}_x^{-1} = \begin{bmatrix} 1/s_1^2 & 0 & . & 0 \\ 0 & 1/s_2^2 & . & 0 \\ . & . & . & . \\ 0 & 0 & . & 1/s_m^2 \end{bmatrix} \text{ and } \mathbf{b} = \mathbf{S}_x^{-1}\mathbf{s}_{xy} = \begin{bmatrix} s_{1y}/s_1^2 \\ s_{2y}/s_2^2 \\ \\ s_{my}/s_m^2 \end{bmatrix}$$

and

$$R^2 = (\mathbf{b}'\mathbf{s}_{xy})^2/s_y^2 = [\sum(s_{iy}/s_i^2)]/s_y^2 = \sum[s_{iy}/(s_is_y)]^2 = \sum r_{iy}^2 .$$

Path Analysis Problem

Data are collected from 116 subjects on their socioeconomic status (SES), IQ, need for achievement (NAch), and UNM GPA. The correlations among these four variables are:

	SES	IQ	NAch	GPA
SES	1.00	.30	.41	.33
IQ	.30	1.00	.16	.57
NAch	.41	.16	1.00	.50
GPA	.33	.57	.50	1.00

Use these data to test the following assumptions (model):

(i) High SES causes high Need for Achievement, but any influence of SES on GPA is

mediated by NAch (i.e., there is no direct causal influence of SES on GPA).
(ii) High NAch causes high GPA.
(iii) High IQ also causes high GPA, but has no direct causal influence on NAch.
(iv) SES and IQ are correlated, but we have no commitment to the sign and/or direction of any causal path(s) between them.

1. Draw the path diagram for this model.

2. Estimate each of the path coefficients in the model.

3. Carry out the Condition 9' and Condition 10 tests for the model.

4. Compute the correlations among the observed variables that are implied by the model.

5. Conduct the overall chi-square test of the model's fit to the data.

6. Decompose each observed correlation (other than $r_{SES,IQ}$) into direct, indirect, spurious, and unanalyzable components.

7. Comment on the overall fit of the model, and on what modifications (if any) of the model would most economically improve its fit to the observed correlations.

8. Indicate at least one change of direction in a single causal path that would convert this to a non-recursive model.

9. Give an example of a fifth variable and its correlations with the original four variables that would constitute a violation of the "relevant variables" condition. Ditto for an example of an omitted variable that would not be relevant (and whose inclusion would thus not affect our estimates of the path coefficients in the present model).

Answers to Path Analysis Problem

The computer run that generated the path-coefficient estimates and provided the Condition 9' and Condition 10 tests was:

```
Set Length = None  Width = 80
TITLE PATH ANALYSIS HOMEWORK
MATRIX DATA VARIABLES = SES IQ NACH GPA/
     CONTENTS = N CORR / FORMAT = FREE FULL
BEGIN DATA
116 116 116 116
1 .30 .41 .330
.30 1 .16 .57
.41 .16 1 .50
```

```
.33 .57 .50 1
END DATA
LIST
REGRESSION MATRIX IN (*)/ VARIABLES = SES TO GPA/
  DEPENDENT = NACH/  ENTER SES / Enter IQ /
  DEPENDENT = GPA/ ENTER IQ NACH/ Enter SES /
```

1. *Draw the path diagram for this model.*

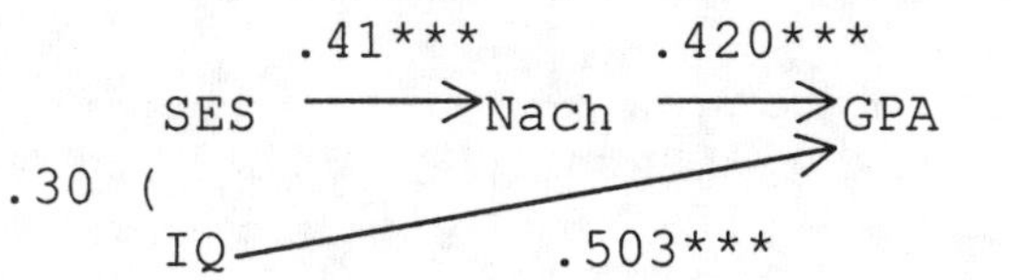

(The question didn't ask for the path estimates yet, but the diagram is a convenient place to "park" them.)

2. *Estimate each of the path coefficients in the model.*
Relevant output from the computer run is:

```
  REGRESSION MATRIX IN (*)/ VARIABLES = SES TO GPA/
    DEPENDENT = NACH/  ENTER SES / Enter IQ /
    DEPENDENT = GPA/ ENTER IQ NACH/ Enter SES /

Equation Number 2    Dependent Variable..   GPA
Block Number  1.  Method:  Enter       IQ       NACH
Multiple R              .70456
R Square                .49641
------------------ Variables in the Equation ------------------
Variable                B          SE B         Beta          T  Sig T

NACH               .419540      .067629      .419540      6.204  .0000
IQ                 .502874      .067629      .502874      7.436  .0000
```

3. *Carry out the Condition 9' and Condition 10 tests for the model.*

<u>Condition 9':</u>

```
Equation Number 1    Dependent Variable..   NACH
Block Number  1.  Method:  Enter       SES
Multiple R              .41000
F =        23.03570        Signif F =  .0000
```

→ **p31 passes.**

Nach & IQ rows from above → *p43, p42 pass.*

<u>Condition 10:</u>

```
Equation Number 1    Dependent Variable..   NACH
Variable(s) Entered on Step Number   2..    IQ
Multiple R              .41183
R Square                .16960
------------------ Variables in the Equation ------------------
```

```
Variable                    B          SE B        Beta           T    Sig T
SES                   .397802       .089863     .397802       4.427    .0000
IQ                    .040659       .089863     .040659        .452    .6518
```

IQ row ⟶ *p32 passes.*

```
---------------------
Equation Number 2     Dependent Variable..    GPA
Block Number  2.   Method:   Enter       SES
Multiple R                .70461
R Square                  .49647
------------------- Variables in the Equation ------------------
Variable                    B          SE B        Beta           T    Sig T
NACH                  .416126       .073580     .416126       5.655    .0000
IQ                    .500663       .070352     .500663       7.117    .0000
SES                   .009189       .076139     .009189        .121    .9042
```

SES row from above ⟶ *p41 passes.*

4. *Compute the correlations among the observed variables that are implied by the model.*
r12 = r12 (unanalyzed) = .30

r13 = p31 = .41 (vs. .41)
r23' = r12 · p31 = .3(.41) = .123 (vs..160)

r14' = r12 · p42 + r13 · p43
= .3(.503) + .41(.420) = .323 (vs..330)
r24' = p42 + r23 · p43
= .503 + .123(.420) = .555 (vs..570)
r34' = r23'·p42 + p43
= .123(.503) + .420 = .482 (vs..500)

4. *Conduct the overall chi-square test of the model's fit to the data.*

$$Q = \frac{\text{Full - model products}}{\text{Reduced - model products}} = \frac{.8304(.50353)}{.8319(.50359)} = .99808;$$

```
Full-model Prods            .8304(.50353)
```

$W = -(116 - 2)\ \ln(.99808) = .219 < \chi^2_{.05}(2) = 5.99$
⟶ can't reject the H_0 of perfect model fit.

6. *Decompose each observed correlation (other than $r_{SES,IQ}$) into direct, indirect, spurious, and unanalyzable components.*
r12 = r12 (unanalyzed) = .30 U

$$\begin{array}{lcccccc}
r13 & = & p31 & = & .41\,D & & . \\
 & & D & & & & \\
r23' & = & r12 \cdot p31 & = & .3(.41) & = & .123\,U
\end{array}$$

(p31•p12 would be I, whereas p21•p31 would be S, so this is all U.)

$$\begin{array}{lcccc}
r14' & = & r12 \cdot p42 & + & r13 \cdot p43 \\
 & & (p12 \cdot p42 \text{ or } p42 \cdot p21) & & p43 \cdot p51 \\
 & & S \quad \text{or} \quad I & & I \\
 & = & .3(.503)\,U & + & .41(.420)\,I \\
 & = & .151\,U & + & .172\,I
\end{array}$$

$$\begin{array}{lcccccc}
r24' & = & p42 & + & r23' \cdot p43 & & \\
 & & D & & r12 \cdot p31 \cdot p43 & & \\
 & & & & (p41 \cdot p12 \text{ or } p41 \cdot p21) & & \\
 & = & & & I \quad \text{or} \quad S & & \\
 & = & .503\,D & + & .123(.420)\,U & = & .503\,D + .052\,U
\end{array}$$

$$\begin{array}{lcccccc}
r34' & = & r23' \cdot p42 & + & p43 & & \\
 & & r12 \cdot p31 \cdot p42 & & D & & \\
 & & (p31 \cdot p12 \cdot p42 \text{ or } p31 \cdot p42 \cdot p21) & & & & \\
 & & (p32 \cdot p42 \text{ or } p31 \cdot p41) & & & & \\
 & & S \quad \text{or} \quad S & & & & \\
 & = & .123(.503)\,S & + & .420\,D & = & .062\,S + .420\,D
\end{array}$$

7. *Comment on the overall fit of the model, and on what modifications (if any) of the model would most economically improve its fit to the observed correlations.*

The fit is excellent. Our overall Q is .998, our chi-square testing the hypothesis of perfect fit is nonsignificant, and our reproduced correlations are quite close to the observed correlations, the worst fit being .123 vs. .160 for r_{23}. There seems little to gain by changing the model in any way.

8. *Indicate at least one change of direction in a single causal path that would convert this to a non-recursive model.*

A very poor question, as it turns out. *Adding* p14 would do it—but that's not just a change of direction. Any change in the paths that are nonzero in our model would have to go through r_{12} to "cycle back," so there's no way to know if the resulting model would be non-recursive or recursive.

9. *Give an example of a fifth variable and its correlations with the original four variables that would constitute a violation of the "relevant variables" condition. Ditto for an example of an omitted variable that would not be relevant (and whose inclusion would*

thus not affect our estimates of the path coefficients in the present model).

Any added variable that affects only one of the variables in the model would clearly not be a relevant variable, and would not alter our estimates of the paths in the current model. I had thought that any variable that does influence (correlate with) two or more of the variables already in the model would be a relevant variable, in that its addition would change the estimate of the magnitude of the causal path between the two variables it correlates with. However, Elaine Raybourn (personal communication) pointed out that if the two existing variables are connected by only a curved, two-headed arrow, the estimate of that correlation already takes into account spurious correlation due to variables not in the model that correlate with them both, so adding an explicit variable that correlates with these two would *not* change our estimate of the total unanalyzed correlation between them.

3
Hotelling's T^2: Tests on One or Two Mean Vectors

This chapter describes a rather direct generalization of Student's t test to situations involving more than one outcome variable. The general approach, as in all of the multivariate techniques we study in this text except factor analysis, is to compute a combined score for each subject that is simply a linear combination of his or her scores on the various outcome measures. The weighting factors employed in obtaining this combined score are those that make the univariate t test conducted on the combined score as large as possible. Just as Student's t test can be used either to test a single sample mean against some hypothetical value or to test the difference between two independent sample means and a hypothesized difference (usually zero) between the means of the populations from which they were drawn, so too is there a single-sample and a two-sample version of Hotelling's T^2 (the statistical procedure discussed in this chapter) corresponding to testing a mean outcome vector or the difference between two vectors of mean scores on dependent measures against corresponding population mean vectors. If the dependent measures are sampled from a multivariate normal distribution, the sampling distribution of Hotelling's T^2 (which is simply the square of the maximum possible univariate t computed on any linear combination of the various outcome measures) has the same shape as the F distribution. Now for the details.

3.1 SINGLE-SAMPLE t AND T^2

Harris and Joyce (1980) had 20 pentads of five students each[1] work independently but in the same room, on anagrams (scrambled-word problems). The partnership that solved the most anagrams received a prize based on the number solved. The main interest was in how these five-person groups decided to divide their potential prize. Data Set 3 in Table 3.1 reports the decisions of these 20 pentads, separated by whether they were asked to

---

[1] The most common sampling unit in studies is the individual subject. The most common label for a set of independently sampled subjects exposed to common conditions is group." However, many other sampling units are employed: classrooms, litters, and-as in the study we are discussing-groups of individuals who interact. The statistical techniques discussed in this chapter can be applied to any study in which the behavior of each of one or two samples of such sampling units is measured in more than one way. Our discussions of general techniques will often use "subjects" as a generic term for "sampling units" and "groups" as a generic term for classes or categories of such sampling units. More specific terms should of course be used when describing a particular study—such as the groups or partnerships or pentads who were subjected in the present study to one of two experimental conditions.

allocate final shares of the prize directly (the outcomes-allocation condition) or, instead, did this indirectly by specifying how much should be subtracted from each person's individual contribution to get his or her final share of the prize (called the expenses-allocation condition). Thus, for instance, the third partnership run in the outcomes-allocation condition decided that Partner 1 should receive \$3.90 of their \$9.50 prize; Partner 2 should receive \$2.86; Partner 3, \$2.60; and Partners 4 and 10, \$.07 each.

The major theory in this area (Adams' 1965 ratio formula) predicts that the mean amounts allocated to the five partners in the outcomes-allocation condition will be \$5.112, \$2.699, \$1.880, \$-.297, and \$-.096 (Partner 4 will be requested to pay 29.7 cents to the rest of the group and Partner 10 will be requested to pay 9.6 cents) for the population from which these data were sampled. These predictions are quite discrepant from the observed sample means. However, before we devote a lot of theoretical effort to explaining this discrepancy, we'd like to be sure that it's not solely due to random variations about the specified population means.

The single-sample version of Student's t test provides a mechanism for testing this null hypothesis with respect to any one of the five means. For example, testing the statistical significance of the difference between m = \$1.277 and the predicted value of \$-.297 yields

$$H_0\colon \mu = -.297$$

$$t = \frac{\overline{X} - \mu_0}{s_{\overline{X}}} = \frac{\overline{X} - \mu_0}{\sqrt{s^2/N}} = \frac{[1.277 - (-.297)]}{\sqrt{.852/10}} = \frac{1.574}{.2919} = 5.392$$

with $N - 1 = 9\ df$.

Because the critical value for a t with 9 degrees of freedom (df) is 3.250 at the $\alpha = .01$ level (see the t table in your introductory statistics text or on your favorite statistics website—perhaps http://fsweb.berry.edu/academic/education/vbissonnette/), we can confidently reject the null hypothesis of a perfect fit between Adams' theory and actual allocations to group members who work on four-interchange problem. We would of course like to examine Adams' predictions for the other four partners as well. We could conduct a separate t test on each of the five outcome measures and compare each to a table of critical values (upper percentiles) of Student's t distribution with 47 degrees of freedom. This, however, leads us straight into the multiple comparison problem. If we adopt a significance level (Type I error probability) of .05 for each t test, the probability of falsely identifying at least one of the five means as significantly different from its predicted value will be somewhere between .05 (the probability if all five measures are perfectly correlated with each other) and .266 [$= 1 - (1 - \alpha)^k = 1 - .95^5$, the experimentwise error rate if the five measures are mutually independent]. We thus need to find a way of directy controlling our α level for the set of five comparisons as a whole. The approach that Hotelling (1931) discovered and that was later subsumed as a special case of Roy's union-intersection principle involves (a) representing any comparison of one or more of the sample means with one or more of the corresponding hypothesized

Table 3.1 Data Set 3: Divisions of Potential Prize, Experiment 3, Harris & Joyce (1980)

Condition	Partners					
	$P1$	$P2$	$P3$	$P4$	$P10$ [a]	Sum
Outcomes	$1.70	$1.70	$1.70	$1.70	$1.70	$8.50
Allocated	1.00	1.00	1.00	1.00	1.00	5.00
	3.90	2.86	2.60	.07	.07	9.50
	2.34	.00	1.62	.00	.00	3.96
	1.30	1.30	1.30	1.30	1.30	6.50
	3.00	3.00	3.00	3.00	3.00	15.00
	4.50	3.50	2.50	1.50	.50	12.50
	2.00	2.00	2.00	2.00	2.00	10.00
	1.70	1.70	1.70	1.70	1.70	8.50
	4.50	4.50	3.50	.50	.50	13.50
Means	$2.594	2.156	2.092	1.277	1.177	9.296
Expenses	$5.00	5.50	1.50	1.00	– 1.50 [b]	11.50
Allocated	7.00	2.00	1.50	– 1.50	.50	9.50
	6.00	5.00	2.50	- 1.50	- 1.50	10.50
	6.00	2.00	1.50	.00	- 1.50	8.00
	7.17	4.17	1.66	.00	- 1.00	12.00
	8.20	2.60	- .20	- 1.70	2.10	11.00
	6.50	6.50	.00	- 1.50	2.00	13.50
	6.50	3.00	2.00	- 1.00	- .50	10.00
	5.50	2.50	4.00	1.00	- 1.50	11.50
	6.50	4.00	1.50	- .50	.00	11.50
Means	$6.437	3.727	1.596	- .570	- .290	10.900

[a] Partners are designated by the number of random interchanges of pairs of letters made in constructing each anagram of that partner's 20-anagram list

[b] A negative entry indicates that the group asked that member to give money back to the group so that the other members would have more to share among themselves.

means as a univariate t ratio computed on a single combined variable that is a linear combination of the original outcome measures, and (b) taking as the test statistic for such comparisons the maximum possible value that such a "combined t ratio" could attain for any selection of coefficients for the linear combining rule and selecting a critical value of that statistic that guarantees occurrence of maximum t ratios exceeding that level only α x 100% of the time when H_0 is correct. The development of this multivariate significance test proceeds as follows:

1. We define a new variable

$$\mathbf{W} = \mathbf{a}_1\mathbf{X}_1 + \mathbf{a}_2\mathbf{X}_2 + \cdots + \mathbf{a}_p\mathbf{X}_p = \mathbf{Xa},$$

where $\mathbf{X}_j$ is an N-element column vector giving each of the N subjects' scores on dependent measure (outcome variable) j; $\mathbf{X} = [\mathbf{X}_1, \mathbf{X}_2, \cdots \mathbf{X}_p]$ is an N x p data matrix whose ith row gives subject i's scores on each of the outcome variables; **a** is a p-element column vector giving the weights by which the dependent measures are to be multiplied before being added together; and **W** is therefore an N-element column vector whose ith entry is subject i's score on the new outcome variable W. In scalar terms, we have

$$W_i = a_1X_{i,1} + a_2X_{i,2} + \cdots + a_pX_{i,p}$$

for each of participants i =1 through N.

2. Our overall null hypothesis is that $\mu_1 = \mu_{10}$; $\mu_2 = \mu_{20}$; ..., $\mu_p = \mu_{p0}$ are all true. If one or more of these equalities is false, the null hypothesis is false. This hypothesis can be expressed in matrix form as

$$\boldsymbol{\mu} = \begin{bmatrix} \mu_1 \\ \mu_2 \\ \cdot \\ \mu_p \end{bmatrix} = \begin{bmatrix} \mu_{10} \\ \mu_{20} \\ \cdot \\ \mu_{p0} \end{bmatrix} = \boldsymbol{\mu}_0, \tag{3.1}$$

and it implies that $\mu_W = \mathbf{a}'\boldsymbol{\mu}_0$.

3. The variance of a linear combination of variables can be readily expressed as a linear combination of the variances and covariances of the original variables, namely

$$s_W^2 = a_1^2s_1^2 + a_2^2s_2^2 + \ldots + a_p^2s_p^2 + 2a_1a_2s_{12} + 2a_1a_3s_{13} + \mathrm{K} + 2a_{p-1}a_ps_{p-1,p}$$
$$= \mathbf{a'Sa},$$

where **S** is the covariance matrix of the outcome variables. Thus the univariate t ratio computed on the combined variable W is given by

$$t(\mathbf{a}) = \frac{\mathbf{a}'\overline{\mathbf{X}} - \mathbf{a}'\boldsymbol{\mu}_0}{\sqrt{\mathbf{a}'\mathbf{Sa}/N}} = \frac{\mathbf{a}'(\overline{\mathbf{X}} - \boldsymbol{\mu}_0)}{\sqrt{\mathbf{a}'\mathbf{Sa}/N}} \tag{3.2}$$

where the notation $t(\mathbf{a})$ emphasizes the fact that the value of t depends on (is a function of) the particular coefficients selected for **a.** Now, we are clearly interested in the absolute magnitude of t, irrespective of sign. Because absolute values are a tremendous inconvenience mathematically, we seek instead to maximize the square of $t(\mathbf{a})$, which is of course equal to

$$t^2 = \frac{N[\mathbf{a}'(\overline{\mathbf{X}} - \boldsymbol{\mu}_0)]^2}{\mathbf{a}'\mathbf{Sa}} = \frac{N\mathbf{a}'(\overline{\mathbf{X}} - \boldsymbol{\mu}_0)(\overline{\mathbf{X}} - \boldsymbol{\mu}_0)'\mathbf{a}}{\mathbf{a}'\mathbf{Sa}} \tag{3.3}$$

It should be fairly clear that multiplication of the coefficients in **a** by a constant will have no effect on $t(\mathbf{a})$. In other words, the elements of **a** are defined only up to a change of units, and we must add a side condition on the possible choices of **a** in order that the vector that maximizes $t^2(\mathbf{a})$ may be uniquely defined. The most convenient such choice is the condition that **a'Sa**, the variance of W, be equal to 1. This reduces our problem to that of maximizing the numerator of $t^2(\mathbf{a})$ subject to the constraint that the denominator be equal to unity. Derivation 3.1 shows that the maximum value of $t^2(\mathbf{a})$ and the particular combining weights that yield this value are given by the solutions for λ and **a**, respectively, in either of Equations (3.4a) or (3.4b):

$$[N(\overline{\mathbf{X}} - \boldsymbol{\mu}_0)(\overline{\mathbf{X}} - \boldsymbol{\mu}_0)' - \lambda\mathbf{S}]\mathbf{a} = \mathbf{0}, \tag{3.4a}$$

whence

$$[N\mathbf{S}^{-1}(\overline{\mathbf{X}} - \boldsymbol{\mu}_0)(\overline{\mathbf{X}} - \boldsymbol{\mu}_0)' - \lambda\mathbf{I}]\mathbf{a} = \mathbf{0}. \tag{3.4b}$$

Thus our problem is reduced to that of finding the characteristic roots and vectors (see Digression 2) of the matrix $N\mathbf{S}^{-1}(\overline{\mathbf{X}} - \boldsymbol{\mu}_0)(\overline{\mathbf{X}} - \boldsymbol{\mu}_0)'$. Note that this matrix will generally ***not*** be symmetric. Many characteristic-root (eigenvalue) functions within statistical packages [e.g., EVAL(X) within SPSS's MATRIX/END MATRIX system] are restricted to symmetric matrices. However, this turns out not to be much of a problem in the present case (although it will reappear in more serious form in chap. 4) because, as Derivation 3.1 shows, the matrix whose characteristic roots we seek is of rank 1. Thus, it has only a single nonzero root, which is equal to the trace of the matrix and can be shown to equal

$$T^2 = N(\overline{\mathbf{X}} - \boldsymbol{\mu}_0)'\mathbf{S}^{-1}(\overline{\mathbf{X}} - \boldsymbol{\mu}_0). \tag{3.5a}$$

Moreover, a solution of Equation (3.4a) or (3.4b) for the discriminant function—that is, for the coefficients **a** that maximize $t^2(\mathbf{a})$ at a value equal to T^2—can be obtained very simply as

$$\mathbf{a} = \mathbf{S}^{-1}(\overline{\mathbf{X}} - \boldsymbol{\mu}_0). \tag{3.5b}$$

Equation (3.5a) should be easy to remember, because it is such a direct generalization of the formula for the univariate single-sample t. Specifically, the univariate t formula can be written as

$$t = \frac{\overline{X} - \mu_0}{s_{\overline{X}}},$$

whence

$$t^2 = \frac{(\overline{X} - \mu_0)^2}{s^2 / N}$$

$$= \frac{N(\overline{X} - \mu_0)(\overline{X} - \mu_0)}{s^2}$$
$$= N(\overline{X} - \mu_0)(s^2)^{-1}(\overline{X} - \mu_0),$$

so that Equation (3.5a) is obtained from the formula for t^2 by replacing s^2 with **S** and $(\overline{X} - \mu_0)$ with $(\overline{\mathbf{X}} - \boldsymbol{\mu}_0)$, keeping in mind the matching up of dimensions of various matrices needed to yield a single number as the end result of applying Equation (3.5).

Yet another procedure for computing T^2 is available. Using the results in Digression 2 on the determinants of partitioned matrices, it can be shown that

$$T^2 = (N-1)\left(\frac{|\mathbf{A} + N(\overline{\mathbf{X}} - \boldsymbol{\mu}_0)(\overline{\mathbf{X}} - \boldsymbol{\mu}_0)'|}{|\mathbf{A}|} - 1\right)$$
$$= \left(\frac{|\mathbf{S} + N(\overline{\mathbf{X}} - \boldsymbol{\mu}_0)(\overline{\mathbf{X}} - \boldsymbol{\mu}_0)'|}{|\mathbf{S}|} - 1\right) \tag{3.6}$$

where $\mathbf{A} = \mathbf{x}'\mathbf{x} = (N - 1)\mathbf{S}$ is the cross-product matrix of deviation scores on the p outcome measures. Either of these two determinantal formulae avoids the need to invert **S**.

4. The statistical significance of the observed value of T^2 must now be determined. T^2 is the square of a univariate t ratio computed on a linear combination of the outcome measures. We know that if scores on X are sampled from a normal distribution, the square of a t ratio whose variance estimate is based on N - 1 degrees of freedom has the same frequency distribution as an F distribution with 1 and N - 1 degrees of freedom. However, this distribution is based on the assumption that a single variable, defined in advance of collecting the data on purely a priori grounds, is being tested. Instead, in the present situation, we have generated a new variable, largely post hoc in nature, that is a linear combination of the various outcome measures whose coefficients were deliberately chosen so as to "inflate" t^2 to the greatest extent possible. Hotelling was able to prove that if scores on $X_1, X_2, \cdots, X_p$ are sampled from a single multivariate normal distribution, the effects of all this capitalization on chance can be exactly compensated for by the simple expedient of multiplying the observed value of T^2 by a constant $(N - p)/[p(N - 1)]$, which is always less than 1.0 (except when p and the constant both equal 1), and then comparing the resulting statistic with the F distribution having p and N - p degrees of freedom. That is,

$$F = \frac{N-p}{p(N-1)}T^2 \tag{3.7}$$

(where p equals the number of dependent variables) is distributed as F with p and N - p

degrees of freedom. This expression assumes that we have available N independently sampled outcome vectors (one for each subject or other sampling unit), whence each variance or covariance estimator is based on $N - 1$ degrees of freedom. If we use instead variance-covariance estimators based on fewer degrees of freedom (e.g., if we compute each deviation score separately about the mean of the *subgroup* to which that subject belongs, rather than about the grand mean across all subjects, whence we have only $N - k$ degrees of freedom, where k is the number of subgroups), the constant multiplier becomes $(df - p + 1)/(p \cdot df)$, where df is the number of degrees of freedom going into each estimate of s^2 or s. This will usually be the number by which each sum of squared deviations or sum of cross products of deviation scores is divided in computing $\mathbf{S}$. In this case, the degrees of freedom for our F test become p and $df - p + 1$.

5. If T^2 exceeds the critical value for the chosen level of significance, we must now specify which variable or combination of variables is "responsible" for the rejection of H_0. The most complete, though seldom the simplest, answer is that that linear combination of the original variables actually used to maximize $t^2(\mathbf{a})$—known as the ***discriminant function***—is the variable (dimension?) on which the observed data differ most sharply from our null hypothesis. We usually, however, wish to test other, more natural or more theoretically interesting, linear combinations of the variables (for example, each original variable by itself) for significant departure from the value implied by our null hypothesis vector. Any of these comparisons that were planned in advance employ the univariate t test and the corresponding critical values in a table of the t distribution, with the alpha levels of the individual tests perhaps adjusted below the traditional .05 level so as to keep α_{exp} under control. Post hoc tests are also conducted by computing the appropriate univariate t ratio on the appropriate combination of the outcome measures, but the resulting value of such a post hoc t is compared with that value of T ($\sqrt{T^2}$) that would have just barely achieved statistical significance. This criterion ensures us that no matter how carefully we search for a combining rule that produces a large t value, we shall have at most an α x 100% chance of making a Type I error (rejecting a true null hypothesis) in *any* comparison.

Conversely, if we are interested primarily in testing the overall significance of the differences between our observed sample means and the hypothesized population means, we may be able to avoid computing a covariance matrix (let alone inverting it) at all. Clearly, T^2 must be greater than or equal to the square of the largest of the univariate t ratios. If, therefore, any one of these univariate t ratios exceeds $\sqrt{T^2_{crit}}$, so will T^2.

If max (t_i^2) < T^2_{crit}, we must of course compute T^2, because it still may exceed T^2_{crit}. Use of this "quick and dirty" test, however, even if successful, foregoes information on the discriminant function, that is, on which particular linear combination of the original variables deviates most drastically from the value implied by our overall null hypothesis.

Example 3.1 Applying all this to our original problem, that of testing the null hypothesis that the vector of mean allocations in the outcomes-allocation condition of Harris and Joyce (1980) arose through random sampling from a population in which the true mean response vector is as predicted by Adams' ratio formula, we obtain

$$\overline{\mathbf{X}}' = [2.594 \quad 2.156 \quad 2.092 \quad 1.277 \quad 1.177],$$
$$\boldsymbol{\mu}_0' = [5.112 \quad 2.699 \quad 1.880 \quad -.297 \quad -.096],$$
$$\boldsymbol{\Delta}' = [-2.518 \quad -.543 \quad .212 \quad 1.574 \quad 1.273]$$

$$\mathbf{X'X} = \begin{array}{c} P1 \\ P2 \\ P3 \\ P4 \\ P10 \end{array} \overset{\begin{array}{ccccc} P1 & P2 & P3 & P4 & P10 \end{array}}{\begin{bmatrix} 82.656 & 68.624 & 62.401 & 30.743 & 26.243 \\ & 62.150 & 53.406 & 29.170 & 25.670 \\ & & 49.354 & 27.152 & 24.652 \\ & & & 23.975 & 22.475 \\ & & & & 21.975 \end{bmatrix}}$$

$$\mathbf{S} = \frac{\mathbf{X'X} - N\mathbf{XX'}}{N-1} = \begin{bmatrix} 1.7075 & 1.4108 & .908 & -.2647 & -.4765 \\ & 1.7407 & .9225 & .1820 & .0327 \\ & & .6211 & .0486 & .0032 \\ & & & .8520 & .8272 \\ & & & & .9024 \end{bmatrix},$$

and

$$\mathbf{S}^{-1} = \begin{bmatrix} 84.669 & 4.359 & -122.346 & -102.913 & 139.231 \\ & 3.658 & -11.163 & -8.498 & 9.999 \\ & & 185.428 & 152.429 & -204.581 \\ & & & 139.134 & -182.114 \\ & & & & 241.975 \end{bmatrix},$$

whence

$$T^2 = N\boldsymbol{\Delta}'\mathbf{S}^{-1}\boldsymbol{\Delta} = 6129.6\text{, and}$$

$$F = [(10-5)/(9 \bullet 5)]T^2 = 681.07$$

with 5 and 5 &, $p < .001$.

[Note that these values of T^2 and F, and the following calculation of the discriminant function coefficients, are based on the **S** matrix as listed in this example, that is, rounded off to four decimal places. If MATLAB uses the **S** matrix as stored internally (i.e., with as much accuracy as that system can muster) in subsequent calculations, a T^2 value of

6293.2 is obtained. This degree of sensitivity to small round-off errors suggests that we may have a nearly singular $\boldsymbol{S}$ matrix. Indeed,

$$\frac{|\mathbf{S}|}{\prod_j s_j^2} = 1.2 x 10^{-4},$$

Cases such as this one used to pose severe—and, unfortunately, difficult to detect—accuracy problems for multivariate computer programs running on 8-bit personal computers.]

Having a value of T^2 in hand, we may return to matrix Equation (3.4a) to compute the vector of coefficients that produced the maximum t^2, whence we have

$$[N(\overline{\mathbf{X}}-\boldsymbol{\mu}_0)(\overline{\mathbf{X}}-\boldsymbol{\mu}_0)' - 6129.6\mathbf{S}]\mathbf{a},$$

that is,

$$\begin{bmatrix} -10{,}402.9 & -8634.0 & -5545.3 & 1582.9 & 2888.7 \\ & -10{,}666.8 & -5655.7 & -1124.1 & -207.4 \\ & & -3806.6 & -294.6 & -16.9 \\ & & & -5197.6 & -5050.4 \\ & & & & -5197.6 \end{bmatrix} \begin{bmatrix} a_1 \\ a_2 \\ a_3 \\ a_4 \\ a_5 \end{bmatrix} = \begin{bmatrix} 0 \\ 0 \\ 0 \\ 0 \\ 0 \end{bmatrix}.$$

As pointed out in the matrix algebra disgression (section D2.12), we solve this homogenous system of equations by solving for a_1 through a_4 as multiples of a_5, namely

$$\mathbf{a}^* = \begin{bmatrix} -10{,}402.9 & -8634.0 & -5545.3 & 1582.9 \\ & -10{,}666.8 & -5655.7 & -1124.1 \\ & & -3806.6 & -294.6 \\ & & & -5197.6 \\ & & & \end{bmatrix}^{-1} \begin{bmatrix} -2888.7 \\ 207.4 \\ 16.9 \\ 5050.4 \\ \end{bmatrix} a_5$$

$$= \begin{bmatrix} -.74574 & .22721 & .77353 & -.32008 \\ & -.52945 & .44338 & .15857 \\ & & -2.06816 & .25688 \\ & & & -.33873 \end{bmatrix} \begin{bmatrix} -2888.7 \\ 207.4 \\ 16.9 \\ 5050.4 \end{bmatrix} a_5,$$

whence $\mathbf{a}'$ = (-5979, .0422, -.8802, - .7489, 1) or any nonzero multiple thereof. In other words—as the reader will wish to verify—computation of the combined variable W = .5979$P1$ + .0422$P2$ - .8802$P3$ - .7489$P4$ + $P10$ for each subject, followed by a t test of the difference between $\overline{W}$ and its predicted value $\mathbf{a}'\mu_0 = 1.642$, will yield a t ratio of $\sqrt{6129.6} = 78.29$. No other combined variable can be found that will discriminate as strongiy as (produce a t ratio whose absolute value is greater than) this particular choice. Alternatively, we could have computed the discriminant function coefficients via Equation (3.5b), yielding

$$\text{a} = \mathbf{S}^{-1}\boldsymbol{\Delta}.,$$

whence a' = [-226.138 -15.977 332.913 283.230 -378.223]. These look rather different from the coefficients computed via Equation (3.4a), but if we multiply through by the constant l/(-378.223) = -.2644, we get a revised coefficient vector of [.5979 .0422 -.8802 -.7488 1.000], as before. Remember that discriminant coefficients are defined only up to a multiplicative constant, because multiplying each of the weights by a constant simply multiplies our new variable (the discriminant function) by a constant, and the univariate t ratio is unaffected by such a change of unit of measurement. Of course, if we want to make computation of the discriminant function coefficients **a** the first step of our analysis and then compute T^2 via $T^2 = N\Delta'\mathbf{a}$, we must use the a vector computed from $\mathbf{a} = \mathbf{S}^{-1}\Delta$ before doing any rescaling of the coefficients.

This discriminant function is not, however, a particularly simple one, so we shall wish to conduct some tests on simpler linear combinations of the variables. For this purpose, we first need to compute t_{crit}. With 5 and 5 degrees of freedom, we would need an F of 5.05 to achieve statistical significance at the .05 level. This corresponds to a T^2 of 9(5)/(10 - 5) times 5.05 = 45.45 = T^2_{crit}, so that $t_{\text{crit}} = 6.742$. The variances for post hoc tests of the differences between observed and predicted means on each original variable are found in the main diagonal of **S**, whereas post hoc tests on more complicated linear combinations are obtained from Equation (3.3). (The one bit of matrix algebra that should be embossed inside your right eyelid is that the variance of a linear combination, $a_1X_1 + a_2X_2 + \cdots + a_pX_p$, is equal to **a'Sa**.) Thus, for instance,

$$t_{P1} = \frac{-2.518}{\sqrt{1.7075/10}} = -6.094;$$

whereas the remaining four individual variables yield t ratios of -1.301, ,851, 5.392 (as before), and 4.23—none of which reach statistical significance by our multivariate criterion. Had we been interested only in these five tests, we could have conducted Bonferroni-adjusted t tests by comparing each of the five ts to $t_{\alpha/9}(9)$, which equals 3.25 for $\alpha_{\text{exp}} = .05$. Thus, the observed mean allocations to Partners 1, 4, and 10 all differ significantly from their predicted values by the Bonferroni criterion but not by the multivariate criterion. However, these are *not* simply five unrelated means; Adams' theory in fact states that the amount allocated should be directly proportional to the partner's individual contribution to the group's performance. An alternative theory (Harris's 1976a linear formula) agrees that the amount allocated will increase linearly with input but suggests that individual subjects or groups will differ in just how strongly they wish to differentiate among outcomes on the basis of differential inputs. We might therefore expect that Adams' predictions would be farther from the observed means, the closer a particular partner's input (performance) is to one of the extremes. As it turns out, subjects given 10-interchange problems performed slightly better than 4-interchange partners, so that P4 and P10 jointly represent one extreme of the input dimension and P1, the other. We therefore would anticipate that Pl - (P4 + P10)/2 would show a significant departure from its predicted mean, and indeed

$$t_{(-2,0,0,1,1)} = \frac{7883}{\sqrt{13.204/10}} = 6.860, p < .05$$

by our post hoc criterion.

Finally, testing a simplified version of our discriminant function, $\mathbf{a}'_{\text{simp}}$ = (1, 0, -1, -1, l), yields

$$t_{(1,0,-1,-1,1)} = \frac{3.031}{\sqrt{.2882/10}} = 17.854,$$

which is also significant. The square of this last t ratio (318.8) may seem like quite a large loss in ability to discriminate observation from prediction as compared to our discriminant function in its full multidecimal glory (which yielded a t^2 of 6,130) until we recognize two points: First, the drop in proportion of variance accounted for is only from $t^2/[t^2 + (N - 1)]$ = .999 for the actual discriminant function to 318.8/327.8 = .973 for $\mathbf{a}_{\text{simp}}$. Second, we have only five more observation vectors than variables in the present problem, so that our optimization procedure is apt to be highly sensitive to chance factors in the data. A sample size this small would ordinarily be recommended only for highly reliable phenomena. On the other hand, we would not want to take as very strong support for Harris' (1976b) linear formula the fact that its predictions for this condition—μ_0 = (2.544, 2.177, 2.129, 1.826, 1.184)—do not differ significantly from the observed means, with T^2 = 4.274 and consequently $F(5,5)$ = .487.

On first examination, situations in which the single-sample T^2 would apply would appear to be rare. Rarely are social or behavioral scientists able to predict the precise magnitude of the effects they study. However, we often do study differences between various treatments administered to the same subject or changes in some effect over time. In such designs (discussed in section 3.8) our null hypothesis is that the mean differences or changes all have the precise population value of zero. The single-sample T^2 thus provides a valuable tool for analyzing such within-subjects or repeated-measures designs—a tool that in fact requires much less stringent assumptions than does the more common Anova approach to such data.

3.2 LINEARLY RELATED OUTCOME VARIABLES

Data Set 4 in Table 3.2 reports the number of mutually cooperative (CC), unilaterally cooperative (CD), and unilaterally competitive (DC') outcomes in a 40-trial Prisoner's Dilemma game for each of 48 subject pairs. (The DD column will be explained shortly.)

These data were gathered in the first of a series of replications and extensions of Deutsch's (1960) article designed to assess the effects of directive versus nondirective motivating instructions on behavior in experimental games (highly structured, two-person, two-choice interaction situations). See Harris, Flint, and Everett (1970) for details.

Ignoring for the moment differences between conditions, we can treat the data as representing 48 subject pairs and compute an overall mean for each outcome measure,

$$(\overline{CC}, \overline{CD}, \overline{DD}) = (11.23, 6.92, 8.33).$$

We would like to be certain that subjects are not merely choosing C or D at random. If they were choosing purely randomly, we would expect 10 CCs in a 40-trial game. We would also expect 10 CDs and 10 DCs. Such a multivariate prediction seems tailor-made for analysis via single-sample T^2.

Example 3.2 Applying the techniques of section 3.1 to the problem of testing the hypothesis that the mean response vector for subjects in the Deutsch replication study arose through sampling from a population in which the true mean response vector was (10, 10, 10), we obtain

$$\overline{\mathbf{X}}' = \frac{1}{48}(539, 332, 400) = (11.229,\ 6.917,\ 8.333), \text{ and}$$

$$\bar{\Delta}' = (1.229, -3.083, -1.667)$$

Because we used the determinantal formula (3.6) to compute T^2, we must return to the matrix equations (3.4) to compute the vector of coefficients that produced the maximum t^2, whence we have

$$\left[N(\overline{\mathbf{X}} - \mu_0)(\overline{\mathbf{X}} - \mu_0)' - 46.10\mathbf{S}\right]\mathbf{a} = [\mathbf{N\Delta\Delta' - (46.10/47)A}$$

$$= \begin{bmatrix} -5024.5 & 654.9 & 545.8 \\ 654.9 & -269.3 & 280.7 \\ 545.8 & 280.7 & -732.5 \end{bmatrix} \begin{bmatrix} a_1 \\ a_2 \\ a_3 \end{bmatrix} = \mathbf{0};$$

whence our discriminant function, in nonnormalized form, is given by $\mathbf{a}' = [.358\ 1.913\ 1]$. (Some scalar-single-symbol-expressions for simple cases are given in section 3.8.)

Table 3.2 Data Set 4: Results of Deutsch Replication 1

Motivating Orientation	Experimenter	Subject Pair	CC	CD	DC	DD
			Number of Responses			
Competitive	DR	A	0	1	7	32
		B	4	9	8	19
	HC	C	11	12	9	8
		D	5	9	7	19
	RC	E	15	11	5	9
		F	8	9	8	15
	RK	G	4	9	10	17
		H	5	8	11	16
	RS	I	2	15	2	21
		J	4	6	6	24
	SI	K	12	14	4	10
		L	1	5	7	27
		Σ	71	108	84	217
Individualistic	DR	M	1	4	11	24
		N	6	5	6	23
	HC	O	9	8	10	13
		P	1	6	8	25
	RC	Q	0	8	6	26
		R	2	6	7	25
	RK	S	17	6	11	6
		T	2	6	6	26
	RS	U	9	2	7	22
		V	9	11	7	13
	SI	W	10	14	9	7
		X	10	4	10	16
		Σ	76	80	98	226
No motivating orientation	DR	Y	10	9	11	10
		Z	3	5	2	30
	HC	AB	11	10	9	10
		BC	5	8	12	15
	RC	CD	6	9	12	13
		DE	9	5	11	15
	RK	EF	11	12	7	10
		FG	3	5	13	19
	RS	GH	11	8	8	13
		HI	7	5	20	8
	SI	IJ	6	10	8	16
		JK	17	7	14	2
		Σ	99	93	127	161

Table 3.2 (continued)

Motivating Orientation	Experimenter	Subject Pair	Number of Responses CC	CD	DC	DD
Cooperative	DR	KL	30	0	10	0
		LM	14	10	11	5
	HC	MN	18	1	17	4
		NO	34	0	4	2
	RC	OP	16	1	20	3
		PQ	40	0	0	0
	RK	QR	31	3	4	2
		RS	39	1	0	0
	RS	ST	19	12	6	3
		TU	2	11	9	18
	SI	UV	37	3	0	0
		VW	13	9	10	8
		Σ	293	51	91	45

To conduct tests on simpler linear combinations of the variables, we first need to compute t_{crit}. With 3 and 45 degrees of freedom, we would need an F of 2.82 to achieve statistical significance at the .05 level. This corresponds to a T^2 of 47(3)/(48 - 3) times F_{crit} = 3.133(2.82) = 8.836, whence $t_{crit} = \sqrt{8.836} = 2.97$. Testing each separate mean outcome against 10.0, we obtain $t_{CC} = .812$, as in our earlier computation;

$$t_{CD} = -\frac{3.08}{\sqrt{739.7/[47(48)]}} = -5.59,$$

which is statistically significant by our experimentwise criterion; and

$$t_{DC} = -\frac{1.67}{.626} = 2.668,$$

which would have been significant had this been an a priori comparison but is not significant by our post hoc criterion. Thus the only single variable of these three that can be reliably discriminated from our hypothesized mean value of 10.0 is *CD*. Note that the correspondence between these individual *t*s and the contribution of the different variables to the discriminant function is far from perfect—a result that should surprise no one who has been through the discussion of multiple regression in chapter 2. The reader has probably guessed that the DD column of Data Set 4 represents a fourth measure of the outcome of the game, namely, the number of mutually competitive choices made during the 40 trials of the game. This conjecture is correct. It is also true that our null hypothesis of random, 50–50 responding implies that μ_{DD} also equals 10.0. Why, then, was DD not included in the preceding analysis? Fundamentally, because it adds no new information beyond that already provided by CC, CD, and DC. Because each pair played 40 trials, CC + CD + DC + DD = 40, whence knowledge of a pair's "scores" on the first three measures permits us to predict with 100% certainty the value of DD for the pair. As might be anticipated (and as you should verify), the true dimensionality (rank) of the complete 4 x 4 covariance matrix is only 3, whence S^{-1} is undefined, and our computational procedures break down.

This problem will crop up whenever there is a perfect linear relationship among the outcome measures—that is, whenever one or more of the dependent variables can be written as a linear combination of the others. In such a case there is a linear dependence among the rows and columns of **A** and **S**, so that $|\mathbf{A}| = |\mathbf{S}| = 0$. Going back to an earlier step in the derivation of T^2, we find that we need a solution to the matrix equation $[N\Delta\Delta' - \lambda\mathbf{I}]\mathbf{a} = \mathbf{0}$. In the case of linear dependence, we cannot use the usual trick of multiplying through by $\mathbf{S}^{-1}$, because it does not exist. However, it can be shown, with a bit of algebra, that if λ and $\boldsymbol{a}$ are solutions to this equation, then λ (same value) and $\mathbf{a}^* = [a_i + c_i a_p]$ are solutions to the matrix equation

$$[N\Delta^*\Delta^{*\prime} - \lambda\mathbf{S}^*]\mathbf{a}^* = \mathbf{0},$$

where the asterisk (*) indicates that that vector or matrix has had its last (*p*th) row and/or column deleted, and where

$$X_p = c_1X_1 + c_2X_2 + \cdots + c_{p\text{-}1}X_{p\text{-}1} + a_0 .$$

We can now multiply through by $\mathbf{S}^{*-1}$ (assuming that **S** is of rank p - 1; if it is of rank p - r, we must delete r of the dependent variables) and proceed to obtain the same formulae as before. Thus, whenever the rank of **S** is less than p, we must delete dependent variables until the reduced system yields a reduced covariance matrix with a nonzero determinant and then apply the usual T^2 formulae to this reduced system of outcome measures. It does not matter which of the measures is deleted, as long as it is one of the dependent variables involved in the perfect linear relationship. For instance, retaining DD but deleting DC in the present example leads to a discriminant function of

$$\begin{aligned} -.442\text{CC} + .913\text{CD} - \text{DD} &= -.642\text{CC} + .913\text{CD} - (40 - \text{CC} - \text{CD} - \text{DC}) \\ &= .358\text{CC} + 1.913\text{CD} + \text{DC}, \end{aligned}$$

which is obviously equivalent to our original discriminant function and which in fact leads to precisely the same value of T^2.

Tatsuoka (1972) suggested that linear dependence in **A** and **S** be handled by first performing a principal component analysis (cf. chap. 6 or section 1.2.11) of the data, then conducting T^2 (or Manova, or multiple regression) on the subjects' scores on the principal components. The number of these principal components (PCs) will equal the rank of the covariance matrix, so that no problem arises in computing the inverse of the covariance matrix of scores on PCs. The resulting discriminant function expressed in terms of scores on PCs can then be converted back to a discriminant function expressed in terms of scores on the original variables by substituting for each PC its definition in terms of the original variables.

The only argument Tatsuoka advanced for preferring this rather involved procedure over simple deletion of one or more original variables is that the resulting discriminant function will usually assign nonzero weights to each of the p original variables. This desideratum can be achieved much more simply, however, by using the variable deletion method to obtain an initial discriminant function and then using the known linear relationship between the deleted variable and the original variables to generate any one of the infinitely many equivalent expressions that assign nonzero weights to all variables. For instance, in the present case,
.358CC + 1.913CD + DC = (.358 + b)CC + (1.913 + b)CD + (1 + b)DC + bDD - 40b
for any choice of b.

A word of caution should be inserted here about the dangers of relying on $|S|$, rather than knowledge of the constraints operating on the data, to reveal linear dependencies among the outcome variabes: A very small error in computing the entries of **S** or **A** can lead to a computed determinant that appears greatly different from zero. For instance, if the entry for $\sum X_{CD}X_{DD}$ in **X'X** is changed from 4637 to 4647, $\sum x_{CD}x_{DD}$ in **x'x** changes from 148.1 to 158.1, and the resulting determinant of **A** is 2.345×10^{10} which is unlikely to be confused with zero. The problem is that the +10 error in this one entry enters into all four-term products involving that entry that are used in computing $|\mathbf{A}|$. Near-dependencies like the example can be made more detectable by dividing the determinant of any crossproduct or covariance matrix by the product of the main diagonal elements (the maximum possible value of its determinant) and casting a suspicious eye on the possibility that $|\mathbf{A}|$ or $|\mathbf{S}|$ is truly zero if this ratio is less than 10^{-2}. (The use of .01 as the criterion is arbitrary.) In the present case, the product of the main diagonal elements of **A** is approximately 12.4 x 10^{12}, and the suggested ratio is approximately $.19 \times 10^{-2}$.

3.3 TWO-SAMPLE t AND T^2

Returning to Data Set 4, we might wish to test the effects of our various instructional sets on the performance of our subjects. For instance, we could ask whether the no motivating orientation (NMO) and cooperative (COOP) conditions differ significantly in the frequency of mutually cooperative outcomes. This calls for the old, familiar t test for the difference between two independent sample means. Thus we have

$$H_0\text{: } \mu_1 = \mu_2 \text{; that is, } \mu_1 - \mu_2 = 0.$$

$$t = [(\overline{X}_1 - \overline{X}_2) - 0] / \sqrt{s_c^2(1/N_1 + 1/N_2)} \text{ with } N_1 + N_2 - 2 \; df; \tag{3.8}$$

where $s_c^2 = (\sum x_1^2 + \sum x_2^2)/(N_1 + N_2 - 2)$ is our best available estimate of the presumed common value of the variance of the two populations from which samples 1 and 2 were drawn. Thus in the present situation,

$$t_{\text{NMO-COOP}} = \frac{8.25 - 24.42}{\sqrt{[(180.25 + 1662.95)/22](2/12)}}$$
$$= \frac{-16.17}{\sqrt{1843.2/132}} = -4.33 \text{ with } 22\ df,\ p < .001.$$

However, our "eyeball analysis" of the data led us to concentrate on the difference in CCs as the most dramatic one; thus we would still like to have an overall test of the difference between the two sample mean vectors. The development of the appropriate multivariate test follows that of the single-sample T^2 point for point. The matrix analog of $\overline{X}_1 - \overline{X}_2$ is $\overline{\mathbf{X}}_1 - \overline{\mathbf{X}}_2$, whereas the matrix analog of $\sum x_i^2$ is $\mathbf{A}_i$, so that s_c^2 becomes

$$\mathbf{S}_c = \frac{\mathbf{A}_1 + \mathbf{A}_2}{(N_1 + N_2 - 2)}.$$

Thus the two-sample T^2 can be computed from any of the following formulae:

$$T^2 = [N_1 N_2 /(N_1 + N_2)](\overline{\mathbf{X}}_1 - \overline{\mathbf{X}}_2)'\mathbf{S}_c^{-1}(\overline{\mathbf{X}}_1 - \overline{\mathbf{X}}_2) \tag{3.9}$$
$$= \text{single nonzero root of } [N_1 N_2 /(N_1 + N_2)](\overline{\mathbf{X}}_1 - \overline{\mathbf{X}}_2)'(\overline{\mathbf{X}}_1 - \overline{\mathbf{X}}_2)\mathbf{S}_c^{-1} \tag{3.10}$$
$$= \frac{|\mathbf{S}_c + N_1 N_2 (\overline{\mathbf{X}}_1 - \overline{\mathbf{X}}_2)(\overline{\mathbf{X}}_1 - \overline{\mathbf{X}}_2)'/(N_1 + N_2)|}{|\mathbf{S}_c|} - 1 \tag{3.11}$$

The application of formula (3.9) to the present comparison proceeds as follows:

$$\begin{array}{cc} \text{NMO} & \text{COOP} \\ \mathbf{X}'\mathbf{X} = \begin{array}{c} \\ \text{CC} \\ \text{CD} \\ \text{DC} \\ \text{DD} \end{array} \begin{array}{c} \begin{array}{cccc} \text{CC} & \text{CD} & \text{DC} & \text{DD} \end{array} \\ \begin{bmatrix} 997 & 803 & 1076 & 1084 \\ 803 & 783 & 949 & 1135 \\ 1076 & 949 & 1557 & 1498 \\ 1084 & 1135 & 1498 & 2673 \end{bmatrix} \end{array} & \begin{array}{c} \\ \text{CC} \\ \text{CD} \\ \text{DC} \\ \text{DD} \end{array} \begin{array}{c} \begin{array}{cccc} \text{CC} & \text{CD} & \text{DC} & \text{DD} \end{array} \\ \begin{bmatrix} 8817 & 784 & 1602 & 517 \\ 784 & 467 & 420 & 369 \\ 1602 & 420 & 1159 & 459 \\ 517 & 369 & 459 & 455 \end{bmatrix} \end{array}; \end{array}$$

$$\overline{\mathbf{X}}' = [99 \quad 93 \quad 127 \quad 161]/12 \qquad\qquad [293 \quad 51 \quad 91 \quad 45]/12;$$

$$\begin{array}{cc} \text{NMO} & \text{COOP} \end{array}$$
$$\mathbf{A} = \begin{bmatrix} 180.25 & 35.75 & 28.25 & -244.24 \\ 35.75 & 65.25 & -35.25 & -112.75 \\ 28.25 & -35.25 & 212.92 & -205.90 \\ -244.24 & -112.75 & -205.90 & 512.93 \end{bmatrix} \begin{bmatrix} 1662.95 & -461.23 & -618.90 & -581.75 \\ -461.23 & 250.25 & 33.25 & 177.75 \\ -618.90 & 33.25 & 468.92 & 117.75 \\ -581.75 & 177.75 & 117.75 & 286.25 \end{bmatrix};$$

$$\mathbf{S}_{\text{pool}} = (\mathbf{A}_1 + \mathbf{A}_2)/22 = \begin{bmatrix} 83.782 & -19.340 & -26.848 & -37.545 \\ -19.340 & 14.205 & -.091 & 2.955 \\ -26.848 & -.091 & 30.992 & -4.007 \\ -37.545 & 2.955 & -4.007 & 36.326 \end{bmatrix}.$$

After deleting *DD* (the fourth row and column of **S**), we have

$$\mathbf{S}^{-1} = \begin{bmatrix} .029432 & .040235 & .025614 \\ .040235 & .12540 & .035223 \\ .025614 & .035223 & .054559 \end{bmatrix}; \; \Delta = \overline{\mathbf{X}}_1 - \overline{\mathbf{X}}_2 = \begin{bmatrix} -16.167 \\ 3.500 \\ 3.000 \end{bmatrix};$$

and

$$\begin{aligned} \mathbf{\Delta}'\mathbf{S}^{-1}\mathbf{\Delta} = & \; [.029432(261.372) + .12540(12.25) + .054559(9) + .19697(93.451)] \\ & + 2[-.040235(56.5845) - .025614(48.5) + .035223(10.5)] \\ = & \; 9.720 - 2(3.149) = 3.422; \end{aligned}$$

whence

$$T^2 = 6(3.422) = 20.532;$$

and

$$F = (N_1 + N_2 - p - 1)T^2/[p(N_1 + N_2 - 2)]$$

$$= (20/66)T^2 = 6.222 \text{ with } p \text{ and } (N_1 + N_2 - p - 1)\ df.$$

that is, with 3 and 20 degrees of freedom, $p < .01$. Thus we can reject the hypothesis that all three population mean differences are zero. In yet another direct analogy to the one–sample T^2, we can conduct all the specific comparisons of the two groups in terms of linear combinations of the outcome measures and be assured that the probability of any one or more of these comparisons leading to a false rejection of the hypothesis of no true difference is at most 01. This is done simply by selecting $\sqrt{T^2_{\text{crit}}}$ as our critical value of $|t|$ for all such comparisons, where T^2_{crit} is that value of T^2 that would have just barely led to rejection of the overall null hypothesis. Thus in the present case we have

$$(20/66)\ T^2_{\text{crit}} = 3.10$$

for significance at the .05 level, whence

$$T^2_{\text{crit}} = 3.3(3.10) = 10.23.$$

Thus our criterion for the statistical significance of any post hoc comparisons is that $|t|$ for that comparison must equal or exceed $\sqrt{10.23} = 3.20$. For instance, testing each separate variable gives us

$$t_{CC} = -16.167\sqrt{6.0/83.782} = -4.33, p < .05;$$

$$t_{\text{CD}} = 2.27,\ ns;\ \ t_{\text{DC}} = 1.32,\ ns;\ \text{and}\ \ t_{DD} = 3.929,\ p < .05.$$

Of course, had we been interested solely in significance tests, we could have computed $\sqrt{T^2_{\text{crit}}}$ and compared it to each of the univariate ts, thereby discovering that t_{CC} exceeds this value and thus, a fortiori, so must T^2. We could therefore omit computation of **S**, $\mathbf{S}^{-1}$, and so on. However, in so doing, we would lose information about the discriminant function, that is, that particular linear combination of our three measures that maximally discriminates between the two groups. For the present data, the discriminant function (ignoring DD) is .842CC + .345CD + .415DC.

3.4 PROFILE ANALYSIS

Our first interpretation of the discriminant function might be that it is essentially CC + CD + DC. However, this equals 40 - DD, so that t^2 for this linear combination is just $t^2_{\text{DD}} = 15.4$, only about 3/4 as large as T^2. Notice, however, that the coefficient for *CC* is roughly twice as large as those for CD and DD, so we might consider the discriminant function to be tapping 2CC + CD + DC—which doesn't seem highly interpretable until we realize that this is equivalent to CC - DD, the difference between the number of mutually competitive and mutually cooperative outcomes during the game. Applying $\mathbf{a}' = (1, 0, 0, -1)$ to the full 4 x 4 **S** matrix and **a** vector yields a squared t ratio of 20.513, very close to T^2. The overall T^2 test for two samples "lumps together" two sources of difference between the two groups' response vectors ("profiles"): a difference in the level of the two curves and differences in the *shapes* of the two curves. Figures 3.1(a) and 3.1(b) illustrate, respectively, a pair of groups that differ only in mean response level and a pair of groups that differ only in shape. The method that analyzes these two sources of difference separately, and in addition provides a simple test of the flatness of the combined or pooled profile for the two groups, is known as ***profile analysis.*** A profile analysis of the response vectors for the two groups involves tests of three separate null hypotheses.

Levels Hypothesis.

H_{01}: The profiles for the two groups are at the same level; that is,

$$\mu_{W,1} = \mu_{\sum X_{i,1}} = \mu_{1,1} + \mu_{2,1} + \cdots + \mu_{p,1} = \mu_{1,2} + \mu_{2,2} + \cdots + \mu_{p,2} = \mu_{W,2};$$

that is,

$$\mu_{W,1} - \mu_{W,2} = 0.$$

This simply tests the hypothesis that the mean of the means (actually, the mean sum) of the separate variables is identical for the two groups. If the parallelism hypothesis is true, then this is equivalent to a test of the hypothesis that the difference between the group means on any variable (this distance being the same for all variables) is truly zero.

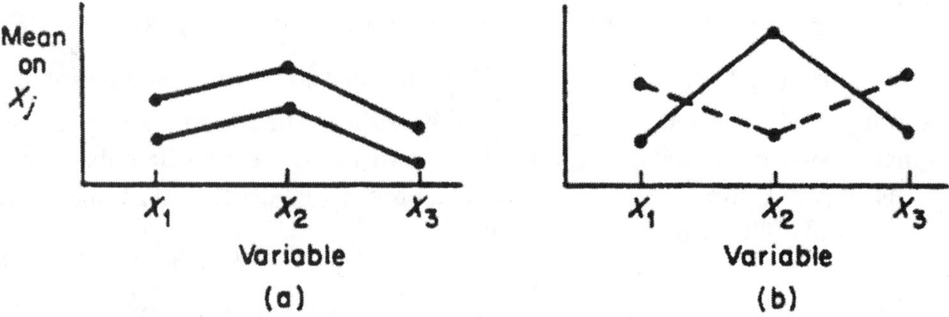

Fig. 3.1 Response vectors for groups differing only in (a) level or (b) slope.

Parallelism Hypothesis

H_{02}: The profiles for the two groups are parallel (have the same shape); that is,

$$\mu_{slope,1} = \begin{bmatrix} \mu_{1,1} - \mu_{2,1} \\ \mu_{2,1} - \mu_{3,1} \\ \text{M} \\ \mu_{p-1,1} - \mu_{p,1} \end{bmatrix} = \begin{bmatrix} \mu_{1,2} - \mu_{2,2} \\ \mu_{2,2} - \mu_{3,2} \\ \text{M} \\ \mu_{p-1,2} - \mu_{p,2} \end{bmatrix} = \mu_{slope,2};$$

that is,

$$\mu_{slope,1} - \mu_{slope,2} = 0.$$

The equivalence of the two forms of the null hypothesis can be seen from the fact that if the two profiles are truly parallel, then the slope of each line segment making up that profile will be the same for both groups.

Flatness Hypothesis

H_{03}: The "pooled profile for the two groups combined is perfectly flat; that is, the combined means are all equal to the same value:

$$\boldsymbol{\mu}_{\text{slope}} = \begin{bmatrix} \mu_1 - \mu_2 \\ \mu_2 - \mu_3 \\ \text{M} \\ \mu_{p-1} - \mu_p \end{bmatrix} = \mathbf{0}$$

(The absence of subscripts on the μ_js indicates that these are *combined* means, pooled across both groups.) The final version of $H_{0,3}$ simply takes advantage of the fact that a flat profile implies that all line-segment slopes are truly zero. These three tests are analogous to the three standard tests of a two-way univariate analysis of variance in which treatments (groups) correspond to rows, and response measures (dependent variables) correspond to columns, whence $H_{0,2}$ corresponds to a test of the row main effect; $H_{0,3}$, to a test of the column main effect; and $H_{0,1}$, to a test of the interaction between rows and columns.

Thus in profile analysis, as in two-way Anova, the "interaction" (parallelism) test takes precedence, with a significant departure from parallelism implying that

(a) the two groups must be compared separately on each outcome measure, because the magnitude and/or direction of the difference between the two groups varies, depending on which variable is considered, and

(b) nonsignificant departures from the equal levels $H_{0,2}$ hypothesis and/or the flatness $H_{0,3}$ hypothesis are essentially noninterpretable, because the significant interaction between groups and measures implies that both are significant sources of variance, the magnitude of their influence depending, however, on the level of the other source of variation in responses.

These three tests are *statistically* independent, but the last two are irrelevant substantively when parallelism is rejected for data displaying a ***crossover interaction*** in which the group difference differs in sign (rather than only in magnitude) from measure to measure.

In order that profile analysis give meaningful results, it is necessary that the various outcomes be measured in meaningful units. Formally, a statement about some empirical finding is meaningful if its truth or falsehood is unaffected by supposedly irrelevant transformations of the data. In the present context, the shape of the pooled profile is affected by a change of origin (addition or subtraction of a constant) of some but not all of the variables, whereas parallelism can be destroyed or guaranteed by changing the units of the different measures. Thus the investigator who uses profile analysis must be able to justify his or her particular choice of origin and units for each measure. (The overall T^2 test for two samples is unaffected by any linear transformation of any of the dependent variables.)

Computationally, the parallelism and flatness hypotheses are handled by converting the original set of p outcome measures to p - 1 *slope* measures, where slope measure $i = X_i - X_{i+1}$ is the difference between two "adjacent" dependent variables. Parallelism is then handled by conducting a two-sample T^2 test on the difference between the slope measures for the two groups, whereas flatness is tested by conducting a single–sample T^2 test of the hypothesis that all of the grand means (combined means, combining the data for both groups) of the various slope measures are equal to zero. The equal levels hypothesis is tested by a univariate t test on the sum of the p original measures. These two transformations (to slope measures and to sums of scores) could be carried out on each single subject, but it is generally much easier to take advantage of the simple relationships between the coefficients used in constructing a linear combination of scores and the mean and variance of that combined score; namely,

$$\overline{X}_W = w_1\overline{X}_1 + w_2\overline{X}_2 + \cdots + w_p\overline{X}_p = \mathbf{w}'\overline{\mathbf{X}} \text{ and } s_W^2 = \mathbf{w}'\mathbf{S}\mathbf{w}.$$

Thus our three significance tests become:

1. $H_{0,1}$: $\mu_{\text{slope},1} = \mu_{\text{slope},2}$ (the interaction between Group and Measure) is tested by

$$T^2 = N_1 N_2 (\overline{\mathbf{X}}_{\text{slope},1} - \overline{\mathbf{X}}_{\text{slope},2})' \mathbf{S}_{\textbf{slope}}^{-1} (\overline{\mathbf{X}}_{\text{slope},1} - \overline{\mathbf{X}}_{\text{slope},2})/(N_1 + N_2),$$

whence

$$F = \{(N_1 + N_2 - p)/[p - 1)(N_1 + N_2 - 2)]\}T^2$$
with $p - 1$ and $N_1 + N_2 - p$ *df*;

where

$$\overline{\mathbf{X}}_{\text{slope},\, i} = \begin{bmatrix} \overline{X}_{1,i} - \overline{X}_{2,i} \\ \overline{X}_{2,i} - \overline{X}_{3,i} \\ \vdots \\ \overline{X}_{p-1,i} - \overline{X}_{p,i} \end{bmatrix} \quad \text{for } i = 1,2;$$

and $\mathbf{S}_{\text{slope}} = \mathbf{S}_x$ computed on the slope measures $= \mathbf{CSC}'$, where

$$\mathbf{C} = \begin{bmatrix} 1 & -1 & 0 & \cdots & 0 \\ 0 & 1 & -1 & \cdots & 0 \\ . & . & . & \cdots & . \\ 0 & 0 & \cdots & 1 & -1 \end{bmatrix}.$$

Alternately,

$$\mathbf{S}_{\text{slope}} = \begin{bmatrix} s_1^2 + s_2^2 - 2s_{12} & s_{12} + s_{23} - s_{13} - s_2^2 & \cdots & s_{1,p-1} + s_{2p} - s_{1p} - s_{2,p-1} \\ & s_2^2 + s_3^2 - 2s_{23} & \cdots & s_{2,p-1} + s_{3p} - s_{2p} - s_{3,p-1} \\ & & \cdots & \vdots \\ & & & s_{p-1}^2 + s_p^2 - 2s_{p-1,p} \end{bmatrix};$$

that is,

the (r, t)th entry of $\boldsymbol{S}_{\text{slope}}$ = the covariance of $(X_r - X_{r+1})$ with $(X_t - X_{t+1})$
$= s_{rt} + s_{r+1,t+1} - s_{r,t+1} - s_{r+1,t}$.

2. H_{02}: $\mu_{w,1} = \mu_{w,2}$ is tested by

$$t = (\overline{X}_{W,1} - \overline{X}_{W,2})/[\sqrt{s_W^2(1/N_1 + 1/N_2)}] \text{ with } N_1 + N_2 - 2 \; df,$$

where $\overline{X}_{w,i} = \overline{X}_{1,i} + \overline{X}_{2,i} + \Lambda + \overline{X}_{p,i}$ for $i = 1,2$ and s_W^2 is the sum of all entries in $\mathbf{S}_c$.

3. H_{03}: $\mu_{slope} = 0$ is tested by

$$T^2 = (N_1 + N_2)\overline{\overline{\mathbf{X}}}'_{slope}\,\mathbf{S}_{slope}^{-1}\,\overline{\overline{\mathbf{X}}}_{slope},$$

whence

$$F = (N_1 + N_2 - p)\,T^2/[(p - 1)\,(N_1 + N_2 - 2)]$$
$$\text{with } p = 1 \text{ and } (N_1 + N_2 - p)\ df,$$

where $\overline{\overline{\mathbf{X}}}_{slope}$ is a column vector of "adjacent" differences in sample grand means.

Finally, note that the results of our three tests in a profile analysis are identical (though, of course, the numerical values of the two sets of discriminant function coefficients change) if we use any set of $(p - 1)$ nonlinearly dependent contrasts among the dependent variables in place of the adjacent-difference (slope) scores.

As an illustration, let us conduct a profile analysis of the difference between the outcomes-allocation and the expenses-allocation conditions for Data Set 3. A profile analysis is quite appropriate for these data for a number of reasons:

1. All five dependent variables are measured in the same units (dollars allocated).
2. The levels test has a meaningful interpretation, because the sum of our five dependent variables is simply the total prize that group would have received had they been the winning group. Because groups were randomly assigned to conditions, this difference should represent random variation in the abilities of the group members and should not be statistically significant.
3. Primary interest is in differences between the two conditions in the distribution of the total prize among the various participants, that is, in the shape of the profile of mean allocations. We would not wish to confound shape differences with differences in the total amount being allocated.
4. The overall shape of the grand mean profile is also of interest, because this is predicted to be a very nearly linear function of mean inputs. (More specifically, allocations are predicted to be linearly related to input within each group, but the variation in total prize from group to group leads to slight departures from linearity in the predicted grand means.) The results of our profile analysis follow:

Parallelism Test (Group x Number of Interchanges Interaction).

We set this up in terms of contrasts that would represent linear, quadratic, cubic, and quartic trends *if* the task difficulty manipulation had led to equally spaced decreases in mean individual contribution going from Pl to P10. (This would be much less tempting if I had computed $\mathbf{S}_{contr}$ or $\mathbf{S}_{slope}$ by hand, in which case the computational simplicity of converting $\mathbf{S}_c$ to $\mathbf{S}_{slope}$ would strongly suggest employing slope measures.)

$$\mathbf{C} = \begin{array}{l} \\ \text{Lin} \\ \text{Quad} \\ \text{Cub} \\ \text{Quart} \end{array} \begin{array}{c} \begin{array}{ccccc} P1 & P2 & P3 & P4 & P10 \end{array} \\ \begin{bmatrix} 2/10^{1/2} & 1/\sqrt{10} & 0 & -.31623 & -.63246 \\ 2/14^{1/2} & -1/\sqrt{14} & -2/\sqrt{14} & -.26726 & .53452 \\ -1/10^{1/2} & 2\sqrt{10} & 0 & -.63246 & .31623 \\ -1/70^{1/2} & -4/\sqrt{70} & 6/\sqrt{70} & -.47809 & .11952 \end{bmatrix} \end{array};$$

$$\overline{\mathbf{X}}_{\text{contr}} = \overline{\mathbf{X}}\mathbf{C}' = \begin{array}{l} \\ \text{Outc} \\ \text{Exp} \end{array} \begin{array}{c} \begin{array}{cccc} \text{Lin} & \text{Quad} & \text{Cub} & \text{Quart} \end{array} \\ \begin{bmatrix} 1.17415 & -.02004 & .10783 & .30968 \\ 5.61336 & 1.5887 & .59040 & .36992 \end{bmatrix} \end{array};$$

$$\mathbf{\Delta}'_{\text{contr}} = \begin{bmatrix} -4.43921 & -1.60891 & -.48257 & -.06024 \end{bmatrix};$$

$$\mathbf{S}_{\text{contr}} = \begin{array}{l} \text{Lin} \\ \text{Quad} \\ \text{Cub} \\ \text{Quart} \end{array} \begin{bmatrix} 1.49123 & -.20840 & .34445 & .18236 \\ & 1.86272 & .42791 & .06691 \\ & & 1.10721 & .49019 \\ & & & .78904 \end{bmatrix};$$

$$\mathbf{a}'_{\text{contr}} = \text{any multiple of } \mathbf{\Delta}'_{\text{contr}} \mathbf{S}^{-1}_{\text{contr}}$$

$$= c\begin{bmatrix} -4.37769 & -2.18122 & 3.14234 & 3.06137 \end{bmatrix}$$

for any nonzero c; and

$$\begin{aligned} T^2_{\text{par}} &= [10(10)/(10+10)]\mathbf{\Delta}'_{\text{contr}} \mathbf{S}_{\text{contr}} \mathbf{\Delta}_{\text{contr}} \\ &= 5[(-4.37769)(-4.43921) + (-2.18122)(-1.60891) + \ldots] \\ &= 106.255, \end{aligned}$$

whence

$$F = 15/[4(18)]T^2 = 22.136 \text{ with 4 and 15 } df, \; p < .01.$$

Thus, we can be confident that the two conditions yield different population patterns of allocation across the five partners. Before beginning post hoc exploration of differences between the groups on particular contrasts among the partners, let's translate our discriminant function into a contrast among the original five measures. Our discriminant function is

-4.43921 Lin - 2.18122 Quad + 3.14324 Cub + 3.06237 Quart

= -4.43921(.63246$P1$ + .31623$P2$ + 0 $P3$ - .31623$P4$ - .63246$P10$)

$$\begin{aligned}
&- 2.18122(\ .53452P1 + .26726P2 + .53452P3 - .26726P4 - .53452P10) \\
&+ 3.14234(- .31623P1 + .63246P2 + \quad 0\ P3 - .63246P4 - .31623P10) \\
&+ 3.06137(\ .11952P1 + .47809P2 + .71714P3 - .47809P4 - .11952P10) \\
&= \quad - 4.5624P1 \ - .2776P2 + 3.3613P3 - 1.4837P4 + 2.9624P10.
\end{aligned}$$

Note that, whereas the coefficients of the optimal linear combination of the contrast scores do not sum to zero, the coefficients of the optimal combination of the original scores do. In other words, any linear combination of the contrast scores implies a contrast among the original measures. Note too that, as the preceding layout suggests, ***a*** (the discriminant function coefficients in terms of the original variables) can be computed from

$$\mathbf{a'} = \mathbf{a'}_{\text{contr}}\mathbf{C}.$$

Interpreting the discriminant function as essentially $P3 + P10 - P1 - P4$ yields

$$\overline{W}_{\text{outc}} = -.602;\ \overline{W}_{\text{exp}} = -4.561;\ s_W^2 = 1.21374; \text{ and}$$

$$t_W^2 = 5(-.602 + 4.561)^2 / 1.21374 = 64.568.$$

Comparing this to our post hoc critical values of $t^2_{\text{crit}} = (72/15)\text{F}_\alpha(4, 15) = 14.69$ at the .05 level, 23.47 at the .01 level, we can see that our simplified discriminant function is statistically significant. It does, however, represent a considerable drop in discriminative ability from that provided by the discriminant function itself (which yields, of course, a squared t ratio equal to $T^2 = 106.255$). In fact, the t^2 for our simplified discriminant function is actually lower than that provided by the LIN contrast all by itself (t^2 = 66.075)! Only by use of a difference between weighted averages can we get a t_W^2 close to T^2; for example, $\mathbf{a'}_{\text{simp}}$ = [-5 0 3 - 1 3] yields a t_W^2 of 104.032. This is a good indication that, for the relatively small sample sizes involved here, our optimization procedure has indeed capitalized on chance and T^2 is apt to considerably overrepresent the t^2 we would obtain were we to compute a univariate t on -4.562$P1$ - .278$P2$ ··· for a new pair of samples.

The squared t ratios for LIN, QUAD, CUB, and QUART are 66.075 ($p < .01$), 6.948 (ns), 1.052 (ns), and .023 (ns), respectively. (Be sure you understand how these t^2s are computed; the required s_W^2 s are given in the main diagonal of $\mathbf{S}_{\text{contr}}$.) The difference between the mean amount allocated to the highest and lowest input partners ($\overline{P1} - \overline{P4}$) is significantly greater for the expenses-allocation than for the outcomes-allocation condition:

$$\begin{aligned}
t_W^2 &= 5(7.007 - 1.317)^2 / [1.25925 + .96159 - 2(-.45175)] \\
&= 5.69^2 / 3.124 = 51.81,
\end{aligned}$$

as is the difference in mean amounts allocated to the partners receiving the easiest and most difficult problems: $t_W^2 = 5(6.727 - 1.417)2/2.27774 = 61.895$.

Finally, the assumption that outcome will be linearly related to input within each group predicts a vector of between-condition differences in mean amounts allocated of (\$3.877, 1.548, -.484, -1.830, -1.505). This prediction, in turn, implies contrast coefficients of **a'** = (3.566, 1.227, -.805, - 2.151, -1.826), which yields a t_W^2 of 75.698 ($p < .01$, slightly larger than that yielded by LIN.

Levels Test (Group Main Effect).

$$\mathbf{S}_c = \frac{(10-1)\mathbf{S}_{outc} + (10-1)\mathbf{S}_{exp}}{10+10-2} = \frac{\mathbf{S}_{outc} + \mathbf{S}_{exp}}{2}$$

$$= \begin{bmatrix} 1.25925 & .50052 & .13400 & -.45175 & .21827 \\ & 2.09221 & .22065 & .06733 & .13291 \\ & & 1.00907 & .36802 & -.66236 \\ & & & .96159 & -.09880 \\ & & & & 1.45503 \end{bmatrix};$$

$$\overline{\mathbf{X}}' = \begin{matrix} \text{Outc} \\ \text{Exp} \end{matrix} \begin{bmatrix} 2.594 & 2.156 & 2.092 & 1.277 & 1.177 \\ 6.437 & 3.727 & 1.596 & -.570 & -.290 \end{bmatrix};$$

$$s_W^2 = 1.25925 + 2.09221 + \cdots + 1.45503 + 2(.50052 + .13400 + \cdots - .09880) = 7.6473;$$
thus

$$t_W = (9.296 - 10.900)/\sqrt{7.63473(1/10 + 1/10)}$$
$$= \sqrt{5(-.604)^2 / 7.63473} = 1.298 \text{ with 18 } df\text{, ns.}$$

As anticipated, the difference in mean potential prize between the two conditions can reasonably be interpreted as due to chance.

Flatness Test.

We've already computed $\mathbf{S}_{contr}$. We lack only

$$\overline{\mathbf{X}}' = \frac{10\overline{\mathbf{X}}'_{outc} + 10\overline{\mathbf{X}}'_{exp}}{20} = \frac{\overline{\mathbf{X}}'_{outc} + \overline{\mathbf{X}}'_{exp}}{2} = [4.5155 \quad 2.9415 \quad 1.8440 \quad .3535 \quad .4435],$$

whence

$$\mathbf{X}'_{contr}\ \overline{\mathbf{X}}\mathbf{C} = [3.39376 \quad .78442 \quad .34912 \quad .33980]$$
$$= \mathbf{\Delta}'_{flatness} \text{ (since the } H_0 \text{ we're testing is that } \mu_{contr} = 0).$$

Continuing on, we get

$$\Delta \mathbf{S}_{\text{contr}}^{-1} = [3.02186 \quad 1.20916 \quad -1.73255 \quad -1.44661 \;],$$

which also gives us our discriminant function coefficients; and

$$T^2 = 20\mathbf{\Delta}'\mathbf{S}_{\text{contr}}^{-1}\mathbf{\Delta} = 202.152,$$

whence

$$F = (15/72)T^2 = 42.115 \text{ with 4 and 15 } df, p < .01.$$

Treating this as essentially $P1 - (P3 + P10)/2$ yields $\mathbf{a}'_{\text{simp}} = (2, 0, -1, 0, -1)$; $t_w^2 = 909.496/4.7673 = 190.778$, $p < .01$. (Our post hoc critical values for exploration of differences among the grand means are the same as for the parallelism test.) The squared t ratios for LIN, QUAD, CUB, and QUART are 154.471 ($p < .01$), 6.606 (ns), 2.202 (ns), and 2.927 (ns). Finally (though we are free to examine any number of post hoc contrasts among the grand means), the hypothesis that the allocations within each group will be linearly related to that group's participants' individual contributions (number of problems solved or amount of money added to potential prize) implies a contrast among the grand means of

$$\mathbf{a}'_{\text{pred}} = [2.463 \quad .932 \quad -.132 \quad -1.674 \quad -1.588],$$

whence

$$t_w^2 = 3037.48/18.7683 = 161.841, p < .01.$$

A few final points should be made before leaving our analysis of Data Set 3. First, the two matrix inversions (of the 5 x 5 $\mathbf{S}_c$ in the overall two-sample case and of the 4 x 4 $\mathbf{S}_{\text{contr}}$ in the profile analysis) are best left to a computer program. Although certainly feasible if carried out carefully via row operations, the opportunities for small but snowballing errors (especially getting a sign wrong) are so numerous as to make the computer's reliability a great advantage. However, once the inverse has been computed, the remainder of the analysis requires only matrix multiplication, which is quite feasible by hand. In fact, post hoc exploration of particular linear combinations of the dependent variables is probably more efficiently carried out, in a problem of this size, via desk–calculator-aided use of Equations (3.2) or (3.3) than by continual rerunning of a "canned" program that computes a score for each subject on each new combination you decide to explore. You could also avoid repetitive restoring by using SAS's PROC MATRIX, although this would require submission of several "batch" runs. The most efficient method would be to use an online, interactive matrix-manipulation language such as MATLAB. Second, the availability of an efficient tool like T^2 for exploration of differences in central tendency should never tempt us to forego examining individual observation vectors. In the present case, for instance, the reader has probably noticed that 6 of the 10 groups in the outcomes-allocation condition assigned equal outcomes to all partners, regardless of differential contributions, whereas none of the expenses-allocation groups did so. Less obvious without having the individual contributions to examine is the fact that 7 of the 10 groups in the expenses-allocation groups recommended equal input-outcome *differences* for all participants (i.e., subtracting 50 cents from each person's individual contribution to atrive at his or her final share of the prize), whereas none of the outcomes-allocation groups recommended such a procedure. These two aspects of the data should of course be described and should be supported by chi-square significance tests, even though these tests were not included in the original strategy for controlling experimentwise alpha.

3.5 DISCRIMINANT ANALYSIS

The clearer the distinction a particular measure makes between two groups, the more useful that measure is apt to be in deciding to which group an as yet unclassified subject belongs. The discriminant function—that linear combination of the original variables that yields the highest possible t ratio for the difference between the two groups—therefore seems a logical candidate for such applications. For instance, if we perform a T^2 analysis of the data from Dick Linn's (1981) study of differences between 16 children classified as hyperactive and 16 not so classified (see section 2.3.4), we obtain a discriminant function of

$$D = .107\text{CPQ} + .041\text{CPTE} + .005\text{LOC} + .026\text{CPTC},$$

yielding

$$\bar{D}_{\text{norm}} = 7.333 \text{ and } \bar{D}_{\text{hyper}} = 14.931.$$

If we now compute a score for each subject on D, we find that the lowest three scores in the hyperactive group are 8.947, 11.798, and 12.085, and the highest three scores in the normal group are 9.753, 9. 199, and 8.689. If we were to reclassify subjects on the basis of their discriminant function scores, considering subject i to be a "predicted hyperactive" if $D_i < 11.132 = (\bar{D}_{\text{norm}} + \bar{D}_{\text{hyper}})/2$, we would wind up classifying all but one of our hyperactive children and none of the normals as "predicted hyperactives." This 94% "hit rate" and 0% "false alarm rate" are not likely to be maintained when this particular discriminant function is applied to a new sample of data. Note that we could increase our (sample) hit rate to 100% by dropping our cutoff point to, say, 8.9—but only at the cost of 2 (12.5%) false alarms. Discriminant analysis includes procedures for taking differential frequencies of occurrence (e.g., the fact that considerably less than half of the total population of children are hyperactive) and differential costs of misclassification (whether misses are more costly than false alarms, or vice versa, and by how much) into account in setting the cutting point.

In particular, if we assume a normally distributed discriminant function, we minimize our probability of misclassifying a subject randomly sampled from the population if we use the cutting point given by Equation (3.12):

$$X_C = \frac{(\mu_{D1} + \mu_{D2})}{2} + \frac{\sigma_D^2}{(\mu_{D1} - \mu_{D2})}\ln(p_2 / p_1) \tag{3.12}$$

where X_C is the dividing line between scores that lead to being classified into group 1 and those that lead to classification into group 2; μ_{D_j} is the population mean of group (or subpopulation) j's scores on the discriminant function; σ_D^2 is the population variance of discriminant-function scores (assumed identical for the two subpopulations); and p_j is the proportion of the total population who are actually members of subpopulation j. If we wish to minimize our expected loss due to misclassifications, we simply substitute $\ln[(p_2/p_1)/(c_2/c_1)]$ for $\ln(p_2/p_1)$ in Equation (3.12), where c_j is the cost of incorrectly classifying an observation from group j as belonging to the other group and $\ln(Y)$ is the natural logarithm (logarithm to the base e) of Y. This has the effect of moving our cutting point closer to the mean of whichever group is relatively rare or costs us relatively little to misclassify, thereby requiring stronger evidence that an observation comes from that group before we so classify it.

We can estimate our misclassification probabilities or costs by substituting sample values for corresponding population values in Equation (3.12)—for example, $\hat{\sigma}_D^2 = s_D^2 =$ **a'S**$_c$**a**—and using the resulting cutting point to classify the scores in our two samples of observations. Note, however, that the cutting point given by Equation (3.12) will *not* minimize the number of errors we make in classifying our sample data unless $p_2/p_1 = n_2/n_1$, that is, unless the relative sizes of our two samples reflect the relative frequencies of occurrence of these two groups or categories in the population. To extrapolate from the classification table based on resorting our sample data to the performance of our discriminant function when applied to scores randomly sampled from the population, we would compute

$$\hat{p}_{\text{mc}} = p_1 \Pr(\text{"Grp2"} | \text{Grp1}) + p_2 \Pr(\text{"Grp1"} | \text{Grp2}), \tag{3.13}$$

where $\hat{p}_{\text{mc}}$ = the probability of misclassifying an observation sampled at random from the population and Pr("Grp j" | Grp i) = Pr("j" | i) is the probability that an observation actually belonging to group i is misclassified as belonging to group j.

To illustrate this point, consider a hypothetical situation where $\mu_{D_1} = 20$, $\mu_{D_2} = 10$, $\sigma_D = 3$, and $p_2/p_1 = 9$ but (for convenience or to increase our statistical power) $n_2 = n_1$. If we ignore the differential population probabilities ("priors") of the two categories, we would use a cutting point of 15, whence (assuming normality), we would have Pr("2"| 1) = Pr(z < -5/3) = Pr("1"| 2) = Pr(t > 5/3) = .055, so that we would, overall, misclassify 5.5% of our $n_1 + n_2$ sample observations. If instead we apply Equation (3.12), we would use a cutting point of 15 + ln(9)(9/10) = 16.98, yielding Pr("2"| 1) = Pr(z < -3.02/3) = .158 and Pr("1"| 2) = Pr(z > 6.98/3) = .001, so that our overall percentage of misclassifications in our pair of equally sized samples would be expected to be

(1/2)(.158) + (1/2)(.001) = .084—worse than our overall error rate ignoring the difference in priors. However, if we draw observations at random out of the *population* and apply the priors-adjusted cutoff of 16.98, we would misclassify 15.8% of the 1/10 of our population who are really from group (subpopulation) 1 and 0.1% of the 9/10 of our population who are really from group 2, for a total misclassification probability of (1/10)(.158) + (9/10)(.001) = .0167. This is a considerable improvement over the .055 misclassification probability ignoring priors. Most discriminant analysis computer programs provide a classification table that makes it easy to compute sample estimates of Pr("2"| 1) and Pr("1"| 2), so that extrapolation to populations having different proportions in categories 1 and 2 is easily accomplished.

Users of discriminant analysis also employ a somewhat different and less standardized terminology in describing their results than do users of T^2, despite the obvious commonality between the two techniques. The two most important differences are that a report of a discriminant analysis is likely (a) to offer a value of Mahalobnis's D^2 statistic ($= \Delta' \mathbf{S}_c^{-1} \Delta = T^2/K$) and/or a chi-square test based on assuming a very large sample size in place of T^2 and its associated F ratio and (b) to report two sets of "classification function" or "likelihood function" (or even "discriminant function") coefficients, one for each group. These coefficients give, for each group, the weights to be applied to a subject's scores on the dependent variables in order to compute the distance of his or her observation vector from the centroid of that group or the likelihood that this observation vector could have been sampled from that group. The subject is then classified into that group for which the distance measure is smaller or the likelihood measure is larger. The resulting classification decisions are identical to those yielded by the single discriminant function from a T^2 analysis, but classification– or likelihood–function coefficients are difficult to interpret substantively, because they do not yield estimates of the subjects' positions on a single underlying dimension.

Huberty (1975) provided an excellent review of discriminant analysis terminology, including a discussion of the computer programs available and of procedures for handling cases in which the assumption of homogeneous covariance matrices (section 3.7.1) is not tenable. Huberty (1984) updated the discussion of computer programs and emphasizes some of the misinterpretations of discriminant analyses.

3.6 RELATIONSHIP BETWEEN T^2 AND MRA

You probably recall that the work of Linn and Hodge (1981) was used in chapter 2 to illustrate stepwise multiple regression analysis (MRA), with CPQ, LOC, CPT–C, and CPT–E used as predictors of GRP, a group-membership variable equal to 1 for each hyperactive child and 0 for each nonhyperactive ("normal") child. How does this relate to the T^2 analysis of these same data just alluded to in section 3.5? As proved in Derivation 3.2, these are equivalent analyses. The F ratio and its associated degrees of freedom for testing the statistical significance of R^2 for the relationship between a dichotomous group-membership variable and a set of dependent variables treated as "predictors" is identical to the F and its degrees of freedom for testing the statistical significance of Hotelling's T^2 for the difference between the two mean vectors. Moreover, $R^2 = T^2/(T^2 + df)$, and the raw-score discriminant function coefficients are identical (to within a multiplicative constant) to the raw-score regression coefficients. [One must be careful, however, in comparing z-score regression weights with standardized discriminant function coefficients, because the former are almost always based on z scores using the overall standard deviation (ignoring groups) as the denominator, whereas the latter are almost always computed using the within-cells standard deviation—the square root of the main diagonal entry of $\mathbf{S}_c$—as the denominator for *z* scores. Once this difference in standard deviations is taken into account, however, the equivalence becomes apparent.]

This relationship between T^2 and MRA can be put to use in a number of ways. For example, if a fully canned T^2 (or Manova) program is not available, a multiple regression program almost certainly will be. Similarly, the tests for the effects of adding predictors to MRA can be translated directly into tests of the effects of adding dependent variables to a T^2 analysis.

For instance, we might be interested in testing the H_0 that the difference in shape between outcomes– and expenses–condition mean profiles (cf. the example in Section 3.5) is entirely attributable to the difference predicted by Harris' (1976b) linear formula. Specifying three additional contrasts orthogonal to (or at least not linearly dependent upon) the predicted contrasts would in itself be a difficult task. However, we bypass this problem by recognizing that T^2 on adding three such contrasts, whatever they are, will equal our overall T^2 for parallelism. We thus have $R^2 = T^2/(T^2 + df) = 75.698/93.698 = .8079$ for the predicted contrast by itself and $R^2 = 106.255/124.255 = .8551$ for the total difference in shape. Thus $F_{incr} = [(.8551 - .8079)/3][(1 - .8551)/15] = 1.629$ with 3 and 15 degrees of freedom, so that our null hypothesis remains tenable.

There is also a relationship between MRA and single-sample T^2 as follows: If our outcome measure, *Y*, is taken to be a constant, that is, the same nonzero number for all subjects, and a "raw score" regression analysis of the relationship between *Y* and a set of *X* scores is conducted, then the resulting *F* and regression coefficients are the same as the *F* and discriminant function coefficients that would have resulted from a single–sample T^2 of the H_0 that all of the X_j have population means of zero. (We can test other hypotheses about μ_x, via raw-score MRA by substituting $(X_j - \mu_{jo})$ for X_j in our regression analysis.) By "raw-score MRA" we mean that β_0 is omitted from our regression model (or, equivalently, that b_0 is forced to be zero), which is equivalent to applying the **x'x**-based formulae of chapter 2 to **X'X** and **X'Y,** the matrices of sums of squares and cross products of *raw* scores, rather than deviation scores. We must also keep in mind that our residual error term will have one more degree of freedom than usual, because we have *not* used a degree of freedom to estimate the mean of each variable. This is merely a computational dodge, but it may prove useful when the discriminant analysis or Manova or T^2 program on your local computer system makes no provision for the single-sample case (as, for example, in the case of SAS's Manova program, PROC GLM).

3.7 ASSUMPTIONS UNDERLYING T^2

3.7.1 The Assumption of Equal Covariance Matrices

The procedure of averaging together the covariance matrices for groups 1 and 2 before carrying out a T^2 analysis of differences between the two groups involves the implicit assumption that the differences between $\mathbf{S}_1$ and $\mathbf{S}_2$ simply represent random fluctuations about a common population covariance matrix Σ. Logically, our null hypothesis includes both the hypothesis that $\mu_1 = \mu_2$ and that $\Sigma_1 = \Sigma_2$. (The second hypothesis is usually thought of as an assumption upon whose correctness the validity of the first hypothesis—the one of real interest to the investigator—depends, rather than as part of H_0.) Rejection of H_0 thus *could* be due to the fact that $\Sigma_1 \neq \Sigma_2$, rather than (or in addition to) a nonnull difference between μ_1 and μ_2. However, studies by statisticians have shown that T^2, like the univariate t, is much more sensitive to differences in means than to differences in variances (and covariances, in the case of T^2). In particular, Ito and Schull (1964) showed that the true significance level of the T^2 test is unaffected by discrepancies between Σ_1 and Σ_2, so long as $N_1 = N_2$ is fairly large. For small and unequal sample sizes, it is handy to have a separate test of the equality of the covariance matrices of the populations from which groups 1 and 2 were drawn. A multivariate analog of Bartlett's test for homogeneity of variance (a standard tool in the kit of the user of univariate analysis of variance) provides such a test. Namely,

$$H_0 : \Sigma_1 = \Sigma_2 = \ldots = \Sigma_k$$

is tested by one of two tests. Each requires calculation of

$$M = n \ln |\mathbf{S}_c| - \Sigma(n_j \ln|\mathbf{S}_j|),$$

where n_j is the degrees of freedom for the variance-covariance estimates in $\mathbf{S}_j$ and $n = \Sigma n_j$. If p (the number of dependent variables) and k (the number of independent groups) are both ≤ 5 and each sample size is ≥ 20, the chi-square approximation can be used, namely

$$\chi^2 = M/C$$

is distributed as χ^2 with $(k - 1)(p)(p + 1)/2$ degrees of freedom, where

$$1/C = 1 - (2p^2 + 3p - 1)/[6(p + 1)(k - 1)](\Sigma\, n_j - n^{-1})$$
$$(= 1 - (2p^2 + 3p - 1)(\,(k + 1)/[6(p + 1)kn] \quad \text{when } n_1 = n_2 = \cdots = n_k).$$

In any other situation, the F approximation must be used, namely

$$F = M/b$$

is distributed as F with f_1 and f_2 degrees of freedom, where

$$b = f/[1 - A_1 - (f_1/f_2)]; \qquad f_1 = \square(k-1)p(p+1);$$
$$f_2 = (f_1 + 2)/(A_2 - A_1^2); \qquad A_1 = 1 - 1/C;$$

and

$$A_2 = (p-1)(p+2)[\sum (1/n_j^2) - 1/n^2]/[6(k-1)]$$
$$[= (p-1)(p+2)(k^2+k+1)/(6k^2n^2) \text{ for equal sample sizes}].$$

The F approximation is excellent so long as none of the covariance matrices is based on fewer than 9 degrees of freedom. For extremely small sample sizes, an extremely complex series solution for the exact distribution of the test statistic M must be employed (cf. Box, 1949).

However, routine application of Box's test as a precondition for performing T^2 analysis is *not* recommended, for at least three reasons. First, as Olson (1974) pointed out, the test is overly powerful in that it is very likely to yield statistically significant evidence against homogeneity of covariance matrices in cases in which violation of this assumption has only very minor effects on the validity of our T^2 critical values. Second, it, like the univariate test from which it was developed, is highly sensitive to violations of the normality assumption, whereas T^2 is highly robust against departures from multivariate normality. Finally, although multivariate analogs of the t-test procedures for the unequal-covariance case are available (see, e.g., Timm, 1975, and Ito, 1969), their validity has only been established for large sample sizes. Thus a decision that violation of this assumption precludes use of the overall T^2 test leaves us, in the small-sample case, with Bonferroni-adjusted Behrens–Fisher or Welch–Aspin t tests (Winer, 1971) on a family of linear combinations specified a priori as our only viable analysis.

3.7.2 Known Covariance Matrix

In some situations we have a priori grounds for specifying, in advance of examining the data, the entries in the population variance-covariance matrix. The first test conducted in this chapter, that of the hypothesis that the grand mean response vector, computed across all four groups, is (10, 10, 10, 10), is an example of such a situation, because the hypothesis of random selection of C or D on each trial with probability = 1/2 implies specific values of the σ_{ij} as well as the μ_j, namely, $\sigma_j = 40(3/16) = 7.5$ and $\sigma_{ij} = 40(-1/16) = -2.5$. (The fact that the observed variances are uniformly larger than these hypothesized values suggests that individual differences in choice probability are inflating the response variabilities. One should be careful in applying formulae for the mean and variance of a multinomial distribution to situations in which the assumption of those formulae that all participants have the same generating probabilities is unlikely to be valid.)

The formulae for T^2 are very easily corrected to known-covariance formulae simply by substituting Σ for $\mathbf{S}_1$ or $\mathbf{S}_2$. The resulting value of T^2 is then referred (without postmultiplication by any constant other than 1.0) to the chi-square table with p degrees of freedom. For instance, in the single-sample case,

$$T^2 = N(\overline{\mathbf{X}}-\boldsymbol{\mu}_0)'\Sigma^{-1}(\overline{\mathbf{X}}-\boldsymbol{\mu}_0)$$

is distributed as chi square with p degrees of freedom; and in the two-sample case,

$$T^2 = N_1N_2(\overline{\mathbf{X}}_1-\overline{\mathbf{X}}_2)'\Sigma^{-1}(\overline{\mathbf{X}}_1-\overline{\mathbf{X}}_2)/(N_1+N_2)$$

is distributed as chi-square with p degrees of freedom. Applying the single-sample formula to our hypothesis of completely random responding from all 48 subject pairs gives

$$\chi^2 = 48\cdot\begin{bmatrix}1.23 & -3.09 & -1.67\end{bmatrix}\begin{bmatrix}.2 & .1 & .1\\ & .2 & .1\\ & & .2\end{bmatrix}\begin{bmatrix}1.23\\-3.09\\-1.67\end{bmatrix}$$

$$= [.2(13.84) + .1(\text{-}6.16)]48 = 48(2.14) = 102.2 \text{ with 3 } df.$$

Actually, had we simply applied the same formula as is used for the estimated covariance-matrix case, keeping in mind that we have an arbitrarily large number of degrees of freedom for the covariance matrix, we would have obtained identical results, because the F distribution with ∞, n_2 degrees of freedom is identical to the distribution of χ^2/n_2 with n_2 degrees of freedom.

3.7.3 The Assumption of Multivariate Normality

The derivation of the distribution of T^2 makes use of the assumption that the vectors of outcome variables are sampled from a multivariate normal distribution. We might suspect from the multitude of empirical sampling experiments that have demonstrated the remarkable robustness of Student's t test (when two-tailed tests are employed) against violations of the normality assumption, that violations of multivariate normality would have similarly small effects on the validity of T^2 tests unless especially "wild" distributions are employed. However, little is known about the robustness of T^2, except that for sufficiently large sample sizes, computed T^2 values do conform to the F distribution, no matter what the shape of the parent population. Empirical sampling studies to determine how large "sufficiently large" is for the T^2 test are as yet too sparse to permit firm generalizations (Olson, 1974).

3.8 ANALYZING REPEATED-MEASURES DESIGNS VIA T^2

Readers familiar with analysis of variance (Anova) will have noticed that the data for Harris and Joyce's (1980) Experiment 3, analyzed via profile analysis in section 3, could have been analyzed instead as a 2 (allocation condition) x 5 (difficulty of anagram set) repeated-measure Anova. The principal results of such an analysis are reported in Table 3.3 (see Winer, 1971, for computational details). Clearly, our test of the main effect of

Table 3.1 Repeated-Measures Anova of Data Set 3

Source	*df*	*SS*	*MS*	*F*
Between-Subjects Effects				
Allocation Condition (*A*)	1	12.864	12.864	1.685
Groups within *A*	18	137.425	7.635	
Within-Subjects Effects				
Task Difficulty (*D*)	4	247.407	61.852	47.124
A × *D*	4	102.365	25.591	19.497
Groups × *D* within *A*	72	94.504	1.31255	

the between-subjects factor is equivalent to the levels test of our earlier profile analysis of these same data, and indeed the two tests yield identical F ratios and degrees of freedom. [Remember that the square of a t ratio is distributed as an F ratio with 1 *df* for the denominator, so that our levels-test t ratio of 1.298 with 18 *df* is exactly equivalent to an $F(1,18)$ of 1.298^2 = 1685.1 Similarly, the Anova test of the main effect of the within-subjects factor (task difficulty) is testing the same null hypothesis as the flatness test, and the test of the allocation condition x task difficulty interaction corresponds to the parallelism test. However, the correspondence in these latter two cases does not extend to identical F ratios. The Anova tests of *within-subjects* effects each have 4 and 72 *df*, whereas the corresponding profile analysis tests yield Fs with 4 and 15 df, and the numerical values of the corresponding F ratios differ.

These differences arise because the two T^2 tests are each based on that contrast among the levels of the within-subjects factor that, when tested against its own variance across sampling units, yields the maximum possible univariate F ratio. The Anova tests of the same effects, however, use an error term—MS for Groups x *D(A),* or, more generally, sampling units x levels of the within-subject factor, nested within levels of the between-subjects factor—that is an average of the variances of the various possible within-subject contrasts.

More specifically, if we construct any set of orthogonal, normalized within-subject contrasts (normalized so that the sum of the squared contrast coefficients for any given contrast is unity and orthogonal in that the sum of the cross products of contrast coefficients for any two contrasts is zero) and then compute the variance across sampling units of each of these p - 1 contrasts, the Anova erorr term will simply be the mean of these p - 1 variances. In our profile analysis of Data Set 3, for instance, we employed just such a set of contrasts in computing $\mathbf{S}_{contr}$. As you can readily verify, the mean of the main diagonal entries of $\mathbf{S}_{contr}$ is (1.491 + 1.863 + 1.107 + .789)/4 = 1.31255, which also equals $MS_{GxD(A)}$, our Anova error term for the within-subjects effects. As you might suspect, the validity of the Anova approach to within-subjects or repeated-measures designs is contingent on these p - 1 contrasts having identical population variances.

The usual univariate F ratios used to test within-subjects effects provide accurate p values (actual Type I error rates close to their nominal alphas) only if the population variance of the difference in response to treatment levels i and j is the same for all possible pairs of levels of the within-subjects factor. Huynh and Feldt (1970) proved that a necessary and sufficient condition for the assumption of homogeneity of treatment-difference variances (the h.o.t.d.v. assumption) to be met is that all normalized contrasts among the levels of the within-subject factor W—that is, all linear combinations $\Sigma c_j T_j$ such that $\Sigma c_j = 0$, where T_j is the response at the jth treatment level of W—have identical population variances, whence testing each such contrast against a single variance estimate ($MS_{S\times W}$) is legitimate. These authors further show that the h.o.t.d.v. assumption is met if and only if $\mathbf{C\Sigma C'} = k\mathbf{I}$, that is, if $\mathbf{C\Sigma C'}$ is proportional to an identity matrix, whence the characteristic roots of $\mathbf{C\Sigma C'}$ are all identical, where $\mathbf{C}$ is a $(p - 1) \times p$ matrix whose rows are any $(p - 1)$ orthogonal vectors of contrast coefficients that have been normalized to unit length. **C** is thus easily constructed by selecting any $(p - 1)$ orthogonal contrasts among the treatment means and dividing each by $\sqrt{\sum c_j^2}$, the square root of the sum of the squares of the coefficients for that contrast (that is, for that row of **C**). For instance, we may always take as the ith row of **C** a row vector having zeros in the first $i - 1$ positions, $p - i$ in the ith position, and -1 in the remaining $p - i$ positions. The h.o.t.d.v. assumption is then tested by computing

$$M = \frac{|\mathbf{CSC'}|}{[tr\,\mathbf{CSC'}/(p-1)]^{p-1}},$$

where tr **A** is the trace of **A** (the sum of its main diagonal elements). In addition, we compute $f = [p(p - 1)/2] - 1$ and $d = 1 - (2p^2 - 3p + 3)/[6(p - 1)df]$, and then compare $-df{\cdot}d{\cdot}\ln M$ to the chi-square distribution having f degrees of freedom. (The term df here is the number of degrees of freedom on which **S**, the sample covariance matrix for the $p - 1$ treatment levels and thus our best estimate of Σ, is based.)

Several design texts, such as Winer (1971) and Kirk (1982), provide a test due to Box (1950) of the assumption of ***compound symmetry***, that is, that all main diagonal elements of Σ are equal to a common value and all of its off-diagonal elements are equal to a second, lower common value. Compound symmetry is, however (as Winer and Kirk both made clear), only a *sufficient* and not a *necessary* condition for the h.o.t.d.v. assumption to be met.

Greenhouse and Geisser (1959) showed that an upper bound to the probability of the observed data given the null hypothesis of no main effect of, or interaction with, the within-subject factor (and thus a conservative test of any effect involving that factor) is provided by dividing both numerator and denominator degrees of freedom for any test involving the within-subjects factor by $p - 1$, where p is the number of response measures per subject, before looking up the critical value for that F ratio in the tables of the F distribution. Computational details of within-subjects univariate Anova, together with expressions for both standard tests (which apply when h.o.t.d.v. holds) and conservative tests, are provided by Winer (1971).

Several authors (e.g., Huynh, 1978; McCall & Appelbaum, 1973; Mendoza, Toothaker, & Nicewander, 1974; Rogan, Keselman, & Mendoza, 1979) have explored the possibility of estimating the magnitude of the departure from the h.o.t.d.v. assumption in the form of a measure, $\hat{\varepsilon}$, that varies between 1.0 (the value assumed in the uncorrected traditional approach) and $1/(p - 1)$ (the value used in the Greenhouse–Geisser correction), and then multiplying numerator and denominator degrees of freedom by this factor. Monte Carlo work has indeed shown that this procedure yields a true alpha level close to the nominal one.

It is important to recognize, however, that these *df*-adjustment procedures correct the positive bias of only the overall test of the within-subjects factor; they do *not* provide a basis for controlling Type I error rates for the specific contrasts that should always follow and clarify the source(s) of a significant overall test.

The usual procedure in the Anova approach to a repeated-measures design is to test each SS_{contr} against the same averaged error term employed in the overall test. But if h.o.t.d.v. is not met, this average *MS* will underrepresent the actual variance of some contrasts and overrepresent (often drastically) the actual variance of other contrasts. Boik (1981) showed that the resulting inflation of Type I error rate (for contrasts with above-average variance) and loss of power (for contrasts with below-average variance) can be quite pronounced even for departures from h.o.t.d.v. that are so mild as to require very little correction to the *df* for the overall test—so mild, in fact, as to be virtually undetectable by the Huynh and Feldt test. Boik (1981, p. 254) concludes that "there is no justification for employing ordinary univariate *F* tests for repeated-measures treatment contrasts." The recommended alternative is to test each contrast against its own variance across sampling units, "that is, use ... $\mathbf{c'Sc}$ rather than $\text{tr}(\mathbf{E})/(k - 1)$ in the denominator of the *F* statistic." If a limited number of a priori contrasts are to be examined, experimentwise error rate is controlled via Bonferroni adjustment of the critical values for these tests; for post hoc contrasts, the T^2-based critical value described and used in section 3.5 should be employed.

The one case in which one could argue for a continuing role for the traditional Anova repeated-measures *F* test is that in which the number of dependent measures (levels of the within-subjects factor) exceeds the degrees of freedom on which each variance estimate is based ($N - 2$ here). In this case, the sample value of T^2 is guaranteed to be arbitrarily large, and thus its critical value must be, as well. The traditional Anova *F* ratio can still be computed, however, because its use of an average of the $(p - 1)$ normalized contrast variances effectively reduces the degrees of freedom needed to look for the "optimal" contrast. Of course, its grossly exaggerated denominator degrees of freedom, $(N - 2)(p - 1)$, will lead to a grossly inflated true alpha level. However, Maxwell and Avery (1982) showed that multiplication of numerator and denominator degrees of freedom by $\hat{\varepsilon}$ produces a relatively unbiased overall test. One should still, of course, base all tests of specific contrasts among the levels of the within-subjects factor on the actual value of s_c^2 computed for scores on that contrast, rather than on the pooled error term. Controlling α_{exp} for such contrasts will be impossible unless the researcher specifies a finite set of all such contrasts of interest to him or her before examining the data and applies Bonferroni-adjusted critical values to these tests. (We know that astute post hoc exploration can uncover one or more contrasts that have zero within-conditions variance and that thus produce an infinitely large F, so the critical value for post hoc tests would have to be infinity.) However, even in this case, the value of the $\hat{\varepsilon}$-adjusted F ratio is questionable, because its statistical significance does not guarantee that any of the specific contrasts will be statistically significant, nor does its failure to achieve statistical significance prove that none of the a priori contrasts the experimenter has specified will be statistically significant. It's probably best to simply consider the overall null hypothesis rejected if one or more of the prespecified contrasts is statistically significant by its Bonferroni-adjusted critical value.

It's time for a concrete example of a within-subjects analysis.

Example 3.2 *Blood Doping.* Brien and Simon (1987) had each of six highly trained male runners run three competitive10-km (10K) races at 9-day intervals. The first run established a baseline level of performance. The second and third races were each run 5 days after one of two infusions: 400 mL of red blood cells (RBCs) that had been withdrawn from the runner himself in two phlebotomies 10 and 18 weeks prior to the baseline run, or 100 mL of a saline solution (roughly the same amount used to suspend the RBCs in the own-blood infusion). Three of the runners (those in the BL-Sal-OB condition) received the saline infusion 5 days prior to the second run and the own-blood infusion 5 days prior to the third run, and the three subjects in the BL-OB-Sal condition received the two infusions in the reverse order. The data appear in Table 3.4.

Tests of specific, single-*df* contrasts are extremely straightforward. The only between-subjects contrast that fits within the main-effects-and-interaction framework (but see Harris, 1994, section 4.5 for alternatives to "sticking with" main effects and interactions in factorial designs) is the test of the main effect of Order, which is simply an independent-means t or F test on the difference between the two groups with respect to

Table 3.2 10K Running Time as Affected by an Infusion of One's Own Blood

		Type of Infusion			
Order	Runner	None	Saline	Own Blood	$\bar{Y}_{runner}$
BL-Sal-OB	1	34.03	34.53	33.03	33.863
	2	32.85	32.70	31.55	32.367
	3	33.50	33.62	32.33	33.150.
Means,Order	1	33.46	33.617	32.303	33.127
BL-OB-Sal	4	32.52	31.23	31.20	31.650
	5	34.15	32.85	32.80	33.267
	6	33.77	33.05	33.07	33.297.
Means,Order	2	33.48	32.377	32.357	32.738
Grand Means		33.47	32.997	32.33	32.932

the simple sum or average of all scores obtained from each runner. In the present case our test of the main effect of order yields

$$t = \frac{.170}{\sqrt{\frac{(.023^2 + .760^2 + ...559^2)}{4}\left(\frac{2}{3}\right)}} = .245 \text{ with } 4\, df.$$

Similarly, a test of a contrast among the levels of the within-subjects factor (i.e., of a contrast among the grand means) is simply a single-sample t test of the H_0 that the population grand mean of the contrast is zero. Thus, for instance, the difference between the OB condition (averaged between the two groups) and the average of the grand means for the BL and SAL conditions is tested by computing a score on [(SAL + BL)/2 – OB] for each of the six runners (yielding 1.250, 1.225, 1.230, .675, .700, and .340) and then computing

$$t = \frac{.90333 - 0}{\sqrt{(.015^2 + .010^2 + ... + .232^2)/(4 \bullet 6)}} = \frac{.90333}{.05816} = 15.534 \text{ with } 4\, df.$$

And an interaction contrast is tested by conducting an independent-means t or F test of the difference between mean score on the within-subjects contrast for the two groups. For instance, the difference between the two orders with respect to the OB versus BL, SAL contrast is based on the six runners' scores on $(BL + SAL)/2 - OB$ = 1.25, 1.225, and 1.230 minutes for the Order-1 runners and .675, .700, .340 minutes for the Order-2 runners, yielding

$$t = \frac{1.235 - .5717}{\sqrt{[(.015^2 + .010^2 + \cdots + .232^2)/4](2/3)}} = \frac{.6633}{.11632} = 5.702 \text{ with } 4\, df.$$

The a priori critical value for the above contrasts is 4.604 at the .01 level and 8.610 at the .001 level, so they are clearly statistically significant on an a priori, per-comparison basis. The fully post hoc critical value is the square root of $[2(4)/(4\text{-}2\text{+}1)]F_\alpha(2,3)$ =

$$\sqrt{(8/3)\begin{pmatrix} 9.552 \\ 30.816 \\ 148.500 \end{pmatrix}} = \begin{pmatrix} 5.047 \\ 9.065 \\ 19.900 \end{pmatrix} \text{ at the } \begin{pmatrix} .05 \\ .01 \\ .001 \end{pmatrix} \text{ level of significance, so both contrasts are}$$

also statistically significant at at least the .05 familywise, fully post hoc level. (The t and F critical values can be obtained online from a number of sites. In particular, the t critical values for this example were obtained from the Southeast Louisiana website, http://www.selu.edu/Academics/Faculty/vbissonnette/tables.htm ; however, the F critical values couldn't be obtained there, as the applets provided don't "work" for *df* this low.)

To complete our catalog of tests that can be conducted without resorting to matrix algebra, we can compute the contrast that accounts for 100% of the variation among the within-subjects grand means (i.e., that has an SS_{contr} identical to the SS for our within-subjects factor) by using contrast coefficients of

$$c_j = \overline{Y}_j - \overline{\overline{Y}} = (33.47 - 32.932,\ 32.997 - 32.932,\ 32.33 - 32.932) = (.538, .065, -.602),$$

which is very close to the BL - OB contrast. The t_{contr} for either or both of these two contrasts can computed as a single-sample t and then compared to the fully post hoc critical value. As the reader will wish to verify, this yields t_{contr} values of 10.759 for the

SS_{Infusion}-maximizing contrast and 9.958 for our simplified (and much more interpretable) version thereof—both statistically significant at the .01 familywise, fully post hoc level.

Similarly, the interaction contrast that accounts for 100% of the difference between the two orders in the pattern of infusion-treatment means (i.e., that has an SS_{contr} identical to the SS for the Order x Infusion interaction) can be computed by subtracting the grand mean of the Order-1 versus Order-2 difference from each of the order-effect differences for the three infusion treatments, that is, by computing

$$c_j = \bar{\Delta}_j - \bar{\bar{\Delta}} = (-.020 - .392,\ 1.240 - .392,\ -.044 - .392) = (-.412,\ .848,\ .-.436),$$

which is very close to the SAL vs. average of BL and OB = (-.5, 1, -.5) contrast. Either or both of these contrasts can of course be tested via an independent-means t-test, with the resulting t being tested against the fully post hoc critical value. As the reader will wish to verify, this yields t_{contr} values of 7.692 for the $SS_{\text{InfusionxOrder}}$-maximizing contrast and 7.572 for our simplified (and much more interpretable) version thereof—both statistically significant at the .05 familywise, fully post hoc level.

We clearly have a carryover effect in this situation, with the runners who received the saline infusion as their second treatment showing no reliable change in their 10K running times as a result of that infusion, but runners who received the saline infusion as their third treatment (after having received an infusion of their own blood nine days before) still benefiting from the increased hematocrit induced by that own-blood infusion (as was verified by hematocrit measures) and thus running almost exactly as fast during their saline-infusion run as they had during their own-blood-infusion run. As a result, the grand mean of the saline-infusion 10K times (averaged across orders) falls halfway between the BL and OB grand means, so that the "linear trend" across infusion levels (i.e., the BL - OB contrast) accounts for nearly 100% of the variation among those three means. On the other hand, order has almost no effect on BL times (of course) or on OB times, but a huge effect on SAL times, with the result that the "Order x (SAL vs. other two)" contrast accounts for nearly 100% of the interaction SS.

Had infusion been a between-subjects effect, all infusion and all infusion x order contrasts would employ the same error term, and maximizing SS_{Infusion} or $SS_{\text{InfusionxOrder}}$ would be equivalent to maximizing the corresponding t_{contr} or F_{contr}. However, because Infusion is a within-subjects factor each contrast is tested against its own normalized variance across the subjects in each group (which is .0542 for the BL - SAL contrast, four times as large as the .0135 normalized variance for the BL and SAL vs. OB contrast). If we wish to know what contrasts maximize the Fs or squared t's for the main effect of infusion and its interaction with order, we must use a bit of matrix algebra to compute Hotelling's T^2 and the associated discriminant function for each of these effects. Using the (.5, .5, -1) and (1, -1, 0) contrasts yields

$$\mathbf{X'}_{\text{contr,Order 1}} = \begin{bmatrix} 1.250 & 1.225 & 1.230 \\ -.50 & .15 & -.12 \end{bmatrix} \text{ and } \mathbf{X'}_{\text{contr,Order 2}} = \begin{bmatrix} .675 & .700 & .340 \\ 1.29 & 1.30 & .72 \end{bmatrix},$$

whence

$$\overline{\mathbf{X}}_{contr} = \begin{bmatrix} 1.235 & -.15667 \\ .572 & 1.10333 \end{bmatrix},\ \overline{\overline{\mathbf{X}}}_{contr} = [.9035 \quad .47333],\ \overline{\Delta}_{contr} = [.663 \quad 1.18997],$$

and

$$\mathbf{S}_{c,contr} = \left(\begin{bmatrix} .000175 & -.00420 \\ -.00420 & .106635 \end{bmatrix} + \begin{bmatrix} .040408 & .066667 \\ .066667 & .110235 \end{bmatrix} \right) / 2 = \begin{bmatrix} .0203275 & .031235 \\ .031235 & .108435 \end{bmatrix}.$$

Thus

$$\mathbf{a}_{Infusion} = \overline{\overline{\mathbf{X}}}_{contr} \mathbf{S}^{-1}_{c,contr}$$

$$= [.9035 \quad .47333] \begin{bmatrix} .10844 & -.03124 \\ -.03124 & .02033 \end{bmatrix} / [(.02033)(.10844) - .03124^2]$$

$$= [.08319\ -.01860]/.0012286 = [67.711\ -15.139]$$

or any multiple thereof, whence

$$\mathbf{d}_{Infusion} = 67.711(½, ½, -1) - 15.139(1,-1,0) = 18.717BL + 48.994SAL - 67.711OB.$$

This full-decimaled optimal contrast yields

$$\max F_{Infusion\text{-}contr} = T^2_{Infusion} = 6\,\overline{\overline{\mathbf{X}}}_{contr} \mathbf{S}^{-1}_{c,contr} \overline{\overline{\mathbf{X}}}'_{contr}$$

$$= 6[67.711\ -15.139] \begin{bmatrix} .9035 \\ .47333 \end{bmatrix} = 324.067,$$

which we test for post hoc statistical significance by comparing $[(4 - 2 + 1)/[2(4)]] \cdot T^2_{Infusion} = 121.525$ to $F_\alpha(2,3) = 19.900$ at the .001 familywise level.

Similarly, the combination of our two basis contrasts that yields the largest F_{contr} for any order x infusion contrast is given by

$$\mathbf{a}_{OrdxInf} = \overline{\Delta}_{contr} \mathbf{S}^{-1}_{c,contr} = [.663 \quad -1.260] \begin{bmatrix} .10844 & -.03124 \\ -.03124 & .02033 \end{bmatrix} / .0012286$$

$$= [.11126\ -.04633]/.0012286 = [90.558\ -37.708],$$

so that the interaction contrast that yields the maximum F_{contr} is any multiple of

$$90.558(.5, .5, -1) - 37.708(1,-1,0) = 7.571BL + 82.987SAL - 90.558OB,$$

and that max $F_{OrdxInf\text{-}contr}$ is given by $(3/2)[.663(90.558) - 1.260(-37.708)] = 161.328$, $p < .01$ by the familywise post hoc critical value.

You could also let a computer program (cf. section 3.9) compute the optimal contrasts. Unfortunately "fully canned" programs such as those in SPSS and SAS provide only **a**, the optimal combination of the basis contrasts. The user must translate **a** into the optimal contrast among the levels of the within-subjects factor via hand calculations or by entering **a** and the definitions of the basis contrasts into a matrix-manipulation program such as MATLAB or SPSS's MATRIX/END MATRIX facility.

Application of T^2 to the analysis of repeated-measures designs is not limited to situations involving a single within-subjects factor. Designs in which the vector of observations obtained from each subject has a factorial structure or in which the same battery of dependent measures is measured under several different treatment conditions for each subject are readily handled by defining appropriate sets of linear combinations of the entries in the observation vector. However, we defer detailed discussion of these cases to chapter 4.

3.9 SINGLE-SYMBOL EXPRESSIONS FOR SIMPLE CASES

In this section, we give the single-symbol (scalar) expressions for the two-variable, three-variable, uncorrelated-variable and proportional-matrices cases for both the single-sample and two-sample T^2 tests. Readers will probably wish to test their understanding of T^2 by deriving these formulae themselves. Readers may also have noticed that T^2 analogs for only two of the three *t*-test procedures usually described in introductory statistics texts have been discussed explicitly in this chapter. The generalizations of the single-sample *t* and the *t* test for independent means have been presented, but there has been no separate section on a generalization of the *t* test for correlated means. The reason for this is that this third application of Student's *t* test is equivalent to a flatness test—which you will recall is a test of the hypothesis that the "profile" of population means of the *p* dependent variables is flat, whence every difference between two "adjacent" means is zero—conducted on a two-variable outcome vector. In this case **S*** is the single entry

$$s_1^2 + s_2^2 - 2s_{12} = s_{\overline{X1}-\overline{X2}} = s_{\overline{d}2},$$

so that

$$T^2_{\text{flatness}} = N\bar{\mathbf{d}}\mathbf{S}^{*-1}\bar{\mathbf{d}}' = Nd^2 / s_d^2$$

is the square of the usual test statistic for the *t* test for correlated means.

$p = 2$

$$T^2 = \frac{K(d_1^2 s_2^2 - 2d_1 d_2 s_{12} + d_2^2 s_1^2)}{s_1^2 s_2^2 - s_{12}^2};$$

when $s_1^2 = s_2^2 = s^2$, then

$$T^2 = \frac{K(d_1^2 + d_2^2 - 2r_{12} d_1 d_2)}{s^2(1 - r_{12}^2)};$$

moreover, in both cases,

$$F = \frac{(df-1)T^2}{2df};$$

$$a_1 = \frac{s_{12}T^2 - Kd_1d_2}{Kd_1^2 - T^2s_1^2}a_2 = \frac{Kd_2^2 - T^2s_2^2}{s_{12}T^2 - Kd_1d_2}a_2 = \frac{s_2^2d_1 - s_{12}d_2}{s_1^2d_2 - s_{12}d_1}a_2$$

$p = 3$

$$T^2 = \frac{K\begin{bmatrix} d_1^2(s_2^2s_3^2 - s_{23}^2) + d_2^2(s_1^2s_3^2 - s_{13}^2) + d_3^2(s_1^2s_2^2 - s_{12}^2) \\ + 2\,d_1d_2(s_{13}s_{23} - s_{12}s_3^2) + d_1d_3(s_{12}s_{23} - s_{13}s_2^2) + d_2d_3(s_{12}s_{13} - s_{23}s_1^2) \end{bmatrix}}{s_1^2s_2^2s_3^2 + 2s_{12}s_{13}s_{23} - s_1^2s_{23}^2 - s_2^2s_{13}^2 - s_3^2s_{12}^2};$$

further; when all variances are equal to a common value, s^2,

$$T^2 = \frac{K\begin{bmatrix} d_1^2(1-r_{23}^2) + d_2^2(1-r_{13}^2) + d_3^2(1-r_{12}^2) \\ +2d_1d_2(r_{13}r_{23} - r_{12}) + d_1d_3(r_{12}r_{23} - r_{13}) + d_2d_3(r_{12}r_{13} - r_{23}) \end{bmatrix}}{s^2\left[1 + 2r_{12}r_{13}r_{23} - r_{12}^2 - r_{13}^2 - r_{23}^2\right]};$$

$$F = \frac{(df-2)T^2}{3df};$$

$$a_1 = \frac{L_{12}L_{23} - L_{22}L_{13}}{L_{11}L_{22} - L_{12}^2}a_3; \qquad a_2 = \frac{L_{12}L_{33} - L_{13}L_{23}}{L_{13}L_{22} - L_{12}L_{23}}a_3;$$

where $L_{ij} = Kd_id_j - T^2s_{ij}$.

Uncorrelated Variables

$$T^2 = K[(d_1/s_1)^2 + (d_2/s_2)^2 + \cdots + (d_p/s_p)^2] = t_1^2 + t_2^2 + \cdots + t_p^2,$$

where t_j is the univariate t computed on X_j alone;

$$F = \frac{(df-p+1)T^2}{p\,df};$$

$$a_j = (d_j/d_p)(s_p^2/s_j^2)a_p.$$

$K\,\Delta\Delta'$ Proportional to S

If $K\,\boldsymbol{\Delta\Delta'} = c\mathbf{S}$, then all linear combinations of the Xj yield the same value of $t^2(\mathbf{a})$, namely c, so that $T^2 = c$ and $\mathbf{a}$ is undefined. However, this can only happen if $r_{ij} = \pm 1$ for every pair of dependent variables.

In the preceding formulae, $K = N$ for the single-sample T^2, and $N_1N_2/(N_1 + N_2)$ for the two-sample T^2; $d_j = \overline{X}_j - \mu_{j0}$ for the single-sample T^2, and $d_j = \overline{X}_{j1} - \overline{X}_{j2}$ (the difference between the sample means for variable j) for the two-sample T^2; and df is the degrees of freedom for each covariance or variance estimate—it will usually be $N - 1$ for the single-sample T^2 and $N_1 + N_2 - 2$ for the two-sample T^2.

3.10 Computerized T^2

3.10.1 Single-Sample and Two-Sample T^2

MATLAB.

```
Sc = [Entries in covariance matrix (either single or pooled
      within-cells), ending each row with a semicolon or
      beginning it on a new line]
Del = [Row-vector of differences - either between the means on
        each measure for the two groups, or between the observed
        and predicted mean on each measure]'
Discr = inv(Sc)*Del
Compute K = N (for the single-sample T**2)
            or N1*N2/(N1+N2) (for the two-sample T**2)
T2 = K*Del'*Discr
```

SPSS MANOVA Program.

Data setup commands (DATA LIST, VAR NAMES, etc.), then

```
Manova list-of-dep-vars by Group (1,2) /
   Print = Cellinfo (Means) Signif (Univ)/
   Discrim Raw Stan
```

(For single-sample T^2, omit "by Group (1,2)" and use a Compute statement to subtract the mean predicted for each measure from that measure before naming it in the dependent-variable list.)

SPSS DISCRIMINANT Program.

```
DISCRIMINANT GROUPS=varname(min,max)
 /VARIABLES=varlist
[/FUNCTIONS={g-1,100.0,1.0**}]
            {nf , cp  ,sig  }
 [/PRIORS={EQUAL**    }]
          {SIZE       }
```

```
          {value list}
[/ROTATE={NONE**    }]
         {COEFF     }
         {STRUCTURE}
 [/STATISTICS=[MEAN  ][COV ][FPAIR][RAW  ]]
              [STDDEV][GCOV][UNIVF][COEFF]
              [CORR  ][TCOV][BOXM ][TABLE]
              [CROSSVALID][ALL]
 [/PLOT=[MAP][SEPARATE][COMBINED][CASES[(n)]][ALL]]
 **Default if the subcommand is omitted.
 Default for /MATRIX OUT and IN is the working data file.
```

3.10.2 Within-Subjects Anova

SPSS MANOVA.

```
Manova dep-variable-list /
   WSFactors name-for-within-subjs-factor /
   Contrast (name-for-within-subjs-factor) = Special (1 1 … 1,
     Coefficients-for-contrast1, contrast2, … contrp-1) /
   Rename = Ave, name-for-contrast1, name-for-contrast2, …,
     Name-for-contrastp-1 /
   WSDesign /
   Print = Cellinfo (Means) Transform Signif (Univ)/
   Pmeans / Discrim (Raw) .
```

For instance, the analysis of Example 3.2 could have been carried out via

```
Title Mixed-Design Anova on Doping Data via Manova .
Data List Free / Runner  BL Sal OB Tot .
Recode Runner (1,2,3=1) (4,5,6=2) into Order .
Value Labels Order (1) BL-SAL-OB (2) BL-OB-SAL .
Begin Data .
1 34.03 34.53 33.03 101.59
2 32.85 32.70 31.55 97.10
3 33.50 33.62 32.33 99.45
4 32.52 31.23 31.20 94.95
5 34.15 32.85 32.80 99.80
6 33.77 33.05 33.07 99.89
End Data .
Manova BL Sal OB by Order (1,2) /
   WSFactor Infusion (3) /
   Contrast (Infusion) = Special (1 1 1, .5 .5 -1, 1 -1 0) /
   Rename = Ave, BLSalVOB, BLvSal /
   WSDesign /
   Print = Signif (Univ) Transform Homogeneity / PMeans /
   Discrim (Raw) .
```

Demonstration Problems

A. *Hypothetical Data.*

Consider the following set of data:

Group	Subject		X_1	X_2	X_3
1	1		3	4	11
	2		9	8	10
	3		3	9	6
		$\bar{\mathbf{X}}_1$	5.0	7.0	9.0
2	4		10	9	11
	5		11	9	10
	6		6	9	12
		$\bar{\mathbf{X}}_2$	9.0	9.0	11.0
		$\bar{\bar{\mathbf{X}}}$	7.0	8.0	10.0

1. Compute a pooled covariance matrix based on deviations about group means (rather than about the grand means). This yields an estimate of the common covariance matrix which is not affected by differences in population means.

2. Using the pooled covariance matrix computed in problem 1, test the null hypothesis that the population grand mean profile is (10, 10, 10).

3. Compute the discriminant function corresponding to the test in problem 2, and compute a score for each participant on this discriminant function. Now conduct a univariate t test on the difference between the mean of this discriminant function and the population mean implied for this combined score by the null hypothesis adopted in problem 2. Compare with your results in problem 2.

4. Now test the null hypothesis that the population mean profiles for groups 1 and 2 are identical, that is, that populations 1 and 2 do not differ in their means on any of the three dependent measures.

5. Compute a score for each subject on the discriminant function corresponding to the test in problem 4. Now conduct a univariate F test on the difference in mean discriminant scores for the two groups. Compare with your results in Problem 4.

6. Using post hoc significance criteria, test the hypotheses that the two groups are sampled from populations having identical means on

(a) X_1, (b) X_2, (c) $X_1 + X_2 + X_3$, (d) $4.0\,X_1 + 2.0\,X_2 + 2X_3$.

The last test uses the observed mean differences as weighting factors.

7. Conduct a profile analysis of these data. Compare the results of the flatness test with the results of the test in Problem 2, which also assumes a flat profile. In addition, test the significance of the difference in grand means on X_1 and X_2.

B. *Some Real Data.*

All of the following refer to Data Set 4.

1. Use Hotelling's T^* to test the significance of the difference between the mean outcome vectors for the NMO and the COMP groups.

2. Compute the discriminant function for the difference between NMO and COMP. Compute each subject-pair's score on the discriminant function D.

3. Conduct a univariate t test on the difference between the NMO and COMP groups in mean D (discriminant function) score. Compare the results of Problem 1.

4. Test the statistical significance of each of the following sources of difference between the NMO and COMP groups:

(a) CC; (b) CD - DC; (c) CC + CD; (d) CC + DD;

(e) a simplified (rounded-off) version of the discriminant function.

Indicate in each case both the conclusion you would reach if this were a post hoc comparison and the conclusion employing an a priori criterion.

C. More Real Data (Red-Shouldered vs. Red-Tailed Hawks).

James Bednarz (1981; Bednarz & Dinsmore, 1981) carried out a number of observations on the nest sites of two species of hawk in Iowa. Included in a set of 20 measures obtained from 12 red-shouldered hawk (RSH) nest sites and 26 red-tailed hawk (RTH) nest sites were four distance measures, namely the distance from the nest site to the nearest building (DB), the nearest road (DRD), the nearest railroad (DRR), and the nearest source of water (DW), all measured in meters. The means and covariance matrix for each group of nest sites follow:

$$\mathbf{S}_{RSH} = \begin{array}{c} \\ DW \\ DB \\ DRD \\ DRR \end{array} \begin{array}{c} \begin{array}{cccc} \text{DW} & \text{DB} & \text{DRD} & \text{DRR} \end{array} \\ \left[\begin{array}{cccc} 14{,}475.0 & 13{,}079.5 & 22{,}327.3 & 24{,}259.1 \\ & 259{,}990.0 & 185{,}273.0 & 78{,}189.4 \\ & & 259{,}418.0 & 93{,}890.9 \\ & & & 207{,}015.0 \end{array}\right] \end{array};$$

$$\overline{\mathbf{X}}'_{RSH} = [142.5 \quad 1000.83 \quad 820.0 \quad 381.667];$$

$$\mathbf{S}_{\text{RTH}} = \begin{matrix} & DW & DB & DRD & DRR \\ DW & 326{,}452 & -4{,}142.0 & -67{,}691.2 & 88{,}803.5 \\ DB & & 47{,}570.0 & 28{,}589.0 & 28{,}802.0 \\ DRD & & & 54{,}110.3 & 6{,}806.0 \\ DRR & & & & 230{,}420.0 \end{matrix};$$

$$\overline{\mathbf{X}}'_{\text{RTH}} = [522.692 \quad 495.0 \quad 308.577 \quad 1247.31].$$

1. Conduct a profile analysis of these data. Be sure to include tests of simplified discriminant functions and of any other interesting combinations of the dependent variables that come to mind.

2. What do the results tell us about Dr. Bednarz' hypothesis that red-tailed hawks tend to choose more accessible nesting sites than do red-shouldered hawks?

Answers

A.

1. $\mathbf{S}_c = \frac{1}{4}\begin{bmatrix} 38 & 6 & 1 \\ 6 & 14 & -11 \\ 1 & -11 & 16 \end{bmatrix} = \begin{bmatrix} 9.5 & 1.5 & .25 \\ 1.5 & 3.5 & -2.75 \\ .25 & -2.75 & 4.0 \end{bmatrix}.$

2. $\overline{\overline{\mathbf{X}}}' = (7, 8, 10)$; $\Delta' = (-3, -2, 0)$;

$$\mathbf{S}_c + N\Delta\Delta' = \mathbf{S}_c + 6\begin{bmatrix} 9 & 6 & 0 \\ 6 & 4 & 0 \\ 0 & 0 & 0 \end{bmatrix} = \begin{bmatrix} 63.5 & 37.5 & .25 \\ 37.5 & 27.5 & -2.75 \\ .25 & -2.75 & 4.0 \end{bmatrix};$$

$$|\mathbf{S}_c| = 49.875; \quad |\mathbf{S}_c + N\Delta\Delta'| = \mathbf{826.50.}$$

Thus, using the determinantal formula,

$$T^2 = (826.5/49.875) - 1 = 15.571.$$

Then

$$F = (4 - 3 + 1)/[3(4)]\, T^2 = T^2/6 = 2.595 \text{ with 3 and 2 } df, \text{ ns.}$$

It is always a good idea to check your work by applying an alternate computational procedure. Using Eq. (3.5), we obtain

$$\mathbf{S}_c^{-1} = \begin{bmatrix} .12907 & -.13409 & -.10025 \\ & .76065 & .53133 \\ . & & .62515 \end{bmatrix},$$

Whence

$$N\Delta'\, \mathbf{S}_c^{-1}\Delta = 6[9(.12907) + 4(.76065) + 12(-.13409)] = 6(2.59515) = 15.5709.$$

Almost identical results are obtained by employing MATLAB. (I initially used EINVH, provided in the second edition of this book; either gives a T^2 value of 15.5710.)

3. It is probably most convenient to work with $[N\Delta\Delta' - \lambda\mathbf{S}]\mathbf{a} = \mathbf{0}$, rather than carrying out the premultiplication by $\mathbf{S}^{-1}$. At any rate, in nonnormalized form, the discriminant function is $.15625X_1 + 1.46874\,X_2 + X_3$. Any other set of coeficients that preserves the ratios of the above figures (that is, their relative magnitude) is correct. The D scores are 17.344, 23.156, 19.687, 25.781, 24.937, 26.156 for subjects 1, ... 6, respectively. The resulting t test gives

$$t = (22.844 - 26.250)/\sqrt{.74493} = -3.946 = -\sqrt{T^2}.$$

Note that as an a priori test (which it certainly is not), this t ratio would have been statistically significant. Recall, too, that we are using a within-groups estimate of σ_c^2.

3. $\Delta' = (4,\ 2,\ 2)$, whence

$$\mathbf{S}_c + (3/2)\Delta\Delta' = \begin{bmatrix} 33.5 & 13.5 & 12.25 \\ & 9.5 & 3.25 \\ & & 10.0 \end{bmatrix},$$

and the determinantal formula gives

$$T^2 = (655.5/49.875) - 1 = 12.143,$$

whence

$$F = T^2/6 = 2.024 \text{ with 3 and } 2df\text{, ns.}$$

5. The resulting t should be equal to $\pm$ the square root of T^2. The discriminant function is $.0170X_1 + .7321\,X_2 + .6810\,X_3$.

6. By far the quickest way to compute these t ratios is to make use of the two-sample version of Eq. (3.2), whence

$$t^2(\mathbf{c}) = \frac{3}{2}(\mathbf{c}'\Delta)^2/(\mathbf{c}'\mathbf{S}_c\mathbf{c}),$$

where $\mathbf{X}_c$ is the set of combined scores. Thus,

$$t^2(1, 0, 0) = (3/2)(16/9.5) = 2.53, \text{ ns};$$
$$t^2(1, 1, 1) = (3/2)(64/15) = 6.4, \text{ ns};$$
$$t^2(1, 0, 0) = (3/2)(4/3.5) = 1.71, \text{ ns};$$
$$t^2(1, 0, 0) = (3/2)(576/188) = 4.596, \text{ ns}.$$

We knew, of course, that none of these comparisons would be significant by the post hoc criterion, but it is worth noting how much better the discriminant function does (on this sample of data) than any of the others. Note, too, how unimportant X_1 is in the discriminant function.

7. (a) **Parallelism:**

$$\mathbf{S}^* = \begin{bmatrix} 10 & -5 \\ -5 & 13 \end{bmatrix}; \quad \mathbf{\Delta}'_{\text{slope}} = (2,0); \quad \mathbf{S}^{*-1} = \frac{1}{105}\begin{bmatrix} 13 & 5 \\ 5 & 10 \end{bmatrix};$$

$$T^2 = (3/2)(52/105) = 78/105; \quad F = 3\,T^2/8 < 1.$$

(b) **Levels:** This is essentially the test in problem 6(c), with

$$F = 6.4 \text{ with 1 and 4 } df, \text{ ns}.$$

(c) **Flatness:**

$$T^2 = 6\begin{bmatrix} 1 & 2 \end{bmatrix}\left(\frac{1}{105}\right)\begin{bmatrix} 13 & 5 \\ 5 & 10 \end{bmatrix}\begin{bmatrix} 1 \\ 2 \end{bmatrix} = 6(73/105) = 4.17;$$

$$F = (3/8)\,T^2 = 1.56 \text{ with 2 and 3 } df, \text{ ns}.$$

This is not the same as the test against (10, 10, 10), because the overall level is unspecified. It is closer to (but not identical to) a test of the hypothesis that μ' = (8.67, 8.67, 8.67). Testing the significance of the difference between the grand means on variables 1 and 2 yields an F of 6(1)2/10 = .6, nonsignificant.

B.

1.

$$\mathbf{A}_c = \begin{bmatrix} 417.17 & 149.75 & 14.25 \\ 149.75 & 226.25 & -79.25 \\ 14.25 & -79.25 & 282.92 \end{bmatrix}; \quad \mathbf{A}_c + 6\mathbf{\Delta\Delta}' = \begin{bmatrix} 449.84 & 132.25 & 64.42 \\ 132.25 & 235.63 & -106.13 \\ 64.42 & -106.13 & 359.96 \end{bmatrix};$$

$$\mathbf{\Delta}' = [-2.333 \quad 1.250 \quad -3.583], \qquad N_1N_2/(N_1 + N_2) = 6;$$

$$|\mathbf{A}_c| = .17355 \times 10^8; \qquad |\mathbf{A}_c + 6\mathbf{\Delta\Delta}'| = .24005 \times 10^8;$$

$$T^2 = 22[(.24005/.17355)-1] = 8.4287; \; F = (22 - 3 + 1)\,T^2 = 2.554 \text{ with 3 and 20 } df, \text{ ns}.$$

2. $D = .75565CC - .68658CD + DC.$

3. $t = \pm\sqrt{T^2} = 2.904.$

4. For post hoc comparisons, $T^2_{crit} = 3.3(3.10) = 10.23$, whence $t_{crit} = 3.20$. Of course, because the overall test failed, we know that none of these specific comparisons will be significant on a post hoc basis. For a priori comparisons, a squared t ratio of $2.074^2 = 4.30$ is required for significance at the .05 level.

(a) $t^2(1, 0, 0, 0) = 132[(2.333)^2/417.17] = 1.722.$
(b) $t^2(0, 1, -1, 0) = 132[(4.833)^2/667.670] = 4.618, \quad p < .05$ if a priori.
(c) $t^2(1, 1, 0, 0) = .164.$ **(d)** $t^2(1, 0, 0, 1) = 2.049.$
(e) $t^2(1, -1, 1, 0) = 8.374.$

Note the (perfectly legitimate) inclusion of DD in test (d), despite its apparently not having been involved in the overall test. This is a result of the linear dependency between DD and the other three outcome measures.

C.

1 and 2.

$$\mathbf{C} = \begin{bmatrix} 3/\sqrt{12} & -1/\sqrt{12} & -.288675 & -.288675 \\ 0 & 2/\sqrt{6} & -1/\sqrt{6} & -.408248 \\ 0 & 0 & 1/\sqrt{2} & -.707107 \end{bmatrix};$$

$$\mathbf{S}_{contr} = \mathbf{C}\mathbf{S}_c\mathbf{C}' = \begin{bmatrix} 221{,}681 & -2{,}160.7 & -51{,}836.9 \\ & 62{,}569.8 & 49{,}527.9 \\ & & 136{,}641.0 \end{bmatrix}.$$

(Recall that a matrix whose below-diagonal entries are omitted is presumed to be a symmetric matrix.)

(a) ***Levels.***

$$t^2 = [12(26)/38]\cdot(2345 - 2573.579)^2/1{,}051{,}292$$
$$= 8.21053(.04970) = .40806 < F_{.05}(1{,}36) = 4.10.$$

Thus we cannot reject the null hypothesis that the average distance from all four "objects" is the same for both species in the population.

(b) ***Flatness.***

$$\bar{\bar{\mathbf{X}}}' = (12\bar{\mathbf{X}}'_{RSH} + 26\bar{\mathbf{X}}'_{RTH})/38$$
$$= [402.632 \quad 654.737 \quad 470.079 \quad 973.947];$$

$$\Delta'_{flat} = \bar{\bar{\mathbf{X}}}'\mathbf{C}' = [-257.172 \quad -54.931 \quad -356.289],$$

whence

$$\mathbf{a}' = \mathbf{\Delta}'\mathbf{S}^{-1}_{\text{contr}} = [-2.14481 \quad 2.4626 \quad -4.31378] \times 10^{-3};$$

and

$$T^2 = 38\mathbf{a}'\mathbf{\Delta} = 74.244.$$

Our significance test is then based on

$$F = (36 - 3 + 1)/(3\cdot36)\ T^2 = 23.362 \text{ with 3 and 34 } df,\ p < .01 .$$

The initial choice of the three contrasts was made on substantive grounds, because buildings, roads, and railroads represent human intrusions as compared to water sources, which are more likely to be "natural" and at any rate represent a survival need. We shall therefore wish to test each of the three original contrasts for significance by dividing 38 times its squared mean by its main-diagonal entry in $\mathbf{S}_{\text{contr}}$, yielding

$$t^2_{(3,-1,-1,-1)} = 38(-257.172)^2/221{,}681 = 11.337$$

for *DW* vs. the other three. Because $t^2_{\text{crit}} = T^2_{\text{crit}} = (108/34)F_\alpha(3, 34) = 9.154$ at the .05 level and 14.036 at the .01 level, we can be fairly confident .that the two species, averaged together, have an overall tendency to choose nest sites closer to water than to human intrusions. However,

$$t^2_{(0,2,-1,-1)} = 38(-54.931)^2/62{,}569.8 = 1.833, \text{ ns}$$

for *DB* vs. *DRD, DRR,* so that we cannot be sure that the overall tendency of these hawks to site their nests closer to buildings than to roads or railroads is true of the population; whereas

$$t^2_{(0,0,1,-1)} = 38(-356.289)^2/136{,}641 = 35.3, p < .01.$$

for *DRD vs. DRR.*

Because our discriminant function for flatness involved nonzero weights for all three of our "basis" contrasts, it is probably going to be easier to find a meaningful interpretation of our significant flatness test that accounts for a large portion of the departure from flatness (i.e., that yields a t^2 close to T^2) if we translate the discriminant function into a contrast among the original four measures. Thus we have

$$\mathbf{a}'_{\text{orig}} = \mathbf{a}'_{\text{contr}}\mathbf{C} = [-1.857 \quad 2.6299 \quad -3.437 \quad 2.664].$$

Interpreting this as essentially *DW* - *DB* + *DRD* - *DRR* yields

$$t^2_{(1,-1,1,-1)} = 38(-755.973)^2/(\mathbf{a}'_{simp}\mathbf{C}\,\mathbf{a}_{simp}) = 38(571{,}495)/330{,}965 = 65.617, p < .01,$$

which is 88.4% as large as T^2. (In terms of r^2 measures, the drop is only from .673 to 646.) Thus a large portion of our departure from flatness can be interpreted as a tendency to find these hawks' nests closer to water and roads than to buildings and railroads—which might for instance, reflect a tendency to avoid noise.

Of course, all of this discussion of the grand mean profile must be qualified by our soon-to-be-gained knowledge that these differences among the four distance measures are somewhat different for red-shouldered hawks (RSHs) than they are for red-tailed hawks (RTHs), as revealed by the parallelism test.

(c) *Parallelism Test.*

$$\begin{aligned}\boldsymbol{\Delta}'_{par} &= [(\overline{\mathbf{X}}_{RSH} - \overline{\mathbf{X}}_{RTH})\mathbf{C}]' = [-373.024 \quad 557.62 \quad 973.731];\\ \mathbf{a}' &= \boldsymbol{\Delta}'_{par}\mathbf{S}^{-1}_{contr} = [-.003350 \quad .038913 \quad .043134]; \text{ and}\\ T^2_{par} &= [12(26)/38)\mathbf{a}'\boldsymbol{\Delta}_{par} = 64.950 \Rightarrow F = (34/108)\,T^2\\ &= 20.447 \quad \text{with} \quad 3 \text{ and } 34\ df, p < .01.\end{aligned}$$

Translating our discriminant function into the original measures, we. find that

$$-.0673\text{DW} + .7589\text{DB} + .3613\text{DRD} - 1.059\text{DRR}$$

is that contrast among our four distance measures on which the two species of hawk show the greatest difference. Interpreting this as essentially (DB + DRD)/2 - DRR (again, perhaps, a noise-avoidance indicant, though restricted this time to the "built environment"), we find that RSHs show a mean of 528.75 m on this measure, actually siting their nests closer to railroads than to buildings or roads, whereas RTHs show the expected tendency, siting their nests an average of 845.5 m farther from the nearest railroad than from buildings and roads; thus

$$t^2_{(0,1,-.5,-.5)} = 8.2105[528.75 - (-845.5)]^2/241{,}522.5 = 64.203,$$

which is statistically significant at the .01 level and which is 98.9% as large as T^2.

We should, of course, also test each of our "basis" contrasts and the simplified discriminant function that seemed to have most to do with the overall shape of the profile of grand means for significant interactions with species. This leads to the following table of comparisons:

Contrast	$\overline{X}_{RSH}$	$\overline{X}_{RTH}$	s_c^2	t^2
(1, -1/3, -1/3, -1/3)	-591.7	−160.9	295,575	5.154
(0, 1, -1/2, -1/2)	400.0	-282.9	93,855	40.802
(0, 0, 1, -1)	438.3	-938.7	273,282	56.973
(.5, -.5, .5, -.5)	-210.0	-455.5	82,741	5.981
(0, .5, .5, -1)	528.75	-845.5	241,522	64.203

Note that the profile analysis breakdown makes no provision for analyzing differences between the two groups on each of the original measures—just as analyzing simple main effects in a factorial Anova requires stepping outside the framework of a main-effects-and-interactions breakdown. We could, however, subsume tests of differences on individual dependent variables and linear combinations thereof (including, but not restricted to, contrasts) by adopting as our critical value $4(36)/(36 - 4 + 1)\ F(4,35) = 12.539$ at the .05 level, 19.197 at the .01 level. None of our decisions thus far would be changed by the increase in critical value to that for an overall T^2 on the differences between the two groups (of which the levels and parallelism tests are special cases), and we could in addition report the fact that the two species differ significantly in the distance they maintain between their nest sites and the nearest building ($t^2 = 18.678$), road ($t^2 = 18.379$), or railroad (27.556) but not in distance to nearest source of water ($t^2 = 5.135$).

Note, too, that we have chosen to interpret the *raw* score discriminant function coefficients and to make all other comparisons in terms of the original units of measurement. This will almost always be the case in a profile analysis, because we should not be doing a profile analysis unless we have the same unit of measurement for all variables. The difference between how close you are to water and how close you are to buildings is quite different from (and more readily interpretable than) the difference between the percentile rank of your nest site in the distribution of closeness to water and its percentile rank (which is in essence what is tapped by the z score) in the distribution of distances from buildings. On the other hand, the complete analysis of these data included measures as wildly different as longest diameter of nest in centimeters, size of the nesting woodlot in hectares, and percentage of branches supporting the nest that fell into diameter class *A*. If you do not have a common unit of measurement, z-score coefficients do give you a kind of common unit (relative standing within the various distributions) and are thus a better basis for interpretation and for testing contrasts. A profile analysis would have been very inappropriate for the analysis of all 20 dependent variables.

Any interpretation of these results must rely heavily on the researcher's expertise in the substantive area within which the study falls and on his or her knowledge of the details of the study's methodology. Under the assumption that woodlots close to buildings and roads are apt to be more "thinned out" and that nest sites close to buildings and roads are thus more readily accessible, the encroachment is consistent with the hypothesis. The fact that RTHs choose nest sites significantly farther from the nearest railroad than do RSHs is a bit harder to reconcile with the hypothesis, unless we argue that railroad right-of-way does not necessarily go along with commercial or residential buildings (except near stations) and therefore need not indicate more accessible nest sites. This would also make the simplified discriminant function for parallelism consistent with the hypothesis.

Somewhat stronger evidence for the hypothesis came from analysis of the other 16 measures, which were more direct indicators of the nature of the site. The species with the higher mean on the simplified discriminant function (the red-tailed hawks) build nests that are easy to get to (high access area), with a low density of surrounding trees (quadrat density), close to the main trunk crotch, with lots of medium size supporting branches, on sloping ground (which might also make for easier access or perhaps simply reflects the tendency to pick scrawny, small-diameter trees of the sort that grow on the topsoil of sloping ground), and so on.

Again, keep in mind that these interpretations are only intended to give you a feel for the way in which a T^2 analysis can, when combined with a knowledge of the substantive area, provide a framework for a much sounder, empirically grounded description of the differences between two groups or conditions. Bednarz' (1981) report of his analysis (including the fact that the discriminant function provided perfect classification of the two species when reapplied to the samples of nest site measures) gives a much more complete description.

4
Multivariate Analysis of Variance: Differences Among Several Groups on Several Measures

The use of Hotelling's T^2 is limited to comparisons of only two groups at a time. Many studies, such as the one that provided the data presented in Data Set 4 (chap. 3), involve more than just two independent groups of subjects. Just as doing a series of six univariate *t* tests to test all the differences on a single variable among four experimental groups would be inappropriate because of the inflation of Type I error this produces, so would a series of six T^2 analyses of the differences in mean response vectors among these same groups be just as inappropriate for the same reason. The answer to the problem of multiple experimental groups in the univariate case is (univariate) analysis of variance (Anova); the solution to the multiple comparison problem when there are two or more outcome measures as well as two or more groups is multivariate analysis of variance (Manova). Before we discuss Manova, it might be well to review the basic rationale and computational formulae of Anova.

4.1 ONE-WAY (UNIVARIATE) ANALYSIS OF VARIANCE

4.1.1 The Overall Test

Let us say that we are interested in assessing differences among the four motivating instruction conditions of Data Set 4 (chap. 3, Table 3.2) in the number of mutually competitive outcomes (DDs) achieved by a pair of subjects in a 40-trial game. The least interesting of the possible true "states of nature" that might underlie these data is the possibility that there are no real differences due to motivating instructions, the differences among the sample means having arisen solely through random sampling from populations having identical means. Thus, before we proceed with any more detailed examination of the means on DD, we must test the null hypothesis that

$$\mu_{COOP} = \mu_{NMO} = \mu_{IND} = \mu_{COMP}$$

More generally, the null hypothesis can be stated as

$$H_0\text{: } \mu_1 = \mu_2 = \ldots = \mu_k$$

where *k* is the number of groups. If this null hypothesis were true, then on the average we would obtain $\overline{X}_1 = \overline{X}_2 = \cdots = \overline{X}_k$, although any particular set of samples would produce pair-wise differences between sample means differing somewhat from zero as a consequence of random fluctuation. Our first task is to select a single statistic that will summarize how far our sample means depart from the equality implied by $H_{0,\text{ov}}$. An attractive candidate, and the one we shall adopt, is

$$s_{\overline{X}}^2 = \sum (\overline{X}_j - \overline{\overline{X}})^2 / (k-1), \tag{4.1}$$

the sample variance of the *k* means. In addition to the obvious reasons for selecting $s_{\overline{X}}^2$ (its sensitivity to any difference between sample means, the algebraic tractability of variances, and so on), it is also equal to one-half the mean squared difference between the $k(k-1)/2$ pairs of sample means. For instance, the variance of the four $\overline{X}$ s for Data Set 3 is $[(18.81 - 13.52)^2 + \cdots + (3.75 - 13.52)^2]/3 = 48.14$, whereas the mean squared pairwise difference is $[(18.81 - 18.09)^2 + (18.09 - 13.41)^2 + \cdots + (18.81 - 3.75)^2]/6 = 96.31$.

The second thing we need to ask is whether the magnitude of the discrepancies among the sample means (as measured by the variance of the *X*s) is consistent with our assumption of identical population means. A basis for comparison is provided by the well-known relationship between the variance of a population and the variance of samples drawn from that population, namely, that the variance of the means is

$$\sigma_{\overline{X}}^2 = \frac{\sigma^2}{n} \tag{4.2}$$

where *n* is the size of the sample on which each mean is based. Thus if we can obtain an estimate of the variance that we assume to be common to the *k* populations from which our samples were drawn, we can thereby also obtain an estimate of how large we would expect the variance of our sample means to be if there were in fact no true differences in population means. [Any true difference between any one or more pairs of population means will, of course, inflate the variance of the sample means beyond the figure computed from Equation (4.2).] We can obtain such an estimate by looking at the variability within each group of subjects and pooling these *k* estimates of σ^2 into a single best estimate, just as we pool our two available estimates in the univariate *t* test for the difference between two independent means. It would then seem informative to compare the ratio of our direct estimate of the variance of sample means drawn from our four populations with our indirect estimate of this same variance arrived at through an examination of the within-group variances in conjunction with the assumption of no true differences among the means of the four populations. In other words, we compute

$$F = \frac{s_{\overline{X}}^2}{s_w^2 / n} = \frac{\text{direct estimate of } \sigma_{\overline{X}}^2}{\text{estimate of } \sigma_{\overline{X}}^2 \text{ assuming } H_0} \tag{4.3}$$

where

$$s_W^2 = \left\{\sum(X_1 - \overline{X}_1)^2 + \sum(X_2 - \overline{X}_2)^2 + \Lambda + \sum(X_k - \overline{X}_k)^2\right\}/(N-k);$$
$$s_{\overline{X}}^2 = \left\{(\overline{X}_1 - \overline{\overline{X}})^2 + (\overline{X}_2 - \overline{\overline{X}})^2 + \Lambda + (\overline{X}_k - \overline{\overline{X}})^2\right\}/(k-1); \tag{4.4}$$

and n is the (common) sample size, that is, the number of subjects in each of our k independent groups, so that the total number of observations is $N = nk$.

Statisticians have computed for us how often F ratios of various sizes would occur if, in fact, the null hypothesis were correct and the additional assumptions of homogeneity of variance and normality of parent populations were met. We can thus compare the value of F (the ratio of the variability we observe among our sample means to the variability we would expect to observe were H_0 correct) we obtain to tabled percentiles of the F distribution. We need to specify two degree-of-freedom parameters for this comparison: the degrees of freedom going into our direct estimate of $\sigma_{\overline{X}}^2$ (k - 1, the number of numbers used in computing $s_{\overline{X}}^2$ minus one degree of freedom "lost" in taking deviations about $\overline{X}$ instead of μ) and the number of degrees of freedom going into our indirect estimate (N - $k = n_1 - 1 + \ldots + n_k - 1$, the sum of the degrees of freedom going into the various group variances). If our computed value of F is greater than the "critical value" tabled for an F with k - 1 and N - k degrees of freedom at the chosen significance level, H_0 is rejected; otherwise, it is not rejected.

Computationally, Equation (4.3) is inconvenient to work with, because each $\overline{X}_j$, as well as $\overline{\overline{X}}$, can be expected to involve decimals and because repetitive subtractions are not easily worked into cumulations of crossproducts on a desk calculator. For computational purposes, it has become traditional to reverse the intuitively obvious comparison procedure, using the variability of the sample means to estimate $\sigma^2 = n\sigma_{\overline{X}}^2$ and comparing this indirect estimate of σ^2 with its direct estimate s^2. The computational version of Equation (4.3) is thus

$$F = \frac{\sum n_j(X_j - \overline{X})^2/(k-1)}{\sum\sum(X_j - \overline{X})^2/(N-k)} = \frac{\left[\sum(T_j^2/n_j) - (\sum\sum X)^2/N\right]/(k-1)}{\left[\sum\sum X^2 - \sum(T_j^2/n_j)\right]/(N-k)}$$
$$= \frac{SS_b/(k-1)}{SS_b/(N-k)} = \frac{MS_b}{MS_w} \tag{4.5}$$

with k - 1 and N - k degrees of freedom, where $T_j = \sum_u X_{ju}$ is the sum of the observations in group j. The results of our one-way Anova are conventionally reported in a summary table (Table 4.1). The computational formulae of Equation (4.5) are, of course, simply raw-score formulae for the two kinds of variances computed in Equation (4.3).

Table 4.1 Summary Table of Anova on Dependent Variable

Source	*df*	*SS*	*MS*	*F*
Between groups	$k-1$	$\sum(T_j^2/n_j)-(\sum\sum X)^2/N$	$SS_b/(k-1)$	MS_b/MS_w
Within groups (error)	$N-k$	$\sum\sum X^2-\sum(T_j^2/n_j)$	$SS_w/(N-k)$	
Total	$N-1$	$\sum\sum X^2-(\sum\sum X)^2/N$		

4.1.2 Specific Comparisons

Rejection of our overall H_0 simply tells us that something other than chance fluctuation is generating the differences among our k group means. It is the purpose of a ***specific comparison procedure*** to allow us to specify in more detail the source of our significant overall F. There are many alternative procedures for conducting such specific comparisons. Rather than repeating the comprehensive analyses of the various approaches by Games (1971) and Kirk (1968), this text focuses on two approaches: Scheffé's contrast method and Bonferroni critical values. Scheffé's contrast method has the following properties:

1. It tests any and all hypotheses of the form $c_1\mu_1 + c_2\mu_2 + \cdots + c_k\mu_k = 0$, that is, all ***linear contrasts*** among the population means. All alternative methods (except the Bonferroni approach to be described next) confine themselves to pairwise comparisons, that is, comparisons involving only two groups at a time. It has been my experience that truly illuminating descriptions of the relationships among three or more groups usually involve more than pairwise comparison of group means. For example, the differences among the four instructional sets of Data Set 3 in terms of mean number of DD outcomes can most naturally be summarized in terms of the difference between the COOP group and the other three (corresponding to the H_0 that $\mu_{COOP} - \mu_{NMO}/3 - \mu_{IND}/3 - \mu_{COMP}/3 = 0$), the difference between the NMO group and the other two ($\mu_{NMO} - \mu_{IND}/2 - \mu_{COMP}/2$), and the difference between the Individualistic and Competitive groups ($\mu_{IND} - \mu_{COMP}$).

2. The procedure supplies significance tests that are easily adjusted—through multiplication by a constant—to handle either a priori or post hoc comparisons. That these two types of comparisons require different significance criteria should be obvious. The probability that the COOP group would score higher than the other three groups by chance is, for instance, considerably lower than the probability that *some* one of the four groups would have the highest sample mean. Similarly the probability that this

prespecified (a priori) difference would achieve a specific magnitude is considerably lower than the probability that *some* after-the-fact (post hoc) contrast would reach that same magnitude.

Each a priori comparison is tested by comparing an obtained F value, defined by

$$F_{\text{contr}} = \frac{\left(\sum c_j \overline{X}_j\right)^2}{\sum (c_j^2/n_j)^2 (MS_\text{w})} = \frac{SS_{\text{contr}}}{MS_\text{w}} = \frac{MS_{\text{contr}}}{MS_\text{w}} \tag{4.6}$$

$$\left\{ = \frac{n\left(\sum c_j \overline{X}_j\right)^2}{(\sum c_j^2)(MS_\text{w})} = \frac{\left(\sum c_j T_j\right)^2}{n(\sum c_j^2)(MS_\text{w})} \text{ when } n_1 = n_2 = \ldots = n_k = n \right\},$$

with the critical value at the chosen level of significance for an F with 1 and N - k degrees of freedom. Each post hoc F_{contr} is compared to k - 1 times the critical value for an F with k - 1 and N - k degrees of freedom. This post hoc procedure has the property that the probability of *any* false rejection of a null hypothesis is at *most* α (the chosen significance level), no matter how many of the infinite number of possible contrasts are computed. Further, the overall F is statistically significant if and only if at least one linear contrast can be found that reaches statistical significance by the post hoc criterion, although it is not at all uncommon to find that no single pairwise difference between any two group means is significant even though the overall F is.

3. When all sample sizes are equal, then any set of k - 1 *independent* contrasts completely partitions the between-groups sum of squares, in the sense that the sum of the $(k - 1)$ SS_{contr}s is equal to SS_b. Scheffe's method thus has enormous descriptive value in specifying the percentage of the total variation among the groups that is attributable to each of k - 1 independent sources of differences among the k groups. Two contrasts are independent if and only if the sum of the cross products of their respective coefficients equals zero, that is, $\mathbf{c}_1'\mathbf{c}_2 = \sum c_{i1} c_{i2} = 0$. It is convenient for descriptive purposes to choose independent contrasts, but this should never get in the way of testing those contrasts among the means that are of greatest theoretical import to the researcher.

To illustrate the Scheffé procedures we have outlined for one-way Anova, let us apply them to the DD measure only in Data Set # 4. We obtain

$$T = \sum T_j = \sum\sum X = 649; \quad \sum\sum X^2 = 12{,}505;$$

$$(\sum T_j^2)/12 = 10{,}509.25; \quad C = 649^2/48 = 8775.02;$$

$SS_{\text{contr 1}} = [217 + 226 + 161 - 3(45)]^2/(12\cdot 12) = 1527.51$

for the COOP group versus the other three groups;

$SS_{\text{contr 2}} = [217 + 226 - 2(161)^2]^2/(12\cdot 6) = 203.5$

for NMO versus COMP and IND; and

$SS_{\text{contr 3}} = [217 - 226]^2/(12 \bullet 2) = 3.38$

for COMP versus IND.

From these results we construct the summary table given by Table 4.2. Note that the

"NMO versus COMP, IND" contrast would be considered statistically significant if it were specified before the data were collected. If, however, the thought of making this particular comparison arose only after seeing the data, it could not be considered statistically significant.

Table 4.2 Summary Table for Effects of Instructions on Frequency of DD Outcomes

Source	*df*	*SS*	*MS*	*F*
Between groups	3	1734.23	578.08	12.745[a]
COOP versus other 3	1	1527.51	(88.1%)[b]	33.7[a]
NMO versus COMP, IND	1	203.35	(11.7%)	4.49
COMP versus IND	1	3.38	(0.2%)	< 1
Within groups	44	1995.75	45.36	
Total	47	3729.98		

[a] $p < .01$ by post hoc criterion.
[b] Figures in parentheses are percentage of SS_b accounted for by each contrast = $100(SS_{contr}/SS_b)$.

The issue of how best to conduct multiple comparisons is still a highly controversial one. In particular, it is clear that the procedure outlined thus far, which entails complete freedom to conduct any a priori test using the same significance level as if that test were the only one conducted, and requiring that post hoc tests employ a critical value that sets the probability that *any* one of the (infinitely many) *possible* comparisons leads to a false rejection of a null hypothesis at α, has some disadvantages. For example, if the researcher simply predicts in advance that every treatment will lead to a distinctly different effect, so that no two of the corresponding population means are identical, then he or she can claim that *any* specific comparison is logically a priori. The Bonferroni procedure, introduced in section 1.1.1, provides a solution to this problem.

The Bonferroni approach adjusts the critical values of the individual tests so as to yield individual Type I error rates α_i that sum to the desired ***setwise error rate*** α_{set}. If the set contains all null hypotheses of any possible interest to the researcher, then $\alpha_{set} = \alpha_{ew}$, the ***experimentwise error rate.*** (See Games, 1971, and/or Kirk, 1982, 1995 for a discussion of the issue of selection of error rate.) If the tests are highly intercorrelated, α_{set} may be considerably lower than $\Sigma\alpha_i$—it does in fact equal α_i when the tests are perfectly correlated. The Bonferroni approach thus provides conservative tests of statistical significance, but it nevertheless often yields less stringent critical values (and thus more powerful tests) than would the Scheffé post hoc criterion, especially when n_t (the total number of comparisons) is small. It also has the advantage of allowing the researcher to set lower α_is for (and thus provide more powerful tests of) the hypotheses of greatest theoretical or practical importance.

The Bonferroni approach has the disadvantage of often requiring critical values for significance levels not included in readily available tables of the percentiles of sampling distributions. The desired critical values will usually be those for the extreme tails of Student's *t* distribution. You are unlikely to find a published table with a column for, e.g., $\alpha = .00833$ (.05/6). However, computer subroutines that generate critical values for any alpha level you specify are now widely available online (e.g., as of this writing,

http://surfstat.newcastle.edu.au/surfstat/main/tables.html

and

http://fsweb.berry.edu/academic/education/vbissonnette/

Alternatively, a FORTRAN program for computation of *z, r, t, F,* and chi-square critical values is available via email (rharris@unm.edu) with the permission of its author (Kevin O'Grady,1981).

The major disadvantage of the Bonferroni approach is its restriction to prespecified sets of comparisons, which thereby reduces its utility for post hoc exploration of the obtained data. Of course, this is not a disadvantage to the researcher who is interested only in testing a limited number of specific hypotheses generated on theoretical grounds or who finds that one of the following two sets of comparisons includes all those he or she would consider meaningful:

1. All pairwise comparisons of the *k* means, for a total of $\binom{k}{2} = k(k-1)/2$ comparisons [the notation $\binom{n}{m}$ represents the number of possible combinations of *m* things selected from a set of *n* things, order ignored].

2. All comparisons of the unweighted means of two subsets of the means, leading to a total of

$$\frac{1}{2}\sum_{i=1}^{k-1}\sum_{j=1}^{k-1}\binom{k}{i}\binom{k-i}{j} = (3^k - 2^{k-1} + 1)/2, \tag{4.7}$$

where *k* is the number of groups. Note that this set includes tests of hypotheses such as $\mu_4 = (\mu_1 + \mu_2 + \mu_3)/3$ but does not include the standard tests for linear, quadratic, cubic, and higher-order trends.

It is especially important to note that n_t, the size of the set of comparisons whose individual α_is sum to α_{set}, is equal to the number of ***possible*** comparisons, not just the number the researcher actually carries out. Thus, for instance, a decision to test the largest mean against the smallest involves an n_t of $k(k-1)/2$, not 1, because it implicitly requires examination of all pairs of means in order to select the largest pairwise difference.

Note too that choice of the Bonferroni approach to controlling the experimentwise error rate for multiple comparisons is inconsistent with the use of the overall *F*, $F_{ov} = MS_b/MS_w$ as a test of the overall hypothesis of no differences among the population means, because this F test considers all sources of differences among the means, including some kinds of comparisons that the researcher has declared to be of no conceivable interest to him or her. Thus the overall null hypothesis for the user of the Bonferroni approach should be rephrased as, "Each of the n_t null hypotheses (in the set of comparisons of interest) is true." This hypothesis is rejected at the α_{ew} level of significance if one or more of the n_t tests leads to rejection of its corresponding null hypothesis at the α_i (usually = α_{ew}/n_t) level of significance.

Finally, note that specialized multiple comparison procedures have been developed for certain sets of prespecified contrasts, such as comparisons of each of the k - 1 experimental groups' means with the single control group's mean (Dunnett's test) or examination of all possible pairwise differences (Tukey's test). These specialized procedures provide more powerful tests than the Bonferroni procedure when they are applicable, because their critical values take into account the nature of the intercorrelations among the various comparisons. The reader is referred to Kirk, 1982, 1995, or Harris, 1994, for details of these specialized procedures. Too recently to appear in either of these texts, Ottaway and Harris (1995) and Klockars and Hancock (1998) developed a critical value (c.v.) for testing all possible subset contrasts. Klockars and Hancock provide tables of this ***subset-contrast critical value*** and Ottaway and Harris showed that this c.v. is very closely approximated by a weighted average of the Scheffé and Tukey HSD critical values in which the Scheffé c.v. gets 70% of the weight.

On the other hand, the Bonferroni approach is more flexible, generalizing readily to situations involving multiple dependent variables. For instance, if the researcher is willing to confine examination to single dependent variables, eschewing all consideration of the emergent variables represented by linear combinations of the original variables, the test of the overall null hypothesis in a situation in which Manova would normally be applied can be accomplished instead by performing p separate univariate analyses, taking $F_{\alpha/p}(df_{eff}, df_{err})$ as the critical value for each such test, where α is the desired significance level for the overall test. Similarly, the researcher could specify a set of $b{\cdot}d$ contrasts encompassing all b of the relevant comparisons among the means on any of the d linear combinations of the dependent measures he or she deemed worthy of consideration, taking $\alpha_{ew}/(b{\cdot}d)$ as the significance level for each such comparison. Such a procedure, however, seems recklessly wasteful of the information that a true multivariate analysis could provide.

The Bonferroni approach can be illustrated with the same data used to illustrate Scheffé procedures, namely the three contrasts among the four means on DD from Data Set 3. From the summary table provided earlier (Table 4.2), we have Fs of 33.7, 4.49, and <1 for tests of

$$\mu_{COOP} = (\mu_{NMO} + \mu_{IND} + \mu_{COMP})/3, \qquad \mu_{NMO} = (\mu_{IND} + \mu_{COMP})/2,$$

and

$$\mu_{IND} = \mu_{COMP},$$

$$\mu_{IND} = \mu_{COMP},$$

respectively. If we had specified these in advance as the only comparisons of any conceivable interest to us, and if we had considered them of equal importance, then we would use a per-comparison error rate α_i of .05/3 = .0167 for each test, yielding a critical value of F_{contr} of 6.09 (using the O'Grady's 1981 program). (This assumes that we have selected an experimentwise error rate of .05.) If, on the other hand, we had actually chosen these comparisons after seeing the data as the most interesting or interpretable of all subset comparisons (cf. set 2 described earlier), then the size of this set would be n_t = $(3^4 - 2^{k+1} + 1)/2 = 25$, and the appropriate critical value for each comparison would be $3.292^2 = 10.84$. Scheffé's post hoc criterion would be 3(2.82) = 8.46, which is smaller than the Bonferroni critical value, despite being based on the infinitely large set of *all* possible contrasts. (Be sure that you understand why this is possible.) Finally, if we anticipate performing the same kind of post hoc exploration of the means separately for each of the four outcome measures, we would thereby be contemplating 100 possible comparisons and would therefore require an F of $3.912^2 = 15.30$ for significance. (As shown later, the Manova equivalent of Scheffé's post hoc criterion, taking into account all possible contrasts on all possible linear combinations of the dependent measures, yields a critical value of 15.44.)

The basic. principle underlying the Bonferroni approach is sufficiently simple that the reader should be able to adapt it quite readily to any multiple comparisons problem he or she encounters in which the size of the set of all "relevant" comparisons can be specified. Because of the conservative nature of the Bonferroni procedure, the reader should, however, consider the possibility that test procedures based on the infinitely large ***set*** of ***all*** possible comparisons may provide less stringent critical values than the Bonferroni approach.

4.2 ONE-WAY MULTIVARIATE ANALYSIS OF VARIANCE

After being exposed to multiple regression and to Hotelling's T^2, it should come as no surprise to the reader that Manova is handled by reducing each subject's scores on p variables to a single number—a simple linear combination of his or her scores on those original p variables. Heuristically, one-way Manova consists of a search for that linear combination of the variables that maximally discriminates among the k groups in the sense of producing the largest possible univariate F ratio, followed by comparison of this largest possible univariate F to a critical value appropriate to such a statistic, taking into account the extreme capitalization on chance involved in finding it.

Formally, it should be fairly clear from the reader's familiarity with the relationship between the variance of a linear combination of variables and the variances of the component variables that

$$F(\mathbf{a}) = MS_{\mathrm{b},V} / MS_{\mathrm{w},V} = \frac{\sum n_j(\bar{V}_j - \bar{\bar{V}})^2/(k-1)}{\sum\sum(V_j - \bar{V}_j)^2/(N-k)} = \frac{\mathbf{a'Ha}}{\mathbf{a'Ea}}\left(\frac{N-k}{k-1}\right), \tag{4.8}$$

where $MS_{b,V}$ is the mean square between groups for variable (dependent measure) V; $MS_{w,V}$ the within-group mean square for dependent variable V;

$$V = a_1X_1 + a_2X_2 + \Lambda + a_pX_p = \mathbf{Xa};$$

and **H** is the between-group SSCP (sums of squares and cross-products) matrix,

$$\begin{bmatrix} SS_{b;1} & SP_{b;1,2} & SP_{b;1,3} & \cdots & SP_{b;1,p} \\ SP_{b;1,2} & SS_{b;2} & SP_{b;2,3} & \cdots & SP_{b;2,p} \\ SP_{b;1,3} & SP_{b;2,3} & SS_{b;3} & \cdots & SP_{b;3,p} \\ \vdots & \vdots & \vdots & \cdots & \vdots \\ SP_{b;1,p} & SP_{b;2,p} & SP_{b;3,p} & \cdots & SS_{b;p} \end{bmatrix},$$

where

$$SP_{b;r,s} = \sum n_j(\bar{X}_{j,r} - \bar{\bar{X}}_r)(\bar{X}_{j,s} - \bar{\bar{X}}_s) = \sum(T_{j,r}T_{j,s}/n_j) - T_rT_s/N$$

and

$$SS_{b;r} = SP_{b;r,r} = \sum n_j(\bar{X}_{j,r} - \bar{\bar{X}}_r)^2 = \sum(T_{j,r}^2/n_j) - T_r^2/N,$$

where $T_{j,r}$ is the sum of observations on variable r in group j and T_r is the sum of all observations in all groups on variable r; and **E** is the within-group covariance matrix, the matrix whose (r, s)th entry is

$$SP_{w;r,s} = \sum\sum(X_{j,r} - \bar{X}_r)(X_{j,s} - \bar{X}_s) = \sum\sum X_rX_s - \sum(T_{j,r}T_{j,s}/n_j).$$

The task is to choose an **a** such that $F(\mathbf{a})$ is maximized. However, maximizing $F(\mathbf{a})$ is equivalent to maximizing **a'Ha/a'Ea**, which is just $F(\mathbf{a})$ with the bothersome constant "stripped off." Further, we know that the univariate F ratio is unaffected by linear transformations of the original data, so that we are free to put a side condition on **a** so that it may be uniquely defined. The most convenient restriction is **a'Ea** = 1, whence, applying the method of Lagrangian multipliers (Digression 1) and the matrix differentiation procedures discussed in Digression 2, we find (Derivation 4.1) that the maximum value of the stripped F ratio and the combining weights that produce this maximum value are given by the largest value of λ and the associated vector, **a**, which together satisfy Equation (4.9) and/or Equation (4.10):

$$[\mathbf{H} - \lambda\mathbf{E}]\mathbf{a} = \mathbf{0}, \tag{4.9}$$

whence

$$[\mathbf{E}^{-1}\mathbf{H} - \lambda\mathbf{I}]\mathbf{a} = \mathbf{0} \tag{4.10}$$

In other words, they are given by the largest characteristic root and associated characteristic vector (eigenvalue and eigenvector) of $\mathbf{E}^{-1}\mathbf{H}$. (Note that $\mathbf{E}^{-1}\mathbf{H}$ is almost always nonsymmetric. Many eigenvalue programs and subroutines require as input a symmetric matrix. Be sure the program you use can handle the "generalized eigenvalue problem.") $\mathbf{E}^{-1}\mathbf{H}$ will in general have a rank greater than or equal to 2 (usually it will equal either k - 1 or p, whichever is smaller), so that $\mathbf{E}^{-1}\mathbf{H}$ will have more than one characteristic root and associated vector. What meaning do the other roots and vectors have? As Derivation 4.1 shows, the *largest* characteristic root is the solution to our maximization problem, with the characteristic vector associated with this largest root giving the linear combining rule used to obtain this maximum possible $F(\mathbf{a})$. The ith largest characteristic root is the maximum possible univariate F ratio obtainable from any linear combination of the p original variables that is uncorrelated with the first i - 1 "discriminant functions," the characteristic vector associated with the ith largest root giving the coefficients used to obtain this maximum possible F.

The distribution of the largest characteristic root (which will, of course, be the basis of our decision as to whether to reject the overall null hypothesis of identical group means on each of the p variables, that is, that $\mu_1 = \mu_2 = \cdots = \mu_k$ where μ_j is a population mean response ***vector*** for condition j) is rather complicated. Tables are, however, available through the efforts of Heck (1960), Pillai (1965, 1967), and Venables (1973, 1974) and are provided in Appendix A, and a FORTRAN program is available at rharris@unm.edu. Three degree-of-freedom parameters must be specified, namely,

$$s = \min(df_{\text{eff}}, p); \quad m = (|df_{\text{eff}} - p| - 1)/2; \quad n = (df_{\text{err}} - p - 1)/2; \tag{4.11}$$

where df_{eff} is the degrees of freedom for the effect being tested (= k - 1 for one-way Manova) and df_{err} is the degrees of freedom for the appropriate error term (= N - k for one-way Manova). When s = 1, $\lambda_{\max}$ has an F distribution (as might have been anticipated from the fact that in such cases Manova reduces either to one-way univariate Anova or to Hotelling's T^2); specifically,

$$s = 1 \rightarrow F = \frac{2n+2}{2m+2}\lambda_{\max} \text{ with } 2m+2 \text{ and } 2n+2 \; df \tag{4.12}$$

$$\left(= \frac{df_{\text{err}} - p + 1}{p}\lambda_{\max} \text{ for } p \geq 2\text{, in which case } df_{\text{eff}} = 1 \right)$$

$$\left(= \frac{df_{\text{err}}}{df_{\text{eff}}}\lambda_{\max} = F_{\text{ov}} \text{ on the single dependent variable when } p = 1 \right).$$

When s = 2 or larger, we must compute $\theta_1 = \lambda_{\max}/(\lambda_{\max} + 1)$ and compare this to the critical value read from the appropriate table in Appendix A. Note that $\theta_1 = SS_{\text{eff,V}}/(SS_{\text{eff,V}} + SS_{\text{err,V}}) = SS_{\text{eff}}/SS_{\text{total}}$ computed on our discriminant function; θ_1 is thus interpretable as a sample r^2 measure—looking ahead to chapter 5, as the squared canonical correlation between group-membership variables representing the effect we are testing and the set of dependent variables.

Specific comparisons following a significant overall greatest characteristic root test are handled by the same reasoning (Roy's union intersection principle) that provided post hoc procedures in Hotelling's T^2, combined with the contrast methods we recommended for one-way univariate Anova. Because λ_{max} represents the most favorable picture of the differences among the groups on all p variables, any univariate F on a specific variable or linear combination of variables that is larger than that value of $F_{max} = (N - k)\,\lambda_{max}/(k - 1)$ that would just barely have led to rejection of the overall null hypothesis must be considered statistically significant at at least the same level employed in testing the overall H_0. The overall null hypothesis is rejected if and only if there exists at least one linear combination of the original variables that would have yielded an F ratio significant by this post hoc criterion. There is thus no need to run any specific post hoc comparison if the overall null hypothesis (based on the greatest characteristic root test) did not achieve statistical significance. A priori tests should of course always be run.

We can perform a Scheffé-type contrast among the means using any single variable or linear combination of variables as the dependent measure. However, the critical value to be used in testing the significance of such a contrast will depend on whether (a) the coefficients for this particular contrast. among the groups and (b) the coefficients for this particular linear combination of the outcome measures were selected on a priori grounds or only after examining the data, as indicated in Table 4.3.

The first column of Table 4.3 needs no further explanation, because it amounts to applying Scheffé's contrast method (discussed in section 4.1) to the linear combination of outcome measures just as if it were the single outcome measure selected for a univariate Anova. The lower right-hand corner represents a relatively straightforward extension of Scheffe's contrast method to the multivariate case. It, like the univariate application, is so designed that the overall test is significant if and only if *some* contrast among the means on *some* linear combination of the outcome measures would yield an F ratio exceeding the specified post hoc critical value. Roy and Bose (1953a, 1953b, summarized in Morrison, 1976) established similar, but more general critical values in the context of simultaneous confidence intervals. The (1 - α)-level interval developed by Roy and Bose (1953a, 1953b) can, when applied to a contrast on a linear combination of variables, be expressed as

$$\mathbf{c}'\overline{\mathbf{X}}\mathbf{a} - \sqrt{\sum \frac{c_j^2}{n_j}(\mathbf{a}'\mathbf{E}\mathbf{a})\lambda_{\text{crit}}} < \mathbf{c}'\boldsymbol{\mu}\mathbf{a} < \mathbf{c}'\overline{\mathbf{X}}\mathbf{a} + \sqrt{\sum \frac{c_j^2}{n_j}(\mathbf{a}'\mathbf{E}\mathbf{a})\lambda_{\text{crit}}}$$

where n is the number of subjects per group; **c** the vector of k contrast coefficients; **a** the vector of p weights defining the linear combination of the dependent measures; **E** the error matrix; $\overline{\mathbf{X}} = [\overline{X}_{i,j}]$ a $k \times p$ matrix of group means on dependent measures; μ a k x p matrix of the corresponding population means; $\mathbf{c}'\boldsymbol{\mu}\mathbf{a}$ therefore the contrast among the

Table 4.3 CriticalValues for Contrasts Performed on Linear Combinations of Variables[a]

		Was this linear combination of the outcome measures selected on a priori grounds? Yes, a priori	No, post hoc
Was this particular contrast among the groups selected on a priori grounds?	Yes, a priori	$F_\alpha(1, df_{err})$	$\frac{p \cdot df_{err}}{df_{err} - p + 1} F_\alpha(p, df_{err} - p + 1)$
	No, post hoc	$df_{eff} \cdot F_\alpha(df_{eff}, df_{err})$	$df_{eff} \cdot F_{crit}$

Note. The entries for a priori tests give the alpha (α) level for tests involving a *single* a priori contrast and/or a *single* a priori linear combination of the dependent measures. Thus, for instance, if you decide to run tests on any number of linear combinations selected on post hoc grounds, and do this for each of five a priori contrasts (using, of course, the upper right-hand entry of Table 4.3 as your critical value for each of these tests), your total experimentwise Type I error rate will be 5·α (or $\sum_{i=1}^{5}\alpha_i$ if you use a different α level for each of the contrasts you explore). Put another way, control of experimentwise α at a given level requires the use of Bonferroni-adjusted critical values when more than one a priori contrast or combination is used; but no such adjustment is needed for multiple post hoc contrasts or combinations.

[a] $F_{crit} = df_{err}\lambda_{crit}/df_{eff} = df_{err}\theta_{crit} / [(1 - \theta_{crit})\, df_{eff}]$, where θ_{crit} is the critical value at the α level of significance of the greatest characteristic root distribution having the degree-of-freedom parameters specified in Equation 4.12.

population means on the specified combined variable for which we are constructing the confidence interval; and $\lambda_{crit} = \theta_{crit}/(1 - \theta_{crit})$ is the α-level critical value of the greatest characteristic root (gcr) of $\mathbf{E}^{-1}\mathbf{H}$; and θ_{crit} the critical value of the gcr of $(\mathbf{H} + \mathbf{E})^{-1}\mathbf{H}$, that is, the critical value tabled in Table A5. By way of illustration, the 95% confidence interval about

$$\mu_{COOP,W} - (\mu_{NMO,W} + \mu_{IND,W} + \mu_{COMP,W})/3,$$

where $W = CC + DC$ (cf. Demonstration Problem A4 at the end of this chapter), is computed as

$$\mathbf{c}'\overline{\mathbf{X}}\mathbf{a} = [384 - (226 + 174 + 155)/3]/12 = 16.583,$$

with degree-of-freedom parameters

$$s = 3;\quad m = (|3 - 3| - 1)/2 = -.5;\quad n = (44 - 3 - 1)/2 = 20,$$

whence

$$\lambda_{\text{crit}} = .260/.740 = .35135$$

and

$$\sum c_j^2 (\mathbf{a'Ea})\lambda_{\text{crit}} / n = (12/9)(2064.85)(.35135)/12 = 80.609,$$

whence

$$16.583 - \sqrt{80.609} < \mu_{\text{COOP},W} - (\mu_{\text{NMO},W} + \mu_{\text{IND},W} + \mu_{\text{COMP},W})/3 < 16.583 + \sqrt{80.609},$$

that is,

$$7.604 < \mathbf{c'\mu a} < 25.561.$$

It is important to understand the degree of protection we obtain with this 95% confidence interval. If we were to run a large number of experiments and follow up each with post hoc confidence intervals of the type that we just constructed, constructing as many as we found interesting (including, if we had the time and the fortitude, all possible such contrasts), then in at most 5% of these experiments would one or more of our statements about the range within which the population values of the various contrasts lie be incorrect. These are thus extremely conservative confidence intervals, in that we can construct as many of them as we please, on whatever grounds we choose, and still be 95% confident (in the sense just described) that ***all*** of these confidence intervals are correct statements.

In the case in which an a priori contrast is to be performed on a combined variable that was constructed only after examining the data (e.g., the discriminant function), the critical value is essentially that for a T^2 analysis performed by treating the a priori contrast as representing the difference between two weighted means: one represented by the positive coefficients in the contrast and the other represented by the negative coefficients. Note that the discriminant function, which maximizes the overall F ratio, will generally not be the combined variable that maximizes the F ratio for this particular contrast among the groups. A separate maximization procedure could be carried out for each contrast, but then the resulting SS_{contr}s would no longer add up to the total sum of squares for the subjects' scores on the discriminant function. This is very similar to the problem arising in higher order Manova (section 4.8) in deciding whether to conduct a separate maximization and greatest characteristic root test for each effect or to select instead the discriminant function from a one-way Manova, which ignores the factorial structure of the groups, followed by a higher order univariate Anova treating subjects' scores on this single discriminant function as the dependent measure.

To give a specific example, consider the problem of following up a Manova on Data Set 3 with various specific contrasts on a variety of linear combinations of the outcome measures. (Remember that here $p = 3$ because of the linear dependence among the four outcome measures .) For these four groups,

$$df_{\text{eff}} = 3;\ df_{\text{err}} = 44;\quad s = 3;\quad m = (|3 - 3| - 1)/2 = -.5;\quad \text{and } n = (44 - 3 - 1)/2 = 20.$$

If we had specified in advance that we would compare $\overline{\text{CD} - \text{DC}}$, the mean value of the combined variable CD - DC, for COOP to $\overline{CD - DC}$ for the other three groups, the resulting F ratio for this contrast on this combined variable would be compared to $F_{.05}(1, 44) = 4.06$. If we had specified CD - DC in advance as a theoretically relevant variable, but decided to compare the COOP and NMO groups only after examining our results, the appropriate critical value would be $3 \cdot F_{.05}(3, 44) = (2.82) = 8.46$. If instead we specified the COOP versus others contrast in advance, but used the (clearly post hoc) discriminant function as our dependent measure, we would compare the F ratio for the contrast to $(3 \cdot 44/42)\ F_{.05}(3, 42) = 8.96$. Finally, if we decided after examining the data to compare the COOP and NMO groups in terms of their means on the discriminant function, the critical value would be 3(44/3)(.260/.740) = 15.44—.260 being the .05-level critical value of the gcr statistic having the specified degree-of-freedom parameters.

If the researcher is uninterested in linear combinations of his or her original variables and will consider only comparisons involving a single dependent measure, he or she may find that the Bonferroni approach described in section 4.1.2 provides more powerful tests of preplanned comparisons than does the gcr-based approach described in this section.

4.3 MULTIPLE PROFILE ANALYSIS

The situations in which we might wish to employ multiple profile analysis are the same as those in which "simple" profile analysis, which was discussed in chapter 3, is appropriate, except that multiple profile analysis can be applied to distinctions among three or more groups. All variables must be expressed in meaningful, nonarbitrary units, and the researcher must be interested in distinguishing between differences in shape and differences in overall level of response, averaged across outcome measures. The three null hypotheses to be tested are the same, and the test procedures are also the same, once allowances for the presence of more than two independent groups have been made. Specifically, the ***parallelism*** hypothesis (that the shapes of the k profiles are mutually identical, that is, that the difference between each pair of adjacent dependent measures is the same for all k groups) is tested by conducting a one-way Manova on slope measures. This can be done either by actually computing each subject's slope scores $d_i = X_i - X_{i+1}$ and proceeding with standard one-way Manova procedures from there, or by using the formulae developed in chapter 3 for the relationship between **S** computed on all p original variables and $\mathbf{S}_{\text{slope}}$ computed on slope scores. The ***levels*** hypothesis is tested by conducting a one-way *univariate* Anova on $W = \sum_p X_i$, the sum of each subject's scores on the p dependent measures. This test is of course most conveniently carried out by employing Formula (4.8) with $\mathbf{a}' = (1, 1, \ldots 1)$. As pointed out in discussing two-group profile analysis in chapter 3, the levels hypothesis is difficult to interpret when the parallelism hypothesis has been rejected. The Flatness hypothesis (that the "pooled" profile is flat, that is, that the p grand means on the various dependent measures are identical) is tested by conducting a single-sample T^2 on the vector of grand mean slope measures (e.g., the vector of differences between adjacent grand means).

Specifically, the hypotheses and their tests are as follows:

Parallelism Hypothesis (Group x Measure Interaction).

$$H_{01}: \begin{bmatrix} \mu_{1,1}-\mu_{1,2} \\ \mu_{1.2}-\mu_{1,3} \\ \mathrm{M} \\ \mu_{1,p-1}-\mu_{1,p} \end{bmatrix} = \begin{bmatrix} \mu_{2,1}-\mu_{21,2} \\ \mu_{2,2}-\mu_{2,3} \\ \mathrm{M} \\ \mu_{2,p-1}-\mu_{2,p} \end{bmatrix} = \Lambda = \begin{bmatrix} \mu_{k,1}-\mu_{k,2} \\ \mu_{k,.2}-\mu_{k,3} \\ \\ \mu_{k,p-1}-\mu_{k,p} \end{bmatrix};$$

that is,

$$\mu_{\text{slope},1} = \mu_{\text{slope},2} = \cdots = \mu_{\text{slope},k},$$

where $\mu_{i,j}$ is the mean on dependent measure i in the population from which group j was sampled, and $\mu_{\text{slope},j}$ is a column vector of the p - 1 differences between means on "adjacent" dependent measures for population j. This hypothesis is tested by computing

$$\lambda_{\text{par}} = \text{greatest characteristic root of } (\mathbf{E}^*)^{-1}\mathbf{H}^*,$$

where **E*** and **H*** are error and hypothesis matrices, respectively, computed on the p - 1 difference scores $(X_1 - X_2)$, $(X_2 - X_3)$, ... ,$(X_{p-1} - X_p)$. Repeating the rather extensive computational procedures required to compute **H** and **E** matrices can be avoided by noting that

$$e^*_{ij} = e_{ij} + e_{i+1,j+1} - e_{i,j+1} - e_{i+1,j}$$

and

$$h^*_{ij} = h_{ij} + h_{i+1,j+1} - h_{i,j+1} - h_{i+1,j}$$

In other words, the (i,j)th entry of the "reduced" or "starred" or "slope" matrix is obtained by adding together the main diagonal entries and then subtracting the off-diagonal entries of the 2 x 2 submatrix whose upper left-hand corner is the (i,j)th entry of the original, p x p hypothesis or error matrix. The degrees-of-freedom parameters for this test are $s = \min(k - 1, p - 1)$; $m = (|k - p| - 1)/2$; and $n = (N - k - p)/2$.

Levels Hypothesis (Group Main Effect).

$$H_{02}: \mu_{w,1} = \mu_{w,2} = \cdots = \mu_{w,k},$$

where $\mu_{w,j}$ is the mean for population j of the sum of all dependent measures and thus is equal to $\mu_{j,1} + \mu_{j,2} + \ldots + \mu_{j,p}$. This hypothesis is tested by comparing

$$F = (SS_{b,W} / SS_{w,W})[(N - k)/(k - 1)],$$

where $SS_{b,W} = \Sigma\Sigma h_{i,j}$ is the sum of all the entries in the original, p x p hypothesis matrix, and $SS_{w,W} = \Sigma\Sigma e_{i,j}$ is the sum of all the entries in **E**, the original p x p error matrix, to the F distribution having k - 1 and N - k degrees of freedom.

Flatness Hypothesis (Measure Main Effect).

$$H_{03}\colon\ \bar{\mu}_1 = \bar{\mu}_2 = \bar{\mu}_3 = \cdots = \bar{\mu}_p,$$

Where $\bar{\mu}_V = (\mu_{1,V} + \mu_{2,V} + \ldots + \mu_{k,V})/k$ is the arithmetic mean of the means on dependent variable V of the k populations from which the k groups were sampled. This hypothesis is tested by computing

$$T^2_{\text{flat}} = N(N-k)\bar{\bar{\mathbf{X}}}'_{\text{slope}}(\mathbf{E}^*)^{-1}\bar{\bar{\mathbf{X}}}_{\text{slope}},$$

where

$$\bar{\bar{X}}_{\text{slope}} = \begin{bmatrix} \bar{\bar{X}}_1 - \bar{\bar{X}}_2 \\ \bar{\bar{X}}_2 - \bar{\bar{X}}_3 \\ \mathrm{M} \\ \bar{\bar{X}}_{p-1} - \bar{\bar{X}}_p \end{bmatrix}$$

is a (p - 1)-element column vector of differences between "adjacent" grand means (averaged across groups); that is,

$$\bar{X}_V = (n_1\bar{X}_{1,V} + n_2\bar{X}_{2,V} + \Lambda + n_k\bar{X}_{k,V})/N.$$

Then

$$F = \frac{(N-k-p+2)}{[(N-k)(p-1)]}T^2_{\text{flatness}}$$

is distributed as F with p - 1 and N - k - p + 2 degrees of freedom.

None of these results are affected if, in place of the transformation to slope measures, any set of p - 1 nonlinearly dependent contrasts among the dependent variables is employed. Use of theoretically or practically relevant contrasts holds out the possibility of obtaining a discriminant function with very simple coefficients. If, however, the optimal combination of the dependent measures is very different from any of the contrasts we employ in computing **H*** and **E***, the process of translating the discriminant function from a linear combination of our chosen contrasts into a contrast among the p original measures will be more complex than if we had based our initial analysis on slope measures.

To illustrate multiple profile analysis, consider Aida Parkinson's (1981) master's thesis.

Example 4.1 Damselfish Territories Parkinson studied differences among three species of damselfish in the nature of the territories they establish and defend. Because locating and measuring these territories involve time consuming and costly scuba diving in the Carribean waters damselfish inhabit, it was feasible to sample only five territories for each species. Table 4.4 presents the mean percentage of each species' territories found to be covered by three kinds of algae, three kinds of rubble, and anemones.

Table 4.4 Mean Percentage Coverage of Damselfish Territories Species

Species	Cerralic algae	Filamental algae	Other algae	Staghom rubble	Sand silt	Lime rubble	Ane-mones
Dusky	0.0%	12.0	10.0	7.2	0.0	42.0	22.4
3-Spot	14.8	12.4	15.2	14.4	0.8	22.4	5.2
Beau-Gregory	32.4	2.0	0.0	38.8	20.4	0.0	0.0

In order to keep our illustrative analysis simple, we take as our dependent variables the three rubble-coverage measures, for which

$$\begin{array}{c} \\ \mathbf{H} = \begin{array}{r} \text{STAG} \\ \text{SAND} \\ \text{LIME} \end{array} \end{array} \begin{array}{c} \begin{array}{ccc} \text{STAG} & \text{SAND} & \text{LIME} \end{array} \\ \begin{bmatrix} 2742.93 & 1881.07 & -3358.13 \\ & 1334.93 & -2175{,}76 \\ & & 4416.53 \end{bmatrix} \end{array}.$$

From the raw data we compute

$$\mathbf{E} = \begin{bmatrix} 3604.8 & -459.2 & -68.8 \\ & 504.0 & -73.6 \\ & & 4723.2 \end{bmatrix}$$

Our profile analysis then proceeds as follows:

Levels test (main effect of species):

$$F = \frac{12\sum\sum h_{ij}}{2\sum\sum e_{ij}} = 6(1168.53/8628.8) = .813, \text{ ns.}$$

Parallelism (species by type of coverage interaction): Using the main diagonals-off diagonals approach, we compute

$$\mathbf{H}^* = \begin{bmatrix} 315.72 & 1718.4 \\ & 10123.2 \end{bmatrix} \text{ and } \mathbf{E}^* = \begin{bmatrix} 5027.2 & -968.0 \\ & 6374.4 \end{bmatrix},$$

whence

$$\mathbf{E}^{*-1}\mathbf{H}^* = \begin{bmatrix} .11817 & .66712 \\ .28752 & 1.6894 \end{bmatrix}$$

and

$$\lambda^2 - 1.80758\lambda + .00783 = 0 \rightarrow \lambda = (1.80758 \pm \sqrt{3.23603})/2$$
$$= (1.80324, .00434),$$

with $\theta = 1.803/2.803 = .643$, $s = \min(2, 2) = 2$, $m = -.5$, $n = (12 - 2 - 1)/2 = 4$. From the gcr tables we find that $\theta_{crit} = .533$ at the .05 level, .658 at the.01 level, whence we can reject parallelism at the .05 level. Our F_{crit} for post hoc tests of particular contrasts among the three measures is 6 times $\theta_{crit}/(1 - \theta_{crit}) = 6.848$ at the .05 level. Our discriminant function is given by the solution to

$$[\mathbf{E}^{*-1}\mathbf{H}^* - 1.80324\mathbf{I}]\mathbf{a} = \begin{bmatrix} -1.68507 & .66712 \\ .28752 & -.11384 \end{bmatrix} \begin{bmatrix} a_1 \\ a_2 \end{bmatrix} = \begin{bmatrix} 0 \\ 0 \end{bmatrix},$$

the two rows of which give us $a_1 = .39590\ a_2$ and $.39594\ a_2$, respectively. Thus our discriminant function for parallelism is .396 (STAG - SAND) + (SAND - LIME) = .396 STAG + .604SAND - LIME. Interpreting this as the difference between percentage coverage of lime rubble and the average of coverage by sand silt and staghorn (coral) rubble yields $\mathbf{a}'_{simp} = (1, 1, -2)$, $F = (12/2)(47{,}682.12/26{,}652.8) = 10.734$, which is quite close to the F of 10.819 yielded by the actual discriminant function and is statistically. significant by our post hoc criterion. The means on this simplified discriminant function are -76.8, -29.6, and 59.2 for the Dusky, 3-Spot, and Beau-Gregory (B-G) damselfish, respectively. Thus the B-G versus Dusky contrast yields a sum of squares of

$$\frac{n\left(\sum c_j \overline{X}_j\right)^2}{\sum c_j^2} = 5[59.2 - (-76.8)]^2 = 46{,}240,$$

which is 97.0% of the total sum of squares (47,682.12) on this simplified discriminant function. This then gives us an F_{contr} of

$$\frac{SS_{contr}/1}{(\mathbf{a}'\mathbf{E}\mathbf{a})/12} = \frac{46{,}240}{26{,}652.8/12} = 20.819,$$

which safely exceeds its fully post hoc critical value of $2 \cdot F_{crit} = 13.6\%$.

Flatness test (main effect of type of coverage): Despite our rejection of the parallelism hypothesis, it might be interesting to see what the "average" damselfish's territory (averaging across species) looks like with respect to the three kinds of rubble. Thus we have

$$\overline{\mathbf{X}}' = (20.133, 7.067, 21.467) \text{ and } \Delta' = (13.066, -14.400).$$

Our discriminant function for flatness is proportional to

$$\Delta'\mathbf{E}^{*-1} = (.00223, -.00192),$$

and

$$T^2 = 15(12)\ \Delta'\mathbf{E}^{*-1}\ \Delta = 10.223,$$

whence

$$F = \frac{12-2+1}{12(2)}T^2 = 4.686 \text{ with 2 and 11 } df, p < .05.$$

Expressing our discriminant function as a contrast among the original measures gives 2.23 (STAG - SAND) - 1.92 (SAND - LIME) = 2.23STAG - 4.15SAND + 1.92 LIME, whence

$$\mathbf{a}'_{\text{simp}} = (1,-2,1) \Rightarrow F = \frac{15(27.466^2)}{13337.6/12} = 10.181,$$

which is quite close to the F yielded by the actual discriminant function (T^2 earlier) and obviously also statistically significant.

This is, of course, only an illustrative analysis. Given that five other measures were computed for each territory (depth and perimeter of territory, number of encrusted and interstitial corals, and number of times a vertical rod thrust through the territory hits something), a profile analysis and/or an overall Manova employing all 12 dependent measures is clearly impossible. (What would the degree-of-freedom parameter, n, be?) The analysis actually recommended was a combination of Bonferroni-adjusted Manovas on logically related subsets of the measures, with a separate profile analysis being conducted on the three algae measures, the three rubble measures, the single number-of-anemones-present measure (the only relevant test here being the levels test, of course), and the five "territory complexity" measures—each of these four profile analyses to be carried out with an alpha level of .025. There is, of course, a great deal of judgment involved in deciding how to apportion alpha among the four analyses and in choosing an acceptable level of familywise α. The important thing is to be aboveboard in reporting these judgments to your readers.

4.4 MULTIPLE DISCRIMINANT ANALYSIS

One of the by-products of a one-way Manova is the set of coefficients that produce the largest possible univariate F ratio when they are taken as the coefficients of a linear function of the original variables. This linear combination of the original p variables is the discriminant function (more accurately, the ***first*** or *primary* ***discriminant function)*** for these k groups, and is a logical candidate for use in classifying some new subject whose classification is unknown into one of the k groups. (Such classification will, however, be fraught with uncertainty if the new subject's scores on the various measures lie outside of the ranges of scores represented in the sample from which the discriminant function was derived. Relationships that are very nearly linear within one range of scores may be distinctly nonlinear when viewed over a wider range.) The new, as yet unclassified, subject's score on the primary discriminant function is computed and compared with the mean scores on this discriminant function for each of the k groups. If the true population proportions of the various kinds of persons who constitute these k groups are identical, and if the "cost" of making an error of classification is identical for all of the various kinds of error that might be made, then the subject is assigned to (given the same classification as) that group whose mean discriminant score is closest to the subject's discriminant score. Where population proportions differ and/or not all errors are equally costly, the "cutoff point" for deciding between group i and group j is moved closer to the mean discriminant score for the more common group or away from the group that is most costly (e.g., the "mentally ill" group or the "habitual criminal" group). This is accomplished by applying Equation (3.12) to the cutting point separating any two adjacent mean discriminant function scores.

There is an added complication in multiple discriminant analysis, in that there are $s = \min(p, k - 1)$ discriminant functions, not just one. The jth discriminant function is that linear combination of the original variables that maximally discriminates among the groups (in the sense that it produces the highest possible univariate F ratio), subject to the condition that it be uncorrelated with the first j -1 discriminant functions. The jth discriminant function is of course given by the characteristic vector associated with the jth largest characteristic root of $\mathbf{E}^{-1}\mathbf{H}$, and it need be considered an indication of a true source of differences among the groups only if it is statistically significant. The first degree-of-freedom parameter for testing the significance of the jth characteristic root is s $= \min(k - j, p - j + 1)$ = the value of s for the first root minus j-1; the other two degree-of-freedom parameters remain the same. (Harris (1973b, 1976b) pointed out that the resulting tests of the significance of the second, third, . . ., sth roots are conservative.] Further, by the logic of these tests, if the jth largest characteristic root is declared nonsignificant, so must all the remaining roots, $\lambda_{j+1}, \lambda_{j+2}, \cdots, \lambda_s$. See section 4.5 for further comments on procedures for determining the number of statistically significant roots.

The existence of multiple discriminant functions opens a Pandora's box of further problems. For one thing, the close connection between the discriminant function and the classification problem in the two-sample case is loosened somewhat when we have three or more groups. Use of the primary discriminant function to classify sampling units does not necessarily lead to the best classification statistics, and a host of alternative classification rules has been suggested, including:

1. Classifying subject i into that group k for which $D_{ik}^2 = (\mathbf{X}_i - \overline{\mathbf{X}}_k)'\mathbf{S}_c^{-1}(\mathbf{X}_i - \overline{\mathbf{X}}_k)$ = the squared Euclidean distance of subject i's vector of scores from the mean vector for group k is minimized.

2. The same classification rule as rule 1 but with $\mathbf{X}_i$ and $\overline{\mathbf{X}}_k$ expressed in terms of scores on the s discriminant functions.

3, 4. Use of $D_{ik}^2 - 2\ln(p_k)$, where p_k is the population proportion of subjects belonging in group k.

5–12. Rules equivalent to rules 1–4 but expressed either as the likelihood $L_{i/k}$ of sampling observation i from population k given multivariate normality or as the corresponding "tail" area;

13–24. Any of rules (1-12) with $\mathbf{S}_k$ substituted for $\mathbf{S}_c$ and so on.

Huberty (1975, 1984) did an excellent job of discussing the relationships among these various classification rules, including the results of Monte Carlo studies of their relative efficiency. The main point is that no single classification rule yields superior classification statistics (e.g., overall percentage of the sample that is correctly classified) for all data sets. Use of the primary discriminant function by itself produces the best classification results in some cases, whereas use of distance measures in the space of all s discriminant functions produces a higher percentage of correct classifications in other cases, and so on. In addition, the reader should keep in mind the caveat of section 3.5 that use of priors (values of p_k) other than N_k/N will almost always lower the overall percentage of correct classification of the sample data, even though use of Equation (3.13) may well show such use to provide greatly improved classification statistics when applied to the population to which the p_k refer.

Having $s > 1$ also means that the D^2 statistic reported by many discriminant analysis programs—which in the two-sample case was simply T^2 divided by the constant $K = N_1N_2/(N_1 + N_2)$—is no longer easily translated into the most appropriate Manova test statistic, the gcr. However, Huberty (1975, p. 563) pointed out that use of the chi square test based on

$$D^2 = \sum_j^k T_j^2 \, ,$$

where T_j^2 is the single-sample T^2 statistic for testing the H_o (that group j's population mean vector is equal to the grand mean vector averaged across groups), is equivalent to using Hotellings's trace statistic, $\Sigma\lambda_i$, as our Manova criterion.

4.5 GREATEST CHARACTERISTIC ROOTS VERSUS MULTIPLE-ROOT TESTS IN MANOVA

The general technique of reducing Manova to a univariate problem by computing a univariate F on the "optimal" linear combination of the p original variables led us naturally to significance tests based on the magnitude of the greatest characteristic root of $\mathbf{E}^{-1}\mathbf{H}$. The original development of multivariate analysis of variance (Wilks, 1932) took the alternative approach of computing the ratio of the "likelihood" of the data under the assumption of population mean vectors identical to the corresponding sample mean vectors, to the likelihood of the observed data under the assumption of no differences among population mean vectors. This leads to ***Wilks' lambda*** statistic, also referred to as the U-statistic (thereby avoiding Greek letters),

$$\Lambda = U = \frac{|\mathbf{E}|}{|\mathbf{H}|+|\mathbf{E}|} = \frac{1}{|\mathbf{E}^{-1}\mathbf{H}+\mathbf{I}|} = \left(\prod_{i=1}^{s} \lambda_i{}^*\right)^{-1},$$

where $\lambda_i{}^*$ is the ith characteristic root of $\mathbf{E}^{-1}\mathbf{H}+\mathbf{I}$. However, $\lambda_i{}^*$ is a solution to the matrix equation

$$[(\mathbf{E}^{-1}\mathbf{H}+\mathbf{I}) - \lambda_i^*\mathbf{I}]\mathbf{a} = \mathbf{0},$$

that is, of

$$[\mathbf{E}^{-1}\mathbf{H} - (\lambda_i^* - 1)\mathbf{I}]\mathbf{a} = \mathbf{0},$$

where $\lambda_i = \lambda_i{}^* - 1$ is the ith characteristic root of $\mathbf{E}^{-1}\mathbf{H}$. Thus greatest characteristic root tests and Wilks' lambda criterion are closely related, with

$$U = \left[\prod_{i=1}^{s}(\lambda_i + 1)\right]^{-1} = \prod_{i=1}^{s}(1-\theta_i) \tag{4.13}$$

where λ_i is defined as above.

Most texts and most computer programs employ Wilks' lambda criteria for significance tests in Manova. The reasons for this preference are at least fourfold:

1. Historical precedent.
2. Although the exact distribution of U is extremely complex, fairly good approximations to this distribution in terms of readily available chi square and F-table entries exist.
3. Under certain circumstances, especially for data in which the successive characteristic roots of $\mathbf{E}^{-1}\mathbf{H}$ are nearly equal, statistical tests based on U are more powerful than greatest characteristic root (gcr) tests.
4. Determinants are easier to compute than are characteristic roots.

In addition, Olson (1974, 1976) reports that Wilks' lambda is often more robust against violations of the assumptions of multivariate normality and homogeneity of covariance matrices than is the gcr statistic. However, in most of these situations a third criterion, the Pillai trace statistic $V = \sum_{i=1}^{s}\theta_i$ (also obviously a multiple-root criterion) is even more robust, so that users concerned about robustness are likely to prefer V over U. Chi-square and F approximations to U, V, and the Hotelling trace statistic (the sum of the λ_i) were provided by Jones (1966) and are sufficiently well known to have been incorporated into the SPSS MANOVA program.

This book, however, concentrates on the gcr approach to significance testing because of its greater heuristic and didactic value and because:

1. It leads directly and naturally to multiple comparison procedures on both an a priori and ad hoc basis.

2. It makes the connection between discriminant analysis and Manova much clearer, whereas texts employing *U* statistics have to introduce gcr procedures as a separate body of techniques for discriminant analysis.

3. The power of gcr tests is greater than that for any of the multiple-root tests when, in the population from which our samples are drawn, the first characteristic root is much larger than for the other roots; and the two approaches have nearly identical power for large sample sizes. (See Morrison, 1976, pp. 223-224, for a summary of these power comparisons.)

4. Closely related to point 2, the source of any statistically significant differences among the groups is much more readily described (via the statistically significant discriminant function or functions and simplified versions thereof) when gcr techniques are employed. It is to this multiple comparison problem—that of identifying the source(s) of significant Manova effects—that we now turn.

4.5.1 "Protected" Univariate Tests

The most common approach (Bock, 1975; Finn, 1974; Gabriel, 1969; Tabachnick & Fidell, 1997) to following up a statistically significant multiple-root test is to perform univariate analyses of variance on each separate dependent measure, with all the potential for compounded Type I error rates that entails and with no hope of detecting crucial *combinations* of variables. As Ryan (1980) and Bird and Hadzi-Pavlovic (1983) point out, the apparent assumption that this ***protected univariate*** approach sets alpha for each of the follow-up tests to at most the alpha employed for the overall test is true only when the overall null hypothesis is true. If the particular contrast on the particular linear combination of measures being examined in a follow-up test has a population value of zero whereas other contrasts on that or other measures differ from zero, the overall probability of falsely rejecting one or more true H_0s under a protected, two-stage procedure can be very much higher than the alpha for the overall test. For instance, the logic used to justify protected tests would permit adoption of the second-stage decision rule: declare $\psi = \Sigma c_j\mu_j \neq 0$ ($\sum c_j \overline{X}_j$ statistically significant, where X is any of the p dependent variables) if and only if $|\sum c_j \overline{X}_j| > 0$. Such a test has a Type I error rate equal to the power of the overall test and is susceptible to a strategy whereby a very weak effect can be given a very high probability of being declared significant simply by including an uninteresting but reliable effect (e.g., having subjects rate the sex of male vs. female stimulus persons or including a cyanide-ingestion group in a clinical study of treatment effects) so as to insure a powerful overall test. In short, our consideration of multivariate test criteria must consider, not just their performance as tests of the overall H_0, but their value as bases for post hoc tests of specific contrasts. Such comparisons strongly favor the gcr test over multiple-root tests.

4.5.2 Simultaneous Test Procedures and Union Intersection

For instance, Gabriel (1969, 1970) recommends, as a general approach to conducting specific comparisons, the use of a ***simultaneous test procedure*** (STP) based on whichever overall test is employed. An STP takes as its critical value that magnitude of the F (or ratio of sums of squares) for contrast that is just large enough to produce sufficient evidence to reject $H_{0,ov}$. Thus, for instance, the U-based STP considers $\Sigma c_j \overline{W}_j$ (where W is any single dependent variable or any linear combination of the p dependent variables) statistically significant (rejects $\sum c_j \mu_{W,j} = 0$) if and only if $1 - SS_{contr,W}/(SS_{contr} + SS_{w,W}) = SS_{w,W}/(SS_{contr,W} + SS_{w,W}) < U_{crit,}$ the critical value of $U = (1 - \theta_1)(1 - \theta_2) \cdots (1 - \theta_s)$. ($SS_{contr,W}$ and $SS_{w,W}$ in the preceding formulae refer to the SS for contrast and for within-cells error, respectively, as computed on the particular linear combination, W, of the dependent measures that is under examination in this specific comparison.)

All STPs are ***coherent***; that is, if any specific contrast is statistically significant, so is the overall test. However, only the gcr-based STP is ***consonant;*** that is, it has the property that if the overall test is statistically significant, there exists at least one contrast on some linear combination of the dependent variables (specifically, the maximized contrast on the first discriminant function) that is statistically significant by the post hoc criterion,

Because of their nonconsonance, each of the STPs based on multiple-root tests can sometimes leave us in the awkward position of having rejected the overall null hypothesis, and having thus concluded that the mean vectors differ, but being unable to specify any particular dimension (linear combination of the dependent variables) along which the groups differ significantly. Moreover, the critical value of F_{contr} for specific contrasts based on multiple-root tests is always larger than the corresponding gcr-based critical value, and the power of gcr-based tests of specific contrasts is therefore higher than that provided by any other STP (Bird & Hadzi-Pavlovic, 1983).

What, then, accounts for the continuing, overwhelming popularity of multiple-root tests? This popularity is partly the result of the fact that most authors have focused entirely on the properties of multivariate criteria as tests of the overall H_0, ignoring the need for specific comparisons to establish ***how*** the groups differ; partly the result of the gcr statistic's poor robustness, mentioned earlier; and partly the result of widespread use of a ***partitioned-U*** procedure that gives the mistaken impression of providing consonance between the overall test and follow-up tests.

4.5.3 Invalidity of Partitioned-U Tests of Individual Roots

The best known of the approximations to U's sampling distribution is Bartlett's (1939) chi-square approximation based on the natural logarithm of U. Specifically,

$$\chi^2 = -m(\ln U), \tag{4.14}$$

where $m = N - k - (p - k)/2$ and ln U is the natural logarithm of U, is distributed as chi square with $p(k - 1)$ degrees of freedom when N is large. This chi-square approximation can be decomposed into additive components, because

$$\chi^2 = -m \ln U = -\sum_{i=1} [m \ln(1+\lambda_i)] = \sum [-m \ln(1-\theta_i)], \tag{4.15}$$

where λ_i is the gcr statistic associated with the ith largest characteristic root—including that associated with the largest root itself. This makes it tempting to speculate that the individual components of U would be chi-square distributed and thus provide significance tests for the individual discriminant functions including that associated with the gcr. In fact, Tatsuoka (1972) and Kendall and Stuart (1961) each asserted that, under the null hypothesis, $-M \ln(1 - \theta_i) = -M \ln(1 + \lambda_i)^{-1}$ is distributed as a chi-square variate having $p + (k - 1) - 2i + 1$ degrees of freedom. However, as pointed out to the present author by W. N. Venables (personal communication, 1974), Lancaster (1963) proved this assertion to be false with respect to λ_1, and Kendall and Stuart (1968) retracted their earlier assertion. Harris (1974) reported that Monte Carlo runs based on the case $s = 10$, $m = 0$, $n = 100$ yielded true Type I error rates for the test of λ_1 based on the first component of $-M \ln U$ of 61% for a nominal α of .05 and 24% for a nominal α of .0l! The problem is that the individual ***population*** roots do indeed generate distributions of $-M \ln(1 + \lambda_i)$ having the specified characteristics, but there is no way of knowing which of the p population roots has generated the ith largest ***sample*** value of λ_i.

This attempt (abortive though it is) to use the individual components of U to test the individual sample values of λ_i does, however, make it clear that those cases in which U leads to rejection of the overall null hypothesis but the gcr test does not are precisely the cases in which no single discriminant function discriminates significantly among the various groups. In such a case, we are in the unenviable state of knowing that *something* beyond random fluctuation is "going on" in our data, but being unable to specify what that "something" is, because no statement about differences among the groups on any single variable or linear combination of variables can be supported by a significance test. This state of limbo can be avoided (ostrich-like?) by concentrating on the gcr approach.

Wilks' lambda is a valid test of the overall null hypothesis for those researchers who are uninterested in saying anything about *how* their groups differ. It could also be used as a preliminary to testing individual discriminant functions via gcr tests—although this would entail a loss of power as compared to simply going directly to the gcr tests of λ_1, λ_2, and so on. What ***is*** objectionable on logical grounds is the procedure many authors of multivariate texts (e.g., Cooley & Lohnes, 1971; Overall & Klett, 1972; Tatsuoka, 1971, 1972; Tatsuoka & Lohnes, 1988) recommend for testing the significance of the individual discriminant functions. (A similar procedure for testing the separate pairs of canonical variates is suggested by each of these authors.) This procedure tests the *residuals* after removing the effects of the first, second, . . ., $(s - 1)$st discriminant functions. This is accomplished by including only the higher values of i in the summation of Equation

(4.15). As soon as the residual after removing $r \geq 1$ characteristic roots fails to achieve significance, it is claimed that each of the first r discriminant functions is statistically significant. In particular, the first discriminant function is never tested directly, with the argument being that if the overall test is significant, then at least the first discriminant function must also be significant. This is analogous to claiming that if we obtain a significant overall F in a one-way Anova, then at least one of the pairwise differences (in particular, the largest mean minus the smallest) must be statistically significant—a claim that is not true.

The SPSS incarnation of the partitioned-U procedure (the "dimension reduction analysis" table it produces by default) now recognizes this logical flaw by labeling the overall U test as a test of roots 1 through s, the test of the first residual as a test of roots 2 through s, and so on, thus making it clear that these do *not* establish the statistical significance of any individual root and its associated discriminant function.

What *can* such dimension reduction analysis do for us? If the residual U_is had the chi-square distributions claimed for them, they might (a) serve as preliminary tests to determine whether further testing of individual roots is worthwhile at each stage or (b), as Mendoza, Markos, and Gonter (1978) have suggested, they might serve to identify the dimensionality of the discriminant function space by the number of successive roots that must be removed before the residual becomes statistically nonsignificant. The latter application looks especially attractive in light of the conservatism (mentioned in section 4.4) of the gcr-based tests of θ_2 against $\theta_\alpha(s - 1, m, n)$, θ_3 against $\theta_\alpha(s - 2, m, n)$, and so on. Unfortunately, the residual U_is do *not* have the claimed distributions, and they appear to be just as conservative as the step-down gcr tests.

Harris (1976b) demonstrated the conservatism of the residual U_i tests in the case in which the overall H_O is true—that is, where all s population roots, θ_i (the population canonical correlations between group-membership and dependent variables), are zero. Mendoza and associates (1978) then followed a long-standing tradition by treating Bartlett's (1939) statement that the chi-square distributions he derived for the residual U_is were strictly valid only if all the preceding θ_is were 1.0 as equivalent to requiring that the deleted roots be merely nonzero. However, to quote from Harris (1985a), who was actually discussing the equivalent application to canonical correlation (chap. 5):

> The results of some Monte Carlo runs suggest that ρ_1 [the square root of θ_1] may have to be quite large (> .4) before the conservatism of the residual-U test becomes negligible. Six simulation runs were conducted, each based on 600 independent random samples of size 117 from a 15-variable multivariate normal population. The six populations employed all had $\rho_2 = \rho_3 = \ldots = \rho_7 = 0$ for the relationship between the first 7 and the last 8 variables. They differed in that ρ_1 = .001, .100, .200, .400, .600, or .990. None of the 600 samples yielded a statistically significant value of U_1 at the .05 level for ρ_1 = .001, .1, or .2, and only 9 samples yielded significance when ρ_1 = .4 ($p < .001$ for the difference between the

> observed number of significant results and the expected number of 30 in each case). The $\rho_1 = .6$ and .99 runs yielded 27 and 33 nominally significant U_1 values, respectively. It might be objected that, because the test of U_1 is carried out only if the overall U test yields significance, we needn't be concerned about the consequences of such low values of ρ_1. This would be incorrect. Because U is based on all 7 roots, it is possible for the overall test to yield significance at the .05 level when r_1^2 [the squared canonical R for the sample] is as low as .0740—a value lower than the minimum value (.0879) observed in the 1200 samples from the $\rho_1 = .001$ and .100 cases of the present runs. Clearly, continued faith in the validity of the partitioned-U procedure for testing residuals is badly misplaced. (p. 690)

The gcr test of θ_2 proved almost identically conservative—as should be expected, because both tests involve acting as if θ_1 were an unbiased estimate of ρ_1^2. One is, then, justified in using either statistic for tests of dimensionality. (I would argue, however, that this is not a terribly useful test, because in any real situation we know a priori that *no* population root is truly zero, the only question being whether we have gathered sufficient data to begin to specify the nature of the corresponding discriminant functions and the differences among our conditions along these dimensions.) However, as soon as you consider the problem of identifying what those dimensions are (i.e., of specifying linear combinations of the dependent variables that differ significantly among your groups or conditions), the gcr tests become an indispensable part of your testing procedure, and any use of residual U_i tests only serves to decrease the power of this search.

Summarizing sections 4.5.1 through 4.5.3 crudely: The gcr approach "speaks up" (rejects the overall null hypothesis) only if it can then go on to provide information about the source(s) of the differences among the groups. Multiple-root tests, on the other hand, speak up even when they have nothing useful to say.

4.54 Simplified Coefficients as a Solution to the Robustness Problem

This leaves us, however, with the problem of the gcr test's sometimes appalling lack of robustness in the presence of violations of multivariate normality and/or homogeneity of covariance matrices. (Olson, 1974, for instance, reports some instances in which the actual alpha for a gcr test of the overall null hypothesis at a nominal alpha of .05 is close to .60.) A very promising answer to this problem is provided by Bird and Hadzi-Pavlovic (1983). These authors pointed out that a statement that the maximized contrast on the discriminant function [obtained by taking as our contrast coefficients $c_j = n_j\,(\overline{D}_j - \overline{\overline{D}})$, where $\overline{D}_j$ is the mean of group j on the discriminant function] is statistically significant is by itself unlikely to be very useful, because our substantive interpretation of the discriminant function and of the contrast coefficients is unlikely to account for the multidecimal details of these coefficients. Instead, we must consider our task incomplete until we have tested the statistical significance of a simplified contrast (usually the difference between the average of one subset of groups vs. the average of another subset of groups) on a simplified discriminant function (usually one in which each coefficient is 0, 1, or -1). The restrictions on **c** and **a** (in H_0 : $\mathbf{c}'\mu\mathbf{a} = 0$) implied by or explicitly stated as a part of this simplification process reduce the size of the set of contrasts in our search and thus also decrease the probability of rejecting our null hypothesis. Bird and Hadzi-Pavlovic provided Monte Carlo data that show that this restricted-comparison procedure provides quite acceptable protection against the inflation of Type I error rate associated with the unrestricted gcr STP when its assumptions are violated, at the cost of minimal loss of power or diminution of actual alpha when assumptions are met.

This introduces a degree of nonconsonance into our test procedure, because it is entirely possible that, when the gcr test just barely exceeds its critical value, our search for simplified **c** and **a** vectors that retain statistical significance may fail. It is best to consider the overall test nonsignificant in such cases, adopting the decision rule that our overall H_0 is rejected if and only if at least one statistically significant *interpretable* specific contrast can be identified. The gcr test of the overall null hypothesis then becomes superfluous except as an indication as to whether there is any point to beginning our search for an interpretable contrast—and also, of course, as a base for the critical value for the post hoc tests involved in that search.

Bird and Hadzi-Pavlovic's data also demonstrate the abysmally low power of multiple-root-based STPs, relative to the gcr-based STP, even under conditions in which the multiple-root test's power as a test of the overall null hypothesis exceeds that of the gcr test by a considerable margin.

4.5.5 Finite Intersection Tests

We pointed out in section 1.1.1 that basing our overall test on whether one or more Bonferroni-adjusted specific comparisons reach statistical significance has the same consonance and coherence properties as do union-intersection tests. Such a procedure can thus be labeled a ***finite intersection test,*** because it requires a priori specification of a finite number of comparisons, as opposed to the infinite number of comparisons (a priori or post hoc) considered by union–intersection tests.

Bonferroni adjustment is, however, based on an inequality whose complete generality carries with it the penalty of being quite conservative when the various comparisons are highly intercorrelated. The Bonferroni bounds on a set of specific comparisons can therefore be improved on and—mathematical complexities permitting—made exact if the specific nature of the intercorrelations among comparisons can be taken into account. For instance, in a univariate Anova in which one plans to conduct all six pairwise comparisons among four experimental conditions (but not more complex contrasts), the Bonferroni inequality yields a critical value for each pairwise *t* ratio of $t_{\alpha_{\exp}/6}(4n-4)$, which equals 3.75 for n = 6 subjects per group and an $\alpha_{\exp}$ of .05. If instead we use the Studentized range test as our overall test and its critical value (which is based on the sampling distribution of the difference between the largest and the smallest of a set of independent sample means) as the critical value for each of our pairwise *t*s (thereby adopting Tukey's, 1953, HSD test procedure), our critical value becomes 3.37. (Harris, 1994, Table 3.2, provides such comparisons for a range of parameter values and also compares Dunnett's test of the difference between a control group's mean and each of the other group means to the corresponding Bonferroni-adjusted test.)

Basing our finite-intersection tests on exact sampling distributions tailored to the specific nature of our a priori set of comparisons can thus yield lower critical values (and therefore higher power) than the corresponding tests based on the general-purpose Bonferroni inequality. P. R. Krishnaiah (e.g., 1965, 1969, 1979) and his colleagues (e.g., Cox, Krishnaiah, Lee, Reising & Schuurman in Krishnaiah, 1980; Mudholkar & Subbaiah, 1980) extended greatly the sets of comparisons for which exact critical values, or at least bounds that should be more precise than the Bonferroni inequality, are available. These researchers have developed, among other things, a T^2 test (a multivariate extension of the Tukey HSD test in which each pair of mean *vectors* is compared) and an F_{max} test (in which an F test is conducted on each of p correlated dependent variables).

For instance, if we wish to restrict our attention to pair-wise comparisons among 6 groups of 6 subjects each in terms of their vectors of means on 2 dependent variables, we could compute the two-sample T^2 for each of the 15 comparisons involved (using, however, the pooled within-cells covariance matrix $\mathbf{S}_c$ for computation of each T^2) and then compare each resulting T^2 to one of the three following critical values:

1. The gcr-based critical value (which controls for *all* contrasts among the mean vectors), calculated from the lower right-hand corner of Table 4.3 as

$$30 \cdot \frac{\theta_{.05}(5,1,13.5)}{1-\theta_{.05}(5,1,13.5)} = 30(.5841/.4359) = 38.823.$$

2. The Bonferroni-adjusted T^2 critical value of

$$\frac{2(30)}{30-2+1} F_{.05/15}(2,29) = (60/29)(6.9884) = 14.46.$$

3. The T^2_{max} critical value. Cox and associates (Cox, Krishnaiah, Lee, Reising, & Schuurman, 1980, Table 29) indicated that the relative efficiency of the gcr test, compared to T^2_{max}, is .62 (computed as the ratio of the two critical values for $\sqrt{T^2}$), so a lower bound on the T^2_{max} critical value would appear to be .615 times the gcr-based critical value, which equals 14.7. Because this critical value cannot logically be larger than the Bonferroni-adjusted critical value, this probably indicates some small discrepancy in the present computation of the gcr value (obtained by running the GCRCMP program in Appendix C for n = 13 and 14) as compared to the value used by Cox et al.

The relative efficiency of these three classes of tests—gcr-based union-intersection test, Bonferroni-based finite intersection test, and more specialized finite-intersection test—seen in the above example is typical of the comparisons I have examined so far. Going to Bonferroni-adjusted tests based on a restricted number of specific comparisons does indeed afford a considerable reduction in critical values as compared to a fully multivariate post hoc critical value. However, the further improvement gained from using one of Krishnaiah's more specialized finite-intersection tests is apt to be so small as to be (as in the preceding example) undetectable in the presence of round-off errors or reliance on various approximations in deriving the more specialized distributions. Given that tables of critical values for such distributions as that of T^2_{max} and of F_{max} are available only for a small set of degree-of-freedom parameters and sometimes (as in the case of F_{max}) only for special cases such as that of equally correlated dependent variables, the reader is probably well advised to employ the more flexible Bonferroni-adjusted finite intersection tests when considering an a priori set of comparisons. However, we can anticipate that specialized finite intersection tests will prove an increasingly attractive alternative to general-purpose Bonferroni adjustment as Krishnaiah's very active research program continues and as computer programs for exact finite-intersection tests become available.

4.6 SIMPLE CASES OF MANOVA

We now give some formulae for simple cases of Manova. Note that when both **H** and **E** are diagonal matrices—which corresponds to having mutually uncorrelated dependent measures—the greatest characteristic root is equal to the largest univariate *F*.

In what follows,

$$e_{ij} = \text{the } (i,j)\text{th entry of } \mathbf{E}; \quad h_{ij} = \text{the } (i,j)\text{th entry of } \mathbf{H};$$

and each λ must be "converted" either to

$$F = \frac{n+1}{m+1}\lambda \text{ with } 2m+2 \text{ and } 2n+2 \; df \text{ (when } s = 1)$$

or to

$$\theta = \lambda / (\lambda + 1)$$

with degrees-of-freedom parameters *s*, *m*, and *n* in order to test for statistical significance.

$p = 2$**:**

$$\lambda = \frac{h_{11}e_{22} + h_{22}e_{11} - 2h_{12}e_{12} \pm \sqrt{(h_{11}e_{22} + h_{22}e_{11})^2 + 4(h_{12}e_{11} - h_{11}e_{12})(h_{12}e_{22} - h_{22}e_{12})}}{2(e_{11}e_{22} - e_{12}^2)}$$

$$= \frac{R_1 + R_2 \pm \sqrt{(R_1 - R_2)^2 + 4h_{12}^2/(e_{11}e_{22})}}{2} \quad \text{when } e_{12} = 0,$$

where R_i is the "stripped" univariate *F* ratio computed on dependent measure *i*, that is,

$$R_i = \frac{k-1}{N-k} F_i.$$

***p* = 3:** λ is obtained from the cubic equation

$$\lambda^3 - (1/d)\left(\sum_{i=1}^{3}\sum_{j=1}^{3} c_{ij}h_{ij}\right)\lambda^2 + (1/d^2)\left(\sum_{i=1}^{3}\sum_{j=1}^{3} g_{ij}\right)\lambda + D = 0$$

where

$$c_{ij} = e_{ik}\,e_{jk} - e_{ij}\,e_{kk}\,; \quad c_{ii} = e_{jj}e_{kk} - e^2_{jk}\,;$$

$$g_{ij} = (h_{kk}h_{ij} - h_{ik}h_{jk})(c_{kk}c_{ij} - c_{ik}c_{jk});$$

$$d = e_{11}e_{22}e_{33} + 2e_{12}e_{13}e_{23} - \sum_{i=1}^{3} e_{ii}e_{jk}^2\,;$$

and

$$D = (1/d^3)\left\{\begin{array}{l} f_1[l_1(h_{22}h_{33} - h_{23}^2) + l_2(h_{22}h_{13} - h_{13}h_{23}) + l_3(h_{12}h_{33} - h_{12}h_{23})] \\ + f_2[l_1(h_{13}h_{23} - h_{12}h_{33}) + l_2(h_{11}h_{23} - h_{12}h_{13}) + l_3(h_{13}^2 - h_{11}h_{33})] \\ + f_3[l_1(h_{12}h_{23} - h_{22}h_{13}) + l_2(h_{12}^2 - h_{11}h_{22}) + l_3(h_{11}h_{23} - h_{12}h_{13})] \end{array}\right\}$$

where

$$f_i = \sum_{j=1}^{3} h_{ij}c_{ij}; \quad l_1 = c_{22}c_{33} - c_{23}; \quad l_2 = c_{22}c_{13} - c_{13}c_{23}; \text{ and } l_2 = c_{33}c_{12} - c_{13}c_{23}.$$

(See Digression 3 for procedures for solving cubic equations.)

When $e_{12} = e_{13} = e_{23} = 0$, the above reduces to

$$\lambda^3 - (\sum R_i)\lambda^2 + \left[\sum_{i=j}\sum\left(R_iR_j\frac{h_{ij}^2}{e_{ii}e_{jj}}\right)\right]\lambda - R_1R_2R_3 - \frac{2h_{12}h_{13}h_{23} - h_{11}h_{23}^2 - h_{22}h_{13}^2 - h_{33}h_{12}^2}{e_{11}e_{22}e_{33}}.$$

If in addition $k = 3$, so that s (the rank of $\mathbf{E}^{-1}\mathbf{H}$) = 2,

$$\lambda = \frac{(\sum R_i)^2 \pm \sqrt{(\sum R_i)^2 - 4S_2}}{2},$$

where

$$S_2 = R_1R_2 + R_1R_3 + R_2R_3 - \sum_i\sum_j \frac{h_{ij}^2}{e_{ii}e_{jj}}.$$

Also, if $k = 2$, then the single nonzero characteristic root is given by

$$\lambda = \frac{k-1}{N-k}T^2 = \sum R_i\,.$$

Outcome Scores and Corresponding Group Means Uncorrelated: If

$$h_{ij} = 0 \quad \text{and } e_{ij} = 0 \text{ for all } (i,j) \text{ pairs for which } i \neq j,$$

then

$$\lambda_i = R_i \quad \text{and } \mathbf{a}_i' = [0, 0, \ldots, 1, \ldots, 0],$$

that is, the characteristic roots are simply the "stripped" univariate *F*s, and the characteristic vector associated with λ_i has zeros everywhere except in the *i*th position. Thus the greatest characteristic root is the largest *F* computed on any *single* outcome measure with the corresponding "discriminant function" simply being that most discriminating single variable. Note, however, that it is not sufficient for **E** alone to be diagonal- - as the rather complex expressions for the $e_{ij} = 0$ cases already described clearly demonstrate. The point is that both **E** *and* **H** are covariance matrices whose off-diagonal entries reflect the intercorrelations among the dependent measures, and each could in fact be converted to an intercorrelation matrix by dividing its (*i, j*)th entry by the square roots of its *i*th and *j*th main diagonal entries: **E** reflects the tendency for an individual subject who scores high on one dependent measure to also score high on other dependent measures, whereas **H** reflects the tendency for those groups that have high *mean scores* on one measure to also have high means on other measures. Thus **E**/(*N* - *k*) is a ***within-groups*** covariance matrix, whereas **H**/(*k* - 1) is a ***between-groups*** covariance matrix. Grice (1966) pointed out that these two sources of correlation are logically—and often empirically—quite distinct. Thus, for instance, the chronic level of anxiety of the subjects who are assigned to a given treatment may be not at all useful in predicting individual differences in response to the treatment, whereas applying different treatments to each of several groups may produce *mean* levels of anxiety that are highly predictive of the ordering of the groups in terms of *mean* responsiveness to the various treatments. All of which is simply to say that the researcher will seldom encounter situations in which both **H** and **E** are uncorrelated—even when the variables have deliberately been submitted to a set of transformations, such as principal components analysis, specifically designed to yield uncorrelated variables. The chances of being able to employ the special formulae for uncorrelated variables (i.e., for **H** and **E** both diagonal) are even slimmer when we consider the usually very small number of degrees of freedom (*k* - 1) on which each of the "covariance" estimates in **H** is based. A sample correlation coefficient based on only two pairs of observations must have a value of either + 1 or - 1 (or be undefined if either s_1^2 or $s_2^2 = 0$), so that if only two groups are employed, each off-diagonal entry in **H** (which is the sum across groups of the cross products of deviations of the means on two measures from the grand means for those two measures) will be either $+\sqrt{h_{ii}h_{jj}}$ or $-\sqrt{h_{ii}h_{jj}}$. When $k > 2$, the expected value of the correlation between the *k* sample means of variable *i* and the *k* sample means of variable *j* is still $\sqrt{1/(k-1)}$, even if the true correlation between population means is zero. Thus, unless k is quite large, it is extremely unlikely that **H** will even come close to being a diagonal matrix.

However, in the case in which a single-degree-of-freedom effect is being tested (as in a 2 x 2 x ··· factorial design), a diagonal **E** matrix *is* sufficient to produce considerable simplification. In this case the single nonzero root λ_1 will equal the sum of the stripped F ratios for the individual dependent variables, and the z-score discriminant function coefficient for dependent variable i will be proportional to the (signed) t test for the effect on that dependent variable.

H *Proportional to* E: If **H** = c**E**, then R(**a**)—the stripped F ratio computed on the linear combination **Xa**—equals c, no matter what our choice of **a** is (with the single exception of **a** = **0**). We thus have s identical values of λ_i, and we can take as discriminant function coefficients for the s identical roots any orthogonal set. Although perfect proportionality is very unlikely for sample data, this case does serve to demonstrate that we can experience a gain from Manova's maximization algorithm (i.e., a maximized F ratio larger than any univariate F ratio) only if the pattern of between-group correlations differs somewhat from the pattern of within-group correlations.

4.7 HIGHER ORDER ANOVA: INTERACTIONS

Anything approaching a comprehensive description of higher order Anova designs—n-way factorial, mixed, hierarchical, and so on—would require an entire textbook unto itself—and indeed several excellent ones (e.g., Winer, 1971, Kirk, 1982, and Harris, 1994, he says modestly) exist. The present section therefore confines itself to a discussion of the major new problem that arises when we move from one-way to higher order univariate analysis of variance: the testing and interpretation of interactions. Such a discussion is necessary for two reasons: (a) The reader needs to be "inoculated" (McGuire & Papageorgis, 1961) against some of the very common but very poor approaches to interpreting interactions he or she will find in the literature; and (b) your growing familiarity with multivariate statistics makes available some tools that can facilitate the interpretation of interactions, even in univariate Anova designs.

By far the most common approach to interpreting a significant interaction is to take the significant overall test as license to describe the table of means in any way one pleases, with no attempt to determine whether one's interpretation (a) is in fact a description of interaction, rather than being confounded with one or both main effects, and (b) represents a statistically reliable aspect of the interaction. Such an approach is closely analogous to the "protected univariate" approach to following up a significant overall Manova test that was excoriated in section 4.5. If the interaction has more than one degree of freedom (*df*), *any* explanation of the interaction should be supported by a test of the single-*df* interaction contrast it implies and by a statement of the percentage of the total SS_{int} accounted for by that contrast. Even in the 2 x 2 (df_{int} = 1) case, it is easy to slip into describing the interaction as essentially one particular cell's being different from the other three; this explanation is confounded with both main effects. If you find that such explanations appeal to you, this is a good sign that it is time to abandon the traditional breakdown into main effects and interactions and instead treat your design as a one-way Anova followed up by whatever contrasts among the *a* (levels of factor *A)* x *b* (levels of *B*) cells seem most meaningful. The critical value for the various post hoc contrasts will in this case be $(ab - 1)F_\alpha (ab - 1, df_{err})$, rather than the usual $[(a - 1)(b - 1)] \cdot F_\alpha([(a - 1)(b - 1)], df_{err})$ appropriate for contrasts required to be orthogonal to both main effects.

As the discussion so far implies, this book extends to interactions its bias toward Scheffé contrast procedures, as opposed to multiple comparison procedures that are restricted to pairwise comparisons. A table of means specifying via subscripts which pairs of means differ significantly from each other is singularly useless in interpreting interactions. An interaction is, at a bare minimum, a difference between two differences (labeled by Johnson, 1976, as a ***tetrad difference*** in conformity to factor analytic terminology) and can therefore be got at only very indirectly by descriptions of pairwise differences, one at a time.

Assuming that you have decided to stick with a main-effects-and-interactions breakdown and are thus confining your explanations to "true" interaction contrasts, you can satisfy this restriction with any set of contrast coefficients that (a) sum to zero across each row and each column and (b) have a sum of cross products of zero with any main-effect contrast. By far the simplest and most reliable way to insure that these conditions are met will be to construct your interaction contrast as a contrast of contrasts, that is, by letting $I_{ij} = a_i b_j$, where I_{ij} is the contrast coefficient to be applied to the mean for cell (*i*, *j*); a_i is the row-*i* coefficient for a contrast among the *a* levels of factor *A;* and b_j is the column-*j* coefficient for a contrast among the *b* levels of *B*. The problems with this approach are twofold: (a) how to choose the row and column contrasts from which to construct the contrast of contrasts; and (b) the fact that there may be no contrast of contrasts that accounts for more than a small portion of SS_{int}.

Your knowledge of characteristic roots and vectors can be of assistance in solving the first problem. As was implicit in Gollob's (1968) Fanova model and as applied explicitly to the present problem by Bradu and Gabriel (1974), Harris (1976c), and Johnson (1976), that contrast of contrasts (c.o.c.) that has the highest possible sum of squares for contrast (and thus accounts for the highest possible portion of the total SS_{int}) can be found by taking the characteristic roots and vectors of **ZZ**' and **Z'Z**, where **Z** is the *a* x *b* matrix of interaction means; that is, $z_{ij} = \overline{Y}_{ij} - \overline{Y}_{i\bullet} - \overline{Y}_{\bullet j} + \overline{Y}_{\bullet\bullet}$. The largest characteristic root of either matrix product gives the SS_{int} for the optimal c.o.c.; the characteristic vector of **ZZ**' gives the row (factor *A*) contrast coefficients; and the characteristic vector associated with the largest root of **Z'Z** gives the column (factor *B*) contrast coefficients. The row- and column-contrast coefficients can then be simplified to facilitate interpretation, and the resulting simplified c.o.c. tested for significance.

With respect to the second problem, it can be shown (personal communication, Kevin Bird) that **Z** matrices exist for which the optimal c.o.c. accounts for only one-third of SS_{int} for a 4 x 4 interaction. Unless the overall test of the interaction yields an F considerably above its critical value, one is going to be left in such a situation with a statistically significant interaction but no statistically significant contrast of contrasts. Most researchers, however, find it very difficult to provide substantive interpretations of interactions that do not imply c.o.c.'s. It would therefore be very useful to have a significance-testing procedure that explicitly takes into account a researcher's commitment to consider only contrasts of contrasts. Such a procedure, using the percentiles of the Wishart distribution, was developed for the large-sample case by Bradu and Gabriel (1974) and by Johnson (1976) and then for the general case by Boik (1986, 1993). Tables are provided in the Boik articles and in Harris (1994).

4.8 HIGHER ORDER MANOVA

To describe the multivariate extensions of each univariate design in detail would require in itself a book at least as long as Kirk (1982). What is done here instead is to:

1. Describe the procedure that would be followed in extending any univariate Anova design to the multivariate (multiple dependent measure) case.
2. Discuss a general class of designs—those involving repeated measures or "within-subjects" designs—that are, as pointed out in chapter 1, commonly treated by univariate Anova techniques although they are usually more properly handled by Manova.

Univariate Anova designs vary considerably in their complexity, in the number of different "error terms" employed in constructing F ratios, and so on. However, the end result of the effort involved in selecting an appropriate univariate Anova model for a particular study will be a ***summary table*** listing each source of variance, its associated degrees of freedom, and which other entry in the table is to serve as the error term (denominator) for the F ratio testing the null hypothesis that this particular component of variance is truly zero in the population. The procedures for extending this design to take into account differences among the groups on all p dependent variables are:

1. Construct a cross-product matrix (more accurately, a matrix of sums of cross products of deviation scores) for each effect you wish to test and each error term you will use, that is, for each row of the univariate Anova summary table. Each rth main diagonal entry of the $\mathbf{H}_i$ matrix corresponding to a particular effect i will be the sum of squares for that effect computed on the basis of subjects' scores on dependent measure r. Each (r,s)th off-diagonal entry in $\mathbf{H}_i$ or $\mathbf{E}_i$ is computed by substituting a product of the subject's score on outcome measures r and s for every occurrence in the single-variable computational formula of a square of the subject's score on a single variable. Thus, for instance,

$$\sum \frac{T_j^2}{n_j} \text{ becomes } \sum \frac{T_{j,r} T_{j,s}}{n_j}$$

and

$$\sum (\overline{X}_j - \overline{\overline{X}})^2 \text{ becomes } \sum (\overline{X}_{j,r} - \overline{\overline{X}}_r)(\overline{X}_{j,s} - \overline{\overline{X}}_s).$$

SPSS's MANOVA and SAS's PROC GLM produce these matrices (upon appropriate request) as a side product on their way to computing characteristic roots and vectors (the next step).

2. For every

$$F \text{ ratio} = \frac{MS_{\text{eff}}}{MS_{\text{err}}} = \frac{SS_{\text{eff}} / df_{\text{eff}}}{SS_{\text{err}} / df_{\text{err}}}$$

computed in the univariate analysis, compute the characteristic roots and vectors of the corresponding matrix product $\mathbf{E}^{-1}_{\text{err}}\mathbf{H}_{\text{eff}}$.

(The characteristic roots of $\mathbf{HE}^{-1}$ are identical to those of $\mathbf{E}^{-1}\mathbf{H}$, but the characteristic vectors are not. A review of the step—premultiplication by $\mathbf{E}^{-1}$—needed to get us from Equation (4.9) to Equation 4.10 shows that EINVH is indeed a more appropriate name for the computer program provided in Appendix B of the second edition of this book than its first-edition name of HEINV. Blush.)

The largest of the characteristic roots (eigenvalues) provides the value of λ_1, which can then be converted to θ_1 via $\theta_1 = \lambda_1/(1 + \lambda_1)$ and compared to the appropriate gcr critical value in Table A.5 (interpolating or running the GCRCMP program in Appendix B if necessary) with degree-of-freedom parameters

$$s = \min(df_{\text{eff}}, p), \quad m = \frac{|df_{\text{eff}} - p| - 1}{2}, \quad n = \frac{df_{\text{err}} - p| - 1}{2}.$$

In cases in which $s = 1$, you may use whatever significance criterion your computer program uses as its default option (e.g., you may employ the "SIGNIFICANCE (BRIEF)" option in SPSS), because all multivariate criteria lead to the same "exact F" in that case.

3. For each source (row) of the summary table that yielded rejection of the overall, multivariate null hypothesis, perform specific post hoc comparisons among the various experimental groups on the various dependent measures and linear combinations thereof. (Of course, specific a priori hypotheses about the pattern of means on a specific variable or combination of variables can always be tested, regardless of whether the corresponding multivariate test yielded statistical significance or not.) This post hoc analysis will almost always include the univariate F ratio for that effect computed separately on each outcome measure—which univariate Fs are of course computed directly from the main diagonal entries of $\mathbf{H}_{\text{eff}}$ and $\mathbf{E}_{\text{err}}$ via the equation

$$F_i = \frac{df_{\text{err}} h_{ii}}{df_{\text{eff}} e_{ii}}.$$

In addition, univariate Fs for the effects of that source on any linear combinations of the outcome measures can be computed via the equation

$$F_i = \frac{df_{\text{err}} h_{ii}}{df_{\text{eff}} e_{ii}}.$$

In particular, each of the characteristic vectors of $\mathbf{E}_{\text{err}}^{-1}\mathbf{H}_{\text{eff}}$ computed in step 2 whose corresponding characteristic root exceeds its criterion for multivariate statistical significance should be interpreted in terms of a simpler, readily verbalized pattern of weights, and that interpretation should be tested by a univariate F ratio on the specified linear combination via the $F(\mathbf{a})$ formula. For instance, $.71X_1 + .49X_2 + .92X_3 + .63X_4$ might be interpreted as representing "essentially" the sum of the four measures, whence $F(\mathbf{a})$ for $\mathbf{a}$ = [l, 1, 1, l] would be computed, whereas $.30X_1 - .82X_2 + .25X_3 + .31X_4$ could be interpreted as the difference between variable two and the average of the other three, implying an **a** of [1, -3 , l, 1]. The post hoc criterion for significance, rather than the critical values of the (univariate) F distribution, should be used in all of these "follow-up" F tests.

The gcr criterion for post hoc F ratios can be computed from the gcr tables in Appendix A via the formula

$$F_{\text{crit}} = \frac{df_{\text{err}}\theta_{\text{crit}}}{df_{\text{eff}}(1-\theta_{\text{crit}})},$$

where θ_{crit} is that value required for significance at the chosen significance level for the applicable degrees of freedom.

4. As the final stage of the analysis, applicable only if $df_{\text{err}} > 1$, each univariate F on a single variable or a combination of variables that is found in step 3 to be statistically significant by the post hoc criterion is further explained through the use of Scheffe'-type contrasts on the differences among the means on that particular original variable or combined variable of the various experimental groups. Again, post hoc significance criteria are preferred, each F_{contr} being compared to $df_{\text{eff}}\,F_{\text{crit}}$, where F_{crit} is the post hoc significance criterion computed for post hoc overall Fs on the effect in step 3.

It has been the author's experience that steps 1–4 provide a highly efficient means of "screening out" the multitude of chance fluctuations that would otherwise crop up if the unfortunately common alternative procedure of simply conducting p different univariate Anova were followed. Moreover, the recommended steps also provide a logical procedure for arriving at a relatively compact summary of the sources of differences among the various groups on the various outcome measures—including descriptions in terms of important linear combinations of these variables that might never have been detected via univariate approaches. There are, however, at least two alternate approaches to conducting a higher order Manova that should be mentioned.

First, if the dependent variables are commensurable, that is, measured in common, meaningful units, the researcher may wish to begin the analysis of each row of the summary table with a profile analysis, rather than a single, overall significance test.

Second, the reader will have noted that the discriminant function used in testing, say, the main effect of factor C is not in general the same discriminant function used in testing the $A \times B$ interaction, as the gcr maximization procedure is free to pick that linear combination of variables that maximizes the univariate F ratio for the particular effect being tested. The researcher may therefore prefer to ignore the factorial structure of his or her groups initially and simply conduct a one-way Manova on the differences among all k experimental groups to test the overall null hypothesis and then perform a higher order *univariate* Anova on the resulting discriminant function or a simpler interpretation thereof—using, of course, gcr post hoc significance levels, rather than the usual F table critical values. Such an approach is roughly equivalent to conducting a test of the total between-group sum of squares in a higher order Anova before beginning the finer-grained tests of main effects and interactions of the various factors. The primary advantage in the multivariate case is that the combined outcome measure used to assess the various main effects and interactions is the same for all such tests. Its primary disadvantage is a loss of power in each of these tests, because the combined measure that maximally discriminates among all k experimental groups may be very different from the combined variable that maximally discriminates among the levels of factor A.

Example 4.2 Eyeball to Eyeball in a Prisoner's Dilemma. It may be helpful to give an example of higher order Manova. In the fall of 1972, students in the Social Psychology Lab course at the University of New Mexico conducted a study of the effect on behavior of subjects being able (or not being able) to see each other's faces and exchanging (or not exchanging) explicit messages about their intentions prior to each trial in a Prisoner's Dilemma game. The design was thus a 2 x 2 factorial with both factors (visual presence or not and communication or not) fixed. (The reader is not necessarily expected to understand the terminology of higher order Anova at this point. Come back to this example after you have had a course in Anova designs other than one-way or after you have read a text in the area.) The summary table for a univariate analysis of, say, the proportion of the 30 trials that yielded mutually cooperative outcomes (the number of such trials, divided by 30) would look like the one shown in Table 4.5.

Table 4.5 Proportion of Mutually Cooperative Choices as f(Contact, Communication)

Source	*df*	*SS*	*MS*	*F*
Visual contact (V)	1	.0122	.0122	.3369
Communication (C)	1	.0270	.0270	.7461
$V \times C$	1	.0159	.0159	.4404
$P(VC)$	72	2.6075	.0362	
Total	75	2.6626		

As can be gathered from this table, there were a total of 76 pairs of subjects, 19 in each of the four experimental conditions. A computer program (BMDX69, to show how long these data have been in hand) was used to generate the cross-product matrix (based on four dependent measures) for each of the effects in Table 4.5 (except the Total row):

$$\mathbf{H}_V = \begin{bmatrix} .01220 & & & \\ .00507 & .00211 & & \\ .00418 & .00174 & .00143 & \\ -.02153 & -.00894 & -.00738 & .03798 \end{bmatrix};$$

$$\mathbf{H}_C = \begin{bmatrix} .02702 & & & \\ .00762 & .00215 & & \\ -.01252 & -.00353 & .00580 & \\ -.02200 & -.00620 & .01020 & .01792 \end{bmatrix};$$

$$\mathbf{H}_{VC} = \begin{bmatrix} .01595 & & & \\ -.02031 & .02586 & & \\ .01350 & -.01719 & .01143 & \\ -.00926 & .01179 & -.00784 & .00537 \end{bmatrix}$$

and

$$\mathbf{E}_{P(VC)} = \begin{bmatrix} 2.60751 & & & \\ -.54722 & .94746 & & \\ -.51449 & -.14619 & .62542 & \\ -1.54555 & -.25407 & .03526 & 1.76471 \end{bmatrix}.$$

Each of these matrices is of course symmetrical, so only the below-diagonal and main-diagonal entries need be presented. The program also output the means for each group on each outcome measure, as summarized in Table 4.6. It may be instructive to show how the cross-product matrices were constructed.

Table 4.6 Mean Proportion of Total Responses Accounted for by Each Outcome

Group	*CC*	*CD*	*DC*	*DD*
Isolated,				
no communication	.14563	.21047	.19479	.44916
communication	.21232	.18421	.20184	.40163
Face-to-face,				
no communication	.19995	.18410	.22800	.38763
communication	.20868	.23163	.18600	.37374

We cannot compute **E** without access to the original scores, but the various **H** matrices require only knowledge of the group mean vectors. Thus, for example, the formula for the sum of squares for the communication main effect for a single variable is $\sum_{j=1}^{2} n_j(\overline{X}_j - \overline{\overline{X}})^2$, where n_j is the number of subjects at the *j*th level of the communication factor (which equals twice the number of subjects in a single group), $\overline{X}_j$ is the mean score on the outcome variable of the subjects at that level of the *C* factor, and $\overline{\overline{X}}$ is the grand mean across all subjects of the dependent measure. Thus, for CC, this sum of squares is $38(.17297 - .19164)^2 + 38(.21050 - .19164)^2 = .027019$, which appears, in rounded-off form, in the first main diagonal position of $\mathbf{H}_c$.

If we now wish to compute, say, the (CC, DC) entry—first column, third row—of $\mathbf{H}_c$, we simply replace each squared deviation term in the previous formula with the product of a deviation score on CC and a deviation score on DC, giving

$$h_{\text{CC,DC}} = 38(.17279 - .19164)(.21139 - .20266) + 38(.21050 - .19164)(.19392 - .20266)$$
$$= -.012507,$$

which is fairly close to the listed value of -.01252. Of course, we would not actually use these deviation-score formulae if we were computing the cross-product matrices "from scratch," but would instead use more convenient (and more accurate) computational versions. When in doubt, however, it is always best to go back to the deviation-score formulae, which minimize the chances of a major goof. Note, too, that it *is* possible to get a negative entry in an off-diagonal cell of one of the **H** or **E** matrices. Your strongly conditioned aversions to negative sums of squares can, however, be safely transferred to the main diagonal entries of these matrices. (Be sure you understand why.) The reader would find it instructive to reconstruct each of the preceding **H** matrices from the mean vectors given earlier.

To carry out our gcr-based significance tests, we can submit the cross-product matrices produced by BMDX69 to a computer program such as MATLAB to obtain the greatest (in this case only) characteristic root and associated vector of $\mathbf{E}^{-1}_{\text{err}}\mathbf{H}_{\text{eff}}$. (More recently developed programs such as BMDP12V, the Manova options within SAS's PROC GLM, and SPSS's MANOVA will, if asked properly, produce the gcr statistics and the associated discriminant function coefficients, along with the **E** and **H** matrices—in a single run.) For example, λ_1 for the *V* main effect is .02502 with $s = 1$, $m = .5$, and $n = 34$; whence $F = (70/3)\,\lambda_1 = .5837$ with 3 and 70 *df.*

The other *F*s yielded by the gcr approach (all with 3 and 70 degrees of freedom) are .4440 for the *C* main effect and 1.0795 for the *V* x *C* interaction. The answer to Demonstration Problem A.6 at the end of this chapter illustrates how the results of univariate tests of each dependent variable and the overall multivariate test can be incorporated into a single summary table.

At this point, we really should stop, because the *F* statistics indicate that the null hypothesis is still a strong contender for each effect. (This example should allay the fears of those who look on multivariate statistics as another of those complicated techniques by which researchers can make *anything* come out significant.) However, we will, for didactic purposes, also report which linear combination of the outcome measures maximally discriminates among the levels of each factor.

The univariate *F* test for the *V* main effect is maximized by choosing

$$.426\text{CC} + .549\text{CD} + .719\text{DC}$$

as our combined variable, whereas

$$.608\text{CC} + .708\text{CD} - .359\text{DC}$$

maximizes the *C* main effect, and

$$.287\text{CC} - .420\text{CD} + .861\text{DC}$$

maximizes the $V \times C$ interaction. If we are troubled by this inconstancy of just what we are talking about in analyzing each effect, we can instead perform a four-group one-way Manova and then perform a univariate analysis on the resulting discriminant function. Thus in the present case we submit

$$\mathbf{H}_{\text{between}} = \mathbf{H}_V + \mathbf{H}_C + \mathbf{H}_{VC}$$

and the same $\mathbf{E}^{-1}$ as before to a program (e.g., MATLAB; see section 5.5.1) that can solve the generalized eigenvalue problem, discovering thereby that for the one-way Manova $\lambda_1 = .0519$ (there are now a second and a third root of .0357 and .0124, which we shall ignore), whence $\theta_1 = .0493$ with $s = 3$, $m = -.5$, and $n = 36$, nonsignificant. (The critical value for θ_1 at the .05 level is .150.) The coefficients of the discriminant function are given by

$$\mathbf{d}' = (.413, .252, .875);$$

that is,

$$D = .413\text{CC} + .252\text{CD} + .875\text{DC}.$$

The univariate Anova on this discriminant function is most simply carried out by realizing that

$$F_{\text{eff}} = 72(\mathbf{d}'\mathbf{H}_{\text{eff}}\mathbf{d})/(\mathbf{d}'\mathbf{E}\mathbf{d}).$$

The results of this Anova are summarized in Table 4.7. The appropriate critical value for

Table 4.7 Summary Table for Anova on Discriminant Function from One-Way Manova

Source	*df*	*SS*	*MS*	*F*
V	1	.00815	Same as *SS*	1.354
C	1	.000828	Same as *SS*	.138
$V \times C$	1	.01106	Same as *SS*	1.837
P(VC)	72	.43354	.00602	

each of these comparisons is $72(3/70)F_{\alpha}(3,70) = 7.344$, rather than the univariate critical value of 2.78. (Be sure you understand why.) Note, too, that the discriminant function derived from the one-way Manova does not do a better job (in terms of the resulting F ratio) than does the discriminant function derived from the higher order Manova for that effect. (Be sure you understand why it *cannot.)* Rao's approximate F statistic does *not* equal the univariate F computed on the discriminant function for that effect. For comparison, the F_{max} for each effect is $72 * \lambda_{crit} = 1.801$, 1.370, and 3.331 for the three rows of Table 4.7.

4.9 WITHIN-SUBJECT UNIVARIATE ANOVA VERSUS MANOVA

An increasingly common experimental design is the ***within-subjects*** (or ***repeated-measures***) design, in which one or more factors of the design are manipulated in such a way that each subject receives all levels of that factor or factors. The primary advantage of this approach is that it provides control for individual differences among the subjects in their responsiveness to the various experimental treatments, and thus provides more powerful tests of the effects of the within-subjects actor(s) when such intersubject variability is high. (It is also necessary that the subject's responses across trials be sufficiently highly correlated to offset the loss of degrees of freedom in the error term.) On the other hand, it also introduces correlations among the means on which the tests of the main effect of and interactions with the within-subjects factor are based. The usual univariate F ratios used to test these effects provide therefore only approximate significance tests, with the resulting p values being accurate only if highly restrictive assumptions about the nature of the correlations among subjects' responses to the various levels of the within-subject factor(s)—the h.o.t.d.v. assumption discussed in section 3.8—are met. These significance tests could be handled more appropriately by treating the subject's responses to the w levels of the within-$\underline{S}$ factor W as a w-element outcome vector and then applying Manova techniques. The main effect of W would then be tested by a flatness test on the grand mean outcome vector (cf. section 4.3 on Multiple Profile Analysis), and interactions with between-participant factors would be reflected in the parallelism test within the profile analysis of each separate term in the Manova summary table. Each statistically significant within-subjects effect should of course be followed up with tests of specific contrasts among the levels of W, using that particular contrast's within-cells variance as the error term. See Keselman, Rogan, and Games (1981) and Keselman (1982) for examples and Boik (1981) for analytic results that clearly demonstrate the inappropriateness of the usual ***univariate Anova*** approach of testing all contrasts against the same pooled error term. (See also our discussion of Boik's findings and recommendations in section 3.8.)

If a battery of q dependent variables is used to measure each subject's response at each of the p levels of W, our overall multivariate test of W will consist of a levels test (single-sample T^2) on the $q(p - 1)$-element vector of scores computed by applying each of our $(p - 1)$ contrasts to each of the q dependent variables, and multivariate tests of interactions with W will consist of parallelism tests (as in profile analysis; see section 4.3) on this same outcome vector. Our follow-up tests of particular contrasts across the levels of W on particular linear combinations of the dependent variables will, however, be complicated somewhat by the fact that the $(p - 1)$ subsets of the $q(p - 1)$-element vector of discriminant function coefficients need not be proportional to each other, even though we would usually wish to impose such a restriction on the maximization process.

Multivariate tests of between-subjects effects will be carried out on the q-element outcome vector obtained by computing the mean or sum of the subject's responses on a particular dependent variable across all levels of W. If we have more than one within-subject factor, tests of any given within-subject effect (main effect or interaction) and of the interaction between between-subjects effects and that particular within-subject effect are carried out on the set of contrast scores that represents that particular within-subject effect.

To make the procedures described in the last two paragraphs a bit more concrete, consider a design involving two between-subjects factors A and B, having three levels each, and two within-subject factors C and D, also having three levels each, with each subject therefore receiving all nine combinations of the levels of these two factors. We either randomize order of presentation of these nine treatment (experimental) conditions separately for each subject or make order of presentation one of the between-subjects factors of our design. Finally, the subject's response to each treatment combination is measured in terms of two different dependent measures. This leads to the following data table:

		C_1						C_2						C_3					
		D_1		D_2		D_3		D_1		D_2		D_3		D_1		D_2		D_3	
		Y_1	Y_2	Y_1	Y_2	Y_1	Y_2	Y_1	Y_2	Y_1	Y_2	Y_1	Y_2	Y_1	Y_2	Y_1	Y_2	Y_1	Y_2
A_1	B_1																		
	B_2																		
	B_3																		
A_2	B_1																		
	B_2									Y_{ijklm}									
	B_3																		
A_3	B_1																		
	B_2																		
	B_3																		

Our analysis of these data (assuming that we wish to maintain our main-effects-and-interactions approach rather than reconceptualizing either the between-subjects or the within-subject component as a 9-level single factor) will involve carrying out four separate Manovas, one on each of the four sets of dependent measures defined in the following table:

		C_1						C_2						C_3					
		D_1		D_2		D_3		D_1		D_2		D_3		D_1		D_2		D_3	
Effects tested	Outcome measures	Y_1	Y_2	Y_1	Y_2	Y_1	Y_2	Y_1	Y_2	Y_1	Y_2	Y_1	Y_2	Y_1	Y_2	Y_1	Y_2	Y_1	Y_2
A, B, AB	Sum-1	1	0	1	0	1	0	1	0	1	0	1	0	1	0	1	0	1	0
	Sum-2	0	1	0	1	0	1	0	1	0	1	0	1	0	1	0	1	0	1
C	*C*11	2	0	2	0	2	0	−1	0	−1	0	−1	0	−1	0	−1	0	−1	0
AC, BC	*C*12	0	2	0	2	0	0	0	−1	0	−1	0	−1	0	−1	0	−1	0	−1
ABC	*C*21	0	0	0	0	0	0	1	0	1	0	1	0	−1	0	−1	0	−1	0
	*C*22	0	0	0	0	0	0	0	1	0	1	0	1	0	−1	0	−1	0	−1
D,	*D*11	2	0	−1	0	−1	0	2	0	−1	0	−1	0	2	0	−1	0	−1	0
AD, BD	*D*12	0	2	0	−1	0	−1	0	2	0	−1	0	−1	0	2	0	−1	0	−1
ABD	*D*21	0	0	1	0	−1	0	0	0	1	0	−1	0	0	0	1	0	−1	0
	*D*22	0	0	0	1	0	−1	0	0	0	1	0	−1	0	0	0	1	0	−1
	*CD*111	4	0	−2	0	−2	0	−2	0	1	0	1	0	−2	0	1	0	1	0
	*CD*112	0	4	0	−2	0	−2	0	−2	0	1	0	1	0	−2	0	1	0	1
CD,	*CD*121	0	0	2	0	−2	0	0	0	−1	0	1	0	0	0	−1	0	−1	0
ACD, BCD	*CD*122	0	0	0	2	0	−2	0	0	0	−1	0	1	0	0	0	−1	0	−1
ABCD	*CD*211	0	0	0	0	0	0	2	0	−1	0	−1	0	−2	0	1	0	1	0
	*CD*212	0	0	0	0	0	0	0	2	0	−1	0	−1	0	−2	0	1	0	1
	*CD*221	0	0	0	0	0	0	0	0	1	0	−1	0	0	0	−1	0	1	0
	*CD*222	0	0	0	0	0	0	0	0	0	1	0	−1	0	0	0	−1	0	1

The first Manova, conducted on Sum-l and Sum-2, involves averaging across all levels of the repeated-measures factors (*C* and *D*) for each of the two dependent measures (Y_1 and Y_2). This, then, yields tests of our between-subjects effects (*A, B,* and the *A* x *B* interaction). The second set of measures (C_{11}, C_{12}, C_{21}, and C_{22}) consists of two orthogonal contrasts across the levels of repeated-measures factor *C* ($2C_1$ - C_2 - C_3 and C_2 - C_3), computed separately for each of the two dependent measures. A 3 x 3 factorial Manova on this set of measures thus gives the interaction between repeated-measures factor C_1 and each of the two between-subjects factors. For instance, testing the *A* main effect for this vector of four contrast scores tests the extent to which the three levels of *A* differ in the pattern of scores on the two C contrasts for each dependent measure. This is, of course, exactly what we mean by an $A \times C$ interaction.

Our test of factor C itself is simply a test of the null hypothesis that the grand mean on each of variables C_{11} through C_{22} is zero in the population. This is accomplished via the flatness test of section 4.3, namely,

$$F = \frac{N(df - p + 1)}{p}\bar{\bar{\mathbf{X}}}'\mathbf{E}^{-1}\bar{\bar{\mathbf{X}}} \text{ with } p \text{ and } df - p + 1 \text{ degrees of freedom,}$$

where df is the degrees of freedom for the within-cells error term, p is the number of contrast measures involved in the analysis (in the present case, 4), and $\bar{\bar{\mathbf{X}}}$ is the vector of grand means (averaged across all subjects, regardless of which AB combination of treatments they received) on the p contrast measures. If the analysis is conducted via a computer program, this will be the test of the CONSTANT effect.

The Manova on measures D_{11}, D_{12}, D_{21}, and D_{22} gives an exactly parallel set of tests for the main effect of repeated-measures factor D (the flatness or CONSTANT test) and of its interactions with the between-subjects factors (tests of the A, B, and AB effects on this set of contrast scores).

Finally, each of measures CD_{111} through CD_{222} is a combination of one of the C-effect contrasts with one of the D-effect contrasts for a particular dependent measure. Each such measure thus represents one component of the $C \times D$ interaction for one of the two dependent measures. For example, CD_{121} represents the combination of our first C-contrast ($2C_1 - C_2 - C_3$) with our second D-contrast ($D_2 - D_3$), computed only on dependent measure 1, thus yielding

$$2(D_2 - D_3)_{C1} - (D_2 - D_3)_{C2} - (D_2 - D_3)_{C3}$$

for Y_1, which equals $2Y_{C1D2,1} - 2Y_{C1D3,1} - Y_{C2D2,1} + Y_{C2D3,1} - Y_{C3D2,1} + Y_{C3D3,1}$.

Under the null hypothesis of no CD interaction, each of these eight contrast scores should have a population mean of zero, so our flatness test of this eight-element grand mean vector provides a test of the CD interaction, whereas each of the other effects tested in our Manova on this set of contrast scores assesses the interaction between one of our between-subjects effects and the $C \times D$ interaction.

Most Manova computer programs (cf. section 4.10) provide an option to read in a matrix of coefficients such as those in the preceding table to define the linear combinations of the elements of the original response vector that are to serve as dependent variable sets for analyses such as this one. Many of the programs also generate the necessary coefficients when given information as to the number of levels in, and a preferred set of contrasts among the levels of, each within-subjects factor in the design.

Example 4.3 Stress, Endorphins, and Pain. Resorting to real data, two researchers from the University of New Mexico (Otto & Dougher, 1985) studied the effects of a stessor (exposure to a snake) and of administration of the drug Naloxone (an endorphin) on tolerance of pain. The two independent (between-subjects) factors were exposure (normal controls vs. snake-phobic controls vs. snake phobics exposed to a snake) and Naloxone (administered, or a placebo injection administered instead). In addition to subjective measures of pain tolerance, three measures of physiological arousal (designated as FAC, SAI, and GSR) were taken at each of three points in time: baseline, after exposure (or not, in the case of controls) to the snake, and after participation in the pain tolerance test. They thus had a 3 **x** 2 X 3 factorial design, with the third factor a repeated-measures factor, and with a three-element vector of outcome (dependent) variables.

Listed here is the SPSS MANOVA command used to carry out this "repeated battery" (see section 5.4.2), or "doubly multivariate" (pp. 532–534 of the SPSS manual; SPSS, Inc., 1983), analysis:

```
MANOVA   FAC1,SAI1,GSR1, FAC2,SAI2,GSR2, FAC3,SAI3,GSR4 BY
          EXPOSURE (0,2)  NALOXONE (0,1)/
         TRANSFORM = SPECIAL ( 1 0 0 1 0 0 1 0 0.
                               0 1 0 0 1 0 0 1 0.
                               0 0 1 0 0 1 0 0 1.
                               1 0 0-1 0 0 0 0 0.
                               0 1 0 0-1 0 0 0 0.
                               0 0 1 0 0-1 0 0 0.
                               1 0 0 1 0 0-2 0 0.
                               0 1 0 0 1 0 0-2 0.
                               0 0 1 0 0 1 0 0-2)/
         RENAME = SUMFAC,SUMSAI,SUMGSR,  STRESFAC,STRESSAI,STRESGSR,
                  PAINFAC,PAINSAI,PAINGSR/
   PRINT = DISCRIM (RAW STAN ALPHA(1.0)) ERROR (SSCP) SIGNIF(HYPOTH)/
   NOPRINT=PARAMETER(ESTIM)/
   METHOD=SSTYPE(UNIQUE)/
   ANALYSIS = (SUMFAC,SUMSAI,SUMGSR/
                STRESFAC,STRESSAI,STRESGSR, PAINFAC,PAINSAI,PAINGSR)
```

The variable names were constructed by combining the abbreviation for the arousal measure with 1, 2, or 3 to represent the time of measurement. However, there was actually an "extra" GSR measurement for each subject, so "GSR4" is the galvanic skin response taken at the same point in the experimental session as FAC3 and SAI3. The names of the transformed scores (RENAME =) were designed to be equally suggestive, with SUM representing a particular arousal measure summed over all three time periods, STRES being the change from baseline to after presentation of the stressor, and PAIN being the difference between the average of the first two periods and level of arousal after undergoing the pain tolerance test. METHOD= SSTYPE(UNIQUE) insured that each effect was tested in the context of the full model (see section 2.7). The ANALYSIS subcommand requested that the program first conduct a Manova on the SUM measures, thereby yielding main effects of exposure, Naloxone, and their interaction; and then perform this same analysis on STRES and PAIN, thereby analyzing the main effect of and interactions with the measurement-period factor.

The results included significant main effects of and a significant interaction between the two between-subjects factors (from the analysis of the SUM measures), significant changes in arousal measures over measurement periods (from the CONSTANT effect in the analysis of the STRES and PAIN measures), and statistically significant differences between the phobic controls and the other two groups in the pattern of their changes in arousal during the experimental session (from the EXPOSURE effect in the analysis of the contrast scores). Because the main purpose of this example was to demonstrate the setup for this sort of analysis, the reader is referred to the authors of the study (ottom@a1.mgh.harvard.edu, dougher@unm.edu) for details.

4.10 Computerized Manova

We've already given an example of a Manova carried out via computer program (see Example 4.3). However, just in case the general principles underlying that example didn't come through via the usual osmotic process, we'll now get explicit about those general principles.

The setup for an SPSS MANOVA analysis is a straightforward generalization of the format for the various T^2 analyses demonstrated in chapter 3. In the general case, our between-subjects factor(s) may have any number of levels (versus the restriction to $k = 2$ for T^2 analyses), we may have more than one within-subjects factor, and we may of course measure subjects' responses to each combination of a level of each of the within-subjects factors in more than one way (p, the size of the dependent-variable battery). As in the T^2 case, we wish to examine the optimal contrasts and/or discriminant functions associated with each within-subjects effect and with each interaction between a between-subjects and a within-subjects effect, and doing so will require some "hand" calculations to get from the discriminant-function coefficients provided by SPSS to the optimal-contrast coefficients they imply.

What is new in this general case is that, for each effect for which the between-subjects portion of the effect has 2 or more degrees of freedom, the four multivariate tests (the greatest-characteristic-root test, Pillai's and Hotelling's trace statistics, and Wilks' lambda) will yield four different (sometimes very different) sampling distributions and associated p values, and the p value for the most useful of the four will not be printed out by SPSS but will have to be looked up in a table or computed via a separate subroutine.

4.10.1 Generic Setup for SPSS MANOVA

```
MANOVA measure list by bsf1(bl1a,bl1b) bsf2 (bl2a,bl2b) ...  /
   WSFACTORS =wsf1 (wl1) wsf2 (wl2) ... /
    CONTRAST (wsf1) = SPECIAL (1 1 1, 1 0 -1, etc.) /
```

```
 CONTRAST (wsf2) = SPECIAL (1 1 1 1, 3 -1 -1 -1, etc.) /
          ...
 RENAME = Ave,Wsf1C1, Wsf1C2, etc. /
WSDESIGN /
 CONTRAST (bsf1) = SPECIAL (1 1 1, 2 -1 -1, etc.) /
 CONTRAST (bsf2) = POLYNOMIAL (2,5,7,12) /
                    ...
PRINT = TRANSFORM CELLINFO (MEANS) SIGNIF (UNIV HYPOTH)
        ERROR (COV) /
DISCRIM = RAW STAN /
OMEANS = TABLES (bsf1, bsf2, bsf1 by bsf2, etc.) /
DESIGN / DESIGN = bsf1(1), bsf1(2), bsf2(1), bsf2(2), bsf2(3),
     bsf1(1) by bsf2(1), bsf1(1) by bsf2(2), etc. /   .
```

In this listing, uppercase terms are MANOVA subcommands and specifications, and lowercase terms are inputs that must be supplied by the user. In particular:

measure list is a list of all $p_{\text{battery}} \cdot wl_1 \cdot wl_2 \ldots$ measures obtained from each sampling unit.

bsf1, bsf2, and so on are the names of variables (already read in) that indicate what level of one of the between-subjects factors was assigned to or measured for that subject, and (*bl1a, bl1b*) provides the lower and upper bounds of the levels of this factor. These values must be consecutive integers and must have been read in for each subject, recoded (via the RECODE) command) from another variable that was read in, or constructed in some other way from the variables that were read in.

wsf1 is the name given to within-subjects factor 1 and is used only within the MANOVA program.

(wl1) gives the number of levels of within-subjects factor 1.

Unlike the between-subjects factors, the within-subjects factor names are *not* names of variables read in for each subject but are internal to the MANOVA program. The levels of each within-subjects factor are inferred from the structure of the *measure list* by counting, with dependent variable moving most slowly, then *wsf1*, then *wsf2*, and so on.

Thus, for instance, if our set of 24 measures on each subject comes from having measured the same battery of 4 dependent variables at each of the 6 combinations of one of the 3 levels of motivation-level with one of the 2 levels of task difficulty, the WSFACTORS = MTVL (3) TD (2) specification tells SPSS that the measure list has the structure

V1AM1TD1, V1M1TD2, V1M2TD1,. . . , V4M3TD1, V4M3TD2.

This is the portion of the MANOVA setup you are most likely to get wrong the first time around, so Rule Number 1 when using MANOVA with within-subjects factors is to *always* (without fail) include a PRINT = TRANSFORM request. This will produce a list of the basis contrasts actually employed by MANOVA, from which any lack of fit between the ordering of your *measures list* and the order in which the WSFACTORS are listed will be immediately apparent.

CONTRAST(wsf1) must come between WSFACTORS and WSDESIGN. Because the

SIGNIF (UNIV) request prints the univariate F for each of the basis contrasts and for its interaction with each of the between-subjects effects, it is important to spell out the particular contrasts in which one is most interested. However, control over the basis contrasts one wishes to employ is lost if the contrasts specified are not mutually orthogonal, because SPSS MANOVA will, in that case, insist on "orthogonalizing" them—usually by retaining one of the contrasts which has been specified (not necessarily the first) and selecting additional contrasts orthogonal to that one. In earlier editions of MANOVA one could get around the totally unnecessary restriction by selecting one of MANOVA's keyword-specified sets of contrasts, such as CONTRAST (wsfactor) = SIMPLE, which tested every other level of the within-subjects factor against the first level. In the version available at this writing, however, MANOVA even "corrects" its own internally generated sets of contrasts. Again, *always* be sure to request PRINT = TRANSFORM to be able to know to what the UNIVARIATE Fs and discriminant function coefficients refer.

No such restriction applies to contrasts among levels of between-subjects factors, although such a restriction would make some sense there, because mutually orthogonal contrasts *do* additively partition SS_{eff} for between-subjects effects but not for within-subjects effects.

Finally, the first, unadorned DESIGN subcommand requests a standard factorial breakdown into main effects and interactions, whereas the second, more involved DESIGN subcommand requests tests of the specific between-subjects contrasts. [MTVL (2), for example, refers to the second single-*df* contrast specified in the CONTRAST subcommand for MTVL.]

4.10.2 Supplementary Computations

Although SPSS's MANOVA program is detailed, it does *not* directly provide any of the following:

1. The maximized F_{eff}.
2. The critical value to which max F_{eff} is to be compared (and thus also the fully post hoc critical value for F_{eff} computed on any linear combination of the measures).
3. F_{eff} for various post hoc linear combinations of the measures
4. In the case of within-subjects factors, the optimal contrast among the levels of that within-subjects factor
5. For doubly multivariate designs, the optimal linear combination of the measures when we restrict attention to contrasts among the levels of the within-subjects factor(s) with respect to a particular linear combination of the dependent-variable battery.

These additional bits of information are easily, if sometimes tediously, computed by hand calculations and/or supplementary computer runs. Examples were given earlier in this chapter, and additional details are given by Harris (1993).

4.10.3 Pointing and Clicking to a Manova on SPSS PC

Don't—at least not as of this writing (summer 2000). The MANOVA program is available only through the syntax window. If you click on "Analyze" and then request "General Linear Model" and then "Multivariate" or "Repeated Measures" (the only menu routes to multivariate analysis of variance) you will be using SPSS's GLM program, which is fine as long as you are either interested only in overall tests or in the limited set of prespecified contrasts that GLM makes available, *and* the between-subjects portion of your design involves only fixed factors.

Taking the second limitation first: SPSS GLM provides the user with the option of specifying which factors are fixed (the default is all factors fixed) and which are random; it then uses this specification to generate the appropriate error terms for the various effects in the model. (A further limitation is that the program will not permit this specification if there are any explicitly defined within-subjects factors; there are various ways to "kluge around" this limitation by computing your basis contrasts outside of the program, but as we shall see, that issue is moot.) Unfortunately, as of July 2000 and versions 8 through 10 of SPSS for PCs, GLM gets it wrong. For instance, in the very simplest case, with a single random factor and a single fixed factor, GLM insists on using the interaction between the two factors as its error term for both the fixed and the random factor; the correct error term for the random factor is simply the within-cells error mean square. (See Harris, 1994, section 4.3, for details.)

The more important limitation (given the infrequency with which researchers admit to using random factors) is that for exploration of specific contrasts (which is of course the ultimate goal of any Manova), GLM insists that you express each desired contrast as a linear combination of the parameters of the model it uses to analyze the data, which means that each contrast requires several lines of combining weights that are very difficult to puzzle out. The instructor and teaching assistants for my department's second-semester graduate statistics course had planned to switch from MANOVA to GLM shortly after the latter became available, and actually did manage to put together some several-page handouts on how to translate main-effect and interaction contrasts into GLM's parameter-matrix weight specifications, but ultimately decided that this was unduly burdensome, given the much more straightforward procedures available in the MANOVA program. It's likely that the SPSS administrators will eventually make available a more user-friendly version of the GLM program, but in its current state it just isn't a practical alternative to MANOVA.

4.10.4 Generic Setup for SAS PROC GLM

For designs where measures for each subject consist entirely of scores on a single dependent variable at every combination of the levels of one or more within-subjects factors (i.e., all factorial Manova designs except for repeated-battery designs), an example setup is

```
PROC GLM;
  CLASS vetgen agegrp
  MODEL a1 b1 a2 b2 a3 b3 a4 b4 a5 b5 a6 b6 a7 b7
   = vetgen agegrp vetgen*agegrp / NOUNI;
  MEANS vetgen agegrp vetgen*agegrp;
  REPEATED moodmeas 7 PROFILE , befaft 2 /
     NOU PRINTM PRINTRV SHORT SUMMARY;
```

```
TITLE '2 rep-measures factors';
```

In this example, a1 through a7 are mood measures obtained after subjects viewed a movie, b1 through b7 are premovie measures, and the battery of seven mood measures are treated as the levels of a within-subjects factor called *moodmeas.* Note that, as in SPSS MANOVA, the first-named within-subjects factor varies most slowly. Thus *moodmeas* and *befaft* correspond to SPSS MANOVA's *wsf*1 and *wsf*2. The CLASS variables *vetgen* and *agegrp* name the between-subjects factors and are equivalent to SPSS MANOVA's *bsfi* factor names. There is no provision for establishing the range of values on each between-subjects factor within GLM, so before entering GLM one must declare as missing all values of the variables that are not to be included in the Manova. Note, too, that SAS indicates interaction effects with an asterisk (*), rather than BY.

The MEANS statement requests tables of means relevant to the specified main effets and interactions. The NOUNI option on the MODEL statement tells GLM not to print univariate Anova summary tables for each of the 14 original measures. If these summary tables *are* wanted, NOUNI should be omitted. *PROFILE* on the REPEATED statement requests that adjacent-difference contrasts (level *i* - level *i*+1) be used as the basis contrasts for the *moodmeas* factor. (Various other options, including POLYNOMIAL contrasts, are available.) The options appearing after the slash on the REPEATED statement and their meanings follow.

NOU: Do not print the univariate-approach Anovas on the within-subjects effects.
PRINTM: Print the contrast coefficients that define the basis contrasts (an important check, although GLM appears to give the contrasts requested without orthogonalizing them).
PRINTRV: Print the eigenvalues and associated discriminant function coefficients. Without this specification one gets only the overall tests, with no indication of what optimal contrast got you there.
SHORT: Print relatively short messages as to how the various multivariate overall tests were computed.
SUMMARY: Provide a univariate Anova summary table for each of the within-subjects basis contrasts.

For doubly multivariate designs one must explicitly construct within-subjects contrasts with respect to each measure in the dependent-variable battery. The SPSS MANOVA "kluge" of specifying a within-subjects factor with a number of levels that is a multiple of the number of measures in the *measures list* does not work here. An example setup is

```
PROC GLM;
   CLASS vetgen agegrp ;
   MODEL b1 b2 b3 b4 b5 b6 b7 a1 a2 a3 a4 a5 a6 a7
     =vetgen agegrp vetgen~agegrp / INT;
   MANOVA H = _ALL_M = b1 - a1, b2 – a2, b3 – a3, b4 – a4,
     b5 – a5, b6 – a6, b7 – a7 PREFIX = befaft / SHORT SUMMARY;
   MANOVA H = _ALL_ M = b1 + a1, b2 + a2, b3 + a3, b4 + a4,
     b5 + a5, b6 + a6, b7 + a7 PREFIX = sum / SHORT SUMMARY;
   TITLE 'Doubly-multiv analysis, own befaft contrasts'
```

In this setup the order in which the measures is named on the MODEL statement is irrelevant. The INT option is highly relevant in that it requests that, in addition to the between-subjects factors and their interactions, the grand mean be tested. Two MANOVA statements are necessary to be able to use the PREFIX option (which labels the basis contrasts as BEFAFT1, BEFAFT2, etc., and the transformed variables used for testing purely between-subjects effects as SUM1, SUM2, etc.). The SUMMARY option on each MANOVA statement requests a univariate Anova summary table for each transformed variable.

Crucial in all of this are the semicolons, which are statement separators in SAS. Their omission is the most common and most subtle error made in setting up SAS analyses.

Demonstration Problems

A. Computations

All of the following problems refer to Data Set 4 (Table3.2), which involved four groups of 12 subject pairs each, with each group receiving a different motivational set (cooperative [COOP], competitive [COMP], individualistic [IND], or no motivating orientation [NMO]) prior to beginning a 40-trial Prisoner's Dilemma game. Six different experimenters were employed, with each experimenter running two complete replications of the experiment. (This experimenter factor is ignored in the first five problems.) The outcome measures of interest here are the number of mutually cooperative (CC), unilaterally cooperative (CD), unilaterally competitive (DC), and mutually competitive (DD) outcomes. Recall that there is a perfect linear relationship among these four measures (their sum must equal 40), so that we can only include three of the four variables in any multivariate test, and $p = 3$ in computing degree-of-freedom parameters for such a test. You may find it instructive to work one or more of the problems first omitting DD and then omitting one of the other measures.

1. Repeat problems B1 and B2 of the Demonstration Problem in chapter 3 using the formulae for Manova.

2. Using Manova, test the null hypothesis of no differences among the population mean outcome vectors for these four experimental conditions.

3. Compute the discriminant function D_i for each subject-pair i.

4. Conduct a one-way Anova on the differences among the four conditions in discriminant function scores. Compare with problem 2. Use Scheffé's contrast method to "decompose" the differences among the groups in mean discriminant function scores into the following three sources of variability:

(a) COOP versus the other three groups;
(b) NMO versus IND and COMP;
(c) IND versus COMP.

What is the appropriate a priori criterion for each of these contrasts? What is the appropriate post hoc criterion?

5. Test the statistical significance of the differences among the four groups in terms of each of the five linear combinations of outcome variables mentioned in problem B4 of the Demonstration Problem in chapter 3. Further decompose the differences among the groups on each of these combined variables into the three sources of variability mentioned in problem 4. What are the appropriate a priori and post hoc criteria for each of these tests?

6. Repeat the analysis of question 4, this time including experimenter (E) as a factor in the design. We thus have a 4 x 6 factorial design, with one complication: Because we are interested in drawing inferences about the impact of experimenters in general, rather than of these six particular experimenters, we must treat E as a *random* factor. As Winer (1971, chap. 5) and Harris (1994, chap. 4) point out, this means that the appropriate error term for the M main effect is the M **x** E interaction, whereas the E main effect and the M x E interaction are each tested against the usual within-cells error term—in the present case, Pairs within ME combinations, or Pairs(ME), or just Within. The summary table for the single dependent variable CC would look as follows:

Source	*df*	*SS*	*MS*	*F*
M	3	2819.73	939.91	20.21
E	5	250.60	50.12	<1
M x E	15	697.65	46.51	<1
Within Pairs(ME)	24	1428.50	59.52	

Note: $F_M = \dfrac{MS_M}{MS_{M \times E}} = 939.91/46.51,$

and $F_E = \dfrac{MS_E}{MS_{\text{within}}} = 50.12/59.52.$

You aren't necessarily expected to be familiar with the details of designs involving random factors, but once you have been given the univariate design you should be able to follow the guidelines of section 4.8 to conduct the corresponding Manova on the three-variable outcome vector (CC, CD, DC). Do so. Be sure to include an interpretation of each statistically significant discriminant function and to test for statistical significance at least one specific contrast on the simplified discriminant function implied by your interpretation.

B. Thought questions

1. It is not uncommon for an overall Manova to indicate signiiicant differences among the various groups, and yet to find that *none* of the obvious comparisons among the groups reach statistical significance by the post hoc criterion. How are we to interpret such a situation?

2. What would the formulae for Manova look like if we knew that our dependent measures were uncorrelated? What does this suggest about the possibility of preceding Manova with a principal component or factor analysis of our dependent measures?

3. In general, Manova produces k - 1 discriminant functions. If two or more of these are statistically significant sources of between-group differences, could we do a better job of discriminating among the groups by employing a linear combination of the various statistically significant discriminant functions? How are we to use the second, third, and so on discriminant functions?

Answers

A.

1. These should lead to the same ultimate F ratio as did the T^2 analyses. In addition, if $F_{max} = (df_{err}/1)\lambda_{max}$ is computed from the Manova results, it will equal T^2. Also, the discriminant function should be the same.

2. $\lambda_{max} = 1.3775$ with $s = 3$, $m = -.5$, $n = 20$; whence $\theta = 1.3775/2.3775 = .5795$, which is compared to $\theta_{crit} = .262$ at the .05 level and .325 at the .0l level, so that we may soundly reject H_0. The other roots are .116 and .054, nonsignificant.

3. $D_i = .832CC_i - .159CD_i + .532DC_i$, which seems close to the number of cooperative responses made by member B of the pair, CC + DC.

4. The overall F ratio = 20.20 with 3 and 44 degrees of freedom, which equals $F_{max} = (44/3)\lambda_{max}$. Applying Scheffé's contrast method to the discriminant function scores, we obtain sums of squares of 1934.3 for COOP versus others, 91.2 for NMO versus the other two, and 10.8 for IND versus COMP. These add to 2036.3, the total SS for the between-groups factor. Note that the COOP versus others contrast accounts for 95.0% of the total between-groups sum of squares. Considering the selection of the discriminant function D as our dependent measure as a post hoc selection, we must use the second column of Table 4.3 to compute significance levels for the contrasts, whence the critical value for

each contrast will be 10.69 if it is considered a priori (which each of these contrasts actually was) or 15.44 if considered post hoc. The obtained Fs for contrast were 57.58, 2.71, and .320, so that the COOP versus others contrast is significant by either criterion, whereas the other two do not reach significance by either criterion.

5. The results of all these analyses can be summarized as follows:

	SS and F for:				
Source	$CD - DC$	$CC + CD$	$CC + DD$	$CC + DC$	CC
Between-groups	208.33(2.06)	1826.39(11.15)	263.00(2.99)	2700.24(19.18)	2819.7(17.4)
COOP versus others	58.78(1.74)	1771.00(32.42)	186.78(6.37)	2475.06(52.74)	2782.6(51.5)
NMO versus other 2	76.06(2.26)	33.35(.61)	68.06(2.32)	210.13(4.48)	36.1(0.7)
IND versus COMP	73.50(2.18)	22.04(.40)	8.17(.28)	15.04(.32)	1.0(.02)
Within-groups	1482.59	2403.58	1289.27	2064.85	2376.7

The critical value for the overall (between-groups) F for each combination of variables is 2.82 if that combination was selected on a priori grounds but 5.15 if it is considered post hoc. Presuming that the selection of these dependent measures is considered a priori (in contrast to the definitely post hoc discriminant function analyzed in Problem 4), the critical value for each specific comparison is 4.06 if the contrast is considered a priori and 8.46 if it is a post hoc comparison of the means.

6. The SPSS MANOVA command used to analyze these data was as follows:

```
MANOVA   FAC1,SAI1,GSR1, FAC2,SAI2,GSR2, FAC3,SAI3,GSR4 BY
           EXPOSURE (0,2)   NALOXONE (0,1)/
         TRANSFORM = SPECIAL ( 1 0 0 1 0 0 1 0 0.
                               0 1 0 0 1 0 0 1 0.
                               0 0 1 0 0 1 0 0 1.
                               1 0 0-1 0 0 0 0 0.
                               0 1 0 0-1 0 0 0 0.
                               0 0 1 0 0-1 0 0 0.
                               1 0 0 1 0 0-2 0 0.
                               0 1 0 0 1 0 0-2 0.
                               0 0 1 0 0 1 0 0-2)/
         RENAME = SUMFAC,SUMSAI,SUMGSR,  STRESFAC,STRESSAI,STRESGSR,
                  PAINFAC,PAINSAI,PAINGSR/
     PRINT = DISCRIM (RAW STAN ALPHA(1.0)) ERROR (SSCP) SIGNIF(HYPOTH)/
     NOPRINT=PARAMETER(ESTIM)/
     METHOD=SSTYPE(UNIQUE)/
     ANALYSIS = (SUMFAC,SUMSAI,SUMGSR/
                 STRESFAC,STRESSAI,STRESGSR, PAINFAC,PAINSAI,PAINGSR)
```

Note the way in which the DESIGN statement specifies the error term for each effect.

You may wish to verify that the pooled within-cells cross-product matrix and the

Hypothesis cross-product matrices are

$$\begin{array}{c} \\ \mathbf{E} = \end{array}\begin{array}{c} \quad\text{CC}\qquad\text{CD}\qquad\text{DC} \\ \begin{bmatrix} 1428.5 & -59.5 & -488.0 \\ & 338.0 & -11.5 \\ & & 549.0 \end{bmatrix} \end{array}; \qquad \begin{array}{c} \\ \mathbf{H}_M = \end{array}\begin{array}{c} \quad\text{CC}\qquad\text{CD}\qquad\text{DC} \\ \begin{bmatrix} 2819.7 & -569.9 & -104.3 \\ & 146.5 & 13.7 \\ & & 89.2 \end{bmatrix} \end{array};$$

$$\mathbf{H}_E = \begin{bmatrix} 250.6 & -14.3 & -8.3 \\ & 68.9 & -8.5 \\ & & 17.4 \end{bmatrix}; \text{ and } \mathbf{H}_{ME} = \begin{bmatrix} 697.6 & -209.3 & -56.0 \\ & 186.25 & -28.3 \\ & & 227.1 \end{bmatrix}$$

The multivariate tests of the *M*, *E*, and *M* x *E* effects are then provided by the characteristic roots of $\mathbf{H}_{\text{ME}}^{-1}\mathbf{H}_{\text{M}}$, $\mathbf{E}_{\text{err}}^{-1}\mathbf{H}_{\text{E}}$, and $\mathbf{E}_{\text{err}}^{-1}\mathbf{H}_{\text{ME}}$, respectively. The only effect that was anywhere near statistical significance was our familiar motivating orientations main effect, with first, second, and third largest thetas being .814, .285, and .192—each with $m = -.5$ and $n = 5.5$, whereas the *s* parameter for the critical value for θ_1, θ_2, and θ_3 is 3, 2, and 1, respectively. Only θ_1 is statistically significant ($p < .01$), with an associated discriminant function of .173CC + .098CD + .042DC. This seems interpretable as CC by itself, and indeed F_{M} for CC = 20.21, which is quite close to the *F* on the discriminant function of (15/3)(.8138/.1862) = 21.85. [An alternative comparison would be to convert the univariate *F* to an $r^2 = 20.21/(20.21 + 15/3) = .802$, as compared to our θ_1 of .814.] This time our Manova has not "paid off" with a new linear combination of the measures that outperforms our best single dependent variable by much. Because we already did an effective breakdown of the *M* main effect on CC in question 5, we know that the Coop versus other three contrast yields an SS_{contr} of 2782.6, accounting for 98.7% of total SS_{M}. However, our F_{contr} changes, because of the new error term, to 2782.6/50.12 = 55.52. This F_{contr} can be compared to the fully post hoc critical value $15\theta_{\text{crit}}/(1 - \theta_{\text{crit}})$ = 15(.705/.295) = 35.85 at the .01 level.

Finally, we can summarize the analysis succinctly in the following table:

		SS For:			(CC, CD,	DC,	DD)		
Source	*df*	CC	. . .	DD	λ_1	θ_1	*s*	*m*	*n*
M	3	2819.7**		1734.2	4.371	.814**	3	−.5	5.5
E	5	250.6		274.6	.253	.202	3	.5	10
M x *E*	15	697.6		523.6	1.027	.507	3	5.5	10
Within	24	1428.5		1197.5					

** $p < .01$ by gcr-based post hoc critical value.

Note that we have omitted the *CD* and *DC* columns and the *F* ratios to conserve space. We could also have included a breakdown into orthogonal contrasts in the univariate portion of the table. The results of the multivariate tests would of course come out exactly the same whether *CC, CD, DC,* or *DD* was the measure deleted from the cross-product matrices.

B.

1. Whatever is going on in our data is simply nonobvious. As long as we have used the gcr test, we know that there will be at least one contrast among the group means for *some* linear combination of the outcome measures that will exceed F_{crit}. In particular, the F for contrast that results when we take the discriminant function as our outcome measure and use contrast coefficients proportional to the difference between each group mean (on the discriminant function) and the grand mean across all groups will exactly equal the stripped F ratio corresponding to λ_{max}, that is, will equal $df_{err} \cdot \lambda_{max}$. If our multivariate test barely reached significance, it may well be that only something very close to this (usually rather arbitrary looking) contrast will reach significance by the post hoc criterion.

If we use likelihood-ratio (U) statistics, however, there is no guarantee that any (even the first) of the discriminant functions is statistically significant, so that we might in fact come up with a significant U statistic but have nothing else to say, that is, be unable to specify any contrast on any linear combination of outcome measures that is statistically significant. Such cases are, of course, those in which the gcr test would nor yield significance and are the basis of the U statistic's generally greater power.

2. See the comments in section 4.7.

3. We could not do any better by using a linear combination of the discriminant functions, because the resulting combined variable will still be a linear combination of the original measures, and if this linear combination had been better than the first discriminant function, it would have been used as the first discriminant function. Put another way, the optimal linear combination of the various discriminant functions will turn out to be one in which all but the first discriminant function receive zero weights. This does not, however, rule out the possibility that a nonlinear combining rule would yield more accurate sorting of subjects into groups than the first discriminant function. For instance, we might put a subject into that group to which he or she is "closest" in the discriminant function space. In other words, we would compute, for each group j, the measure

$$d_{uj} = \sqrt{(d_{1u} - \bar{d}_{1j})^2 + (d_{2u} - \bar{d}_{2j})^2 + \cdots + (d_{su} - \bar{d}_{sj})^2}\,,$$

where d_{iu} is subject u's score on discriminant function i, and $\overline{d_{ij}}$ is the mean of group j on discriminant function i. We would then put the subject into that group for which d_{uj} was smallest. It would be interesting to explore the conditions under which this procedure does indeed provide better discrimination among the groups than does the first discriminant function.

5
Canonical Correlation: Relationships between Two Sets of Variables

In chapter 2 we discussed multiple regression as a means of measuring the degree of relationship between a set of predictor variables and a single predicted (outcome) variable. In this chapter we shall discuss a technique known as ***canonical correlation*** for assessing the degree of relationship between a set of p predictor variables and a set of q outcome variables. (Actually, canonical correlation is a perfectly symmetric technique in which the distinction between the predictor set and the outcome set is not mirrored by any difference in statistical treatment of the two sets, and the distinction thus need not be made.) Our approach to multiple regression was to obtain a "combined predictor," a linear combination of scores on the original p predictor variables X_1, X_2, ... , X_p, and correlate this combined variable with our single Y score. In canonical correlation, we obtain two linear combinations: one combination of the p predictor variables and one combination of the q outcome measures Y_1, Y_2, ... , Y_4. Naturally we take as our coefficients for these linear combinations those vectors **a** and **b** (of length p and q, respectively) that make the Pearson product–moment correlation between the two combined variables $u = \mathbf{Xa}$ and $v = \mathbf{Yb}$ as large as possible. The value of the maximum possible Pearson r is known as the ***canonical correlation*** R_c between the two sets of variables, and u and v are known as ***canonical variates,*** and **a** and **b** constitute the two sets of ***canonical coefficients.***

5.1 FORMULAE FOR COMPUTING CANONICAL *R*s

As should be obvious to those who have studied chapters 2 through 4, the correlation between u and v is given by

$$r(\mathbf{a},\mathbf{b}) = \frac{(\mathbf{a}'\mathbf{S}_{xy}\mathbf{b})}{\sqrt{(\mathbf{a}'\mathbf{S}_x\mathbf{a})(\mathbf{b}'\mathbf{S}_y\mathbf{b})}}, \qquad (5.1)$$

where $\mathbf{S}_x$ and $\mathbf{S}_y$ are the variance-covariance matrices for the left- and right-hand set of measures (the Xs and the Ys), respectively, and $\mathbf{S}_{xy}$ is a matrix each of whose rows gives the covariances of one of the Xs with each of the Ys. The matrices involved in Equation (5.1) are, then, submatrices of the overall covariance matrix of all $p + q$ variables under study:

$$\mathbf{S} = \begin{bmatrix} \mathbf{S}_x & \mathbf{S}_{xy} \\ \mathbf{S}_{xy} & \mathbf{S}_y \end{bmatrix}.$$

In addition to being the starting point of efforts to derive R_c Equation (5.1) provides a convenient basis for exploration of relationships between specific linear combinations of the variables in the two sets-for example, the correlation between simplified canonical variates. To eliminate the mathematically inconvenient square root, we seek a method of maximizing r^2, knowing that this will maximize the absolute value of r as well. (Note that the sign of r is arbitrary because this sign can be reversed by multiplying either u or v by -1.) To simplify our "maximanda" further, while assuring the uniqueness of **a** and **b** in the face of the well-known invariance of the Pearson r under linear transformations of either of the variables being correlated, we establish the side conditions that $\mathbf{a'S}_x\mathbf{a}$ (the variance of our linear combination of predictor variables) = $\mathbf{b'S}_y\mathbf{b}$ (the variance of the linear combination of outcome variables) = 1.

As Derivation 5.1 points out, the maximum value of $r^2(\mathbf{a, b})$ and the two sets of combining weights that yield this maximum value are given by the largest value of θ and the associated vectors **a** and **b** that simultaneously satisfy Equations (5.2):

$$[\mathbf{S}_{xy}\mathbf{S}_y^{-1}\mathbf{S'}_{xy} - \theta\,\mathbf{S}_x]\,\mathbf{a} = \mathbf{0}$$

and (5.2)

$$[\mathbf{S'}_{xy}\,\mathbf{S}_x^{-1}\mathbf{S}_{xy} - \theta\,\mathbf{S}_x]\,\mathbf{a} = \mathbf{0}.$$

The solutions for θ in Equations (5.2) are the characteristic roots of the matrices

$$\mathbf{S}_x^{-1}\mathbf{S}_{xy}\mathbf{S}_y^{-1}\mathbf{S'}_{xy} \quad \text{and} \quad \mathbf{S}_y^{-1}\mathbf{S'}_{xy}\,\mathbf{S}_x^{-1}\mathbf{S}_{xy}\ .$$

The characteristic vector associated with the gcr of the first matrix gives the canonical coefficients for the left-hand set of variables, and the characteristic vector associated with the gcr of the second matrix gives the canonical coefficients for the right-hand set of variables, and the gcr of either matrix gives the squared canonical correlation between the two sets of variables.

The largest characteristic root (i.e., the square of the canonical correlation between the two sets) is tested for significance by comparing it directly (no division by 1+λ is required) to the gcr tables of Appendix A with degree-of-freedom parameters

$$s = \min(p, q); \quad m = (|p - q| - 1)/2; \quad \text{and} \quad n = (N - p - q - 2)/2. \qquad (5.3)$$

There are, in general, s nonzero roots of $\mathbf{S}_x^{-1}\mathbf{S}_{xy}\mathbf{S}_y^{-1}\mathbf{S'}_{xy}$. Just as with Manova, the "extra" roots have an intepretation quite similar to our interpretation of the largest root. The pair of characteristic vectors associated with the ith largest characteristic root represents linear combinations of the *X*s and the *Y*s, respectively, that correlate as highly with each other as possible, subject to the conditions that they be uncorrelated with the first i - 1 canonical variates. These successive roots thus describe s uncorrelated ways in which the two sets of measures relate to each other, that is, the s "factors" they have in common.

The investigator will of course wish to concentrate his or her interpretive efforts on those pairs of canonical variates that he or she is sure represent more than just random fluctuation in the data. The procedure for testing the *i*th canonical R and associated pair of canonical variates for statistical significance is identical to that for testing the first, largest canonical R, except that the first degree-of-freedom parameter s_i now equals min($p - i + 1$, $q - i + 1$). The other two parameters, m and n, remain the same. However, as pointed out in section 4.4, the tests of all but the first of these canonical Rs are conservative.

As might be suspected from our knowledge of the Pearson r, the value of the canonical R between two sets of variables is unaffected by linear transformations of the variables within either set. In particular, precisely the same canonical R (or canonical Rs) is (are) obtained when the variables are first transformed to standard-score form, which of course results in the substitution of correlation matrices for the covariance matrices in the formulae we have developed so far. The numerical values of **R**-derived canonical coefficients will be different from the **S**-derived coefficients, but the difference simply reflects the same combined variable described either in terms of raw scores (**S**-derived coefficients) or in terms of standard scores (**R**-derived coefficients).

5.1.1 Heuristic Justification of Canonical Formulae

The formulae just given for canonical ***R*** and the canonical coefficients look—and in fact are, as shown when we discuss explicit formulae for simple cases—quite complicated. However, the product matrices whose characteristic roots and vectors must be obtained are highly patterned. The matrix used to obtain the canonical variates for the left-hand set of variables (the X variables) begins with the inverse of the matrix of covariances of the Xs, $\mathbf{S}_x^{-1}$, and the matrix used to obtain the canonical variates for the right-hand set of variables begins with $\mathbf{S}_y^{-1}$. Both products involve both inverses ($\mathbf{S}_x^{-1}$ and $\mathbf{S}_y^{-1}$) and both between-set covariance matrices ($\mathbf{S}_{xy}$ and $\mathbf{S}'_{xy} = \mathbf{S}_{yx}$) that make up the complete $(p + q)$ x $(p + q)$ covariance matrix, with inverses and between-set covariance matrices alternating in such a way as to maintain conformability of the dimensions, that is, in such a way that the indices also alternate (assuming $\mathbf{S}'_{xy}$ to be rewritten as $\mathbf{S}_{yx}$).

More importantly, the products "make sense " in terms of the multiple regression formulae with which we are already familiar. For instance, if we single out for attention the *j*th column of $\mathbf{S}_{xy}$, $\mathbf{s}_{xy_j}$, then

$$\mathbf{s'}_{\mathbf{xy}_j}\mathbf{S}_{\mathbf{x}}^{-1}\mathbf{s'}_{\mathbf{xy}_j}$$

is clearly equal to $s_j^2(R^2_{y_j \cdot x})$, that is, to that portion of the variation among scores on Y_j that is predictable from the optimum linear combination of the Xs, in other words, to the variance of the predicted scores on Y_j due to the (linear) effects of the Xs. Thus

$\mathbf{S}'_{xy}\mathbf{S}_x^{-1}\mathbf{S}_{xy}$ represents the *q* x *q* covariance matrix of the *q* residual variates. (The algorithms in computer programs for stepwise multiple regression, discussed in section D4.4.3, involve computation of just such matrices.) Thus $\mathbf{S}_y^{-1}\mathbf{S}'_{xy}\mathbf{S}_x^{-1}\mathbf{S}_{xy}$ is a matrix analog of $(s_y^2 R^2_{y \cdot x})/s_y^2 = R^2_{y \cdot x}$, and it is indeed equal to that value when there is only one *Y* variable.

As Derivation 5.2 shows, the coefficients of the *i*th left-hand canonical variate $\mathbf{Xa}_i$ are exactly proportional to the regression coefficients for predicting $\mathbf{Yb}_j$ (the *i*th right-hand canonical variate) from scores on the *X*s. Similarly, the matrix equation defining the canonical variates for the set of *X* variables implies an equality between canonical *R* and the multiple *R* resulting from predicting a particular linear combination of the *X* variables (one of the left-hand canonical variates) from knowledge of scores on the *Y* variables. Thus the process of finding characteristic roots and vectors of $\mathbf{S}_y^{-1}\mathbf{S}'_{xy}\mathbf{S}_x^{-1}\mathbf{S}_{xy}$ and $\mathbf{S}_x^{-1}\mathbf{S}_{xy}\mathbf{S}_y^{-1}\mathbf{S}'_{xy}$ amounts to a partitioning of the variance of each set predicted by knowledge of scores on variables in the other set (i.e., the variance shared by the two sets of variables) into portions attributable to s different pairs of linear combinations of the variables in the two sets.

Beginning with a complex set of *pq* between-set correlations whose interpretation should take into account the *p*(*p* - 1)/2 + *q*(*q* - 1)/2 within-set correlations, we arrive at a new set of s = min(*p*, *q*) pairs of variables—each a simple linear combination of the original variables in its set—having the following extremely simple correlation matrix:

$$
\begin{array}{c|cccccc|cccccc|}
 & u_1 & u_2 & \cdot & \cdot & u_{s-1} & u_s & v_1 & v_2 & \cdot & \cdot & v_{s-1} & v_s \\
\hline
u_1 & 1 & 0 & \cdot & \cdot & 0 & 0 & R_1 & 0 & 0 & \cdot & 0 & 0 \\
u_2 & 0 & 1 & \cdot & \cdot & 0 & 0 & 0 & R_2 & 0 & \cdot & 0 & 0 \\
\cdot & \cdot & \cdot & \cdot & \cdot & \cdot & \cdot & 0 & 0 & \cdot & \cdot & \cdot & \cdot \\
\cdot & \cdot & \cdot & \cdot & \cdot & \cdot & \cdot & \cdot & \cdot & \cdot & \cdot & 0 & 0 \\
u_{s-1} & 0 & 0 & \cdot & \cdot & 1 & 0 & 0 & 0 & \cdot & \cdot & R_{s-1} & 0 \\
u_s & 0 & 0 & \cdot & \cdot & 0 & 1 & 0 & 0 & \cdot & \cdot & 0 & R_s \\
\hline
v_1 & R_1 & 0 & 0 & \cdot & 0 & 0 & 1 & 0 & \cdot & \cdot & 0 & 0 \\
v_2 & 0 & R_2 & 0 & \cdot & 0 & 0 & 0 & 1 & \cdot & \cdot & 0 & 0 \\
\cdot & 0 & 0 & \cdot & \cdot & \cdot & \cdot & \cdot & \cdot & \cdot & \cdot & \cdot & \cdot \\
\cdot & \cdot & \cdot & \cdot & \cdot & 0 & 0 & \cdot & \cdot & \cdot & \cdot & \cdot & \cdot \\
v_{s-1} & 0 & 0 & \cdot & \cdot & R_{s-1} & 0 & 0 & 0 & \cdot & \cdot & 1 & 0 \\
v_s & 0 & 0 & \cdot & \cdot & 0 & R_s & 0 & 0 & \cdot & \cdot & 0 & 1 \\
\end{array}
$$

where R_i is the *i*th largest coefficient of canonical correlation.

5.1.2 Simple Cases of Canonical Correlations

In this section, we give explicit scalar (single-symbol as opposed to matrix form) formulae for "simple" cases in which canonical analysis might be appropriate. Note that because of the extreme complexity of the single-symbol expressions that result from multiplying together four matrices, two of which are inverses of the original correlation matrices, the p or $q = 3$ case has not been included. The reader will find it much easier to employ the matrix operations for computing R_c and the canonical variates, rather than trying to "plug into" single-symbol expressions for any case beyond $p = q = 2$.

For $q = 1$,

$$R_c^2 = \mathbf{s}'_{xy}\,\mathbf{S}_x^{-1}\mathbf{s}_{xy} = R_{Y\bullet X}^2$$

is the squared multiple correlation between the single Y score and the p X scores;

$$\mathbf{a} = \mathbf{S}_x^{-1}\mathbf{s}_{xy}$$

gives the regression coefficients for predicting Y from the Xs; and **b** is undefined, because multiplication of the single Y score by any constant has no effect on its correlation with any other variable.

A parallel expression holds, of course, for $\boldsymbol{p} = 1$.

For $p = q = 2$,

$$R_c^2 = [m_s \pm \sqrt{m_s^2 - 4(r_{13}r_{24} - r_{14}r_{23})^2}\,]/2,$$

where

$$\begin{aligned} m_s &= \frac{\begin{bmatrix} r_{13}^2 + r_{14}^2 + r_{23}^2 + r_{24}^2 - 2r_{13}r_{14}r_{34} - 2r_{23}r_{24}r_{34} \\ -2r_{13}r_{23}r_{12} - 2r_{14}r_{24}r_{12} + 2r_{12}r_{34}(r_{13}r_{24} + r_{14}r_{23}) \end{bmatrix}}{(1 - r_{12}^2)(1 - r_{34}^2)} \\ &= [R_{1\bullet 3,4}^2 + R_{2\bullet 3,4}^2 - 2r_{12}C_1]/(1 - r_{12}^2) \\ &= [R_{3\bullet 1,2}^2 + R_{4\bullet 1,2}^2 - 2r_{34}C_2]/(1 - r_{34}^2); \\ C_1 &= [r_{13}r_{23} + r_{14}r_{24} - r_{34}(r_{13}r_{24} + r_{23}r_{14})]/(1 - r_{34}^2); \\ C_2 &= [r_{13}r_{14} + r_{23}r_{24} - r_{34}(r_{13}r_{24} + r_{23}r_{14})]/(1 - r_{12}^2); \end{aligned}$$

and

$$R_{i\bullet j,k}^2 = [r_{ij}^2 + r_{ik}^2 - 2r_{ij}r_{ik}r_{jk}]/(1 - r_{ik}^2)$$

is the squared multiple correlation of variable i predicted from variables j and k

$$a_1/a_2 = (r_{12}R^2_{2\bullet 3,4} - C_1)/[R^2_{1\bullet 3,4} - r_{12}C_1 - (1-r^2_{12})R^2_c]$$
$$= [(1-r^2_{12})R^2_c + r_{12}C_1 - R^2_{2\bullet 3,4}]/(C_1 - r_{12}R^2_{1\bullet 3,4})$$

and

$$b_1/b_2 = (r_{34}R^2_{4\bullet 1,2} - C_2)/[R^2_{3\bullet 1,2} - r_{34}C_2 - (1-r^2_{34})R^2_c]$$
$$= [(1-r^2_{34})R^2_c + r_{34}C_2 - R^2_{4\bullet 1,2}]/(C_2 - r_{34}R^2_{3\bullet 1,2})$$

When $r_{12} = r_{34} = 0$ (i.e., when the variables within each set are uncorrelated), these expressions become

$$R^2_c = \tfrac{1}{2}[r^2_{13} + r^2_{14} + r^2_{23} + r^2_{24} \pm \sqrt{(r^2_{13} + r^2_{14} + r^2_{23} + r^2_{24})^2 - 4(r_{13}r_{24} - r_{14}r_{23})^2}\,];$$
$$a_1/a_2 = (r_{13}r_{23} + r_{14}r_{24})/(R^2_c - r^2_{13} - r^2_{14})$$
$$= (R^2_c - r^2_{23} - r^2_{24})/(r_{13}r_{23} + r_{14}r_{24});$$

and

$$b_1/b_2 = (r_{13}r_{14} + r_{23}r_{24})/(R^2_c - r^2_{13} - r^2_{14})$$
$$= (R^2_c - r^2_{13} - r^2_{14})/(r_{13}r_{14} + r_{23}r_{24});$$

Variables within Each Set Uncorrelated. For $\mathbf{R}_x = \mathbf{R}_y = \mathbf{I}$,

R_c is the characteristic root of $\mathbf{R}_{xy}\mathbf{R}'_{xy}$ or of $\mathbf{R}'_{xy}\mathbf{R}_{xy}$, where of course

$$\mathbf{R}_{xy}\mathbf{R}'_{xy} = \begin{bmatrix} \sum_{j=1}^{q} r^2_{x_1y_j} & \sum_{j=1}^{q} r_{x_1y_j}r_{x_2y_j} & \cdots & \sum_{j=1}^{q} r_{x_1y_j}r_{x_py_j} \\ \sum_{j=1}^{q} r_{x_1y_j}r_{x_2y_j} & \sum_{j=1}^{q} r^2_{x_2y_j} & \cdots & \sum_{j=1}^{q} r_{x_2y_j}r_{x_py_j} \\ \vdots & \vdots & \vdots & \vdots \\ \sum_{j=1}^{q} r_{x_1y_j}r_{x_2y_j} & \sum_{j=1}^{q} r_{x_2y_j}r_{x_py_j} & \cdots & \sum_{j=1}^{q} r^2_{x_py_j} \end{bmatrix}$$

and $\mathbf{R}'_{xy}$ is a q x q matrix whose (i, j)th entry is the sum of the cross products of columns i and j of $\mathbf{R}_{xy}$. The characteristic vectors of $\mathbf{R}_{xy}\mathbf{R}'_{xy}$ yield the canonical variates for the left-hand set of variables, and the characteristic vectors of $\mathbf{R}_{xy}'\mathbf{R}_{xy}$ yield the right-hand canonical variates.

5.1.3 Example of Canonical Analysis

Data collected in the fall of 1970 by Professor Douglas P. Ferraro (Ferraro & Billings, 1974a,1974b) on drug use among undergraduates at the University of New Mexico serve to illustrate the use of canonical analysis in uncovering interesting relationships between two sets of variables. The items on this anonymous questionnaire included a number of questions on general drug use (from cigarettes to "hard" narcotics such as heroin), as well as a set of questions explicitly directed toward marijuana usage and a number of questions tapping general background variables (age, sex, and so on). These data thus afford the opportunity to investigate (among other things) the extent to which various constellations of background variables are predictive of drug use. Focusing on marijuana for the moment, Table 5.1 reports the correlations between eight background variables (sex, age, class in school, parents' annual income, degree of religious activity, political activity, degree of agreement with parents on Vietnam, and degree of agreement with parents on premarital sex and abortion) and six questions pertaining to marijuana usage (how often the respondent had smoked marijuana, the age when he or she had first smoked it, how frequently he or she currently smoked it, whether the respondent would use it in the future if he or she had the chance, whether he or she feels pot use should be restricted to those 18 years of age or older, and whether he or she feels marijuana use should be legalized). The between-set correlations reported in Table 5.1 are of course insufficient as the basis for a canonical analysis, because we must consider the within-set correlations as well. Table 5.1 is provided so as to give the reader some feeling for the interpretive chore involved in the usual approach of examining these 48 correlations one at a time.

Table 5.2 reports the six coefficients of canonical correlation, together with the corresponding pairs of canonical variates. The overall null hypothesis is hardly in doubt. Because these data came from the responses of 2675 students to the anonymous questionnaire, we have $m = (|8 - 6| - 1)/2 = ½$ and $n = (2675 - 16)/2 = 1329.5$, with s being 6 for R_c^2, 5 for R_2^2, and so on. Taking $n = 1000$ as the basis for a conservative critical value, we see from the tables in Appendix A that each of the first three coefficients of canonical correlation is statistically significant at the .01 level. The first pair of canonical variates seem to be identifying a tendency for young, politically active, but religiously inactive, respondents who disagree with their parents on the topics of premarital sex and abortion to have tried marijuana at an early age, to have used marijuana frequently in the past, to be in favor of legalizing its use, and to be willing to try marijuana again in the future should the opportunity present itself. Of course, this first canonical correlation is only .551, with the specified linear combination of the background variables accounting

Table 5.1 Correlations of Background Variables with Marijuana Questions

Background variables	Marijuana questions					
	Nmarij	FstMarij	NowUseMj	WldUseMj	RestrMj	LegalMj
Sex	−.186	.174	−.171	.144	.034	−.079
Age	−.080	.202	−.083	.057	−.107	−.052
Class	−.005	.144	−.026	−.008	−.052	.043
Pannuinc	.101	−.102	.109	−.084	.009	.052
ReligAct	.389	−.319	.325	−.349	.100	.329
YrPolAct	−.150	.117	−.136	.125	−.033	−.132
AgrPVN	−.174	.134	−.162	.206	−.067	−.195
AgrPSex	−.261	.246	−.200	.272	−.072	−.308

Note: The questions and the available responses were as follows:
Sex: 0 = male; 1 = female.
Age: 0 = 17 or younger; 1 = 18; 2 = 19; and so on.
Class: 0 = freshman; 1 = sophomore; ... 4 = graduate student.
Pannuinc (Parents' annual income): 0 = under \$5000; 1 = \$5,000 - 10,000; ... 4 = over \$25,000.
ReligAct ("How active are you in your religion?"): 0 = very active; 1 = active; 2 = barely active; 3 = inactive.
YrPolAct ("How active are you politically?"): 0 = very active; ... 3 = inactive.
AgrPVN ("Do you agree with your parents on Vietnam ?"): 0 = strongly disagree; 1 = generally disagree; 2 = do not know; 3 = generally agree; 4 = strongly agree.
AgrPSex ("Do you agree with your parents on premarital sex and abortion?"): 0 = strongly disagree; ... 4 = strongly agree.
NMarij ("How many times have you used marijuana?"): 0 = never; 1 = 1 time; 2 = 2-4 times; 3 = 5-14 times; 4 = 15-30 times; 5 = more than 30 times.
FstMarij ("When did you first use marijuana?"): 1 = elementary school, grades l-6; 2 = junior high school, grades 7-9; 3 = high school, grades 10-12; 4 = college, freshman year; ... 7 = college, senior year; 8 = have never used.
NowUseMj ("How often do you currently use marijuana?"): 0 = do not use marijuana; 1 = less than once per month; 2 = l-4 times per month; 3 = 1-3 times per week; 4 = 4-6 times per week; 5 = 7 or More times per week.
WldUseMj ("If you had the opportunity in the future, would you use marijuana?"): 0 = yes; 1 = no.
RestrMj ("If marijuana were legalized, should it be restricted to those 18 years of age or older?"): 0 = yes; 1 = no.
LegalMj ("Do you feel that society should legalize marijuana at this time?"): 0 = no; 1 = undecided; 2 = yes.

for only about 30% of the total variation in the specified combination of the responses to the marijuana questions. The second canonical *R* of .344 seems to suggest primarily that older, junior-senior respondents have used marijuana extensively in the past but began this use late in their academic career. (As Herbert H. Blumberg, personal communication, 1983, pointed out, both tendencies may be an artifact of the greater number of years older

Table 5.2 Canonical Analysis of Background Variables versus Marijuana Questions

i:	1	2	3	4[a]	5[a]	6[a]
R_i^2:	.3035	.1182	.0254	.0062	.0042	.0008
R_i:	.551	.344	.159	.079	.065	.028
Sex	−.188	.185	*.719*			
Age	−*.309*[b]	*.473*	−*.322*			
Class	.045	*.616*	.155			
Pannuinc	.085	−.078	−*.385*			
ReligAct	*.623*	.180	−.117			
YrPolAct	−*.258*	.031	.011			
AgrPVN	−.194	−.123	−.023			
AgrPSex	−*.395*	.002	−*.504*			
Nmarij	*.295*	*1.815*	−*.581*			
FstMarij	−*.185*	*1.920*	−.228			
NowUseMj	.061	−.508	−*.662*			
WldUseMj	−*.244*	−.209	−.157			
RestrMj	.067	−.147	*.491*			
LegalMj	*.414*	.270	*.723*			

a. Coefficients have been omitted since R_i^2 is nonsignificant.
b. Italicized coefficients are those that were emphasized in developing substantive interpretations of the canonical variates.

students have had "to accumulate heavy and late use." We should keep constantly in mind that the relationships identified in canonical analysis of measured—as opposed to manipulated—variables are *correlational* relationships, subject to a host of possible alternative explanations.) The third R_i of .159 indicates that young females from low-income families who disagree with their parents on the issues of premarital sex and abortion tend not to have used marijuana much in the past nor to be using it at present, but nevertheless to be more in favor of legalizing marijuana for everyone, 18 or not, than most other respondents.

It must be kept in mind that there is some loss in shared variance when we go from the canonical variates as actually computed by the canonical analysis to the "simplified" canonical variates implied by the above verbal descriptions. For instance, the correlation between

$$(\text{-Age} + \text{ReligAct} - \text{YrPolAct} - \text{AgrPSex})$$

and

$$(\text{Nmarij} - \text{FstMarij} + \text{LegalMj} - \text{WldUseMj})$$

is only .539, versus .551 for the first pair of canonical variates. Similarly, the second and third pairs of "simplified" canonical variates yield correlations of .285 (vs. .344) and.148 (vs. .159). In each case, however, the simplified canonical variates still yield correlations that exceed the critical values for the canonical correlations.

At any rate, the three sources of relationship between the two sets of variables uncovered by the canonical analysis seem to be tapping important trends that would have been difficult to uncover through examination of the individual pairwise correlations. The first canonical correlation, with its suggestion that "activist" students who have "tuned out" religion are especially likely to use marijuana, is particularly interesting in light of theories that predict that drug users should be primarily dropouts who have been discouraged by the difficulty of political reform and have turned their interests inward. [For another example of the use of canonical correlation, see Doob and MacDonald (1979), whose study was described briefly in Example 1.5 (chap. l)].

5.2 RELATIONSHIPS TO OTHER STATISTICAL TECHNIQUES

All of the techniques we have talked about in preceding chapters can be seen as special cases of canonical correlation. When one of the two sets of variables consists entirely of group membership dummy variables, one-way Manova results. When $s = 1$, conducting a canonical analysis on the two "sets" of data is equivalent to conducting a multiple regression of the set having two or more variables as predictors of the single variable in the other set. Within a canonical analysis in which both groups have two or more variables, two multiple regression analyses are "imbedded." If we compute u (the canonical variate for the left-hand set of variables) for each case and then conduct a multiple regression analysis of the relation between u and the right-hand set of variables, the resulting set of regression coefficients will be identical to the coefficients that define u. Similarly, a multiple regression analysis of v predicted from the p variables of the left–hand set produces u as the best combination of those p variables for predicting v. This suggests an iterative (and unfortunately very inefficient) procedure for finding u and v by regression analyses involving successive "guesses" as to the coefficients defining u and v.

For those readers who are not easily satisfied by the hand-waving approach to statistics, a proof of the relationship between canonical analysis and Manova is given in Derivation 5.3, which may of course be skipped without serious damage to the reader's understanding of multivariate statistics. The relationship between canonical analysis and multiple regression when $s = 1$ is detailed in section 5.1.2 and can easily be derived from the formulae given in chapter 2. The relationship between canonical analysis and multiple regression in the more general case was presented mathematically in Derivation 5.2 and is demonstrated in the demonstration problems at the end of this chapter. Figure 1.1 from chapter 1 displays these relationships graphically.

Knapp (1978) gave examples of still other techniques (e.g., chi-square contingency table analysis and tests of differences among correlated proportions) as special cases of Canona and argued for its use as a "general parametric significance-testing system." See, however, Isaac & Milligan (1983), who pointed out that Knapp (along with

Holland, Levi, & Watson, 1980) misinterprets the overall likelihood-ratio chi-square test of section 5.3 as a test of the first canonical variate. Isaac and Milligan also pointed out that some of Knapp's suggested applications imply new approaches to commonly used significance tests—approaches that have not as yet been subjected to systematic analytic or Monte Carlo exploration by mathematical statisticians. Most of these differences from current approaches appear to involve substituting an *F* test for a chi-square test, which is equivalent to replacing an assumption of infinite degrees of freedom (i.e., of known population variance) with the finite and often quite small degrees of freedom associated with a sample variance estimate. For instance, McNemar's test for the difference between two correlated proportions (e.g., the proportion of students with correct answers to two different items) is equivalent to a *z* ratio for the difference between .5 and the proportion of nonmatching students (those answering only one of the two items correctly) who are more "successful" on the first item. The canonical correlation approach to this same test reduces to a *t* test of the hypothesis that the mean difference between dichotomized scores on the two items is zero in the populations. (The *t* test uses the observed sample variance of the difference score, whereas the *z* test uses the known population variance of a dichotomous variable with generating probability of exactly .5.) Because there are a number of reasons for expecting the *t*-test approach to have greater power than the *z*-ratio approach when the null hypothesis is not true, I hope that mathematical statisticians will take the discrepancy between the two approaches (and similar differences between other current approaches and their canonical-correlation-derived alternatives) as an issue worthy of study, rather than as a basis for rejecting the Canona-based approaches.

One final, parenthetical note on Isaac and Milligan (1983): These authors correctly cited Mendoza, Markos, and Gonter's (1978) claim that the step-down gcr procedure recommended in sections 4.4, 4.5.3, and 5.1 is more conservative than partitioned-*U* tests in establishing the number of statistically significant canonical correlations. However, Mendoza and associates' conclusion was based on their Monte Carlo runs, which showed differences between the two procedures' rejection rates that were (a) very small and (b) not tested for statistical significance. My own runs, reported in Harris (1985a) and summarized earlier in section 4.5.3, also showed trivial, nonsignificant differences between the conservatism of the gcr-based test of the *i*th root and the partitioned-*U* test of the residual from the first *i* roots.

Finally, Browne (1979) pointed out that a particular variety of factor analysis (interbattery FA) is closely related to Canona. In particular, "interbattery factor loadings are merely rescaled canonical loadings" (p. 82). This would seem to imply that all other techniques are various special cases of factor analysis—despite the fact that this book represents factor analysis as a purely within-set technique and thus omits it entirely from Figure 1.1 in chapter 1. If we consider any computation of the characteristic roots and vectors of any covariance matrix or of any function of covariance matrices to be an instance of factor analysis, this is clearly true. Indeed, inter-battery factor analysis begins with extraction of the eigenvalues and eigenvectors of a between-set correlation matrix $\mathbf{R}_{xy}$ that has been corrected for the correlations among the variables within each of the two sets (batteries) of measures—that is, with the **R**-based equivalent of Equations (5.2). However, all other varieties of FA of which I am aware begin with a single, uncorrected, overall correlation matrix. This importation of between-set considerations into factor analysis is laudable, but whether that thereby makes all between-set techniques (such as those discussed in chaps. 2 through 5) special cases of factor analysis is a matter of choice of terminology.

5.3 LIKELIHOOD-RATIO TESTS OF RELATIONSHIPS AMONG SETS OF VARIABLES

Just as there are likelihood-ratio tests as competitors to the λ_{max} tests in Manova, so are there likelihood-ratio tests of the overall null hypothesis of no relationship between two sets of variables. Specifically,

$$H_0\colon \mathbf{S}_{xy} = \mathbf{0}$$

is tested by comparing

$$\chi^2 = -(N - \frac{p+q}{2} - \frac{3}{2})\ln\left(\frac{|\mathbf{S}|}{|\mathbf{S}_x||\mathbf{S}_y|}\right)$$

$$= -(N - \frac{p+q}{2} - \frac{3}{2})\sum_{i=1}^{s}\ln(1 - R_i^2), \qquad (5.4)$$

where R_i is the ith coefficient of canonical correlation, with the chi-square distribution having pq degrees of freedom, where **S** is the $(p + q)$ x $(p + q)$ covariance matrix of all the variables, ignoring the division into subsets. This test makes good intuitive sense, because the ratio of determinants involved can be rewritten (using the formulae for determinants of partitioned matrices given in Digression 2) as $|\mathbf{S}_y - \mathbf{S}'_{xy}\mathbf{S}_x^{-1}\mathbf{S}_{xy}|/|\mathbf{S}_y|$, that is, the ratio of the "generalized variance" of the set of X variables with the effects of the Y variables "partialed out" to the "generalized variance" of the "raw" Y variables. The concept of generalized variance is discussed in the next chapter. Another interpretation of Equation (5.4), given in the second line, is based on the product of the portions of variance left unexplained by the first, second, ..., sth canonical correlations. The big advantage of Equation (5.4) over the gcr test is that Equation (5.4) is readily generalizable to a test of the hypothesis of no relationships between any pair of k sets of variables, as shown in the next section.

The last expression in Equation (5.4) makes clear the fact that the likelihood-ratio test of the significance of canonical correlation, like the corresponding tests for Manova, essentially involves a pooled test of all sources of linear relationship between the two sets of variables. The likelihood-ratio test therefore provides a test of the overall null hypothesis that $\mathbf{S}_{xy} = 0$, but it must not be interpreted (as, unfortunately, many authors do, as was discussed in section 4.5) as a test of the statistical significance of R_c^2.

Nor may the individual components of $-M \ln U$, $-M \ln(1 - R_i^2)$, be treated as chi-square variates having $p + q - (2i - 1)$ degrees of freedom, Tatsuoka (1971c) notwithstanding. Nor may the residual after "removing" the effects of the first i variates,

$$-M \sum_{j=i+1}^{s} \ln(1 - R_j^2),$$

be treated as a chi-square variate having $(p - i)(q - i)$ degrees of freedom, unless we acknowledge that this test (a) is quite conservative unless the roots removed have near-unity population values and (b) does ***not*** establish the statistical significance of any single R_i and its associated pair of canonical variates. See section 4.5 of this book, Lancaster (1963), and Harris (1976b) for further details on the bases for these proscriptions against partitioned-U tests. Recall, at a minimum, that for $s = 10$ (a not unusually high value for practical applications of Canona), and for a sample size over 200, the partitioned-U test of R_c^2 yields true Type I error rates of 61% for tests at a nominal α of .05 and 24% for tests at the nominal .01 level.

You may also wish to review the discussion in section 4.5 of the relative power and robustness of the gcr versus multiple-root tests (Wilks' lambda and the Hotelling and Pillai trace statistics) as tests of the overall null hypothesis (for which, if it is one's sole interest, the trace statistics are often superior) and as beginning points for identification of particular dimensions along which the two sets of measures are related (for which gcr-based specific comparisons are uniformly more powerful). These points apply equally well to canonical analysis.

5.4 GENERALIZATION AND SPECIALIZATION OF CANONICAL ANALYSIS

A number of variations on the procedures described in this chapter have been proposed. For instance, Horst (196la, 1961b) provided computational algorithms for a situation in which m sets of variables are to be related by selecting a linear transformation of the variables within each set such that the sum of the $m(m - 1)/2$ intercorrelations between combined variables is as large as possible. Carroll (1968) approached Horst's problem by maximizing the largest characteristic root of the matrix of intercorrelations among the m canonical variates. Roy (1957) discussed the canonical correlation between two sets of *residual* scores (each set having been corrected for the linear effects of the same third matrix). Also Meredith (1964) proposed that canonical analysis be based on covariance matrices that have been "corrected for" unreliability, in the spirit of factor analytic procedures (cf. chapter 7). Significance tests for these methods have not been developed, but they may nevertheless be useful descriptively. The reader is referred to the papers just cited for computational procedures, which are not presented in detail in this book. Instead, a maximum-likelihood significance test for the null hypothesis of no relationships between any pair of m sets of variables—which logically precedes either Horst's or Carroll's analyses or pairwise canonical analyses in accordance with the procedures of this chapter—will be presented. Then the consequences of requiring that the canonical variates for the left- and right-hand sets of variables be identical, will be explored. We also discuss ***rotation*** of canonical variates as an aid to interpretation, the use of the ***redundancy coefficient*** as a measure of total variance two sets of measures have in common, and (finally) some of the statistical techniques that are not available in canonical analysis—a frustratingly large set.

5.4.1 Testing the Independence of *m* Sets of Variables

Wilks (1935) provided a test of the m-set case that is a straightforward generalization of the test described in Equation (5.4). Box (1949) subsequently improved the accuracy of the test by modifying the multiplicative constant. Specifically, we test

$$H_0\text{: } \Sigma_{ij} = \mathbf{0} \quad \text{for i} \neq \text{j}$$

where Σ_{ij} is a submatrix containing the population covariances of the variables in set i with those in set j, by comparing

$$\begin{aligned} \chi^2 &= -C_1 \ln \frac{|\mathbf{S}|}{|\mathbf{S}_1||\mathbf{S}_2|\Lambda|\mathbf{S}_m|} \\ &= -C_1 \ln \frac{|\mathbf{R}|}{|\mathbf{R}_1||\mathbf{R}_2|\Lambda|\mathbf{R}_m|} \end{aligned}, \tag{5.5}$$

with the chi-square distribution having $f = C_2/2$ degrees of freedom, where p_i is the number of variables in set i; $\mathbf{S}$ is the covariance matrix of all Σp_i variables; $\mathbf{S}_i$ is the p_i x p_i covariance matrix of the variables in set i and

$$C_1 = N - 1 - \frac{3C_2 + 2C_3}{6C_2};$$

$$2C_2 = \left(\sum p_i\right)^2 - \sum p_i^2;$$

$$C_3 = \left(\sum p_i\right)^3 - \sum p_i^3;$$

with $C_1 = N - 3/2 - \dfrac{p(m+1)}{3}$ when all m submatrices are of size p. A demonstration of this test may be in order.

Example 5.2 *Consistency of Behavior across Different Experimental Games* Ron Flint, in his University of New Mexico doctoral dissertation study (1970), had each subject play each of four different kinds of experimental games: Prisoner's Dilemma (PD), Chicken, Battle of Sexes (BS), and Apology. Half of the subjects (the fixed subjects) had the same "partner" in all four games, and the other subjects (the mixed subjects) had a different partner for each game. Crossed with this mixed-fixed partner variable was the payoff procedure. One of two payoff procedures was employed: the *regular* payoff of one mil per point earned or the *ranked payoff* of seven graduated "prizes" (50 cents, $1, $1.50, $2.00, $2.50, $3.00, or $3.50), depending on subject's ranking in terms of total points earned across all four games, relative to other subjects occupying the same role (row player or column player). To summarize, the four experimental conditions were:

FREG: fixed partner, regular payoff procedure,
FRANK: fixed partner, ranked payoff procedure,
MREG: mixed partners, regular payoff procedure,
MRANK: mixed partners, ranked payoff procedure.

The main focus of the research was on the consistency of individual differences in behavior across games and the amount of attenuation in between-game correlations produced by "mixing" partners. Canonical analysis of the relationship between behavior in each pair of games seems a "natural" for this problem. It might, however, be useful to have an overall test of the null hypothesis of no consistency whatever (zero intercorrelations) between behavior in one game and in any other game. We demonstrate application of the Wilks–Box test of Equation (5.5) to testing this overall null hypothesis for the behavior of the 20 subject–pairs in the FREG condition, measuring this behavior in terms of four dependent variables: the by now nauseatingly familiar CC, CD, and DC (see Data Set 4), plus ρ_o, the correlation (phi coefficient, really) between the two players' choices across the 30 trials of each game.

The 16 x 16 correlation matrix for these measures is given in Table 5.3. The overall determinant of this 16 x 16 matrix is $|\mathbf{R}| = .39865 \times 10^{-11}$, and the four within-game submatrices have determinants $|\mathbf{R}_{PD}| = .40933$, $|\mathbf{R}_{Ch}| = .11227$, $|\mathbf{R}_{BS}| = .18821 \times 10^{-1}$, and $|\mathbf{R}_{Apol}| = .80276 \times x\ 10^{-2}$. Further, because we have same-sized matrices, we can use the simpler formula for C_1 yielding $C_1 = 20 - 1.5 - 4(5)/3 = 11.8333$ and

$$\chi^2 = -11.8333 \ln \frac{.39865 \times 10^{-11}}{(.40933)(.11227)(.018821)(.0080276)}$$

$$= -11.8333 \ln(5.7415 \times 10^{-7})$$

$$= -11.8333(-14.37) = 170.05 \text{ with 96 degrees of freedom, } p < .01.$$

We can thus be confident that there is some degree of consistency in these subject-pairs' behavior across

Table 5.3 Correlation Matrix for Game OutcomeVariables, Flint (1970)

	Prisoner's Dilemma				*Chicken*				*Battle of the Sexes*				*Apology*			
	CC	*CD*	*DC*	ρ_0	*CC*	*CD*	*DC*	ρ_0	*CC*	*CD*	*DC*	ρ_0	*CC*	*CD*	*DC*	ρ_0
CC	1.0	−.263	−.489	.439	−.450	.672	.230	−.267	−.168	.320	−.010	.114	−.043	−.188	−.049	.240
CD		1.0	−.087	−.441	.795	−.472	−.747	.448	.533	−.154	−.587	−.051	−.563	.605	.237	−.500
DC			1.0	−.370	.212	−.433	−.034	.121	.003	−.249	.227	−.260	−.114	−.145	.458	−.351
ρ_0				1.0	−.507	.502	.379	.176	−.538	.463	.330	.138	.231	−.158	−.223	.280
CC					1.0	−.689	−.752	.400	.447	−.188	−.379	−.281	−.589	.569	.401	−.657
CD						1.0	.233	−.308	−.342	.374	.110	.241	.430	−.375	−.539	.626
DC							1.0	−.427	−.484	.034	.609	.018	.355	−.564	.056	.382
ρ_0								1.0	.153	−.023	−.203	.116	−.112	.117	.093	−.192
CC									1.0	−.804	−.876	.569	−.254	.234	.134	−.246
CD										1.0	.518	−.619	.016	.087	−.192	.132
DC											1.0	−.580	.213	−.262	.087	.045
ρ_0												1.0	.326	−.259	−.283	.332
CC													1.0	−.652	−.765	.785
CD														1.0	.169	−.738
DC															1.0	−.700

the four kinds of game. To get an (admittedly inflated) estimate of the *magnitude* of this consistency, we can compute the six pairwise canonical correlation coefficients, yielding squared canonical *R*s for pairs of games:

	PD	Ch	BS	Apol
PD	1	.8000	.7510	.7235
Ch		1	.7454	.7254
BS			1	.5813

For each of these pairwise R_c^2 s, we have $s = \min(4,4) = 4$; $m = (|4 - 4| - 1)/2 = -.5$; and $n = (20 - 4 - 4)/2 = 5$, yielding $\theta_{crit} = .708$ at the .05 level and .788 at the .01 level. If we wish to control our experimentwise alpha at .05, we would require a Bonferroni–adjusted critical value of (4,-.5,5), which program GCRCMP reveals to be -.792. Only the PD versus Chicken canonical *R* exceeds this critical value—though that fact would be enough to reject our overall null hypothesis had we wished to use this Bonferroni-adjustment procedure as an alternative to the pleasures of Equation (5.5). Overall, the pairwise s were lower for the Mixed conditions than for the Fixed conditions (mean of .541 vs. .695), thus *suggesting* that between-game consistency is enhanced by keeping the same partner (even though row- and column-players were separated by a partition and thus did not know whether or not they were paired with the same partner as they moved from booth to booth). However, I have no idea of how to test this difference for statistical significance.

Finally, we should point out how very useful repeated-battery Canona (next section) would be for these data. For instance, the canonical correlation of .894 between behavior in Prisoner's Dilemma and behavior in Chicken is based on the correlation between $-.417z_{CC} + 1.287z_{CD}, + .405z_{DC} + .094z_{DD},$ and $.405z_{CC} - .646\, z_{CD} - .607\, z_{DC} + .222\, z_{DD}$. The pairs of canonical variates for the other comparisons show a similarly distressing tendency to differ. It would make so much sense to restrict our analysis to "matching" canonical variate coefficients for each pair of games—and perhaps to the same coefficients within every game.

5.4.2 Repeated-Battery Canona

One common situation in which the question of the relationship between two or more sets of variables arises, occurs when an experimental treatment or a battery of tests is administered repeatedly to the same group of subjects. The canonical correlation between any two administrations of the test battery could then be considered a generalized reliability coefficient, and the canonical correlation between any two repetitions of the treatment (i.e., any two trials) would afford a measure of the stability of behavior under those conditions. Finally, if different treatments were administered on different occasions, but the set of dependent measures used to assess treatment effects was the same on all occasions, the canonical R between any two such occasions would be a measure of the generalizability of the subjects' behavioral tendencies (predispositions, traits) across experimental conditions. (Note that this is the obverse of the usual interest in establishing *differences* in behavior as a function of conditions.)

In all of these examples, it would seem natural to require that the *same* linear combinations of each set of variables be employed in computing canonical correlations, rather than leaving the computing algorithm free to pick different canonical variates for the two sets. Thus, in the two-set case, we would seek to maximize

$$r^2(\mathbf{a}) = (\mathbf{a'S}_{xy}\mathbf{a})^2 / [(\mathbf{a'S}_x\mathbf{a})\,(\mathbf{a'S}_y\mathbf{a})], \tag{5.9}$$

rather than the square of the expression in Equation (5.1). In the m-set case, we would adopt Horst's criterion (mentioned earlier) of maximizing the sum of the $m(m - 1)/2$ intercorrelations between pairs of combined variables, but with the added restriction that the same combining coefficients be used within each set. I (Harris, 1973a) developed expressions for the special case in which the various $p \times p$ main diagonal submatrices of the population covariance matrices are identical, but not for the more general case.

The solution under the assumption of homogeneity of main-diagonal submatrices requires taking the greatest characteristic root and associated characteristic vector of $(\overline{\mathbf{S}})^{-1}\overline{\mathbf{S}}_{ij}$(s), where $\overline{\mathbf{S}}$ is the simple average of the m main-diagonal submatrices and $\overline{\mathbf{S}}_{ij}$ is the simple average of the $m(m - 1)/2$ above-diagonal submatrices. (DeGroot & Li,

1966, arrived at an identical expression for the more restrictive case where $m = 2$ and where the two population mean vectors are identical.) The sampling distribution of the resulting correlation measure has not been derived.

As the second edition of this book was going to press, I discovered a paper that provides solutions to the descriptive aspects of repeated-battery Canona. Meredith and Tisak (1982) show that the single linear combination of the variables in a repeated battery that has the highest possible average between-occasion covariance, subject to the constraint that its average variance across conditions equal 1.0, can be obtained via the following steps:

1. Compute

$$\mathbf{P} = \frac{1}{m}\sum_{i=1}^{m}\mathbf{R}_{ii} \quad \text{and} \quad \mathbf{Q} = \frac{1}{m}\sum_{i=1}^{m}\sum_{j=1}^{m}\mathbf{R}_{ij},$$

where m = the number of occasions on which the battery of p measures is assessed, and $\mathbf{R}_{ij}$ is the matrix of correlations between scores on occasion i and scores on occasion j. ($\mathbf{R}_{ii}$ is of course the matrix of correlations among the p variables on a single occasion i.)

2. Compute any factorization of **P**; that is, compute **F** such that $\mathbf{P} = \mathbf{FF}'$. One very simple approach, for instance, is to compute the eigenvalues and eigenvectors of **P** and then take $\mathbf{F} = \mathbf{AD}^{1/2}$, where **D** is a p x p matrix whose jth column is the characteristic vector associated with the eigenvalue in the jth main-diagonal element of **D**. (We show in chapter 6 that this is one aspect of principal components analysis.)

3. Compute $\Gamma = \mathbf{F}^{-1}\mathbf{Q}(\mathbf{F}^{-1})'$.

4. Compute the largest eigenvalue of Γ and its associated eigenvector. If the computer program you are using doesn't automatically normalize the eigenvectors, normalize it yourself by dividing each coefficient a_{1j} by $\sqrt{\sum a_{1j}^2}$, the square root of the sum of the squared coefficients. Label this normalized characteristic vector associated with the largest characteristic root of Γ as $\mathbf{u}_1$.

5. The p weights defining the repeated-battery canonical variate are then given by the elements of $\mathbf{w}_1 = (\mathbf{F}^{-1})'\,\mathbf{u}_1$.

These steps can be demonstrated with respect to the experimental game data of Example 5.3 (Flint, 1970) in the form of the MATLAB (Moler, 1982) commands used to compute the repeated-battery canonical variate for (CC, CD, DC, ρ_0) across the four different games that subjects played:

```
R = <1.0    -.263 -.489 ... -.188 -.049  .240
     -.263   1.0 -.087 ...  .605  .237 -.500
                 :
      .240 -.500 -.351 ... -.738 -.700  1.0 >
P = ( R(1:4,1:4) + R(5:8,5:8) + R(9:12,9:12) + R(13:16,13:16) )/4
Q = 4*P + 2*( R(1:4,5:8) + R(1:4,9:12) + R(1:4,13:16) )
```

```
Q = Q + 2*( R(5:8,9:12) + R(5:8,13:16) + R(9:12,13:16) )
<V,D> = EIG(P)        [This step sets D = diagonal matrix of eigenvalues, while the
F = V*SQRT(D)                                 columns of V give the eigenvectors.]
GAMMA = INV(F)*Q*INV(F')
<U,PHI> = EIG(GAMMA)
U1 = U(1:4,4)         [Column 4 selected because the 4th eigenvalue was the largest one.]
DEN = SQRT(U1'*U1)
U1 = U1/DEN           [This normalizes u1.]
W1 = INV(F')*U1
DW1 = KRON(I(4),W1)
SW1 = DW1'*R*DW1      [This computes the covariance matrix of the m scores on w1.]
MKR = SQRT(DIAG(DIAG(SW1)))
RW1 = INV(MKR)*SW1*INV(MKR)   [This computes the matrix of between-occasion
                                correlations for the (first) canonical variate.]
```

The results of this MATLAB run included:

$$\mathbf{P} = \begin{bmatrix} 1 & -.6020 & -.7205 & -.54825 \\ & 1 & .20825 & -.5265 \\ & & 1 & -.51925 \\ & & & 1 \end{bmatrix};$$

$$\mathbf{Q} = \begin{bmatrix} 1.8860 & .4300 & -2.2280 & -\ .0010 \\ & 4.1300 & -2.6030 & -\ .3140 \\ & & 6.8060 & -2.1670 \\ & & & 1 \end{bmatrix};$$

$$\mathbf{F} = \begin{bmatrix} -.2504 & .3283 & .0989 & .9052 \\ -.1679 & -.1843 & .6523 & -.7158 \\ -.2127 & -.0469 & -.5965 & -.7725 \\ -.0717 & -.5773 & -.1034 & .8068 \end{bmatrix}; \quad \text{and}$$

$$\mathbf{\Gamma} = \begin{bmatrix} 3.9220 & 0.3471 & .3177 & 1.5609 \\ 1.3572 & 8.0076 & .0104 & -2.2760 \\ 1.7735 & -1.1346 & 9.7760 & 1.3688 \\ -.1528 & -1.1938 & 1.4907 & 2.4448 \end{bmatrix};$$

so that the largest root of $\Gamma = 10.3833$; associated eigenvector is

$$[.0976 \;\; -.1511 \;\; 1.0648 \;\; .2208]';$$

normalized vector is

$$\mathbf{u}_1 = [.0885 \;\; -.1371 \;\; -.9660 \;\; .2003]';$$

and

$$\mathbf{w}_1 = [-.0619 \;\; .6781 \;\; -.8985 \;\; .0589]';$$

leading to

$$\mathbf{S}_{W_1} = \begin{bmatrix} 1.3489 & .4507 & .5998 & .6175 \\ & .9824 & .5713 & .5494 \\ & & .6211 & .4301 \\ & & & 1.4076 \end{bmatrix} \text{ and } \mathbf{R}_{W_1} = \begin{bmatrix} 1 & .3915 & .6553 & .5195 \\ & 1 & .7313 & .5415 \\ & & 1 & .4997 \\ & & & 1 \end{bmatrix}.$$

Thus we have found that $-.0619z_{CD} + .6721z_{CD} - .8985z_{DC} + .0589z_{p0}$ has a higher average between-game covariance than any other linear combination of the four measures in our battery with an average variance of unity. (Note that the trace of $\mathbf{S}_{W_1}$ is indeed 4.000.) Given the restriction to average variance of unity, this linear combination probably also maximizes the average between-game (more generally, between-condition or between-occasion) correlation—but this has not yet been proved. At any rate, the average between-game correlation yielded by this repeated-battery canonical variate is .556. Our canonical variate seems quite close to $z_{CD} - z_{DC}$, which represents the greater tendency for column player to double-cross row player (leading to a DC outcome) than vice versa. Given that CD and DC can be expected to have nearly identical variances, this would also be close to CD - DC, which in these games is directly proportional to the amount by which column player's winnings exceed row player's winnings. This suggests that there is a tendency for the same player of a continuing partnership to come out ahead across all four games. To demonstrate that not a great deal of between-game correlation is sacrificed in going to this simplified, highly interpretable variable, the between-game correlations for $z_{CD} - z_{DC}$ were computed, yielding

$$\mathbf{R}_{\text{CD}-\text{DC}} = \begin{bmatrix} 1 & .3961 & .6279 & .5108 \\ & 1 & .6899 & .4910 \\ & & 1 & .4961 \\ & & & 1 \end{bmatrix}$$

and an average between-game correlation of .531—fairly close to the corresponding average of .556 for the full-decimal repeated-battery canonical variate.

On the other hand, both of these averages are well below the average of our six pairwise canonical variates, namely, .848. (At this late stage in your growing familiarity with the logic of multivariate statistics, it should be very clear to you why each of the six correlations in $\mathbf{R}_{W_1}$ must be lower in absolute value than the corresponding unrestricted canonical correlation between those two games.) I suspect that this degree of drop in average correlation in going from unrestricted canonical correlations to repeated-battery correlations is typical and that the critical value for a repeated-battery correlation should therefore be much lower than for an unrestricted canonical correlation. Once again, however, significance tests for repeated-battery Canona have not yet been developed—Meredith and Tisak (1982, p. 63) "reserve this topic for subsequent development."

As in ordinary Canona, repeated-battery Canona does not stop with the identification of a single repeated-battery canonical variate. A total of p such variates, each having an average between-occasion covariance of zero with the preceding repeated-battery canonical variates, is obtained. For instance, the second largest eigenvalue of Γ for the present data is 8.261, with an associated $\mathbf{w}_2$ of [.462 .019 - .618 - 1.1361' . All p repeated-battery canonical variates can be obtained simultaneously from

$$\mathbf{W} = (\mathbf{F}^{-1})' \, \mathbf{U} \, .$$

Finally, Meredith and Tisak also considered an alternative criterion for selecting a repeated-battery canonical variate, namely, that the largest eigenvalue of the resulting $\mathbf{S}_{W_1}$—that linear combination of the m scores on $z_{W1} = W_{11}z_1 + W_{12}z_2 + \cdots + W_{1p}z_p$ that accounts for the largest possible proportion of the between-occasion variance—be maximized. This has the advantage of considering the possibility that a highly relevant battery of measures may be negatively correlated between some pairs of measurement occasions or conditions but positively correlated for others. However, finding such a battery requires an iterative procedure in which a trial vector of p combining weights for the battery items is selected, the largest eigenvalue of the resulting m x m covariance matrix is computed, the optimal battery for this linear combination of the measurement occasions is computed, and so on. There is, moreover, no guarantee that this iterative procedure will converge on the optimal battery rather than settling down at a "plateau."

Clearly, then, there is still considerable mathematical work to be done on repeated-battery Canona. Meredith and Tisak have, however, brought us much closer to a workable procedure.

5.4.3 Rotation of Canonical Variates

This section is marked (#) as expendable in a first reading because it employs techniques that will not be presented in detail until section 6.5. However, many readers will have some familiarity with the notion of rotation within the context of factor analysis. If you are not a member of this group, you may wish to delay reading this section until after you read section 6.5.

The initial transformation of our $(p + q)$ variables into 2·min(p, q) variables having the extremely simple pattern of intercorrelations described in section 5.1.1, although very elegant, has at least two potential disadvantages:

1. The "lion's share" of the shared variance between the two sets is usually accounted for by the first pair of canonical variates, with each successive pair yielding a considerable lower R_i^2 than the preceding ones. This is a very useful property when one is trying to condense the between-set relationships into as few dimensions as possible. However, for any given number of conceptual dimensions, it may be reasonable to expect the contributions of the underlying conceptual variables to the between-set relationships to be nearly equal.

2. The canonical variates are not always easy to interpret substantively, especially when each original variable contributes importantly to more than one of the statistically significant canonical variates. (Several colleagues have assured me that this greatly

understates the difficulty of interpreting canonical variates. However, given the large number of possibilities provided by simple averages of, and differences between simple averages of, various subsets of the variables, I seldom find it difficult to find a simplified canonical variate that can be justified on theoretical or logical grounds. If you're having trouble coming up with interpretations, try rereading section 1.3.1.}

Both of these disadvantages could be overcome if the statistically significant canonical variates were subjected to additional transformations such that each transformed (rotated) variate was (a) a linear combination of the $d \leq s$ canonical variates retained from the canonical analysis and thus also a linear combination of the original variables, and (b) uncorrelated with the other transformed canonical variates except for its single "matched" variable from the other set of measures, and if, further, (c) the coefficients relating rotated canonical variates to original variables assumed a theoretically or readily interpretable pattern (e.g., with each original variable receiving a relatively large *z*-score weight in the definition of only one of the rotated canonical variates).

We cannot meet all of these desiderata simultaneously. However, the same sorts of rotation techniques discussed in section 6.5 guarantee that desideratum (a) can be applied to the *s* *X*-based (left-hand) canonical variates and to the *s* *Y*-based (right-hand) canonical variates (*cvs*) separately, so that we can retain zero intercorrelations among the rotated *cvs* for each set while using desideratum (c) to select one from the potentially infinite number of such sets of rotated *cvs*. The resulting rotated *cvs* will not, however, show the simple pattern of between-set correlations yielded by the canonical analysis; each rotated *X*-based canonical variate will have nonzero correlations with each of (rather than just one of) the rotated *Y*-based canonical variates. This may, however, be a small price to pay for the gains in interpretability provided by the rotation process—especially because it is often found that the correlations between the two sets of rotated *cvs* are still heavily concentrated into $d \leq s < s^2$ correlations between "matching" rotated *cvs*. Interested readers are referred to section 6.5 for the specific formulae needed to rotate each set of *cvs* to a pattern of coefficients satisfying the varimax criterion of interpretability (by far the most common criterion), to Cliff and Krus (1976) for examples of the advantages of rotation, and to Wilkinson (1977) for procedures for and examples of rotation to a theoretically specified pattern of coefficients.

Rotation of canonical variates poses no new significance-testing problems, as long as it is treated as a descriptive tool facilitating the post hoc search for specific between-set relationships that meet the gcr-based post hoc critical value. However, a few points need to be kept in mind:

First, most users of the rotation procedure will, as an unfortunate carryover from common factor analytic practice, choose their rotation (and base their subsequent interpretations) on the basis of the resulting loadings of each original variable on (its zero-order correlation with) each rotated canonical variate, rather than on the rotated canonical variate coefficients (the raw-score or *z*-score weights actually used to compute a subjects's score on a given rotated canonical variate). Computer programs (such as SPSS's MANOVA) that provide the option of rotating canonical variates typically permit only loadings-based criteria. However, the loading of a particular variable on a particular rotated (or unrotated) canonical variate only tells us how well that single variable could substitute for the canonical variate if we had to "make do" with this single variable. It need tell us nothing about the role of that variable in defining the sort of person who receives a high score on the canonical variate when his or her scores on all of the variables in the set are known. Interpretation of canonical variates (rotated or not) should therefore be based on the canonical variate coefficients and not on the loadings. This being the case, it seems clear that our rotation process should seek a simple pattern of canonical variate coefficients, rather than of canonical variate loadings.

[If your goal is to find that single variable that relates most strongly with the other set, this would be best accomplished by examining the q multiple correlations between your q Y measures and the set of p X variables. This is efficiently accomplished via the equation

$$[R^2_{Y_i \bullet X}] = \mathbf{R}'_{xy} \mathbf{R}_x^{-1} \mathbf{R}_{xy},$$

and is known as ***multivariate multiple regression analysis.*** Note that, as the absence of R_y^{-1} indicates, multivariate multiple regression does not take the correlations among the criteria into account (as indeed we don't want it to, given the limited goal under discussion) and is simply an efficient way of carrying out several separate MRAs.]

Second, the set of defining weights yielded by rotation of the canonical variates is not the stopping point of our analysis. Each of the rotated *cvs* must be given a substantive interpretation and the between-set correlations based on the simplified weights implied by our substantive interpretation must be examined.

Third, because the pattern of correlations between the set of rotated *cvs* is no longer the simple one displayed on page 203, all d^2 between-set correlations must be computed and examined (rather than only the d correlations between "matching" *cvs*). This is also true of simplified versions of unrotated *cvs*, but the intercorrelations of "nonmatching" canonical variates are apt to be considerably more nonzero after rotation.

Example 5.3 *A Canonical Cautionary.* In the spirit of Lord's (1953) "On the statistical treatment of football numbers," I put together a hypothetical data set and accompanying scenario (Harris, 1989) that I hoped would make obvious to all readers the superiority of canonical variate coefficients over structure coefficients as bases for interpreting canonical variates. To quote myself,

> The beefy-breasted bowery bird (BBBB) was in very serious trouble. Its last known natural nesting habitat (the alleys behind bars in the Bowery) was scheduled for elimination [in] an urban renewal project. Its only hope for avoiding extinction was the development of an in-captivity breeding program ....
>
> Unfortunately, BBBBs are both strictly monogamous and very picky about their mates. Because no zoo could afford to "host" more than one pair ... of BBBBs (their staple food, stale Manhattan Mountain red wine, was becoming as rare and expensive as the eucalyptus leaves koalas insist on), it was crucial ... that only compatible birds be paired. (p. 17)

Into this gap steps ethologist Bird E. Watcher, who recognizes that Canona could be used to identify what characteristics of male BBBBs female BBBBs find attractive and vice versa. He therefore takes four measures on each of 100 happily mated BBBB couples: MTL (length of male's head from tip of beak to base of skull), MSL (male skull length, from base of beak to base of skull), FTL (female total head length), and FSL (female skull length. Measurements on the 100 pairs fit neatly into the five clusters shown in Figure 5.1a, yielding the correlation matrix displayed at the top of Figure 5.1b. A

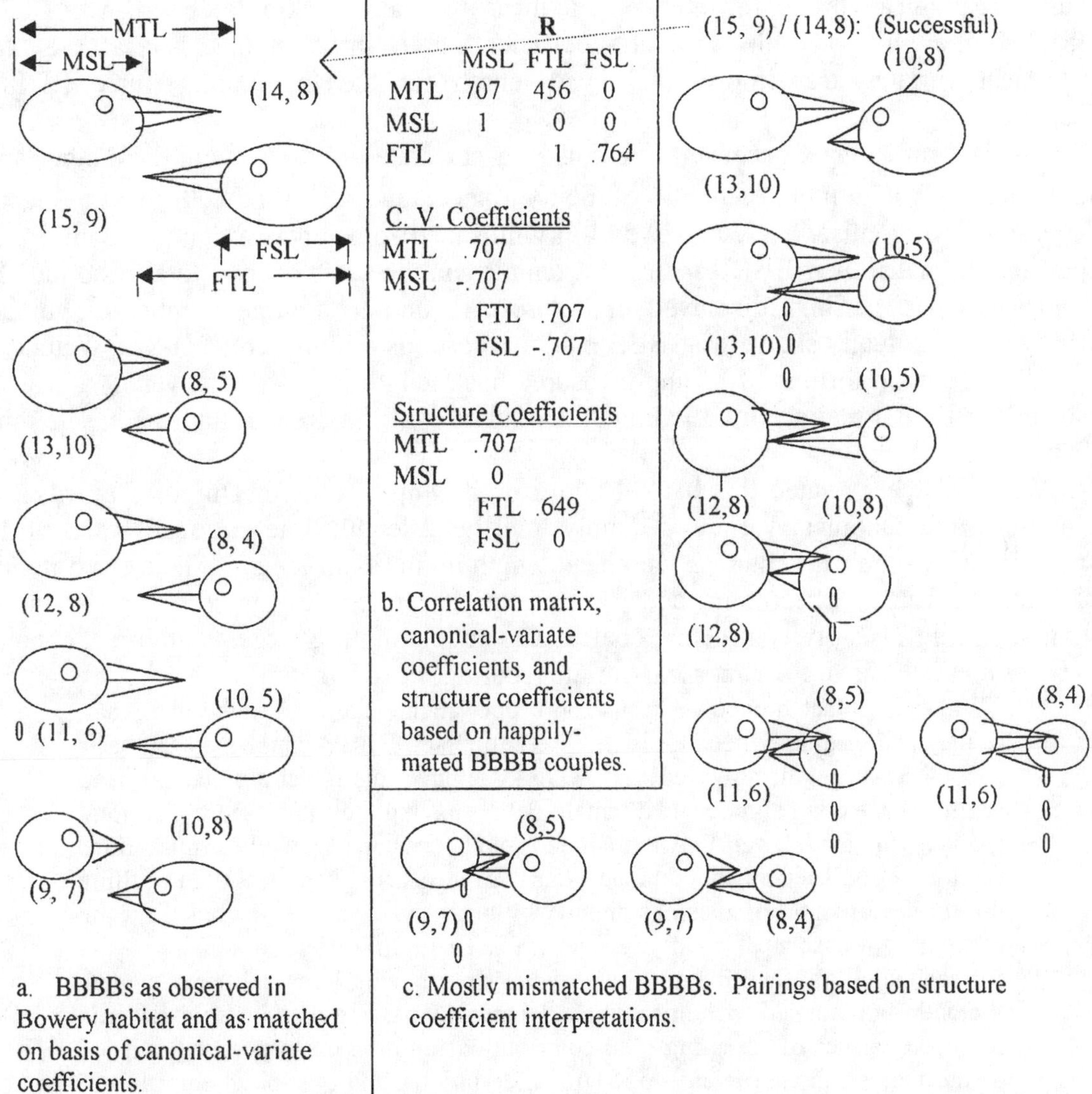

a. BBBBs as observed in Bowery habitat and as matched on basis of canonical-variate coefficients.

b. Correlation matrix, canonical-variate coefficients, and structure coefficients based on happily-mated BBBB couples.

c. Mostly mismatched BBBBs. Pairings based on structure coefficient interpretations.

Fig. 5.1 Naturally occuring versus Canona-based pairings of Beefy-Breasted Bowery Birds (BBBBs). Figure prepared from data and analyses thereof in Harris (1989)

canonical analysis conducted on the relationship between the two male measures and the two female measures leads to a canonical R of 1.00 between .707MTL - .707MSL and

.707FTL - .707FSL—in essence, between the male and female beak lengths. Indeed, the reader can readily verify that for every happily mated BBBB the male's beak length (MTL - MSL) is identical to the female's beak length.

Unfortunately Professor Watcher makes the mistake of taking his analysis to a statistical consultant (Professor Loh Ding) who, like most of his colleagues, is convinced that canonical variates should be interpreted on the basis of the structure coefficients, rather than on the scoring coefficients. In the present case MSL has a zero correlation with the canonical variate for the two male measures and FSL correlates zero with FTL - FSL, so the only possible structure-coefficient-based interpretation is that the two canonical variates are simply total head length for the male and female BBBB, respectively.

This is a bit strange, because MTL and FTL correlate only .456 with each other and thus share only a bit more than 20% of their variance, rather than the 100% that the R_c of 1.0 seemed to promise, but Watcher dutifully informed zookeepers around the world that they should match their BBBB males and females on the basis of their total head length. This led to the pairings displayed in Figure 5.1c and to a large number of either frustrated or mutually skewered BBBB pairs because (as a junior colleague of Watcher's who had the good fortune of having measured beak length directly discovered) a crucial part of the BBBB mating ritual is mutual stimulation of the area around the base of the beak.

When Watcher related this baleful tale to Loh Ding, Professor Ding explained that, although each canonical variate *is* simply total head length, one *measures* each bird's score on that canonical variate as total head length minus skull length. [This corresponds to a quite commonly recommended "compromise" strategy of interpreting canonical variates on the basis of the structure coefficients but computing scores on those canonical variates on the basis of the canonical-variate (scoring) coefficients.]

> Watcher contacted once again his zoo-keeper clientele and informed them that, as they had originally been told, they were to match BBBB couples on the basis of total head length; however, they were to achieve this matching of total head length by selecting a male and a female BBBB having the same "score" on (total head length - skull length). Although one could occasionally overhear mutterings among the zoo keepers about the peculiarity of measuring beak length but calling it total head length, or musings about why birds that had been matched on the conceptual variable of total head length seemed to differ so much on the corresponding physical measure while invariably having identical beak lengths, Watcher soon quelled such misgivings by carefully passing on Ding's wise words about the virtues of assessing total contributions and the corresponding evils of partial regression coefficients, operational defintions, and dust-bowl empiricism. Most important, of course, the clarified matching system worked splendidly.... When BBBB males and females were matched on this measure, they got along splendidly, and 100% of BBBB couples so matched produced offspring. (pp. 22-23)

This cautionary tale is followed in the paper by an explicit discussion of its moral(s) and of various counterarguments, including a debunking of the myth that multiple regression analysis arbitrarily chooses one of two or more highly correlated predictors to assign a large regression weight while assigning near-zero weights to the rest of the predictors in that cluster. The main point is simply that each of the various proposals for interpreting canonical variates or assessing the importance of the contributions of various original measures to the canonical variate falls flat on its face when applied to the BBBB scenario—except of course for interpreting and scoring the canonical variates on the basis of the canonical-variate coefficients.

5.4.4 The Redundancy Coefficient

A shortcoming of R_c^2 as a measure of the relationship between two sets of measures is illustrated by the following hypothetical between-set correlation matrix:

$$\begin{array}{c} \\ X_1 \\ X_2 \\ \vdots \\ X_{20} \end{array} \begin{array}{c} \begin{array}{ccccc} Y_1 & Y_2 & Y_3 & \cdots & Y_{10} \end{array} \\ \begin{bmatrix} 0 & 1.0 & 0 & \cdots & 0 \\ 0 & 0 & 0 & \cdots & 0 \\ & \vdots & & & \\ 0 & 0 & 0 & \cdots & 0 \end{bmatrix} \end{array}$$

In this example, the only between-set relationship of any kind arises from the perfect correlation between X_1 and Y_2. It is intuitively obvious that our set of 20 X-measures and our set of 10 Ys have almost no variance in common; yet $R_c^2 = 1.0$. Stewart and Love (1968) proposed an alternative or supplementary measure of the overall degree of relationship between two sets of measures that seems more reasonable than R_c^2 in this and many other situations. Their ***redundancy coefficient*** $r_{Y \bullet X}^2$ for the proportion of Y-variation accounted for by the Xs is simply the sum of the squared loadings of each Y on each of the s Y-based canonical variates, divided by q, the number of Ys in our right-hand set. A similar definition holds for $r_{X \bullet Y}^2$. For the example just given, then, $r_{Y \bullet X}^2 = 1/10$, whereas $r_{X \bullet Y}^2 = 1/20$. Note that the two redundancy coefficients are *not* symmetric unless $p = q$.

There was initially a lively debate over the properties of the redundancy coefficient(s). (See Gleason, 1976, for a useful summary and DeSarbo, 1981, for further extensions of redundancy analysis.) The most important point to keep in mind about the redundancy coefficient(s) is that it is interpretable as the proportion of the variance in one set of measures that is accounted for by the measures in the other set only if we are referring to standardized (z-score) measures. Because any z_Y or z_X has a variance of 1.0, this ensures that $r_{Y \bullet X}^2$ will differ from $r_{X \bullet Y}^2$ by a factor of (p/q), whether or not this is intuitively reasonable. If we wish to talk about the proportion of explained variance in the metric of the original measures (assuming that such a common metric exists), we must modify our procedure for computing $r_{Y \bullet X}^2$ by multiplying each squared loading by the variance of Y_j before carrying out our sum, and then dividing by the sum of the variances of all the Ys. Thus, for instance, if $s_{Y2}^2 = 100$ and all other s_Y^2 s = 2, "raw" $r_{Y \bullet X}^2 = 100/118$ rather than 1/10. Clearly the redundancy coefficient is very severely affected by changes of unit and can be given a meaningful interpretation only if a common metric is established within each set of measures. In many cases conversion to z scores (the conversion implicitly assumed by Stewart & Love, 1968) will be the most reasonable way of establishing such a common metric. In others (e.g., where the q Ys are responses to the same dependent variable on each of several trials or percentage of items on various MMPI subscales that are answered in the abnormal direction) a common metric will already exist, and conversion to z scores would destroy meaningful information conveyed by differential variances.

There is a large literature on various aspects of redundancy analysis. Van den Wollenberg (1977) pointed out that if one is really interested in the overall predictability of one set of measures from another, then one should select linear combinations of measures that directly maximize between-set redundancy, rather than measuring the (lower) redundancy of canonical variates. Cramer and Nicewander (1979) suggested a number of "symmetric, invariant measures of multivariate association," each of them some function of the s canonical R_is. Muller (1981) showed that redundancy analysis is equivalent to predicting each of the measures in one set from a separate multiple regression analysis of its relationship to the rotated principal components (chap. 6) of the variables in the other set. This demonstrates again that redundancy assesses total predictability of the separate variables in one set from the various dimensions underlying the other set of variables, whereas canonical correlation measures the maximum possible mutual predictability of (association between) any single dimension of each of the two sets.

Bruce Thompson asserted that the equivalence between the redundancy coefficient and the average of the multiple R^2 also shows that the redundancy coefficient "is not truly multivariate" (1991, p. 89) and is instead an "averaged univariate statistic" (1990, p. 19) that therefore does not, as true multivariate methods do, "honor the [multivariate] reality to which the researcher is purportedly trying to generalize" (1990, p. 80) . True. However, Thompson (1990, 1991) also recommends interpreting canonical variates primarily on the basis of the structure coefficients, whose relative magnitudes are, as we saw in the Canonical Cautionary (example 5.3) totally insensitive to the correlations among the variables that define a given canonical variate. They are also purely—indeed, not even averaged—univariate statistics telling us how well we can reproduce scores on the canonical variate from a single original variable, and they thus do not honor the multivariate reality that *combinations* of variables are almost always (and often quite substantially) more strongly related to a given criterion than are single variables. It seems strange to be very concerned about ignoring the correlations among the variables in the outcome-variable set, but not at all concerned about ignoring the correlations among the variables in the predictor-variable set.

Finally, DeSarbo (1981) provided a method for selecting variates that provides a compromise between maximizing measures of association and measures of redundancy.

5.4.5 What's Missing from Canonical Analysis?

The fact that Canona is a straightforward generalization of MRA suggests that we should have available Canona counterparts for each of the measures and significance tests discussed in chapter 2. Unfortunately, this is not the case. In particular, none of the following three extrapolations from multiple regression analysis is currently (2001) available:

1. *An unbiased estimate of the population value squared canonical correlation,* ρ_c^2. We know that R_c^2 has an expected value larger than max $(p, q)/(N - 1)$, because every sample value of R_c^2 must exceed the largest R^2 between any single variable in one of our sets of measures and the variables in the other set. We also have some limited Monte Carlo evidence, from the same runs that were the basis for Harris (1976b), that even $(p + q)/(N - 1)$ is an underestimate of the expected value of R_c^2.

Thompson (1990) attempted to derive an approximation formula via multiple regression analyses on extensive Monte Carlo runs, but was able to achieve an R_c^2 between the sample bias in R_c^2 (population ρ_c^2 - mean sample R_c^2) and the best two predictors thereof (ratio of number of subjects to $p + q$ and total sample size) of only .523 to .787 for the 11 different sets of Monte Carlo runs he examined. His Monte Carlo results did, however, provide evidence that the degree of positive bias in R_c^2 is actually quite small as long as a ratio of N to $(p+q)$ of 10 or more is employed and/or the true squared population R_c^2 is .4 or higher. He also reported that extending Wherry's (1931) correction to the Canona situation—i.e., taking

$$1-\left(\frac{N-1}{N-p-q-1}\right)(1-R_c^2)$$

as the estimate of ρ_c^2—leads to estimates that respond to variations in N, $(p+q)$, and the true value of ρ_c^2 in the same way that the Monte Carlo results indicate such an estimate should. The Wherry formula can't be fully accurate, however, since we know the null distribution of R_c^2 depends on three parameters, two of which—s and m—Wherry's correction collapses into the single parameter, $p+q$. Moreover, Thompson's results do *not* support his claim that the Wherry formula overcorrects R_c^2, so that "the researcher can be reasonably certain [ρ_c^2] will lie between the adjusted [R_c^2] and the calculated [R_c^2]" (p. 29). For instance, for the case where $p = q = 6$, $N = 3$, and $\rho_c^2 = 0$ Thompson's Table 3 reports a mean sample R_c^2 of .447; the Wherry correction yields for that case an estimate of .158, so the average researcher investigating the relationship between two nearly unrelated sets of variables and following Thompson's advice would mistakenly assume a ρ_c value between .40 and .67. More generally, in over one-fourth of the parameter combinations explored by Thompson the Wherry correction *over*estimated ρ_c^2. Thompson's Monte Carlo investigations have gotten us closer to but have nevertheless left us short of the goal of unbiased estimation of ρ_c^2.

2. *A test of the increment to R_c^2 yielded by adding k variables to one of the two sets of measures.* Clearly the magnitude of the increment to R_c^2 must be greater than the corresponding increment to $R_{Y \bullet X}^2$, because in the Canona case our optimization routine is free to consider new combinations of the *outcome* variables as well as new combinations of the predictor variables. We thus *cannot* use the MRA test as a sound basis for deriving an appropriate significance test for Canona.

To illustrate the problem, I wrote a "quickie" program that drew 200 random samples of size 20 from a four-variable multivariate normal population in which all between-set correlations (r_{13}, r_{14}, r_{23}, and r_{24}) were zero, and computed for each sample (a) R_c^2 for X_1 and X_2 versus X_3 and X_4; (b) R_c^2 (actually multiple R^2) for X_1 versus X_3 and X_4; (c) the difference between these two R_c^2 s, ΔR_c^2; (d) $F_1 = \dfrac{\Delta R_c^2}{(1-R_c^2)/(N-2-1)}$; and (e) $F_2 = \dfrac{\Delta R_c^2}{(1-R_c^2)/(N-4-1)}$. F_1 and F_2 represent two attempts to generalize the MRA test of ΔR^2 to Canona. However, in the 200 iterations, F_1 exceeded $F_{.05}(1,17)$ 24 times (effective α of .12) and exceeded its .01-level critical value 7 times. Both rejection rates were significantly higher than their nominal alphas. F_2 exceeded its .05-level critical value 18 times, and it exceeded $F_{.01}(1,15)$ three times. (The first $\alpha_{\text{effective}}$ is significantly greater than .05, but the departure from the nominal alpha of .01 is not statistically significant.) The discrepancy from nominal alpha of the F_2 test in this example might be considered tolerable—for example, not much greater than the discrepancy most researchers are willing to tolerate in assessing robustness of various statistics. However, this particular example is probably one of the more "benign" situations; we can expect the discrepancy to be much greater in situations involving more variables in either set, for instance.

3. *A test of the statistical significance of a given canonical variate coefficient.* This is equivalent to the $k = 1$ case of the test in (2); thus it suffers from the same problem of a shift in outcome-measure weights as well as predictor-variable weights when the new

variable is added.

Limitations 2 and 3 are especially nettlesome when we consider that, just as MRA is the appropriate technique for unequal-*n* Anova, so unequal-*n* Manova is most appropriately conceptualized as a special case of Canona in which one of the two sets of measures consists of group-membership variables or between-group contrast scores. The appropriate test for the overall effect of a multiple-*df* factor is the increment to R_c^2 that results when all of the group-membership (or contrast) variables representing that factor are added last—and we have just seen that no such test exists. On the other hand, the fact that this test *does* exist in Manova (it is, for instance, routinely employed in SPSS MANOVA and in SAS's PROC GLM) suggests that extending the Manova test to the more general case of Canona might prove to be a trivial task for a friendly mathematical statistician. In particular, as Derivation 5.3 demonstrates, the Manova test is equivalent to testing the statistical significance of the gcr of $(\mathbf{S}_y - \mathbf{P}_{reduced})^{-1}(\mathbf{P}_{full} - \mathbf{P}_{reduced})$, where $\mathbf{P}_{full} = \mathbf{S}'_{xy}\mathbf{S}_x^{-1}\mathbf{S}_{xy}$ based on all group-membership variables and $\mathbf{P}_{reduced}$ is the corresponding matrix product when all group-membership variables *except* the ones representing the effect being tested are included in $\mathbf{S}_x$ and $\mathbf{S}_{xy}$. Could it be that this test is equally valid when the *X*s are not group-membership or contrast scores? A set of 200 Monte Carlo runs for the $p = q = 2$ case suggests so, because it yielded 7 (vs. the expected 10) declarations of significance at the .05 level and 3 (vs. the expected 2) at the .01 level, with neither departure from expectation being statistically significant. More comprehensive analytic and/or Monte Carlo work is needed.

5.5 Computerized Canonical Correlation

5.5.1 Matrix-Manipulation Systems

5.5.1.1 *MATLAB.*

```
R = [entries of overall correlation matrix, with 1st p rows and columns
      representing the X set of variables and the next (last) q rows and
      columns representing the Y set of variables]
Rxx = R(1:p,1:p)
Rxy = R(1:p,p+1:p+q)
Ryy = R(p+1:p+q,p+1:p+q)
[CVX, LX] = eig(inv(Rxx)*Rxy*Inv(Ryy)*Rxy')
[CVY, LY] = EIG(Inv(Ryy)*Rxy'*Inv(Rxx)*Rxy)
```

For instance, the canonical correlation(s) and canonical variates for the relationship between behavior in Prisoner's Dilemma and behavior in Chicken (Example 5.2, section

5.4.1) can be obtained via

```
R =[1      -.263 -.489  .439  -.450   .672   .230  -.267
    -.263  1     -.087 -.441   .795  -.472  -.747   .448
    -.489 -.087  1     -.370   .212  -.433  -.034   .121
     .439 -.441 -.370  1      -.507   .502   .379   .176
    -.450  .795  .212 -.507  1      -.689  -.752   .400
     .672 -.472 -.433  .502  -.689  1       .233  -.308
     .230 -.747 -.034  .379  -.752   .233  1      -.427
    -.267  .448  .121  .176   .400  -.308  -.427  1     ];
Rxx = R(1:4,1:4);
Rxy = R(1:4,5:8);
Ryy = R(5:8,5:8);
[CVX,LX] = eig(inv(Rxx)*Rxy*inv(Ryy)*Rxy')
CVX =

   -0.2945    0.5969   -0.7901    0.0403
    0.9094    0.1331   -0.4252    0.4662
    0.2862    0.7912    0.2677    0.2349
    0.0663   -0.0039    0.3510    0.8520


LX =

    0.8000         0         0         0
         0    0.0015         0         0
         0         0    0.2947         0
         0         0         0    0.4826

[CVY,LY] = eig(inv(Ryy)*Rxy'*inv(Rxx)*Rxy)

CVY =

    0.4055   -0.7420   -0.6434   -0.3653
   -0.6463   -0.2928   -0.7293   -0.5250
   -0.6073   -0.6031    0.0944   -0.4022
    0.2216    0.0064    0.2130   -0.6551

LY =

    0.8000         0         0         0
         0    0.0015         0         0
         0         0    0.2947         0
         0         0         0    0.4826
```

Thus we know that the (first) squared canonical R between the two sets of variables is .800, and $r_2^2 = .4826$. (Note that the eigenvalues—listed along the main diagonal of the second matrix specified on the left side of the *eig* command—are not ordered with respect to size.) Moreover, the first pair of canonical variates is -.2945*CC* + .9094*CD* + .2862*DC* + .0663ρ_0 for Prisoner's Dilemma versus .4055*CC* - .6463*CD* - .6073*DC* + .2216ρ_0 for Chicken, and that the coefficients for the pair of canonical variates associated with r_2^2 are given in the *fourth* (not second) column of CVX and CVY—that is, the columns of the first-named matrix "match up" with the main-diagonal entries of the second-named matrix when you run the *eig* command.

Note, too, that command and matrix names are case sensitive (at least on the version running on UNM's Ultrix system), so that typing RXX instead of Rxx, or EIG instead of eig, will get you an error message in place of the output you were seeking; and that ending a command with a semicolon suppresses printing of the output of that command.

Of course we could have entered a raw-data matrix into MATLAB and then used MATLAB commands to compute the correlation matrix, as was demonstrated towards the end of chapter 2 (section 2.4.3.1).

5.5.1.2 *SAS PROC MATRIX and SPSS Matrix/End Matrix.*

Both of these matrix-manipulation systems have eigenvalue/eigenvector functions. However, SPSS and SAS thropugh Version 7 give accurate results only for symmetric matrices, and the matrix products whose eigenvalues must be computed for Canona are seldom symmetric. For instance, the matrix product whose eigenvalues are given by LX in the example of section 5.5.1.1 is

```
 0.3218   -0.0927   -0.2254    0.1205
-0.2535    0.7045    0.0724   -0.1294
-0.1534    0.1728    0.0883    0.0214   .
 0.0380    0.0476   -0.0344    0.4642
```

Fortunately, both systems warn the user about this restriction. For instance, an attempt to use SPSS's matrix command EVAL to obtain the eigenvalues of the above matrix yields the message "Source operand non-symmetric for EVAL" and a refusal to carry out the operation. But this nonetheless leaves the user without the desired eigenvalues and eigenvectors. For this we can invoke a procedure described by Bock (1975, pp. 91–92):

To find the eigenvalues and eigenvectors of $\mathbf{E}^{-1}\mathbf{H}$, where **E** and **H** are square, symmetric matrices, first find **C** (the Choleski factorization of **E**) such that **CC'** = **E**. (SPSS's CHOL function carries out this factorization for you.) Then compute the eigenvalues and eigenvectors of the square, symmetric matrix $\mathbf{C}^{-1}\mathbf{H}(\mathbf{C}^{-1})'$. These eigenvalues are the same as the eigenvalues of $\mathbf{E}^{-1}\mathbf{H}$. The eigenvectors of $\mathbf{E}^{-1}\mathbf{H}$ are then given by pre-multiplying the matrix of eigenvectors of $\mathbf{C}^{-1}\mathbf{H}(\mathbf{C}^{-1})'$ by $(\mathbf{C}^{-1})'$. In the Canona application $\mathbf{r}_{xy}\mathbf{R}_x^{-1}\mathbf{r}'_{xy}$ takes the role of **H** and $\mathbf{R}_y$ takes the role of **E** when computing the canonical variates of the **Y** set of measures, $\mathbf{r}'_{xy}\mathbf{R}_y^{-1}\mathbf{r}_{xy}$ takes the role of **H,** and $\mathbf{R}_x$ takes the role of **E** when computing the canonical variates of the **X** set. The eigenvalues are the same for both computations and of course equal the squared cannonical correlations.

For the PD-versus-Chicken example just analyzed, the SPSS Matrix commands would be:

```
Matrix .
Compute R = {1,-.263,-.489, .439,-.450, .672, .230,-.267;
    -.263,  1  , -.087, -.441, .795,-.472, -.747,  .448;
    -.489, -.087,  1  , -.370, .212,-.433, -.034,  .121;
     .439, -.441,-.370,  1   ,-.507, .502,  .379,  .176;
    -.450,  .795, .212, -.507,  1  ,-.689, -.752,  .400;
     .672, -.472,-.433,  .502,-.689,  1  ,  .233, -.308;
     .230, -.747,-.034,  .379,-.752, .233,   1  , -.427;
    -.267,  .448, .121,  .176, .400,-.308, -.427,   1 } .
Compute Rxx = R(1:4,1:4) .
Compute Rxy = R(1:4,5:8) .
Compute Ryy = R(5:8,5:8) .
Print Rxx .
Print Rxy .
Print Ryy .
  /* First, eigenvalues and vectors for X variables.  */
Compute Hx = Rxy*inv(Ryy)*T(Rxy).
Print Hx / Title = "Rxy*inv(Ryy)*Rxy'".
Compute Err = Rxx .
Compute H = Hx .
Compute CT = Chol(Err) .
Compute C = T(CT) .
Compute RescldH = Inv(C)*H*T(Inv(C)) .
Compute RescldH = (Rnd(1.E9*Rescldh) )/1.E9 .
Call Eigen(RescldH, RescldVs, Lambdas) .
Print Lambdas / Title = "Eigenvalues of Inv(Err)*H." .
Print RescldVs / Title = "Rescaled eigenvectors." .
Compute Vecs = T(Inv(C))*RescldVs .
Print Vecs / Title = "Eigenvectors of inv(Err)*H) for X vars." .
   /* Second,eigvals and eigvecs for Y variables.  * /
Compute Hy = T(Rxy)*inv(Rxx)*Rxy.
Print Hy / Title = "Rxy'*inv(Rxx)*Rxy".
Compute Err = Ryy .
Compute H = Hy .
Compute CT = Chol(Err) .
Compute C = T(CT) .
Compute RescldH = Inv(C)*H*T(Inv(C)) .
Compute RescldH = (Rnd(1.E9*Rescldh) )/1.E9 .
Call Eigen(RescldH, RescldVs, Lambdas) .
Print Lambdas / Title = "Eigenvalues of Inv(Err)*H." .
Print RescldVs / Title = "Rescaled eigenvectors." .
Compute Vecs = T(Inv(C))*RescldVs .
Print Vecs / Title = "Eigenvectors of inv(Err)*H) for Y  vars." .
End Matrix .
```

These commands yield results identical (to at least four decimal places) to those obtained from MATLAB. [The two statements in the preceding listing that include "`Rnd(1.E9*RescldH)`" round off the entries in the matrix product to 9 decimal places; this is necessary because roundoff errors in its calculation would otherwise lead SPSS's *Eigen* function to "think" it's asymmetric and thus to refuse to compute its eigenvalues and eigenvectors.]

The SAS PROC IML function corresponding to SPSS's CHOL is ROOT. And for Version 8 and later versions of IML there is a GENEIGN subroutine: "CALL GENEIGEN(**V**,**L**,**E**,**H**)" takes matrices **E** and **H** as input and puts the eigenvectors of $\mathbf{E}^{-1}\mathbf{H}$ into the columns of matrix **V** and the eigenvalues into the main-diagonal entries of matrix **L**.

5.5.2. SAS PROC CANCORR

The primary emphasis throughout this book has been on SPSS. However, the SAS program for canonical correlation is much more straightforward and easier to use than is SPSS's only alternative (namely, tricking MANOVA into doing the Canona). The basic setup for a SAS PROC CANCORR analysis is:

```
OPTIONS Nocenter LS = 72;
DATA dsname ;
   TITLE 'Title of your choosing';
   LABEL
      VarName1 = 'Label for 1st variable'
      VarName2 = 'Label for 2nd variable'
                 ...
       VarName3 = 'Label for (p+q)th variable';
  INPUT  List of variable names in order they appear
        in the input dataset ;
   CARDS;
    The data, in freefield format.
;
PROC CANCORR S C REDUNDANCY
     VPREFIX=Rootname for X variables
     WPREFIX=Rootname for Y variables
     VNAME= 'Label for X set' WNAME= 'Label for Y set' ;
     VAR Names of X variables;
     WITH Names of Y variables;
     RUN;
```

For example, the setup for a Canona of the relationship between high-school test scores and college grades (taken from an example in the SAS Sample Library—SAS Institute, 1986) is

```
OPTIONS Nocenter LS = 72;
DATA GRADES ;
   TITLE 'HIGH SCHOOL TEST SCORES AND COLLEGE GRADES';
   LABEL
      MAT_TEST = 'MATH       TEST,  HIGH SCHOOL'
      VER_TEST = 'VERBAL     TEST,  HIGH SCHOOL'
      CRE_TEST = 'CREATIVITY TEST,  HIGH SCHOOL'
      MAT_GRAD = 'MATH       GRADE, COLLEGE'
      ENG_GRAD = 'ENGLISH    GRADE, COLLEGE'
      SCI_GRAD = 'SCIENCE    GRADE, COLLEGE'
      HIS_GRAD = 'HISTORY    GRADE, COLLEGE'
HUM_GRAD = 'HUMANITIES GRADE, COLLEGE';
  INPUT  MAT_TEST VER_TEST CRE_TEST
         MAT_GRAD ENG_GRAD SCI_GRAD HIS_GRAD HUM_GRAD ;
   CARDS;
    27     62    46   2.5   2.0   2.9   1.2   1.4
    21     38    76   2.4   1.3   2.3   1.0   2.1
                              …

    77     59    28   1.0   3.9   0.0   3.0   1.6
    31     43    81   2.4   1.3   3.5   2.0   2.6
;
PROC CANCORR S C REDUNDANCY
     VPREFIX=TEST WPREFIX=GRADE
     VNAME= 'HIGH SCHOOL TEST SCORES'  WNAME= 'COLLEGE GRADES' ;
     VAR MAT_TEST VER_TEST CRE_TEST;
     WITH MAT_GRAD ENG_GRAD SCI_GRAD HIS_GRAD HUM_GRAD;
     RUN;
```

PROC CANCORR's output is very clearly labelled, as illustrated by the following selection:

```
                           …
HIGH SCHOOL TEST SCORES AND COLLEGE GRADES
3
                    01:55 Thursday, November 18, 1999

Canonical Correlation Analysis

                             Adjusted        Approx        Squared
          Canonical          Canonical       Standard      Canonical
         Correlation        Correlation       Error       Correlation
   1       0.786866          0.706830        0.087371       0.619159
   2       0.558817          0.446715        0.157775       0.312276
   3       0.298422          0.199275        0.208985       0.089055
                                  …
Test of H0: The canonical correlations in the
                current row and all that follow are zero
```

```
        Likelihood
           Ratio        Approx F        Num DF        Den DF      Pr > F

    1    0.23858896     1.5211            15      33.52812      0.1531
    2    0.62647846     0.8561             8            26      0.5642
    3    0.91094451     0.4562             3            14      0.7171

Multivariate Statistics and F Approximations
S=3    M=0.5    N=5

Statistic                         Value        F  Num DF Den DF   Pr > F
Wilks' Lambda               0.238588963  1.52111  15   33.5281  0.1531
Pillai's Trace              1.02048998   1.44347  15        42  0.1723
Hotelling-Lawley Trace      2.177597886  1.54851  15        32  0.1461
Roy's Greatest Root         1.625764614  4.55214   5        14  0.0113

NOTE: F Statistic for Roy's Greatest Root is an upper bound.

HIGH      SCHOOL     TEST      SCORES      AND      COLLEGE      GRADES
4
                  01:55 Thursday, November 18, 1999

Canonical Correlation Analysis

Raw Canonical Coefficients for the HIGH SCHOOL TEST SCORES

                 TEST1         TEST2
MAT_TEST    0.01412363    0.05253417    MATH TEST, HIGH SCHOOL
VER_TEST    0.06372086   -0.05505824    VERBAL TEST, HIGH SCHOOL
CRE_TEST    0.03910690    0.00789695    CREATIVITY TEST, HIGH SCHOOL


                   TEST3
MAT_TEST    0.0035547921  MATH        TEST,  HIGH SCHOOL
VER_TEST    0.0161546665  VERBAL      TEST,  HIGH SCHOOL
CRE_TEST   -0.045814414   CREATIVITY TEST,  HIGH SCHOOL


Raw Canonical Coefficients for the COLLEGE GRADES

                  GRADE1         GRADE2

MAT_GRAD      0.00125904     1.03998703   MATH    GRADE, COLLEGE
ENG_GRAD      1.13355171     0.31902355   ENGLISH GRADE, COLLEGE
SCI_GRAD      0.72522584     0.28512302   SCIENCE GRADE, COLLEGE
HIS_GRAD     -0.07106458     0.59285090   HISTORY GRADE, COLLEGE
HUM_GRAD      0.65047132     0.07338639   HUMANITIES GRADE, COLLEGE
```

```
              GRADE3
MAT_GRAD      0.188283972      MATH        GRADE, COLLEGE
ENG_GRAD      1.683241585      ENGLISH     GRADE, COLLEGE
SCI_GRAD      0.279315214      SCIENCE     GRADE, COLLEGE
HIS_GRAD     -0.218593883      HISTORY     GRADE, COLLEGE
HUM_GRAD     -0.565547293      HUMANITIES GRADE, COLLEGE
```

Note that CANCORR reports tests of the partitions of Wilks' lambda, which was not spoken kindly of in section 4.5.3. However, the labeling makes it clear that these are *not* tests of individual roots.

More serious is CANCORR's provision of an upper bound on an F approximation to (and thus a lower bound on the p value associated with) the gcr statistic. Strictly speaking, the p value provided by the approximation *is* a lower bound—but so grossly liberal as to be not only useless, but misleading. For instance, for the case just given ($\theta_1 = .619$ for s, m, and $n = 3$, .5, and 5), the actual p value is .153, versus the approximate p of .011. An approximation that's off by a factor of 14 is not worth reporting, and the unwary reader may be tempted to take it seriously as providing at least the right order of magnitude. For instance, I recently reviewed a manuscript of an extensive Monte Carlo study comparing the gcr statistic to alternatives that used the CANCORR approximation to the gcr p-value, with the result that the "finding" of excessive liberalism of the gcr test was useless, because it could have been due entirely to the gross liberality of the CANCORR approximation.

5.5.3. Canona via SPSS MANOVA

It's been many versions since SPSS had a program devoted to canonical correlation. However, it's possible to carry out the initial stages of a canonical analysis via a run through the Manova program—once to get the canonical variates for the X variables and once to get them for the Y variables. These runs omit any BY phrase and name one set of variables in the dependent-variable list and the other as covariates (i.e., in the WITH phrase). Thus, for instance, the analysis of high-school test scores versus college grades that we set up for SAS analysis in section 5.5.2 would be set u p in SPSS MANOVA as

```
SET LENGTH = None  WIDTH = 80 .
TITLE High School Test Scores And College Grades .
DATA LIST FREE / Mat_Test Ver_Test Cre_Test
         Mat_Grad Eng_Grad Sci_Grad His_Grad Hum_Grad .
VAR LABELS
      Mat_Test  Math Test,  High School /
      Ver_Test  Verbal Test,  High School /
      Cre_Test  Creativity Test, High School /
      Mat_Grad  Math Grade, College /
      Eng_Grad  English Grade, College /
      Sci_Grad  Science Grade, College /
```

```
        His_Grad  History Grade, College /
        Hum_Grad Humanities Grade, College   .
BEGIN DATA .
    27     62     46    2.5    2.0    2.9    1.2    1.4
    21     38     76    2.4    1.3    2.3    1.0    2.1
                               ...
    77     59     28    1.0    3.9    0.0    3.0    1.6
    31     43     81    2.4    1.3    3.5    2.0    2.6
END DATA .
MANOVA Mat_Test Ver_Test Cre_Test
     WITH Mat_Grad Eng_Grad Sci_Grad His_Grad Hum_Grad /
   PRINT = CELLINFO(MEANS) SIGNIF(UNIV EIGEN) ERROR(COV COR) /
   DISCRIM(RAW STAN ALPHA(1.0))    .
```

The output from a Canona carried out via SPSS MANOVA is not as clearly labeled as that from a PROC CANCORR run. For instance, the output from the above analysis included (as does any Canona-via-SPSS run) a page of tests of and "canonical variates" relevant to "EFFECT .. CONSTANT"—which tests the almost always meaningless hypothesis that the population means of the variables in the dependent-variable list are all exactly zero (after adjusting for their relationships to the variables in the WITH phrase). Also, the actual canonical variates and tests of their intercorrelations are found in the "EFFECT .. WITHIN CELLS Regression" section. Finally, the multiple regression of each dependent variable predicted from the "covariates" is given, but the regressions of the variables in the WITH phrase as predicted by the variables in the dependent-variable list are *not* given.

5.5.4. SPSS Canona from Correlation Matrix : Be Careful

One advantage of SPSS over SAS as a tool for Canona is the option of inputting to SPSS MANOVA the matrix of correlations among your $p + q$ variables, rather than the raw data. This is especially useful when you are reanalyzing data that were reported in a journal that, like most, is willing to publish the correlation matrix but not the raw data. For instance, a setup for the high-school-to-college example we've been examining would be

```
SET LENGTH=NONE
TITLE 'HIGH SCHOOL TEST SCORES AND COLLEGE GRADES'
SUBTITLE CORRELATION MATRIX INPUT
MATRIX DATA VARIABLES = MATHHS  VERBALHS CREATVHS MATHCOL ENGLCOL
     SCICOL HISTCOL HUMANCOL / FORMAT = FREE FULL  /
     CONTENTS = N MEANS STDDEVS CORR /
BEGIN DATA
20  20     20     20     20     20     20     20
41.9 56.55 50.9  2.04   2.72   1.845 2.255  2.440
21.93 13.98 18.52 1.172 .835   1.270 1.145   .974
```

```
 1.0000   .5065   -.3946       .2643       .0823       .1942
               -.0492      .1779
  .5065 1.0000    -.4002      -.2195       .3421      -.0083
                .0161      .4450
 -.3946 -.4002   1.0000        .0004      -.1585       .1826
               -.0645      .2680
  .2643 -.2195     .0004      1.0000      -.6159       .6326
               -.6109     -.3077
  .0823   .3421   -.1585      -.6159      1.0000      -.8008
                .6161      .1828
  .1942 -.0083     .1826       .6326      -.8008      1.0000
               -.6805      .0249
 -.0492   .0161   -.0645      -.6109       .6161      -.6805
               1.0000      .1032
  .1779   .4450    .2680      -.3077       .1828       .0249
                .1032     1.0000
END DATA
VAR LABELS
      MATHHS   'MATH TEST,  HIGH SCHOOL' /
      VERBALHS  'VERBAL TEST,  HIGH SCHOOL'  /
                         ...
       HISTCOL  'HISTORY GRADE, COLLEGE' /
      HUMANCOL  'HUMANITIES GRADE, COLLEGE'/
MANOVA MATHHS VERBALHS CREATVHS WITH MATHCOL ENGLCOL
   SCICOL HISTCOL HUMANCOL /
   MATRIX = IN(*)/
   PRINT = CELLINFO(MEANS) SIGNIF(UNIV EIGEN) ERROR(COV COR) /
   DISCRIM(RAW STAN ALPHA(1.0))    .
```

However, as Harris (1999b) pointed out, recent versions of SPSS (both "batch" and PC versions) require (though they don't inform the user of this requirement) that the variables be listed in the same order in the MANOVA statement that they were in the MATRIX INPUT statement. Thus, for instance, if we keep the same MATRIX INPUT statement as in the above listing but reverse in our MANOVA command the order in which the two sets of variables are named (i.e., if we use

```
MANOVA MATHCOL ENGLCOL SCICOL HISTCOL HUMANCOL
                   WITH MATHHS VERBALHS CREATVHS  / ),
```

MANOVA actually carries out a Canona of the relationship(s) between MATHHS, VERBALHS, CREATVHS, MATHCOL, and ENGCOL (the first five variables named in the MATRIX INPUT statement) versus SCICOL, HISTCOL, and HUMANCOL (the last three variables named in the MATRIX INPUT statement). Even the multiple regressions are mislabeled: what is labeled as the regression of MATHCOL on MATHHS, VERBALHS, and CREATVHS is actually the multiple R between MATHHS and SCICOL, HISTCOL, and HUMANCOL. This is not a mistake that's easy to catch, unless you're trying to demonstrate to your students the symmetry of Canona with respect to labeling of the two sets of variables, which is how I first noticed the problem. (One more demonstration of the constant interplay between research and teaching.)

Demonstration Problems and Some Real Data Employing Canonical Correlation

1. Given the following data:

Subject	X_1	X_2	Y_1	Y_2	Z_1	Z_2
1	1	0	4	2	11	25
2	1	0	1	3	14	25
3	1	0	3	5	12	26
4	1	0	2	2	11	30
5	0	1	2	2	13	25
6	0	1	4	5	12	30
7	0	1	5	4	12	27
8	0	1	5	5	15	28

(a) Conduct a canonical analysis on the relationship between variables $Y_{1\text{-}2}$ and variables $Z_{1\text{–}2}$. Be sure to include in your analysis the two sets of canonical coefficients and the canonical correlations between the two sets of variables. Do not worry about significance tests.

(b) Compute the two canonical variates for each of the eight subjects. Designate by Y_i^* the left-hand canonical variate for subject *i*, and by Z_i^* the right-hand canonical variate for subject *i*.

(c) Calculate the correlation between the two canonical variates, that is, compute the Pearson product–moment correlation between Y^* and Z^*. Compare this number with your findings in part (a).

(d) Do a multiple regression of Z^* on Y_1 and Y_2. Report the betas and the multiple correlation between Z^* and Y_1 and Y_2. Compare these results with those of part (a).

(e) Do a multiple regression of Y^* predicted from Z_1 and Z_2. Report the regression coefficients and the multiple correlation. Compare these results with those of part (a).

2. Still working with the data of problem 1:

(a) Conduct a canonical analysis of the relationship between variables $X_{1\text{–}2}$, on the one hand, and variables Y_1, Y_2, Z_1, and Z_2 on the other hand. Report both sets of canonical coefficients, the canonical correlation between the two sets of variables, and the appropriate test statistic (F or greatest root) for testing the overall null hypothesis.

(b) Treating subjects 1 through 4 as members of one group and subjects 5 through 8 as members of a second group, conduct a one-way Manova on the differences among the groups in mean response vectors. Report the discriminant function .and the appropriate statistic for testing the overall null hypothesis.

(c) Label the discriminant function obtained in part (b) as $(c_{y1}, c_{y2}, c_{z1}, c_{z2})$. Calculate for each subject the single score, $D_i = c_{y1}Y_{1i} + c_{y2}Y_{2i} + c_{z1}Z_{1i} + c_{z2}Z_{2I}$. Now conduct a t test on the difference in D between the two groups.

(d) Using Hotelling's T^2, conduct a test of the overall null hypothesis of no difference between the two groups in overall mean response vectors.

(e) Compare and contrast the results obtained in parts (a)–(d).

1. As part of a cooperative study of differences between patients classified as clearly paranoid and patients classified as clearly nonparanoid in behavior in a binary prediction experiment and in the Prisoner's Dilemma game, each patient who volunteered for the study went through a psychiatric interview and also took a 400-question "personality inventory" containing, among other things, items from several scales of the MMPI. As a result, we have available both interview (the psychiatrist's ratings of suspiciousness of the study and the overall degree of paranoia) and paper-and-pencil (scores on the Suspicion, Paranoia, and Schizophrenia scales of the MMPI) measures of paranoia. The matrix of intercorrelations among these five measures is reported in the tabulation.

	MMPI Scales			Interviewer Ratings	
	Susp	Par	Schiz	Susp′	Par′
Susp	1.0	.49[a]	.63[a]	.24	.12
Par		1.0	.63[a]	.42[a]	.40[a]
Schiz			1.0	.34[b]	.14
Susp'				1.0	.54[a]
Par'					1.0

$N = 44$

[a] $p < .05$ } by standard test for statistical significance
[b] $p < .01$ } of a single correlation coefficient.

Conduct and interpret a canonical analysis of these data.

4. How are we to interpret the "extra" set of canonical coefficients we obtained in problem 3?

<u>Note:</u> You may use any mixture of hand and computer computations you find convenient.

Answers

1. **(a) Canonical analysis on Y_1 and Y_2 versus Z_1 and Z_2**

$$\mathbf{S}_{11} = \frac{1}{7}\begin{bmatrix} 15.5 & 8.0 \\ 8.0 & 14.0 \end{bmatrix},\quad \mathbf{S}_{11}^{-1} = \frac{7}{153}\begin{bmatrix} 14.0 & -8.0 \\ -8.0 & 15.5 \end{bmatrix},\quad \mathbf{S}_{22} = \frac{1}{7}\begin{bmatrix} 14 & -4 \\ -4 & 32 \end{bmatrix},$$

$$\mathbf{S}_{22}^{-1} = \frac{7}{432}\begin{bmatrix} 32 & 4 \\ 4 & 14 \end{bmatrix},\quad \mathbf{S}_{12} = \frac{1}{7}\begin{bmatrix} 0 & 6 \\ 5 & 7 \end{bmatrix},\quad \text{and} \quad \mathbf{S}_{12}' = \frac{1}{7}\begin{bmatrix} 0 & 5 \\ 6 & 7 \end{bmatrix},$$

whence

$$\frac{1}{153}\begin{bmatrix} 0 & 6 \\ 5 & 7 \end{bmatrix}$$

$$\mathbf{S}_{11}^{-1}\mathbf{S}_{12} = \begin{bmatrix} 14 & -8 \\ -8 & 15.5 \end{bmatrix} = \frac{1}{153}\begin{bmatrix} -40 & 28 \\ 77.5 & 60.5 \end{bmatrix},$$

and

$$\frac{1}{153}\begin{bmatrix} 0 & 5 \\ 6 & 7 \end{bmatrix}$$

$$\mathbf{S}_{22}^{-1}\mathbf{S}_{12}' = \begin{bmatrix} 32 & 4 \\ 4 & 14 \end{bmatrix} = \frac{1}{153}\begin{bmatrix} 24 & 188 \\ 84 & 118 \end{bmatrix}.$$

<u>Note:</u> As pointed out in section D.2.3, "hand" multiplication of matrices is facilitated by positioning the postmultiplying matrix immediately above and to the right of the premultiplying matrix.

Putting the last two matrix products together in two different orders gives us

$$\frac{1}{66{,}096}\begin{bmatrix} 24 & 188 \\ 84 & 118 \end{bmatrix}$$

$$\mathbf{S}_{11}^{-1}\mathbf{S}_{12}\mathbf{S}_{22}^{-1}\mathbf{S}_{12}' = \begin{bmatrix} -40 & 28 \\ 77.5 & 60.5 \end{bmatrix} = \frac{1}{66{,}096}\begin{bmatrix} 1392 & -4216 \\ 6942 & 21{,}709 \end{bmatrix}$$

and

$$\frac{1}{66{,}096}\begin{bmatrix} -40 & 28 \\ 77.5 & 60.5 \end{bmatrix}$$

$$\mathbf{S}_{22}^{-1}\mathbf{S}_{12}'\mathbf{S}_{11}^{-1}\mathbf{S}_{12} = \begin{bmatrix} 24 & 188 \\ 84 & 118 \end{bmatrix} = \frac{1}{66{,}096}\begin{bmatrix} 13{,}610 & 12{,}046 \\ 5{,}785 & 9{,}491 \end{bmatrix}.$$

We now have two matrices, either of whose largest characteristic root will give us the square of the maximum canonical correlation. Always calculate **A** for both matrices as a check on your calculations. Using the formula for the characteristic roots of a 2 x 2 matrix, we compute the following:

(a) Using $\mathbf{S}_{11}^{-1}\mathbf{S}_{12}\mathbf{S}_{22}^{-1}\mathbf{S}_{12}'$,

$$\lambda = \frac{(a+d) \pm \sqrt{(a-d)^2 + 4bc}}{2}, \quad \text{where the matrix is} \quad \begin{bmatrix} a & b \\ c & d \end{bmatrix},$$

whence

$$\lambda = \frac{23{,}101 \pm \sqrt{295{,}710{,}601}}{2(66{,}096)} = \frac{23{,}101 \pm 17{,}196.2}{2(66{,}096)} = (.30484, .04467).$$

(b) Using $\mathbf{S}_{22}^{-1}\mathbf{S}_{12}'\mathbf{S}_{11}^{-1}\mathbf{S}_{12}$, we obtain

$$\lambda = \frac{23{,}101 \pm \sqrt{295{,}710{,}601}}{2(66{,}096)} = (.30484, .04467),$$

which agrees quite well with the first calculation. The maximum canonical correlation is thus $\sqrt{.30484} = .55212$. (We are only keeping all those decimal places for comparison with subsequent calculations.)

Now we want to obtain the canonical coefficients. The coefficients for the *Y* variables are given by the matrix product that begins with the inverse of those variables. Thus,

$$[\mathbf{S}_{11}^{-1}\mathbf{S}_{12}\mathbf{S}_{22}^{-1}\mathbf{S}_{12}' - \mathbf{I}\lambda]\mathbf{a} = \mathbf{0},$$

or

$$\left\{\begin{bmatrix} 1392 & -4216 \\ 6942 & 21709 \end{bmatrix} - 66{,}096\,\lambda\,\mathbf{I}\right\}\mathbf{a} = \mathbf{0}$$

or

$$\begin{bmatrix} 1392 - 20{,}148.5 & -4216 \\ 6942 & 21{,}709 - 20{,}148.5 \end{bmatrix}\begin{bmatrix} a_{11} \\ a_{12} \end{bmatrix} = \mathbf{0}.$$

Here again we have a check on our calculations. The top and bottom row of the preceding matrix each define the ratio between a_{11} and a_{12}. These ratios should, of course, agree. Thus we obtain

$$a_{11} = \frac{4216}{18{,}756.5}a_{12} = -.2248a_{12}$$

and

$$a_{11} = \frac{1560.5}{6942} a_{12} = -.2248 a_{12}$$

Fortunately, the two answers agree closely. The coefficients for the right-hand variables are given by

$$\begin{bmatrix} 13{,}610 - 20{,}148.5 & 12{,}046 \\ 5785 & 9491 - 20{,}148.5 \end{bmatrix} \begin{bmatrix} b_{11} \\ b_{12} \end{bmatrix} = \mathbf{0};$$

$$b_{11} = \frac{12{,}046}{6538.5} b_{12} = 1.8423 b_{12} \quad \text{and} \quad b_{11} = \frac{10{,}657.5}{5785} b_{12} = 1.8423 b_{12}.$$

We could normalize these vectors to unit length, but there is no need to do so for present purposes, because division of each of two variables by a (possibly different) constant does not affect the correlation between them. Thus we will take

$$\mathbf{a}'_1 = (-.2248, 1) \quad \text{and} \quad \mathbf{b}'_1 = (1.8423, 1)$$

as the characteristic vectors defining the left- and right-hand canonical variates, respectively.

(b) and (c) Calculation of individual canonical variates and the correlation between them:

Left-hand variate (variable for Ys) = Y_2 - .2248Y_1
Right-hand canonical variate = Z_2 + 1.8423Z_1
These calculations are straightforward, but as a means of checking your answers, the two canonical variates for each subject are presented in the following tabulation:

Subject	Y^*	$Z^*/10$
1	1.1008	4.5265
2	2.7752	5.0792
3	4.3256	4.8108
4	1.5504	5.0265
5	1.5504	4.8950
6	4.1008	5.2108
7	2.8760	4.9108
8	3.8760	5.5635

$$r_{Y^*Z^*} = \frac{\sum Y^* Z^* - \frac{\left(\sum Y^*\right)\left(\sum Z^*\right)}{N}}{\sqrt{\left(\sum Y^{*2} - \frac{\left(\sum Y^*\right)^2}{N}\right)\left(\sum Z^{*2} - \frac{\left(\sum Z^*\right)^2}{N}\right)}}$$

$$= \frac{112.3265 - (22.1552)(40.0231)/8}{\sqrt{\left(72.5431 - \frac{(22.1552)^2}{8}\right)\left(200.8789 - \frac{(40.0231)^2}{8}\right)}}$$

$$= .5220,$$

which is relatively close to the largest canonical correlation computed in part (a).

(d) Multiple regression of Z* on Y_1 and Y_2: $\mathbf{B} = \mathbf{S}_{22}^{-1}\mathbf{s}_{12}$, where $\mathbf{S}_{22}$ is the covariance matrix of the predictor variables (Y_1 and Y_2) and $\mathbf{s}_{12}$ is the column vector of covariances of Y_1 and Y_2 with Z*. We must therefore calculate these two correlations, giving

$$s_{Y_1Z^*} = \frac{(\sum Y_1)(\sum Z^*)}{8} = 130.6753 - \frac{(26)(20.0231)}{8} = .76002 \text{ and}$$

$$s_{Y_2Z^*} = 141.7023 - \frac{(28)(40.0231)}{8} - 1.6211, \text{ whence}$$

$$\mathbf{B} = \frac{\frac{1}{432}\begin{bmatrix} 5.000 \\ 5.6512 \end{bmatrix}}{\begin{bmatrix} 14 & -8 \\ -8 & 15.5 \end{bmatrix}} = \frac{1}{432}\begin{bmatrix} 182.6048 \\ 99.1168 \end{bmatrix}.$$

Note that b_1/b_2 = -.2246, which is quite close to the -.2248 of part (a). We next compute

$$\text{multiple } R = \sqrt{\mathbf{s}_{12}'\mathbf{B}/s_2^2} = \sqrt{.197445/.6478} = .55208,$$

which is quite close to the largest canonical r calculated in part (a).

(e) Multiple regression of Y^* on Z_1 and Z:

$$s_{Z_1Y^*} = 281.9400 - (100)(22.1552)/8 = 5.0000$$

and

$$s_{Z_2Y^*} = 603.8416 - (216)(22.1552)/8 = 5.6152,$$

whence

$$\mathbf{B} = \frac{\frac{1}{432}\begin{bmatrix} 5.000 \\ 5.6512 \end{bmatrix}}{\begin{bmatrix} 14 & -8 \\ -8 & 15.5 \end{bmatrix}} = \frac{1}{432}\begin{bmatrix} 182.6048 \\ 99.1168 \end{bmatrix}.$$

Note that b_1/b_2 = 1.8423, which is quite close to the 1.8423 of part (a).

$$\text{Multiple } R = \sqrt{\frac{\mathbf{s}'_{12}\mathbf{B}}{s_{Y^*}^2}} = \sqrt{\frac{3.4101}{11.1865}} = .55212.$$

Note: One of the values of going through a problem such as this by hand is that you then have a "test problem" to aid in checking out your computer program. When these data were analyzed using a locally developed program, the computer agreed to five decimal places with the figures quoted here for **A** and the canonical correlation. It did nor agree even to two decimal places, however, with the appropriate values for the canonical coefficients. The clincher came when the program's canonical coefficients were used to compute Y^* and Z^*, yielding a correlation lower than that given by the hand-computed coefficients. (Why is this such a crucial test?) We thus had a problem with this program. This problem turned out to have been caused by a simple mistake in punching the statements of an SSP subroutine used by this program onto IBM cards, and was subsequently corrected. ("SSP" stands for IBM's Scientific Subroutine Package, a very widely distributed package of subroutines useful in "building" programs.)

P.S. The comparisons with the computer program should be done starting with the intercorrelation matrix, because this is what the computer actually does.

Thus, for the current problem,

$$\mathbf{R}_{11} = \begin{bmatrix} 1 & .54308 \\ .54308 & 1 \end{bmatrix}, \mathbf{R}_{22} = \begin{bmatrix} 1.0 & -.18898 \\ -.18898 & 1.0 \end{bmatrix}, \mathbf{R}_{12} = \begin{bmatrix} 0 & .26941 \\ .35714 & .33072 \end{bmatrix}$$

$$\mathbf{R}_{11}^{-1}\mathbf{R}_{12}\mathbf{R}_{22}^{-1}\mathbf{R}_{12}' = \frac{1}{6.79884}\begin{bmatrix} .143189 & -.456309 \\ .678647 & 2.233037 \end{bmatrix},$$

whence

$$\lambda_1 = .30488, \quad \lambda_2 = .04472,$$

and

$$\begin{bmatrix} -.28382 & -.067116 \\ .099818 & .0235638 \end{bmatrix}\begin{bmatrix} a_{11} \\ a_{22} \end{bmatrix} = 0 \Rightarrow a_{11} = -.23647 a_{21}.$$

On dividing a_1 and a_2 by their respective standard deviations, we obtain as raw-score canonical coefficients $\mathbf{a}_1 = (-.2247, 1)$, just as we had before.

2.

(a) Canonical analysis of group membership variables versus others. The problem with applying our usual formulae to the variables as defined is that

$$\mathbf{R}_{11} = \begin{bmatrix} 1 & -1 \\ -1 & 1 \end{bmatrix} \Rightarrow |\mathbf{R}_{11}| = 0.$$

so that $\mathbf{R}^{-1}$ is undefined. We thus must delete one of the two redundant (because linearly related) variables, whence

$$\mathbf{S}_{11} = s_1^2 = 2; \qquad \mathbf{S}_{11}^{-1} = 1/2; \qquad \mathbf{S}_{12} = [-3 \quad -2 \quad -2 \quad -2];$$

$$\mathbf{S}_{22} = \begin{bmatrix} 15.5 & 8.0 & 0 & 6.0 \\ & 14.0 & 5.0 & 7.0 \\ & & 14.0 & -4.0 \\ & & & 32.0 \end{bmatrix}; \quad \mathbf{S}_{22}^{-1} = \frac{1}{10}\begin{bmatrix} .97555 & -.62604 & .21825 & -.01868 \\ & 1.41060 & -.57911 & -.26358 \\ & & .98063 & .20834 \\ & & & .39970 \end{bmatrix};$$

and

$$\mathbf{S}_{11}^{-1}\mathbf{S}_{12} = [-1.5 \quad -1 \quad -1 \quad -1]; \quad \mathbf{S}_{22}^{-1}\mathbf{S}'_{12} = .4\begin{bmatrix} -.51840 \\ .18558 \\ -.46862 \\ -.15822 \end{bmatrix}; \quad \mathbf{S}_{11}^{-1}\mathbf{S}_{12}\mathbf{S}_{22}^{-1}\mathbf{S}'_{12} = .48754.$$

Clearly the characteristic root of this matrix is its single entry, so that

$$\lambda = R_c^2 = .48754$$

and the characteristic vector for the group membership variables is any arbitrary scalar, and the single entry 1.0 when normalized.

$$\mathbf{S}_{22}^{-1}\mathbf{S}'_{12}\,\mathbf{S}_{11}^{-1}\mathbf{S}_{12} = .2\begin{bmatrix} -.51840 \\ .18558 \\ -.46862 \\ .15822 \end{bmatrix}[-3 \quad -2 \quad -2 \quad -2] = \begin{bmatrix} .31104 & .20736 & .20736 & .20736 \\ -.11135 & -.07423 & -.07423 & -.07423 \\ .28117 & .18745 & .18745 & .18745 \\ .09493 & .06329 & .06329 & .06329 \end{bmatrix}$$

Because the first column of this matrix is exactly 1.5 times each of the other three columns, the matrix is clearly of rank 1, and

$$\lambda = \text{trace} = .31104 - .07423 + .18745 + .06329 = .48755,$$

which checks rather closely with our previous figure. The canonical correlation between the group membership variables and the others is thus given by $\sqrt{.48755} = .6977$.

The parameters for our gcr test of the significance of this correlation are

$$s = 1,\ m = (|1 - 4| - 1)/2 = 1,\ \ n = (8 - 5 - 2)/2 = 1/2,$$

whence

$$F = \frac{3/2}{2}\frac{\lambda}{1-\lambda} = \frac{3}{4}\frac{.48755}{.51245}$$
$$= .7136 \text{ with 4 and 3 } df, \text{ ns.}$$

The solution for the right-hand canonical variates proceeds as follows:
Using the first two rows of $\mathbf{S}_{22}^{-1}\mathbf{S}'_{12}\,\mathbf{S}_{11}^{-1}\mathbf{S}_{12} - \lambda\mathbf{I}$ gives us:

$$(.31104 - .48755)b_1 + .20736(b_2 + b_3 + b_4) = 0 \rightarrow \ b_1 = 1.17478(b_2 + b_3 + b_4)$$
$$-.11135\, b_1 - .56178\, b_2 +\text{-}\ .07423(b_3 + b_4) = 0 \rightarrow \ b_2 = -(.20504/.69259)(b_3 + b_4)$$
$$= -.29605(b_3 + b_4),$$

whence

$$b_1 = 1.17478(1 - .29605)(b_3 + b_4) = .82699(b_3 + b_4).$$

Using the third row gives:

$$.28117\, b_1 \quad + .18745 b_2 \quad - .30010 b_3 + .18745 b_4 = 0,$$

$$.28117(.82699)(b_3 + b_4) + .18745(-.29605)(b_3 + b_4) - .30010 b_3 + .18745 b_4 = 0,$$

whence

$$b_3 = (.36448/.12207)\, b_4 = 2.96157\, b_4,$$

which tells us that

$$(b_3 + b_4) = 3.96157\, b_4 \rightarrow b_2 = 3.96157(-.29605) b_4 \text{ and } b_1 = 3.96157(.82699)\, b_4,$$

and thus

$$\mathbf{b}' = (3.2762, -1.1728, 2.9616, 1).$$

Using the fourth equation as a check,

$$.09493\, b_2 + .06329(b_2 + b_3) - .42426 b_4 = .00004,$$

which is reassuringly close to zero.

(b) One-way Manova on *P* versus *NP*:

$$\mathbf{T} = \begin{bmatrix} 10 & 12 & 48 & 106 \\ 16 & 16 & 52 & 110 \end{bmatrix}; \quad \mathbf{G}' = \begin{bmatrix} 26 & 28 & 100 & 216 \end{bmatrix};$$

$$\mathbf{H} = \frac{\mathbf{T}'\mathbf{T}}{4} - \frac{\mathbf{GG}'}{8} = \begin{bmatrix} 4.5 & 3 & 3 & 3 \\ 3 & 2 & 2 & 2 \\ 3 & 2 & 2 & 2 \\ 3 & 2 & 2 & 2 \end{bmatrix}; \quad \mathbf{E} = \begin{bmatrix} 11 & 5 & -3 & 3 \\ 5 & 12 & 3 & 5 \\ -3 & 3 & 12 & -6 \\ 3 & 5 & -6 & 30 \end{bmatrix}.$$

Cheating a bit by using MATLAB we find that the largest characteristic root is given by $\lambda = .95141$, with associated characteristic vector (the discriminant function) given by

$$(.7041, -.25071, .63315, .21377),$$

or

$$D = 3.2765 Y_1 - 1.1728\, Y_2 + 2.9618\, Z_1 + Z_2.$$

Note that $\lambda/(1 + \lambda) = .48755$, precisely equal to the R_c^2 of part (a), and that the discriminant function is almost identical to the right-hand canonical variate of part (a).

The parameters for the gcr test of the significance of the differences between the two groups are the same as for part (a).

(c) *t* Test on discriminant scores:

Subject	$D - 14$	Subject	$D - 14$
1	.60912	5	.47460
2	.15663	6	1.55899
3	.00350	7	1.86880
4	.27715	8	3.73131

$$t = \frac{\overline{X}_1 - \overline{X}_2}{\sqrt{\dfrac{(\sum x_1^2 + \sum x_2^2)(1/2)}{6}}} = \frac{1.6468}{\sqrt{\dfrac{.1986546 + 5.502477}{12}}}$$

$$= \frac{1.6468}{\sqrt{.475094}} \Rightarrow t^2 = 5.7802,$$

and

$$t = 2.389 \text{ with } 6\ df, p > .05\ .$$

(d) Hotelling's T^2 on difference between two groups:

$$T^2 = \frac{N_1 N_2}{N_1 + N_2}(\overline{\mathbf{X}}_1 - \overline{\mathbf{X}}_2)\mathbf{S}^{-1}(\overline{\mathbf{X}}_1 - \overline{\mathbf{X}}_2)$$

$$= \frac{n}{2}\left(\frac{\mathbf{T}_1 - \mathbf{T}_2}{n}\right)' \bullet 2(\text{n - 1})\mathbf{E}^{-1}\left(\frac{\mathbf{T}_1 - \mathbf{T}_2}{n}\right) = \frac{n-1}{n}(\mathbf{T}_1 - \mathbf{T}_2)'\mathbf{E}^{-1}(\mathbf{T}_1 - \mathbf{T}_2);$$

$$T^2 = (1 - \frac{1}{n}(6,4,4,4)\frac{1}{10}\begin{bmatrix} 1.3950 & -.77621 & .59749 & .10936 \\ & 1.4644 & -.71486 & -.30942 \\ & & 1.3235 & .32409 \\ & & & .43878 \end{bmatrix}\begin{bmatrix} 6 \\ 4 \\ 4 \\ 4 \end{bmatrix}$$

$$= (3/4)[36(1.3950) + 48(-.77621 + .59749 + .10936)$$
$$+ 32(-.71486 -.30942 + .32409) + 16(1.4644 + 1.3235 + .43878)]$$
$$= (3/4)(7.611152) = 5.7084,$$

which is quite close to the 5.7082 we calculated as the square of the univariate t for the difference between the two groups in mean "discriminant score."
(Be sure you understand why we do not compare T^2 directly with Student's t distribution.)

(e) Comparisons: The canonical coefficients for the right-hand variables (the *Y*s and *Z*s) arE practically identical to the coefficients of the Manova-derived discriminant function. They are also identical with the T^2-derived discriminant function, although you were not required to calculate this last function.

When these coefficients are used to obtain a single combined score for each subject, and a standard univariate t test is conducted on the difference between the two groups on the resulting new variable, the square of this is the same as the value of Hotelling's T^2 statistic. It is also equal to 6 (= N - 2) times the largest characteristic root (in fact, the only nonzero root) obtained in the Manova, and it is further equal to $6\lambda/(1 - \lambda)$, where λ is the largest root obtained in Canona with membership variables included.

The significance tests conducted in parts (a), (b), and (d) arc identical. The reader may wish to verify that exactly the same results are obtained if the "dummy variable" approach is taken in part (a).

3. To quote from Harris, Wittner, Koppell, and Hilf (1970, p. 449):

> An analysis of canonical correlation (Morrison, 1967, Chapter 6) indicated that the maximum possible Pearson r between any linear combination of the MMPI scale scores and any linear combination of the interviewer ratings was .47.* The weights by which the standard scores of the original variables must be multiplied to obtain this maximum correlation arc -.06, 1.30, and -.05 for Susp, Pa, and SC, respectively, and .73 and .68 for Susp' and Pa'.
>
> Morrison (1967) provides two tests of the statistical significance of canonical *R:* a greatest characteristic root test (p. 210) and a likelihood-ratio test (p. 212). By the first test, a canonical *R* of .49 is significantly different from zero at the .05 level. By the second test, the canonical *R* between the original measures is statistically significant: .... Both tests must be interpreted very cautiously . . ., since they each assume normally distributed variables.

Note that in this paper I fell into the very trap mentioned in section 5.3: interpreting the likelihood-ratio test as testing the statistical significance of the correlation between the first pair of canonical variates.

4. The other set of canonical coefficients obtained (because $s = 2$) represent a second linear combination of the MMPI scales that correlates with a second linear combination of the interviewer ratings as highly as possible for two combinations that must be uncorrelated with the first pair of canonical variates. However, this correlation isn't very high, because our second characteristic root is only .0639. We therefore needn't bother with interpreting this second correlation.

6
Principal Component Analysis: Relationships within a Single Set of Variables

The techniques we have discussed so far have had in common their assessment of relationships between two sets of variables: two or more predictors and one outcome variable in multiple regression; one predictor (group membership) variable and two or more outcome variables in Hotelling's T^2; two or more group membership variables and two or more outcome variables in Manova; and two or more predictors and two or more outcome variables in canonical correlation. Now we come to a set of techniques (principal component analysis and factor analysis) that "look inside" a single set of variables and attempt to assess the structure of the variables in this set independently of any relationship they may have to variables outside this set. Principal component analysis (PCA) and factor analysis (FA) may in this sense be considered as logical precursors to the statistical tools discussed in chapters 2 through 5. However, in another sense they represent a step beyond the concentration of earlier chapters on observable relationships among variables to a concern with relationships between observable variables and unobservable (latent) processes presumed to be generating the observations.

Because there are usually a large number of theories (hypothetical processes) that could describe a given set of data equally well, we might expect that this introduction of latent variables would produce a host of techniques rather than any single procedure for analyzing data that could be labeled uniquely as "factor analysis." This is in fact the case, and this multiplicity of available models for factor analysis makes it essential that the investigator be prepared to commit himself or herself to some assumptions about the structure of the data in order to select among alternative factor analysis procedures. (This is true of all statistical techniques. For instance, one needs to decide whether the dependent variables being measured are the only variables of interest or if instead there may be interesting linear combinations of them worth exploring before deciding whether to apply Manova or to simply perform Bonferroni-adjusted Anovas on the factorial design. It is perhaps a positive feature of PCA-FA techniques that they make this need to choose among alternative procedures harder to ignore.)

The particular procedure discussed in this chapter—principal component analysis, or PCA—is a "hybrid" technique that, like factor analysis, deals with a single set of variables. Unlike the factors derived from FA, however, principal components are very closely tied to the original variables, with each subject's score on a principal component being a linear combination of his or her scores on the original variables. Also unlike factor analytic solutions, the particular transformations of the original variables generated by PCA are unique. We turn next to a description of the procedures that provide this uniqueness.

Note by the way, that PCA is *principal* (primary, most important) component analysis, not (many careless authors notwithstanding) *principle* component analysis, which is presumably a technique for developing taxonomies of lifestyles, axioms, and morals.

6.1 DEFINITION OF PRINCIPAL COMPONENTS

A principal component analysis (PCA) of a set of p original variables generates p new variables, the *principal components,* PC_1 PC_2, ... PC_p, with each principal component being a linear combination of the subjects' scores on the original variables, that is,

$$\begin{aligned} PC_1 &= b_{1,1}X_1 + b_{1,2}X_2 + \cdots + b_{1,m}X_m = \mathbf{Xb}_1; \\ PC_2 &= b_{2,1}X_1 + b_{2,2}X_2 + \cdots + b_{2,m}X_m = \mathbf{Xb}_2; \\ &\vdots \qquad\qquad \vdots \qquad\qquad \vdots \\ PCm &= b_{m,1}X_1 + b_{m,2}X_2 + \cdots + b_{m,m}X_m = \mathbf{Xb}_m; \end{aligned} \tag{6.1}$$

or, in matrix form, **PC** = **XB**, where each column of **B** contains the coefficients for one principal component. (Note that **PC** is a single matrix, ***not*** a product of two matrices. At the risk of offending factor analysts who consider PCA a degenerate subcategory of FA, we can symbolize the matrix of scores on the PCs as **F**. We have more to say about notation in section 6.1.1.)

The coefficients for PC_1 are chosen so as to make its variance as large as possible. The coefficients for PC_2 are chosen so as to make the variance of this combined variable as large as possible, subject to the restriction that scores on PC_2 and PC_1 (whose variance has already been maximized) be uncorrelated. In general, the coefficients for PC_j are chosen so as to make its variance as large as possible subject to the restrictions that it be uncorrelated with scores on PC_1 through PC_{j-1}. Actually, an additional restriction, common to all p PCs, is required. Because multiplication of any variable by a constant c produces a new variable having a variance c^2 times as large as the variance of the original variable, we need only use arbitrarily large coefficients in order to make s^2 for the PC arbitrarily large. Clearly, it is only the relative magnitudes of the coefficients defining a PC that is of importance to us. In order to eliminate the trivial solution, $b_{i1} = b_{i2} = \cdots = b_{ip}$ $= \infty$, we require that the squares of the coefficients involved in any PC sum to unity, that is, $\sum_i \mathbf{b}_{ij}^2 = \mathbf{b}_j'\mathbf{b}_j = 1$. The variance of a component (linear combination of variables) that satisfies this condition can be termed that component's *normalized* variance (just as the variance across subjects of a contrast among the means of a within-subjects factor is know as the normalized variance of that contrast—cf. chap. 4).

Retrieving from the inside of your left eyelid, where it should be permanently inscribed, the relationship between the combining weights of a linear combination and the variance of that combination, it should be apparent (cf. section D2.3 if it is not) that the variance of PC_i is given by

$$\begin{aligned} s^2_{PC_i} &= [1/(N-1)]\sum(b_{i,1}x_1 + b_{i,2}x_2 + \Lambda + b_{i,1}x_m)^2 \\ &= \mathbf{b}'_i\,\mathbf{S}_x\mathbf{b}_i\,. \end{aligned} \tag{6.2}$$

Derivation 6.1 shows that maximizing Equation (6.2), subject to our side condition, requires solving for λ and ***b*** in the matrix equation

$$[\mathbf{S}_x - \lambda\mathbf{I}]\mathbf{b} = \mathbf{0},$$

that is, it requires finding the characteristic roots and vectors (eigenvalues and eigenvectors) of $\mathbf{S}_x$. The result is ***p*** characteristic roots, some of which may be zero if there is a linear dependency among the original variables. Each root is equal to the variance of the combined variable generated by the coefficients of the characteristic vector associated with that root. Because it is this variance that we set out to maximize, the coefficients of the first principal component will be the characteristic vector associated with the *largest* characteristic root. As it turns out, PC_2 is computed via the characteristic vector corresponding to the second largest characteristic root, and in general PC_i has the same coefficients as the characteristic vector associated with the *i*th largest characteristic root λ_i.

Finally, $\mathbf{b}_i'\mathbf{b}_j$—which is the covariance between PC_i and PC_j—is zero for $i \neq j$. Thus the orthogonality of (lack of correlation among) the characteristic vectors of $\mathbf{S}_x$ is paralleled by the absence of any correlation among the PCs computed from those sets of coefficients.

6.1.1 Terminology and Notation in PCA and FA

As we progress through our discussion of PCA and FA, the number of matrices we have to consider will multiply rapidly. So far we've considered only the matrix of coefficients used to compute scores on the PCs from scores on the *X*s. In section 6.2 we consider the coefficients for computing *X*s from PCs. More than a few minutes of exposure to reports of factor analyses will force you to consider the ***loadings*** of the *X*s on (i.e., their correlations with) the factors. As hinted earlier, many factor analysts consider PCA to be such a degenerate subcategory of FA that they take umbrage at the use of a common terminology. However, giving in to this bit of ideology would quickly exhaust the alphabet as we search for unique, single-letter labels for our matrices, so for purposes of the discussion in his chapter and in chapter 7, we adopt the following common notation for both PCA and FA:

A ***factor*** is any variable from which we seek to (re)generate scores on or the correlations among our original variables. A ***component*** is any such combination that is computable as a linear combination of the original measures. (As we show later, factors can usually only be estimated from the observed scores.)

A ***principal component*** is any component that meets the definition given earlier in section 6.1, that is, that has the property of being uncorrelated with the other p - 1 PCs and having maximum variance among all linear combinations that are uncorrelated with those PCs having lower indices than (i.e., "extracted" before) this one.

$_N\mathbf{X}_p$ is an N x p matrix of the scores of the N sampling units (usually subjects) on the p original variables. This is the familiar data matrix, **X**, we have been using throughout this book and, as has been the case, we usually omit the dimensioning subscripts. This is true for the other matrices defined below as well, especially because we often need to use subscripts to indicate further distinctions among matrices of the same general type.

$_N\mathbf{Z}p$ is an N x p matrix of the N subjects' z scores on the p original variables.

$_N\mathbf{F}_m$ is an N x m matrix of the N subjects' scores on the $m \leq p$ factors (or components). **PC** is that special case of **F** in which the factors are all principal components. $\mathbf{F}_z$ (almost always the matrix under consideration in FA) is that special case of **F** in which the scores are z scores. We also occasionally use the alternative label **B** for **F**, and h_{ij} in place of f_{ij} to represent subject i's score on factor j. This occurs when we wish to emphasize the hypothetical (latent) aspect of factors.

$_p\mathbf{B}_m$ is the ***factor*** (or component) ***weight matrix,*** a matrix each of whose columns gives the coefficients 'needed to compute (or estimate) a subject's score on a given factor from his or her set of scores on the original variables. The use of **B** for this matrix should remind you of regression coefficients, and indeed the most common method of estimating scores on the factors is via multivariate regression analysis of each of the factors predicted from the set of Xs. **B** is usually labeled in the output of FA programs as the *factor score coefficient matrix.*

$_p\mathbf{P}_m$ is the factor (component) ***pattern matrix*** (or simply the "factor pattern"), a matrix each of whose rows gives the coefficients for computing a subject's deviation score on that X—that is, $x_{i,u} = X_{iu} - \overline{X}i$, sampling unit u's deviation score on original variable X_i —from that unit's (often unknown) scores on the p factors. It is the existence of this matrix that forced us to use the inelegant **PC** to represent scores on the PCs. Further, the very strong tradition of thinking of a pattern matrix (**P**) as defining original variables in terms of factors forces us to use a different term for **B**. I would be very grateful for any suggestion of a good single-word designation for **B**. (The factor "revelation," perhaps?) One further indulgence with respect to **P**: The individual elements of this matrix are labeled as a_{ij}s—primarily so as to avoid confusion with p, the number of X_is in **X**, but also because of a certain alphabetical closeness to the b_js that play for computation of factors from Xs the role that the a_js play for computation of original variables from scores on the factors.

${}_p\mathbf{R}_m$ is the *structure* matrix, a p x m matrix each of whose rows gives the m correlations between that original variable and each of the factors. These correlations are known as the *loadings* of the variables on the factors. We also label this matrix as $\mathbf{R}_{xf}$, in a manner that should remind you of the $\mathbf{R}_{xy}$ matrix of chapter 5. Although the emphasis among factor analysts is usually on reading this matrix row-wise (i.e., as defining original variables in terms of factors), the symmetry of the correlation coefficient insures that we can also take each column of this matrix as giving the correlations of that factor with the p original variables. This is very useful information, of course, in using the regression approach to estimating or computing factor scores from original scores. This matrix is also commonly labeled as **S**, but 1 prefer to avoid this notation because of its confusion with $\mathbf{S}_x$ the matrix of variances of and covariances among the Xs.

Finally (for now), we use $\mathbf{R}_x$ as before to symbolize the p x p matrix of correlations among the Xs and (as is necessary when oblique factors are examined) $\mathbf{R}_f$ to represent the m x m matrix of correlations among the factors.

For any of the above matrices, we can use a caret (^) to indicate an estimate of that matrix and a subscript "pop" (or the corresponding Greek letter) to designate a matrix based on population parameters.

Much of the above represents notational "overkill" in the case of PCA, because so many of the matrices are equivalent to each other in PCA, but getting our notation straight now is intended to avoid confusion with the very necessary distinctions we'll have to make when we start rotating PCs or considering the more general case of factor analysis.

6.1.2 Scalar Formulae for Simple Cases of PCA

We now present formulae that can be used to obtain principal components of 2 x 2, 3 x 3, and uncorrelated covariance matrices. These formulae were derived via procedures discussed in Digression 2.

2 x 2 Covariance Matrix. Where the covariance matrix is given by

$$\begin{bmatrix} a & c \\ c & b \end{bmatrix},$$

the characteristic equation is given by

$$\lambda^2 - (a+b)\lambda + (ab - c^2) = 0;$$

whence

$$\lambda = \left[\frac{(a+b)+\sqrt{(a-b)^2+4c^2}}{2}, \frac{(a+b)-\sqrt{(a-b)^2+4c^2}}{2} \right],$$

and the characteristic vectors are $[-c/(a-\lambda), 1]$ in nonnormalized form, whence

$$\mathbf{b}_1' = (c, \lambda_1 - a)/\sqrt{c^2+(\lambda_1-a)^2},$$

$$\mathbf{b}_2' = (-c, a-\lambda_2)/\sqrt{c^2+(a-\lambda_2)^2}$$

in normalized form.

If $a = b$, then the characteristic roots are $a + c$ and $a - c$, with corresponding characteristic vectors in normalized form of $(\sqrt{.5}, \sqrt{.5}) = (707, .707)$ and $(\sqrt{.5}, -\sqrt{.5}) = (707, -.707)$. Which root and vector correspond to the first principal component depends on the sign of c. If c is positive, the sum of the two variables is the largest principal component; if c is negative, the difference accounts for a greater portion of the variance.

3 X 3 Covariance Matrix. The covariance matrix

$$\begin{bmatrix} a & b & c \\ b & d & e \\ c & e & f \end{bmatrix},$$

yields the characteristic equation

$$\lambda^3 - (a+d+f)\lambda^2 + (ad+df+af-b^2-c^2-e^2)\lambda - (adf-ae^2-fb^2-dc^2+2bce) = 0,$$

which is rather difficult to develop further algebraically. See Digression 3 for procedures for solving cubic equations.

If, however, $a = d = f$, the characteristic equation becomes

$$(\lambda-a)^3 - (b^2+c^2+e^2)\ (\lambda-a) - 2bce = 0\,.$$

The "discriminant" of this equation is $b^2c^2e^2 - (b^2+c^2+e^2)^3/27$, which is always zero (when $b = c = e$) or negative (when the off-diagonal entries are not all equal). (Can you see why?) Thus a trigonometric solution is called for, namely,

$$(\lambda - a) = K\cos(\phi/3), K\cos(120^{\circ}+\phi/3), K\cos(240^{\circ}+\phi/3),$$

where

$$\cos\phi = \sqrt{27}bce/(b^2+c^2+e^2)^{3/2},$$

and

$$K = 2\sqrt{(b^2+c^2+e^2)/3} = \sqrt{(b^2+c^2+e^2)}/.86603.$$

Any "scientific-model" calculator (including ones currently retailing for $10) provides one- or two-key computation of the trigonometric functions involved in this solution.

If any one of the covariances (say, b) is equal to 0, then $\cos\phi = 0$, whence $\phi = 90°$, $\phi/3 = 30°$, and $(\lambda - a) = (.86603K, -.86603K, 0)$, whence

$$\lambda = a + \sqrt{c^2 + e^2}, a - \sqrt{c^2 + e^2}, a.$$

The characteristic vector corresponding to the largest characteristic root is then given by

$$[c/\sqrt{2c^2 + 2e^2}, e/\sqrt{2c^2 + 2e}, \sqrt{.5}\,]$$

in normalized form. The other two principal components are given by

$$[-c/\sqrt{2c^2 + 2e^2},\ -e/\sqrt{2c^2 + 2e},\ .707] \text{ and } [e/\sqrt{c^2 + e^2},\ -c/\sqrt{c^2 + e^2},\ 0].$$

In the equicovariance (or equicorrelation) case, where $a = d = f$ and $b = c = e$, we have $\lambda = a + 2b$, $a - b$, $a - b$, and the corresponding characteristic vectors are $(3^{-1/2}, 3^{-1/2}, 3^{-1/2}) =$ (.86603, .86603, .86603) and any two vectors orthogonal to (uncorrelated with) the first, for example, $\left[\frac{2}{\sqrt{6}}, -\frac{1}{\sqrt{6}}, -\frac{1}{\sqrt{6}}\right]$ and $(0, 2^{-1/2}, -2^{-1/2})$.

If you wish to employ the iterative procedure (cf. section 4.6) for finding the characteristic vectors in the 3 x 3 case, it may be helpful to note that

$$\begin{bmatrix} a & b & c \\ b & d & e \\ c & e & f \end{bmatrix}^2 = \begin{bmatrix} a^2 + b^2 + c^2 & (a+)b + ce & (a+f)c + be \\ & b^2 + d^2 + e^2 & (d+f)e + bc \\ (symmetric) & & c^2 + e^2 + f^2 \end{bmatrix},$$

that is, each main diagonal entry in the squared matrix is the sum of the squares of the entries in the corresponding row (or column) of the original matrix. The off-diagonal entry in a given row or column is calculated by adding the product of the other two off-diagonal entries in the original matrix to the product of the entry in the same position in the original matrix times the sum of the diagonal entry in the same row and the diagonal entry in the same column. Note that the square of a matrix with equal main diagonal entries will not in general have equal main diagonal entries.

Uncorrelated m x m Matrix. The uncorrelated matrix

$$\begin{bmatrix} a_{11} & 0 & \cdots & 0 \\ 0 & a_{22} & \cdots & 0 \\ \vdots & \vdots & . & \vdots \\ 0 & 0 & \cdots & a_{mm} \end{bmatrix}$$

has characteristic roots equal to $a_{11}, a_{22}, \ldots, a_{mm}$. The characteristic vector corresponding to the root $\lambda_j = a_{jj}$ has coefficients that are all zero except for the jth position, which equals unity.

Equicovariance Case for m x m Matrix. If every main diagonal entry is equal to a common value s^2, and every off-diagonal entry is equal to a common value s^2r (for a positive value of r), then the greatest characteristic root (gcr), $\lambda_1 = s^2[1 + (m - 1)r]$, and the corresponding normalized characteristic vector is $\left[1/\sqrt{m}, 1/\sqrt{m}, \ldots, 1/\sqrt{m}\right]$. Each remaining root is equal to $s^2(1 - r)$, with vectors equal to any set that are orthogonal to the first characteristic vector. If r is negative, then
$s^2[1 + (m - 1)r]$ is the smallest root.

6.1.3 Computerized PCA

MATLAB.

```
Sx = [Entries of Sx, spaces between correlations; each
      row begun on a new line or separated from previous
      row by a semicolon.]
[V,D] = eig(Sx)
```

or (if you want the correlation-based PCs)

```
Rx = [Entries of Rx ]
[V,D] = eig(Rx)
```

Each column of **V** will contain one of the eigenvectors; **D** will be a diagonal matrix containing the eigenvalues, in the same order as the columns of **V** but not necessarily in descending numerical magnitude. Thus you must scan **D** for its largest entry and then look in the corresponding column of **V** for the first principal component; and so on.

SPSS Factor, Syntax Window. If working from raw data, read the data into the Data Editor as usual, then

```
FACTOR VARIABLES = Variable-list
  / METHOD = {Correlation**, Covariance}
  / PRINT = INITIAL
  / ROTATION = NOROTATE   .
```

If you wish to obtain the PCs of the correlation matrix you may of course omit the METHOD subcommand, because that's the default.

If you are working from an already-computed covariance or correlation matrix,

```
MATRIX VARIABLES = Variable-list /
  / FORMAT = FREE UPPER  / N = Sample size
```

```
  / CONTENTS = CORR
FACTOR /MATRIX = IN {COR = *, COV = *}
  / METHOD = {Correlation**, Covariance}
  / PRINT = INITIAL
  / ROTATION = NOROTATE   .
```

SPSS Factor, Point-and-Click.

```
Statistics
   Data Reduction . . . Factor …
```

Select variables to be factored by highlighting in the left window, then clicking on the right arrow to move the variable names to the right window. Ignore the row of buttons at the bottom of the panel and click "OK" to take defaults of an unrotated PCA on the correlation matrix with all eigenvalues reported and with the PC coefficients reported for components having above-average eigenvalues.

To analyze the covariance matrix, click the "Extraction" button and choose "Analyze covariance matrix".

To report a specific number of components to be displayed, click the "Extraction" button and fill in the blank after choosing "Extract Number of Factors ___."

6.1.4 Additional Unique Properties (AUPs) of PCs

So far we have focused on the (normalized) variance-maximizing properties of principal components. Other authors (especially those who focus more on factor analysis) emphasize instead the following property:

AUP1: *PC_1 is that linear combination of the p original variables whose average squared correlation with those original variables is maximized.*

If those original variables are individual items that are being considered for inclusion in a personality scale or an attitude scale, then any component can be considered as defining a total scale score, and the loading of each original variable on the component can be thought of as a *part–whole correlation*, which is often used as a criterion in deciding which items to retain in and which to drop from a scale. PC_1 represents, then, that (weighted) scale score that relates most strongly to its consituent items, that is, for which the average squared part–whole correlation is highest.

Each subsequent PC provides that component (weighted scale score) with the highest possible average squared correlation with the original variables, subject to being uncorrelated with the preceding PCs.

AUP2: *Both the principal components and the vectors of combining weights that define them are mutually uncorrelated*; that is,

$$r_{\text{PC}i,\text{PC}j} = 0 \text{ for all } i \neq j$$

and

$$r_{b_i,b_j} = 0 \text{ for all } i \neq j, \text{ which is true if and only if } \mathbf{b}_i'\mathbf{b}_j = 0.$$

The second version of this property of principal-component weights ($\mathbf{b}_i'\mathbf{b}_j = 0$) is usually described as *orthogonality* of the two sets of weights, so we can rephrase AUP2 as:

Principal components are both uncorrelated and defined by mutually orthogonal sets of combining weights.

Note that it is the *combination* of the PCs' being uncorrelated and the **b**s' being orthogonal that is unique to PCs, not either of those two properties by itself. As is commonly known and as is shown in section 6.5, the PCs can be subjected to a procedure known as "orthogonal" rotation, which, for any given angle of rotation, produces another set of components that are also mutually uncorrelated. What is not as commonly known is that (a) such "orthogonally" rotated PCs are related to the original variables by vectors of combining weights that are *not* mutually orthogonal, and (b) it is possible to rotate PCs in a way that retains the orthogonality of the defining weights—but in that case the rotated PCs are no longer mutually uncorrelated with each other. (I have in fact been unable to find any published mention of these two facts prior to this edition of this book, although it has been presented in conference papers, e.g., Harris & Millsap, 1993.) We return to the difference between uncorrelated-components rotation and orthogonal-weights rotation in section 6.5.5.

AUP3: *The sum of the squared loadings of the original variables on any given PC equals the normalized variance of that PC.*

This equivalence would appear to make the fact that PCs are optimal both in terms of squared loadings and in terms of normalized variance self-evident, and has also led every author I've checked to assume that, because the sum across all components of the sums of their squared loadings remains constant under "orthogonal" rotation (rotation that preserves the zero correlations among the components), so must the normalized variances. However, the appearance of obviousness is misleading because the equivalence of sums of squared loadings and normalized variances holds *only* for the *principal* components, and not for other sets of linear combinations of the original variables.

AUP 4: Joliffe (1986) proved that *the sum of the squared multiple correlations between the first m* ***R****-derived principal components and the p original variables is higher than for any other set of m mutually uncorrelated components.* **R**-derived PCs thus reproduce scores on the original variables more efficiently than do any other uncorrelated linear combinations of those original variables.

With all these unique properties of principal components in mind, let's proceed to examine interpretations (section 6.2) and use of (section 6.3) principal components.

6.2 INTERPRETATION OF PRINCIPAL COMPONENTS

It has already been demonstrated that the PCs of a set of variables are uncorrelated with each other, and that they are hierarchically ordered in terms of their variances, with the *i*th PC having the *i*th largest variance. The variance of a variable is of course a measure of the extent to which subjects differ in their scores on that variable. It is thus reasonable to interpret PC_1 as that linear combination of the original variables that maximally discriminates among our subjects, just as the discriminant function derived in Manova maximally discriminates among the various groups. PCA is thus a sort of "internal discriminant analysis."

It is also true that the PCs partition the total variance of the original variables into *p* additive components. From the general properties of characteristic roots (section D2.12), we know that

$$\sum_j \lambda_j = \sum_j s^2_{PC_j} = \text{trace } \mathbf{S}_x = \sum_i s^2_i .$$

It is customary to describe this last property of PCA as indicating that PC_j "accounts for" a certain proportion—$s^2_{PC_j} / \sum_i s^2_i$—of the variation among the original variables. It would not, of course, be difficult to construct a set of *m* variables that also had the property that their variances summed to $\sum_i s_i^2$, but that had no relationship at all to those original variables. For instance, we could use a random number generator to construct *m* unit-normal distributions, and then multiply each by $\sum_i s_i^2 /m$. We thus need to provide some additional justification for the phrase *accounts for*, that is, for the assertion that PCA in some way explains, or describes the structure of, the variation among the original variables.

The strongest support for this way of talking about PCs comes from the fact that knowledge of our subjects' scores on the *m* principal components, together with knowledge of the coefficients defining each PC, would be sufficient to reproduce the subjects' scores on the original variables perfectly. Just as the PCs are defined as linear combinations of the original variables, so the original variables can be defined as linear combinations of the PCs. In fact, the coefficients that must be used to generate X_j in the equation

$$X_j = a_{j,1}PC_1 + a_{j,2}PC_2 + \cdots + a_m PC_m \tag{6.3}$$

are simply the weights variable *j* receives in the various linear compounds that define the PC*s*, that is, $a_{j,k} = b_{k,j}$ for all *k*, *j*. Because we can reproduce the score made by each subject on each original variable, we can, a fortiori, reproduce any measure or set of measures defined on the original variables—such as, for example, the covariance matrix or the intercorrelation matrix of the original variables. The latter two "reproductions" do not require detailed knowledge of each individual's score, but only the values of the characteristic roots and vectors. Specifically, the covariance of two original variables X_i and X_j is given by the formula

$$s_{ij} = \frac{\sum[(b_{1,i}pc_1 + b_{2,i}pc_2 + \cdots + b_{m,i}pc_m)(b_{1,j}pc_1 + b_{2,j}pc_2 + \cdots + b_{m,j}pc_m)]}{N-1} \tag{6.4}$$
$$= b_{1,i}b_{1,j}\lambda_1 + b_{2,i}b_{2,j}\lambda_1 + \cdots + b_{m,i}b_{m,j}\lambda_m;$$

whence

$$\mathbf{S}_x = \mathbf{BDB}',$$

where **B** is an m x m matrix whose jth column contains the normalized characteristic vector associated with λ_j; **D** is a diagonal matrix whose jth main diagonal entry is λ_j ; and PC_i is a deviation score on the ith principal component.

To illustrate this reproducibility property, consider a very simple example in which we seek the principal components of the system of variables consisting of the first and second variables from the Linn and Hodge (1981) study of hyperactivity (see section 2.3.4): CPT-C and CPT-E, the number of correct responses and the number of errors, resepectively, in the selective-attention task. We'll use the data from the 16 non-hyperactive children, which yields a covariance matrix of

$$\mathbf{S}_x = \begin{bmatrix} 112.829 & -59.238 \\ -59.238 & 72.829 \end{bmatrix},$$

whence (taking advantage of the computational formulae provided in section 6.1.2),

$$\lambda = \left[185.658 \pm \sqrt{40^2 + 4(59.238^2)}\right]/2 = 155.352, 30.306;$$
$$c^2 + (\lambda_1 - a)^2 = 59.238^2 + 42.523^2 = 5317.346;$$
$$\mathbf{b}_1' = (-59.238, 42.523)/72.920 = (-.812, .583);$$

and

$$\mathbf{b}_2' = (59.238, 82.523)/\sqrt{59.238^2 + 82.523^2} = (59.238, 82.523)/101.583$$
$$= (.583, .812).$$

The results of a PCA are typically reported in the form of a ***coefficient matrix***, that is, a matrix **B** whose columns contain the normalized coefficients of the principal components (in other words, the normalized eigenvectors of $\mathbf{S}_x$), or in the form of the equivalent table. For the present analysis, the coefficient matrix (in tabular form) would be

	PC_1	PC_2
X_1	−.812	.583
X_2	.583	.812
$\lambda_j = s^2_{PC_j}$	155.352	30.306

It is also common practice to list the characteristic root for each PC at the bottom of the column containing the coefficients that define that PC. (Both of these "common practices" are becoming somewhat diluted by the ease of performing PCA via a general-purpose factor analysis program, together with the strong bias of such programs toward display of the structure matrix, rather than the coefficient matrix, and toward provision of the eigenvalues of $\mathbf{R}_x$ rather than of $\mathbf{S}_x$ unless specifically directed otherwise.)

The PCs are defined in terms of the original variables as follows:

$$-.812X_1 + .583X_2 = PC_1; \quad .583X_1 + .812X_2 = PC_2 .$$

However, as the form in which these two equations are listed is meant to suggest, we can as readily consider the equations as a pair of simultaneous equations to be solved for X_1 and X_2, whence

$$\begin{gathered} -.473X_1 + .340X_2 = .583PC_1 \\ .473X_1 + .659X_2 = .812PC_2 \\ \Downarrow \\ X_2 = .583PC_1 + .812PC_2 \end{gathered}$$

(from adding the preceding two equations), and

$$X_1 = (.812PC_2 - .659X_2)/.473 = -.812PC_1 + .584PC_2.$$

Note that the coefficients used in the linear combination of PCs that "generates" or "reproduces" a particular original variable bear a close relationship to (and are in fact identical to, except for round-off error) the coefficients in that row of the coefficient matrix. In other words, as long as we normalize the columns of **B** to unit length (coefficients whose squares sum to 1.0), **B** and **P** are identical in PCA. Put another way, if we read the columns of **B**, we are using it as a coefficient matrix, whereas read row-wise it serves as a factor (component) pattern matrix.

This relationship, which was stated without proof a few paragraphs earlier, is proved in Derivation 6.2. We demonstrate it by conducting multiple regression analyses of each of variables X_1 and X_2 from scores on the two principal components of this two-variable system. Let's let SPSS do the calculations:

```
TITLE Predicting CPTC, CPTE from PCs of same; Linn's
   Hyperactivity Data, Controls .
DATA LIST FREE/LOC CPQ CPTC CPTE HYPER .
VAR   LABELS   LOC   Locus   of   Control   /  CPQ   Connors   Parent
Questionnaire /
    CPTC Attention task corrects / CPTE Attention task errors /
    HYPER Hyperactivity    .
VALUE LABELS HYPER (0) Non-hyper control (1) Hyperactive  .
BEGIN DATA
      14.     36.     189.     11.      0.
      17.     41.     183.     15.      0.
```

```
     11.      26.     194.      2.      0.
      8.       5.     198.      2.      0.
     13.      11.     182.     19.      0.
     10.      30.     196.      2.      0.
     11.       4.     199.      0.      0.
     17.      12.     177.     13.      0.
     16.      33.     183.     10.      0.
     14.       9.     174.     29.      0.
     18.       4.     191.      5.      0.
     16.      36.     158.     12.      0.
      9.       7.     193.      5.      0.
     24.      19.     181.     24.      0.
     20.      33.     193.      3.      0.
     19.      10.     182.      5.      0.
End Data .
Compute PC1 = -.812*CPTC + .583*CPTE
Compute PC2 =  .583*CPTC + .812*CPTC
Regression   Variables = CPTC CPTE PC1 PC2   /
    Descriptives = Defaults Corr Cov /
    Dep =  CPTC CPTE / Enter PC1 PC2    .
```

The output from this run includes the following:

```
Dependent Variable: Attention task corrects
Model Summary
                                   Std.Error
               R       Adjusted of the
Model   R      Square  R Square Estimate
1      1.000   1.000   1.000    2.082E-07
a  Predictors: (Constant), PC2, PC1

Coefficients
                 Unstandard          Standard
                 ized                ized
                 Coeffi              Coeffi
                 cients              cients         t          Sig.
Model            B          Std.     Beta
                            Error
1     (Constant)1.061E-14  .000                           .000 1.000
      PC1        -.813      .000     -.953     -181305254.684   .000
      PC2         .583      .000      .302       57495359.598   .000
a  Dependent Variable: Attention task corrects

Dependent Variable: Attention task errors

Model Summary
```

Model	R	R Square	Adjusted R Square	Std. Error of the Estimate
1	1.000	1.000	1.000	1.36 E-07
1	1.000	1.000	1.000	1.36 E-07

a Predictors: (Constant), PC2, PC1

Coefficients

Model		Unstandardized Coefficients B	Std. Error	Standardized Coefficients Beta	Sig.
Model		B	Std. Error	Beta	
1	(Constant)	6.233E-14	.000	.000	1.000
1	(Constant)	6.233E-14	.000	.000	1.000
	PC1	.583	.000	.852	.000
	PC1	.583	.000	.852	.000
	PC2	.813	.000	.524	.000
	PC2	.813	.000	.524	.000

a Dependent Variable: Attention task errors

Thus, as we had predicted, the original observations are indeed perfectly reproduced to within round-off error in the computations, and the regression coefficients for accomplishing this perfect prediction are identical to the scoring coefficients in the row of the normalized scoring-coefficient matrix corresponding to a given original variable.

This, then, is one sense in which the PCs "account for" variation among the original variables: the multiple R between the PCs and each original variable is 1.0, and thus knowledge of scores on the PCs is sufficient to exactly reproduce scores on the original variables.

How unique is this contribution? Are there other linear combinations of the original variables that would also have a perfect multiple correlation with each of the original variables? The answer is of course "Yes." One conspicuous example takes as the coefficients for the jth linear compound a vector of all zeros except for the ith entry. More generally, *any* linearly independent set of vectors would provide a perfect multiple correlation. Thus this property, by itself, does not represent an advantage of principal components over any other set of linear transformations of the original variables.

What, then, makes PCs "better" than other linear compounds as explanations of the data? Are PCs, for instance, more likely to reflect the "true," underlying sources of the variability among the Xs? A simple hypothetical example shows that this is not the case.

Example 6.1 *Known Generating Variables* Assume that scores on two observable variables, X_1 and X_2, are generated by a pair of underlying variables, H_1 and H_2, with X_1 being equal to the sum of the underlying processes, and $X_2 = H_1 + 3H_2$. [This implies that $H_1 = (3X_1 - X_2)/2$ and $H_2 = (X_1 - X_2)/2$.] Assume further that H_1 and H_2 are uncorrelated and that each has a unit normal distribution (and thus a mean and standard deviation of zero and one, respectively). Such a state of affairs would produce an observed covariance matrix of

$$\begin{bmatrix} 2 & 4 \\ 4 & 10 \end{bmatrix}$$

and a factor (component) coefficient matrix of

	PC_1	PC_2
X_1	.383	−.235
X_2	.922	.974
s^2_{PC}	11.655	.345

Thus the linear compounds suggested by PCA as explanations of the data are

$$.383X_1 + .922X_2, \text{ and } .974X_1 - .235X_2,$$

as contrasted to the true generating variables' relationships to X_1 and X_2 (in normalized form), $.707X_1 - .707X_2$ and $.948X_1 - .316X_2$. Looked at from another angle, PCA suggests that X_1 and X_2 are related to underlying processes by $X_1 = .383\text{PC}_1 - .235\text{PC}_2$, and $X_2 = .922\text{PC}_1 + .974\text{PC}_2$ as contrasted to the "true" relationships of $X_1 = H_1 + H_2$ and $X_2 = H_1 + 3H_2$. Thus there is no guarantee that PCA will uncover the true generating processes. The advantages of PCA must lie elsewhere. It is to these advantages we now turn.

6.3 USES OF PRINCIPAL COMPONENTS

The advantages of PCA are (at least) twofold.

6.3.1 Uncorrelated Contributions

The first advantage of PCA is that the principal components are *orthogonal* to each other, as the vectors in the trivial (identity) transformation $H_j = X_j$ are not. This is of benefit in two ways. First, the comparisons we make among subjects with respect to their scores on PC_i are uncorrelated with (do not provide the same information as) comparisons based on their scores on PC_j. Second, any subsequent analysis of the relationship between the variables in the set on which the PCA was conducted and some other variable or set of variables can use subjects' scores on the PCs, rather than their scores on the original variables, without any loss of information. This will have the effect of greatly simplifying the computations involved in these subsequent analyses, as well as providing an automatic "partitioning" of the relationships into those due to correlations between PC_1 and the set of outcome variables, those due to correlations between PC_2 and the set of outcome variables, and so on. For instance, a multiple regression analysis conducted on the relationship between a set of uncorrelated variables (such as m PCs) and some outcome variable yields an $R^2 = \sum_i r_{iy}^2$, the sum of the squares of the separate Pearson *r*s between *y* and each of the uncorrelated predictor variables, and *z*-score regression coefficients each of which is simply equal to r_{iy}. Also, if neither the means nor the within-group scores on some set of outcome variables are correlated with each other, the test statistic resulting from a multivariate analysis of variance conducted on these uncorrelated variables will simply equal $(k - 1)/(N - k) \cdot \max_i(F_i)$, where F_i is the univariate *F* ratio obtained from a univariate Anova conducted on the single outcome measure X_i. Interpretation of the results of these analyses is also facilitated by the partitioning, in multiple regression, of the overall R^2 into *m* uncorrelated sources of predictability, and in the case of Manova, by the partitioning of the differences among the groups into their differences on each of *m* uncorrelated variables. This gain in interpretability of results will, however, be illusory if the uncorrelated variables do not themselves have simple substantive interpretations. This raises the *labeling problem,* the very persistent and usually difficult task of finding substantive interpretations of some set of hypothetical latent variables that have been derived through PCA or FA. This traditionally and commonly involves an examination of the factor structure, that is, of the correlation between each original variable and each latent variable, although we argue (in section 7.6) that interpretations should be based on the factor-score coefficients. Because the partitioning and computational advantages of employing PCs accrue as well to *any* orthogonal set of linear compounds of the original variables, researchers will naturally prefer to employ a set of linear compounds whose factor structure, or factor pattern, takes on a very simple form. Empirically, the factor structures resulting from PCA tend not to have "simple structure"; thus many researchers prefer to treat the results of a PCA as merely an intermediate step that provides one of many possible sets of uncorrelated latent variables. These various possible sets of orthogonal linear compounds are all related by a simple set of transformations known as *orthogonal rotations* (discussed in more detail in sections 6.5.1 and 6.5.2), so that PCA can indeed serve as a starting point for the search for simple structure. If such a procedure is followed, however, the researcher automatically loses the second major advantage of PCA, its hierarchical ordering of the latent variables in terms of the percentage of variance for which each accounts.

6.3.2 Computational Convenience

The principal components resulting from PCA, unlike the latent variables represented by any other set of linear compounds of the original variables, are designed to account for the highest percentage of the variation among the variables with as few PCs as possible. The first PC accounts for a larger portion of the variance of the system than any other linear compound (except a simple multiple of itself), and each successive PC accounts for as high a percentage of the total variance as is possible while remaining uncorrelated with all preceding PCs. If, as is often the case, the first few PCs account for some large percentage (80%? 90%?) of the total variance, the researcher may be willing to sacrifice completeness of description (which may well be illusory anyway, because PCA blithely partitions all of the variance despite the fact that a fair portion of the system's total variance may reflect sampling fluctuation or relationships to hypothetical processes unique to a given variable) for compactness of description by, essentially, throwing away subjects' scores on the remaining PCs. (This should not be taken as suggesting that the PCs associated with larger roots are error free. They, just like the low-variance PCs, are subject to random variation. Results we discuss in section 7.7 indicate that the *uniqueness* of the various variables—that portion not held in common with other variables—is indeed more concentrated in the early components than in the smaller-variance ones. However, uniqueness consists both of error variance and of reliable variance that is simply not shared with other variables. A variable that is uncorrelated with other dependent measures may be an important dimension of differences among your experimental conditions.)

Thus, even if the PCA factor structure is to be subjected to a series of rotations in search of simple structure, the PCA will prove useful to the researcher in providing some evidence on the total number of hypothetical, latent variables he or she needs to include in a final description of the structure of the original set of variables. It must be kept in mind, however, that this reduction in the dimensionality of the system may cost more in terms of the accuracy of subsequent analyses of the data than is immediately apparent from the small percentage of variance left unaccounted for by the "discarded" PCs. The linear compound that maximizes differentiation among the subjects in terms of their scores on the original variables (i.e., PC_1) will not in general be the same as the linear compound that correlates most highly with some criterion variable or that maximally discriminates between various experimental or clinical groups. Thus, for instance, it is possible that PC_4 may correlate more highly with some criterion variable than does PC_1, so that its deletion from the full set of PCs would reduce R^2 by a considerably greater percentage than its share of the total variation among the Xs. We see a dramatic (diabolical?) demonstration of this fact in Example 6.4.

6.3.3 Principal Component Analysis as a Means of Handling Linear Dependence

The computational procedures for multiple regression, T^2, Manova, and canonical analysis all require the inversion of the covariance matrix of one or more sets of variables. When there is a linear relationship among the variables in the set—that is, when the covariance matrix **S** is of lower rank than the number of variables —$|\mathbf{S}| = 0$, and $\mathbf{S}^{-1}$ is therefore undefined. In previous chapters, we have recommended that such a situation be handled by deleting one or more of the variables involved in this linear dependence until the number of remaining variables equals the rank of **S**. A more elegant approach to this same problem would be to perform a PCA of the variables in each set involved, and then perform the multiple regression or other analysis of between-set relationship on the PCs, the number of which will match the rank of the original covariance matrix. The resultant measure of relationship (R^2 or R_c) and test of significance (T^2, gcr, or F) will be identical, no matter which approach (deletion of variables or preliminary PCA) is taken and no matter which variables from among the subset of linearly dependent variables are eliminated in the deletion procedure.

A decision between the two approaches must therefore be made on one of two grounds: (a) computational convenience, or (b) interpretability of the resulting regression weights, discriminant function, or canonical coefficients.

As far as computational convenience is concerned, the crucial question is whether several analyses involving this set of variables, or only one, is to be performed. The additional work of computing PCs is worthwhile only if several analyses are to be carried out on these same data, thus leading to savings of computational effort every time the inverse or determinant of $\mathbf{S}_x$, the variance–covariance matrix for the *Xs*, is required in a subsequent analysis.

In terms of interpretability, the crucial point is that *each* of the following sets of variables contains all of the information available in the full set of p measures: the s PCs; any s "factors" resulting from rigid or oblique rotation of the s PCs; or the s variables in the reduced set of outcome measures remaining after any $p - s$ of the variables involved in the linear dependence have been deleted. Thus any set of regression coefficients, any discriminant function, or any canonical variate obtained by beginning the analysis with any one of these sets of s variables can be translated without loss of information into a linear combination of any of the other sets of variables.

By way of illustration, consider the Manova you conducted in the Demonstration Problem in chapter 4 on differences among the COMP, NMO, COOP, and IND groups in their behavior in the Prisoner's Dilemma game. (Cf. Data Set 4, Table 3.2.) In that study, the four outcome measures—CC, CD, DC, and DD—were constrained to add to 40 for each subject. This linear dependence was handled by deleting DD, leading to a gcr of 1.3775 and a first discriminant function of $D_1 = .832\text{CC} - .159\text{CD} + .532\text{DC}$. However, because CC + CD + DC = 40 - DD, an infinite number of equally valid discriminant functions, each leading to exactly the same gcr, can be obtained from the expression $D_1 = (.832 - c)\text{CC} - (c + .159)\text{CD} + (.532 - c)\text{DC} + c\text{DD}$, where c is any constant whatever. In particular, if $c = .832$, we obtain a discriminant function in which CC receives no direct weight in the discriminant function, all of its contributions having been "channeled through" or "absorbed into" the other three variables. As the reader can readily confirm, this is precisely the discriminant function we would have obtained had we applied the Manova procedure after deleting CC, rather than DD. Similarly, the expressions that result by taking $c = -.159$ or $c = .532$ are those that would have followed deletion of CD or DC, respectively. Thus there is no basis for the advantage sometimes claimed for preliminary PCA over deletion of variables, namely, that PCA yields an expression that involves *all* of the original variables, because such an expression is readily obtained from the deletion procedure simply by taking into account the nature of the linear relationship among the measures.

It might be argued that PCA provides a means of discovering the fact of linear dependence and of specifying the nature of the dependence. This is true, because a perfect linear relationship among the p variables will result in only $s < p$ nonzero eigenvalues of $\mathbf{S}$, and the rows of the "factor pattern" resulting from PCA provide p

equations relating the p variables perfectly to only s unknowns; this set of equations can be solved in various ways to display the linear dependence among the original measures. However, the comments in chapter 3 on the inadvisability of using the determinant of **S** as a test for linear dependence apply to PCA, too. It is highly unlikely that random sampling from any multivariate population will yield a sample covariance matrix having a determinant of zero or fewer than p nonzero characteristic roots, *unless* a linear dependence has been "built into" the set of variables by the researcher (or by his or her supplier of tests). Even in the latter case, round-off errors or relatively small mistakes in transcribing or keypunching the data will lead to a nonzero determinant or to a full complement of p PCs. However, this state of affairs *is* more likely to be detected upon examination of the PCA, because most of us have much stronger intuitive feeling for "proportion of variance accounted for" figures (for example, the last eigenvalue divided by the sum of all the characteristic roots) than for "near-zero" determinants—which may be 10^8 or larger for "near-miss" covariance matrices.

At any rate, a preliminary PCA on the **E** matrix from the Demonstration Problem in chapter 4 yields

$$PC_1 = .757CC - .013CD - .099DC - .646DD;$$
$$PC_2 = .398CC - .334CD - .635DC + .571DD;$$

and

$$PC_3 = .134CC - .799CD + .580DC + .085DD.$$

(A PC_4 was computed but it accounted for only .0002% of the total variance.)

Note that we chose to perform the initial PCA on **E**, rather than on **H** or on the total covariance matrix, ignoring groups, **H** + **E**. Where the separation of the subjects into groups is, as in the present case, on the basis of some set of experimental manipulations, we would not wish to run the risk of confusing ways in which the variables are alike in their responsiveness to experimental treatments (as reflected in the **H** matrix) with their tendency to covary within each group as a consequence of common genetic or developmental roots (as reflected in **E**. If, however, the groups are established on the basis of some organismic variable such as sex, occupational status, or broad age groupings, whether PCA is to be applied to **E** or to **H** + **E** will depend on the population to which the experimenter wishes to generalize his or her results. If the analysis is to be relevant to sub-populations that are homogenous with respect to the variable used to assign subjects to groups, the PCA should be applied to **E**. If instead it is to be relevant to a more general population in which the current "independent" variable is allowed to vary naturally, **H** + **E** should be analyzed. None of these decisions will have any effect on the ultimate outcome of the Manova, because either set of PCs contains all of the information contained in the original variables, so there can be no difference in results. However, performing the PCA on **E** will yield the greatest computational savings in the subsequent Manova.

At any rate, we compute $\mathbf{H}_{PC}$ for the three principal components via the formula $\mathbf{H}_{PC} = \mathbf{B'HB}$, where **B** is the factor weight matrix for the PCA (each column representing one of the PCs) and **H** is the original 4 x 4 hypothesis matrix. Of course, $\mathbf{E}_{PC}$ is a diagonal matrix having the variances of the PCs on the main diagonal. The result is a gcr of 1.3775, as had been the case for the original analysis deleting DD, and a first discriminant function of

$$\begin{aligned} D_1 &= .725\ PC_1 + .959\ PC_2 + .686\ PC_3 \\ &= .665CC - .577CD + .289DC - .371DD \\ &= 1.042CC - .200CD + .666DC \end{aligned}$$

in nonnormalized form. After normalizing so as to achieve a sum of squared coefficients equal to unity, we have

$$D_1 = .832CC - .159CD + .532DC$$

as before. Similarly, we could convert our original discriminant function in terms of CC, CD, and DC into one involving PC_1, PC_2, and PC_3 by substituting in expressions for each original variable as a function of the three PCs.

Given the tedium of computing $\mathbf{H}_{pc}$ and the complete equivalence of the gcr and discriminant function produced by the deleted-variable and initial-PCA approaches to handling linear dependence, the initial-PCA approach would seem ill advised *unless* at least one of the following is true:

1. The researcher is interested in the results of the PCA in their own right, and would thus perform the PCA anyway.
2. The researcher plans several other analyses of the same set of data.
3. The researcher expects the PCs or some rotation thereof to have such a clear and compelling substantive interpretation that the discriminant function expressed as a function of the PCs is more meaningful to him or her than any of the discriminant functions based on the original variables.

6.3.4 Examples of PCA

Example 6.2: *Components of the WISC–R*. The *W*echsler *I*ntelligence *S*cale for *C*hildren–*R*evised (WISCR–R; Wechsler, 1974) is one of the most oft-used (and most often factor-analyzed) measures of intelligence. It is made up of 12 sub-tests: 6 subscales based primarily on processing of verbal and numerical information (Information, Similarities, Arithmetic, Vocabulary, Comprehension, and Digit Span) and 6 performance-based (Picture Completion, Picture Arrangement, Block Design, Object Assembly, Coding, and Mazes). The SPSS syntax-window setup given next provides the matrix of correlations among the twelve subscales of the WISC–R for the norming sample of 2200 children (1100 males and 1100 females, evenly distributed across several age groups):

```
Title
MATRIX DATA VARIABLES =
   INFO       SIMIL     ARITHM    VOCAB     COMPR      DIGSPAN  COMPL
ARRANGEM
   DESIGN  ASSEMBLY CODING  MAZES  /
   CONTENTS = N CORR / FORMAT = FREE FULL
BEGIN DATA
203 203 203 203 203 203 203 203 203 203 203 203
```

```
1   .62 .54 .69 .55 .36 .40 .42 .48 .40 .28 .27
.62 1   .47 .67 .59 .34 .46 .41 .50 .41 .28 .28
.54 .47 1   .52 .44 .45 .34 .30 .46 .29 .32 .27
.69 .67 .52 1   .66 .38 .43 .44 .48 .39 .32 .27
.55 .59 .44 .66 1   .26 .41 .40 .44 .37 .26 .29
.36 .34 .45 .38 .26 1   .21 .22 .31 .21 .29 .22
.40 .46 .34 .43 .41 .21 1   .40 .52 .48 .19 .34
.42 .41 .30 .44 .40 .22 .40 1   .46 .42 .25 .32
.48 .50 .46 .48 .44 .31 .52 .46 1   .60 .33 .44
.40 .41 .29 .39 .37 .21 .48 .42 .60 1   .24 .37
.28 .28 .32 .32 .26 .29 .19 .25 .33 .24 1   .21
.27 .28 .27 .27 .29 .22 .34 .32 .44 .37 .21 1
END DATA .
```

We analyze the correlation matrix—by far the most common choice. Indeed, the typical article reporting the results of a principal component analysis or of a factor analysis provides only the correlation matrix—not even the standard deviations or variances from which the covariance matrix could be (approximately, given the two-decimal-place reporting of the correlations) computed. This in turn means that we must rely on entering Factor program subcommands in the syntax window; input of the correlation matrix is not available via the point-and-click route. The commands:

```
Factor  Matrix = In (Cor = *) /  Print = Defaults FScore
     / Criteria = Factors (4)
     / Rotate = NoRotate   .
```

yield, in part, the following output:

Factor Analysis

Total Variance Explained

	Initial Eigen-values			Extraction Sums of Squared Loadings		
Component	Total	% of Variance	Cumulative %	Total	% of Variance	Cumulative %
1	5.398	44.980	44.980	5.398	44.980	44.980
2	1.160	9.666	54.646	1.160	9.666	54.646
3	.969	8.075	62.722	.969	8.075	62.722
4	.746	6.220	68.942	.746	6.220	68.942
5	.653	5.439	74.381			
6	.618	5.150	79.531			
7	.527	4.394	83.925			
8	.514	4.282	88.207			
9	.426	3.548	91.755			
10	.378	3.148	94.903			
11	.340	2.835	97.738			
12	.271	2.262	100.000			

Extraction method: Principal component analysis

Examining the "% of Variance" column shows that the drop in percentage of variance explained by the successive prinicipal components drops very slowly after the first four PCs. This is usually a sign that those latter (in this case, eight) PCs are largely analyzing error variance. In addition, most authors who have conducted PCAs or FAs on the WISC-R have concluded that either two or three factors beyond the first, general factor are sufficient. I therefore asked the Factor program to print out the loadings/scoring coefficients for only the first four PCs, namely

```
Component Matrix
            Compo
            nent
            1        2        3           4
INFO        .774     -.243    -.202       -2.026E-02
SIMIL       .777     -.160    -.252        1.854E-02
ARITHM      .679     -.338     .165       -.215
VOCAB       .806     -.267    -.239        5.858E-02
COMPR       .730     -.143    -.316        9.284E-02
DIGSPAN     .509     -.400     .464       -.380
COMPL       .652      .350    -.131       -9.522E-02
ARRANGEM .629      .255    -6.547E-02  .222
DESIGN      .758      .302     .135       -2.691E-02
ASSEMBLY .648      .467    3.341E-02    3.333E-02
CODING      .464     -.166     .561        .633
MAZES       .511      .431     .330       -.287
Extraction Method: Principal Component Analysis.
a 4 components extracted.
```

Note that the "Component Matrix" reported here is actually the structure/pattern matrix—that is, it reports the simple correlation between each original variable and each principal component. However, the columns of loadings are columnwise proportional to the normalized scoring coefficients, so little would be gained by having SPSS print out a separate scoring-coefficient matrix. For instance, our "FSCORE" designation on the "PRINT =" subcommand above did yield printout of the "Component score coefficient" matrix. Within the first column of that matrix the entries for INFO and ARITHM are .143 and .126, respectively, for a ratio of .143/.126 = 1.135, as compared to the ratio of their loadings as reported in the "Component matrix" here, namely .774/.679 = 1.140. (Because the loadings and scoring coefficients are reported to three significant digits, we can expect only three significant digits of accuracy in the ratio of any two of them, so the two ratios are indeed identical to within round-off error, as are all other ratios of pairs of loadings as compared to ratios of pairs of scoring coefficients.)

Note, too, that the sum of the squares of the loadings in each column equals (to within round-off error) the eigenvalue for that PC, which in turn equals the normalized variance of the PC. Put algebraically,

$$\sum_i r^2_{X_i,\mathrm{PC}_j} = \lambda_j = s^{2(norm)}_{\mathrm{PC}_j}\,.$$

These are thus not normalized coefficients as we have defined them in this chapter. (Neither are the "Component score coefficients" reported by SPSS. The sum of the squares of the SPSS-reported scoring coefficients in each column equals $1/\lambda_j$, the reciprocal of the eigenvalue of the PC that column represents.) This difference in normalization criterion is unimportant for computing scores on PCs from scores on original variables, but needs to be taken into account if we want to compute scores on original variables from scores on PCs. For this latter task we know that the loadings provide the *z*-score regression coefficients—that is, the combining weights for predicting *z*-scores on original variables from *z*-scores on PCs. But, as just pointed out, the loadings aren't normalized scoring coefficients, so how does this jibe with our statement in section 6.2 that, for unrotated PCs, the rows of the loading/pattern/structure matrix provide the coefficients for predicting original variables from PCs "as long as we normalize the columns of **B** to unit length"? The "out" here is that scores on the PCs are *not* *z*-scores. Any set of *z*-scores has a variance of 1.0, whereas each PC has a variance equal to its λ_j. (After all, we chose the combining weights for each PC so as to yield maximum, not just average, variance.) The entries in the loading matrix can be used to predict *z*-scores on original variables from *z-scores* on the PCs, and the entries in the (normalized) scoring-coefficient matrix can be used to predict *z*-scores on the original variables from *raw scores* on the PCs.

Now, how about interpretation of these first four PCs? The first PC is very straightforward. Every one of the 12 original variables receives a positive weight, ranging from .464 to .806, in computing scores on PC_1, so it is pretty obviously a general factor representing some property (general intelligence?) that all 12 subscales have in common.

PC_2 is almost as straightforward. Seven of the subscales receive negative weights and five of them receive positive weights. The primary decision we have to make is where to "draw the line" between weights that are large enough to warrant including the corresponding subscale in our interpretation of PC_2 and those that are small enough to warrant ignoring the corresponding subscale in our interpretation. There are two primary standards to be considered: Among users of PCA and FA a rule of thumb has emerged that structure coefficients greater than .3 in absolute value are important, whereas those below this value are not. Alternatively, we could employ the same sort of "scree test" we have used in previous chapters, namely, considering the most positive coefficient and all other positive coefficients that are more than one-third to one-fourth as large as that most positive coefficient "important" (nonignorable), considering the most highly negative coefficient and all other negative coefficients whose absolute values are more than one-third to one-fourth as large as that most negative coefficient important, and ignoring ("zeroing out") all the rest. Using the .3 criterion would lead us to interpret PC_2 as whatever would lead someone to have a high score on the Picture Completion, Design, Assembly, and Mazes subscales but a low score on the Arithmetic and Digit Span subscales, or as the difference between whatever is being tapped by the first four subscales and whatever is being tapped by the latter two subscales.

If, on the other hand, we apply the scree test we would retain all five of the subscales that receive positive coefficients in PC_2, because their values range from .255 to .467, a range of only 1.8; we would also include all seven of the negative-coefficient subscales, because those coefficients range from -.166 to -.400, a ratio of only 2.4. PC_2 would then be interpreted as a tendency to score high on Picture Completion, Object Arrangement, Design, Assembly, and Mazes whereas getting (relatively) low scores on Information, Similarities, Arithmetic, Vocabulary, Comprehension, Digit Span, and Coding. Researchers and/or clinicians intimately familiar with the nature of each subscale would be best qualified to provide a substantive interpretation of this profile of subscale scores and to name the dimension underlying it, but it is probably not too farfetched to consider it as a tendency to perform better on concrete tasks (the positive-coefficient subscales) than on more abstract, verbally or numerically presented tasks.

PC_3 appears to represent whatever would lead one to have high scores on Coding, Mazes, and Digit Span but low scores either on Comprehension by itself or on that subtest as well as on Information, Similarities, and Vocabulary. PC_4, on the other hand, appears to be either the simple difference between Coding and Digit Span or a tendency to be high on both Coding and Object Arrangement but low on Digit Span, Arithmetic, and Mazes. I won't attempt a substantive interpretation of either of these last two PCs.

I do, however, point out some of the features of these interpretations and of the pattern/structure/scoring-coefficient matrix from which they were derived that make them unpopular with most factor analysts. First, the very hierarchical ordering (in terms of proportion of original *z*-score variance explained) of the PCs that make the principal-components solution so useful for estimating how many underlying dimensions are necessary to account reasonably well for the correlations among the original variables becomes, in the eyes of most researchers, a liability once we've determined how many components to examine; researchers and users of test batteries such as this one tend to think in terms of underlying factors that are of roughly equal importance. Second, many of the 12 subtests (in fact, all of them if we use the scree-test criterion, rather than the .3 rule of thumb) enter into the description of (have high loadings on) more than one of the PCs, which makes it difficult to come up with substantive interpretations that make the conceptual uniqueness of (and empirical absence of correlation among) the dimensions clear. Third, many of the original variables also load in different directions on different PCs; if each of these subscales and the components derived therefrom are tapping aspects of intelligence, how can, for example, Digit Span indicate (or at least be associated with) both high (PCs 1 and 3) and low (PCs 2 and 4) intellectual capacity?

The second and third objections could be seen (as I do) as in large part a reflection of social and behavioral scientists' general difficulty in understanding the distinction between a collection of univariate results and a truly multivariate result involving a linear combination of measures as a variable in its own right. (For example, the fact that being *better* at spatial than at verbal tasks is an important aspect of one's intellectual aramentarium doesn't at all imply that skill at verbal tasks is a sign of low intelligence.)

The reader is invited to review section 1.3's general discussion of interpretation of linear combinations. Nonetheless, it's hard to deny that interpretation of our set of underlying dimensions would be much easier if (a) each original variable loaded substantially on one and only one component and (b) the various components accounted for roughly equal proportions of the z-score variance of the original variables. This sort of ***simple structure*** is the primary goal of factor (or component) *rotation*, to which we turn in section 6.5.

Assuming instead that we find the principal-component solution congenial in its own right, there is the additional problem of quantifying how good our verbal/substantive interpretations of the PCs are. For example, the PCs themselves are uncorrelated; how close to true is this of the linear combinations of variables implied by our simplified descriptions of the PCs? And (for example) the first PC has a normalized variance of 5.40, 45.0% as large as the total z-score variance of the 12 subscales. How close to 5.40 is the normalized variance of the simple sum or average of all 12 subscales (which is the linear combination of PC_1 implied by our interpretation thereof)? We turn later to this general question of quantifying the goodness/badness of our interpretations of components, in section 6.3.5.

Example 6.3 Attitudes Toward Cheating. A questionnaire seeking to ascertain their attitudes toward cheating was sent to a random sample of University of New Mexico (UNM) students. The even-numbered questions 12 through 22 asked how often the respondent engaged in various practices (copying another student's assignments; copying someone's paper during a test; using crib notes during a test; plagiarizing papers; handing in someone else's research paper with his or her consent; stealing someone else's paper to hand in as your own), and the odd-numbered questions, 13 through 23 asked how often the "typical UNM student" engaged in each of these same activities. Responses to these 12 questions were subjected to a PCA based on the intercorrelation matrix, with the resulting factor structure listed partially in Table 6.1. This table reports only those PCs whose associated characteristic roots were greater than 1.0 (the average value for all 12). The remaining characteristic roots were:

Root:	.831	.731	.630	.527	.451	.380	.325	.229
Percent:	7.1	5.9	5.2	4.4	3.8	3.0	2.6	2.0

Thus the characteristic roots after the third show a very slow decline. We would certainly expect a distinction between the even-numbered variables (how often the respondent cheats in various ways) and the odd-numbered questions to emerge from any factor analysis of these scores, and this is one of the criteria for judging the adequacy of the PCA. There is a hint of this distinction in this "component structure" in that the loadings on the first, general component are uniformly higher for odd-numbered than for even-numbered variables, and in that, with one exception, loadings on component 2 are positive for all odd-numbered questions but negative for all even-numbered questions. The remaining two PCs are more difficult to interpret. Making the arbitrary decision to concentrate on loadings of .5 or better leads to a description of PC_3 as representing the difference between responses to questions 14 and 16 versus question 22 (and possibly question 20), which might be considered a difference between "public" and "private" cheating. Essentially, PC_4 taps only responses to question 22.

Table 6.1 PCA on Questions 12–23 of Cheating Questionnaire

Question (activity)	PC_1	PC_2	PC_3	PC_4	Sum of squared correlations with PC_1–PC_4
	Correlation of resp to question with				
12 (Copy assignment?)	.321	−.693	−.161	.008	.609
14 (Copy test paper?)	.400	−.354	.502	.467	.754
16 (Crib notes during test)	.161	−.522	.535	−.025	.585
18 (Plagiarize?)	.332	−.689	−.155	−.377	.750
20 (Use other's paper?)	.342	−.338	−.481	.177	.495
22 (Stolen other's paper?)	.191	−.142	−.524	.615	.710
13 } How often typical UNM student does these things	.635	.171	.135	.035	.452
15	.788	.068	.294	.201	.753
17	.730	.264	.272	−.203	.718
19	.702	−.100	−.296	−.455	.797
21	.679	.410	−.181	−.060	.666
23	.624	.411	−.168	.144	.608
Eigenvalue	3.455	1.936	1.420	1.085	7.896
% Variance accounted for	28.8	16.2	11.8	9.1	69.8

Example 6.4 *Fat, Four-eyed, and Female, Again.* At the end of section 6.3.2 you were promised (threatened with?) a demonstration of the dangers of PCA as a preliminary to other statistical techniques. We use data collected by Mary B. Harris (Harris, Harris, & Bochner, 1982) for this purpose. However, the study, which was used as our first example of Manova (Example 1.3), is also such an excellent demonstration of the potential gain from performing Manova (instead of Bonferroni-adjusted univariate Fs), and of the deficiencies of loadings on the Manova discriminant function as a means of interpreting that emergent variable, that I won't be able to resist sermonizing on those issues as well.

As you recall, "Chris Martin" was described to 159 Australian psychology students as either a man or a woman in his or her late twenties who, among other things (such as having dark, curly hair and preferring casual clothes), is either overweight or of average weight and was either said to wear glasses or not to wear glasses. Each subject rated this stimulus person on 12 adjective pairs. Two of the scales (masculinity and femininity) were replaced by the combined variable SEXAPP = masculinity minus femininity for males and the reverse for females. All items were then scored such that a high rating indicated a favorable rating, leaving us with the 11 dependent variables listed in Table 6.2. A 2 x 2 x 2 Manova was conducted on these data, yielding statistically significant main effects of wearing glasses and of being overweight, but no significant main effect of Chris Martin's gender and no statistically significant interactions. (See the full report for discussion of possible reasons for the surprising lack of evidence for a sex-role stereotype.)

We focus on the obesity main effect, which yielded a λ_1 of 3.095 (maximized F ratio of 184.76), which, with $s = 1$, $m = 4.5$, and $n = 69$, exceeds by a considerable margin its .0l-level critical value of $\lambda_{crit} = (5.5/70)F_{.01}$ (11, 140) = .1809 and $F_{crit} = 150(.1809) = 27.135$. Table 6.2 reports the univariate F for the obesity main effect, the raw and z-score discriminant function coefficients (defining z scores in each case on the basis of the pooled within-cells standard deviation), and the loading on (simple correlation with) the discriminant function for each of the dependent variables. Before moving on to a

Table 6.2 Manova Test of Obesity Main Effect

Dependent variable[a]	Univariate $F(1, 150)$	Discriminant function coefficients		Loading on discriminant function
		Raw		
X_1 (Assertive)	2.410	.07008	.08143	−.114
X_2 (Active)	91.768	*−.59177*	*−.66512*	*−.705*
X_3 (Intelligent)	1.048	.15989	.14298	−.075
X_4 (Hardworking)	8.352	.00051	.00053	−.213
X_5 (Outgoing)	.796	*.31325*	*.40301*	−.066
X_6 (Happy)	4.091	−.14862	−.19514	−.149
X_8 (Attractive)	53.166	*−.66652*	*−.61317*	*−.536*
X_9 (Popular)	4.300	*.46336*	*.41961*	−.1525
X_{10} (Successful)	2.553	−.05199	−.04830	−.118
X_{11} (Athletic)	63.805	*−.37275*	*−.45078*	*−.588*
SEXAPP[b]	7.707	−.03223	−.06535	−.204

Note. Adapted by this book's author from Harris, Harris, & Bochner (1982). Non-italicized coefficients were ignored ("zeroed out") in interpreting the discriminant function.

[a] Label in parentheses appeared on favorable end of scale.

[b] Masculininty - femininity for males, vice-versa for females.

comparison with a Manova performed on principal component scores, we should use the entries in Table 6.2 to review a number of points about the interpretation of Manova discriminant functions.

Those z-score discriminant function coefficients whose absolute magnitudes seemed large enough to warrant inclusion in our simplified discriminant function are italicized in Table 6.2. Because all of the variables retained had a common metric (7 = very well described by the favorable adjective of the pair, 1 = very well described by the unfavorable end of the scale), we prefer to base our choice of simplified weights and our interpretation of the resulting new variable ($X_5 + X_9 - X_2 - X_8 - X_{11}$) on the raw-score coefficients. Our interpretation is that the stereotype of the overweight stimulus person is that of someone who is (relative to the average-weight person) outgoing and popular but inactive, unattractive, and unathletic—or, perhaps more accurately, someone who is more outgoing and popular than one would expect on the basis of his or her relatively low levels of physical attractiveness, activity, and athleticism.

By comparison, an attempt to base the interpretation on the loadings (by far the more popular approach) would miss this interesting partern of ratings and simply report (correctly but uninformatively) that the overweight stimulus person is rated low on activity, attractiveness, and athleticism. Perhaps because of the social value usually placed on these latter three characteristics, the overweight person is seen as (nonsignificantly) less popular and less outgoing than the average-weight stimulus person. Clearly what the loadings are doing is simply looking at each original variable by itself, with no consideration given to its relationship to the other measures. This is made especially clear by the fact that the loadings are directly proportional to the univariate *t*s (and the squared loadings are thus directly proportional to the univariate *F*s), as the reader can verify. In other words, the loadings do not provide an interpretation of the discriminant function at all, but simply a repeat (although in a more subtle form and after a great deal more computational effort) of the results of the univariate *F* tests. Keep this in mind when we begin our discussion of factor loadings versus factor score coefficients as the appropriate basis for interpreting factors in PCA or FA.

Rather than analyzing our 11 variables directly, we could instead precede our Manova with a PCA of the within-cells correlation matrix and then use subjects' scores on the resulting 11 PCs or some subset thereof as our dependent variables for a Manova. [There are at least three other possible bases for our initial PCA: analysis of the within-cells covariance matrix (proportional to **E**) or analysis of the total covariance matrix or the total correlation matrix (each based on **H** + **E**). Use of either total matrix, ignoring experimental conditions, would forfeit the computational convenience of having a diagonal **E** matrix. Whether to use a correlation or a covariance matrix depends primarily on whether the various dependent variables can be considered to have been measured on a common scale. These are of course important questions if the PCA is of substantive interest rather than simply being a preliminary step in our Manova.]

The preliminary PCA yields eigenvalues of 3.095, 1.517, 1.207, .996, 832, .767, .629, .584, 531, .456, and .381, with the first three PCs accounting for 52.9% of the total within-cells *z*-score variance and the first nine accounting for 92.4%. Table 6.3 reports, for each of the 11 PCs, its univariate *F* for the main effect of obesity, the *z*-score coefficients used in computing scores on that PC from *z* scores on the original variables, and its raw-score and *z*-score discriminant function coefficients.

Any subsequent Manova is indeed made very simple by this preliminary PCA. Because $s = 1$ and the PCs are mutually uncorrelated, the maximized *F* ratio for the optimal linear combination of all 11 PCs, or any subset thereof, is simply equal to the sum of the univariate *F*s for the PCs included in the analysis, with the *z*-score discriminant function coefficients being directly proportional to the univariate *t*s (or the appropriately signed square roots of the univariate *F*s). For instance, if all 11 PCs are "retained" in our Manova, the resulting maximized *F* ratio for the obesity main effect is $47.54 + 14.30 + \cdots + 58.73 = 184.77$ (identical, within rounding error, to the *F* obtained in our Manova of the original variables), and the resulting discriminant function turns out to be entirely equivalent to our original-variable-based discriminant function, once the relationship between each PC and the 11 original variables has been taken into account.

Table 6.3 PCA-Based Manova of Obesity Main Effect

Discriminant function			Univariate	z score principal component weights										
PC	Raw	z score	*F*(1, 150)	x_1	x_2	x_3	x_4	x_5	x_6	x_8	x_9	x_{10}	x_{11}	*SEXAPP*
1	−.164	−.507	47.54	.223	.356	.306	.257	.322	.199	.355	.378	.346	.247	.267
2	.183	.278	14.30	−.229	−.292	.421	.334	−.362	.318	−.031	−.038	.233	−.487	.217
3	.054	.065	.79	.584	.183	−.062	.200	.130	.331	−.505	−.435	.015	−.074	.078
4	.001	.001	.00	−.148	.129	−.053	−.500	.133	.304	−.037	.084	−.352	−.177	.657
5	−.125	−.104	2.00	−.254	−.183	−.423	.164	−.004	.710	−.038	.206	−.037	.312	−.222
6	−.717	−.550	55.78	−.292	.472	.102	.426	−.360	−.093	−.057	−.168	−.433	.331	.168
7	.099	.062	.71	−.321	−.183	−.093	−.040	.085	−.209	−.388	−.178	.529	.423	.402
8	−.228	−.133	3.27	.364	.078	.023	−.338	−.719	.050	−.166	.354	.169	.213	.026
9	−.163	−.086	1.38	.232	−.130	−.657	.355	−.173	−.213	.301	.059	.037	−.191	.405
10	−.085	−.039	.27	.296	−.590	.281	−.012	−.029	.060	.300	−.261	−.307	.443	.179
11	1.479	.564	58.73	.091	−.278	.115	.284	.203	−.223	−.504	.601	−.332	−.017	.075

Interpretation is another matter. Although only 3 of the PCs need be considered in our simplified discriminant function, the linear combination of these 3 PCs has no meaning until we have given a substantive interpretation to each of the PCs, and this interpretation would have to consider all 11 original variables for PC_1, 6 for PC_2 and 3 for PC_3. Alternatively, we could translate our simplified combination of the PCs into the linear combination of the z_xs and then simplify *those* weights—but that would be much more roundabout than simply doing our Manova on the *X*s in the first place.

More important for our present purpose, what happens if we decide to submit only a subset of our PCs to the subsequent Manova? If, for instance, we keep only the first three PCs (accounting for 52.9% of the variance the PCA took into account), our maximum *F* now becomes only 62.63, a 66.2% loss in the magnitude of *F*. Even if we retain the first nine PCs, thus "throwing away" only 7.6% of our interindividual (within-groups) variance, the resulting max *F* is only 125.76, which represents a 31.9% drop.

The main point is that the linear combinations of the *X*s that are most important in accounting for within-groups variance (i.e., the PCs with high associated eigenvalues) may be very different from those combinations that are important in discriminating among groups (i.e., the *X*-based discriminant functions). Indeed, the variance-maximizing criterion we set for our PCA is antithetical to our attempt in Manova to maximize the ratio of between-groups to within-groups variance, and it is only the fact that in a PCA (or FA) of sample data, the low-eigenvalue PCs are usually based disproportionately on error variance and thus also tend to have low between-group variance that keeps preliminary PCA or FA from being a disaster.

Monte Carlo data collected as part of Kevin Bird's dissertation at the University of New South Wales indicate that the actual Type I error rate for Manova statistics based on retaining the first m components (or factors) is very close to the nominal alpha if we compute our critical values as if we had only m dependent variables, rather than the full p we started with. Bird (1975) was thus on safe ground in endorsing such treatment of Manova results based on PCs. On the other hand, this happy circumstance arises only because there are numerous cases like the present one in which the reduction of dimensionality hurts our subsequent analysis more than the percentage of variance accounted for by our factors would suggest.

I would therefore recommend that researchers go directly to a Manova (or T^2 or MRA) based on the p original variables *unless* one of the following two circumstances applies:

1. There are strong reasons for expecting the principal components or factors that emerge from the preliminary analysis to have very clear-cut interpretations, so that they aid in interpreting the subsequent Manova discriminant function.

2. The number of original variables is so high, relative to the degrees of freedom for "error," that the power of a Manova based on all p original variables is unacceptably low, and some reduction of dimensionality is therefore essential.

6.3.5 Quantifying Goodness of Interpretation of Components

As pointed out in section 1.3.1, almost any verbal interpretation of an emergent variable (e.g., a PC) implies a particular linear combination of the original variables from which the latent variable was derived. (The particular combining weights defining that linear combination are usually considerably simpler than the multi-decimal weights that define the absolutely optimal—e.g., normalized-variance-maximizing—emergent variable, but simple explanations sometimes generate multi-decimal linear combinations, as for example when linear trend is tested for unequally spaced levels of an independent variable.) A necessary condition for an interpretation of a PC to be considered adequate is, therefore, that the simplified linear combination of the original variables implied by that interpretation mimic "fairly closely" the properties of the PC it interprets.

The two most obvious properties of a PC that we would want our interpretation to mimic are the normalized variance of the PC (its variance divided by the sum of the squares of the combining weights that define it, which of course equals its eigenvalue) and its zero correlations with the other PCs. As a first step in deciding whether your interpretations of a set of PCs are adequate you should therefore compute, for each of the PCs you feel are sufficiently important to retain for further study, the normalized variance of that simplified PC (to be compared to the eigenvalue for the PC it describes) and its correlations with the other simplified PCs (to be compared with the zero correlations of the "actual" PC with the other actual PCs). If you are uncertain as to which of two or more interpretations of a given PC to adopt, you can carry out these comparisons for each of the alternative interpretations and use the alternative interpretations' relative goodness of fit to the actual PC's properties as a (*not* the only) criterion in choosing among the alternative interpretations. (Another, and usually more important, criterion is the construct validities of the alternative interpretations, i.e., how well each fits into an internally consistent and empirically validated or at least empirically testable theoretical framework.)

I should point out, however, that this prescription is *not* one you will find in other textbooks nor one that is adhered to by very many researchers. Instead, most authors simply examine the pattern of the correlations of the original variables with a given PC (which is directly proportional to the pattern of the scoring coefficients—although this is no longer true after rotation of the PCs as described in section 6.5), give the PC a verbal label that would be consistent with the difference between a person with high scores on the original variables that have substantial positive loadings and low scores on the original variables that have substantial negative loadings and a person who has the mirror-image pattern of scores, and leave it at that. This is roughly equivalent to conducting an analysis of variance, determining that the overall *F*-ratio is statistically significant, and then using that as *carte blanche* to describe the pattern of the means any way you wish without supporting your descriptions of aspects of the differences among the means with tests of the contrasts (or pairwise differences) they imply—not an uncommon practice itself, but one that a large majority of researchers, textbook authors, and journal editors find unacceptable.

Assuming that you find the arguments for measuring goodness of fit of explanations compelling, let's demonstrate the process by applying it to the WISC–R PCs derived in Example 6.2. Because we don't have scores for individual subjects on the 12 subscales, we use the formulae developed earlier for the relationship between the combining weights of a linear combination and its variance, together with the relationship between two sets of combining weights and the correlation between the linear combinations they represent, to obtain the normalized variances of and correlations among the four PCs we decided to retain. Recall that PC_1 was interpreted as "general intelligence" (which is where most reports of interpretations would stop), which implies (more accurately, was based on) a simplified PC (PC_{simp1}) of (1,1,1,1,1,1,1,1,1,1,1,1) or any constant multiple thereof. PC_2 was interpreted as "whatever would lead someone to have a high score on the Picture Completion, [Object Arrangement], Design, Assembly, and Mazes subscales but a low score on the seven other subscales, or as the difference between whatever is being tapped by [these two discrete and exhaustive subsets of the subscales]" and was later given the verbal label of "a tendency to perform better on concrete tasks ... than on more abstract, verbally or numerically presented tasks". Thus PC_{simp2} is defined by the combining weights (to be applied to *z*-scores on the WISC–R subscales) of (-1/7, -1/7, -1/7, -1/7, -1/7, -1/7, 1/5, 1/5, 1/5, 1/5, -1/7, 1/5). I was less certain as to the interpretation to give to PC_3. PC_{simp3a} = (-1/4, -1/4, 0, -1/4, -1/4, 1/3, 0, 0, 0, 0, 1/3, 1/3) —the average of (*z* scores on) Coding, Mazes, and Digit Span minus the average of Completion, Information, Similarities, and Vocabulary, and PC_{simp3b} = (0, 0, 0, 0,-1, 1/3, 0, 0, 0, 0, 1/3, 1/3)—the Coding, Mazes, Digit Span average, minus Completion.

Similarly, two alternative interpretations of PC_4 were considered: PC_{simp4a} = (0, 0, -1/3, 0, 0, -1/3, 0, 1/2, 0, 0, 1/2, -1/3)—Coding and Arrangement versus Digit Span, Arithmetic, and Mazes—and PC_{simp4b} = (0, 0, 0, 0, 0, -1, 0, 0, 0, 0, 1, 0)—the difference between Coding and Digit Span.

The MATLAB commands I used to carry out these calculations were as follows:

```
r = [1   .62 .54 .69 .55 .36 .40 .42 .48 .40 .28 .27
     .62 1   .47 .67 .59 .34 .46 .41 .50 .41 .28 .28
     .54 .47 1   .52 .44 .45 .34 .30 .46 .29 .32 .27
     .69 .67 .52 1   .66 .38 .43 .44 .48 .39 .32 .27
     .55 .59 .44 .66 1   .26 .41 .40 .44 .37 .26 .29
     .36 .34 .45 .38 .26 1   .21 .22 .31 .21 .29 .22
     .40 .46 .34 .43 .41 .21 1   .40 .52 .48 .19 .34
     .42 .41 .30 .44 .40 .22 .40 1   .46 .42 .25 .32
     .48 .50 .46 .48 .44 .31 .52 .46 1   .60 .33 .44
     .40 .41 .29 .39 .37 .21 .48 .42 .60 1   .24 .37
     .28 .28 .32 .32 .26 .29 .19 .25 .33 .24 1   .21
     .27 .28 .27 .27 .29 .22 .34 .32 .44 .37 .21 1  ]
c = [ 1, 1, 1, 1, 1, 1, 1, 1, 1, 1, 1, 1
     -1/7,-1/7,-1/7,-1/7,-1/7,-1/7, 1/5, 1/5, 1/5, 1/5,-1/7, 1/5
     -1/4, -1/4,  0,-1/4,-1/4, 1/3,  0,  0,  0,  0, 1/3, 1/3
      0  ,    0,  0,   0 , -1, 1/3,  0,  0,  0,   0, 1/3, 1/3
      0  ,    0,-1/3,  0,   0,-1/3,  0, 1/2, 0,   0, 1/2,-1/3
      0, 0, 0, 0, 0, -1, 0, 0, 0, 0, 1, 0 ]
covars = c*r*c'
normlzn = diag(diag(c*c'))
sqrtnorm = sqrt(normlzn)
normdvars = diag(inv(sqrtnorm)*covars*inv(sqrtnorm))
stddevs = sqrt(diag(diag(covars)))
corrs = inv(stddevs)*covars*inv(stddevs)
```

These commands yielded the following results (edited for readability):

	PC_{simp1}	PC_{simp2}	PC_{simp3a}	PC_{simp3b}	PC_{simp4a}	PC_{simp4b}
Normalized Variance :	5.2733	1.1210	1.0586	0.7150	0.7367	0.7100
Eigenvalue of corresponding PC :	5.398	1.160	.969	.969	.746	.746

Correlations among simplified PCs:

1.0000	-0.0463	-0.2917	-0.1936	-0.0222	-0.0295
-0.0463	1.0000	0.2760	0.1905	0.1419	0.0774
-0.2917	0.2760	1.0000	0.8135	-0.0706	0.0498
-0.1936	0.1905	0.8135	1.0000	-0.0792	-0.0029
-0.0222	0.1419	-0.0706	-0.0792	1.0000	0.6998
-0.0295	0.0774	0.0498	-0.0029	0.6998	1.0000

Max |correlation|

```
with other PC:      .292     .276     .292      .194     .142     .077
                                   or .194* or .191*
```

*Top figure holds if PC_{simp3a} adopted; bottom figure if PC_{simp3b} adopted.

These results are not as helpful in deciding between pairs of alternative interpretations as we might have wished. For both pairs the two comparisons go in opposite directions: The simplified PC that more closely matches the actual PC's normalized variance has a higher correlation with a PC other than the one it's based on than does the alternative simplified PC. In the case of the two alternative interpretations of PC_4 it matters very little which interpretation we adopt, and we should probably let substantive considerations dictate the choice. I'm not an intelligence-theory expert, so I'll simply let parsimony (simple difference versus difference between subsets of size 2 and 3), together with the fact that a .142 correlation based on 203 cases is statistically significant, whereas a .077 correlation is not, persuade me to adopt PC_{simp4b}. The differences between PC_{simp3a} and PC_{simp3b} are more substantial: Interpretation *a* yields a normalized variance that's less than 10% larger than PC_3's eigenvalue, whereas interpretation *b* falls short on this dimension by over 30%; but adopting PC_{simp3a} would lead to a maximum correlation between any two simplified PCs of .292, whereas adopting PC_{simp3b} would reduce that maximum correlation to .194—still statistically significant, but representing less than 4% of the variance shared by any two simplified PCs. Giving priority once again to parsimony, together with the intuitive judgment that match between normalized variances is less important for PCs beyond the first, leads me to choose PC_{simp4b}.

Our tentative choice of interpretations (simplified PCs 1, 2, 3b, and 4b) thus leaves us with a set of four simplified PCs that are nearly orthogonal to each other and that retain the same rank ordering with respect to normalized variance as do the actual PCs, but with a substantial enough correlation between PC_{simp3b} and simplified PCs 1 and 2 and a substantial enough undershooting of PC_3's normalized variance that we might be tempted to consider a third interpretation somewhere between 3a and 3b.

Finally, this example clearly demonstrates the amount of extra work involved in quantifying the goodness of fit of your interpretations of PCs to the properties of those PCs—but I believe that this extra work is worth the corresponding reduction in the risk of misleading your readers as to just how well your interpretations represent the PCs.

6.4 SIGNIFICANCE TESTS FOR PRINCIPAL COMPONENTS

So far we have discussed PCs on a purely descriptive level. We have thus required no distinction between population and sample PCs and no assumptions about the nature of the distributions of the various variables. However, in using the PCs as an aid in describing the "structure" underlying a system of variables, or in "explaining" the covariance among these variables, we are usually hopeful that this description or explanation will be relevant to the *population* of observations from which our particular *sample* of observations was drawn. The number of inferential tests that have been developed for PCs is rather small.

Further, significance tests available for PCs derived from a *correlation* matrix $\mathbf{R}_x$ are even more problematical than those based on $\mathbf{S}_x$. At first glance this may seem strange, because $\mathbf{R}$ is simply that special case of $\mathbf{S}_x$ where the covariances are computed on standard scores rather than raw scores. Because investigators are free to begin their PCA with whatever transformation(s) of their data seem most meaningful to them—raw scores, logarithms, squares, cube roots—it seems somewhat churlish to tell them that if they happen to choose transformation to standard scores they cannot use many of the significance tests that would otherwise be available to them. However, there *is* a major difference between using, say, a logarithmic transformation on the data to take into account multiplicative relationships among variables, and converting all scores to standard scores.

Let X_i be a single observation in a sample of N observations from some population, $\overline{X}$ and s being the sample mean and standard deviation of the N observations. Then $\log(X_i)$ involves the single variable X_i , whereas $z_i = (X_i - \overline{X})/s$ involves the *three* variables X_i , $\overline{X}$, and s. Large values of $\log(X_i)$ can arise only through large values of X_i, whereas large values of z_i can arise if X_i is large, $\overline{X}$ is small, or s is small. Another way of phrasing this difference is to point out that we could specify, in advance of collecting any data, what value of $\log(X_i)$ would replace each value of X_i that might be observed in the data, whereas we have no idea what value of X_i will lead to a transformed score of z_i until we have collected all of the data and computed s.

Somewhat more formally: If X_i is normally distributed, then so is $a X_i + b$, where a and b are preselected constants; and the distribution of x_i^2 has the same shape as a chi-square distribution with one degree of freedom. However, $z_i = x_i/s_i$ has a much more complex distribution, which was derived by Cramer (1946), who also showed that a nonlinear transformation of z_i has the t distribution. The distribution of correlation-based PCs *is* therefore more complex than the distribution of covariance-based PCs. However, when no test appropriate to some situation is available for PCs obtained from PCA of an intercorrelation matrix, but is available for covariance-matrix-derived PCs, the latter may be used as a rough guide under the rationale that you could have simply taken standard scores as your basic measures.

A second caution must be noted with respect to the significance tests that follow: They are all based on the assumption of very large samples. I am unaware of Monte Carlo studies to establish just how large a sample is required for validity. Certainly $N - m$ should be greater than 30; a referee suggested that "$N - m > 100$ would probably be more appropriate." Small-sample confidence intervals and p values can be expected to be larger than these given here, although how much larger is not generally known. The following summary of available tests is patterned closely after that provided by Morrison (1976, section 7.7).

6.4.1 Sampling Properties of Covariance-Based PCs

1. L_i (the ith *sample* characteristic root) is distributed independently of $\mathbf{b}_i$ (the ith sample characteristic vector).

2. $\sqrt{df}(L_i - \lambda_i)$ is distributed as a normal distribution with mean zero and variance equal to $2\lambda_i^2$ and is distributed independently of the other characteristic roots.

3. $H_0: \lambda_{q+1} = \lambda_{q+2} = \ldots = \lambda_{q+r}$, that is, the hypothesis that the r intermediate characteristic roots are truly identical in terms of population values, can be tested by comparing

$$-df \sum_{q+1}^{q+r} \ln(L_j) + (df)r \ln\left[(L_j)/r\right]$$

with the desired critical value of the chi-square distribution having $r(r + 1)/2 - 1$ degrees of freedom. The term "*df*" here and in test 2 refers to the number of degrees of freedom going into each estimate of an s_j, and it will usually be equal to the number of observations minus 1. Where, however, deviation scores are taken about each of several group means in obtaining a "pooled" covariance matrix that is independent of differences in group means on the different variables, $df = N - k$, where N is the number of subjects and k is the number of groups; and "ln" stands for the natural logarithm, that is, the logarithm to the base e = 2.78. . . . This test, developed by Anderson (1963), makes good intuitive sense, because it can be rewritten as

$$\chi^2 = (df)\ln\left(\bar{L}^r / \prod_j L_j\right)$$

where $\bar{L}$ is the mean of the r roots being tested for equality and $\prod_j L_j$ is the product of these roots. If the sample values are all equal, then $L_j = 1$ for all the relevant j, whence $\bar{L}^r$ identically equals $\prod_j L_j$. The farther the L_js depart from equality, the greater will be the ratio between the rth power of their mean and their product. (The logarithm of 1 is, of course, zero.)

4. The *confidence interval* about a given λ_i is given by Anderson (1963) as

$$\frac{l_i}{1 + z_{\alpha/2}\sqrt{2/df}} < \lambda_i < \frac{l_i}{1 - z_{\alpha/2}\sqrt{2/df}},$$

where $z_{\alpha/2}$ is the (1 - α/2) percentile of the normal distribution, and the confidence interval constructed is at the (1 - α) level of confidence. Thus, for instance, the 95% confidence interval for the first PC derived from a covariance matrix each of whose elements had 72 degrees of freedom would be between $.755L_1$ and 1.488 $[= 1/(1 - 1.96\sqrt{2/df})]$ times the observed value of the first PC.

5. Ho: $\beta_i = \beta_{i0}$ —that is, the null hypothesis that the characteristic vector associated with the distinct root L_j is equal to a specific "null hypothesis" vector—is tested by comparing

$$df[L_i \beta'_{i0} \mathbf{S}^{-1} \beta_{i0} + (1/L_i)\beta'_{i0} \mathbf{S}^{-1} \beta_{i0} - 2]$$

with the chi-square distribution having m - 1 degrees of freedom. This test makes good intuitive sense, because it is essentially equal to the number of degrees of freedom going into each covariance estimate, multiplied by $[w + (1/w) - 2]$, where w is the ratio between the variance of the hypothesized linear combination of the original variables—$\beta_{i0}'\mathbf{S}\,\beta_{i0}$— and the variance of PC_i - L_j. If the two are equal, then $w = 1$ and $w + (1/w) = 2$. Any inequality, however, produces a value of $w + (1/w) > 2$. (For instance,1/2 + 2 = 2.5; 3 + 1/3 = 3.33; and so on.) This is a very useful test in determining whether the substantive interpretation of a particular principal component is viable, or whether the discrepancy between the coefficients that would support the interpretation perfectly and those actually obtained in the PCA cannot reasonably be attributed to chance fluctuation. For instance, PC_1, obtained from the PCA of the data for game 1, FREG condition (Example 6.1), was interpreted as "essentially" equal to (CC + DD) - (CD + DC), which implies coefficients of .5 for CC and DD and - .5 for CD and DC. The test described in this paragraph can be used to quantify the "essentially" by assigning a specific value to the probability that the differences between these "ideal" coefficients and the ones actually obtained are due solely to "chance" fluctuation.

6.4.2 Sampling Properties of Correlation-Based PCs

1. H_0: The last p - 1 characteristic roots are equal, which is equivalent to

$$H_0 : \rho_{ij} = p \text{ for all } i \neq j,$$

can be tested by comparing

$$(df/\hat{\lambda}^2)\left[\sum_i \sum_j (r_{ij} - \bar{r})^2 - \hat{\mu} \sum_{k-1} (\bar{r}_k - \bar{r})^2\right]$$

with the chi-square distribution having $(p - 1)(p - 2)/2$ degrees of freedom; where

$$\bar{r}_k = \left(\sum_{i \neq k} r_{ik}\right) / (p-1)$$

is the mean of the correlations in row k of $\boldsymbol{R}$;

$$\bar{r} = (\sum_{i>j}\sum r_{ij})/[p(p-1)/2]$$

is the mean of all the $p(p - 1)/2$ correlations in **R**; $\hat{\lambda} = 1 - \bar{r}$; and

$$\hat{\mu} = (p-1)^2(1-\hat{\lambda}^2)/[p-(p-2)\hat{\lambda}^2].$$

This test developed by Lawley (1963) essentially involves a comparison of the variability of row means with the variability among all the correlation coefficients, ignoring which variables are involved.

2. *Ho*: The first *m* principal components provide a perfect description of the population factor structure, with the remaining PCs having population values of zero; this is tested by comparing

$$\chi^2 = -(df - 2p/3 - m/3 - 5/6)\ln(|\mathbf{R}_{rep}|/|\mathbf{R}|) \qquad (6.5)$$

with the chi-square distribution having $[(p - m)^2 - m - p]/2$ degrees of freedom, where $\mathbf{R}_{rep}$ is the correlation matrix "reproduced" from the loadings on the first *p* PCs only, and **R** is the observed correlation matrix. The ratio between $|\mathbf{R}_{rep}|$ and $|\mathbf{R}|$ will of course approach 1 (and the natural logarithm of this ratio will approach zero) as the goodness of fit of the reproduced correlations to the observed correlations increases. The natural logarithm of the ratio of the two determinants can be approximated for very large samples by the sum of the squares of the discrepancies between observed and reproduced correlations. This test was actually developed (by Lawley, 1940, and Bartlett, 1954) for testing the adequacy of the fit of the maximum-likelihood factor solution involving *m* common factors, and significant findings must therefore be interpreted cautiously, because both minres and maximum likelihood factor solutions (chapter 7) provide better fits than selection of the same number of PCs from a PCA. A nonsignificant result, however, definitely indicates that including additional PCs is unnecessary to obtain a fit that is "perfect" within the limits of sampling fluctuation.

Note, further, that this is a meaningful test only if its degrees of freedom are positive or zero, that is, only if *m* is an integer less than or equal to $(2p + 1 - \sqrt{8p+1})/2$. This tells us, for instance, that we cannot even test the hypothesis that a single component explains the correlations among 3 variables, because $[2p + 1 - \sqrt{8p+1}]/2 = [7 - \sqrt{25})/2 = 0$—which makes sense when we recognize that there are only 3 correlations to be accounted for and 3 unknown loadings of the original variables on the single component, so we can always reproduce the correlations among 3 variables perfectly with a single component (which will *not* in general be the first principal component). We can, however, test the hypothesis that zero components are sufficient to explain the correlations among 3 variables—that is, that the 3 are mutually uncorrelated in the population from which we are sampling.

3. H_0: $\rho_{ij} = 0$ for all *i, j*—that is, there is no structure to be explained—was tested in chapter 2.

6.5 ROTATION OF PRINCIPAL COMPONENTS

As pointed out earlier, the uniqueness of the coefficients derived in PCA is achieved by requiring a descending order of importance (percentage of variance accounted for) among the PCs. This will generally be quite useful in providing evidence as to how many latent variables need be assumed in order to account for a sizeable percentage of the variance in the system of original variables. Indeed, Kaiser (1960a) has argued that a whole host of criteria involving both statistical and practical considerations suggest the number of PCs having associated characteristic roots greater than $\sum s_i^2/p$ (= 1 when the PCA is performed on **R**) as the best single criterion for the number of factors to be assumed in any analysis of structure, whether PCA or one of the multitude of factor analysis models we discuss in chapter 7. However, this hierarchical structure may (and usually does) correspond quite poorly to the researcher's preconceptions (theoretical commitments?) as to the nature of the latent variables that might be producing the intercorrelations among the observable variables. The researcher might, for instance, feel that all latent variables should be of roughly equal importance, or he or she might feel that each original variable should be "produced" by a relatively small number of latent "determinants." Once the variance-maximization criteria of PCA are abandoned, an infinitude of alternative sets of uncorrelated linear combinations of the original variables become available to the researcher. Fortunately, each of these alternative "factorizations" is related to the others (including PCA) by a set of operations known equivalently as *orthogonal transformations* or *rigid rotations*. To provide a concrete frame-work within which to develop definitions of and formulae for these techniques, let us consider an earlier example in somewhat greater detail.

Example 6.1 revisited *Known generating variables* Assume that our observable variables, X_1 and X_2, are generated as:

$$X_1 = H_1 + H_2 \text{ and } X_1 = H_1 + 3H_2,$$

where $r_{H_1,H_2} = 0$, and each latent variable has a unit normal distribution. Then

$$s_1^2 = \sum(h_1 + h_2)^2 = 2;\ s_{12} = \sum(h_1 + h_2)(h_1 + 3h_2) = 1 + 3 = 4;$$
$$s_2^2 = \sum(h_1 + 3h_2)^2 = 10;\ \text{and}\ s_{12} = 10;$$

that is,

$$\mathbf{S} = \begin{bmatrix} 2 & 4 \\ 4 & 10 \end{bmatrix},$$

whence, applying PCA to **S**, we obtain

$$\lambda = [12 \pm \sqrt{8^2 - 4(4^2)}]/2 = 6 \pm 5.655 = 11.655,\ .346.$$

Conducting a PCA of **S** yields the following pattern and structure matrices:

	S-derived factor pattern		**S**-derived factor structure[a]	
	PC_1	PC_2	PC_1	PC_2
X_1	.3828	−.9239	.9239	−.3828[a]
X_2	.9239	.3828	.9974	.0721
s^2_{pc}	11.656	.344	1.848	.152

[a] $r_{X_i PC_j} = b_{ij}\sqrt{\lambda_i}/s_i$.

Further,

$$\mathbf{R} = \begin{bmatrix} 1 & \sqrt{.8} \\ \sqrt{.8} & 1 \end{bmatrix} = \begin{bmatrix} 1 & .894 \\ .894 & 1 \end{bmatrix},$$

whence, via PCA, we obtain

	R-derived weight matrix		**R**-derived factor structure	
	PC_1	PC_2	PC_1	PC_2
X_1	.707	−.707	.973	−.231
X_2	.707	.707	.973	.231
$\sum r^2_{X_i PC_j}$	1.893	.107	1.893	.107

In addition to these two PCA-derived matrices, we have the "true" structure, which we determine as

$$h_1 + h_2 = x_1; \quad h_2 = (x_2 - x_1)/2;$$

$$h_1 + 3h_2 = x_1; \quad h_1 = x_1 - h_2 = (3x_1 - x_2)/2;$$

$$r_{x_1,h_1} = \sum x_1 h_1 / \sqrt{x_1^2 \cdot \sum h_1^2} = (3s_1^2 - s_{12}) / \sqrt{s_1^2(9s_1^2 - 6s_{12} + s_2^2)} - 1/\sqrt{2}$$

$$= .707;$$

and so on, whence we obtain

	"True" weights			"True" structure	
	H_1	H_2		H_1	H_2
X_1	.9486	−.707		.707	.707
X_2	−.3162	.707		.3162	.9486
			$\sum r^2_{X_i H_j}$:	.6	1.4

6.5.1 Basic Formulae for Rotation

Each of the three factor structures can be plotted on a graph having the latent variables as axes (dimensions) (see Figure 6.1). However, examination of these figures shows (allowing for some numerical inaccuracies) that the relationship between variables X_1 and X_2 remains constant, with the cosine of the angle between them being equal to r_{12} = .894. The only differences among the three factor structures lie in the orientation of the two reference axes. We thus ought to be able to find a means of expressing the effects of these "rotations." In the general case of an observed variable plotted as a function of two latent variables L_1 and L_2 and then redescribed after rotation through an angle θ in terms of L_1* and L_2*, we have Figure 6.2.

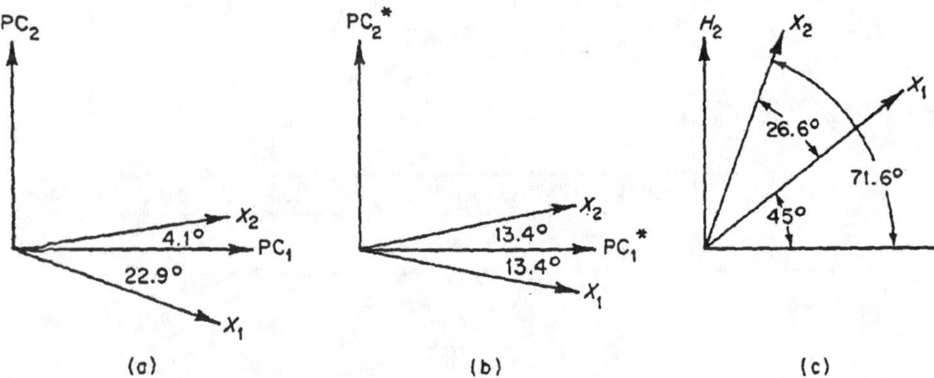

Figure 6.1 Factor structures, Example 6.1

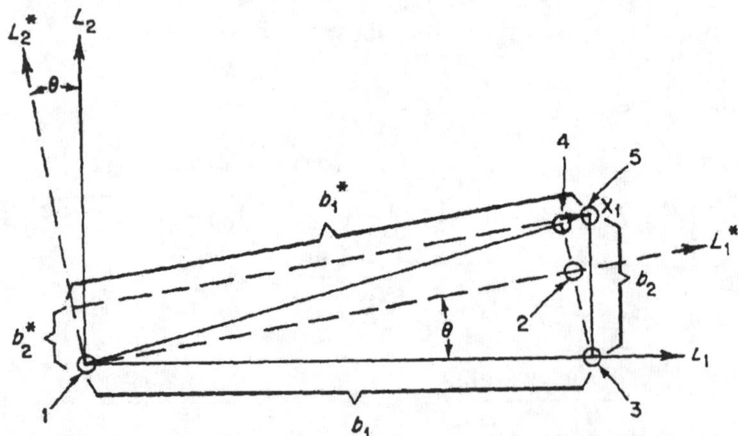

Figure 6.2 Rotation, general case

(Note that in plotting a variable, the length of the vector connecting it to the origin is equal to its "communality," that is, the sum of the squares of its loadings on the factors of the factor structure, and is equal to 1 when all PCs are retained.) By consideration of the triangle formed by points 1, 2, and 3 and the 234 triangle, we see that $b_1^* = \overline{12} + \overline{45} = (\cos\theta)b_1 + (\sin\theta)b_2$ (where $\overline{ij}$ is the length of the line segment connecting points i and j; and $b_2^* = \overline{34} - \overline{23} = (\cos\theta)b_2 - (\sin\theta)b_1$.

Put in matrix form, this becomes

$$[b_1^* \;\; b_2^*] = [b_1 \, b_2] \begin{bmatrix} \cos\theta & -\sin\theta \\ \sin\theta & \cos\theta \end{bmatrix} = [b_1 \, b_2]\mathbf{T}.$$

The reader may readily verify that the second two factor structures can indeed be obtained from our PCA of the covariance matrix by rotations of -9.5° and 67.9°, respectively, that is, by application of transformation matrices

$$\begin{bmatrix} .973 & .226 \\ -.226 & .973 \end{bmatrix} \text{ and } \begin{bmatrix} .384 & .923 \\ -.923 & .394 \end{bmatrix}.$$

We could of course pick any other value of θ that gives us a "desirable" factor structure. For example, we might wish to have a zero loading of one variable on one of our hypothetical variables and zero elsewhere. This would be accomplished by a θ of -22.9° and a transformation matrix **T** of

$$\begin{bmatrix} .9206 & .3884 \\ -.3884 & .9206 \end{bmatrix},$$

which leads to a factor structure of

	"Simple" Factor Structure	
	H_1^*	H_2^*
X_1	1.000	.000
X_2	.891	.454
$\sum r^2_{X_i H_i}$	1.794	.206

Note, incidentally, that in general **TT'** = **T'T** = **I,** which is the defining property of an orthogonal matrix.

When we move to three or more dimensions, it should be fairly clear that we can rotate any pair of axes while holding all other axes fixed, thus changing the loadings of the original variables only on that particular pair of latent variables. The matrix expression of this process of pairwise rotation of reference axes is

$$\mathbf{T} = \mathbf{T}_1 \cdot \mathbf{T}_2 \cdot \mathbf{T}_3 \cdots \cdot \mathbf{T}_f$$

where each orthogonal matrix on the right is simply an *m x m* identity matrix, with the exception of the entries

$$t_{ii} = \cos\theta_{ij}, \quad t_{ji} = \sin\theta_{ij}, \quad t_{ij} = -\sin\theta_{ij}, \quad \text{and} \quad t_{jj} = \cos\theta_{ij},$$

where θ_{ij} is the angle through which axes *i* and *j* are rotated; and **T** is the single transformation matrix that, when premultiplied by the original factor structure **P**, yields the new factor structure **P***, which is the result of the *r* [usually equal to $p(p - 1)/2$, where *p* is the number of PCs or factors retained] pair-wise rotations of axes. If in fact some of the original PCs are "ignored" and thus kept fixed throughout the various rotations, the rows and columns corresponding to these "dropped" dimensions can be deleted, making **T** and each component $\mathbf{T}_i$ only *m* x *m* $(m < p)$. For any large number of "retained" PCs, the process of rotating to obtain a factor structure that is easy to interpret is quite time-consuming, because $m(m - 1)/2$ pairs of axes must be considered. The researcher pursuing the "intuitive-graphical" approach to rotation will find a calculator with trigonometric functions extremely helpful.

6.5.2 Objective Criteria for Rotation

This section draws heavily on Harman's (1976, chapter 14) summary of analytic methods. The indeterminacy we have seen in the positioning of our reference axes, and thus in the configuration of "loadings" (correlations) of the observable variables on the latent variables, has been (probably needlessly) a sore point with factor analysts for many years. To a mathematician there is no indeterminacy, because the configuration of vectors representing the relationships among the variables is the same, regardless of how we choose to orient the axes we use in obtaining a numerical expression of this geometric configuration. However, most psychologists are at least as interested in describing the axes (dimensions) themselves as in describing the relations among the original variables. Concern over the issue of rotation of axes was probably heightened by the extravagant claims of early factor analysts that factor analysis yields information about *the* "true" sources of intercorrelations among the observable variables, that is, about the true, fundamental dimensions underlying responses in some situation or class of situations. A plethora of versions of the truth is rather damaging to such claims; it was therefore important to these pioneers in the field to develop principles for selecting just one of the infinite number of possible solutions as ***the*** solution.

We know now, of course, that these claims were based on a fundamental misconception about the nature of scientific explanations. PCA and FA, like any other techniques for the construction of models for the explanation of empirical data, can at best establish the plausibility of one of many alternative explanations of a given set of data. However, there is no harm in choosing the positions of our reference axes in such a way as to make the task of providing substantive interpretations of the hypothetical variables as simple as possible, so long as we recognize that factor structures having "simple structure" are no more (nor less) valid than any of the other factor structures that are obtainable by rotation of our initial PCA. No consideration of simple structure will, for instance, generate the "true" structure underlying Example 6.1. (We might, however, postulate—as Einstein did and as is done implicitly whenever we use maximum-likelihood statistical techniques—that Nature does not play tricks on us and that all "true" natural laws are simple ones. This would then justify putting greater faith in "simple structure" solutions than in alternative factor structures.)

There are (not necessarily unreasonably or even unfortunately) many kinds of simplicity. In order to make the search for simple structure an objective process, rather than a subjective art, it is necessary to specify exactly what criteria for or measures of simplicity are to be employed. The first person to attempt an explicit definition of "simple structure" was Thurstone (1947). He phrased his criteria (as have all subsequent researchers) in terms of the loadings within the $p \times m$ rotated factor structure, as follows:

1. Each row of the factor structure should contain at least one zero. (This sets as a minimum criterion of simplicity of description of the variables the criterion that not all of the hypothetical variables should be required to describe a given original variable.)
2. Each column of the factor structure should contain at least m zeros. (This sets a minimum condition on the simplicity of our description of a given latent variable in terms of the original variables.)
3. Every pair of columns should contain several original variables whose loadings vanish in one column but not in the other. (This criterion aids in distinguishing between the two hypothetical variables in terms of their relationship to the original variables.)
4. If the number of factors (or retained PCs) is four or more, every pair of columns of the factor structure should contain a large number of responses with zero loadings in *both* columns. (This is useful in distinguishing this particular pair of latent variables from the other latent variables.)
5. For every pair of columns, only a small number of original variables should have nonzero loadings in both columns. [This is essentially a rephrasing of condition 4.]

There are two major drawbacks to Thurstone's criteria:

1. They can almost never be satisfied by any set of real data (although they can often be rather closely approximated).
2. No objective measures are provided of how far we are from satisfying the criteria.

Thurstone's criteria have, however, provided a general framework for other authors' search for numerically specifiable criteria for simple structure.

Around 1953 or 1954, several authors, starting from slightly different conceptualizations of simple structure, arrived independently at the *quartimax* criterion. To see how these apparently different criteria led to precisely the same results, we need to consider the fact that the ***communality*** of each original variable (the portion of its variance it shares with other variables in the system or with the underlying factors—there's some disagreement over definitions, which is addressed in chapter 7—which is equal to the sum of the squares of that variable's loadings on the various hypothetical variables as long as these latent variables are uncorrelated) remains constant as the reference axes are rotated. If the rotation is from an initial PCA, the communalities will of course be 1.0 for all variables, but latent variables generated by factor analysis, or factor structures based on only some of the PC*s* from a PCA, yield communalities less than 1.0. In a graphical plot of a factor structure, the communality of a variable is mirrored by the length of its vector, which is clearly unaffected by the orientation of the reference axes. Formally, then

$$h_j^2 = \sum L_{ji}^2 = \text{constant for each variable } X_j, \tag{6.6}$$

where L_{ji} is the correlation of variable j with latent variable i, that is, it is the entry in the jth row, ith column of the factor structure. However, if this is constant across rotations, then so is the sample variance of the communalities, namely,

$$\begin{aligned} \sum (h_j^2 - \overline{h_j^2})^2 &= \sum\sum (h_j^2)^2 - \left(\sum h_j^2\right)^2 / m \\ &= \sum\sum L_{ji}^4 + 2\sum_{\substack{u=1 \\ u>v}}^{p}\sum_{v=1}^{p}\left(\sum L_{ju}^2 L_{jv}^2\right) - m(\overline{h_j^2}) \\ &= Q + CP \text{ - a constant;} \end{aligned} \tag{6.7}$$

that is, the variance (actually, sum of squared deviations from the mean communality) of the communalities is decomposable, in a manner reminiscent of the decomposition of total variance in Anova into within- and among-group sums of squares, into a term Q, which is the sum of the fourth powers of the loadings of the original variables on the latent variables, and a term CP, which is the sum (over all pairs of columns of the factor structure) of the sums of cross-products of the loadings in each pair of columns. Because the sum of Q and CP is a constant, maximization of Q (which is what was proposed by Ferguson, 1954; Newhaus and Wrigley, 1954; and Saunders, 1953) is equivalent to minimization of CP (which was proposed by Carroll, 1953). Any rotation scheme that attempts to minimize CP or maximize Q is known as a *quartimax* rotation.

As Harris (1985b) showed algebraically, maximization of Q can be seen as desirable from at least three different viewpoints:

1. It maximizes the variance of all *mp* squared loadings, thus tending towards loadings that are close to zero or to one.
2. It maximizes the sum of the variances of the squared loadings within each row, so that each variable tends to have a near-unity loading on one or two latent variables and near-zero loadings on all other latent variables, simplifying the description of that variable's ***factorial composition***.
3. It maximizes the kurtosis of the distribution of squared loadings, which will be high to the extent that there is a sharp differentiation between high and low loadings within the factor structure. (Kurtosis is, for a unimodal distribution, a measure of its "peakedness," that is, the tendency for observations to cluster both tightly around the mean and broadly in the tails.)

Minimization of CP, on the other hand, drives the average correlation between pairs of columns of squared loadings towards zero, so that we can more readily differentiate between two pairs of latent variables by examining the pattern of their loadings on the original variables.

Despite the multiplicity of desirable properties of quartimax rotations, however, the quartimax criterion runs far behind the popularity of the varimax criterion for rotation. The reason for this is most clearly seen through consideration of the definition of Q in terms of the variances of the loadings within rows of the matrix. Maximizing Q is thus perfectly compatible with a factor structure in which a large portion of the variance is accounted for by a general factor on which each and every variable has a very high loading—which is precisely the feature of the factor structure produced by PCA that most researchers find unacceptable and that led to consideration of the possibility of rotating this initial solution in the first place. Kaiser (1956, 1958) proposed maximizing instead the variance of the loadings within *columns* of the matrix. This of course precludes the emergence of a strong general factor and tends to produce a factor structure in which each latent variable contributes roughly the same amount of variance.

Specifically, the raw varimax criterion is

$$V^* = (1/p^2)\sum_i^m \left[p\sum_j^p L_{ji}^4 - (\sum_j^p L_{ji}^2)^2 \right] \qquad (6.8)$$

Note that the sum of the squared loadings within a given column of the matrix is not constant across rotations, although (for uncorrelated latent variables) the sum across all columns of these column sums of squared loadings is both constant and the same as the sum of the squared communalities.

This criterion is called the raw varimax criterion because Kaiser proposed a correction to each loading in the factor structure before carrying out the rotation. The raw varimax procedure tends to give equal weight to variables having very low communalities and those having near-unity communalities. Kaiser found this undesirable, and corrected for this tendency by dividing each loading within a given row of the factor structure by the square root of the communality of that variable. After the rotation process has been completed, the effects of this "Kaiser normalization" are removed by multiplying each loading in the rotated structure by the square root of the communality of the variable described in that row. A varimax rotation employing Kaiser normalization of the loadings is referred to as normal varimax or simply varimax rotation.

Computationally, varimax rotation is simplified by the fact that the angle of rotation for any given pair of reference axes that produces the maximum increase in V (the normalized varimax criterion) can be solved for explicitly, rather than requiring trial-and-error search. Specifically, for the latent variables r and s, the optimal angle of rotation, ϕ_{rs}, is given by

$$\tan(4\phi) = \frac{4\left[p\sum d_{rs}c_{rs} - \left(\sum c_{rs}\right)\left(\sum d_{rs}\right)\right]}{p\sum d_{rs}^2 - \left(\sum d_{rs}\right)^2 - 4p\sum c_{rs}^2 + 4\left(\sum c_{rs}\right)^2}. \qquad (6.9)$$

where

$$d_{rs} = v_{jr}^2 - v_{js}^2$$

is the difference between the squared normalized loadings of variable j on factors r and s;

$$c_{rs} = v_{jr}v_{js}$$

is the product of the loadings of variable j on the two factors; and

$$v_{ji} = L_{ji} / \sqrt{h_j^2}.$$

The right-hand side of Equation (6.9) seems to be related to the correlation between d_{rs} and $2c_{rs}$, but I've been unable to move from this astute observation to a convincing intuitive rationale for the reasonableness of Equation (6.9). (Note that each summation is over the p variables from $i = 1$ through $i = p$.) Because of the cyclic nature of tan ϕ, there will be two choices of 4ϕ that satisfy Equation (6.9). Which to select is determined by the sign of the numerator and of the denominator of Equation (6.9) as indicated in Table 6.4. (Table 6.4 is derived from a consideration of the second derivative of V, the

Table 6.4 Quadrant Within Which 4ϕ Must Fall as Function of Signs of Numerator and Denominator of Equation (6.9)

		Sign of numerator	
		+	−
Sign of denominator	+	$0° \le 4\phi \le 90°$	$270° \le 4\phi \le 360°$
	-	$90° \le 4\phi \le 180°$	$180° \le 4\phi \le 270°$

normalized criterion, with respect to ϕ.) A "large iteration" of the varimax rotation scheme requires that latent variable (factor) 1 be rotated with factor 2; the new factor 1 is then rotated with original factor 3; and so on, until all $m(m - 1)/2$ pairs of factors have been rotated. Kaiser has shown that each large iteration increases the value of V, and that V has an upper bound of $(m - 1)/m$, so that the rotation scheme eventually converges. In practice, rotations are continued until the difference between V for the large iteration just completed and the value of V after the preceding large iteration falls below some criterion or until the rotation angles recommended for the next large iteration all fall below some minimum value.

Kaiser presented evidence that normalized varimax rotation satisfies the very desirable criterion of "factorial invariance," that is, the independence of the definition of a set of latent variables in terms of a battery of original variables from the "contaminating" effects of the particular other variables included in the set on which the initial factor solution is obtained.

More specifically, Kaiser proved mathematically that when the vectors representing the original variables fall into two collinear clusters (i.e., two factors provide sufficient explanation of the variation among the original variables), the angle of rotation satisfying Equation (6.9) is independent of the number of variables. Although he was unable to provide a mathematical proof beyond the two-factor case, he did provide a rather convincing empirical demonstration of the invariance property for a case with 4 factors and 24 variables, which he adds in several stages, beginning with 6 variables.

Incidentally, an expression similar to Equation (6.9) is available for quartimax rotation, namely,

$$\tan(4\phi_q) = 4\sum c_{rs} d_{rs} / \left(\sum (d_{rs}^2 = 4c_{rs}^2\right), \qquad (6.10)$$

where the subscript on ϕ serves as a reminder that this is the angle of rotation for the quartimax rotation scheme, and where the appropriate quadrant for 4ϕ is given by Table 6.4.

Finally, Lawley and Maxwell (1963) outlined least-squares methods for rotating to factor structures that come as close as possible to an a priori pattern of ones and zeros. The practical significance of Lawley and Maxwell's work was greatly increased by Joreskog's (1967, 1969, 1973, 1978) development of an efficient (although initially very user-*un*friendly) computer program, LISREL, for such ***confirmatory*** factor analysis. We discuss LISREL and other programs for confirmatory factor analysis in chapter 7.

6.5.3 Examples of Rotated PCs

To illustrate the very simplest case of rotated principal components, we apply Equations (6.9) and (6.10) to the "known-generator" case, Example 6.1. With just two hypothetical variables, these formulae produce the desired solution in just one step, with no iteration being required. Because it does not matter which of the several structures generated by rotation of the initial PC solution we use, we choose to work with the intuition-based "simple structure" solution, because the first row (loadings of X_1) is so simple. Table 6.5 gives the required computations. Based on these intermediate computations, we have as the angle of rotation for a quartimax solution,

$$\tan(4\phi) = 2(.4755)/.6910 = 1.3763,$$

Table 6.5 Intermediate Calculations for Quartimax and Varimax Rotation

j	$2L_{j1}L_{j2} = 2c_j$	$L_{j1}^2 - L_{j2}^2 = d_j$	$4c_j^2$	d_j^2	$2c_j d_j$	$d_j^2 - 4c_j^2$
1	0	1	0	1	0	1
2	.8090	.5878	.6545	.3455	.4755	−.3090
Sum	.8090	1.5878	.6545	1.3455	.4755	.6910

whence

$$4\phi_q = 53.96° \text{ and } \phi_q = 13.49°,$$

whence

$$\sin\phi_q = .23345 \text{ and } \cos\phi_q = .97237,$$

so

$$\mathbf{T} = \begin{bmatrix} .972 & -.233 \\ .233 & .972 \end{bmatrix}$$

whence we have

	Quartimax Factor Structure	
	H_1	H_2
X_1	.972	$-.233$
X_2	.972	.233
$\sum r^2_{X_i H_j}$	1.890	.109

This is identical (except for computational error in the third digit) to the correlation-derived (although not to the covariance-derived) PCA, as can be confirmed by applying Equation (6.10) to that factor structure. This might lead us to suspect that the PCA of a two-variable system will always have optimally simple structure in the quartimax sense if it is based on the intercorrelation matrix.

Applying Equation (6.9), we obtain

$$\tan(4\phi) = \frac{2[2(.4755) - (1.5878)(.8090)]}{[2(1.3455) - (1.5878^2 - 2(.6545) + (.8090)^2} = \frac{-2(.33353)}{-.48463} = 1.3764,$$

whence

$$4\phi = -126.04°; \ \phi = \text{-}31.51°; \ \sin\phi = \text{-}.52250; \text{ and } \cos\phi = .85264;$$

whence

$$\mathbf{T} = \begin{bmatrix} .853 & .5225 \\ -.5225 & .853 \end{bmatrix},$$

whence we have

	Varimax Factor Structure	
	H_1	H_2
X_1	.8526	.5225
X_2	.5225	.8526
$\sum r^2_{X_i H_j}$	1.0000	1.0000

The varimax rotation has certainly produced a level contribution of factors, although the resulting structure seems in this case to be less interpretable than the quartimax (or PCA) solution. In terms of the varimax simplicity criterion, the varimax structure produces a value of $[(.727 - .5)^2 + (.273 - .5)^2] \times 2 = .206$, as compared with values of .0207, .0000, and .0424 for the **S**-derived PCA, the **R**-derived PCA, and the intuitive simple solution, respectively.

As an example of rotation of a larger factor structure, PC_1 through PC_4, (those having eigenvalues greater than 1.0) from the PCA of the cheating questionnaire (Example 6.3) were subjected to a varimax rotation, yielding the rotated structure of Table 6.6. For comparison purposes, the entire 12-component factor structure was also subjected to

Table 6.6 Varimax Rotation of PC_1–PC_4, Cheating Questionnaire

Question	PC_1	PC_2	PC_3	PC_4
12	−.047	−.661	.299	.282
14	.181	−.017	.835	.157
16	−.075	−.292	.661	−.238
18	−.011	−.856	.132	−.004
20	.111	−.421	−.019	.552
22	.039	−.014	.014	.841
13	.643	−.048	.187	.023
15	.727	−.047	.463	.089
17	.792	−.096	.159	−.238
19	.588	−.643	−.193	−.007
21	.785	−.032	−.184	.120
23	.725	.096	−.098	.251
Σr^2	3.112	1.871	1.597	1.315
% Total	25.9	15.6	13.3	10.9

varimax rotation, producing as the first four factors, the structure described in Table 6.7. Considering first the results of rotating only the four most important PCs, we find that varimax rotation has accentuated the difference between one's own cheating and perception of others' cheating. Now the first and most important (26% of total variance) factor involves only others' cheating, the even-numbered questions having near-zero correlations with factor 1. The other factors are not as easy to label. In general, perceptions of others' cheating have quite low loadings on factors 3 and 4. Another way of saying this is that reports of one's own cheating behavior are more "factorially complex" than perceptions of others' cheating. At any rate, based on examination of the variables having loadings greater than .5 (in absolute value) on a given factor, factor 2 (own copying of assignments, own plagiarizing of another's paper, others' plagiarizing) seems to reflect abstention from cheating on assignments (except that the low loading of variable 13 does not fit this description); factor 3 seems to reflect one's own cheating on tests (although the .46 loading for variable 15 suggests that we might be able to include others' cheating on exams in the factor), and factor 4 seems to reflect one's own frequency of particularly flagrant cheating (handing in another's paper and, especially,

Table 6.7 First 4 PCs After Varimax Rotation of All 12 PCs, Cheating Questionnaire

Question	PC_1	PC_2	PC_3	PC_4
12	−.049	−.224	.903	.085
14	.031	−.068	.956	.020
16	−.036	−.102	.166	−.020
18	−.031	−.935	.073	.032
20	.070	−.089	.050	.113
22	.045	−.028	.018	.984
13	.123	−.005	.077	.009
15	.120	−.025	.283	.017
17	.217	−.038	.060	−.056
19	.237	−.252	−.037	.015
21	.889	.037	.039	.059
23	.228	.038	.021	.114
Σr^2	0.876	1.017	1.051	1.010
% Total	8.2	8.5	8.8	8.4

stealing someone else's paper).

An initial glance at the results of rotating all 12 PCs (Table 6.7) as compared with rotation of only the first four PCs (Table 6.6) suggests that the two are quite different; however, closer inspection reveals that each variable that has the highest loading on that factor in one matrix also has the highest loading on that factor in the other matrix. Further, the rank order among the variables having high loadings is quite consistent across the two structures. In general, the pattern that results from rotating the full 12-component structure is quite similar to the pattern that results from rotating only the first 4 components, except that in the former the highest loading within a given column is greatly accentuated at the "expense" of other loadings in the column. The extent of this accentuation is indicated by the fact that within each column of the full 12-factor matrix (only the first 4 columns of which are reproduced in Table 6.7) there is one loading that is at least three times as large in absolute magnitude as the next largest loading, the smallest of these "largest loadings" being .825.

There is a strong tendency for rotation of a complete component structure to lead to a "matching up" of factors with original variables, as indicated in Table 6.8, a list of largest loadings. This tendency will of course obscure more subtle patterns such as those offered as interpretations of the results displayed in Table 6.9. Note, too, the extremely level contributions of the 12 factors. In fact, the factor we suspect to be the most meaningful (factor 1) has the third lowest sum of squared loadings (total contribution to variance accounted for) of the 12 factors. Probably, too, if we wished to use this structure to estimate the communalities of the variables, we should simply sum the loadings for a given variable *without* its loading on the factor on which it has its highest loading (which is really just measuring itself). These are going to be rather low communalities. Such are the results of overfactoring.

Table 6.8 Large Loadings for Cheating Questionnaire

Factor:	1	2	3	4	5	6	7	8	9	10	11	12
Largest loading on factor:	.889	.935	.956	.984	−.970	−.945	−.969	.944	−.825	.919	.876	.885
Next largest loading:	.237	−.252	.283	.114	−.141	−.251	−.173	.231	−.221	.230	.234	.213
Variable having highest loading:	21	18	14	22	20	13	16	12	15	23	19	17
Sum of $(\text{loading})^2$:	.988	,1.017	1.051	1.010	1.019	1.061	1.014	1.004	.846	1.040	.949	1.000

Note that we could have, had we chosen to do so, listed all 12 hypothetical variables (4 rotated PCs and 8 unrotated PCs) in a single factor structure after we had rotated the first 4 PCs. Applying rotations to the first four PCs only has not in any way affected the relationship between the original variables and the last 8 PCs, and we can at any time "restore" perfect reproducibility of the original intercorrelation matrix by adding the unrotated PCs to our "pre-rotation" equations. The difference between rotation of the full PCA factor structure and rotation of only the "important" PCs is that the latter procedure must attempt to simplify the descriptions of all p PCs, whereas the former can concentrate on simplifying the descriptions of those m PCs that account for most of the variance and are thus most worth describing.

6.5.4 Individual Scores on Rotated PCs

In section 6.2 we showed, through multiple regression analysis of the relationship between PCs and original variables, that scores on PCs would be perfectly "predicted" from scores on the original variables and that the coefficients of the regression equations for computing scores on PC_i were simply the entries in column i of the factor pattern (or the entries in that column of the factor structure multiplied by s_j/λ_i). The situation after rotation of the PCs is not quite as simple. It is still true that the multiple correlation between each rotated PC and the original variables is 1.0. (The reverse is *not* true. A given original variable will generally *not* be perfectly reproducible from scores on the rotated PCs. How can this be? Remember that we generally do *not* rotate all of the PCs.) However, the entries of the factor loading matrix (i.e., of the factor structure) may be quite different from the weights given the different original variables in the formulae relating rotated PCs to original variables. Harking back to our known-generator problem, for instance, we note that all four correlations between an original variable and a latent variable are positive, whereas the formulae that would be used to compute h_1 and h_2 from scores on x_1 and x_2 are $h_1 = 3x_1 - x_2$ and $h_2 = x_2 - x_1$. This discrepancy should come as no surprise after our experience with multiple regression and the great difficulty in predicting the regression coefficient a variable will receive on the basis of its simple correlation with the outcome variable.

In the known-generator problem we started with the weight matrix (whose columns give the coefficients to be used in computing each latent variable from scores on the original variables) and constructed the factor structure (whose entries are latent variable–original variable correlations) from this information. Generally, of course, we have to work in the opposite direction. It would therefore be helpful to have available a set of formulae for computing the weight matrix for rotated PCs from the factor structure.

One approach to developing such formulae is to consider the original variables to be a set of predictor variables from which we wish to "predict" scores on one (and eventually all) of the rotated PCs. We saw in chapter 2 that all the information needed to develop a regression equation for predicting standard scores on Y from z scores on a set of Xs were the intercorrelations among the predictors and between each predictor and Y. In our present case, the predictor–predicted correlations for predicting a particular rotated PC are simply the loadings in the corresponding column of the rotated factor structure.

Thus, for instance, the regression coefficients for predicting z scores on H_1, (the first of the "true" generating variables) from Example 6.1 are obtained by computing

$$\mathbf{b}_{h_1} = \mathbf{R}^{-1} r_{x_1 h_1} = \begin{bmatrix} 1 & \sqrt{.8} \\ \sqrt{.8} & 1 \end{bmatrix}^{-1} \begin{bmatrix} .707 \\ .3162 \end{bmatrix} = \frac{1}{.2} \begin{bmatrix} 1 & -.894 \\ -.894 & 1 \end{bmatrix} \begin{bmatrix} .707 \\ .3162 \end{bmatrix} = \begin{bmatrix} 2.121 \\ -1.580 \end{bmatrix}.$$

For H_2, we have

$$\mathbf{b}_{h_2} = \frac{1}{.2} \begin{bmatrix} 1 & -.894 \\ -.894 & 1 \end{bmatrix} \begin{bmatrix} .707 \\ .9486 \end{bmatrix} = \begin{bmatrix} -.707 \\ 1.580 \end{bmatrix}.$$

Because the procedures for computing $\mathbf{b}_{h_1}$ and $\mathbf{b}_{h_1}$ have in common premultiplication by $\mathbf{R}^{-1}$, we could have solved for the two sets of coefficients somewhat more compactly by writing

$$\mathbf{B}_\mathbf{h} = [\mathbf{b}_{h_1} \; \mathbf{b}_{h_2}] = \mathbf{R}^{-1}\mathbf{F} = \mathbf{R}^{-1}[\mathbf{r}_{\mathbf{x},h_1} \; \mathbf{r}_{\mathbf{x},h_1}], \tag{6.11}$$

that is, simply by premultiplying the (rotated) factor structure by $\mathbf{R}^{-1}$.

Note that the coefficients we obtain are those employed in the z-score regression equation, that is,

$$z_{h_1} = 2.121\, z_1 - 1.580\, z_2 = (2.121/\sqrt{2})X_1 - (1.580/\sqrt{10}X_2 = 1.5X_1 - .5X_2;$$

and

$$z_{h_2} = -.707\, z_1 + 1.580\, z_3 = -.5X_1 + .5X_2.$$

Because we generated X_1 and X_2 in the first place on the basis of their relationships to two hypothetical variables having unit normal distributions, we know that these are also the raw-score formulae for H_1 and H_2. More generally, we would have to multiply both sides of the expression relating h_i to the Xs by s_{h_i} to obtain an expression for H_i, the raw score on hypothetical variable *i*; that is,

$$\mathbf{H}_i = s_{h_i} \mathbf{X}\, \mathbf{b}_{h_i} ,$$

where $\mathbf{H}_i$ is a column vector of scores on hypothetical variable (rotated PC) *i*. However, we *cannot* ordinarily determine s_{h_i}—nor, fortunately, do we need to know the absolute magnitudes of subjects' raw scores on the hypothetical variables.

Note, too, that

$$R_{h_1 \bullet \mathbf{x}} = .707(2.125) + (.3162)(-1.580) = 1..000,$$

and

$$R_{h_2 \bullet \mathbf{x}} = .707(.707) + (.949)(1.580) = .999,$$

as we might have expected from the fact that there was an exact relationship between PCs and Xs before rotation. We show when we discuss factor analysis that this same basic multiple regression technique can be used to *estimate* factor scores on the basis of the factor structure and **R**, but the resulting R^2 values will be less than unity.

The multiple-regression approach to deriving scores on rotated PCs is a straightforward application of what is hopefully a very familiar technique. However, it also requires the inversion of the full p x p matrix of correlations among the original variables, thus negating to a large degree one of the purported advantages of PCs, their computational convenience. Given the perfect relationship between rotated PCs and original variables, we might suspect that there would be some computationally simpler method of computing scores on the rotated PCs. In particular, because "rotation*' simply involves a set of linear transformations of the intercorrelations between PCs and Xs, we might expect rotation to have the effect of carrying out a similar set of transformations on the PCs themselves. As it turns out (and as we shortly prove), this is almost correct. Rigid rotation of a PCA-derived factor structure—that is, postmultiplication of **F** by $\mathbf{T} = \mathbf{T}_1 \cdot \mathbf{T}_2 \cdot \cdots \cdot \mathbf{T}_r$—has the effect of postmultiplying $\mathbf{Z}_{PC}$ —the N x m matrix of standard scores on the m PCs—by **T** as well, and thus of postmultiplying Z_x—the matrix of standard scores on the Xs—by $\mathbf{B}_z \cdot \mathbf{T}$, where $\mathbf{B}_z$ is the standard-score weight matrix from the original PCA. In other words,

$$\mathbf{Z}_h = \mathbf{Z}_{pc} \cdot \mathbf{T} = \mathbf{Z}_X \cdot \mathbf{B}_z \cdot \mathbf{T}; \tag{6.12}$$

where the matrices are as defined above. Thus, returning to Example 6.1 (the known-generator problem), we have, for scores on the "true" hypothetical variables (whose correlations with the original variables are obtained by rotating the **S**-based PC axes through an angle of -67.5°),

$$\begin{bmatrix} .3827 & .9239 \\ -.9239 & .3827 \end{bmatrix}$$

$$\mathbf{Z}_h = \mathbf{Z}_{pc} \cdot \begin{bmatrix} .3827 & .924 \\ -.924 & .3827 \end{bmatrix} = \mathbf{Z}_X \cdot \begin{bmatrix} .1586 & -2.2277 \\ .8558 & 2.0639 \end{bmatrix}$$

$$= \mathbf{Z}_X \cdot \begin{bmatrix} 2.119 & -.760 \\ -1.579 & 1.581 \end{bmatrix} = \mathbf{X} \cdot \begin{bmatrix} 1.498 & -.499 \\ -.499 & .500 \end{bmatrix}.$$

Note that the entries of the standard-score weight matrix are obtained by multiplying each entry in the **S**-derived factor pattern by $s_i / \sqrt{\lambda_j}$. Be sure you understand why.

The proof of Equation (6.12) is given in Derivation 6.3. This proof includes proof of the fact that rotation of the factor structure through an angle θ does *not* lead to the same transformation being applied to the raw-score weight matrix. Note, too, that the rotated factor structure and the rotated weight matrix are quite different. As we argued in the case of Manova (cf. Example 6.4), and as we argue again in the case of factor analysis (section 7.6), it is the weight matrix (i.e., the factor score coefficients defining how each person's score on that factor can be computed from his or her scores on the original variables) that should be used in interpreting just what it is that our latent variables are tapping. Once we have inferred an interpretation of each factor from its "operational definition" as provided by its column of the weight matrix, we can then use the factor structure to reinterpret each original variable in terms of the various factors that go into determining scores on that variable, by looking at the corresponding row of the factor structure.

This suggests that if our goal is to simplify the task of coming up with meaningful interpretations of our factors, we should apply the criteria for simple structure to the weight matrix, rather than to the factor structure. (If, however, our primary concern is with the simplicity of our interpretation of the original variables in terms of the underlying factors, the usual criteria for rotation are entirely appropriate, although we should examine the resulting *weights* to know what it is we're relating the original variables to.) Thus, for instance, we could seek an angle of rotation that would lead to X_2 having a coefficient of zero in the definition of H_1, whence we require (in the case of our known-generator problem) that $.8558 \cos\theta + 2.0639 \sin\theta = 0$, whence $\tan\theta = -.41465$, $\theta = -22.52°$, and

$$\mathbf{Z}_h = \mathbf{Z}_X \bullet \begin{bmatrix} 2.757 & -1.9972 \\ .000 & 2.234 \end{bmatrix} = \mathbf{X} \cdot \begin{bmatrix} 1.950 & -1.414 \\ .000 & 1.117 \end{bmatrix}.$$

In section 7.6, we address the questions of whether the criteria for simple structure are equally appropriate as criteria for simple weights and how to talk your local computer into rotating to simple factor score coefficients. For the moment, however, let me try to buttress the position that factor score coefficients should be the basis for interpreting (naming) factors by resorting to a case in which we know the factors that are generating our observed variables.

Example 6.5 *A factor fable*. An architect has become reenamored of the "box" style of architecture popular in the 1950s. She applies this style to designing residences in the Albuquerque area (where annual snowfall is sufficiently low to make flat roofs practical). After designing a few hundred homes for various clients, all as some variant of a (hollow) rectangular solid, she detects a certain commonality among the designs. She suspects that it may be possible to characterize all of her designs in terms of just a few underlying factors. After further consideration, she indeed comes up with three factors that appear to characterize completely all of the homes she designs: frontal size (*f*, the perimeter of the house's cross section as viewed "head on" from the street), alley size (*a*, the perimeter of the cross section presented to a neighbor admiring the house from the side), and base size (*b*, the perimeter of the house around the base of its outside walls). Figure 6.3 illustrates these measurements.

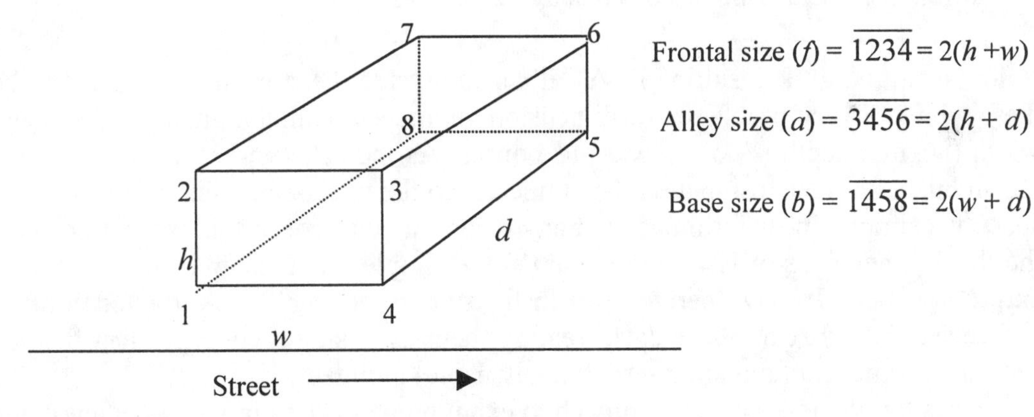

Figure 6.3 Architectural dimensions of houses

However, the architect is not completely satisfied with these measures as a basis for designing homes. For instance, even after a client has specified his or her preferences for *f*, *a*, and *b* it takes a bit of experimenting with rectangles having the required perimeters to find three that will fit together.

The problem is that her three factors appear to be neither logically independent nor perfectly correlated. Having heard about PCA's ability to extract orthogonal dimensions from measurements having a complex pattern of intercorrelations (and being one of those "dust bowl empiricists" who prefers to keep her explanatory factors closely tied to observable features), she decides to submit measurements on a representative sample of 27 houses to a PCA. The 27 houses selected for measurement are shown schematically in

Figure 6.4, and the measures of f, a, and b for these 27 houses are presented in the first 3 columns of Table 6.9.

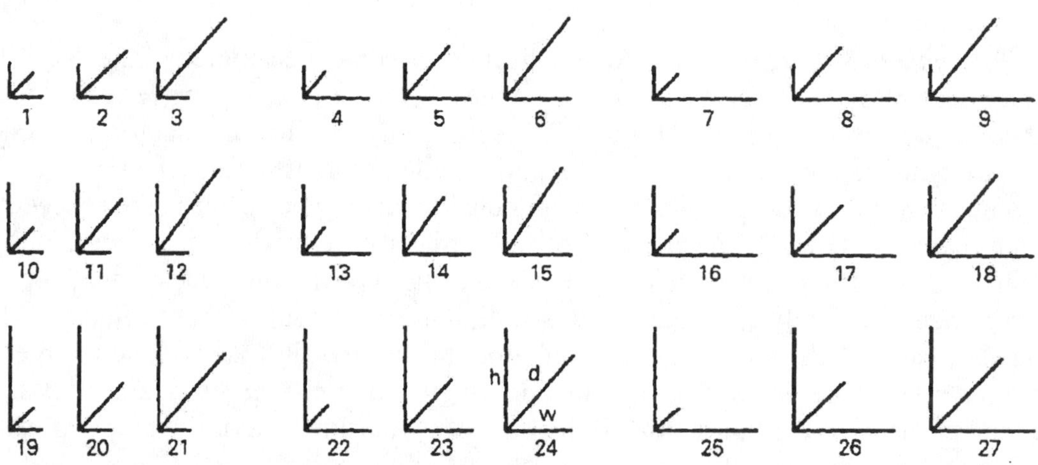

Figure 6.4 **Schematic representation of 27 houses**

Before examining the resulting PCA, let's step outside the story for a moment to belabor the obvious: a very usable set of dimensions is indeed available, namely the height (h), width (w), and depth (d) of the various houses. As indicated in Figure 6.3, the measures the architect has been using are related linearly to the usual three dimensions of Cartesian space via simple linear formulae, and any technique that doesn't allow us to discover (a) the three dimensions of h, w, and d and (b) the simple relationship of f, a, and b to h, w, and d is suspect. It may, therefore, be instructive to "try out" PCA, rotated components, and (eventually) FA on these data, treating them as posing a concept identification task for our hypothetical (and unrepresentatively naive) architect,

This is by no means a new approach to examining FA. It is in fact a blatant plagiarism of L. L. Thurstone's (1947) box problem, Kaiser and Horst's (1975) addition of measurement error and additional sampling units to Thurstone's "data," Gorsuch's (1983) box plasmode, and probably dozens of other reincarnations. What is (as far as I know) unique to this application is the use of data that fit the rotated-components model perfectly. Each of our measured variables is a perfect linear combination of two of the underlying dimensions, and all communalities are thus 1.0. Any failure to uncover the true situation therefore cannot be attributed to imperfect measurement (as in Gorsuch's deliberately rounded-off measures), the presence of "noise" variables (e.g., Gorsuch's longest inner diagonal and thickness of edge), or nonlinear relationships (e.g., squared height in both Gorsuch's and Thurstone's examples). It will be seen, nevertheless, that attempting to interpret our components in terms of loadings leads to just such failure. Back to our story.

The correlations among f, a, and b have, for this systematically constructed (designed) set of houses, a very simple form, namely,

$$\mathbf{R}_x = \begin{bmatrix} 1 & .5 & .5 \\ & 1 & .5 \\ & & .5 \end{bmatrix}$$

Table 6.9 Scores on Observed and Derived Variables for27 Houses

	Measures			Components			Implied by loadings		
House	f	a	b	$f+a-b$	$f-a+b$	$-f+a+b$	$f+a$	$f+b$	$a+b$
1	2	2	2	2	2	2	4	4	4
2	2	3	3	2	2	4	5	5	6
3	2	4	4	2	2	6	6	6	8
4	3	2	3	2	4	2	5	6	5
5	3	3	4	2	4	4	6	7	7
6	3	4	5	2	4	6	7	8	9
7	4	2	4	2	6	2	6	8	6
8	4	3	5	2	6	4	7	9	8
9	4	4	6	2	6	6	8	10	10
10	3	3	2	4	2	2	6	5	5
11	3	4	3	4	2	4	7	6	7
12	3	5	4	4	2	6	8	7	9
13	4	3	3	4	4	2	7	7	6
14	4	4	4	4	4	4	8	8	8
15	4	5	5	4	4	6	9	9	10
16	5	3	4	4	6	2	8	9	7
17	5	4	5	4	6	4	9	10	9
18	5	5	6	4	6	6	10	11	11
19	4	4	2	6	2	2	8	6	6
20	4	5	3	6	2	4	9	7	8
21	4	6	4	6	2	6	10	8	10
22	5	4	3	6	4	2	9	8	7
23	5	5	4	6	4	4	10	9	9
24	5	6	5	6	4	6	11	10	11
25	6	4	4	6	6	2	10	10	8
26	6	5	5	6	6	4	11	11	10
27	6	6	6	6	6	6	12	12	12

As we know from the equi-correlation formulae of section 6.1.2, the first PC will be a general factor (the sum or average of f, a, and b) and will account for $[1 + 2(.5)]/3 = 2/3$

of the variance. In fact, we have a factor structure and a normalized coefficient matrix of

$$\mathbf{R}_{xf} = \begin{bmatrix} .8165 & .5 & .2867 \\ .8165 & -.5 & .2867 \\ .8165 & 0 & -.5774 \end{bmatrix} \text{ and } \mathbf{B} = \begin{bmatrix} .577 & .707 & .408 \\ .577 & -.707 & .408 \\ .577 & 0 & -.816 \end{bmatrix}.$$

Either matrix informs the architect that she may use the simple sum or average of all three measures (labeled, perhaps, as "overall size"), $f + a + b$—which we know to also equal $2(h + w + d)$—as one dimension; the difference between frontal size and alley size ("covert size"?), $f - a$—which also equals $2(w - d)$—as the second dimension; and the difference between b and the average of the other two dimensions ("living area dominance"?), $b - (f + a)/2$—which also equals $w + d - 2h$—as the third dimension.

This, as the reader will wish to verify, does indeed establish three uncorrelated factors—but the solution still lacks the simplicity of the ***h, w, d*** system we had in mind. For instance, it would still be a difficult process to construct a house from specifications of the desired "overall size," "covert size," and "living area dominance."

Rather than taking you through a series of varimax, quartimax, or other rotations, let's cheat and move directly to the solution we're really searching for, namely,

$$h = (f + a - b)/2,\ w = (f - a + b)/2, \text{ and } d = (-f + a + b)/2.$$

The resulting structure and weight matrices are

$$\mathbf{R}_{xf} = \begin{bmatrix} .707 & .707 & 0 \\ .707 & 0 & .707 \\ 0 & .707 & .707 \end{bmatrix} \text{ and } \mathbf{B} = \begin{bmatrix} 1 & 1 & -1 \\ 1 & -1 & 1 \\ -1 & 1 & 1 \end{bmatrix}.$$

Assuming that our architect happens on this particular rotation, she is now faced with the task of interpreting the resulting rotated components. Let's consider this task with respect to the first of the rotated components—the one we know is "actually" the height of the house. What she has to do is to consider what kind of house would get a high score on both f and a, but a low score on b. As an aid to this consideration, she can calculate a score for each house on $f + a - b$ and then sort the 27 houses in terms of these scores. When she does, she finds that houses 1 through 9 (see Figure 6.4) get a score of 2; houses 10 through 18 get a score of 4; and houses 19 through 27 each get a score of 6. (See also the fifth column of Table 6.9.) It probably won't take her long to recognize that it is the *height* of the house that this dimension is tapping.

This will be the case ***unless*** she listens to factor analysts, who claim that it is the first column of $\mathbf{R}_{xf}$ she should be examining, not **B.** After all, the loadings are more stable under replication (just as r_{iy}s are more stable than regression coefficients in MRA), they are always bounded between 0 and 1, they are likely to remain the same as we add other correlated variables to our search, and they don't involve the terrible mistake (common to "dust-bowl empiricists" such as I) of tying latent variables to observable measures. If she heeds all this advice, she will conclude that b has not a whit to do with the first factor (it has a loading of zero on that factor) and that what she should be looking for is what it is about a house that would lead it to have a high score on both f and a. She recognizes that one way to see which houses are high (or moderate or low) on both frontal size and alley size is to compute a score for each house on $f + a$. (Indeed, several textbook authors have recommended estimating factor scores by adding up scores on variables with high loadings on the factor—the *salient loadings* approach to estimating factor scores.) Sorting the seven houses on the basis of $f + a$ gives the scores listed in the eighth column of Table 6.9 and yields the clusters of identically scored houses displayed in Figure 6.5.

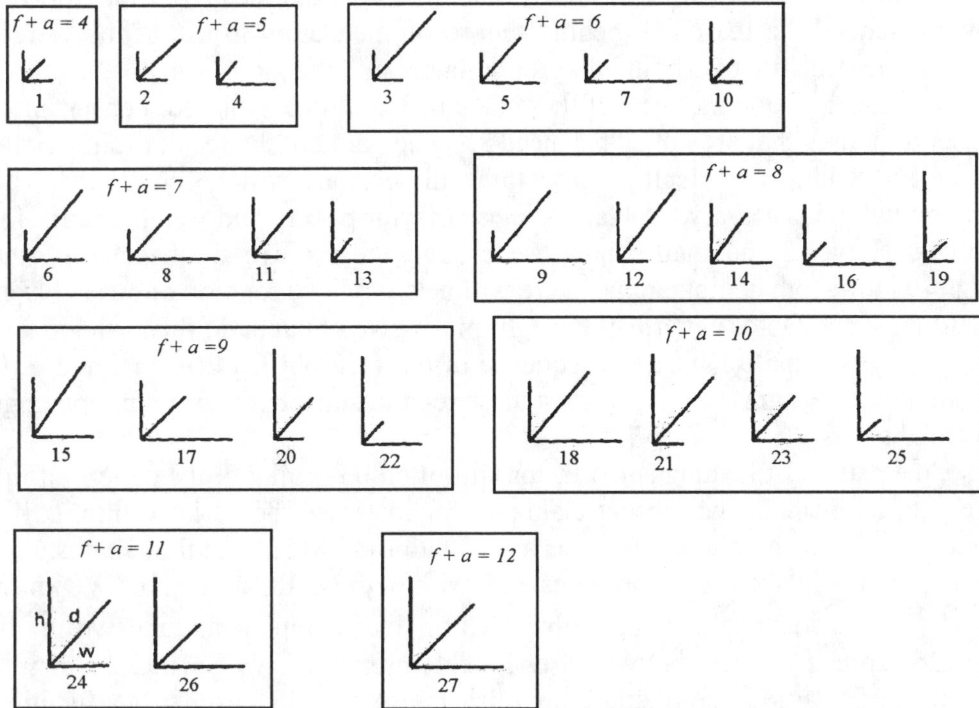

Fig. 6.5 Same 27 houses, sorted on the basis of the loadings-based interpretation of Factor 1. The odds against inferring "height" as the underlying dimension generating this rank ordering are enormous—as they should be because it is *not* height, but $f + a = 4h + 2w + 2d$, that is being examined. This is not to say that the loadings are "wrong." They tell us quite correctly that f and a each have a .707 correlation with h, whereas b has a zero correlation with h; moreover, if we wanted to estimate the height of these houses with no information other than frontal size, the loading of frontal size on the height factor tells us just how well we would do (namely, 50% of the variance). Finally, once we have correctly identified height, width, and depth as our three factors, the loadings now provide the weights by which to interpret the original variables in terms of the factors. Thus they tell us, quite correctly, that frontal area is directly proportional to height + width, and so on. It would be just as wrong to attempt to use the rows of **B** to interpret the "factorial composition" of the original variables as it is to use the columns of $\mathbf{R}_{xf}$ to interpret the factors in terms of their relationship to the original variables. (Note,

however, that our use of $\mathbf{R}_{xf}$ to interpret the original variables as functions of the underlying factors is really based on the identity of this matrix with $\mathbf{P}_z$, the matrix whose rows indicate how to compute each z_x as a linear combination of scores on the factors. When we consider nonorthogonal factors, this identity breaks down, and the sorts of arguments we've been pursuing in this example suggest that we should prefer the pattern matrix **P** over the structure matrix $\mathbf{R}_{xf}$ for this purpose when the two matrices are no longer identical. There actually are factor analysts who concur with this preference—Gorsuch [1983], for instance, reported that Thurstone and his students generally prefer **P**, and Gorsuch himself recommends that both **P** and $\mathbf{R}_{xf}$ be examined—but they do not seem to have extended the reasoning behind this recommendation to use of the weight matrix **B** versus the structure matrix in interpreting factors.)

There are a number of other aspects of these data that could be explored. For instance, if we use a set of houses that are not a "balanced" set (i.e., are not the result of factorially combining height, width, and depth), these three dimensions will be correlated, and oblique rotation will be necessary to obtain a "clean" factor pattern and weight matrix for these variables. However, no matter how we change the relative frequencies of the various height-width-depth combinations, there will be an oblique rotation of our PC axes that will yield the same factor pattern and weight matrix we obtained in this orthogonal–factors case. This is a simple logical consequence of the fact that f ***is*** $2(h + w)$ and h ***is*** $f + a - b$, no matter how hard we try to distort these relationships by interpreting loadings rather than weights.

Of course, the pattern of loadings on a factor will often be similar to the pattern of the factor score coefficients. In the present example, for instance, the variable implied by interpreting the loadings on height $(f + a)$ has a correlation of .8165 (and thus shares 67% of its variance) with height. (It also correlates .408 with both width and depth.) We know that $\mathbf{R}_{xf}$ and **B** yield identical interpretations of principal components, and we might suspect that the degree to which the two bases for interpretation diverge (i.e., the degree to which the loadings-based interpretations are misleading) would be primarily a function of how far the final solution has been rotated from the principal components or principal factors solution. Analytical and "ecological" studies of the bounds on and parameters determining the loadings–weights discrepancy would be helpful in establishing how much revision is needed of past, loadings-based interpretations of factor analyses.

The moral of our factor fable should be clear: when our observed variables can indeed be generated as linear combinations of underlying factors, factor (including component) analysis can identify these factors—provided that we base our interpretations on the factor score coefficients, rather than on the loadings.

6.5.5 Uncorrelated-Components Versus Orthogonal-Profiles Rotation

In section 6.1.4 I asserted that the PCs of a correlation matrix have the largest sum of normalized variances attainable by any set of mutually uncorrelated linear combinations of z scores on the original variables. This being the case, it follows that any of the "orthogonal" rotations (more about the quote marks later) " we've carried out should have yield rotated PCs with lower total normalized variance than the pre-rotation sum. The normalized variance of a component **Xb** = $b_1z_1 + b_2z_2 + \cdots + b_pz_p$ is easily computed via

$$s_{rpc}^{2(\text{norm})} = \mathbf{b'Rb}/(\mathbf{b'b}) = \left(\sum\sum b_i b_j r_{ij}\right)/\left(\sum b_i^2\right),$$

where the *rpc* index stands for "rotated principal component". Applying this formula to the known-generator problem (Example 6.1), we find that the normalized variances of the original two variables are 1.794 and .206, for a total of 2.000. (The sum of the normalized variances of all *p* PCs of a correlation matrix will always equal the number of variables, since any set of *z* scores has unity variance.) Rotation to the "true" structure yielded $rpc_1 = 2.121z_1 - 1.580z_2$ and $rpc_2 = -.707z_1 + 1.580z_2$, with resulting normalized variances of 1/7 + 1/3 = .476—less than a quarter of the pre-rotation total. Taking a real data set, namely, the twelve WISC–R subscales of Example 6.2, the first four PCs yielded a total normalized variance (and sum of squared loadings across those four PCs) of 5.398 + 1.160 + .969 + .746 = 8.273. After varimax rotation, the twelve subscales still have squared loadings on the four rotated PCs that sum to 8.273, but their normalized variances are now 1.712, 1.528, .997, and .849, for a total of 5.086, 61.5% of the pre-rotation total. (Don't expect any help from SAS or SPSS in obtaining these normalized variances. However, most computer programs impose on rotated PCs [as they do on the original PCs] the side condition that **b'Rb** = 1; if that's the case for the program you're using, then the normalized variance of any given component will simply equal the reciprocal of its squared scoring coefficients—though the scoring coefficients will have to be asked for explicitly, since they are not a part of the default output.)

Millsap (in Harris & Millsap, 1993) proved that this "leakage" of normalized variance under "orthogonal" rotation (those quote marks again) of the components is a completely general phenomenon, always holding whenever the PCs of a correlation matrix are rotated by any angle not a multiple of 90°. This seemed to contradict a theorem proved by Joliffe (1986) that the sum of the normalized variances of any set of components of a given set of variables is the same for any choice of mutually orthogonal vectors of combining weights—and aren't the vectors of combining weights that define rotated PCs mutually orthogonal?

Well, no! "Orthogonal" rotation, as almost universally practiced, yields *rpc*s that are uncorelated with each other (i.e., whose vectors in the Euclidean space whose axes are the original variables are perpendicular to each other, and thus orthogonal in the space of the original variables), but whose vectors of combining weights are *not* uncorrelated with each other (and are thus *not* orthogonal to each other in combining-weight space). Put another way, if we interpret the vector of combining weights for a given *rpc* as defining that profile of *z* scores that would lead to a relatively high score on that *rpc*, then those profiles will be somewhat redundant with each other. For instance, the two *rpc*s from the "true"-structure rotation of the known-generator variables are close to mirror images of each other, one being approximately $z_1 - z_2$ and the other approximately $z_2 - z_1$, and the two sets of combining weights indeed correlate -.873 with each other. (That *scores* on the *rpc*s those combining weights define are nevertheless uncorrelated with each other is attributable to [a] the "approximately" in the above descriptions and [b] the fact that rpc_1 accounts for over twice as much of the individual-difference variance as does rpc_2.) The redundancies among the varimax-rotated components of the WISC–R are less dramatic, but the correlations among the combining-weight vectors of the four rotated PCs nevertheless range from -.094 to -.488.

What if we would like what we say about the profile of scores that characterizes one *rpc* to be nonredundant with what we say about what leads to a high score on any other *rpc* – i.e., what if we would prefer to keep our *rpc*-defining weight vectors orthogonal to each other? In that case, as shown by Harris and Millsap (1993), we could rotate our PCs (transform them to components we find easier to interpret) by applying an orthogonal transformation matrix of our choice to the matrix of factor scoring coefficients (first normalized to unit length by dividing each column of the scoring-coefficient matrix by the square root of the sum of the squares of its initial entries). Moreover, Joliffe's formerly pesky theorem referred to earlier tells us that the resulting components will no longer suffer the leakage of normalized variance that components rotated under the constraint of remaining uncorrelated do. For instance, if we apply to the known-generator problem's (normalized) scoring-coefficient matrix the same transformation matrix we used to accomplish a varimax rotation of its structure matrix, we get a scoring-coefficient matrix for the rotated PCs of

$$\begin{bmatrix} .707 & -.707 \\ .707 & .707 \end{bmatrix} \begin{bmatrix} .853 & .5225 \\ -.5225 & .853 \end{bmatrix} = \begin{bmatrix} .973 & -.234 \\ .234 & .973 \end{bmatrix}.$$

In other words, our rotated PCs are $rpc_1 = .973z_1 + .234z_2$ and $rpc_2 = -.234z_1 + .973z_2$. The sum of cross-products of the two vectors of combining weights—and therefore the correlation between the two sets of combining weights—equals zero, and the normalized variances are 1.407 and .593, for a total of 2.000 equal to the pre-rotation total. On the other hand, the correlation between the two *rpc*s is

$$[.973 \quad .234] \begin{bmatrix} 1 & .894 \\ .894 & 1 \end{bmatrix} \begin{bmatrix} -.234 \\ .973 \end{bmatrix} / \sqrt{1.407(.593)} = .874$$
$$\quad [1.182 \quad 1.104] \quad .798$$

and the sum of the squared loadings of original variables on *rpc*s is given by the sum of the squares of the entries in

$$\begin{matrix} & rpc_1 & rpc_2 \\ z_1 & .996 & .826 \\ z_2 & .931 & .992 \end{matrix},$$

the post-rotation structure matrix, and thus equals 1.859 (for rpc_1) + 1.666 (for rpc_2),for a total of 3.525, considerably higher than the pre-rotation total of 2.000.

A similar pair of rotations applied to the first two PCs of Harman's (1967) oft-analyzed Eight Physical Variables example yielded, for uncorrelated-variables rotation, a decrease of total normalized variance by about 20% and defining vectors that now correlated -.44; and, for orthogonal-profiles rotation, *rpc*s that correlated +.44 and a total sum of squared loadings that increased by about 20%.

Thus you may (must?) choose between a rotation that keeps the rotated components uncorrelated with each other, keeps the sum of the squared loadings across the rotated components constant, but "leaks" normalized variance and yields redundant defining profiles, and a rotation that keeps the profiles defining the rotated components nonredundant and keeps the sum of the normalized variances constant but yields correlated components having a very different (higher?) sum of squared loadings than the original PCs.

I am far from understanding all the implications of this difference for practical component analysis—this is, after all, marked as an optional section. I know of no one who has used orthogonal-profiles rotation in a published report—though there is a close analogy to trend analysis of repeated-measures designs, where the nearly universal preference is to employ trend contrasts whose defining contrast coefficients are mutually orthogonal, rather than trend contrasts yielding uncorrelated scores for individual participants. At a bare minimum, however, the results Roger Millsap and I have obtained so far suggest that (a) there are more ways of rotating components than had heretofore been dreamed of in your philosophy, Horatio; (b) we don't really have a firm grip on how most appropriately to compare variances of components; and (c) we should either come to grips with problem (b) or abandon maximization of normalized variance as the definitional criterion for principal components.

Demonstration Problems

A. Do the following analyses "by hand." (Use of a calculator is all right, as is doing part C first.)

1. Conduct a PCA on the covariance matrix of the predictor variables employed in the demonstration problems in chapter 2. Report the factor pattern and the factor structure separately.

2. Compute each subject's score on each of the PCs computed in problem 1. Now compute the sample variance on scores on PC_1 , PC_2, and PC_3. To what previously computed statistics do these variances correspond?

3. Using the subjects' scores on the PCs, compute all of the correlations among the PC*s* and between the PCs and the original variables. To what do these correlations correspond?

4. "Interpret" the PCs obtained in problem 1. For each interpretation (e.g., "PC*2* is simply the sum of X_1, X_2, and X_3") specify the "ideal" characteristic vector implied by

that interpretation and conduct a significance test to determine whether the characteristic vector that is actually obtained can reasonably be interpreted as being a random fluctuation from that ideal characteristic vector.

5. Test the hypothesis that the second and third characteristic vectors are identical except for sampling fluctuation.

6. Use the coefficients defining PC_1 alone to "reproduce" the covariance matrix. Test the null hypothesis that the discrepancies between observed and "reproduced" covariances are due solely to random fluctuation.

7. Repeat problem 6 on the basis of PC_1 and PC_2.

8. Repeat problem 6 on the basis of PC_1, PC_2, and PC_3.

9. Why must the significance tests in Problems 4 through 8 be interpreted with extreme caution?

10. Conduct a multiple regression analysis (MRA) of the relationship between the outcome variable and scores on PC_1, PC_2, and PC_3. How does this compare with the results of the MRA you performed before on the relationship between Y and X_1 - X_3?

11. Conduct a MRA of the relationship between the outcome variable and scores on PC_1 and PC_2. How does this compare with the results of problem 10 and with the results of the MRA you performed before on the relationship between Y and $X_1 - X_2$?

12. Conduct a MRA of the relationship between Y and PC_1. How does this compare with problems 10 and ll?

13. Use graphical methods to rotate the factor structure obtained in problem 1 to what you consider the most easily interpretable structure.

14. Do a varimax rotation of PC_1, PC_2, and PC_3. Compare with problem 13.

15. Do a varimax rotation of PC_1 and PC_2. Compare with problems 13 and 14.

16. Compute a score for each subject on each of the three hypothetical variables obtained in problem 13. Compute the variance of each hypothetical variable. Could you have predicted these variances in advance?

17. Compute a score for each subject on each of the two hypothetical variables obtained in problem 15. Compute the variances. Could you have predicted them in advance?

B. Repeat the analyses of part A, beginning with the correlation matrix.

C. Perform as many of the analyses in parts A and B as possible using available computer programs.

Answers

A.

1. Note that the first covariance-derived principal component accounts for 64% of the variation in *raw* scores but only 50% of the variation in *standard scores*.

	Factor pattern				Factor structure		
	PC_1	PC_2	PC_3		PC_1	PC_2	PC_3
X_1	−.2643	−.3175	.9107		−.7006	−.6208	.3521
X_2	.1211	.9259	.3579		.1741	.9819	.0750
X_3	.9568	−.2049	.2062		.9873	−.1560	.0310
λ_j	17.567	9.559	.3737	ΣL^2	1.4959	1.3739	.1306
% accounted	.639	.348	.0136	% variance accounted for	.499	.458	.0435

2. On computing the variance of each column of numbers, we find that (except for round-off errors) the variance of principal component i is equal to the eigenvalue associated with that principal component.

	Score on		
Subject	PC_1	PC_2	PC_3
A	.6911	.4629	6.9503
B	−.5949	3.9679	6.7009
C	7.7962	−1.2623	6.2144
D	3.0550	−2.3211	7.7638
E	8.7871	4.6106	7.4511

3. The correlations between PCs are all zero, whereas the correlations between *X*s and PCs are given by the entries in the factor structure.

	X_1	X_2	X_3	PC_1	PC_2	PC_3
X_1	1	−.70502	−.58387	−.7005	−.6298	.3521
X_2		1	.02111	.1741	.9819	.0750
X_3			1	.9873	−.1560	.0310
PC_1				1	0	0
PC_2					1	0
PC_3						1

4. First, taking the extremely simple hypotheses that $\mathbf{B}_1' = (0\ 0\ 1)$, $\mathbf{B}_2' = (0\ 1\ 0)$, and $\mathbf{B}_3' = (1\ 0\ 0)$, we have

$$\chi^2 = 4[16.5/17.567 + 17.567(171/1004) - 2] = 7.725 \text{ with } 4\ df, \text{ ns} \quad \text{for the first } H_0;$$

$$\chi^2 = 4[8.5/9.559 + 9.559(435/1004) - 2] = 12.123 \text{ with } 4\ df, \quad p = .05 \quad \text{for the second } H_0;$$

and

$$\chi^2 = 4[2.5/.7373 + .3737(2243/1004) - 2] = 22.099 \text{ with } 4\ df, \quad p < .001 \quad \text{for the third } H_0.$$

Because the major discrepancy occurs in our description of PC_3, we might be tempted to modify only that description and retest for significance. However, this is not really legitimate, because we know that PCs are uncorrelated, and we thus cannot change our description of PC_3 without also changing our predictions for PC_1 and PC_2 so as to keep them orthogonal to PC_3. This suggests that we should really consider ourselves as constructing a hypothesized factor pattern. That then suggests that we could simply see what covariance matrix is implied by our hypothesized factor pattern, and test that against the observed covariance matrix. The significance test described in section 6.4.2, however, applies to correlation matrices, not covariance matrices (although we recommended simply substituting $\mathbf{S}$ and $\hat{\mathbf{S}}$ for $\mathbf{R}$ and $\hat{\mathbf{R}}$ when you wanted to test that sort of hypothesis). The correlation matrix implied by our initial descriptions is simply a 3 x 3 diagonal matrix, that is, it corresponds to the prediction that all off-diagonal correlations are equal to zero in the population. However, we have a test for that, namely,

$$\chi^2 = (\text{complicated constant})\cdot\ln(1/|\mathbf{R}|) = -1.5\ \ln(.1789) = 2.58 \text{ with } 3\ df, \text{ ns.}$$

This seems inconsistent with the results of our tests of the separate predictions until we realize that those predictions involve not only predictions about interrelationships among the original variables but also hypotheses about the variances of the original variables. If we test the hypothesis, for instance, that the variance of variable 1 is truly .3737 in the population (which is what our hypothesized factor pattern implies), then we obtain

$$\chi^2 = 4(2.5)/(.3737) = 26.76 \text{ with } 4\ df, p < .01.$$

Of course, if the assumption of no nonzero intercorrelations were true, then the population value of the third characteristic root would be 2.5, not .3737, and our "reproduced" value of s_i would be identically equal to its observed value. At any rate, trying out different sets of hypothesized PCs leads finally to the conclusion that we must use coefficients that reflect the relative magnitudes of the actual PC coefficients if we are to avoid significant chi-square values, especially for PC_3 . Remember, however, that these tests are all large-sample approximations and probably grossly underestimate the true probabilities of our hypotheses.

5. $q = 1, r = 2$; H_0: $\lambda_2 = \lambda_3$. Thus

$$\begin{aligned}\chi^2 &= -4[\ln(9.559) + \ln(.3737)] + 4(2)\ln[(9.559 + .3737)/2] \\ &= 4\ln\{4.966^2/[9.559(.3737)]\} = 4(1.932) \\ &= 7.728 \text{ with } 2\ df,\ p < .05.\end{aligned}$$

Thus we can reject the hypothesis that PC_2 and PC_3 have identical eigenvalues in the population.

6. The reproduced covariance matrix is **PDP'**, where **P** in this case is merely a column vector consisting of the coefficients of PC_1. Thus,

$$\hat{\mathbf{S}} = \begin{bmatrix} -.2642 \\ .1211 \\ .9568 \end{bmatrix} [17.567][-.2643\ \ .1211\ \ .9568] \begin{bmatrix} 1.227 & -.562 & -4.442 \\ & .258 & 2.036 \\ & & 16.082 \end{bmatrix}.$$

Using the test in Equation (6.5), p. 263 with $|\mathbf{R}| / |\hat{\mathbf{R}}|$ replaced with $|\mathbf{S}| / |\hat{\mathbf{S}}|$ yields

$$\chi^2 = -(5 - 3/3 - 2/3 - 11/6)\ln(.00013/16.5) = 17.63$$

with. $[(3\text{-}1)^2 - 3 - 1]/2 = 0$ degrees of freedom.

Zero degrees of freedom?

It becomes a little clearer why the hypothesis that one PC is sufficient can be so soundly rejected when we consider the predicted correlation matrix, namely,

$$\hat{\mathbf{R}} = \begin{bmatrix} 1 & -.999 & -1 \\ & 1 & .999 \\ & & 1 \end{bmatrix}$$

Unity correlations are by definition perfect, and any deviation from perfection permits rejection of the null hypothesis.

7. Based on PC_3 and PC_2:

$$\hat{\mathbf{S}} = \begin{bmatrix} -.2643 & -.3175 \\ .1211 & .9259 \\ .9569 & -.2049 \end{bmatrix} \begin{bmatrix} 17.567 & 0 \\ & 9.559 \end{bmatrix} \begin{bmatrix} -.2643 & .1211 & .9569 \\ -.3175 & .9259 & -.2049 \end{bmatrix} = \begin{bmatrix} 2.191 & -3.372 & -3.821 \\ & 8.452 & .222 \\ & & 16.482 \end{bmatrix};$$

$$\hat{\mathbf{R}} = \begin{bmatrix} 1 & -.783 & -.635 \\ & 1 & .019 \\ & & 1 \end{bmatrix};$$

$$\chi^2 = -1.5\ln(|\hat{\mathbf{R}}|/|\mathbf{R}|) = 5.87 \text{ with } (9+3)/2 = 6\ df, \text{ ns.}$$

8. Perfect reproduction, $\chi^2 = 0$.

9. The very small sample size.

10. In terms of z scores, $\hat{z}_y = .87053\, z_{pc_1} + .44984\, z_{pc_2} + .12319\, z_{pc_3}$;

$$R^2 = \sum r^2_{y,pc_i} = .97536, \text{ whence } R = .9876.$$

This matches perfectly (well, almost perfectly) the results from the MRA using the original scores. We can get back to the regression equation based on original scores by substituting into the regression equation the definition of z_{pc_i} in terms of the original scores.

11.

$$\mathbf{S} = \begin{bmatrix} 17.567 & 0 \\ & 9.560 \end{bmatrix};\ \mathbf{b} = \begin{bmatrix} .210 \\ .147 \end{bmatrix};\ \text{and } \mathbf{b}_z = \begin{bmatrix} .871 \\ .450 \end{bmatrix}.$$

Thus $R^2 = .961$, whence $R = .980$.

12. $R^2 = .758$; $R = .871$.

13. The three pairwise plots are shown in Fig. 6.6.

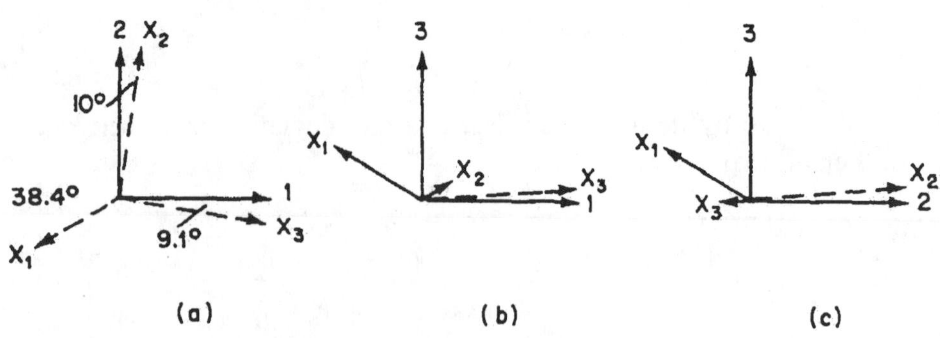

Fig. 6.6 Rotation of factor structure from Problem 1

Try $\theta_{12} = -6.6^\circ$, $\theta_{13} = +2.8^\circ$, and $\theta_{23} = +1.8^\circ$. Thus,

$$\mathbf{T}_{12} = \begin{bmatrix} .993 & .115 & 0 \\ -.115 & .993 & 0 \\ 0 & 0 & 1 \end{bmatrix}; \mathbf{T}_{13} = \begin{bmatrix} .999 & 0 & -.049 \\ 0 & 1 & 0 \\ .049 & 0 & .999 \end{bmatrix}; \mathbf{T}_{23} = \begin{bmatrix} 1 & 0 & 0 \\ 0 & 1.0 & -.031 \\ 0 & .031 & 1.0 \end{bmatrix};$$

$$\mathbf{T} = \mathbf{T}_{12} \cdot \mathbf{T}_{13} \cdot \mathbf{T}_{23} = \begin{bmatrix} .992 & .113 & -.053 \\ -.115 & .993 & -.025 \\ .049 & .031 & .999 \end{bmatrix};$$

whence the rotated structure is

$$\mathbf{FT} = \begin{bmatrix} -.606 & -.685 & .404 \\ .063 & .997 & .041 \\ .999 & -.042 & -.017 \end{bmatrix}.$$

The varimax criterion (sum of variances of squared loadings within each column) is .8938 for this rotated structure, as compared with .9049 before rotation!

14. Because each variable's communality is 1.000, Kaiser normalization is superfluous (because it requires that we divide the loadings within each row by the square root of the communality of that variable). The calculations for the first large iteration went as follows:

For PC_1 and PC_2:

$d = L_1^2 - L_2^2$	$c = L_1 L_2$
.10576	.43532
−.93404	.17087
.94983	−.15397

$$\tan 4\theta_{12} = \frac{4[3(-.26141) - (.12065)(.45222)]}{3(1.78580) - (.12065)^2 - 4(3)(.24241) + 4(.4522)^2}$$
$$= -3.355/3.252 = -1.03173,$$

whence

$$\theta_{12} = -11.475°,$$

whence

$$\mathbf{T}_{12} = \begin{bmatrix} .98 & .199 & 0 \\ -.199 & .98 & 0 \\ 0 & 0 & 1 \end{bmatrix},$$

and the new structure is

H_1	H_2	PC_3
−.5635	−.7480	.352
−.0248	.9970	.075
.9983	.0434	.031

For H_1 versus PC_3:

$d = L_1^2 - L_2^2$	$c = L_1 L_2$
.19363	−.19835
−.00501	−.00186
.99564	.03095

$$\tan 4\theta_{1*3} = \frac{4[3(-.00758) - (1.18426)(-.16926)]}{3(1.02882) - (1.18426)^2 - 12(.04030) + 4(.16926)^2}$$

$$= -.71084/1.31499 = .54057,$$

whence

$$\theta_{1*3} = 7°5.91',$$

whence

$$\mathbf{T} = \begin{bmatrix} .99233 & 0 & -.12358 \\ 0 & 1 & 0 \\ .12358 & 0 & .99233 \end{bmatrix},$$

and the new structure is

H'_1	H_2	H_3
−.5157	−.7480	.4189
−.0153	.9970	.0775
.9945	.0434	−.0926

For H_2 versus H_3:

$$\tan 4\theta_{2*3*} = -.78368/.48735 = 1.60804; \quad \theta_{2*3*} = 14°31.9';$$

$$\mathbf{T} = \begin{bmatrix} 1 & 0 & 0 \\ 0 & .96801 & -.24091 \\ 0 & .25091 & .96801 \end{bmatrix},$$

and the new structure is

H'_1	H'_2	H'_3
−.516	−.619	.593
−.015	.985	−.175
.995	.019	−.101

After this first large iteration, the varimax criterion is now 1.078 (versus .905 initially). Because it ultimately converges to 1.083 (we cheated by looking at the computer's work), it hardly seems worth continuing the process by hand. The final varimax-rotated structure is

H''_1	H''_2	H''_3
−.582	−.460	.671
−.025	.974	−.223
.980	.002	−.197

15. Varimax rotation of PC_1 and PC_2 was the first step of problem 14. This produces a varimax criterion of .916, as compared with .905 for the original structure and .875 for the intuitive solution. However, this assumes either that we do not wish to employ Kaiser normalization, or that we wish to apply the varimax criterion to the entire factor structure (retain all three PCs), even though we are rotating only two of the PCs. If we retain only the two PCs and do use Kaiser normalization, we obtain

$$\tan(4\theta) = -2.0532/1.0880 = 1.887; \quad \theta = 15°31';$$

and the new structure is

H_1	H_2
−.509	−.786
−.095	.993
.993	.114

16. Regression approach:

$$\mathbf{B}_h = \mathbf{R}^{-1}\mathbf{F} = \begin{bmatrix} 5.635 & 3.911 & 3.209 \\ & 3.715 & 2.206 \\ & & 2.828 \end{bmatrix} \begin{bmatrix} -.606 & -.685 & .404 \\ .063 & .997 & .041 \\ .999 & -.042 & -.017 \end{bmatrix} = \begin{bmatrix} .041 & -.095 & 2.38 \\ .068 & .932 & 1.69 \\ 1.019 & -.118 & 1.34 \end{bmatrix}$$

"Direct" approach:

$$\mathbf{Z}_h = \mathbf{Z}_{PC} \cdot \mathbf{T} = \mathbf{Z}_x \cdot \begin{bmatrix} -.09971 & -.16237 & 2.35550 \\ .08424 & .87311 & 1.70691 \\ .92729 & -.26920 & 1.37015 \end{bmatrix} \begin{bmatrix} .992 & .113 & -.053 \\ -.115 & .993 & -.025 \\ .049 & .031 & .999 \end{bmatrix} = \mathbf{Z}_x \cdot \begin{bmatrix} .035 & -.099 & 2.35 \\ .067 & .929 & 1.68 \\ 1.018 & -.120 & 1.37 \end{bmatrix}$$

The variance of each z_{h_i} is 1.0. The variance of raw scores on the hypothetical variables (rotated PCs) is impossible to determine.

17.

$$\mathbf{R}^{-1}\mathbf{F} = \mathbf{R}^{-1} \cdot \begin{bmatrix} -.509 & -.786 \\ -.095 & .993 \\ .993 & .114 \end{bmatrix} = \begin{bmatrix} -.053 & -.180 \\ -.153 & .866 \\ .965 & -.009 \end{bmatrix} = \mathbf{B}_h,$$

or

$$\mathbf{Z}_{PC} \cdot \mathbf{T} = \begin{bmatrix} -.09971 & -.16237 \\ .08424 & .87311 \\ .92729 & -.26920 \end{bmatrix} \begin{bmatrix} .96355 & .26752 \\ -.26752 & .96355 \end{bmatrix} = \begin{bmatrix} -.053 & -.183 \\ -.152 & .864 \\ .966 & -.011 \end{bmatrix}.$$

Note that the multiple regression method requires (even after $\mathbf{R}^{-1}$ has been found) a total of 18 multiplications, whereas the "direct" approach requires only 12. This discrepancy equals $p^2m - pm^2 = pm(p - m)$, where p is the number of original variables and m is the number of rotated PCs.

B. Computations based on correlation matrix

1.

	Factor pattern				Factor structure		
	PC_1	PC_2	PC_3		PC_1	PC_2	PC_3
	.705	.004	.702		.9757	.0043	.2191
	−.546	−.637	.535		−.7574	−.6311	.1674
	−.456	.770	.440		−.6326	.7622	.1374
Eigenvalue	1.927	.979	.094	ΣL^2	1.927	.979	.094
% accounted	64.2	32.6	3.1				

2. Note that the scores on the PCs are nor z scores, even though they are linear combinations of true z scores. Once again the variance of each column of scores is equal

to the eigenvalue associated with that PC.

	Score on		
Subject	PC_1	PC_2	PC_3
A	.9684	−.3480	−.0660
B	.1861	−1.3900	−.1760
C	−.5190	1.1930	−.3840
D	1.4510	.6612	.3399
E	−2.0900	−.1210	.2856

3. The correlations between PCs are still zero. The correlations between original scores and PCs are the entries in the ***R***-derived factor structure.

4, The obvious set of hypothesized PCs is

	PC_1	PC_2	PC_3
	2/6	0	1/3
$\mathbf{P}_0 =$	−1/6	−.707	1/3
	−1/6	.707	1/3

that is, the hypotheses that PC_1 is the difference between z_1 and z-scores on the other two original variables; that PC_2 is the difference between z_2 and z_3; and that PC_3 is the simple sum of the three original variables. These hypotheses lead to predicted variances of 1.929, .979, and .151, respectively, and to values of $\beta'_o \mathbf{S}^{-1}\beta_o$ (actually $\beta'_o \mathbf{R}^{-1}\beta_o$ in the present case, of course) of .905, 1.075, and 10.10. The end result is a chi-square value of 2.976, .204, and 2.208 for each test, all with four degrees of freedom and none statistically significant. Using $\mathbf{P}_o$ to reproduce the intercorrelation matrix yields an $\tilde{\mathbf{R}}$ of

$$\begin{bmatrix} 1 & -.592 & -.592 \\ & 1 & -.070 \\ & & 1 \end{bmatrix}$$

to which we apply our overall significance test, obtaining

$$\chi^2 = 1.5 \ln(.246/.179) = .481 \text{ with 3 } df, \text{ ns.}$$

5. $\chi^2 = 4 \ln\{.5365^2/[(.979)(.094)]\} = 4.58$ with 2 df, ns.

6. Same as Problem A6.

7. Using PC_1 and PC_2 yields

$$\hat{\mathbf{R}} = \begin{bmatrix} 1 & -.737 & -.613 \\ & 1 & -.002 \\ & & 1 \end{bmatrix};$$

whence

$$\chi^2 = -\ 1.5\ \ln(.079/.179) = 3.26 \text{ with 6 } df, \text{ ns.}$$

8. Perfect reproduction.

9. Small sample sizes and use of **R** matrix.

10. The correlations between the PCs and Y are - .9266, .2257, and .2594, respectively. These are also the z-score coefficients for predicting z_y from z scores on the PCs.

$$R^2 = .9266^2 + .2257^2 + .2594^2 = .9678;\ R = .988.$$

These values of R^2 and R match both Problem Al0 and the results of the Demonstration Problem in chapter 2. The. coefficients for the regression equation are different here, but comparability is restored when the definitions of the PCs in terms of the original variables are substituted.

11. Keeping only PC_1 and PC_2 , we simply drop the term corresponding to PC_3 from our regression equation, thereby decreasing R^2 to .901 and R to .950. This is a greater loss than from dropping the third covariance-derived PC, but less than dropping the third original variable.

12. R^2 = -850, R = .922. We get more "mileage" out of the first **R**-derived PC (in this particular example, although nor as a general rule) than out of the first **S**-derived PC or out of the first original variable.

13–15. Because the **R**-derived factor structure is the same as the covariance-derived factor structure, except for a rigid rotation, any objective rotation scheme (such as varimax) will produce the same final rotated factor structure regardless of whether we start with **R** or **S.** The same should be true of graphical-intuitive solutions (at least approximately), except in so far as one or the other starting point gives different insights into the possibilities for simple structure. As a check on your work, the results of the first large iteration in the varimax rotation scheme are given below.

$\theta_{12} = +30°10'$			$\theta_{2*3} = 12°12'$			$\theta_{1*3*} = 45°$		
PC_1^*	PC_2^*	PC_3	PC_1^*	PC_2^*	PC_3^*	PC_1^{**}	PC_2^*	PC_3^{**}
.846	−.487	.219	.846	−.429	.317	.822	−.429	−.374
−.972	−.165	.167	−.972	−.126	.199	−.547	−.126	.828
−.164	.977	.137	−.164	.984	−.072	−.167	.984	.065

16, 17. By the same reasoning as in the answer to Problems 13-15, we would expect the expressions for the relationships between original variables and *z* scores on hypothetical variables to be the same as if we had begun with the covariance matrix—we do, after all, "wind up" with the same factor structure. However, being too lazy to perform the rotations by hand, the author was led to recognize a problem that is apt to confront a user whose computer program does not compute factor scores as an option. Namely, if such a user wishes to employ the "direct approach" to computing scores on rotated PC*s,* he or she will not in general know the angles of rotation used by the program in satisfying, say, a varimax criterion, and thus will not know **T**. However, we have

$$\mathbf{F^* = FT},$$

where **F*** is the rotated factor structure; whence

$$\mathbf{F'F^* = F'FT};$$

whence

$$\mathbf{T = (F'F)^{-1}(F'F^*)},$$

so that we can compute **T** knowing only the original and rotated factor structures. This may sound like scant comfort, because it involves inverting a matrix. However, **F'F** is only p x p and is a diagonal matrix (this latter property is generally not true of **F*'F*),** so that its inverse is simply a diagonal matrix whose main diagonal entries are the reciprocals of the characteristic roots of **S** or of **R**. To illustrate this on the graphically rotated PC*s:*

$$\mathbf{F'F} = \begin{bmatrix} -1.271 & -1.397 & .374 \\ .719 & -.664 & -.037 \\ .015 & .011 & .093 \end{bmatrix}; \quad \mathbf{T} = \begin{bmatrix} -.660 & -.725 & .194 \\ .735 & -.678 & -.038 \\ .160 & .117 & .990 \end{bmatrix};$$

$$\mathbf{Z}_h = \mathbf{Z}_x \begin{bmatrix} .508 & .994 & 2.290 \\ -.393 & -.644 & 1.745 \\ -.329 & .778 & 1.435 \end{bmatrix}; \quad \mathbf{T} = \mathbf{Z}_x \begin{bmatrix} .034 & -.102 & 2.37 \\ .066 & .927 & 1.68 \\ 1.018 & -.121 & 1.33 \end{bmatrix},$$

as was obtained in Problem A16.

7
Factor Analysis: The Search for Structure

Several times in the previous chapter we mentioned the possibility of retaining only the first few PCs—those accounting for a major percentage of the variation in the original system of variables—as a more economical description of that system. The PCs having low associated eigenvalues are likely to be describing error variance, anyway, or to be representing influences (causal factors?) that affect only one or a very few of the variables in the system. However, the basic PCA model leaves no room for error variance or for nonshared (specific) variance. Models that *do* provide explicitly for a separation of shared and unique variance lead to the statistical techniques known collectively as ***factor analysis*** (FA).

Excellent texts on computational procedures and uses of factor analysis are available (cf. especially Gorsuch, 1983; Harman, 1976; and Mulaik, 1972). In addition, Hinman and Bolton (1979) provide informative abstracts of 30 years of factor analytic studies. This chapter is therefore confined to a brief survey of the various types of factor analytic solution, together with an attempt to add a bit of insight into the bases for preferring FA to PCA, or vice versa.

7.1 THE MODEL

$$z_{ij} = a_{j1}h_{i1} + a_{j2}h_{i2} + \ldots + a_{jm}h_{im} + d_j u_{ij} = \sum_{k=1}^{m} a_{jk} f_{ik} + d_j u_{ij}, \qquad (7.1)$$

where z_{ij} is subject i's z score on original variable j; h_{ik} is subject i's z score on hypothetical variable k (and *can* thus be alternatively symbolized as $f_{z,ik}$); u_{ij} is subject i's z score on the ***unique*** (nonshared) variable associated with original variable j; and a_{jk} and d_j are coefficients that must usually be estimated from the data. Written in matrix form this becomes

$$\mathbf{Z} = \begin{bmatrix} z_{11} & z_{12} & \cdots & z_{1p} \\ z_{21} & z_{22} & \cdots & z_{2p} \\ \vdots & \vdots & \ddots & \vdots \\ z_{N1} & z_{N2} & \cdots & z_{Np} \end{bmatrix} = [\mathbf{z}_1 \quad \mathbf{z}_2 \quad \ldots \quad \mathbf{z}_p] = [\mathbf{H} \mid \mathbf{U}] \left[\frac{\mathbf{P}}{\mathbf{D}} \right], \qquad (7.2)$$

where [**A** | **B**] represents a combined matrix obtained by appending the columns of the matrix **B** immediately to the right of the columns of the matrix **A**; **P** (the common-factor pattern) is an m x p matrix giving the common-factor coefficients a_{jk}; **D** is a diagonal matrix whose jth main diagonal entry is d_j; $\mathbf{H} = [\boldsymbol{h}_1\ \boldsymbol{h}_2 \ldots \boldsymbol{h}_p] = \mathbf{F}_z$ is an N x p matrix giving each subject's score on the p ***common factors***; and **U** is an N x m matrix giving each subject's score on the m unique factors. It is assumed that the unique factors are uncorrelated with each other.

Please be patient with a couple of inelegances in the above notation: the use of a_{jk} to represent the individual elements of **P** and the omission of a z subscript on **H** or the entries therein to remind us that we're talking about z scores. Because factor analysts almost always analyze correlation matrices (although the growing body of researchers who see factor analysis as one special case of the analysis of covariance structures à la Joreskög, 1978, are an exception), and because the standard deviations of the hypothetical variables in factor analysis are almost always unknowable, it seems a relatively harmless indulgence to assume (in this chapter, although we did not in chapter 6) that h_js are deviation scores expressed in standard-deviation units. Please also be aware that this notation overlaps imperfectly with that used by other authors. For instance, Gorsuch (1983) used **S** as the label for the factor structure matrix and w_{fi} and w_{if} for the elements of **P** and **B**, respectively. Be sure to check the author's definition of his or her terms before trying to compare results, equations, and so on, between texts or articles—or even between two editions of, say, this book.

At first glance, this would appear to be a much less parsimonious model than the PCA model, because it involves $p + m$ hypothetical (latent) variables, rather than just p. However, the sole purpose of the p unique factors is to permit the elimination of unique variance from the description of the relationships among the original variables, for which description only the m (usually much less than p) common factors are employed. We shall thus need expressions for the common factor portion of the model, namely,

$$\hat{\mathbf{Z}} = \begin{bmatrix} \hat{z}_{11} & \hat{z}_{12} & \cdots & \hat{z}_{1p} \\ \hat{z}_{21} & \hat{z}_{22} & \cdots & \hat{z}_{2p} \\ \vdots & \vdots & \ddots & \vdots \\ \hat{z}_{N1} & \hat{z}_{N2} & \cdots & \hat{z}_{Np} \end{bmatrix} = \mathbf{HP}; \tag{7.3}$$

where the carets ("hats") emphasize the fact that the entries of $\hat{\mathbf{Z}}$ are only approximations to or estimates of the entries in $\boldsymbol{Z}$.

In PCA, the latent variables were always uncorrelated (although we could have considered ***oblique*** rotation of the PCs). In some forms of FA, however, this is no longer true. In the case of correlated factors, we must provide the additional information contained in the ***interfactor correlation matrix***,

$$\mathbf{R}_h = \left[r_{h_i h_j} \right] = \mathbf{R}_f = \left[r_{f_i f_j} \right] \tag{7.4}$$

This information is often provided implicitly in the factor structure $\mathbf{R}_{xf} = \mathbf{PR}_f$, which gives the correlations between original variables and factors. This relationship implies that $\mathbf{R}_{xf}$ and $\mathbf{P}$ are identical in the case of orthogonal factors, in which case $\mathbf{R}_f = \mathbf{I}$. That this is true can be seen by considering the multiple regression equation for "predicting" z scores on X_i from z scores on a set of uncorrelated predictors, namely,

$$\hat{z}_i = r_{X_i h_1} z_{h_1} + r_{X_i h_2} z_{h_2} + \ldots + r_{X_i h_p} z_{h_p},$$

which, by comparison with Equation (7.2), shows the equivalence between $\mathbf{P}$ and $\mathbf{R}_{xf}$ in this case.

Note the difference between **P** in the present context and **B**, the weight (factor score coefficient) matrix discussed in chapter 6. **P** is *not,* in particular, equal to $\mathbf{B}_z$, the standard-score weight matrix used in Equation (6.12). In FA, the emphasis has traditionally been on describing the original variables in terms of the hypothetical variables (although this emphasis should be reversed during the initial phase of labeling the factors), so that **P** refers to the coefficients used to "generate" z scores on the Xs from (unknown and usually unknowable) z scores on the hypothetical variables. Weight matrices might then be referred to as X-to-f patterns, whereas, by tradition, factor patterns are f-to-X patterns.

Actually, the reproduced (estimated) scores of the subjects on the original variables are seldom obtained. Instead, the emphasis in FA is traditionally on reproducing the correlations among the original variables. Of particular relevance to this goal. are the following matrix equations:

$$\mathbf{R} = [1/(\mathbf{N}-1)]\mathbf{Z}'\mathbf{Z} = \mathbf{R}_r + \mathbf{D}'\mathbf{D} = \begin{bmatrix} h_1^2 & r_{12} & \cdots & r_{1p} \\ r_{12} & h_2^2 & \cdots & r_{2p} \\ \vdots & \vdots & \ddots & \vdots \\ r_{1p} & r_{2p} & \cdots & h_p^2 \end{bmatrix} + \begin{bmatrix} d_1^2 & & & \\ & d_2^2 & & \\ & & \ddots & \\ & & & d_p^2 \end{bmatrix},$$

and

$$\hat{\mathbf{R}}_r = \mathbf{P}\mathbf{R}_h\mathbf{P}' = \mathbf{R}_{xf}\mathbf{P}' = \mathbf{P}\mathbf{R}'_{xf} \qquad (7.5)$$

(= **PP'** for orthogonal factors).

Implicit in the above equations is that $\mathbf{R}_r$, the ***reduced correlation matrix***, is simply the matrix of correlations among the Xs with h_js the ***communalities*** of the variables, substituted for unities on the main diagonals. The communalities are sometimes thought of as shared or common portions of the variances of the z's and are sometimes simply treated as parameters needed to make the factor model come out right. (More about this in the next section.) Note that under either interpretation h_j^2 bears no necessary relationship to the square or the variance of h_j, one of the m hypothetical variables or factors—each of which, being a z score, has unit variance. Use of h_j^2 to represent communalities is so strongly traditional in factor analysis that we shall have to put up with yet another bit of notational awkwardness.

The goal of FA is to select values for the communalities h_j^2 and the pattern coefficients a_{jk} that will do the best job of reproducing the off-diagonal elements of **R**. Many of the concepts and computational techniques required in factor analysis were introduced in the chapter on PCA. The additional considerations that arise in FA are essentially three in number:

1. The problem of estimating communalities;

2. The goal of reproducing **R** as opposed to maximizing percentage of variance accounted for.

3. The difficulties of describing nonorthogonal factors.

These differences have some important consequences for the usefulness of FA as compared with PCA. For one thing, the introduction of communalities adds yet another dimension along which factor "solutions" may differ. Whereas any factorization of the complete **S** or **R** matrix is related to all other factorizations by a rigid rotation, this is true of factorizations of $\mathbf{R}_r$ only if they are based on identical communalities. For another, the relationship between original scores and factor scores is no longer a perfect one. The fact that only part of the variance of a given variable is accounted for by our system of common factors introduces uncertainty into our reproduction of r_{ij}. Factor scores must therefore be estimated, usually through multiple regression methods, with the multiple R being less than unity. However, the "direct approach" to computing scores on rotated PCs (see section 6.5.4) can be generalized to computing estimates of scores on rotated factors. Our original estimates are

$$\mathbf{F} = \mathbf{Z}\mathbf{R}_x^{-1}\,\mathbf{R}_{xf};$$

our estimates of scores on rotated factors would therefore be

$$\mathbf{H}^* = \mathbf{Z}\mathbf{R}_x^{-1}\,\mathbf{R}_{xf} = \mathbf{Z}\mathbf{R}_x^{-1}(\mathbf{R}_{fx}\mathbf{T}) = \mathbf{Z}(\mathbf{R}_x^{-1}\mathbf{R}_{fx})\mathbf{T} = \mathbf{H}\mathbf{T}\,,$$

where **T** is the transformation matrix that accomplishes the rotation of the original factor structure $\mathbf{R}_{fx}$ to the new factor structure $\mathbf{R}^*_{fx}$.

7.2 COMMUNALITIES

As pointed out in Digression 2, the rank of a matrix is closely related to the dimensionality of the system of variables involved in that matrix. If the rank of the matrix **A** is p, then p columns can be found such that any column of **A** can be written as a linear combination of these columns. The linear relationships between variances of variables and variances of linear combinations of these variables imply that if a correlation matrix is of rank m, any of the original variables can be expressed as a linear combination of m

latent variables. This implication is further supported by examination of the Laplace expansion of the determinantal equation used to compute the characteristic roots of **R.** If the rank of **R** is *m*, then there will be exactly *m* nonzero and *p* - *m* zero characteristic roots.

Now, it is extremely unusual—except in the case in which the researcher has deliberately included "derived" measures defined as linear combinations of other variables in the system—for a sample covariance or correlation matrix to have anything other than full rank *p*. However, if the main-diagonal entries are adjusted to values other than unity, it may be possible to bring the values of the determinant of $\mathbf{R}_r$—and more generally the values of all (*m* + 1) x (*m* + 1) or larger minor determinants of $\mathbf{R}_r$—very close to zero. It could then be argued—and is so argued by factor analysts—that the apparent full rank of **R** was an artifact of including in the analysis error variance and specific variance, neither of which can have anything to do with the relationships among different variables in the system. These relationships can thus be described in terms of *m* (hopefully a much smaller number than *p*) latent "generating" variables.

(Note again that $\mathbf{R}_r$ is the correlation matrix with communalities substituted for the original 1.0 entries on the main diagonal. In other words, $\mathbf{R}_r = \mathbf{R} - \mathbf{D}(1 - h_i^2)$, where $\mathbf{D}(c_i)$ is a diagonal matrix whose *i*th main diagonal entry is c_i.)

There are three general procedures for estimating communalities:

1. Through a direct theoretical solution of the conditions necessary to achieve minimum possible rank.
2. Through an iterative process in which a set of communalities is assumed, a FA of the reduced matrix is carried out and used to compute a new set of communalities that are substituted into $\mathbf{R}_r$ for a second solution, and so on, until the communalities that are assumed before a given FA is carried out and those computed on the basis of the resulting factor pattern are consistent;
3. Through various empirical approximations.

7.2.1 Theoretical Solution

To obtain a theoretical solution for the communalities needed to achieve rank *m*, algebraic symbols for the communalities are inserted in the main diagonal of **R** and all determinants of minors of order (*m* + 1) x (*m* + 1) are set equal to zero. Some of these equations will involve the unknown communalities and can be used to solve for them whereas others will involve only off-diagonal elements and will thus establish conditions on the observed correlation coefficients that must be satisfied in order that rank *m* be obtained. Actually, this latter class of conditions can always be obtained through the requirement that the various available solutions for a given communality produce the same value of h_j^2. A number of authors (e.g., Harman, 1976) have, by consideration of the number of determinants of a given order existing within a matrix of particular size, established the minimum rank that can be achieved by adjustment of the communalities, irrespective of the values of the off-diagonal elements, and the number of linearly independent conditions on these off-diagonal elements that are needed to achieve a yet lower rank. Thus, for instance, Harman (1976, p. 72) stated that "three variables can always be described in terms of one common factor," whereas 54 linearly independent conditions on the off-diagonal entries of a 12 *x* 12 correlation matrix must be satisfied in order to achieve rank 1.

To see how this works, consider the three-variable case. If all 2 x 2 minors are to vanish, then we must have

$$c_1c_2 = r_{12} = 0; \ c_1c_3 = r_{13;} \text{ and } c_2c_3 = r_{23;}$$

whence

$$c_3/c_2 = r_{13}/r_{12} \text{ and } c_2/c_1 = r_{23}/r_{13};$$

whence

$$c_1 = |r_{12}\, r_{13}/\, r_{23}|; \ c_2 = |r_{12}\, r_{23}/\, r_{13}|; \ c_3 = |r_{13}\, r_{23}/\, r_{12}|.$$

Applying these results to our PCA and MRA Demonstration Problem data, we find that cl = (.705)(.524)/.021, which equals approximately 23. However, communalities greater than 1.0 cannot exist for real data, so we must conclude that it is ***not*** possible to achieve rank = 1 for these data in a meaningful way. Instead, we can ask only that the rank be 2, for which it is necessary that

$$|\mathbf{R}_r| = c_1c_2c_3 + 2r_{12}r_{13}r_{23} - c_1r_{23}^2 - c_2r_{13}^2 - c_3r_{12}^2 = 0.$$

For these data, $r_{12}r_{13}r_{23}$ and r_{23} are both very small, so that an approximate solution (really, a family of solutions) is provided by the equation

$$c_1c_2c_3 = c_2r_{13}^2 + c_3r_{12}^2 = .497c_2 + .341c_1,$$

whence

$$c_1 = .497/c_3 + .341/c_2.$$

Thus, for instance, $c_2 = c_3 = 1$ and $c_1 = .838$ would serve as one satisfactory set of communalities, and $c_1 = c_2 = 1$ and $c_3 = .755$ or $c_1 = c_2 = c_3 = .916$ would also yield a reduced correlation matrix of rank 2. Cases in which communalities producing a desired rank *m* can be found, but only if one or more communalities are greater than unity, are called ***Heywood*** cases, and indicate either that more than *m* factors are needed to describe the system or that nonlinear relationships among the variables exist. At any rate, the conditions listed by Harman as necessary in order that a rank of 1 be attainable, namely, that

$$\text{rank 1 attainable} \Leftrightarrow r_{js}\, r_{jt}/\, r_{st} = h_j^2 = \text{constant for all } s,\ t \neq j \qquad (7.6)$$

should be supplemented by the requirement that $0 < h_j^2 < 1$. A similar set of conditions for attaining rank 2 have been developed, namely,

$$h_j^2 = \frac{r_{aj}\begin{vmatrix} r_{jb} & r_{jd} \\ r_{cb} & r_{cd} \end{vmatrix} - r_{cj}\begin{vmatrix} r_{jb} & r_{jd} \\ r_{ab} & r_{ad} \end{vmatrix}}{\begin{vmatrix} r_{ab} & r_{ad} \\ r_{cb} & r_{cd} \end{vmatrix}} = \text{constant} \qquad (7.7)$$

for all a, b, c, d different from each other and $\neq j$ (again subject to the restriction that this constant value lie between zero and unity). Actually, these conditions need not be met exactly to make it reasonable to assume a rank of 1 or 2, because these will be subject to a certain amount of error variability. If the various values of h_j^2 are fairly close to each other, they may simply be averaged together and the analysis continued. Spearman and Holzinger (1925, sic) have provided formulas for the sampling error of "tetrad differences" of the form $r_{jt}\, r_{su} - r_{ju}\, r_{st}$ which must all be zero within sampling error if Equation (7.6) is to hold. Kenny (1974) proved that this tetrad is zero if and only if the second canonical correlation between variables j and s, as one set, and variables t and u, as the other set, is zero. Each tetrad difference can thus be tested by computing and testing for statistical significance the appropriate second canonical R via the step-down gcr test of section 5.1 (although Kenny recommended the second component of the partitioned-U test), with Bonferroni-adjusted critical values to take into account the multiplicity of tests.

Ultimately, the decision as to whether a particular assumption about the rank of **R** is tenable should rest on the significance of the difference between the observed and reproduced correlations resulting from use of those communality estimates in a factor analysis.

The expressions for 4 x 4 and higher order determinants are sufficiently complex that no one has as yet worked out closed-form expressions for the conditions necessary to attain rank 3 or higher. Instead, it is recommended that minimum-residuals or maximum-likelihood procedures, which require that the number of factors be specified and then produce appropriate communalities as a by-product of the analysis, be employed where a rank higher than 2 is suspected.

7.2.2 Empirical Approximations

A very wide variety of rough-and-ready approximations to communalities have been suggested for use in those cases (e.g., principal factor analysis) where communalities must be specified before the analysis can begin. It can be shown, for instance, that

$$R^2_{j\bullet\text{oth}} \leq h_j^2 \leq r_{jJ},$$

where $R^2_{j\cdot oth}$ is the square of the multiple correlation between variable j and the m - 1 other variables, and where r_{jJ} is the reliability with which variable j is measured. Thus the squared multiple correlation can be used to set a lower bound on h_j^2 and the reliability of the variable sets an upper bound on h_j^2. This $R^2_{j\cdot oth}$ is readily obtainable as 1.0 minus the reciprocal of the jth main diagonal entry of $\mathbf{R}_x^{-1}$. However, r_{jJ} is itself a theoretical construct that can only be approximated by, for example, the correlation between two successive measurements of the same group on that variable. This test–retest reliability could be deflated by real changes in the subjects between the two measurements or it might be inflated artificially by memory effects.

Other approximations—$\max_j \sum r_{ij}^2$, $\max_{i,k}(r_{ij} r_{jk} / r_{ik})$, sum of the squared loadings on those PCs having eigenvalues greater than 1.0—have been suggested, but none has been shown to be superior to the others in generating communalities that reduce the rank of **R** as much as possible.

7.2.3 Iterative Procedure

Under this procedure, a rank m is assumed and an initial guess as to communalities is made. A principal factor analysis of the reduced correlation matrix is conducted (i.e., the characteristic roots and vectors are obtained), and the sum of the squared loadings of each variable on the first m factors is obtained. These are taken as the new communality estimates and are "plugged into" the main diagonal of **R**. A PFA is conducted on this new $\mathbf{R}_r$ and the process continued until each communality inserted into $\mathbf{R}_r$ matches the sum of the squared loadings of that variable on the first m factors derived from a PFA of $\mathbf{R}_r$. Wrigley (1958) found that this method converges for all the matrices he tried, with the convergence being fastest when $R^2_{j\cdot oth}$ was taken as the initial estimate of h_j^2 and especially poor when unities were used as initial estimates. Gorsuch (1983, p. 107), however, reported personal experience with failures to converge—usually where the iteration procedure couldn't "decide" between two variables with nearly equal communalities. He also reviewed the still sparse empirical literature on this iterative procedure.

7.2.4 Is the Squared Multiple Correlation the True Communality?

Earlier we remarked that $R^2_{j\cdot oth}$ is a lower bound on the communality of a given variable. (See Dwyer, 1939, for a proof.) Guttman (1956) argued that $R^2_{j\cdot oth}$ should be considered the best possible estimate of communality, because it represents the maximum proportion of X_i's variance that can be predicted from its linear relationship to the other variables. [As we know, however, sample values of $R^2_{j\cdot oth}$ systematically overestimate the true population value $\rho^2_{j\cdot oth}$. Gorsuch (1983) suggests that this might compensate for $R^2_{j\cdot oth}$'s understatement of communality. The discussion in this section will bypass this estimation problem by focusing on $\rho^2_{j\cdot oth}$.] Should we, then, consider $\rho^2_{j\cdot oth}$ the "true" communality, that is, that measure that uniquely embodies the conceptual definition of communality?

The primary argument against treating $\rho^2_{j\cdot oth}$ as *the* communality (as Gorsuch, 1983, puts it), and the basis of proofs that $\rho^2_{j\cdot oth}$ is merely a lower bound, is the fact that there are many matrices for which it is possible to achieve lower rank of the reduced correlation matrix by using values of h_j^2 greater than $\rho^2_{j\cdot oth}$. This could, however, be taken as indicating that the reduced rank achieved in such cases is merely a mathematical illusion achieved by assuming a proportion of shared variance that is unattainable. Cases in which minimum rank is attainable only if one or more communalities exceed their corresponding values of $\rho^2_{j\cdot oth}$ would, in this view, be a form of a Heywood case.

Minimum rank does, however, have a concrete, operational consequence. If $\mathbf{R}_r$ has rank *m*, it is possible to reproduce $\mathbf{R}_r$ perfectly from only *m* latent variables. This holds even if some h_j^2s exceed unity, although in that case we would implicitly be using imaginary numbers having negative variances as the scores on some of our underlying factors. This might be quite realistic in substantive areas (e.g., analysis of electrical circuits) in which imaginary (complex) numbers model important properties. In most applications, however, negative variances are taken as a sign that minimum rank is *not* attainable without violating fundamental side conditions of the factor analysis model. Similarly, then, use of communalities greater than $\rho^2_{j\cdot oth}$ can be justified only if it can be shown that the resulting factors do not violate the side condition imposed by our fundamental definition of communality as the proportion of that variable's variance it shares with other variables. There are at least three arguments suggesting that X_j might indeed have a larger proportion of shared variance than its $\rho^2_{j\cdot oth}$:

1. There may be *nonlinear* relationships producing shared variance not tapped by $\rho^2_{j\cdot oth}$. This would, however, be inconsistent with the fundamental assumption (Equation 7.1) of linear relationships among factors and *X*s and thus among the *X*s themselves.

2. The regression variate (linear combination of the other *X*s) whose squared correlation with X_j is reported by $R^2_{j \cdot oth}$ maximizes that correlation only among linear combinations whose coefficients are constrained to be *unbiased* estimates of the corresponding population regression coefficients. Perhaps use of techniques such as ridge regression (see section 2.8) would allow us to achieve a proportion of shared variance greater than $R^2_{j \cdot oth}$. (How this would apply to a population correlation matrix is unclear.) This possibility does not appear to have been explored.

3. Finally, Gorsuch (1983, p. 103) states that "the population $R^2_{j \cdot oth}$ gives only the overlap with the other variables, and not the overlap the variable might have with the factors in addition to that indicated from the overlap with the other variables." The most obvious instance in which this might occur would be where one or more factors are ***singlets,*** each defined by a single original variable. However, that portion of *Xj*'s overlap with itself that is unrelated to the other *X*s must surely be considered ***specific*** variance, which is defined (e.g., Harman, 1976, or Comrey, 1973) to be a portion of that variable's uniqueness, rather than a part of its communality. More generally, any variance that *Xj* shares with a given factor but not with the other *X*s would appear to fit the conceptual definition of specificity. Further, if the factors each have a squared multiple correlation of 1.0 with the *X*s, then each is expressible as a linear combination of the *X*s, and any squared correlation between *Xj* and a linear combination of the factors can therefore be reexpressed as a squared correlation between *Xj* and a linear combination of the *X*s and thus cannot exceed $R^2_{j \cdot oth}$ except through its correlation with itself.

This last point only shows that the use of $\rho^2_{j \cdot oth}$ as our definition of communality establishes an internally consistent system. If communalities other than $\rho^2_{j \cdot oth}$ or unity are employed in analyzing a population correlation matrix, the resulting factors will have squared multiple correlations with the *X*s that are less than unity, and it might therefore be possible to construct sets of factor scores that (a) would not directly involve *Xj* itself, but (b) would yield a squared multiple correlation with *Xj* greater than $\rho^2_{j \cdot oth}$. A demonstration or proof that this is indeed possible would be very useful—although we would need to supplement such a demonstration with a justification for considering the excess over $\rho^2_{j \cdot oth}$ as part of that variable's communality, rather than part of its specificity.

7.3 FACTOR ANALYSIS PROCEDURES REQUIRING COMMUNALITY ESTIMATES

7.3.1 Principal Factor Analysis

Principal factor analysis (PFA) differs computationally from principal component analysis (PCA) only in that the main diagonal entries of **R** are replaced by communalities. (PFA can be carried out on the covariance matrix, in which case the *j*th main diagonal entry would be $h_j^2 s_j^2$. However, factor analysts almost invariably take **R**, rather than **S**, as their starting point.) The first principal factor (PF_1) accounts for as large a percentage of the *common* variance as possible, with its associated eigenvalue being equal to the sum of the squared loadings. By way of illustration, we can conduct a PFA of the PCA Demonstration Problem data (chapter 6). First, taking that set of values for the communalities for which Σh_j^2 is as large as possible while ensuring that $\mathbf{R}_r$ has rank 2, we find that

$$\mathbf{R}_r = \begin{bmatrix} .821 & -.705 & -.584 \\ & 1.000 & .021 \\ & & 1.000 \end{bmatrix};$$

whence

$$|\mathbf{R}_r| = 0, \; S_2 = 1.803, \text{ and } S_1 = 2.821;$$

whence our three eigenvalues are 1.8415, .9795, and 0.0, with corresponding eigenvectors

$$\begin{bmatrix} -.669 \\ .570 \\ .475 \end{bmatrix}, \begin{bmatrix} -.005 \\ -.639 \\ .769 \end{bmatrix}, \text{ and } \begin{bmatrix} .745 \\ .516 \\ .426 \end{bmatrix};$$

whence the factor structure and pattern are

	PF_1	PF_2	PF_3
X_1	−.906	−.006	X
X_2	.775	−.632	
X_3	.648	.761	
$\sum r^2_{X_i PF_j}$	1.842	.980	0

One might suppose that, because the same basic method (extraction of eigenvalues and eigenvectors) is used in PFA as in FA, and because we are examining an unrotated solution, PFA would show the same basic properties as PCA. In particular, we would expect the eigenvectors to be directly proportional to the columns of **B**—that is, to provide weights by which the principal factors can be defined in terms of the original variables. This is not the case, even though our particular choice of a communality for X_1 preserves the perfect multiple correlation between each of the two factors and the Xs. (The squared multiple correlation between X_1 and the other two variables is .821, so that using .821 as our communality estimate for X_1 and 1.0 for the other two preserves the deterministic relationship between each factor and the Xs.)

As the reader will wish to verify, in the present case,

$$\mathbf{B} = \mathbf{R}_x^{-1}\mathbf{R}_{xf} = \begin{bmatrix} -.0004 & .0000 \\ -.7621 & .6481 \\ -.6322 & -.7753 \end{bmatrix}$$

and

$$\mathbf{R}'_{xf}\,\mathbf{R}_x^{-1}\mathbf{R}_{xf} = \begin{bmatrix} R^2_{f1X} & 0 \\ 0 & R^2_{f2X} \end{bmatrix} = \begin{bmatrix} 1 & 0 \\ 0 & 1 \end{bmatrix}.$$

Thus, examination of the eigenvectors would tempt us to conclude that PF_1 is whatever it is that leads subjects to have a high score on z_2 and z_3 but a low score on z_1, whereas examining the factor score coefficients would tell us instead (and correctly) that a high scorer on PF_1 is someone with low scores on both z_2 and z_3.

That it is the **B**-based interpretation that is correct is indicated by the fact that the correlation of z_1 with $-.0004z_1 - .7621z_2 - .6322z_3$ (the factor definition given by the first column of $\mathbf{B}_z$) is .906 as advertised in $\mathbf{R}_{xf}$, and the correlation of z_1 with $-.669\,z_1 + .570\,z_2 + .475\,z_3$ (the eigenvector-implied "factor") is .973. (The fact that this comes out as a positive correlation, rather than negative, is rather arbitrary, because the eigenvector routine attends only to squared correlations.) Thus the eigenvectors of $\mathbf{R}_r$ cannot be interpreted as defining the factors. The loadings can, however, be obtained from the eigenvectors by multiplying each by $\sqrt{\lambda_j}$. This of course implies that the loadings, being directly proportional to the eigenvector coefficients, are just as misleading as a basis for interpreting the factors.

Note again that this discrepancy between eigenvectors (or loadings) and factor score coefficients occurs despite the fact that this is an unrotated principal factor solution.

7.3.2 Triangular (Choleski) Decomposition

From the point of view of most factor analysts, the primary value of principal factor analysis is that it provides a convenient way of obtaining an initial solution that can subsequently be rotated to obtain a more compelling or more easily interpreted set of loadings. Any other procedure that factored the initial matrix into orthogonal hypothetical variables capable of "reproducing" $\mathbf{R}_r$ would do as well as PFA's eigenvector approach for this purpose. One such alternative procedure for which computational algorithms are readily available is the method of ***triangular (or Choleski) decomposition.*** This procedure, like PFA, yields as a first factor a general factor on which all original variables have nonnegligible loadings. The second factor, however, has a zero correlation with the first original variable; the third factor has zero correlations with the first two *original* variables; and so on. In other words, the factor structure is a truncated lower triangular matrix—lower triangular because all above-diagonal loadings are precisely zero and truncated because, unless $\mathbf{R}$ rather than $\mathbf{R}_r$ is factored, there will be fewer columns than original variables. The method is also known as the ***square root*** method of factoring, because (a) the resulting "Choleski matrix" has the property that $\mathbf{C'C} = \mathbf{R}_r$ in direct analogy to $\sqrt{x}$, which has the property that $\sqrt{x} \cdot \sqrt{x} = x$; and (b) solving the scalar expressions for the triangular-solution loadings involves taking a lot of square roots. (See Harman, 1976, or Mulaik, 1972) for computational details.)

Triangular decomposition analysis (TDA) would clearly be a desirable factor analytic procedure if the researcher's theory led him or her to predict a hierarchical factor structure like the one produced by TDA. However, most researchers employing factor analysis find such a structure unacceptable for much the same reasons PFA is not popular as a final solution. Because the TDA factor pattern for a given set of communalities is obtainable via a rigid rotation of the PFA factor pattern, and because computational convenience is no longer a relevant consideration, the primary reason for continued interest in triangular decomposition is the role of Choleski decomposition in providing a solution to the generalized eigenvalue problem (cf. section 5.5.1.2).

7.3.3 Centroid Analysis

The centroid method was developed by Thurstone (1947) as a computationally simpler approximation to PFA. This simplicity derives from the fact that the *n*th centroid factor is the unweighted sum of the residual variates, that is, of the measures corrected for the effects of the first n - 1 centroid factors. However, because processing times for computerized PFAs of even rather large numbers of original variables are now measured in seconds, there is little need for centroid analysis unless the researcher finds himself or herself away from a computer. Any sufficiently ancient factor-analysis textbook (or this same section of the second edition of this book) can provide the computational details.

7.4 METHODS REQUIRING ESTIMATE OF NUMBER OF FACTORS

The primary goal of factor analysis is usually thought to be to reproduce as accurately as possible the original intercorrelation matrix from a small number of hypothetical variables to which the original variables are linearly related. The factor analytic methods discussed so far have not explicitly incorporated this goal into the factoring process; the goal is represented only in the process of selecting communalities, and even here the emphasis is on the approximate rank of the reduced correlation matrix, with no clear cut specification of how much error in reproduction of **R** is to be tolerated. Each of two as yet relatively unused factor analytic procedures establishes a measure of goodness of fit of the observed to the reproduced correlation matrix and then selects factor loadings in such a way as to optimize this measure. The minimum-residuals, or ***minres***, method seeks to minimize the sum of the squared discrepancies between observed and reproduced correlations, whereas the ***maximum-likelihood*** approach finds the factor structure that maximizes the likelihood of the observed correlation matrix occurring, each likelihood being conditional on that particular proposed structure's being the population structure.

Each of these methods has the added advantage of providing a natural base for a significance test of the adequacy of the factor solution. An additional advantage of the maximum-likelihood method is that it readily generalizes to ***confirmatory factor analysis***, that is, to situations in which the researcher specifies in advance of performing the analysis certain patterns that the final factor structure should fit. This theoretical commitment is usually expressed in the form of certain loadings that are preset to zero or to unity, with the maximum-likelihood method being used to estimate the remainder of the loadings and then to test one or more of the null hypotheses that

1. The a priori reproduced correlation matrix $\mathbf{R}_{ap}$, implied by the factor loadings obtained under the specified side conditions, is the population matrix (If this hypothesis is true, then the differences between the off-diagonal elements of **R** and of $\mathbf{R}_{ap}$ are all due to sampling fluctuations.)
2. The post hoc factors derived with no restrictions on the loadings (other than that they lie between zero and unity, inclusive) imply a correlation matrix that is in fact the population correlation matrix.
3. The a priori restrictions on the factor structure are correct in the population. Therefore, no significant improvement in goodness of fit will be observed when factor loadings are estimated after dropping these restrictions. Thus a test of H_0: "a priori restrictions hold" against H_1: "no a priori restrictions" will not show significance.

The test used for each of these hypotheses follows the same form as Equation (6.7), with the numerator determinant being the determinant of the correlation matrix reproduced under the most restrictive assumptions in each case. See the work of Mulaik (1972, pp. 381–382) for details. Confirmatory factor analysis is discussed more fully in section 7.9 below.

The disadvantages of minres and maximum-likelihood factor analysis are that (a) each requires prior commitment as to the number of factors that underlie responses on the original variables, and (b) the computational procedures for each are extremely complex. The first disadvantage is actually no more serious than the problem of knowing when to stop adding variables to a prediction equation in stepwise multiple regression, because the significance test provided by Equation (6.7) can be applied repeatedly for 1,

2, 3, … factors, stopping when hypothesis (b) fails to be rejected. The problem, as in stepwise multiple regression, is that the Type I error rate for the statistical test in Equation (6.7) applies to *each* test and not to the overall probability of unnecessarily adding a factor in a sequential process such as just described. However, Joreskög (1962, cited by Mulaik, 1972) pointed out that this latter probability is less than or equal to the significance level of each separate test. Cattell and Vogelmann (1977) reviewed the findings prior to 1977 on alternative approaches to determining the number of factors to be included in an analysis; they then provided, as their title indicates, "a comprehensive trial of the scree and KG [Kaiser–Guttman: retain factors with eigenvalues greater than 1.0] criteria."

At the time the second edition of this book was published, the computational problem was still a very serious one. Until shortly before then, only iterative procedures requiring large amounts of computer time and not carrying any guarantee of convergence to a final solution were available. However, Joreskög (1967, 1970) published procedures for maximum-likelihood factor analysis (procedures that are also straightforwardly adaptable to minres factor analysis) that are known to converge and that do this much more rapidly than any of the previously available techniques. These techniques have been incorporated into the nationally distributed LISREL program (Joreskög & Sörbom, 1982). Maximum-likelihood and minres techniques, together with the much broader class of covariance structure analyses discussed by Joreskög (1978), are now readily available. However, "readily available" does not necessarily mean "inexpensive."

Neither maximum-likelihood nor minres procedures would ordinarily be used unless at least four variables were being analyzed, because a 3 x 3 correlation matrix can always (except for Heywood cases) be perfectly reproduced by a single factor. Moreover, where a theoretical solution for communalities producing perfect fit with a specified number of factors can be obtained, there is again no need for minres or maximum-likelihood procedures. In fact, it can be shown that *a PFA performed on a reduced correlation matrix having the same communalities as those produced by a minres or maximum-likelihood solution yields a factor structure that is identical (except possibly for an orthogonal rotation) to that minres or maximum-likelihood solution.* Thus minres and or maximum-likelihood solutions are apt to be sought only when fairly large numbers of original variables are involved and therefore only in situations in which desk-calculator computation is impractical and use of a computer is essential.

In addition to LISREL, maximum-likelihood (ML) factor analysis is now available both in SAS PROC FACTOR and in SPSS FACTOR. However, only exploratory ML factor analysis is provided in either program. Even though Joreskög's work led to much more efficient algorithms than previously available for ML methods, a great deal of searching of parameter spaces is still necessary, with the probability of converging on a local, rather, than an overall, maximum ever present. Martin and McDonald (1975) estimated that Heywood cases (communalities greater than 1.0) occur about 30% to 40% of the time when using maximum-likelihood algorithms for exploratory factor analysis.

7.5 OTHER APPROACHES TO FACTOR ANALYSIS

All of the approaches mentioned so far can be considered members of the PFA "family" of procedures. TDA can be accomplished either directly or via orthogonal rotation of a PFA solution; centroid analysis is merely a computationally convenient approximation to PFA; and minres and maximum-likelihood solutions are PFA applied to reduced correlation matrices having "optimal" communalities inserted. An approach to factor analysis that is *not* simply a special case of or orthogonal rotation of a PFA, but that follows the basic model in Equations (7.1) and (7.2), is ***multiple group*** factor analysis. This method differs from the preceding ones in at least two ways: (a) It requires specification of both the number of factors and the communalities, and (b) it yields oblique factors as the initial solution. (This is also true for maximum-likelihood FA under some circumstances.)

The basic procedure is somewhat reminiscent of centroid analysis in that composite variables, each equal to the unweighted sum of some subset of the original variables, form the basis of the analysis. The researcher specifies, either on theoretical grounds or via more formal methods of cluster analysis, which variables "go together" to form a particular cluster of measures of essentially the same underlying concept. Each factor is then simply the unweighted sum of all the variables in a particular cluster, with the resultant factor structure simply being a matrix of correlations of original variables with these composite variables.

It is of course highly unlikely that these composite variables will be uncorrelated with each other. As with any other oblique solution, the researcher may wish to submit $\mathbf{R}_f$ (the matrix of correlations among the factors) to PCA in order to generate a pair of orthogonal reference axes. Because I have personally had little experience with oblique rotations, I have chosen not to discuss at any length the problem of whether the factor pattern, the factor structure, or projections of either onto orthogonal reference vectors are the most appropriate bases for intepreting the factorial composition of the original variables. Given the discussion in this chapter and chapter 6 of the loadings versus weights issue in interpreting factors, the reader can guess my probable bias toward interpreting pattern coefficients. Gorsuch (1983) recommended that both the pattern and the structure be considered, and he suggested that reference vectors and primary vectors played a primarily historical role as a means of easing the computational labor of oblique rotations. Good introductions to these issues are provided by Gorsuch (1983), Jennrich and Sampson (1966), and Harris and Knoell (1948).

A wide variety of other approaches to the analysis of within-set variability have been taken that do not fit Equations (7.1) and (7.2). Among the more important of these are ***cluster analysis*** (Tryon & Bailey, 1970) and ***image analysis*** (Guttman, 1953). Cluster analysis, instead of assuming the same linear relationship between observed and hypothetical variables for all subjects, seeks out different profiles of scores on the various measures that are typical of different clusters of individuals. This is often closer to the researcher's actual aims than is the continuous ordering of all subjects on the same underlying dimensions. ("Subjects" and "individuals" of course refer here to whatever sampling units the researcher employs.) Image analysis differs from PFA and related approaches primarily in that it permits the possibility of correlations among the various unique factors, restricting the common factors to explanation of that portion of the variance of each variable that is directly predictable via linear relationships from knowledge of the subjects' scores on the other measures, that is, to the variable's ***image*** as "reflected" in the other variables in the set. This amounts to taking the population value of $R^2_{j\cdot oth}$ (the squared multiple correlation between variable j and the other variables) as the *exact* value of variable j's communality, rather than as a lower bound on h_j^2.

Mulaik (1972, chapter 8) and Pruzek (1973), in a more complete form, pointed out several interesting relationships among image analysis, maximum-likelihood factor analysis, and ***weighted component analysis.*** [In weighted component analysis, each original variable is rescaled by multiplying it by the best available estimate of the square root of its error variance before applying PCA to the matrix of variances and covariances of the resealed data.] Equivalently, we may apply PCA to the matrix $\mathbf{E}^{-1}\mathbf{R}\mathbf{E}^{-1}$, where $\mathbf{E}$ is a diagonal matrix whose *i*th diagonal entry is the error variance of variable *i.* Image analysis is interpretable as weighted component analysis in which 1 - $R^2_{j\cdot oth}$ is taken as the estimate of the error variance of variable *j.* Maximum-likelihood FA is also interpretable as a special case of weighted component analysis. Another very important point made by Mulaik and Pruzek is that image analysis, unlike other approaches to FA, yields completely *determinate* factor scores, that is, factors each of which have a multiple correlation of unity with scores on the original variables.

The most promising development in recent years is, however, the increasing availability of and use of confirmatory (as opposed to purely exploratory) factor analysis. Confirmatory factor analysis is discussed more fully in section 7.8 below.

7.6 FACTOR LOADINGS VERSUS FACTOR SCORING COEFFICIENTS

By far the most common procedure for interpreting (naming) the factors resulting from a principal components analysis or a factor analysis is to single out for each factor those variables having the highest loadings (in absolute value) on that factor. The highly positive loadings then help to define one end of the underlying dimension, and the highly negative loadings (if any) define the opposite end. I believe that this practice should be supplanted in most cases by examination instead of the linear equation that relates subjects' scores on that factor to their scores on the original variables. For principal component analysis these two procedures yield identical results, because the PCA factor pattern can be read either rowwise to obtain the definitions of the variables in terms of the principal components or columnwise to define the principal components in terms of scores on the original variables. As soon, however, as communalities are introduced or the pattern is rotated, this equivalence between the variable-factor correlations (that is, the loadings) and the factor score coefficients (the coefficients in the linear equation relating factors to scores on original variables) is lost.

7.6.1 Factor Score Indeterminacy

For purposes of the present discussion, assume that the factor score coefficients are obtained via multiple regression techniques, which leads to a multiple R of 1.0 in the case of unity communalities but less than unity as the communalities depart from unity. This makes clear the close connection between the problem of interpreting or naming factors and the problem of interpreting a multiple regression equation. Moreover, if communalities other than unities are employed, the multiple correlation between scores on a given factor and the *X*s drops below 1.0, so that factor scores must be estimated, rather than computed from the data in hand. (As a number of authors have pointed out, this situation can be interpreted instead as involving perfectly determinate factor scores—but an infinitely large choice among equivalent sets of such scores, each having the same ability to reproduce the correlations in $\mathbf{R}_r$)

The classic papers on factor score indeterminacy are those of Guttman (1955, 1956), in which he pointed out that if the multiple correlation between a factor and the original variables drops as low as .707 (= $\sqrt{.5}$), it is possible to find two sets of factor scores that are uncorrelated with each other but that are both consistent with the factor loadings! He further shows that PFA leads to determinate factor scores if $h_j^2 = R^2_{j \cdot oth}$ -- which is of course the basic assumption of image analysis—and a very large number of variables are factored.

There has been renewed interest in the indeterminacy issue over the past few years, with (as Steiger, 1979a pointed out) much of the discussion being repetitive of arguments of the 1950s, but also with some interesting new insights into the problem forthcoming. McDonald and Mulaik (1979) reviewed the literature as of that date. Steiger (1979a, p. 165) pointed out that there are essentially two clusters of positions on the indeterminacy issue:

1. One group of researchers "stress the fact that factor analysis (even if it could be performed on population correlation matrices) does not uniquely identify its factors—rather, it identifies a range of random variables that can all be considered factors." For instance (as Steiger, 1979b, pointed out), if the multiple correlation between a factor and the observable variables is .707, it is possible to construct a set of scores that are uncorrelated with each and every one of the variables in the set that was analyzed, and yet have that external variable correlate anywhere from -.707 to + .707 with scores on the factor.

2. Another group of researchers feel that the "factors underlying a domain of variables can be reasonably well-identified [primarily through regression estimates], and factor indeterminacy should be viewed as a lack of precision which stems from sampling a limited number of variables from the domain of interest" (Steiger, 1979a, p. 166).

To these two positions we should of course add a third position:

3. Researchers who feel that the indeterminacy problem should be avoided by employing techniques—usually PCA (see next section) or image analysis—that provide factors that can be computed deterministically from scores on the original variables.

7.6.2 Relative Validities of Loadings-Derived Versus Scoring-Coefficient-Derived Factor Interpretations

Much of this literature has ignored a second, and in my opinion more crucial, source of indeterminacy—the strong tradition of interpreting factors in terms of loadings, rather than in terms of factor score coefficients. As the Factor Fable of section 6.5.5 demonstrated, even when scores on the underlying factors are deterministic linear combinations of the original variables, and all communalities are 1.0 (as in rotated components analysis), interpretation of the factor in terms of the loadings may imply a variable that rank orders the subjects (or other sampling units) very differently from the way in which the factor itself does. In section 7.3.1 we saw, in the context of a principal factor analysis with determinate factor scores, that the linear combination of the original variables implied by interpreting the loadings on PF_1 was very different from PF_1 identified (correctly) by the entries in the first column of **B**. The loadings-derived linear combination did not in fact have the correlation with X_1 that X_1's loading on PF_1 said it should have. The **B**-defined PF_1 did have the correct correlation with X_1.

The reader who has gotten to this point "honestly" (i.e., by suffering through all previous chapters) has had the advantage of seeing this issue recur in each of the techniques involving choosing optimal linear combinations of sets of variables. Interpreting a regression variate on the basis of the loadings of the predictors on it turns out to be equivalent to simply examining the zero-order correlation of each predictor with Y, and thus throws away any information about correlations among the predictors. Interpreting a discriminant function in T^2 or in Manova with a single-degree-of-freedom effect in terms of loadings of the dependent variables on the discriminant function is equivalent to simply examining the univariate Fs. Thus, we might as well omit the Manova altogether if we're going to interpret loadings rather than discriminant function coefficients. We have repeatedly seen the very misleading results of interpreting loadings rather than score coefficients in analyses in which we can directly observe our variables; it therefore seems unlikely that loadings will suddenly become more informative than weights when we go to latent variables (factors) that are not directly observable.

Fortunately, we needn't limit ourselves to speculation about the generalizability of the superiority of scoring coefficients over structure coefficients to the case of factor analysis. James Grice, in his 1995 University of New Mexico NM dissertation (subsequently published as Grice and Harris, 1998), conducted a Monte Carlo study in which he drew samples of varying sizes from a population of 76,000 observation vectors that yielded nine population correlation matrices. These nine matrices (drawn from seven published sample matrices and two artificially constructed matrices) yielded varimax-rotated factor structures that were of low, medium, or high factorial complexity (roughly, the average number of factors on which a given variable has substantial loadings), with the loadings within the cluster of variables having substantial loadings on a given factor displaying either low, medium, or high variability.

On the basis of the identity (to within a columnwise scaling factor) of loadings and scoring coefficients in the unrotated-PC case, together with the earlier finding (Wackwitz and Horn, 1971) of very small differences between loading-based and regression-based factor-score estimates in PFAs based on samples from a structure matrix having perfect simple structure with identical loadings within clusters, it was hypothesized that regression-based factor scores (whether based on the exact scoring coeffients or on 0,1,-1 simplifications of the scoring-coefficient vectors) would outperform loadings-based factor scores (whether employing exact loadings or unit-weighted "salient loadings" as weights) when factorial complexity and/or loadings variability were moderate or high, but that this difference would be smaller and perhaps even slightly reversed for the low-factorial-complexity, low-loadings-variability condition.

The particular properties examined were ***validity*** (the correlation between estimates of scores on a given factor and the true population scores on that factor for the cases

Table 7.1 Mean Validity, Univocality, and Orthogonality of Regression and Loading Estimates for Three Levels of Complexity

Factorial Complexity	Factor Scores: True	Regression-Based	Loadings-Based	Difference	Difference Between Squares
		Validity			
Low	1.0	.871	.836	.035*	6.09%
Medium	1.0	.781	.718	.062*	9.76
High	1.0	.846	.724	.121*	19.26
		Univocality			
Low	.070	.112	.255	-.143*	-8.45%
Medium	.069	.142	.365	-.223*	-11.54
High	.069	.119	.312	-.193*	-8.43
		Orthogonality			
Low	.070	.090	.390	-.300*	-26.29%
Medium	.069	.153	.719	-.566*	-50.52
High	.069	.132	.664	-.532*	-44.16

Note. Validity = correlation with true scores; univocality = correlation with true scores on factors other than the one being estimated); and orthogonality = average correlation between estimates of pairs of factors. Entries in "True" column represent validity, univocality, and orthogonality in the population of 76,000 cases. From Grice (1995), with permision.
*p < .007.

sampled), ***univocality*** (the tendency for estimates of scores on a given factor to correlate highly with true scores on *other* factors), and ***orthogonality*** (the average correlation between pairs of estimated factors). As shown in Table 7.1, factor-score estimates based on factor-scoring (regession) coefficients were substantially superior to those based on the loadings on all three measures, even for the three low-factorial-complexity population matrices.

Thus, sets of scores estimated on the basis of the loadings correlate more poorly with true scores on the factors they're estimating, correlate more highly with factors they're supposed to correlate zero with, and lead to estimated factors whose intercorrelations are higher (even though they're estimating orthogonal factors) than do regression-coefficient-based estimates. The difference is most dramatic with respect to orthogonality: Loadings-based factor estimates that share 30% to 57% of their variance versus regression-based estimates' average (for any given level of factorial complexity) of less than 2.5% shared variance. One can easily envision a researcher developing a salient-loadings factor-estimation procedure based on one sample of data, assuming that this procedure yields uncorrelated factor estimates for that original population, applying this scoring algorithm to a sample of data gathered under different experimental conditions, and then taking the .5 or above average correlation between the resulting estimates as proving that the original factor solution no longer applies and that an oblique solution is necessary. But the superiority of regression-based estimators with respect to univocality and validity is also substantial, especially at moderate to high levels of factorial complexity.

Of course, as pointed out in chapter 6, any substantive interpretation of a factor that is based on the pattern of the loadings will imply a relative spacing of individual scores on that factor that is very nearly the same as the spacing of the loadings-based factor-score estimates, so that loadings-based interpretations suffer from the same deficiencies, relative to regression-based interpretations, that the corresponding sets of factor-score estimates do. Researchers who follow the fairly common practice of interpreting (and labeling) their factors on the basis of the loadings, but scoring them on the basis of the factor-scoring coefficients, thus put themselves in the same position as the zookeepers of the Canonical Cautionary (Example 5.3) who match their birds on the basis of beak length but describe what they're doing as matching on the basis of total head length.

7.6.3 Regression-based Interpretation of Factors is Still a Hard Sell

The reader should keep clearly in mind, however, that the great majority of factor analysts (and journal referees and editors) would assume as a matter of course that it is the loadings, rather than the factor score coefficients, that define a given factor. A referee of this book has assured me that there are "unpublished papers by Kaiser and McDonald" that thoroughly demolish the use of factor score coefficients to interpret factors. However, my own impression is closer to that of Kevin Bird (personal communication, 1983), who suggested that factor score coefficients have "simply never been taken seriously in the factor analysis literature as a basis for interpretation."

Some authors have come tantalizingly close. For instance, proponents of the view that the nonuniqueness of factors poses serious dilemmas for FA often point out how important it is to be able to uniquely rank order subjects in terms of their scores on the factor if we are to infer what it is tapping—but they then proceed without comment to base their interpretations on the loadings, thereby developing hypothetical variables that may rank order the subjects very differently than does the factor they're seeking to interpret. Similarly, Gorsuch (1983) argues for the value of considering the factor pattern (i.e., the coefficients that would allow us to compute scores on original variables from scores on the factors) in interpreting the factorial composition of the original variables, but he didn't see the parallel argument for interpreting factors in terms of the way we compute them from scores on the variables. He also pointed out (in his section 12.2.1) that taking the pattern coefficients (which are, in the orthogonal-factors case, equal to the loadings) as weights for computing or estimating factor scores is an "indefensible approximation"—but he failed to see that this might apply to interpretation, as well as computation, of the factor scores.

We can probably take comfort in the likelihood that factors and loadings-based interpretations of them only rarely rank order the cases in dramatically different ways. Recall, for instance, that the Factor Fable of section 6.5.5 yielded a loadings-based interpretation of height that correlated with actual height at about a .8 level. Similarly, PF_1 and the loadings-based interpretation of it (section 7.3.1) correlate -.707. This may be an adequate level of similarity for substantive areas in which a .5 correlation may be cause for celebration.

One implication of turning to an emphasis on factor score coefficients is that if our goal is to describe the nature of the variables underlying the covariance among our observed variable, then our criterion for rotation should be, not the simplicity of the factor *structure*, but the simplicity of $\mathbf{B}_z$ the p x m matrix of regression coefficients predicting z scores on the factors from z scores on the original variables. One can use Equation (6.12) or the equivalent factor-analytic formula of section 7.1 to carry out hand rotation to an intuitively satisfying, simple pattern of factor score coefficients. If input of the factor-loading matrix is available as an option in the computer program you're using (as it is in SPSS), you can "trick" the program into applying the varimax or some other criterion to $\mathbf{B}_z$, by inputting that matrix to the program. However, it is questionable whether the varimax criterion is nearly as compelling a measure of simplicity when applied to $\mathbf{B}_z$ as it is when applied to $\mathbf{F}_z$.

The problem is that the z-score regression weights for predicting a given factor from z scores on the original measures tend to be much larger for factors with low associated eigenvalues. Because it is only the relative magnitudes of the coefficients defining a given factor that matter in interpreting that factor, we would ideally like to normalize the coefficients within each column of $\mathbf{B}_z$ to unit length before applying the varimax criterion. Kaiser normalization, however, normalizes the rows of the matrix submitted to the rotation algorithm. It would seem to be fairly simple to make this change so as to implement rotation to simple factor score cofficients as an option in all factor analysis programs—but I know of no program that offers this option.

One could argue reasonably for retaining simplicity of the loadings as the important criterion in rotation. The loadings are (for orthogonal factors) the regression coefficients relating the original measures (deterministically) to the underlying factors. If your ultimate goal (once the factors have been interpreted) is to have as simple a model as possible for the variables in terms of the factors, it would be reasonable to rotate to simple structure as usual. However, *interpretation* of the factors must still be based on the factor-score coefficients, not on the loadings.

Of course, the whole issue of interpretation may be finessed if selection of variables is made on the basis of their a priori specified relations to underlying conceptual variables, especially if the techniques of confirmatory factor analysis are used to test the adequacy of your a priori model. Harris (1999a), however, pointed out that if the conceptual variables were initially derived from loadings-based interpretation of factors yielded by an exploratory factor analysis, the researcher may mistakenly take confirmation of the factor structure he or she "fed into" the CFA program as also confirming the (mistaken) interpretations of the conceptual variables, thus perpetuating the original mistake.

7.7 RELATIVE MERITS OF PRINCIPAL COMPONENT ANALYSIS VERSUS FACTOR ANALYSIS

7.7.1 Similarity of Factor Scoring Coefficients

Before launching into a general discussion, let us return to our PFA of the PCA demonstration problem (chapter 6) data for some examples. We already have the **R**-derived PCA from this demonstration problem and the factor structure from a PFA based on communalities $h_1^2 = .821$, $h_2^2 = h_3^2 = 1.0$. On first attempting a PFA based on equal communalities, it was discovered that solution of the theoretical expression for the rank 2 communalities in that case could *not* ignore the $2r_{12}r_{13}r_{23}$ and r_{23}^2 terms. Solving the complete cubic equation, $c^3 + .01738 - .83840c = 0$ (cf. Digression 3), led to the value $c = .905$ (rather than the .916 previously estimated). The factor structure, the Kaiser-normalized factor structure (which is what is submitted to rotation in most applications of varimax rotation), and the z_x-to-z_f factor pattern for each of these three analyses are compared in Tables 7.2 through 7.4. Note that all three factor structures are highly similar, with the major effect of lowering variable 1's communality (PFA) having been to reduce its loadings on the hypothetical variables, with a slight additional tendency to raise the loadings of variables 2 and 3. Both of these effects can be seen as arising from the requirement that the sum of the squared loadings in a given row must equal the communality for that variable.

The similarity in the factor structures for the PCA and PFA_2 is even greater. Each column of the PFA_2 factor structure is directly proportionate to the corresponding column of PCA, with the "reduction factor" being .975 (= $\sqrt{1-.095/1.926}$) and .950 (which equals $\sqrt{1-.095/.979}$) for loadings on factor 1 and factor 2, respectively. Thus the pattern of the loadings is identical for the two analyses, and we gain nothing by our insertion of communalities, except perhaps a feeling of having virtuously "chastised" ourselves for having error variance in our data. As the equivalences in parentheses in the sentence before last suggest, this close match between PCA and equal-communalities PFA is no freak coincidence. In fact, *for any equal-communalities PFA, the* jth *column of the factor structure is equal to the* jth *column of the unity-communalities (R-based PCA) factor structure multiplied by* $\sqrt{1-(1-c)/\lambda_j}$, *where* c *is the common communality and* λ_j *is the eigenvalue associated with the* jth *PC.* This is proved in Derivation 7.1 (Appendix C).

Table 7.2 Comparison of Factor Structures for PCA Versus Two PFAs of Same Data

	PCA (PC_1 and PC_2)			PFA_1			PFA_2		
	PC_1	PC_2	h_j^2	f_1	f_2	h_j^2	f_1	f_2	h_j^2
z_1	.976	.004	.952	−.906	.004	.821	.951	.004	.905
z_2	−.757	−.631	.972	.775	−.632	1.000	−.738	−.600	.905
z_3	−.633	.762	.981	.648	.762	1.000	−.617	.724	.905
Σr_{xf}^2	1.926	.979	2.905	1.842	.979	2.821	1.831	.884	2.715

Next, note that Kaiser normalization makes the structures even more similar, tending to "undo" the initial effects of inserting communalities. In fact, the Kaiser-normalized factor structure for the equal-communalities PFA is ***identical*** to the Kaiser-normalized factor structure for the **R**-derived PCA of the same data! Thus what is currently the most popular procedure for analytic rotation, the procedure that Kaiser claims satisfies the long-sought goal of factor invariance, essentially amounts to ignoring any differences among the variables in communality. Of course, we "shrink" the loadings back to nonunity communalities after completing the rotation—but that is like saying that we are to append a note reminding ourselves that not all of the variance for a given variable is shared variance. A referee has pointed out that the equal-communalities model crudely discussed earlier was introduced by Whittle (1952, as cited in Pruzek and Rabinowitz, 1981) and analyzed extensively by Pruzek and Rabinowitz (1981). See this latter paper especially if you're interested in equal-communalities analyses beyond their use here in discussing PCA versus FA.

Table 7.3 Comparison of Kaiser-Normalized Factor Structures

	PCA			PFA$_1$			PFA$_2$		
	PC_1	PC_2	h_j^2	f_1	f_2	h_j^2	f_1	f_2	h_j^2
z_1	1.000	.004	1.000	−1.000	.005	1.000	1.000	.004	1.000
z_2	−.768	−.640	1.000	.775	−.632	1.000	−.768	−.640	1.000
z_3	−.639	.769	1.000	.648	.762	1.000	−.639	.769	1.000
Σr_{xf}^2	1.998	1.001	3.000	2.021	.980	3.000	1.998	1.001	3.000

Table 7.4 compares the relationship between factor scores and original variables for the three analyses. As was pointed out in section 7.1, the major cost of PFA as compared to PCA (ignoring the difference in computational effort) is the resulting loss of contact between original variables and hypothetical variables, as reflected in the drop in $R^2_{i \cdot oth}$,

Table 7.4 Comparison of Factor-Score Coefficients

	PCA		PFA$_1$		PFA$_2$	
	PC_1	PC_2	f_1	f_2	f_1	f_2
z_1	.705	.004	.000	.000	.494	.004
z_2	−.546	−.637	.762.	.648	−.383	−.612
z_3	−.456	.770	.632	.776	−.320	.740
$R^2_{f \cdot x}$	1.000	1.000	1.000	1.000	.948	.903

(the squared multiple correlation between each factor and the original variables) from unity to lower values when nonunity communalities are inserted. As Table 7.4 makes clear, the magnitude of this drop is generally considerably smaller than the "average" percentage reduction in communality. In fact, for the present data, lowering the communality of variable 1 to .8209 had *no* discernible effect on $R^2_{j \cdot oth}$. (This is due to the fact that the communality selected for variable 1 is identically equal to its squared multiple R with the other two variables, which Guttman [1956] showed to be a sufficient condition for determinate factor scores.) In the equal-communalities case, the reduction in $R^2_{j \cdot oth}$ is directly proportional to the decrease (from unity) in communality and inversely proportional to the magnitude of the characteristic root associated with that particular factor, that is, $R^2_{j \cdot oth} = 1 - (1 - c)/\lambda j$, where c is the common communality and

λ_j is the jth characteristic root of **R**, the "full" correlation matrix. Thus, for those PCs that are retained (usually those having eigenvalues greater than unity), the reduction in $R^2_{j \bullet oth}$ is less than the reduction in h_j^2. It is, moreover, concentrated in the less important factors, thus being consistent with our intuitive feeling that the low-variance PCs are more likely to be describing nonshared variance than are the first few PCs. (Note that we did not say "*confirming* our intuition" because it is far from clear in what sense inserting communalities "removes" nonshared variance.) This relatively small effect on $R^2_{j \bullet oth}$ of inserting communalities is typical of PFA. For instance, Harman (1976, p. 355) gave an example of a PFA of an eight-variable correlation matrix in which two factors were retained and in which the mean communality was .671, but $R^2_{f_1 \bullet x} = .963$ and $R^2_{f_2 \bullet x} = .885$. Harman also, incidentally, provided a general formula for the magnitude of the squared multiple correlation between factor j and the original variables, namely,

$$R^2_{f_j \bullet x} = 1 - \mathbf{r}'_{x \bullet f_j} \mathbf{R}^{-1}\mathbf{D}(1 - h_j^2)\mathbf{r}'_{x \bullet f_j}. \qquad (7.8)$$

Deciding how best to characterize the degree of indeterminacy in a given factor analysis has become a minor cottage industry within the statistical literature. (See, e.g., McDonald, 1974 & 1977, Mulaik, 1976, and Green, 1976.) However, *any* loss of contact between the original variables and the hypothetical variable is to be regretted if it does not gain us anything in terms of the simplicity of our description of the factor loadings or of the x-to-f factor pattern—which it certainly does *not* in the equal-communalities case.

It therefore seem compelling to the author that *PCA is greatly preferable to PFA unless strong evidence exists for differences among the original variables in communalities.*

Table 7.4 does suggest that two quite similar factor structures may yield x-to-f factor patterns that differ greatly in their interpretability. For instance,

$$\text{PC}_1 = .705z_1 - .546z_2 - .456z_3,$$

and f_1, the first factor from PFA_1, is given by

$$f_1 = .762z_2 + .632z_3.$$

In other words, f_1 does not involve x_1 at all, even though x_1 has a loading of -.906 on f_1. This reinforces the emphasis of section 7.6 on interpreting factor-score coefficients, rather than factor loadings. Note, however, that the expressions for PC_1 and f_1 are not as discrepant as they appear at first, because we have

$$z_1 \approx -.906f_1 = -.690z_2 - .573z_3,$$

whence

$$PC_1 \approx .705z_1 - .546z_2 - .456z_3 = -1.032z_2 - .860z_3 = -1.20f_1.$$

Nevertheless, the task of *interpreting* f_1 as initially expressed will probably be considerably simpler than that of interpreting PC_1 even though the subjects' scores on the two will be highly similar. In general, then, inserting communalities ***may*** lead to more readily interpretable factors if communalities are unequal.

This is, however, a considerably weaker claim for the virtues of FA than is implied by the usual statement that FA "removes" nonshared variance that PCA leaves. The altogether too tempting inference that the relationships revealed by FA are more "reliable" than those contained in a PCA of the same data, because error variance has been "removed" along with all the other nonshared variance, must of course be strongly resisted. In fact, the opposite is probably closer to the truth.

Mulaik (1972), working from classic reliability theory, showed that weighted component analysis (cf. section 7.5) in which the weighting factor for each variable is equal to its true variance (the variance of true scores on that measure, which is equal to its observed variance minus error of measurement) minimizes the proportion of error variance to be found in the first (or first two or first three ...) PCs. This procedure is equivalent to performing a PFA with reliabilities in the main diagonal of **R**. However, the uniqueness of a variable (1 minus its communality) is equal to the sum of its error variance and its specificity (the latter being its reliable, but nonshared, variance, that is, that variance attributable to factors that are not common to any of the other variables). It seems quite likely that in most situations in which FA is employed, the variables would be nearly homogeneous with respect to their reliabilities, but heterogeneous with respect to their communalities. In such cases, if accurate communality estimates are obtained, relatively more of the total error variance in the system will be retained in the first m principal factors than in the first m PCs. (Of course, factors that are as nearly error-free as possible would be obtained by inserting reliabilities, rather than either unity or communalities, in the main diagonal of **R**. However, obtaining accurate estimates of reliabilities is at least as difficult as determining communalities.) A systematic survey of the relative homogeneity of communalities and reliabilities in "typical" situations would help to resolve this issue. In the meantime, glib reference to FA's "removal" of nonshared variance is apt to be misleading.

7.7.2 Bias in Estimates of Factor Loadings

The discussion in the previous section focused on the similarity of the factors derived from FA versus PCA. Velicer and Jackson (1990) and many of the other authors in the special issue of *Multivariate Behavioral Research* in which their article appears emphasized instead the difference between PCA and FA in the absolute values of the eigenvalues and eigenvector coefficients derived therefrom—in essence, if an m-factor model holds perfectly in the population, how well does FA versus PCA perform in estimating the parameters of this true model? The general conclusion is that retaining the first m components of a PCA leads to components whose normalized variances and defining coefficients are systematically biased estimates of the population parameters of the underlying factor model. Indeed, the first m components do not constitute an internally consistent model, in that the sum of the squared loadings of each original variable on the first m components doesn't equal the communality that would have to be "plugged into" the reduced correlation matrix to yield eigenvectors that match the PCs—whence the iterative procedure that most FA programs use to estimate communalities. Thus if the focus is on estimating the parameters of a factor model, regardless of whether individual subjects' scores on those factors can be accurately determined (cf. the factor indeterminacy problem discussed earlier), FA is clearly preferable to PCA. The reader would be well advised to consult the articles in the *MBR* special issue just mentioned to counterbalance the present author's dust-bowl-empiricist bias toward PCA.

7.8 COMPUTERIZED EXPLORATORY FACTOR ANALYSIS

MATLAB. For PFA you could enter the reduced correlation or covariance matrix (i.e., the correlation or covariance matrix with communalities replacing 1.0's or variances on the main diagonal) and then use the Eig function described in section 6.1.3 to get the characteristic roots and vectors of that reduced matrix. However, the resulting eigenvalues will almost always include some negative values (corresponding to imaginary—in the mathematical sense of involving the square roots of negative numbers—factor loadings and factor variances) and will almost always yield for each original variables a sum of squared loadings that does not match, as it should, the communality that was "plugged in" for that variable in the main diagonal of the reduced matrix. To obtain an internally consistent PFA solution you would thus need to carry out the series of iterations between successive communality estimates that was described in section 7.3.1. Other factor-analytic solutions will prove even more tedious to carry out via "hands-on" matrix-algebraic manipulations, so you'll almost certainly find it worthwhile to resort instead to a "fully canned" program such as SPSS's FACTOR command.

SPSS Factor, Syntax Window.

```
FACTOR VARIABLES = Variable-list
  / METHOD = {Correlation**, Covariance}
 [/PRINT=[DEFAULT**] [INITIAL**] [EXTRACTION**] [ROTATION**]
         [UNIVARIATE] [CORRELATION] [COVARIANCE] [DET] [INV]
         [REPR] [FSCORE] [SIG] [ALL]]
 [/PLOT=[EIGEN] [ROTATION [(n1,n2)]]]
 [/DIAGONAL={value list}]
            {DEFAULT** }
 [/CRITERIA=[FACTORS(n)] [MINEIGEN({1.0**})] [ITERATE({25**})]
                                  {n    }           {n    }
            [{KAISER**}] [DEFAULT**]]
             {NOKAISER}
 [/EXTRACTION={PC**       }] [/ROTATION={VARIMAX**    }]
              {PA1**      }             {EQUAMAX      }
              {PAF        }             {QUARTIMAX    }
              {ALPHA      }             {OBLIMIN({0})}
              {IMAGE      }                      {n}
```

```
              {ULS        }              {PROMAX({4}    }
              {GLS        }                     {n}
              {ML         }              {NOROTATE      }
              {DEFAULT**}                {DEFAULT**     }
 [/SAVE=[({ALL}[rootname])]]
          {n  }
```

In this listing (which is an abbreviated version of the command syntax provided by the Help function on SPSS for PC), terms enclosed in curly brackets are alternative choices for a single entry, whereas series of terms enclosed in square brackets are alternative entries, any one or more of which may be chosen. A double asterisk indicates the default option if no specific option is selected. And "n" indicates a number that the user is to enter.

Under the Extraction subcommand, "PC" and "PA1" both stand for PCA; "PAF" stands for PFA; "ULS" stands for minres ("unweighted least squares") FA; "ML" stands for maximum-likelihood FA; and "IMAGE" stands for Kaiser's (1963) image analyes. "ALPHA" and "GLS" stand for extraction methods that have not been discussed explicitly in this chapter. Moreover, the alternatives offered the user will be somewhat different from one factor-analysis program to the next, and a given abbreviation can refer to somewhat different methods in different programs (e.g., "ALPHA" in SPSS versus in SAS's PROC FACTOR). If you are unfamiliar with some of the options offered, either ignore them or read the user manual and/or the references provided by the manual for details. However, a very large percentage of all exploratory factor analyses are either PCA or FA, usually followed by varimax rotation of the factors generated by the initial extraction.

Under the Rotation subcommand, "Promax" indicates a rotation method that begins with an orthogonal, varimax rotation and then attempts to find an oblique rotation thereof that comes closer to oblique simple structure. Promax is actually a more general technique for approximating a "target" factor pattern, but the implementation in SPSS FACTOR uses the initial varimax solution as its target pattern. (See Cureton, 1976, Gorsuch, 1970, Hakstian, 1974, Hendrickson & White, 1964, and Trendafilov, 1994, for further discussion of promax rotation.)

If working from an already-computed covariance or correlation matrix (read in via the MATRIX DATA command demonstrated in Example 6.2, section 6.3 above),

```
FACTOR /MATRIX = IN {COR = *, COV = *}
  / METHOD = {Correlation**, Covariance}
  / PRINT = …
                   etc.
```

SPSS Factor, Point-and-Click

```
Statistics
   Data Reduction . . . Factor …
```

Select the variables to be factored by highlighting them in the left window, clicking on the right arrow to move the variable names to the right window. Then click on the "Extraction" button at the bottom of the panel to choose from among the options described in the previous listing; click on the "Rotation" button to register your choice of rotation method; click the "Scores" button to bring up a menu that permits you to request a display of the scoring-coefficient matrix and/or to ask that individual cases' scores on the factors be saved to the data-editor window; click on the "Descriptives" button to request display of the individual variables' mean, standard deviation, and so on and/or various aspects of the correlation matrix; and then click "OK" to begin the analysis. Alternatively, if you would like to see, add to, or modify the SPSS commands that are generated by your point-and-click selections, click on "Paste" (rather than on "OK"), make any modifications or additions to the commands that subsequently appear in the syntax window, then choose the "Run All" option from the menu that pops up when you click the "Run" button in the taskbar at the top of the syntax window.

Example 7.1 *WISC-R Revisited.* Applying the preceding syntax listing to the WISC-R data used in chapter 6 to demonstate PCA, the following setup was used to carry out a couple of rotated orthogonal factor analyses:

```
TITLE WISC-R FA: 4-FACTOR PFA, ROTATED .
SET WIDTH = 80  Length = None .
MATRIX DATA VARIABLES =
           INFO, SIMIL, ARITHM, VOCAB, COMPR, DIGSPAN, COMPL,
   ARRANGEM, DESIGN, ASSEMBLY, CODING, MAZES
   / FORMAT = FREE FULL / CONTENTS = N CORR    .
BEGIN DATA    .
220 220 220 220 220 220 220 220 220 220 220 220
1   .62 .54 .69 .55 .36 .40 .42 .48 .40 .28 .27
.62 1   .47 .67 .59 .34 .46 .41 .50 .41 .28 .28
.54 .47 1   .52 .44 .45 .34 .30 .46 .29 .32 .27
.69 .67 .52 1   .66 .38 .43 .44 .48 .39 .32 .27
.55 .59 .44 .66 1   .26 .41 .40 .44 .37 .26 .29
.36 .34 .45 .38 .26 1   .21 .22 .31 .21 .29 .22
.40 .46 .34 .43 .41 .21 1   .40 .52 .48 .19 .34
.42 .41 .30 .44 .40 .22 .40 1   .46 .42 .25 .32
.48 .50 .46 .48 .44 .31 .52 .46 1   .60 .33 .44
.40 .41 .29 .39 .37 .21 .48 .42 .60 1   .24 .37
.28 .28 .32 .32 .26 .29 .19 .25 .33 .24 1   .21
.27 .28 .27 .27 .29 .22 .34 .32 .44 .37 .21 1
END DATA    .
FACTOR
   MATRIX  IN (CORR = *) /
   PRINT EXTRACTION  ROTATION FSCORE /
   CRITERIA = FACTORS (4) /
  EXTRACTION = PAF / Rotation = NoRotate /
```

```
ROTATION = VARIMAX /  Extraction = ML / Rotation = NoRotate /
Rotation = Varimax  .
```

The first analysis is a PFA (designated "PAF", for "principal axis factoring", in SPSS), with the factor structure/pattern/scoring coefficients printed before a subsequent varimax rotation. The second analysis employs the maximum-likelihood method for the initial extraction of orthogonal factors, followed by a varimax rotation. Both methods specify that four factors are to be extracted, regardless of the distribution of the eigenvalues. Output included the following:

```
Analysis Number 1
Extraction   1 for analysis    1, Principal Axis Factoring (PAF)

PAF   attempted to extract   4 factors.

More than   25 iterations required. Convergence =       .00186
```

[The iterations referred to here are to obtain a set of internally consistent communality estimates – see section 7.2.3.]

```
Factor Matrix:
              Factor  1      Factor  2      Factor  3      Factor  4
INFO            .75228        -.21091        -.07799        -.05229
SIMIL           .75423        -.14133        -.16052        -.02763
ARITHM          .66149        -.24296         .32961        -.19356
VOCAB           .81048        -.29996        -.19766         .09148
COMPR           .69934        -.14003        -.20812         .02511
DIGSPAN         .46722        -.17585         .32490         .03852
COMPL           .61192         .25108        -.09842        -.12148
ARRANGEM        .57971         .16308        -.07045         .09721
DESIGN          .74145         .33496         .11661        -.03060
ASSEMBLY        .61635         .39745        -.03483         .00061
CODING          .42024        -.02792         .24485         .25126
MAZES           .46071         .25917         .10599         .02870
```

[Note that this is the factor structure/pattern matrix. Further, unlike the PCA situation, the structure matrix is *not* identical to the factor-scoring-coefficient matrix, which is printed a bit later in the output.]

```
Final Statistics:

Variable  Communality  Factor   Eigenvalue    Pct of Var    Cum Pct
                     *
INFO         .61923  *    1      4.96562         41.4          41.4
SIMIL        .61536  *    2       .69174          5.8          47.1
ARITHM       .64270  *    3       .42908          3.6          50.7
VOCAB        .79429  *    4       .14055          1.2          51.9
COMPR        .55263  *
```

```
DIGSPAN       .35627   *
COMPL         .46193   *
ARRANGEM      .37707   *
DESIGN        .67647   *
ASSEMBLY      .53907   *
CODING        .30047   *
MAZES         .29148   *
```

[Note that most criteria for number of factors would have recommended retaining only the first factor. Theoretical considerations led us to retain four, representing the Spearman model of intelligence, consisting of a single general factor and a number of specific factors.]

```
Skipping  rotation   1 for extraction   1 in analysis  1

Factor Score Coefficient Matrix:

              Factor  1     Factor  2     Factor  3     Factor  4

INFO            .12418      -.15237       -.06694       -.11709
SIMIL           .13306      -.09994       -.17316       -.07096
ARITHM          .13450      -.25108        .48251       -.36468
VOCAB           .24438      -.42975       -.35633        .29214
COMPR           .09855      -.05269       -.20121        .02318
DIGSPAN         .05476      -.09078        .28569        .07285
COMPL           .08898       .19972       -.10189       -.17745
ARRANGEM        .07458       .12179       -.04170        .12378
DESIGN          .18677       .41435        .18167       -.03390
ASSEMBLY        .10981       .33534       -.04318        .00843
CODING          .04940      -.01376        .20565        .30224
MAZES           .05971       .13699        .07482        .04608
```

[Note that, e.g., the ratio between INFO's and SIMIL's loadings on PF1 is .9974, whereas the ratio between their scoring coefficients in the formula for estimating scores on PF1 is .9893.]

```
Covariance Matrix for Estimated Regression Factor Scores:

              Factor  1    Factor  2    Factor  3    Factor  4
Factor  1       .92742
Factor  2     -.01960       .63748
Factor  3     -.01537       .03113       .49112
Factor  4     -.00435      -.00923      -.05520       .22068
```

[The main diagonal entries of this matrix give the squared multiple correlation between each factor and the original variables. The fact that these are substantially below zero—and only .221 for the fourth factor – demonstrates the factor indeterminacy problem. The off-diagonal entries give the correlations among the *estimated* sets of factor scores; even though the factors being estimated are perfectly uncorrelated, their least-squares estimators are not—although the two most highly intercorrelated factor estimates share only 5.5% of their variance.]

```
VARIMAX   rotation   2 for extraction   1 in analysis  1
      - Kaiser Normalization.

VARIMAX converged in 7 iterations.

Rotated Factor Matrix:
```

	Factor 1	Factor 2	Factor 3	Factor 4
INFO	.63751	.28845	.35567	.05568
SIMIL	.65373	.34174	.26005	.05985
ARITHM	.35308	.22574	.68343	-.00002
VOCAB	.79180	.23795	.28220	.17632
COMPR	.64626	.30437	.18477	.09056
DIGSPAN	.20218	.14038	.50971	.18942
COMPL	.33242	.57733	.12752	-.04304
ARRANGEM	.34309	.46770	.10988	.16896
DESIGN	.25355	.71516	.29626	.11387
ASSEMBLY	.23048	.68527	.09287	.08793
CODING	.15173	.20872	.31628	.36585
MAZES	.11452	.48071	.17883	.12373

```
Factor Transformation Matrix:
```

	Factor 1	Factor 2	Factor 3	Factor 4
Factor 1	.65314	.60045	.43213	.16166
Factor 2	-.48399	.79399	-.36773	-.01068
Factor 3	-.58185	.00955	.77974	.23102
Factor 4	.02467	-.09464	-.26468	.95937

[This matrix gives each varimax-rotated factor as a linear combination of the unrotated principal factors.]

```
Factor Score Coefficient Matrix:
```

	Factor 1	Factor 2	Factor 3	Factor 4
INFO	.19091	-.03598	.08849	-.10609
SIMIL	.23428	.00561	-.02199	-.08551
ARITHM	-.08037	-.07948	.62320	-.21397
VOCAB	.58215	-.22553	-.09153	.24204
COMPR	.20752	.01323	-.10107	-.00775
DIGSPAN	-.08473	-.04337	.26052	.14571
COMPL	.01636	.22782	-.06747	-.18153

```
ARRANGEM        .01708      .12937     -.07783      .11987
DESIGN         -.18509      .44608      .07897      .03521
ASSEMBLY       -.06525      .33098     -.11176      .01229
CODING         -.07328     -.00791      .10676      .34560
MAZES          -.06970      .14097      .02157      .06968
```

[After varimax rotation the differences between the factor pattern/structure and the factor revelation (scoring-coefficient-matrix) are much more pronounced.]

```
Covariance Matrix for Estimated Regression Factor Scores:

              Factor  1    Factor  2    Factor  3    Factor  4
Factor  1      .75464
Factor  2      .09832       .72185
Factor  3      .14076       .07991       .57344
Factor  4      .06998       .05805       .05830       .22677
```

[Varimax rotation has led to more nearly equal factor indeterminacies: the average of the 4 squared multiple R's between the 4 factors and the 12 original variables is .5692 both before and after rotation, but the range of multiple R's is considerably lower after rotation. On the other hand, the factor-score estimates are now more highly intercorrelated, although none of the estimates of rotated factors share more than about 14% of their variance.]

```
Extraction   2 for analysis   1, Maximum Likelihood (ML)

>Warning # 11382
>One or more communality estimates greater than 1.0 have been
> encountered during iterations.  The resulting improper solution
➤ should be interpreted with caution.

ML    extracted   4 factors.   12 iterations required.
```

[As pointed out in section 7.4, this is a quite common result of an attempt to use the maximum-likelihood method. Because the final solution had a maximum communality of .9990, we're probably OK to proceed to interpret the ML solution.]

```
Test of fit of the    4-factor model:

Chi-square statistic: 6.8554,  D.F.: 24, Significance:      .9998
```

[This tells us that, with communalities estimated to maximize fit, four factors fit the observed correlation matrix to within chance variation.]

Factor Matrix:

	Factor 1	Factor 2	Factor 3	Factor 4
INFO	.54341	.53684	-.17289	.00708
SIMIL	.47376	.60276	-.13686	-.04132
ARITHM	.99948	-.00577	.00020	-.00024
VOCAB	.52408	.64845	-.30970	.02468
COMPR	.44350	.56790	-.17556	-.10118
DIGSPAN	.45143	.19398	-.02986	.31884
COMPL	.34301	.50671	.23855	-.15731
ARRANGEM	.30302	.50071	.13999	.00378
DESIGN	.46334	.55090	.39585	.04255
ASSEMBLY	.29324	.55104	.39247	-.04466
CODING	.32166	.25009	.06374	.31078
MAZES	.27210	.34620	.30182	.06409

Final Statistics:

Variable	Communality		Factor	SS Loadings	Pct of Var	Cum Pct
INFO	.61343	*	1	2.88151	24.0	24.0
SIMIL	.60821	*	2	2.72918	22.7	46.8
ARITHM	.99900	*	3	.65863	5.5	52.2
VOCAB	.79167	*	4	.24352	2.0	54.3
COMPR	.56026	*				
DIGSPAN	.34397	*				
COMPL	.45607	*				
ARRANGEM	.36214	*				
DESIGN	.67668	*				
ASSEMBLY	.54566	*				
CODING	.26666	*				
MAZES	.28910	*				

Skipping rotation 1 for extraction 2 in analysis 1

Factor Score Coefficient Matrix:

	Factor 1	Factor 2	Factor 3	Factor 4
INFO	.00140	.16851	-.16265	.01329
SIMIL	.00120	.18668	-.12704	-.07649
ARITHM	.99467	-.69999	.07429	-.17056
VOCAB	.00250	.37769	-.54062	.08591
COMPR	.00100	.15671	-.14519	-.16689
DIGSPAN	.00068	.03588	-.01655	.35252
COMPL	.00063	.11304	.15949	-.20977
ARRANGEM	.00047	.09525	.07982	.00430

```
DESIGN          .00143       .20675       .44527       .09546
ASSEMBLY        .00064       .14717       .31415      -.07129
CODING          .00044       .04138       .03161       .30738
MAZES           .00038       .05909       .15440       .06539
```

[This initial ML solution is related to a PFA with the same communalities to within an orthogonal rotation. However, because the unrotated PFA solution yields near-maximum similarity between the factor structure and the factor revelation, these two matrices may be quite different in the initial ML solution – as they are here.]

Now, how about an example of an oblique rotation? Following is the setup for an oblimin-rotated PFA. I used only three factors because allowing them to be uncorrelated permits their intercorrelations to reflect what is thought by most researchers feel is a general factor influencing all 12 subscales to some degree, whereas in an orthogonal-factors model a separate (indeed, most likely, the first principal) factor is needed to represent this general factor.

```
Title Oblique Rotation of WISCR Subscales
MATRIX DATA VARIABLES =
   INFO  SIMIL   ARITHM  VOCAB   COMPR   DIGSPAN COMPL   ARRANGEM
   DESIGN  ASSEMBLY CODING  MAZES  /
   CONTENTS = N CORR / FORMAT = FREE FULL
BEGIN DATA
203 203 203 203 203 203 203 203 203 203 203 203
1   .62 .54 .69 .55 .36 .40 .42 .48 .40 .28 .27
.62 1   .47 .67 .59 .34 .46 .41 .50 .41 .28 .28
                    …
.28 .28 .32 .32 .26 .29 .19 .25 .33 .24 1   .21
.27 .28 .27 .27 .29 .22 .34 .32 .44 .37 .21 1
END DATA
Subtitle Oblimin-rotated PAF; delta = 0, -1.5, 0.4
   /*  Max permissible delta = 0.8    */
Factor / Width = 72 / Matrix = In (Cor = *)
     / Print = Defaults FScore Inv Repr
    / Criteria = Factors (3) Delta (0) / Extraction = PAF
    / Rotation = Oblimin
  / Analysis = All  / Print = Defaults FScore
    / Criteria = Factors (3) Delta (-1.5)  Iterate (50)
    / Extraction = PAF
    / Rotation = Oblimin
  / Analysis = All  / Print = Defaults FScore
    / Criteria = Factors (3) Delta (0.4)  Iterate (50)
    / Extraction = PAF
   / Rotation = Oblimin
```

The degree of correlation among the factors after oblimin rotation is determined in a very nonlinear fashion by the value of the delta parameter, whose default value is zero and which can range between negative infinity and .8.

We aren't particularly interested in the initial, orthogonal solution, so I start the listing of the output just as the oblimin rotation is about to begin:

```
OBLIMIN   rotation   1 for extraction   1 in analysis  1 - Kaiser
Normalization

OBLIMIN converged in 9 iterations.

Pattern Matrix:
              Factor  1     Factor  2     Factor  3

INFO             .65776        .02650        .16963
SIMIL            .68370        .10442        .04782
ARITHM           .25574        .02147        .54257
VOCAB            .86696       -.06872        .09256
COMPR            .71190        .07969       -.03373
DIGSPAN          .03230       -.01264        .62349
COMPL            .21576        .54614       -.07947
ARRANGEM         .24284        .42800       -.01710
DESIGN          -.00505        .73972        .15452
ASSEMBLY         .02662        .75191       -.07219
CODING           .03176        .15498        .34585
MAZES           -.08360        .52623        .11942
```

[No more "Factor Matrix" labeling; after oblique rotation the pattern, structure, and scoring-coefficient matrices are all different. This pattern matrix is read by rows and tells us what linear combination of the three factors most closely reproduces z-scores on the original variable listed in a given row—i.e., the factorial composition of each original variable.]

```
Structure Matrix:

              Factor  1     Factor  2     Factor  3

INFO             .77297        .53525        .56308
SIMIL            .77881        .56974        .49478
ARITHM           .58336        .45479        .70107
VOCAB            .87609        .53709        .55993
COMPR            .74388        .52293        .41732
DIGSPAN          .38467        .31632        .63592
COMPL            .52262        .64626        .31517
ARRANGEM         .50945        .57643        .33482
```

```
DESIGN           .56217          .81281          .51712
ASSEMBLY         .47063          .73344          .31476
CODING           .33188          .34640          .44080
MAZES            .32541          .53124          .33111
```

[The structure matrix is neither fish nor fowl. It tells us what the correlation between each original variable and each rotated factor is, but this information is insufficient to tell us either how to estimate scores on original variables from scores on factors (which is what the pattern matrix does) or how to estimate scores on factors from scores on original variables (which is what the factor revelation does).]

```
Factor Correlation Matrix:

              Factor  1     Factor  2     Factor  3

Factor  1      1.00000
Factor  2       .64602       1.00000
Factor  3       .57826        .49415       1.00000
```

[With the default value of delta = 0 we get factors that share an average of about 33% of their variance. We could of course submit this matrix of correlations among the 3 factors to a subsequent PFA to see if we could, e.g., account for the correlations among the factors reasonably well on the basis of a single, higher-order factor that might, e.g., represent general intelligence. I'll hold off on that until the section on confirmatory factor analysis.]

`Factor Score Coefficient Matrix:` [The *factor revelation*, in the author's biased terminology.]

```
             Factor  1     Factor  2     Factor  3
INFO           .18158        .02512        .11108
SIMIL          .20235        .05729        .02358
ARITHM         .05744        .00661        .35695
VOCAB          .41698       -.01604        .09920
COMPR          .17280        .05234       -.03895
DIGSPAN       -.00793        .00345        .32512
COMPL          .05464        .16671       -.05113
ARRANGEM       .04676        .12192       -.00611
DESIGN         .01271        .38278        .14599
ASSEMBLY       .02152        .26943       -.04923
CODING        -.00073        .03248        .13612
MAZES         -.00368        .12439        .05586
```

[This matrix is read by columns and tell us how to estimate scores on factors from scores on the original variables. It is thus the appropriate starting point for the concept-identification task of inferring what the factor represents on the basis of the observable characteristics (pattern of scores on the subscales) of someone who gets a high versus a low score on that factor.]

```
Covariance Matrix for Estimated Regression Factor Scores:

                   Factor  1    Factor  2    Factor  3
Factor  1            .89048
Factor  2            .63659       .83303
Factor  3            .58787       .49380       .69084
```

[Moderate factor-score indeterminacy: average squared multiple R between factors and original variables = .80478. Intercorrelations among factor-score estimates are higher than the actual correlation between the two factors in some cases, lower in others.]

The output from this delta = 0 rotation is sufficient to illustrate most of the important points about oblique rotations. But let's look at the inter-factor correlation matrix and the covariance matrix for estimated factor scores for the other two values of delta. First, for delta = -1.5 we have:

```
Factor Correlation Matrix:
                   Factor  1    Factor  2    Factor  3
Factor  1           1.00000
Factor  2            .47433      1.00000
Factor  3            .44316       .38722      1.00000

Covariance Matrix for Estimated Regression Factor Scores:
                   Factor  1    Factor  2    Factor  3
Factor  1            .84900
Factor  2            .49699       .80858
Factor  3            .49294       .41113       .66520
```

For delta = +0.4 we get much more highly intercorrelated factors:

```
Factor Correlation Matrix:
                   Factor  1    Factor  2    Factor  3

Factor  1           1.00000
Factor  2            .86074      1.00000
Factor  3            .86144       .81661      1.00000

Covariance Matrix for Estimated Regression Factor Scores:
                   Factor  1    Factor  2    Factor  3
```

```
Factor  1         .92114
Factor  2         .81207          .88042
Factor  3         .80590          .75328          .82122
```

How do we decide which value of delta to employ? It's a matter of personal preference, any theoretical considerations that might help judge how highly correlated the factors "should" be, and/or how interpretable the resulting factors are. Factor intercorrelations in the .8 range seem a bit extreme, but as shown in the next section a confirmatory factor analysis left free to estimate the factor intercorrelations indeed came up with an estimate in that range.

We now turn to CFA.

7.9 CONFIRMATORY FACTOR ANALYSIS

7.9.1 SAS PROC CALIS

Section 7.4 mentioned (in the context of determining number of factors) the use of LISREL as a vehicle for confirmatory factor analysis (CFA) This was indeed the first practical solution to the computational complexity of the high-dimensional parameter search needed to find that set of parameters that provide optimal fit to the data. However, the successive versions of LISREL have lagged behind alternative programs in terms of user-friendliness. We focus on SAS's PROC CALIS, though any structural equations modeling (SEM) program (e.g., EQS or AMOS) could be used. We "save" discussion of those programs for chapter 8.

The basic setup for CFA is as follows:

```
TITLE "Confirmatory FA of …";
DATA datasetname (TYPE = CORR);
  _Type_ = 'CORR';
  Input _NAME_ $
      Names of original variables ;
  Cards;
Name-of-1st-variable     1st row of correlation matrix
Name-of-2nd-variable    2nd row of correlation matrix
                  …
Name-of-pth-variable     1st row of correlation matrix
;
PROC CALIS DATA=datasetname  METHOD=ML  EDF=df;
   Title2 "Confirming 3 description of model being tested Using
Factor Statements";
   Title3 "  F1 = factorname1, F2 = factorname2, …";
```

```
Title4 "  HEYWOOD specifies non-neg Uniquenesses";
Title5 "       N = Gives number of factors ";
FACTOR HEYWOOD   N = number-of-factors;
MATRIX _F_
   [,1] = entries in first col of hypothesized pattern matrix,
   [,2] = entries in second col of hypothesized pattern matrix,
                                 ...
   [,N] = entries in last col of hypothesized pattern matrix;
```

In this listing,

df is the degrees of freedom on which each variance estimate is based, and usually equals the number of subjects (cases) – 1, but would of course equal number of subjects – *k* (the number of independent groups) if the correlation matrix being analyzed is based solely on within-cells information;

N is the number of factors in your hypothesized model; and

the entries in each row of the hypothesized pattern matrix are alphanumeric parameter names for loadings to be estimated and specific numerical values (usually zero) for those designated by the model.

In addition, if you wish to permit (or specify) an oblique solution, you enter the inter-factor correlation matrix (below- and on-main-diagonal elements only) as the _P_ matrix:

```
MATRIX _P_
   [1,] = 1.,
   [2,] = correlation between factors 1 & 2,   1.,
   [3,] = r betw F1 & F3, r betw F2 & F3   1.;
          and so on through N rows.
```

Example 7.1 Revisited: *Model Comparisons Galore.* To illustrate, consider the following confirmatory FA(s) of the WISC–R data:

```
OPTIONS Nocenter LS = 72;
TITLE "WISC-R Confirmatory FA, Using PROC CALIS";
Data WISC (TYPE = CORR);
  _Type_ = 'CORR';
  Input _NAME_ $
          INFO SIMIL ARITHM VOCAB COMPR DIGSPAN COMPL
   ARRANGEM DESIGN ASSEMBLY CODING MAZES ;
  Cards;
INFO      1   .62 .54 .69 .55 .36 .40 .42 .48 .40 .28 .27
SIMIL     .62 1   .47 .67 .59 .34 .46 .41 .50 .41 .28 .28
ARITHM    .54 .47 1   .52 .44 .45 .34 .30 .46 .29 .32 .27
                      ...
MAZES     .27 .28 .27 .27 .29 .22 .34 .32 .44 .37 .21 1
;
PROC CALIS DATA=WISC  METHOD=ML  EDF=220;
```

```
   Title2  "Confirming  3  Orthogonal  Factors  Using  Factor
Statements";
   Title3 "  F1 = Verbal, F2 = Spatial, F3 = Numeric";
   Title4 "   HEYWOOD specifies non-neg Uniquenesses";
   Title5 "      N = Gives number of factors ";
   FACTOR HEYWOOD   N = 3;
   MATRIX _F_
      [,1] = A1 A2 0. A4 A5 0. 0. 0. 0. 0.  0.  0.,
      [,2] = 0. 0. 0. 0. 0. 0. B7 B8 B9 B10 0.  B12,
      [,3] = 0. 0. C3 0. 0. C6 0. 0. 0. 0.  C11 0. ;
PROC CALIS DATA=WISC  METHOD=ML  EDF=220;
   Title2 "Trying 4 Orthogonal WISC-R Factors ";
   Title3 "  F1 = Verbal, F2 = Spatial, F3 = Numeric";
   Title4 "   HEYWOOD specifies non-neg Uniquenesses";
   Title5 "      N = Gives number of factors ";
   FACTOR HEYWOOD   N = 4;
MATRIX  _F_
      [,1] = A1 A2 0. A4 A5 0. 0. 0. 0. 0.  0.  0.,
      [,2] = 0. 0. 0. 0. 0. 0. B7 B8 B9 B10 0.  B12,
      [,3] = 0. 0. C3 0. 0. C6 0. 0. 0. 0.  C11 0.,
      [,4] = D1 D2 D3 D4 D5 D6 D7 D8 D9 D10 D11 D12;
RUN;
PROC CALIS DATA=WISC  METHOD=ML  EDF=220;
   Title2 "4 Orthogonal Factors, Equal Loadings within Clusters";
   Title3 "  F1 = Verbal, F2 = Spatial, F3 = Numeric";
   Title4 "   HEYWOOD specifies non-neg Uniquenesses";
   Title5 "      N = Gives number of factors ";
   FACTOR HEYWOOD   N = 4;
   MATRIX  _F_
      [,1] = AA AA 0. AA AA 0. 0. 0. 0. 0.  0.  0.,
      [,2] = 0. 0. 0. 0. 0. 0. BB BB BB BB  0.  BB,
      [,3] = 0. 0. CC 0. 0. C6 0. 0. 0. 0.  CC  0.,
      [,4] = D1 D2 D3 D4 D5 D6 D7 D8 D9 D10 D11 D12;
RUN;
PROC CALIS DATA=WISC  METHOD=ML  EDF=220;
   Title2 "Confirming 3 Oblique Factors Using Factor Statements";
   Title3 "  F1 = Verbal, F2 = Spatial, F3 = Numeric";
   Title4 "   HEYWOOD specifies non-neg Uniquenesses";
   Title5 "      N = Gives number of factors ";
   FACTOR HEYWOOD   N = 3;
   MATRIX  _F_
      [,1] = A1 A2 0. A4 A5 0. 0. 0. 0. 0.  0.  0.,
      [,2] = 0. 0. 0. 0. 0. 0. B7 B8 B9 B10 0.  B12,
      [,3] = 0. 0. C3 0. 0. C6 0. 0. 0. 0.  C11 0. ;
   MATRIX  _P_
      [1,] = 1.,
      [2,] = RF12   1.,
      [3,] = RF13   RF23   1.;
```

```
RUN;
PROC CALIS DATA=WISC  METHOD=ML  EDF=220;
   Title2 "3 Oblique Factors, Equally Intercorrelated";
   Title3 "  F1 = Verbal, F2 = Spatial, F3 = Numeric";
   Title4 "   HEYWOOD specifies non-neg Uniquenesses";
   Title5 "      N = Gives number of factors ";
   FACTOR HEYWOOD   N = 3;
   MATRIX _F_
      [,1] = AA AA 0. AA AA 0. 0. 0. 0. 0.  0.  0.,
      [,2] = 0. 0. 0. 0. 0. 0. BB BB BB BB  0.  BB,
      [,3] = 0. 0. CC 0. 0. CC 0. 0. 0. 0.  CC  0.;
   MATRIX _P_
      [1,] = 1.,
      [2,] = RF   1.,
      [3,] = RF   RF   1.;
RUN;
```

The preceding input requests tests of six different models, all having in common the same specification as to which variables have non-zero loadings on the Verbal, Spatial, and Numeric factors: a model that assumes that those three factors are (a) orthogonal and (b) sufficient to account for the intercorrelations among the 12 subscales; one that adds a fourth, general factor (assessing general intelligence, *g*, perhaps?) orthogonal to the three specific factors; a four-orthogonal-factors model that also assumes that the variables defining any one of the three specific factors load equally on that factor; a model that once again assumes only the three specific factors but permits them to be correlated with each other; a model that assumes three oblique factors but assumes in addition that the three are equally intercorrelated; and the same model with equal loadings within the variables loading on each factor.

The output from these runs is voluminous, but most of it consists of echoing back the parameters specified by the user, reporting the default values assumed by the model, and spelling out the details of the parameter search. The important output from the first CFA requested includes:

```
WISC-R Confirmatory FA, Using PROC CALIS
Confirming 3 Orthogonal Factors Using Factor Statements
  F1 = Verbal, F2 = Spatial, F3 = Numeric
   HEYWOOD specifies non-neg Uniquenesses
      N = Gives number of factors
18:28 Saturday, February 12, 2000

Covariance Structure Analysis: Maximum Likelihood Estimation

Fit criterion . . . . . . . . . . . . . . . . . .      1.1803
Goodness of Fit Index (GFI) . . . . . . . . . . .      0.8393
GFI Adjusted for Degrees of Freedom (AGFI). . . .      0.7679
Root Mean Square Residual (RMR) . . . . . . . . .      0.2815
```

```
Parsimonious GFI (Mulaik, 1989) . . . . . . . . . .      0.6867
Chi-square = 259.6642       df = 54     Prob>chi**2 = 0.0001
Null Model Chi-square:      df = 66                 1080.8576
RMSEA Estimate  . . . . . . 0.1316  90%C.I.[ 0.1158, 0.1478]
Probability of Close Fit  . . . . . . . . . . . .      0.0000
ECVI Estimate . . . . . . . 1.4122  90%C.I.[ 1.1959, 1.6649]
Bentler's Comparative Fit Index . . . . . . . . . .    0.7973
Normal Theory Reweighted LS Chi-square  . . . . .    252.7263
Akaike's Information Criterion. . . . . . . . . .    151.6642
Bozdogan's (1987) CAIC. . . . . . . . . . . . . .    -85.8366
Schwarz's Bayesian Criterion. . . . . . . . . . .    -31.8366
McDonald's (1989) Centrality. . . . . . . . . . .      0.6279
Bentler & Bonett's (1980) Non-normed Index. . . .      0.7523
Bentler & Bonett's (1980) NFI . . . . . . . . . .      0.7598
James, Mulaik, & Brett (1982) Parsimonious NFI. .      0.6216
Z-Test of Wilson & Hilferty (1931). . . . . . . .     10.7870
Bollen (1986) Normed Index Rho1 . . . . . . . . .      0.7064
Bollen (1988) Non-normed Index Delta2 . . . . . .      0.7997
Hoelter's (1983) Critical N . . . . . . . . . . .          63
```

[New fit indices are invented on what seems like a daily basis. In general they divide into tests of statistical significance of the discrepancies between the hypothesized model and the data (e.g., the chi-square measures) and measures—usually on a zero-to-one scale—of the closeness of the fit between model and data. This latter category further subdivides into "power" measures of overall fit (e.g., the GFI) and "efficiency" measures that give credit for parsimony (e.g., the AGFI). Subsequent printouts will focus on just a few of the more commonly used fit indices.]

```
Estimated Parameter Matrix _F_[12:3]
Standard Errors and t Values
Lower Triangular Matrix

            FACT1                   FACT2                   FACT3

INFO        0.7798      [A1 ]       0.                      0.
            0.0594    13.1264       0.          0.          0.        0.
SIMIL       0.7776      [A2 ]       0.                      0.
            0.0595    13.0745       0.          0.          0.        0.
ARITHM      0.                      0.                      0.7047    [C3 ]
            0.          0.          0.          0.          0.0980  7.1906
VOCAB       0.8778      [A4 ]       0.                      0.
            0.0564    15.5691       0.          0.          0.        0.
COMPR       0.7423      [A5 ]       0.                      0.
            0.0606    12.2563       0.          0.          0.        0.
```

```
DIGSPAN     0.                      0.                       0.6386   [C6 ]
            0.         0.           0.           0.          0.0933 6.8467
COMPL       0.                      0.6518        [B7 ]      0.
            0.         0.           0.0659      9.8879       0.       0.
ARRANGEM    0.                      0.5808        [B8 ]      0.
            0.         0.           0.0676      8.5926       0.       0.
DESIGN      0.                      0.8100        [B9 ]      0.
            0.         0.           0.0624     12.9851       0.       0.
ASSEMBLY    0.                      0.7326        [B10]      0.
            0.         0.           0.0641     11.4367       0.       0.
CODING      0.                      0.                       0.4541   [C11]
            0.         0.           0.           0.          0.0815 5.5703
MAZES       0.                      0.5302        [B12]      0.
            0.         0.           0.0687      7.7177       0.       0.
```

[In addition to providing estimates of the loadings that were left as free parameters, this output also presents the equivalent of path analysis's Condition 9 tests—i.e., it provides *t* tests of the null hypotheses that each loading is zero in the population.]

```
Covariance Structure Analysis: Maximum Likelihood Estimation

Standardized Factor Loadings

                   FACT1         FACT2          FACT3

INFO             0.77982       0.00000        0.00000
SIMIL            0.77762       0.00000        0.00000
ARITHM           0.00000       0.00000        0.70466
VOCAB            0.87782       0.00000        0.00000
COMPR            0.74227       0.00000        0.00000
DIGSPAN          0.00000       0.00000        0.63860
COMPL            0.00000       0.65176        0.00000
ARRANGEM         0.00000       0.58077        0.00000
DESIGN           0.00000       0.81004        0.00000
ASSEMBLY         0.00000       0.73259        0.00000
CODING           0.00000       0.00000        0.45412
MAZES            0.00000       0.53019        0.00000
```

[Had we begun this analysis with a covariance matrix, rather than a correlation matrix, this matrix would have converted the loadings to reflect the relationships between each original variable and underlying factors having standard deviations of 1.0. As is, this matrix provides a compact summary of the factor pattern that most closely fits our hypothesized model.]

```
Squared Multiple Correlations
```

Parameter		Error Variance	Total Variance	R-squared
1	INFO	0.391888	1.000000	0.608112
2	SIMIL	0.395314	1.000000	0.604686
3	ARITHM	0.295338	1.000000	0.704662
4	VOCAB	0.229439	1.000000	0.770561
5	COMPR	0.449028	1.000000	0.550972
6	DIGSPAN	0.592187	1.000000	0.407813
7	COMPL	0.575211	1.000000	0.424789
8	ARRANGEM	0.662707	1.000000	0.337293
9	DESIGN	0.343835	1.000000	0.656165
10	ASSEMBLY	0.463319	1.000000	0.536681
11	CODING	0.793778	1.000000	0.206222
12	MAZES	0.718903	1.000000	0.281097

[The squared multiple correlations are, for orthogonal factors, simply the sum of the squared loadings on the various factors. For oblique models, multiple regression analysis must be employed.]

The logic of model testing is the mirror image of the usual null hypothesis significance testing (NHST) in that the researcher is usually "pulling for" a nonsignificant test of overall fit. Fortunately most researchers who use CFA to test a specific model recognize that no model is perfect and therefore take nonsignificance (correctly) as indicating only that there are insufficient data to be confident of the direction in which the various assumptions of that model depart from reality.

The real forte of CFA is in the comparison of alternative models. There is in fact a readymade tool for such comparisons: For any two ***nested*** models (two models, one of which is a special case of the other), the difference between their respective chi-square tests of overall fit is itself distributed as a chi-square with degrees of freedom equal to the difference beween their respective degrees of freedom (which equals the number of parameters in the more general model that are set to a priori values in the more restricted model). We therefore now proceed now to compare a number of other models to this first, three-orthogonal-factors model and to each other. Condensed output from those other models is as follows:

```
          Title2 "Trying 4 Orthogonal WISC-R Factors ";
          MATRIX _F_
             [,1] = A1 A2 0. A4 A5 0. 0. 0. 0. 0.  0.  0.,
             [,2] = 0. 0. 0. 0. 0. 0. B7 B8 B9 B10 0.  B12,
             [,3] = 0. 0. C3 0. 0. C6 0. 0. 0. 0.  C11 0. ,
             [,4] = D1 D2 D3 D4 D5 D6 D7 D8 D9 D10 D11 D12;

Fit criterion . . . . . . . . . . . . . . . . .      0.1119
```

```
Goodness of Fit Index (GFI) . . . . . . . . . . .       0.9818
GFI Adjusted for Degrees of Freedom (AGFI). . . .       0.9662
Root Mean Square Residual (RMR) . . . . . . . . .       0.0234
Parsimonious GFI (Mulaik, 1989) . . . . . . . . .       0.6248
Chi-square = 24.6078       df = 42      Prob>chi**2 = 0.9851
Null Model Chi-square:     df = 66                 1080.8576
```

Test of improvement in fit by using a fourth factor:
Difference chi-square = 259.66 - 24.61 = 235.05 with 54 - 42 = 12 *df*, $p < .001$.

```
Title2 "4 Orthogonal Factors, Equal Loadings within
Clusters";
        MATRIX _F_
           [,1] = AA AA 0. AA AA 0. 0. 0. 0. 0.  0.  0.,
           [,2] = 0. 0. 0. 0. 0. 0. BB BB BB BB  0.  BB,
           [,3] = 0. 0. CC 0. 0. CC 0. 0. 0. 0.  CC  0.,
           [,4] = D1 D2 D3 D4 D5 D6 D7 D8 D9 D10 D11 D12;
```

```
Fit criterion . . . . . . . . . . . . . . . . . .       0.1807
Goodness of Fit Index (GFI) . . . . . . . . . . .       0.9705
GFI Adjusted for Degrees of Freedom (AGFI). . . .       0.9549
Root Mean Square Residual (RMR) . . . . . . . . .       0.0312
Parsimonious GFI (Mulaik, 1989) . . . . . . . . .       0.7499
Chi-square = 39.7469       df = 51      Prob>chi**2 = 0.8731
Null Model Chi-square:     df = 66                 1080.8576
```

Test of H_0 of equal loadings within clusters, first 3 factors:
Difference chi-square = 39.75 - 24.61 = 15.14 with 9 *df*, $p = .087$.

```
Confirming 3 Oblique Factors Using Factor Statements

Fit criterion . . . . . . . . . . . . . . . . . .       0.1672
Goodness of Fit Index (GFI) . . . . . . . . . . .       0.9730
GFI Adjusted for Degrees of Freedom (AGFI). . . .       0.9587
Root Mean Square Residual (RMR) . . . . . . . . .       0.0316
Parsimonious GFI (Mulaik, 1989) . . . . . . . . .       0.7519
Chi-square = 36.7795       df = 51      Prob>chi**2 = 0.9329
Null Model Chi-square:     df = 66                 1080.8576

Interfactor Correlation Matrix

               FCOR1                  FCOR2                FCOR3
FCOR1          1.0000                 0.7424   [RF12]      0.7902 [RF13]
               0.         0.          0.0433  17.1380      0.0521 15.1571
```

```
FCOR2         0.7424    [RF12]     1.0000                 0.6729 [RF23]
              0.0433   17.1380     0.           0.        0.0638 10.5398

FCOR3         0.7902    [RF13]     0.6729    [RF23]       1.0000
              0.0521   15.1571     0.0638   10.5398       0.            0.
```

Test of H_0 that factor intercorrelations are all zero:
Difference chi-square = 259.662 -36.780 = 222.882 with 54 – 51 = 3 *df*, $p < .001$.

```
 3 Oblique Factors, Equally Intercorrelated
  F1 = Verbal, F2 = Spatial, F3 = Numeric
Unequal Loadings within Clusters  22:11 Wednesday, August 9, 2000
      N = Gives number of factors

Covariance Structure Analysis: Maximum Likelihood Estimation

Fit criterion . . . . . . . . . . . . . . . . . . .          0.1849
Goodness of Fit Index (GFI) . . . . . . . . . . . .          0.9703
GFI Adjusted for Degrees of Freedom (AGFI). . . . .          0.9563
Root Mean Square Residual (RMR) . . . . . . . . . .          0.0346
Parsimonious GFI (Mulaik, 1989) . . . . . . . . . .          0.7792
Chi-square = 40.6679       df = 53      Prob>chi**2 = 0.8924
Correlations among Exogenous Variables
---------------------------------------------------------------
Row & Column             Parameter                    Estimate
---------------------------------------------------------------
   2       1     FCOR2        FCOR1          RF         0.743521
   3       1     FCOR3        FCOR1          RF         0.743521
   3       2     FCOR3        FCOR2          RF         0.743521
```

Test of H_0 that factor intercorrelations are all equal to some common value:
Difference chi-square = 40.668 - 36.780 = 3.888 with 53 - 51 = 2 *df*, $p = .143$.

```
3 Oblique Factors, Equally Intercorrelated  12:37 Tuesday, July
25, 2000
  F1 = Verbal, F2 = Spatial, F3 = Numeric
   Equal Loadings Within Clusters
      N = Gives number of factors
Covariance Structure Analysis: Maximum Likelihood Estimation
Fit criterion . . . . . . . . . . . . . . . . . . .          0.3220
Goodness of Fit Index (GFI) . . . . . . . . . . . .          0.9489
GFI Adjusted for Degrees of Freedom (AGFI). . . . .          0.9357
Root Mean Square Residual (RMR) . . . . . . . . . .          0.0797
Parsimonious GFI (Mulaik, 1989) . . . . . . . . . .          0.8914
Chi-square = 70.8315       df = 53      Prob>chi**2 = 0.2068
Null Model Chi-square:     df = 66                 1080.8576
```

Correlations among Exogenous Variables

Row & Column		Parameter			Estimate
2	1	FCOR2	FCOR1	RF	0.755893
3	1	FCOR3	FCOR1	RF	0.755893
3	2	FCOR3	FCOR2	RF	0.755893

Test of difference between model assuming equal factor intercorrelations and equal within-factor loadings versus model assuming equal factor intercorrelations but allowing unequal loadings of original variables on the three factors:
Difference chi-square = 70.832 - 40.668 = 30.164 with 62 – 53 = 9 *df*, $p < .001$.

Considering sets of nested models, it's clear that a fourth orthogonal factor provides large and statistically significant improvement in fit over a three-orthogonal-factor model and in fact leaves us with nonsignificant departures from the former model. Moreover, assuming that the subscales that define (have nonzero loadings on) each of the specific factors have identical loadings yields only a slight, statistically significant deterioration in fit. Relaxing the orthogonality requirement in the three-factors model also provides large and statistically significant improvement in fit, once again leading to a model that provides a nonsignificant test of overall fit, and assuming that the factor intercorrelations are identical does not affect the fit substantially or statistically significantly, but assuming equal loadings within clusters does.

The correlations among the three oblique factors could themselves be submitted to a secondary factor analysis; because the correlations among three variables can always be perfectly accounted for (barring a Heywood-case communality outside the 0-to-1 range) by a single factor, this model is equivalent to assuming that a fourth, higher-order factor (the *g* factor that was extracted as a fourth, general factor in the four-orthogonal-factors model) is operating.

We're thus left with two rather viable models: (a) four orthogonal factors, one of them being a general factor and the loadings within clusters of specific-factor-defining variables being equal; and (b) three correlated factors whose essentially equal intercorrelations are accounted for by a single second-order general factor. Neither of these two models is a special case of the other, so we don't have a formal test of their relative fit – and comparisons are complicated by the fact that the four-orthogonal-factors model has only 51 degrees of freedom, whereas the three-equally-correlated-factors model has 53 *df.* The AGFI (Adjusted Goodness of Fit Index) is intended to adjust for differences in parsimony, as is the Parsimonious Fit Index, but so are a number of other indices, and there's no consensus on a single way of handling comparison of non-nested models with different *df.* It's clear, though, that the two models provide quite similar degrees of fit to the observed correlations: GFI of .9705 versus .9703, root mean square residual of .031 versus .035, and p value for test of perfect-fit hypothesis of .873 versus .892.

It's not unusual to find this sort of near equivalence between a model with a general factor and m specific factors, all mutually orthogonal, on the one hand, and a model with m equally correlated factors. In fact, it is easily shown that if each of the m oblique factors is indeed generated as a weighted average of a specific and the general factor [i.e., oblique factor $i = a_i$ (specific factor i) $+ c_i$ (general factor)], with all m+1 factors being mutually uncorrelated, and scores on the p original variables display perfect simple structure with respect to the three oblique factors [i.e., a unique subset of the original variables has nonzero loadings on each of the m oblique factors] then the correlations among the original variables can also be perfectly reproduced by an m+1 orthogonal-factors model in which one of the orthogonal factors is a general factor on which all original variables have nonzero loadings and the other three orthogonal factors show the same pattern of zero versus nonzero loadings as do the three oblique factors. (I believe that there is a published proof of the algebraic equivalence of these two models, although I was unable to retrieve the reference to it as this book was going to press.) This makes sense, in that both models are saying that each original variable is affected both by the general factor and by one of the m specific factors—it's simply a matter of personal preference as to whether the general factor is represented as a separate factor orthogonal to the specific factors or as a "higher order" factor generating the correlations among a set of oblique factors. It is somewhat ironic, however, that a large majority of factor analysts find an orthogonal general factor distasteful but a higher-order factor appealing.

It is important to recognize that, although the two models may be equally good at accounting for the correlations among the original variables, the specific factors they represent are *not* the same. If a set of data fit these two models closely, then regression-based estimates of scores on the specific factors in the orthogonal-factors model (a) will be computed as the average of the variables with high loadings on that factor, minus the average of the scores on all variables and (b) will represent ability on that specific factor, relative to (over and above) the general factor. On the other hand, regression-based estimates of the factors in the correlated-factors model (a) will be computed as the average of the variables with high loadings on that factor and (b) will represent a combination of ability on that specific factor and ability on (i.e., will be uncorrected for) the general factor.

I should point out that the assumptions discussed here also imply that, in the orthogonal-factors representation, the loadings on the general factor of the original variables within a cluster associated with a given specific factor will be directly proportional to their loadings on that specific factor. We would therefore expect to get an even closer match between the two models as applied to the WISC–R data above if we could impose this proportionality constraint, rather than assuming equality of the 12 loadings on the general factor. Such a constraint is more readily handled within a structural equations model, so we defer the attempt to the next chapter.

8 The Forest Revisited

There are at least four issues that cut across the specific materials of Chapters 2 through 7, but that also require some acquaintance with the material in those chapters for meaningful discussion. The present chapter discusses (a) the applicability of the multivariate techniques discussed so far to data having "weaker" than interval scale properties; (b) the robustness of multivariate significance tests in the presence of violations of the assumptions (especially multivariate normality) used to derive them; (c) how nonlinear relationships among or between variables can be detected and how to handle nonlinearity when it is detected; and (d) the utility of the very general techniques provided for testing the multivariate general linear hypothesis and for modeling structural equations as alternatives to the more specialized procedures described in Chapters 2 through 7. Finally, some suggestions for further study are offered.

8.1 SCALES OF MEASUREMENT AND MULTIVARIATE STATISTICS

As was mentioned briefly in Chapter 1, a number of psychologists, most notably S. S. Stevens (e.g., 1951, 1968), have taken the position that the majority of the most common and most powerful statistical procedures -- including all of those we have discussed in this *Primer* -- are meaningless unless the researcher can establish that his or her data were generated through measurement processes having at least interval scale properties. For all other data (that is, almost all data gathered by psychologists, sociologists, or political scientists), other statistical procedures designed specifically for ordinal or nominal data must be employed -- if they are available. Such nominal- and ordinal-scale versions of multivariate statistical procedures are not in general available, except in the limited sense that the entries in a correlation matrix can be Spearman rank-order correlations or phi coefficients, instead of Pearson *r*s, and any of the multivariate procedures discussed in this Primer can take a correlation matrix as its starting point. This substitution of nominal or ordinal coefficients for the "interval-scale" Pearson *r* does not eliminate the problem that, from the point of view of Stevens's adherents, the optimization criteria (for example, percent of variance accounted for or ratio of among-group to within-group variability), as well as the fundamental operation of computing various linear combinations of scores on the original variables, are meaningless. Thus if Stevens's strictures are adhered to, multivariate statistical procedures of any sort would be available for and applicable to only a very small percentage of the data collected by social and behavioral scientists.

That I do not accept Stevens's position on the relationship between strength of measurement and "permissible" statistical procedures should be evident from the kinds of data used as examples throughout this Primer: level of agreement with a questionnaire item, as measured on a five-point scale having attached verbal labels; dichotomous variables such as sex converted to 0-l numeric codes; and the qualitative differences among the treatments administered k experimental groups converted to k - 1 group membership 0–l variables. The most fundamental reason for this willingness to apply multivariate statistical techniques to such data, despite the warnings of Stevens and his associates, is the fact that the validity of statistical conclusions depends only on whether the numbers to which they are applied meet the distributional assumptions (usually multivariate normality and homogeneity of covariance matrices from group to group) used to derive them, and not on the scaling procedures used to obtain the numbers. In other words, statistical tests are "blind" as to how the numbers "fed" to them were generated. Moreover, we have such strong mathematical and empirical evidence of the robustness of statistical procedures under violation of normality or homogeneity of variance assumptions (cf. section 8.2) that the burden of proof must be presumed to be on the shoulders of those who claim that a particular set of data can be analyzed only through "nonparametric" (a better term would be "distribution free") statistical techniques.

The various alternative measures of correlation that have been proposed for "non-interval-scale" data provide especially convincing demonstrations of the low returns obtained by fretting over whether data have "truly" interval scale properties. Most introductory statistics texts provide special computational formulae for Spearman's ρ (or r_s) for use when two ordinal variables are to be related; the point biserial correlation coeficient r_{pb} and the biserial correlation r_b for the relationship between a dichotomous, 0–1 variable and an interval-scale measure; and the phi coefficient ϕ or the tetrachoric correlation r_{tet} when two dichotomous variables are to be related. (Cf. Glass and Stanley, 1970, for computational formulae.) These various measures of relationship fall into two groups.

The special formulae for r_s, r_{pb}, and ϕ turn out to yield precisely the same numerical values as would be obtained by blindly applying the formula for Pearson's product-moment coefficient of correlation r to the ranked or dichotomized data. (If the original data are not in the form of ranks or 0–1 measures, r computed on the original data will generally differ from r_s, r_{pb}, or ϕ.) Moreover, the significance levels for r_s, r_{pb}, and ϕ, computed so as to take into account the special nature of the sets of data being related, are so close to the corresponding significance levels for Pearson's r as to be practically indistinguishable. Table 8.1 lists these significance levels for a few values of N - 2, the degrees of freedom on which the sample estimate of the corresponding population correlation is based. This table provides, by the way, an excellent example of the robustness of Pearson's r under violations of the assumption of normality. A set of ranks has a rectangular distribution, and a set of 0–1 scores has a binomial distribution, and yet ignoring the departures from normality introduces very little error in the resultant significance test. Thus the only reason for using the special computational formula for r_s, r_{pb}, or ϕ, rather than simply using the Pearson r formula, is computational convenience. For most problems in which multivariate statistical procedures are required, a computer

Table 8.1 Representative Critical Values for Measures of Association

$df = N - 2$	r	r_s	r_{pb}	ϕ
5	.754	.786	.754	.740
10	.576	.591	.576	.566
15	.482	.490	.482	.475
20	.423	.428	.423	.418
25	.381	.385	.381	.378

Note: All critical values are for $\alpha = .05$.

program will be used for computing correlations, so that computational convenience is actually on the side of Pearson's r, because most computer programs employ r rather than r_S, r_{pb}, or ϕ. One can only pity the not entirely apocryphal doctoral students whose advisors send them in search of a computer program that can compute r_S, r_{pb}, or ϕ because their data have "only" ordinal or dichotomous properties.

The computational formula for r_S also illustrates the complete bankruptcy of the argument sometimes made by Stevens's followers that level of measurement determines the kinds of arithmetic operations that can be carried out on the data. For instance, addition and subtraction are said to be meaningless for ordinal data, and multiplication and division of pairs of interval-scale numbers is verboten. Following these proscriptions would successfully eliminate r_S, because it relies on squared differences between the rankings.

The remaining two commonly used correlation measures—the biserial correlation r_b and the tetrachoric coefficient of correlation r_{tet} --are "wishful thinking" measures. Each is the correlation measure that would have been obtained if the subjects' "true" scores on what the researcher assumes are the underlying bivariate normal dimensions had been correlated, that is, if these true scores had not been "distorted" from their "truly" normal distribution by floor effects, ceiling effects, the crudity of the available (pass-fail) scale, or a discrepancy between fact and assumption. Use of these measures is based on the untestable assumption that true relationships always yield normally distributed variables.

There is a third, much less commonly used, set of measures of association that do involve more than a renaming of or wishful adjustment of Pearson's r. Kendall's tau τ is, like r_S, a measure of the association between two sets of ranks. However, τ has the straightforward property (not shared by r_S) of being equal to the difference between the probability that two randomly selected subjects will be ordered identically on both variables, and the probability that their relative positions on one variable will be the reverse of their relative positions on the other variable. Along similar lines, Goodman and Kruskal(1954) developed measures of association for two-way contingency tables (bivariate frequency distributions for nominal data) that are equal to the gain in probability of correctly predicting a randomly selected subject's category on the second variable that results from knowing his or her category on the first variable. Thus τ and the Goodman-Kruskal measures have probabilistic interpretations that are quite different from the "shared variance" interpretation on which Pearson's r, r_s, r_{pb}, ϕ, r_b, and r_{tet} are based. The researcher may find these probabilistic interpretations more congenial in some situations and may therefore prefer τ or the Goodman-Kruskal measures. Such a preference carries with it, however, the penalties of computational inconvenience and inability to integrate the resulting measure into multivariate statistical procedures.

It is hoped that this discussion of alternatives to Pearson's r has convinced the reader that searching out statistical procedures and measures of relationship especially designed for nominal or ordinal data is usually a waste of time. This is not to say, however, that the researcher may simply ignore the level of measurement provided by his or her data. It is indeed crucial for the investigator to take this factor into account in considering the kinds of theoretical statements and generalizations he or she makes on the basis of significance tests. The basic principle is this: Consider the range of possible transformations, linear or otherwise, that could be applied to the data without changing any important properties or losing any of the useful information "stored" in the numbers. Then restrict generalizations and theoretical statements to those whose truth or falsehood would not be affected by the most extreme such permissible transformation. This is, as the reader who has encountered Stevens's classification of strength of measurement into nominal, ordinal, interval, and ratio will recognize, precisely the logic Stevens employed in proscribing, for example, arithmetic means as measures of central tendency for data that have only ordinal properties. Data with only ordinal scale properties may be subjected to any monotonic (order-preserving) transformation without affecting any important properties. Therefore any two sets of ordinal data for which it is true that $\overline{X}_1 > \overline{X}_2$, but that the largest single observation falls in group 2 or the smallest single observation falls in group 1, can be converted to a set of transformed data in which $\overline{X}_1' < \overline{X}_2'$ simply by making the largest observation many orders of magnitude larger than any other observation (this of course does not affect its rank in the distribution) or by making the smallest observation many orders of magnitude smaller than any other observation. For instance, it is perfectly legitimate to compute the two sample means for

Group	Original data	$\overline{X}$	Transformed data	$\overline{X}'$
1	1, 7, 3, 2, 2	3.0	1, 70, 3, 2, 2	15.6
2	4, 6, 6, 3, 6	5.0	4, 6, 6, 3, 6	5.0

either the original or the transformed data and to use "interval scale" statistical tests such as the t test to determine whether the two sets of numbers could have arisen through random sampling from populations of numbers having identical means. However, any generalization from the statistical tests to theoretical statements about the true effects of whatever conceptual independent variable differentiates the two groups, that is, on the conceptual dependent variable the experimenter was attempting to measure, is meaningless (not verifiable) if indeed transformations like the one we applied ("multiply the largest value by 10 and leave the others unchanged") are permissible—as they must be if the original data have *only* interval-scale properties, because in that case any monotonic transformation leaves all important properties unchanged. However, most researchers (and most readers of research reports) would find this sort of transformation completely unacceptable and would feel that it does lose (more strongly, distort) meaningful information contained in the data. For instance, assume the original numbers represented each subject's stated level of agreement (on a scale ranging from 0 to 10) with the proposition that "Men should be barred from any job for which there is a qualified woman candidate until the sex ratio for persons holding such jobs reaches unity." Probably no psychologist would claim that the difference between subjects responding "6" and "7" in antimasculinism (assuming this to be the conceptual dimension that the question was designed to tap) is precisely the same as the difference in antimasculinism between subjects who responded "2" and "3." Thus the measurement process would readily be admitted not to guarantee interval-scale measurement. On the other hand, the difference in antimasculinism represented by the difference in responses of "6" and "7" is almost certainly nowhere near 10 times as great as the conceptual difference between responses of "2" and "3." In other words, the measurement process—as is quite commonly the case—provides some information about relative magnitudes of differences among individuals on the property or dimension being measured, but not complete or fully reliable information. Perhaps such measurement processes should be labeled as having "reasonably ordinal" properties, because only reasonable monotonic transformations of the resulting numbers are permissible. The conditions for reasonableness of a transformation cannot be easily summarized mathematically, but they probably include at least the following two conditions:

1. The transformation must be a priori, that is, specified (or at least specifiable) before the actual data obtained in the study are examined.
2. The transformation must preserve the ordering not only of all pairs of observations actually obtained in the study but also of all pairs of observations that might have been obtained.

These two conditions are most easily met by continuous functions that can be expressed in closed form, such as square root, logarithmic, and arcsine transformations—and indeed, empirical sampling studies have shown that such "reasonable" or "well-behaved" transformations of data have miniscule effects on the results of normal-curve-based statistical tests.

Any readers who still feel that they must have proof of interval-scale properties before applying MRA, T^2, and so on to their data should take two papers and retire to their inner sanctum to read Lord's (1953) humorous discussion of "the statistical treatment of football numbers" and Gaito's (1980) review of subsequent analyses strongly supportive of Lord's statement that "the numbers do not know where they came from."

On the other hand, if the researcher is committed to constructing a theoretical framework or a set of empirical operations that depends only on ordinal-scale measurement of the concepts, then statements must be confined to those that are meaningful under *any* monotonic transformation, not just reasonable ones. For example, many sociologists have concerned themselves with the issue of the *shape* of various societies' class structure. Does a bar graph of the number of persons in that society falling into the upper, middle, and lower classes have a pyramidal, rectangular, or inverted pyramidal shape? These same sociologists often claim that any given indicator of social status (e.g., income) provides a continuous, ordinal measure of true status, with the dividing line between the classes being a matter of arbitrary decision on the researcher's part. However, if this is true, then the question of shape is a meaningless one, because the percentage of subjects falling into the various classes (and thus the shape of the frequency distribution) can be manipulated at will simply by moving the (arbitrary) dividing lines. If a pyramidal shape is desired, put the dividing line between the upper and middle classes at an annual income of $250,000 and the lower-middle dividing line at $15,000; for an inverted pyramid, put the lines at $2000 and $50; and so on. Statistical tests such as chi-square goodness-of-fit tests can certainly be used to indicate whether any given resultant frequency distribution differs significantly from a rectangular one, but under the researcher's assumptions about strength of measurement, such tests tell us more about his or her preconceptions about a particular society than about any empirical property of that society. The point is that the issue of strength of measurement comes into play ***not*** in determining what statistical tests are "permissible," ***nor*** in deciding whether the results of such tests are valid, but rather in deciding what sorts of meaningful (i.e., verifiable) theoretical statements about relationships among the conceptual variables being tapped by the experimenter's measurements can be made. (The class structure example just cited also illustrates that Stevens's classification of levels of measurement as nominal, ordinal, interval, or ratio provides not even an ordinal scale of "level of measurement," because there is no question in the researcher's mind that ordinal measurement of social status is possible, but considerable doubt as to whether nominal measurement is possible! See Brown's (1965) *Social Psychology* text for a full discussion of the problem of providing nonarbitrary definitions of class boundaries.)

There are some multivariate statistical procedures that are more likely than others to provide answers to questions that are not theoretically meaningful even for "reasonably ordinal" data. The most striking example is provided by ***profile analysis*** of data undergoing T^2 or Manova analysis. If the researcher would find it reasonable or acceptable to change the unit or origin of individual variables within the set of outcome measures, the shape of the grand mean profile and parallelism or lack thereof of the various group profiles would become largely a matter of the arbitrary choices of unit and origin for each of the scales. Thus profile analysis, although valid statistically, would have little theoretical value for the researcher. Reasonable monotonic transformations applied to every variable in the vector of outcome measures would not, however, seriously affect the results of a profile analysis. The reader familiar with univariate Anova will recognize the parallels between the present discussion of profile analysis and the issue of the interpretability of interaction effects. The general conclusion for both Anova and profile analysis is that interactions that involve an actual *reversal* of the direction of a difference (called ***crossover*** interactions) are unaffected by monotonic transformations, whereas interactions that do not involve a crossover but merely a change in the absolute magnitude of a difference may be made to "disappear" by judicious selection of a transformation.

Two additional areas in which the experimenter needs to give especially careful thought to whether the results of his or her multivariate statistical tests may be greatly affected by arbitrary scaling decisions are when:

1. Attempting to determine the presence or absence of significant nonlinearity in the relationships among variables (cf. section 8.3).
2. Assessing the relative importance of the individual variables in a set as contributors to multiple R, R_c, T^2, the gcr obtained from a Manova, or a principal component on the basis of *raw-score* regression, discriminant function, or characteristic vector coefficients.

8.2 EFFECTS OF VIOLATIONS OF DISTRIBUTIONAL ASSUMPTIONS IN MULTIVARIATE ANALYSIS

All of the significance tests outlined in this book were derived under at least the assumption of multivariate normality – that is, that the observed data vectors are independent random samples from a population in which any linear combination of the variables in the data vector is normally distributed. In addition, T^2 and Manova tests involve the assumption that the populations from which the different groups' data were sampled all have identical covariance matrices. These two assumptions are almost certainly not valid for any real set of data—and yet they are *nearly* valid for many sets of data. Moreover, the fact that a particular assumption was used in deriving a test does not mean that violation of that assumption invalidates the test, because the test may be quite robust under (i.e., insensitive to) violations of the assumptions used to derive it. We have extremely strong evidence—both mathematical and empirical—that the univariate tests of which the tests in this book are direct generalizations and that were derived under the assumptions of (univariate) normality and homogeneity of variance are in fact extremely robust under violation of those assumptions (cf., e.g., Boneau, 1960; Donaldson, 1968; Lindquist, 1953; Norris & Hjelm, 1961; and Winer, 1971).

The major exceptions to this statement occur for very small and unequal sample sizes or for one-tailed tests. I strongly feel that one-tailed statistical tests are almost never appropriate in a research situation (and in very few applied situations), because they require that the researcher be willing to respond to a fantastically large effect in the direction opposite to his or her prediction in exactly the same way that he or she would to a miniscule difference among the experimental groups. (However, see Part 4 of the book edited by Steiger, 1971, and the "Counterarguments" section of chapter 2 of Harris, 1994 for opposing views on this issue.) We are therefore left with only the unequal-n restriction on our general statement about the robustness of normal-curve-based univariate statistical tests. As a general guideline, normal-curve-based tests on a single correlation coefficient can be considered valid for almost any unimodal X and Y population for any number of observations greater than about 10; and normal-curve-based F or t tests can be considered valid for even U-shaped population distributions so long as two-tailed tests are used; the ratio between the largest and smallest sample variance is no greater than about 20 to 1; the ratio between the largest and smallest sample size is no greater than about 4; and the total degrees of freedom for the error term (s_c^2 or MS_W) is 10 or more. When the violations of assumptions are close to these bounds, a test at the nominal 5% level might actually have close to a 10% probability of yielding a false rejection of Ho. For even grosser violations of the assumptions, transformation of the data so as to meet the assumptions more closely (e.g., converting the observations to ranks based on all the observations in all groups, a procedure referred to as Kruskal-Wallis Anova) or use of a specialized test procedure (e.g., Welch's t^* test for the difference between two sample means arising from populations with grossly different variances) may be called for. The book by Winer (1971) is a good reference for such transformations and tests.

Unfortunately, the situation is not quite so clear-cut for the multivariate tests described in this book. It is easy to present an intuitively compelling argument that tests based on Roy's union-intersection principle (e.g., the significance tests provided for multiple regression and gcr tests for T^2, Manova, and canonical analysis) should display the same kind of robustness as the univariate tests. In each case, the multivariate test involves computing the largest possible value of a univariate test statistic (r, t, or F) that could result from applying the univariate statistical procedure to any linear combination of the several variables, and the relevant distribution is thus the distribution of the maximum value of such optimized univariate statistics. If the input to this maximum-value distribution (i.e., the distribution of the univariate statistic) is valid despite violation of a particular assumption, then so should be the resulting multivariate distribution. Put another way, we know that the univariate statistic computed on any a priori combination of the variables would follow the normal-curve-based univariate distribution quite well despite violations of the assumptions, so any nonrobustness must arise solely in the adjustments needed to correct for the fact that our linear combinations was deliberately picked to provide as large a t (or r or F) as possible—an unlikely source of nonrobustness. However, there is many a slip between intuitive conviction and mathematical or empirical proof, and the latter two are in somewhat short supply where the issue of robustness of multivariate significance tests is concerned. What is known includes the following:

1. Significance tests on the overall significance of a multiple correlation coefficient and on individual regression coefficients are unaffected by the use of cluster sampling, a departure from simple random samplings in which a random sample of large units (e.g., census tracts or mental hospitals) is selected from the population and then simple random sampling (or additional layers of cluster sampling) is carried out within each macro unit (Frankel, 1971). However, Cotton (1967) pointed out that most experimental research

employs volunteer (or at least self-selected) subjects from an available pool, which constitutes neither random sampling nor cluster sampling, so that the most useful studies of robustness would be studies of the "behavior" of various statistics under conditions of random assignment to treatments of a very nonrandom sample of subjects.

2. A multivariate generalization of the central limit theorem assures us that for sufficiently large sample sizes vectors of sample means have a multivariate normal distribution (Ito, 1969). The catch comes in trying to specify how large "sufficiently large" is.

3. The true significance levels for Hotelling's T^2 match the nominal significance levels, despite very nonhomogenous covariance matrices, as long as $N_1 = N_2$. Large discrepancies in both sample sizes and covariance matrices can lead to correspondingly large discrepancies between the true and nominal significance levels (Ito & Schull, 1964).

4. The power of the test of homogeneity of covariance matrices reported in section 3.6.1 appears to be low when nominal and actual significance levels of the overall test in Manova are quite close and high when the true significance level is considerably higher than the nominal significance level (Korin, 1972). Thus the multivariate test of homogeneity of covariance matrices must be taken somewhat more seriously than Bartlett's test in the univariate situation, with the latter test having high power to detect departures from homogeneity that are too small to have any appreciable effect on the overall *F* test.

5. Monte Carlo data (especially those provided by Olson, 1974) demonstrate an appalling lack of robustness of Manova criteria—especially the gcr test—under some conditions. These results—together with Bird and Hadzi-Pavlovic's evidence that the gcr test's robustness is greatly improved if we confine our attention to interpretable contrasts and combinations of measures—are discussed in section 4.5. At a minimum, however, data such as Olson's suggest that we can no longer be optimistic about the robustness of a multivariate technique until this has been thoroughly investigated for that particular statistic.

6. Nonparametric alternatives to the tests described in this book are under development for use when the data show evidence of gross violation of multivariate normality or homogeneity of covariance matrices (cf., e.g., Cliff, 1996; Eye, 1988; Harwell, 1989; Krishnaiah, 1969; and Sheu & O'Curry, 1996).

The issue of robustness of mutivariate tests has been the focus of a great deal of effort by both mathematical and applied statisticians recently, and large strides are still to be expected. Especially promising has been a growing recognition that these investigations must extend beyond overall, omnibus tests to the specific comparisons that are the ultimate goal of our statistical analyses. Kevin Bird and his co-workers (Bird, 1975; Bird & Hadzi-Pavlovic, 1983; and ongoing research on constrained interactions) seem especially to have a feeling for the most important gaps needful of filling through analytic and Monte-Carlo work.

8.3 NONLINEAR RELATIONSHIPS IN MULTIVARIATE STATISTICS

All of the statistical procedures discussed in this book analyze only the linear relationships among the variables within a set or between the variables of different sets. By this we mean that the general formula relating one variable to another variable or to a set of other variables involves only the variables themselves (no logarithms, squares, square roots, and so on), connected only by plus or minus signs (no multiplicative combinations). The exception to this is that independent, group-membership variables are treated as though they were nominal variables, with no attempt to specify the form of their functional relationship to other variables—except through specific contrasts (à la Scheffé) used to specify the source of significant overall effects. Thus the multivariate procedures in which the question of linearity arises are:

1. Those involving combinations of continuous predictor variables, such as the predicted score in multiple regression, the canonical variate in canonical analysis, and the factor score in PCA or factor analysis.
2. Those involving combinations of outcome variables, such as the discriminant function in T^2 or Manova, the canonical variate in canonical analysis, and the principal component in PCA.

(Note that the canonical variate is mentioned under both categories, as is appropriate to the symmetry of canonical analysis, and that considering the hypothetical variables—principal components, factors—in PCA and FA as the independent variables is the traditional, but not the necessary, way of looking at these analyses.) Of these various multivariate procedures, multiple regression is the one whose users are most apt to plan specifically for, test for, and adjust for nonlinearity, although there is certainly just as great a likelihood of nonlinear relationships obscuring the results of the other procedures.

With respect to linearity, we are in much the same situation as with normality and with interval scale measurement: almost no variables have a truly linear relationship over their full range, and yet a great many have a nearly linear relationship over the ranges apt to be included in a given study. Thus, for instance, Weber's law (that any two physical stimuli having a given *ratio* of physical energies will yield the same *difference* in resulting subjective sensations) is known to break down for very low or very high levels of physical energy, but is nevertheless a very useful empirical law over the broad middle range of physical intensity. Moreover, most kinds of nonlinear relationships that can be anticipated on theoretical grounds can be incorporated into a linear model by use of appropriate transformations.

To illustrate this last point, consider a study by Steele and Tedeschi (1967) in which they attempted to develop an index based on the payoffs used in an experimental game that would be predictive of the overall percentage of cooperative choices C made in that game by a pair or pairs of subjects. They had each of 42 subject pairs play a different two-person, two-choice game, with the payoffs for the various games being generated randomly. Four payoffs, labeled as T, R, P, and S, were used in each of the 42 games. On theoretical grounds, the authors expected the best index to be a ratio of some two of the following measures: T, R, P, S, $T - R$, $T - P$, $T - S$, $R - P$, $R - S$, $P - S$, or a square root or logarithmic transformation of such a ratio. This unfortunately yields 135 "candidates" for best index. The authors selected among these (and 75 other possible indices generated unsystematically) by correlating them all with C, then using the 15 having the highest correlation with C as predictor variables in a multiple regression analysis, and finally comparing the squared regression weights each of these 15 attained in the best predictor equation. Rather than detailing the many ways in which these procedures betray a lack of understanding of the goals of multiple regression analysis, let us consider a more efficient approach to the authors's initial question. The stipulation that the index should involve the ratio of 2 of the 10 measures just specified can be formulated somewhat more generally as

$$C = \prod_{i=1}^{10} c_i^{\beta_i} \tag{8.1}$$

where $c_1 = T$; $c_2 = R$; . . . ; $c_{10} = P - S$; and β_i is an exponent associated with term *i*. Thus, for instance, the index $(T - R)/(T - S)$ is obtained from Equation (8.1) by taking $\beta_5 = 1$, $\beta_7 = -1$, and all other $\beta_i = 0$. This looks very far from a linear relationship. However, taking logarithms of both sides of Equation (8.1) gives

$$\log C = \sum_{i=1}^{10} \beta_i \tag{8.2}$$

Thus we can apply standard linear multiple regression procedures to Equation (8.2), letting $Y = \log C$ and $X_i = \log c_i$. The b_i that optimally fit Equation (8.2) will provide optimal exponents for Equation (8.1) as well. If we now ask for the optimal weights under the assumption that the ideal index is logarithmically related to the ideal ratio, that is, that

$$C = \log\left(\prod_{i=1}^{10} c_i \beta_i\right) = \sum_{i-1}^{10} \beta_i \log c_i,$$

we see at once that standard multiple regression analysis again applies with the same predictor variables as before, but this time with "raw" cooperation percentage as the predicted variable. If, finally, we assume that the ideal index is the square of a ratio of differences, we need only take $\log(C^2) = 2 \log C$ as our predicted variable. The transformations suggested here may of course lead to distributions sufficiently nonnormal to be a problem in significance testing.

This example is not meant to imply that all situations involving nonlinearity in which the precise nature of the relevant functions can be spelled out on a priori grounds can be reduced by appropriate transformations to linear form. For instance, had we wanted to find the best function of the payoffs to use as a predictor of a whole set of measures of the outcomes of a game (e.g., CC, CD, and DC as defined in Data Set 4, Table 3.2), where the function of the payoffs is to be multiplicative and only linear combinations of the outcome measures are to be considered, we could *not* reduce the problem to the usual (linear) canonical correlation procedures. However, most a priori specifications can be reduced to linear combinations of transformed variables. If not, there is little that can be done beyond taking the linear formulae (or polynomial expressions—see later discussion) as approximations to the true combining rule.

Where no a priori specification of the nature of any nonlinearity can be made, several options are open to the investigator, including

1. Proceeding with the usual linear analysis on grounds either of convenience or of the ease of interpreting linear combinations, recognizing that measures of relationship (multiple *R*, canonical *R*, T^2, gcr in Manova, or the percent of variance accounted for by PC_1) may be lower than if transformed variables or additional, nonlinear terms were used.
2. Performing an initial linear analysis, followed by a test for nonlinearity and a modified analysis if and only if the test reveals statistically reliable nonlinearity.
3. Performing an analysis employing a polynomial model (squares, cubes, and so on, and cross products of the original variables), which is highly likely to provide a close approximation to *any* function, followed by deletion of terms in the model that do not make a statistically significant contribution.
4. Converting the continuous variables into nominal, group-membership variables and performing factorial Anova instead of multiple regression, or Manova instead of canonical analysis.

Approach 1 needs no further explanation, except to point out that it is a particularly compelling choice in the case of PCA or FA, where the "predictor" variables are unobservable anyway. Procedure 2 is an extension to other multivariate techniques of the "step-up" variety of stepwise multiple regression procedures that are employed by or an option in many MRA programs, where the variables to be added are squares, cubes, and cross products of the original variables. Formal significance tests for such stepwise procedures were presented in Table 2.2. Similar tests for T^2, Manova, and Canona are less well known, although Dempster (1969) applied stepwise procedures to T^2 and cited the work of Pothoff and Roy (1964) for the application to Manova. For all of these techniques, the most intuitively meaningful measure of the improvement provided by addition of one or more terms is the increase in the square of the correlation between the augmented set of variables and the other variable or set of variables in the situation. For multiple regression and canonical analysis this measure is clearly multiple R^2 and R_c, respectively. Because T^2 and Manova are (as was shown in chap. 4) special cases of Canona, the corresponding measure for these techniques is the change in $\theta = \lambda/(1 + \lambda)$, where λ is the gcr of $\mathbf{E}^{-1}\mathbf{H} = SS_{a,D}/SS_{w,D}$ (the "stripped" *F* ratio computed on the discriminant function D) in the general case of Manova and equals $T^2/(T^2 + N_1 + N_2 - 2)$ in the special case of Hotelling's T^2. For each of these two techniques, θ represents the square of the canonical correlation between the group membership variables and the outcome measures, and thus provides a simple, readily interpretable measure of the importance of the additional terms.

Such formal procedures may well be preceded by "eyeball tests" of the curvilinearity of relationships. It was pointed out in chapter 2 (cf. especially the demonstration problem) that the best visual check of the linearity assumption in multiple regression is provided by a plot of Y versus $\hat{Y}$ or versus $Y - \hat{Y}$ (i.e., against the predicted or the residual scores on Y). The primary basis for this conclusion is the fact that curvilinearity of the relationship between Y and a given predictor variable X_i does *not* necessarily imply curvilinearity in the relationship between Y and $\hat{Y}$—which is all that is necessary in multiple regression analysis. By similar arguments it can be seen that a plot of C_x versus C_y, where C_x and C_y are the canonical variates of the left- and right-hand sets of variables, is the best overall visual check for curvilinearity in canonical analysis.

Approach 3 is practical only if the investigator has available a very large number of observations, because the total number of variables involved if k original variables, their squares, their cubes, and the cross-products of all these variables are included is $3k + 3k(3k - 1)/2$—which increases quite rapidly with k and has the potential for exhausting the degrees of freedom for the analysis quite rapidly. (It is important, however, if one is interested in reasonable estimates of the linear term, to use powers and cross products of deviation scores.)

Approach 4 has the major problem of being rather sensitive to the particular cutting points established for each variable's set of categories, as well as yielding some very small subgroups of subjects having particular combinations of scores on the various measures, unless a large amount of data is available.

8.4 THE MULTIVARIATE GENERAL LINEAR HYPOTHESIS

All of the significance tests described in chapters 2 through 5—including specific comparisons explaining statistically significant overall measures—can be represented as tests of a single, very general hypothesis: the ***multivariate general linear hypothesis*** (mgl hypothesis). The mgl hypothesis can be stated as

$$\mathbf{C\beta M = 0}, \tag{8.3}$$

where

$$\beta = \begin{bmatrix} \beta_{1,1} & \beta_{1,1} & \cdots & \beta_{1,1} \\ \beta_{1,1} & \beta_{1,1} & \cdots & \beta_{1,1} \\ \mathrm{M} & \mathrm{M} & \mathrm{O} & \mathrm{M} \\ \beta_{1,1} & \beta_{1,1} & \cdots & \beta_{1,1} \end{bmatrix}$$

is a q x p matrix giving the parameters relating the q predictor variables (usually group-membership variables) to the p outcome measures via the multivariate general linear model (mgl model),

$$\mathbf{Y} = \mathbf{X}\beta + \varepsilon. \qquad (8.4)$$

where **Y** is an N x p matrix of scores of the N subjects on the p outcome measures and is of rank t; **X** is an N x q matrix of scores on the predictor variables, is usually referred to as the design matrix, and is of rank r; ε is an N x p matrix of residual scores, that is, of errors of prediction, which, for purposes of significance testing, are assumed to be sampled from a multivariate normal population having mean vector (0, 0, ... , 0) and covariance matrix Σ of full rank p. Thus β is essentially a matrix of population regression coefficients. Referring back to Equation (8.3), **C** is a g x q ($g \leq r$, the rank of **X**) matrix whose jth row specifies a linear combination of the q parameters that relate the predictor variables to the jth outcome measure, and **M** is a p x u matrix ($u \leq t$, the rank of **Y**) whose ith column specifies a linear combination of the p parameters that relate the outcome measures to the ith predictor variable. Thus, for instance, taking

$$\mathbf{C} = \begin{bmatrix} 1 & 0 & 0 & -1 \\ 0 & 1 & 0 & -1 \\ 0 & 0 & 1 & -1 \end{bmatrix} \text{ and } \mathbf{M} = \begin{bmatrix} 1 & 0 & 0 & \cdots & 0 \\ 0 & 1 & 0 & \cdots & 0 \\ 0 & 0 & 1 & \cdots & 0 \\ \vdots & \vdots & \vdots & \ddots & \vdots \\ 0 & 0 & 0 & \cdots & 1 \end{bmatrix},$$

$$\text{with } \mathbf{X} = \begin{bmatrix} 1 & 0 & 0 & 1 \\ 1 & 0 & 0 & 1 \\ \vdots & \vdots & \vdots & \vdots \\ 1 & 0 & 0 & 1 \\ 0 & 1 & 0 & 1 \\ \vdots & \vdots & \vdots & \vdots \\ 0 & 1 & 0 & 1 \\ 0 & 0 & 1 & 1 \\ \vdots & \vdots & \vdots & \vdots \\ 0 & 0 & 1 & 1 \end{bmatrix}$$

(thus setting $g = 3$, $q = 4$, $u = p$), yields a test of the hypothesis that

$$\begin{bmatrix} \beta_{1,1} - \beta_{4,1} \\ \beta_{2,1} - \beta_{4,1} \\ \beta_{3,1} - \beta_{4,1} \end{bmatrix} = \begin{bmatrix} \beta_{1,2} - \beta_{4,2} \\ \beta_{2,2} - \beta_{4,2} \\ \beta_{3,2} - \beta_{4,2} \end{bmatrix} = \cdots = \begin{bmatrix} \beta_{1,p} - \beta_{4,p} \\ \beta_{2,p} - \beta_{4,p} \\ \beta_{3,p} - \beta_{4,p} \end{bmatrix} = \begin{bmatrix} 0 \\ 0 \\ 0 \end{bmatrix},$$

where $\hat{\beta}_{ij} - \hat{\beta}_{4j}$ (the least-squares estimates of the corresponding population parameters) will be equal to $\overline{Y}_{i,j} - \overline{\overline{Y}}$; β_1 to β_3 thus correspond to μ_1 to μ_3; and β_4 corresponds to $\overline{\mu} = (\mu_1 + \mu_2 + \mu_3)/3$. In other words, these choices of matrices yield the overall null hypothesis for a three-group, p-outcome-measure Manova. (Note that the direct equivalent of Equation (8.3) is a statement of equality between a 3 x p matrix of differences between βs whose columns are the column vectors listed above and a 3 x p matrix consisting entirely of zeros.) Similarly, taking **C** as before but letting

$$\mathbf{M} = \begin{bmatrix} 2 \\ -1 \\ -1 \\ 0 \\ \cdot \\ \cdot \\ \cdot \\ 0 \end{bmatrix}$$

yields the null hypothesis that none of the three groups has a population mean difference other than zero between Y_1 and the average of Y_2 and Y_3. On the other hand, letting **M** be an identity matrix and letting

$$\mathbf{C} = [1 \quad -1 \quad 0 \quad 0]$$

yields the null hypothesis that groups 1 and 2 do not differ in their population mean outcome vectors.

If the rows of **C** meet an ***estimability condition*** (which is always fulfilled by any ***contrast*** among the columns of β, that is, by any set of coefficients that sum to 0), then the null hypothesis stated in Equation (8.3) is tested by computing the greatest characteristic root (gcr) of $\mathbf{E}^{-1}\mathbf{H}$, where

$$\mathbf{E} = \mathbf{M'Y'[I - X_1(X_1'X_1)^{-1}X_1']YM}; \tag{8.5a}$$

and

$$\mathbf{H} = \mathbf{M'Y'X_1(X_1'X_1)^{-1}C'[C(X_1'X_1)^{-1}C']^{-1}C(X_1'X_1)^{-1}X_1'YM}, \tag{8.5b}$$

where $\mathbf{X}_1$ is an N x r submatrix of **X** that has the same rank as **X**. (Such a submatrix is called a ***basis*** for **X**, because the subjects' scores on the remaining $q - r$ independent variables can be determined via linear equations from their scores on these first r measures.) $\mathbf{C}_1$ is, according to Morrison (1976, p. 176), **C** partitioned in conformance with our partitioning of **X**. This gcr is then compared to the appropriate critical value in the gcr tables of Appendix A with degree-of-freedom parameters

$$s = \min(g, u); \quad m = (|g - u| - 1)/2; \quad n = (N - r - u - 1)/2. \qquad (8.6)$$

This is a very broad model that subsumes overall tests, profile analysis, specific comparisons, and a wide range of other tests. One approach to teaching multivariate statistics would be to begin with the mgl hypothesis and present all other significance tests as special cases thereof. This book eschews that approach for both didactic and practical reasons. On didactic grounds, the multivariate general linear hypothesis is simply too abstract to be readily digestible until the user has built up a backlog of more specific procedures. Having developed some understanding of and experience with multiple regression, T^2, Manova, and canonical analysis, the user can be impressed with and perhaps put to use the elegance and generality of the mgl model; without such a background, the main reaction is apt to be one of awe, fear, or bewilderment. On the practical level, Equations (8.5a) and (8.5b) are much more complex than the expressions we had to deal with in chapters 2 through 5. Those two equations simply do not lend themselves to "hand" exploration of even the simplest cases. Even when a computer program for the mgl model is available, the very generality of the model means that the user must devote considerable labor to selecting the **C**, **M**, and design matrices for even the simplest hypotheses, and this procedure must be repeated for each hypothesis to be tested. In a case in which a canned program for a specific procedure exists, the user will generally find it much more convenient to use it rather than resorting to a mgl hypothesis program.

That said, we must nevertheless recognize that SAS PROC GLM and SPSS MANOVA have been given such generality that they are really mglm programs "in disguise," with construction of the appropriate design, contrast, and so on matrices being triggered by simple commands reminiscent of the specific techniques that are special cases of the mglm.

For the researcher who wishes to be explicit about his or her use of the mgl model, Morrison (1976, chap. 5) provides listings of the **C**, **M**, and design matrices for many of the most often used statistical procedures. Pruzek (1971), in a paper that incorporated many of the suggestions for "reform" of statistics that were discussed in section 1.0, illustrated the utility of the mgl model as a unifying tool in discussing methods and issues in multivariate data analysis.

However, I agree with one of the referees of this text that it seems a bit of a letdown to have finally arrived at the most general model in the armament of multivariate statisticians and then to be simply passed along to other authors. If you agree, you might wish to consider the following example. If not (note the "#"), you may simply skip over the example to the paragraph beginning "The main point. . . ."

Example 8.1 *Unbalanced MANOVA via the Multivariate General Linear Model.* Let's at least examine one application of the multivariate general linear model in detail: testing the A main effect for the highly unbalanced 2 x 2 Manova used as an example in exploring the relationship between Canona and Manova in Derivation 5.3. Following Morrison (1976, p. 187), we give the design matrix here. The first three rows represent the three cases in the A_1B_1 condition, the next five rows represent the A_1B_2 condition, and so on—except that ellipses (dots) are used in the A_2B_1 condition to represent another eight rows of the same pattern. Columns 1 and 2 represent α_1 and α_2; the next two columns represent β_1 and β_2; columns 5 through 8 represent the four $\alpha\beta_{ij}$ parameters; and column nine represents μ in the usual factorial model for each dependent measure m,

$$Y_{ijm} = \mu + \alpha_{im} + \beta_{jm} + \alpha\beta_{ijm} + \varepsilon_{ijm} :$$

$$\mathbf{X} = \begin{bmatrix} \multicolumn{2}{c}{\alpha_i} & \multicolumn{2}{c}{\beta_j} & \multicolumn{4}{c}{\alpha\beta_{ij}} & \mu \\ 1&0&1&0&1&0&0&0&1 \\ 1&0&1&0&1&0&0&0&1 \\ 1&0&1&0&1&0&0&0&1 \\ \hline 1&0&0&1&0&1&0&0&1 \\ 1&0&0&1&0&1&0&0&1 \\ 1&0&0&1&0&1&0&0&1 \\ 1&0&0&1&0&1&0&0&1 \\ 1&0&0&1&0&1&0&0&1 \\ \hline 0&1&1&0&0&0&1&0&1 \\ \vdots&\vdots&\vdots&\vdots&\vdots&\vdots&\vdots&\vdots&\vdots \\ 0&1&1&0&0&0&1&0&1 \\ \hline 0&1&0&1&0&0&0&1&1 \\ 0&1&0&1&0&0&0&1&1 \end{bmatrix}$$

Our first task is to find a basis for **X,** that is, r columns of **X** from which all 9 - r other columns can be generated as linear combinations of the basis columns. Knowing that among-group differences have 3 degrees of freedom, we're tempted to assume **X** has a rank of 3; this ignores, however, the grand mean (the "constant effect"), so that **X** actually has a rank of 4. (If in doubt, let your local matrix manipulation program compute the eigenvalues of **X'X**; exactly 4 of them will be nonzero.) Because the columns within each set (the first two representing the A main effect; the third and fourth the B main effect; and the fifth through eighth the AB interaction effect) sum to a unit vector (column 9), we know they're linearly dependent, so we take one column from each set, yielding an $\mathbf{X}_1$ of

$$\mathbf{X}_1 = \begin{bmatrix} 1\,1\,1\,1 \\ 1\,1\,1\,1 \\ 1\,1\,1\,1 \\ ---- \\ 1\,0\,0\,1 \\ 1\,0\,0\,1 \\ 1\,0\,0\,1 \\ 1\,0\,0\,1 \\ 1\,0\,0\,1 \\ ---- \\ 0\,1\,0\,1 \\ \vdots\,\vdots\,\vdots\,\vdots \\ 0\,1\,0\,1 \\ ---- \\ 0\,0\,0\,1 \\ 0\,0\,0\,1 \end{bmatrix}.$$

Because we are interested in the overall effect for both measures, **M** is simply a 2 x 2 identity matrix.

$$\mathbf{X}_1'\mathbf{X}_1 = \begin{bmatrix} 8 & 3 & 3 & 8 \\ & 13 & 3 & 13 \\ & & 3 & 3 \\ & & & 20 \end{bmatrix}$$

and

$$(\mathbf{X}_1'\mathbf{X}_1)^{-1} = \begin{bmatrix} .7 & .5 & -.7 & -.5 \\ & .6 & -.6 & -.5 \\ & & 1.133 & .5 \\ & & & .5 \end{bmatrix}.$$

$\mathbf{X}_1(\mathbf{X}_1'\mathbf{X}_1)^{-1}\mathbf{X}_1'$ is a 20 *x* 20 matrix (more generally, N x N, which must surely strain the limits of available matrix algebra programs for large data sets) that I choose not to display, as is of course the identity matrix minus that matrix. The end result is that

$$\mathbf{E} = \begin{bmatrix} 60 & 33 \\ 33 & 60 \end{bmatrix},$$

which is identical to SPSS MANOVA's result. So far, this is all straightforward, although we might feel a bit uneasy about what appears to be (1, 0), rather than (1, 0, -l), coding of our "predictors" (see section 2.7).

Next we need to select the appropriate contrast for the *A* main effect. In terms of the unpartitioned **C** matrix, this is simply [l -1 0 0 0 0 0 0 01. Thus, following Morrison's stipulation, we drop the same columns of **C** that we dropped from $\mathbf{X}_1$, yielding $\mathbf{C}_1$ = [1 0 0 0]. Here again we feel uneasy, because this appears to correspond to a test of the hypothesis that $\alpha_1 = 0$. This time our fears are justified, because carrying through the matrix multiplications of Equation (8.5b) yields

$$\mathbf{H} = \begin{bmatrix} 22.857 & -11.429 \\ & 5.714 \end{bmatrix}$$

What does this correspond to? It's definitely not the **H** matrix computed by MANOVA for this effect, nor is it the **H** matrix MANOVA would have computed had we used a sequential testing procedure in which the *A* main effect was tested without correcting for confounds with *B* or *AB*. (Try it.) Nor does the intermediate matrix, $(\mathbf{X}_1'\mathbf{X}_1)^{-1}\mathbf{X}_1'\mathbf{Y}$—whose columns should contain parameter estimates for our two dependent variables—seem to correspond to such estimates. That matrix would be

$$\begin{array}{c} \\ A_1B_1 \\ A_1B_2 \\ A_2B_1 \\ A_2B_2 \end{array} \begin{array}{c} \begin{array}{cc} \bar{Y}_1 & \bar{Y}_2 \end{array} \\ \begin{bmatrix} -4 & 2 \\ -4 & 4 \\ 6 & -4 \\ 10 & 6 \end{bmatrix} \end{array}$$

[The $\overline{\mathbf{Y}}$ matrix, the matrix of unweighted means for the levels of ***A***, and the matrix of weighted means (essentially ignoring levels of **B**) are, respectively,

$$\begin{bmatrix} 8 & 8 \\ 6 & 8 \\ 6 & 10 \\ 10 & 6 \end{bmatrix}, \begin{bmatrix} 7 & 8 \\ 8 & 8 \end{bmatrix}, \text{ and } \begin{bmatrix} 6.75 & 8.00 \\ 6.67 & 9.33 \end{bmatrix}.$$

The problem would appear to be in our choice of **C**. However, I'm unable to think of any linear combination of α_1, β_1, $\alpha\beta_{11}$, and μ that will be equal to $\alpha_1 - \alpha_2$ —at least not without putting additional side conditions on the parameters of the model and thus effectively changing the design matrix. Apparently, simply dropping the same columns from **C** as one does from **X** is not, in the factorial design, sufficient to produce a reasonable $\mathbf{C}_1$, nor in fact does (0, 1) coding provide any clear way to come up with a linear combination of the basis parameters to test the effects that are of interest to us.

The usual side conditions are that

$$\sum_i a_i = \sum_j \beta_j = \sum_i \alpha\beta_{ij} = \sum_j \alpha\beta_{ij} = 0.$$

This then implies that $\alpha_2 = -\alpha_1$, $\beta_2 = -\beta_1$, $\alpha\beta_{12} = -\alpha\beta_{11}$, and so on, leading to an $\mathbf{X}_1$ of

$$\mathbf{X}_1 = \begin{bmatrix} 1 & 1 & 1 & 1 \\ 1 & 1 & 1 & 1 \\ 1 & 1 & 1 & 1 \\ 1 & 1 & 1 & 1 \\ \\ 1 & -1 & -1 & 1 \\ 1 & -1 & -1 & 1 \\ 1 & -1 & -1 & 1 \\ 1 & -1 & -1 & 1 \\ 1 & -1 & -1 & 1 \\ \\ -1 & 1 & -1 & 1 \\ \vdots & \vdots & \vdots & \vdots \\ -1 & 1 & -1 & 1 \\ \\ -1 & -1 & 1 & 1 \\ -1 & -1 & 1 & 1 \end{bmatrix}.$$

Further calculations give us $\mathbf{C}_1 = [\,1\ 0\ 0\ 0]$—as before, but more meaningfully this time and also $\mathbf{X}'_1\mathbf{X}_1$ and $(\mathbf{X}'_1\mathbf{X}_1)^{-1}$ matrices

$$\begin{bmatrix} 20 & -10 & 6 & 4 \\ & 20 & -4 & 6 \\ & & 20 & -10 \\ & & & 20 \end{bmatrix} \text{ and } \begin{bmatrix} .07083 & .03333 & -.01667 & -.00417 \\ & .07083 & -.00417 & -.01667 \\ & & .07083 & .03333 \\ & & & .07083 \end{bmatrix}.$$

$\mathbf{C}_1(\mathbf{X}'_1\mathbf{X}_1)^{-1}\,\mathbf{C}'_1$ is simply a 4 x 4 matrix consisting entirely of all zeroes except for a (1,1) entry of .07083. $(\mathbf{X}'_1\mathbf{X}_1)\,\mathbf{X}'_1\mathbf{Y}$ equals

$$\begin{bmatrix} -.5 & 0 \\ -.5 & 1 \\ 1.5 & -1 \\ 7.5 & 8 \end{bmatrix}$$

These entries *do* match up with appropriate parameter estimates when we recall the equivalence of regression estimates in the full 2 x 2 model and in an unweighted-means analysis. Thus, for instance, 7.5 is the unweighted mean of the 4 means on Y_1 [(8 + 6 + 6 + 10)/4], and +1—the (2, 1) entry of the above matrix—comes from $\hat{\beta}_1$ for Y_2 = [(8 + 10)/2 - (8 + 6)/2]/2 = (9 - 7)/2 = 8. Putting all these intermediate calculations together gives us

$$\mathbf{H} = \begin{bmatrix} 3.5294 & 0 \\ 0 & 0 \end{bmatrix},$$

which matches both the **H** matrix computed by MANOVA for the A main effect and the univariate sums of squares in unweighted-means analyses of Y_1 and Y_2 (which should, of course, appear on the main diagonal of **H**). Thus, for instance, the unweighted average on Y_2 of the A_1B_1 and A_1B_2 means identically equals the average of the two means for level 2 of A [(8 + 8)/2 - (10 + 6)/2 = 8 - 8 = 0, so that sum of squares should equal zero. For Y_1, our two unweighted averages of means are (8 + 6)/2 = 7 and (6 + 10)/2 = 8, so that our $SS_A = n_h[2(7 - 7.5)^2 + 2(8 - 7.5)^2] = 3.528*(2)(.25 + .25) = 3.529$. [$n_h = 4/(1/3 + 1/5 + 1/10 + 1/2)$.] Because the **H** and **E** matrices match those used in Derivation 5.3, our mglm analysis will yield the same greatest characteristic root and the same significance test results as did the SPSS MANOVA run.

The main point of Example 8.1 is that, even though all the analyses we discussed in chapters 2 through 5 can be seen as special cases of the mglm, it really doesn't provide a very convenient "grand model." It is much easier to keep a sound intuitive feeling for what you're doing if you instead think of Canona as the most general case. However, the fact that all of the hypotheses contained in the mgl hypothesis are supposed to be testable, whether or not **X** consists entirely of discrete variables, suggests that some of my concerns about "missing" significance tests in Canona might be alleviated by more diligent study of their equivalents in the multivariate general linear hypothesis.

8.5 STRUCTURAL EQUATION MODELING

8.5.1 General Approach and Some Examples

A technique that has a legitimate claim to being a relatively user-friendly generalization and extension of the techniques we've discussed so far is ***structural equations modeling*** (SEM), which essentially combines path analysis (section 2.9) with factor analysis (chap. 7). The factor analytic (measurement) portion of the model expresses the relationship between the observable ("manifest") variables and a usually much smaller number of factors (latent variables), and the path analytic portion of the model expresses the relationships among the latent variables. It would seem that the factors will have to be oblique if the path portion of the model is to be at all interesting – but we'll challenge that assumption toward the end of this section.

Example 8.2 *Path Analysis of Scarr and Weinberg (1986) via SEM* We begin with an analysis that involves only observable ("manifest") variables, namely the path analysis we carried out via multiple-regression analyses back in chapter 2 (Example 2.5). The setup for this analysis in CALIS LinEqs is as follows:

```
OPTIONS Nocenter LS = 72;
TITLE "Scarr Model, PROC CALIS";
```

```
Data SCARR (TYPE = CORR);
  _Type_ = 'CORR';
  Input _NAME_ $
        WAIS YRSED DISCIPL POSCTRL KIDSIQ;
  Cards;
WAIS     1.0 .560 .397 .526 .535
YRSED    .560 1.0 .335 .301 .399
DISCIPL .397 .335 1.0 .178 .349
POSCTRL .526 .301 .178 1.0 .357
KIDSIQ  .535 .399 .349 .357 1.0
;
PROC CALIS DATA=SCARR  METHOD=ML  NOBS=105 PALL;
   Title2 "Replicating SCARR PA via SEM ";
   Title3 " Using LineEQS within CALIS";
   LinEqs
      DISCIPL      = p32 YRSED + D3,
      POSCTRL     = p42 YRSED + D4,
   KIDSIQ     = p51 WAIS + p53 DISCIPL + p54 POSCTRL + D5;
   STD YRSED WAIS = 2*1.0 , D3 - D5 = U3 - U5;
RUN;
```

The output from this run includes:

```
Number of endogenous variables = 3
Manifest:      DISCIPL    POSCTRL    KIDSIQ

    Number of exogenous variables = 5
Manifest:      WAIS       YRSED
Error:         D3         D4         D5

Predicted Model Matrix

               WAIS        YRSED     DISCIPL      POSCTRL        KIDSIQ
WAIS          1.0000      0.5600      0.1876       0.1686       0.46030
YRSED         0.5600      1.0000      0.3350       0.3010       0.31905
DISCIPL       0.1876      0.3350      1.0000       0.1008       0.25447
POSCTRL       0.1686      0.3010      0.1008       1.0000       0.19751
KIDSIQ        0.4603      0.3191      0.2545       0.1975       0.93584
Determinant = 0.3783 (Ln = -0.972)
```

[Strangely, the correlation of KIDSIQ with itself is reproduced as .936, indicating a failure in the optimization procedure. CALIS notices this and issues a warning a page or two later.]

```
Scarr Model, PROC CALIS       14:24 Saturday, August 19, 2000   6
Replicating SCARR PA via SEM
 Using LineEQS within CALIS
```

```
Covariance Structure Analysis: Maximum Likelihood Estimation

Fit criterion . . . . . . . . . . . . . . . . . .         0.3194
Goodness of Fit Index (GFI) . . . . . . . . . . .         0.8929
GFI Adjusted for Degrees of Freedom (AGFI). . . .         0.5985
Root Mean Square Residual (RMR) . . . . . . . . .         0.1233
Parsimonious GFI (Mulaik, 1989) . . . . . . . . .         0.3572
Chi-square = 33.2221       df = 4       Prob>chi**2 = 0.0001
Null Model Chi-square:     df = 10                  134.3211
```

[Notice how close the overall chi-square test is to the 32.285—also with 4 *df*—obtained in our earlier, MRA-based analysis.]

```
Residual Matrix
                 . . .

Average Absolute Residual = 0.07446
Average Off-diagonal Absolute Residual = 0.1053
WARNING: Fitting a correlation matrix should lead to
    insignificant diagonal residuals.
         The maximum value 0.0641608139 of the diagonal residuals
    may be too high for a valid chi**2 test statistic and
    standard errors.

Rank Order of 7 Largest Residuals

   POSCTRL,WAIS    DISCIPL,WAIS    KIDSIQ,POSCTRL    KIDSIQ,DISCIPL
     0.35744000      0.20940000        0.15948785        0.09452855

     KIDSIQ,YRSED  POSCTRL,DISCIPL       KIDSIQ,WAIS
       0.07994875       0.07716500        0.07469787
```

[This printout of the reproduced correlation matrix, the differences between observed and reproduced correlations, and singling out of the largest discrepancies is certainly more convenient than the hand calculations we used in chapter 2.]

```
Manifest Variable Equations

 DISCIPL =     0.3350*YRSED   + 1.0000 D3
 Std Err       0.0924 P32
 t Value       3.6259

 POSCTRL =     0.3010*YRSED   + 1.0000 D4
 Std Err       0.0935 P42
 t Value       3.2189
```

```
KIDSIQ  =    0.1662*DISCIPL + 0.1116*POSCTRL + 0.4103*WAIS
 Std Err      0.0827 P53        0.0824 P54        0.0835 P51
 t Value      2.0104            1.3542            4.9158
                                                          + 1.0000 D5
```

[The above path-coefficient estimates are identical, to 3 decimal places, to our earlier, MRA-based estimates. The Condition 9' tests provided by CALIS LinEqs, however, are slightly inflated: 3.626 vs. 3.608, 3.219 vs. 3.203, 2.010 vs. 1.855, etc. This is to expected, given that the LinEqs tests are asymptotic. Where available, the MRA-based tests are more accurate, especially with small sample sizes. As the SAS/STAT online User's Guide (SAS Institute, Inc., 1999) puts it, these tests are "valid only if the observations are independent and identically distributed, the analysis is based on the nonstandardized sample covariance matrix **S**, and the sample size *N* is sufficiently large (Browne 1982; Bollen 1989b; Jöreskog and Sörbom 1985)" .]

```
Lagrange Multiplier and Wald Test Indices _PHI_[5:5]
Symmetric Matrix
Univariate Tests for Constant Constraints
```

[These are tests of the implicit assumptions that disturbance terms are uncorrelated with each other and with the exogenous manifest variables. because "freeing up" these correlations amounts to confessing that you've omitted one or more relevant variables from your model, this should be the last place you look for possible modifications to your model.]

```
Lagrange Multiplier and Wald Test Indices _GAMMA_[3:2]
General Matrix
Univariate Tests for Constant Constraints
-------------------------------------------
|  Lagrange Multiplier   or   Wald Index   |
-------------------------------------------
|  Probability  | Approx Change of Value |
-------------------------------------------
```

[The Gamma matrix holds paths from exogenous to endogenous variables. The Lagrange Multiplier tests are single-df chi-square tests of the paths our model claims are zero (the ones not labeled in brackets).]

```
                        WAIS                          YRSED
DISCIPL          7.484                      13.147   [P32]
                 0.006   0.305

POSCTRL         21.287                      10.361   [P42]
                 0.000   0.521

KIDSIQ          24.166   [P51]               1.639
                                             0.200   0.135
```

```
Lagrange Multiplier and Wald Test Indices _BETA_[3:3]
Univariate Tests for Constant Constraints
-----------------------------------------
| Lagrange Multiplier  or  Wald Index   |
-----------------------------------------
| Probability  | Approx Change of Value |
-----------------------------------------
```

[The Beta matrix holds the path coefficients for relationships among the exogenous variables. This output thus provides the rest of our Condition 10 tests.]

```
                DISCIPL              POSCTRL              KIDSIQ

DISCIPL         SING                 0.767                3.018
                .       .            0.381   0.085        0.082   0.380

POSCTRL         0.767                SING                 11.619
                0.381   0.087        .       .            0.001   0.750

KIDSIQ          4.042   [P53]        1.834   [P54]        SING
                                                          .       .
```

[If these p values are compared to our earlier, MRA-based Condition 10 test p values, these prove a bit liberal. Again, not surprising.]

Example 8.3 *All Three Colleges in the Faculty Salary Example* Towards the end of section 2.9 (as a prologue to Example 2.6) we indicated that the obvious assumption with respect to the relationship between the two group-membership variables (g.m.v.s) representing which of the three colleges the faculty member had been hired into was that of reciprocal causation—which thereby rules out the usual MRA solution for path coefficients. The appropriate model can, however, be set up within CALIS LinEqs (or any other SEM program) and turned over to the tender mercies of the maximum-likelihood search for parameter estimates—provided that our model is identified. With two separate path coefficients from g.m.v. 1 to g.m.v. 2 and from g.m.v. 2 to g.m.v. 1 we have 7 parameters and only 6 correlations from which to estimate them. However, assuming these two path coefficients to be equal reduces the number of parameters to 6 and makes our model just-identified. The LinEqs setup is as follows:

```
OPTIONS Nocenter LS = 72;
TITLE "3-College Salaries, PROC CALIS";
Data FACSAL (TYPE = CORR);
  _Type_ = 'CORR';
  Input _NAME_ $
        MEngvEd MedvEng BeingFm Salary;
  Cards;
MengvEd     1.0000      .1581     -.6383      .7403
MedvEng      .1581     1.0000     -.0118      .7357
BeingFm     -.6383     -.0118     1.0000     -.2584
Salary       .7403      .7357     -.2584     1.0000
;
```

```
PROC CALIS DATA=FACSAL  METHOD=ML  NOBS=260 PALL;
   Title2 "Reciprocal Caus'n  PA via SEM ";
   Title3 " Using LineEQS within CALIS";
   LinEqs
      MEngvEd   = p13 BeingFm + p121 MedvEng + D1,
      MedvEng   = p23 BeingFm + p121 MEngvEd + D2,
      Salary    = p41 MEngvEd + p42 MedvEng + p43 BeingFm + D4;
```

[Note that the equality of the two paths between MEngvEd and MedvEng is implicit in our having given them the same label in the above equations]

```
STD BeingFm = 1.0 , D1 D2 D4 = U1 U2  U4;
```

[Note that we are *not* setting the Femaleness variable to 1.0 for each subject, nor giving the disturbances new names. Rather, we're setting BeingFm's *variance* to 1.0 and indicating that the *variances* of the disturbances are to be left as parameters (labeled U1, U2, and U4) to be estimated from the data.]

```
RUN;
```

The output from this run included the following:

```
Number of endogenous variables = 3
Manifest:      MENGVED   MEDVENG   SALARY

     Number of exogenous variables = 4
Manifest:      BEINGFM
Error:         D1          D2          D4

College Salaries, PROC CALIS   08:21 Sunday, August 20, 2000   6
Reciprocal Caus'n  PA via SEM
 Using LineEQS within CALIS

Covariance Structure Analysis: Maximum Likelihood Estimation

Fit criterion . . . . . . . . . . . . . . . . .          0.0000
Goodness of Fit Index (GFI) . . . . . . . . . . .          1.0000
GFI Adjusted for Degrees of Freedom (AGFI). . . .           .
Root Mean Square Residual (RMR) . . . . . . . . .          0.0000
Manifest Variable Equations
```

[Because this is a just-identified model, we fit the observed correlations perfectly – as would any just-identified model. We are, however, very interested in the estimates of and significance tests of the path coefficients that yield this perfect fit.]

```
 MENGVED =      0.0954*MEDVENG -  0.6372*BEINGFM +  1.0000 D1
 Std Err        0.0297 P121       0.0470 P13
 t Value        3.2101          -13.5469
```

```
MEDVENG =     0.0954*MENGVED +  0.0491*BEINGFM +  1.0000 D2
Std Err       0.0297 P121       0.0643 P23
t Value       3.2101            0.7640

SALARY  =     0.8160*MENGVED +  0.6099*MEDVENG +  0.2696*BEINGFM
Std Err       0.0107 P41        0.0082 P42        0.0106 P43
t Value      76.2309           74.0087           25.5096
                                                   +  1.0000 D4
```

Two of the preceding path coefficient estimates are questionable with respect to Condition 9': The underrepresentation of females in Medicine, relative to Engineering, is small and statistically nonsignificant, and the reciprocal-influence paths between our two g.m.v.s are statistically significant, but smaller than the customary effect-size cutoff for *z*-score path coefficients of .10. This latter is actually a desirable outcome, in that it suggests that we might do a reasonable job of fitting these data with a model that assumes that the "choice" between Medicine and Engineering is independent of (based on different considerations than) the choice between those two fields versus Education. Such a model would finesse one of the questionable aspects of our original model, namely its assumption that the disturbances of the two g.m.v.s are uncorrelated. (This is the major reason that reciprocal causation usually requires additional "instrument variables" we can assume are correlated with each g.m.v. but not with its disturbance term.) With nonzero paths between them, its highly likely that the indirect path from each disturbance through the other g.m.v. would lead to a nonzero correlation with the other g.m.v.'s disturbance. However, uncorrelated g.m.v.s make uncorrelated disturbances quite reasonable.

Dropping P121 from our model specification (i.e., setting the reciprocal influences of the two g.m.v.s on each other to zero) drops our GFI only to .981 (although statistically significantly so), increases our root mean square residual from zero to .083, and leaves unchanged the conclusion that the -.26 overall correlation between salary and being female is the result of a moderately positive direct effect of femaleness on salary that is overwhelmed by the strong, indirect effect of females' being less likely to be hired into one of the two high-paying colleges.

Example 8.4 *Increment to Canonical* R^2 *via CALIS LinEqs?* The plan for this analysis was to demonstrate that SEM could be used to carry out Canonas and thus also to test the significance of increments to R_c^2 resulting from adding one or more variables to one or both sets of variables—i.e., to reduce by one (subject to the asymptotic nature of the SEM tests) the count of "what's missing" from our Canona armamentarium (section 5.4.5). However, the translation of Canona into SEM proved more difficult than anticipated. I knew, of course, that the ability of the canonical variates to account for the between-set correlations is not unique—any rotation of the canonical variates that preserves their zero within-set intercorrelations does just as well in this respect. This is closely analogous to the situation in factor analysis. Because any rotation of a set of factors leaves the reproduced correlation matrix unchanged, SEM cannot distinguish among alternative rotations of a given factor solution and thus cannot be used to carry out exploratory rotations of factors. However, the Canona orientation is unique in that any rotation away from the original solution yields nonzero correlations between the non-matching pairs of canonical variates—for example, rotated canonical variate 1 will now have a nonzero correlation with rotated canonical variate 3. I therefore hoped that incorporating the restriction to zero correlations between non-matching pairs of canonical variates into my SEM model would serve to reproduce the Canona orientation. That didn't happen.

The model I tried was set up in CALIS LinEqs (using the same high-school-versus-college data we employed throughout section 5.5) as follows:

```
TITLE2 'Now Trying Calis LinEqs ';
PROC CALIS DATA=GRADES  METHOD=ML PALL;
```

[PALL asks for all available ouput—a bit of overkill, because the only non-default output I was very interested in was the scoring coefficients, but CALIS had ignored the more specific PLATCOV request in previous runs.]

```
   Title3 "  F1 - F3 are High School canonical variates";
   Title4 "  F4 - F6 are College canonical variates ";
    LinEqs
      MAT_TEST  = H11 F1 + H12 F2 + H13 F3 + E1,
      VER_TEST  = H21 F1 + H22 F2 + H23 F3 + E2,
      CRE_TEST  = H31 F1 + H32 F2 + H33 F3 + E3,
      MAT_GRAD  = C14 F4 + C15 F5 + C16 F6 + E4,
      ENG_GRAD  = C24 F4 + C25 F5 + C26 F6 + E5,
      SCI_GRAD  = C34 F4 + C35 F5 + C36 F6 + E6,
      HIS_GRAD  = C44 F4 + C45 F5 + C46 F6 + E7,
      HUM_GRAD  = C54 F4 + C55 F5 + C56 F6 + E8,
            F4  = rc1 F1 + D1,
            F5  = rc2 F2 + D2,
            F6  = rc3 F3 + D3 ;
```

[It might be more realistic to add equations for F1 through F3 as functions of F4 through F6, but that shouldn't affect the estimates of canonical *r*s, structure coefficients, and scoring coefficients, and leaving F1–F3 as exogenous variables made it possible to set their variances to 1.0 so that rc1, rc2,and rc3 would indeed represent the canonical correlations and H11, C25, etc. would represent structure coefficients.]

```
   STD E1 - E8 = U1 - U8, F1 - F3 = 3*1.;
   BOUNDS 0. <= U1 - U8  <=1. , .0 <= rc1 rc2 rc3 <= 1.;
```

The first problem with the results of this run was that it yielded a message that "LEVMAR optimization could not be completed. More iterations than 50 or more function calls than 125 may be needed." In other words, the optimization routine hadn't converged. I tried giving LinEqs starting values for paths between canonical variates and observed variables that were close to the structure coefficients reported by PROC

CANCORR [this is done by putting the starting value in parentheses immediately after the parameter's label—e.g. "··· + H13(-.527) F3 ···"], but I still got the LEVMAR-optimization error message. I then upped the limits on itnerations and function calls (by adding "maxiter = 250" between ML and PALL) for both of these runs. This time both runs converged, taking a total of 4.32 and 4.11 seconds of cpu time, respectively (as compared to 0.13 cpu seconds for CANCORR). Unfortunately, the two solutions (using default initial values and initial values close to the Canona solution we already had in hand) differed substantially from each other and from the actual canonical correlations and canonical variates: The analytically obtained (PROC CANCORR) R_c^2's were .787, .559, and .298; but the estimates for rc1, rc2, and rc3 in the default-initial-values run were .880, .631, and .012; and the corresponding estimates for the CANCORR-matching-initial-estimates run were .822, .759, and .467.

My first thought was that the LinEqs solutions might represent rotations of the Canona solutions—but that wouldn't account for obtaining higher correlations between the factors (supposedly representing the canonical variates) than the maximum possible value of .787. Clearly there is some constraint that canonical analysis imposes on its emergent variables that the equations I "plugged into" LinEqs do not. The factors derived by LinEqs dutifully match the constraints of zero intercorrelations among themselves and between non-matching pairs of factors; of direct effects only on the variables in one of the two subsets; and of the factors actually being components, i.e. deterministic linear combinations of the observed variables, as revealed by the squared multiple correlations of 1.0 reported for F4, F5, and F6. (F1, F2, and F3 were treated as exogenous and thus didn't have their squared multiple correlations reported.) However, they do *not* satisfy the restriction that the canonical variates for a given set of variables should be a linear combination solely of the variables in that set. For instance, in the first of the two LinEqs runs the regression equation for scores on F1 (representing canonical variate 1 for the high-school measures) was `0.02631*MAT_TEST -.38753*VER_TEST -.48380* CRE_TEST + .09217* MAT_GRAD -.05290* ENG_GRAD -.61887* SCI_GRAD - .00613* HIS_GRAD -.24783* HUM_GRAD`. For a well-behaved canonical variate the last five coefficients should all be zero. Because we didn't succeed in requiring that LinEqs define each factor as a function only of the variables within a single set of observed variables, it is free to come up with linear combinations of all eight observed variables that correlate more highly with each other than can linear combinations that satisfy the Canona restriction.

Unfortunately, I don't know of any way within LinEqs or any other SEM program to impose constraints on the scoring coefficients. Appropriately enough for a system whose focus is on testing a priori models, SEM focuses on how observed variables are generated from emergent (latent) variables, rather than the other way around.

8.5.2 SEM Is Not a General Model for Multivariate Statistics

Because I couldn't coax a Canona solution out of LinEqs, testing the effects of removing an observed variable from the system (setting each of its loadings on the canonical variates to zero) as a way of testing the increment to R_c^2 yielded by adding that variable isn't possible. More generally, SEM is clearly not a general model within which all of the techniques we've discussed so far can fit. Even where there is a fit with respect to the restrictions and the fit criteria between SEM and one of the "classic" techniques, several considerations—the asymptotic nature of the significance tests within SEM; uncertainty as to whether the SEM optimization routine may have gotten "stuck" in a local maximum and the resulting need to try runs with different start values; the inability to impose on SEM optimization criteria other than overall fit between observed and reproduced covariances; and the host of options, output labeling, simplicities of model specification, etc. strongly suggest that preference should be given to computing algorithms and programs designed specifically for MRA, T^2, Manova, PCA, or exploratory FA. This matches Bentler's (1995) recommendation and is consistent with Bentler and Wu's (1995) provision within their EQS program of separate programs basic statistics such as Anova, MRA, and exploratory FA.

Let's finish up our discussion of SEM with an example of a technique for which it clearly is the method of choice, confirmatory factor analysis.

Example 8.5 *Higher-Order Confirmatory Factor Analysis via SEM: WISC–R One More Time*. Towards the end of chapter 7 we mentioned that the correlations among a set of three oblique factors could be accounted for by assuming that each loads on a single, general, higher-order factor. This implies a relationship among four latent variables and could thus be modeled via SEM. One way to analyze this model is via the LinEqs subsystem of PROC CALIS. The setup for one such analysis—corresponding to the three-equally-correlated-factors CFA of the WISC–R that we carried out via CALIS Factor in section 7.9—is as follows:

```
PROC CALIS DATA=WISC  METHOD=ML  EDF=220;
   Title2 "Trying 3 Oblique, 1 2nd-Order WISC-R Factors";
   Title3 "  F1 = Verbal, F2 = Spatial, F3 = Numeric";
   Title4 "   Equal Loadings on General Factor"
    Title5 "    Using LinEqS Statement";
   LinEqs
      INFO       = X1 F1 + E1,
      SIMIL      = X2 F1 + E2,
      ARITHM     = X3 F3 + E3,
      VOCAB      = X4 F1 + E4,
      COMPR      = X5 F1 + E5,
      DIGSPAN    = X6 F3 + E6,
      COMPL      = X7 F2 + E7,
```

```
      ARRANGEM  = X8 F2 + E8,
      DESIGN    = X9 F2 + E9,
      ASSEMBLY  = X10 F2 + E10,
      CODING    = X11 F3 + E11,
      MAZES     = X12 F2 + E12,
      F1 = X13 F4 + E13,
      F2 = X14 F4 + E14,
      F3 = X15 F4 + E15;
  STD F4 = 1.0,
      E1 - E12 = U14-U25,
      E13-E15 = 3*1.0;
  BOUNDS 0 <= U14-U25;
RUN;
```

Some of the output from this run follows:

```
Number of endogenous variables = 15

Manifest:     INFO      SIMIL     ARITHM    VOCAB     COMPR
              DIGSPAN   COMPL     ARRANGEM  DESIGN    ASSEMBLY
              CODING    MAZES
Latent:       F1        F2        F3


     Number of exogenous variables = 16
Latent:       F4
Error:        E1        E2        E3        E4        E5
              E6        E7        E8        E9        E10
              E11       E12       E13       E14       E15

WISC-R Confirmatory FA, Using PROC CALIS                          43
Trying 3 Oblique, 1 2nd-Order WISC-R Factors
  F1 = Verbal, F2 = Spatial, F3 = Numeric
   Using LinEqs System          18:28 Saturday, February 12, 2000

Covariance Structure Analysis: Maximum Likelihood Estimation

Fit criterion . . . . . . . . . . . . . . . . . .      0.1672
Goodness of Fit Index (GFI) . . . . . . . . . . .      0.9730
GFI Adjusted for Degrees of Freedom (AGFI). . . .      0.9587
Root Mean Square Residual (RMR) . . . . . . . . .      0.0316
Parsimonious GFI (Mulaik, 1989) . . . . . . . . .      0.7519
Chi-square = 36.7795      df = 51      Prob>chi**2 = 0.9329
```

These are identical to the corresponding fit statistics for the three oblique factors, general interfactor correlations model analyzed via CALIS Factor in chapter 7—as they should be, given that the models are saying the same thing. Similarly, the manifest variable equations in standardized form (which give the estimates of the factor pattern coefficients) are identical for the two solutions. The primary difference between the two outputs is that the CALIS LinEqs output reports the loadings (pattern coefficients) of each of the specific factors on the general, higher-order factor but does not explicitly report the interfactor correlations, whereas the reverse is true of the Factor-statement output.

However, these two sets of estimates are readily translated into each other: The loadings of the Verbal, Spatial, and Numeric factors on the general factor are .9337, .7951, and .8463, respectively. But we know that that the correlation between any two of the specific factors must equal the product of their loadings on the general factor (because that is the only factor they have in common), which implies an interfactor correlation matrix of

$$\mathbf{R}_{ff} = \begin{bmatrix} 1.000 & .7424 & .7902 \\ & 1.000 & .6729 \\ & & 1.000 \end{bmatrix},$$

which is exactly what our earlier, CALIS Factor analysis reported. Alternatively, we could subject this $\mathbf{R}_{ff}$ to a single-factor FA; when we do so we obtain a GFI of 1.000, a chi-square of 0.000, and the same loadings of specific factors on the general factor that CALIS LinEqs reported.

What about the four-orthogonal-factors model? As pointed out toward the end of chapter 7, the assumption that each oblique specific factor just identified is actually being generated by the joint effects of a general factor and a "pure" specific factor uncorrelated with the general factor implies that a four-orthogonal-factor model should also fit the data, but with the additional restriction that the loadings on the general factor of the variables in each cluster of variables associated with a given specific factor should be directly proportional to their loadings on that specific factor. Moreover, the constant of proportionality should be equal to the ratio between the loading of the specific factor on the general factor and its loading on the corresponding "pure" factor (i.e., to the square root of the specific factor's uniqueness). There is no way to separate unique and error variance for these factors, but the uniqueness can't be greater than the complement of the square of the specific factor's loading on the general factor. Thus the constant of proportionality must be at least $.934/\sqrt{1-.934^2} = 2.614$ for Verbal, $.795/\sqrt{1-.795^2} = 1.310$ for Spatial, and 1.834 for Numeric.

We can use the LINCON (for "linear constraints") statement within CALIS to require that the loadings on the orthogonal factors satisfy proportionality conditions or any other linear constraint of the form a·[loading i] + b·[loading j] = or <= or >= c, where a, b, and c are numeric constants. Thus, for instance, the condition that D3 be 3 times as large as C3 would be represented in LINCON as "D3 - 3 * C3 = 0". Ideally we would like to leave the constant of proportionality as an parameter to be estimated from the data, but this is not possible within LINCON, so we approximate that situation by setting them to specific numerical values close to the lower bound imposed by the estimated loading on the general factor. A set of CALIS commands to accomplish this is as follows:

```
PROC CALIS DATA=WISC  METHOD=ML  NOBS=221 DFREDUCE=3 ;
```

["DFREDUCE=3" informs CALIS that we shouldn't get full credit for having specified all 12 of the general-factor loadings as deterministic linear functions o f the other loadings, because we used the output from the higher-order FA to estimate the three constants of proportionality.]

```
   Title2 "Trying 4 Orthogonal WISC-R Factors, LinEqs ";
   Title3 "  F1 = Verbal, F2 = Spatial, F3 = Numeric";
   Title4 "   HEYWOOD specifies non-neg Uniquenesses";
   Title5 "   g loadings cluster-wise propl to specif loadings ";
   LinEqs
      INFO       = A1 F1 + D1 F4 + E1,
      SIMIL      = A2 F1 + D2 F4 + E2,
      ARITHM     = C3 F3 + D3 F4 + E3,
      VOCAB      = A4 F1 + D4 F4 + E4,
      COMPR      = A5 F1 + D5 F4 + E5,
      DIGSPAN    = C6 F3 + D6 F4 + E6,
      COMPL      = B7 F2 + D7 F4 + E7,
      ARRANGEM   = B8 F2 + D8 F4 + E8,
DESIGN     = B9 F2 + D9 F4 + E9,
      ASSEMBLY   = B10 F2 + D10 F4 + E10,
      CODING     = C11 F3 + D11 F4 + E11,
      MAZES      = B12 F2 + D12 F4 + E12;
  STD F1 - F4 = 4*1.0,
      E1 - E12 = U14 - U25;
  BOUNDS 0 <= U14 - U25 <= 1.;
    LINCON D1 - 3 * A1 = 0, D2 - 3 * A2 = 0,
           D4 - 3 * A4 = 0, D5 - 3 * A5 = 0,
           D7 - 1.5 * B7 = 0,
           D8 - 1.5 * B8 = 0, D9 - 1.5 * B9 = 0,
           D10 - 1.5 * B10 = 0, D12 - 1.5 * B12 = 0,
           D3 - 2 * C3 = 0, D6 - 2 * C6 = 0,
           D11 - 2 * C11 = 0 ;
RUN;
```

(This is, by the way, the counterexample I promised when at the beginning of this section I made the overly general statement that to be at all interesting the measurement-model portion of a structural equations model needed to involve oblique factors.)

The most compact evidence that CALIS indeed imposed the proportionality conditions specified in the above is actually provided by a portion of the output from the CALIS Factor analysis that corresponds to the above run, namely

```
Standardized Factor Loadings

              FACT1         FACT2         FACT3         FACT4

INFO        0.25374       0.00000       0.00000       0.76121
SIMIL       0.25199       0.00000       0.00000       0.75597
ARITHM      0.00000       0.00000       0.33917       0.67833
VOCAB       0.27309       0.00000       0.00000       0.81927
COMPR       0.23665       0.00000       0.00000       0.70996
DIGSPAN     0.00000       0.00000       0.25297       0.50594
COMPL       0.00000       0.37298       0.00000       0.55948
ARRANGEM    0.00000       0.34209       0.00000       0.51313
DESIGN      0.00000       0.45355       0.00000       0.68032
ASSEMBLY    0.00000       0.39472       0.00000       0.59207
CODING      0.00000       0.00000       0.20818       0.41636
MAZES       0.00000       0.29353       0.00000       0.44030
```

Note that the FACT1 (Verbal) loadings here are exactly one-third of the corresponding FACT4 (*g*) loadings; the FACT2 (Spatial) loadings are 1/1.5 times the corresponding loadings on *g*, and the Numeric loadings are one-half the corresponding general-factor loadings. Within these bounds CALIS LinEqs was nevertheless able to come up with a pretty good fit to the observed correlations:

```
Covariance Structure Analysis: Maximum Likelihood Estimation

Fit criterion . . . . . . . . . . . . . . . . . .      0.1784
Goodness of Fit Index (GFI) . . . . . . . . . . .      0.9703
GFI Adjusted for Degrees of Freedom (AGFI). . . .      0.9546
Root Mean Square Residual (RMR) . . . . . . . . .      0.0489
Parsimonious GFI (Mulaik, 1989) . . . . . . . . .      0.7498
Chi-square = 39.2389        df = 51       Prob>chi**2 = 0.8851
```

Recalling that the overall chi-square for the model with four orthogonal factors and unrestricted general-factor loadings was 24.608 with 42 *df*, our difference chi-square testing the null hypothesis that in the population the proportionality restrictions are satisfied perfectly is 39.239 - 24.608 = 14.631 with 9 *df*, $p = .102$. Thus we don't have enough data to be confident about specifying which loadings depart in which direction from the proportionality conditions implied by the Spearman model for these data.

The other important comparison is to the higher-order-factor model, which yielded a chi-square of 36.780 with 51 *df* and a GFI and RMR of .973 and .032 (vs. the .970 and .049 yielded by the preceding model). Some of the difference in root mean square residual (RMR) between these two models may be due to the fact that it includes the (inevitably perfect) fit between observed and reproduced correlations among the three oblique factors in that model. At any rate, there is little to choose between them.

Finally (with respect to our now beaten-to-death WISC-R example), keep in mind that, although these last two models are (essentially) equivalent in terms of their fit to the correlations, the meaning of the specific factors and the factor-score estimates they yield for individual cases are very different, as revealed in the two scoring-coefficient matrices (obtained only by asking nicely by adding the PALL option at the beginning of the CALIS command); the following listing has been edited to get the output from both models on the same page:

Latent Variable Score Regression Coefficients

	Four-orthogonal-factors model				Higher-order-factor model			
	FACT1	FACT2	FACT3	FACT4	FACT1	FACT2	FACT3	FACT4
INFO	0.183	-.140	-.092	0.154	0.610	0.078	0.158	0.161
SIMIL	0.177	-.136	-.089	0.150	0.592	0.075	0.153	0.156
ARITHM	-.190	-.131	0.457	0.146	0.209	0.075	0.777	0.156
VOCAB	0.272	-.211	-.139	0.221	0.941	0.120	0.243	0.249
COMPR	0.140	-.106	-.070	0.121	0.465	0.059	0.120	0.123
DIGSPAN	-.089	-.062	0.217	0.074	0.091	0.033	0.340	0.068
COMPL	-.069	0.219	-.032	0.057	0.072	0.289	0.053	0.054
ARRANGEM	-.056	0.179	-.026	0.048	0.058	0.233	0.043	0.044
DESIGN	-.130	0.429	-.063	0.101	0.148	0.590	0.108	0.110
ASSEMBLY	-.081	0.255	-.037	0.065	0.087	0.345	0.063	0.065
CODING	-.064	-.044	0.156	0.054	0.063	0.023	0.234	0.047
MAZES	-.042	0.134	-.019	0.037	0.044	0.175	0.032	0.033

8.5.3 Other User-Friendly SEM Programs

Any of the examples discussed in this section on SEM could have been run on EQS, after which CALIS's LinEqs statement was modeled. There are a few minor differences in syntax, such as EQS's use of paired variable names to represent covariances—for example, (V1,V3) for the covariance between variables 1 and 3. But any user who is familiar with CALIS's Factor and LinEqs statements should experience a great deal of positive transfer in learning EQS, and users will find the EQS documentation at least as easy to use as SAS's online *User's Guide.* My reliance on CALIS for the examples in sections 6.5 and in this section was due to my being quite new to EQS at the time I was preparing the examples and having temporary problems getting EQS to read already-prepared data files. EQS also shares with Amos (Arbuckle, 1995, and http://www.spss.com/amos/overview.htm) an option to draw the path diagram for your model and have the program generate the appropriate syntax (equations, etc.) to test that model.

One last note on SEM—a partial retraction, actually: In the 1985 edition of this book I pointed out the similarity between SEM's sometime partitioning of the effect of each variable into direct and indirect effects and the very similar partitioning—cf. section 2.6—of multiple R^2 into b_{zi}, r_{iy} products that have no known operational definition and can produce negative estimates of a variable's total contribution to R^2. However, the SEM partitioning is of r_{ij}, not of squared correlations, and the components do seem to make sense as measures of the relative magnitudes of various factors influencing the correlation between any two variables—*provided* that the user remembers that these components are *not* in a correlation metric and must therefore be interpreted relative to each other, rather than with respect to an absolute, -1 to +1 scale.

8.6 WHERE TO GO FROM HERE

The primary advice to the novice is: Get thee to some data. By far the best way to consolidate your understanding of the techniques discussed in this book is to apply them to data that are important to you. As suggested in chapter 1, statistical understanding resides in the fingertips, be they pushing a pencil, operating a pocket calculator, hitting keyboard keys, or clicking a mouse button. Secondary advice, once you have had some experience with the techniques discussed in this book, is to use reading of other texts and of journal articles to extend your knowledge along one or more of three lines: greater mathematical sophistication, greater variety of techniques, and acquaintance with opposing points of view.

More sophisticated treatments of multivariate statistics are available in most other texts in this area, especially Morrison (1976, 1990) and Anderson (1958, 1984). However, readers who encountered matrix algebra for the first time in this book will probably find these two texts much more comprehensible if they first strengthen their knowledge of matrix algebra by reading, for example, Horst (1963, 1983) or Searle (1966, 1982)—or by taking a course on linear algebra, finite math, or introductory mathematical statistics.

The Morrison and Anderson texts are more comprehensive than this *Primer* in their coverage of statistical techniques, as are the books by Press (1972, 1982) and Overall and Klett (1972, 1983). Tatsuoka's (1971; Tatsuoka & Lohnes, 1988) excellent and quite readable texts cover very nearly the same areas as this book, although he emphasizes geometric interpretations much more than I have in this book. Finally (with respect to comprehensive multivariate texts), the various editions of the justifiably popular Tabachnick and Fidell (e.g., 1997) text emphasize "how to" much more than "whence cometh" and provide many excellent examples of applications to social science research, with a 966-page [sic] fourth edition "hot off the presses" (2001) as I get camera-ready copy of this book ready.

The reader will probably wish to supplement the rather cursory treatment of factor analysis in this book with a text dealing solely with FA, such as Harman (1976), Mulaik (1972), Gorsuch (1983), Comrey (1973, 1992) and/or Loehlin (1998). Loehlin's text also discusses path and structural-equations models, and provides a natural "bridge" to these two scantily-covered (in this book) areas. Most of the references cited in section 2.9.7 apply to SEM as well as to path analysis. In addition, the first eleven pages of Bentler's (1995) *EQS Structural Equations Program Manual* provide a balanced, although highly condensed summary of issues in SEM (including criticisms thereof) and provide hundreds of references to discussions of those issues, and Byrne (1994) provides an excellent introduction to SEM as implemented on EQS. Byrne (2001—recent enough to provide me with an excuse for not yet having read it) promises to do the same for Amos.

Moving on to procedures not covered in the present text: ***Cluster analysis*** involves finding sets of variables or subjects that "hang together" and is often what researchers who turn to factor analysis really have in mind. Tryon and Bailey (1970) and Späth (1980) are good starting points. The major problem you will encounter is that there are many times more clustering algorithms than there are factor-analysis models, with frequently low agreement among the clusters yielded by alternative algorithms and little consensus among researchers as to the preferred algorithm. Because clusters are discrete (a variable is or is not a member of a given cluster), cluster analysis is a subset of ***categorical*** or ***discrete multivariate analysis***, for which see Bishop, Fienberg, and Holland (1978). With respect to one particular choice between categorical and "classical" techniques—that among loglinear models, logistic models, and factorial Anova with arcsin-transformed or "raw" proportion as the dependent variable—I modestly recommend Harris's (1994, chap. 8) discussion of the different metrics for "zero interaction" implicit in the different approaches.

Finally I wish to mention two rather recent (or at least recently popular) developments that seem very promising and that I hope to get better acquainted with "soon": ***latent growth-curve modeling*** for handling data sets where different cases are measured at different points in time (Bock, 1991; McArdle, 1996; Meredith & Tisak, 1990), and ***hierarchical linear modeling*** for situations where the coefficients of a linear model are themselves affected by other variables, which was initially developed to take the different levels of sampling of cases into account but which also seems to be a "natural" alternative for representing interaction and moderation effects (Little, Schnabel, & Baumert, 2000; McArdle & Hamagami, 1996; Singer, 1998) .

Many of the techniques mentioned in this section rely heavily on parameter-space searches, rather than on the analytic solutions that characterize the classic techniques discussed in the first 6.5 chapters of this book. I hope, however, that the thorough acquaintance with classic techniques this book has attempted to provide will be of help in anticipating the properties that the solutions generated by these cpu-intensive searches should have and in recognizing when something (e.g., a Heywood case or entrapment in a local maximum) may have gone awry.

8.7 SUMMING UP

Well, what have we learned .about multivariate statistics? A lot of computational details, certainly—but that's a matter of trees, not forest. What are the underlying themes (morals, principles)? Primarily the ones laid out in chapter 1, but with the added emphasis and further understanding gained from seeing them applied to successively more general techniques:

1. Multivariate statistics have two contributions to make: (a) They allow for control over the inflation of Type I error rate attendant upon analyzing *sets* of measures. (b) They allow for the identification of linear combinations of measures that reveal more about

between-set (or, in the case of PCA and FA, within-set) relationships than does any single variable.

2. Contribution (b) is frankly exploratory. If you prefer to deal strictly with single variables or a priori linear combinations thereof, contribution (a) is obtainable via the use of Bonferroni-adjusted critical values at considerably less cost to the power of your tests than is entailed in the "fully multivariate" approach.

3. The gain in predictive or discriminative power from taking the multivariate approach --that is, the difference between the performance of our optimal combination (regression variate or discriminant function or canonical variate) and the best single original variable is often rather dramatic.

4. We haven't really taken advantage of this gain, however, until we have carried out specific comparisons on a simplified, interpretable version of the optimal combination and have demonstrated (a) that it is statistically significant, too, when compared against the same critical value we used for the optimized statistic and (b) that it accounts for almost as much of the between-set variance as does the multi-decimal version.

5. This process is often very rewarding in terms of suggesting new, emergent variables worthy of further study in their own right. (See especially Examples 1.2, 1.3, and 6.4.) Of course, we throw away this benefit if we attempt to interpret our optimal combination in terms of its correlations with single original variables; this often amounts to simply repeating the message provided by separate univariate analyses, so that we might as well have skipped the multivariate analysis altogether.

6. As points 4 and 5 imply, our assessment of the virtues of alternative multivariate techniques must focus on how naturally they lead to specific comparisons and on their performance as bases for specific comparisons. Relative power or robustness of the omnibus test of the overall null hypothesis is of minor concern if such comparisons would lead us to choose a technique whose omnibus test bears no clear relationship to specific comparisons or that provides low power for these specific comparisons—which are, after all, the ultimate "payoff" of our analysis.

Finally, I hope that the *Primer of Multivariate Statistics* has managed to convey the utility of and the basic simplicity of multivariate statistics, while providing enough mathematical and computing skills to enable you actually to compute and interpret such statistics. Since we began our examination of the trees with a social psychologist's favorite Shakespearean quote, let's mark our exit from the forest with the single-word phrase with which a mathematician (Moler, 1982) chose to end every MATLAB session:

Adios.

Digression 1

Finding Maxima and Minima of Polynomials

A problem that crops up time after time in statistics is that of finding those values of $b_0, b_1, ..., b_m$ that cause a criterion function

$$C = b_0 + a_1b_1 + a_2b_2 + \cdots + a_mb_m + c_1b_1^2 + c_2b_2^2 + \cdots + c_mb_m^2$$

to take on its largest (or smallest) possible value. The *b*s here are unknowns, while the *a*s and the *c*s are known constants (usually sums or sums of squares of data values). This is a problem for which the tools of differential calculus are ideally suited. We shall of course barely tap the wealth of techniques available in that branch of mathematics.

D1.1 DERIVATIVES AND SLOPES

The problem of maximizing (or minimizing) C is approached by setting the *derivatives* of C with respect to $b_0, b_1, b_2, ..., b_m$ equal to zero and then solving the resultant set of $m + 1$ equations in m unknowns. This approach makes good intuitive sense, since the derivative of any function $f(x)$ with respect to x—$df(x)/dx$—is essentially a measure of how rapidly $f(x)$ increases as x increases, that is, of the *rate of change* of $f(x)$, or the number of units of change produced in $f(x)$ by a unit change in x. If the function is a continuous one that takes on its maximum value at $x = x^*$, then $f(x)$ must decrease on either side of x^*, that is, for any change up or down in the value of x. Thus for values of x slightly less than x^*, the derivative of $f(x)$ must be positive; for values of x slightly greater than x^*, the derivative of $f(x)$ must be negative; and at x^* the derivative must temporarily equal zero. (A similar argument establishes that df/dx must be equal to zero at the function's minimum, too. Higher order derivatives can be used to tell whether a maximum or a minimum has been located, but the

Note. There has been little change since 1985 in how functions are maximized, how matrix operations are carried out, and so on, so Digressions 1–3 and Appendices A and C have been, with the exception of a few minor modifications, copied directly from the second edition of this book. Please forgive the resulting inconsistencies with the more up-to-date layout of the newly composed sections of this book.

context will usually make this clear, anyway, and for purposes of this text, you may assume that whichever extremum—maximum or minimum—is sought by the derivative procedure has been found.) The only "trick," then, is to be able to compute the derivatives of C. This requires application of a very few, intuitively sensible, rules.

Rule 1 *The derivative of a constant is zero.* A constant is any expression that does not include within it the variable with respect to which the derivative is being taken. Thus, for instance, $d(a_1b_1)/db_2$ is equal to zero, since a_1b_1 is not modified in any way by (has a rate of change of zero with respect to) changes in b_2.

Rule 2 $d(cx)/dx = c$. A plot of cx against x will of course be a straight line whose slope (increase in vertical direction divided by increase in horizontal direction for any two points on the line) is equal to c. Put another way, if y is equal to a constant times x, each unit of increase in x is magnified (or shrunk, if $c < 1$) by c in its effect on y, with c units of change in y, for every one unit change in x.

Rule 3 $d(cx^2)/dx = 2cx$. This is the most "difficult" of the rules, since $f(x)$ plotted against x no longer forms a straight line, but a "positively accelerated" curve, for which the rate of change in $f(x)$ is itself dependent on the value of x. For values of x close to zero, small changes in x produce very small changes in x^2 (and thus in cx^2), while the same increment applied to a large value of x makes a relatively large change in x^2. For instance, $(.2)^2 - (.1)^2$ is only .03, but $(25.1)^2 - (25.0)^2 = 5.01$. Note, too, that the ratio of the change in x^2 to the change in x is $.03/.10 = 2(.15)$ in the first case and $5.01/.1 = 2(25.05)$ in the second case—exactly equal to twice the mean of the values x takes on during the change. The derivative of a function for a given value of the differentiating variable (the variable with respect to which the derivative is taken, that is, the variable whose effects on some function we wish to assess) is essentially the *instantaneous* rate of change or, put geometrically, the slope of a straight line tangent to the curve representing the function at a point directly above the specified value of the differentiating variable. As can be seen from a plot of $f(x) = x^2$ against x, this slope is close to zero when x is zero, but rapidly increases as x increases.

Rule 4 *The derivative of a sum is equal to the sum of the derivatives, that is,*

$$d[f(x) + g(x) + \cdots]/dx = df(x)/dx + dg(x)/dx + \cdots.$$

or

$$d\left[\sum f_j(x)\right]/dx = \sum[df_j(x)/dx].$$

This rule makes excellent intuitive sense, since it essentially asserts that net rate of change is equal to the sum of the separate rates; for example, the total rate at which a reservoir into which each of three streams brings 2000 gallons per minute and out of which the city takes 4000 gallons per minute, with an evaporation loss of 1000 gallons per minute, is 2000 + 2000 + 2000 − 4000 − 1000 = 1000 gallons per minute. As you might suspect from the frequency with which summation signs occur in statistics, this is an oft-used rule.

Application of Rules 1–4 As an example of the application of these rules, consider a situation (for example, univariate regression) in which we want to select values of b_0 and b_1 that will make $E = \Sigma(Y - b_0 - b_1X)^2$ as small as possible. First, in order to get the criterion function into a form in which Rules 1–4 can be applied, we must carry out the squaring operation, yielding

$$E = \sum(Y^2 + b_0^2 + b_1^2X^2 - 2b_0Y - 2b_1XY + 2b_0b_1X).$$

We next apply Rule 4, to give

$$\begin{aligned} dE/db_0 &= \sum d(Y^2 + b_0^2 + \cdots + 2b_0b_1X)/db_0 \\ &= \sum[d(Y^2)/db_0 + d(b_0^2)/db_0 + \cdots + d(2b_0b_1X)/db_0] \end{aligned}$$

and

$$dE/db_1 = \sum[d(Y^2)/db_1 + d(b_0^2)/db_1 + \cdots + d(2b_0b_1X)/db_1].$$

Now we can apply Rules 1–3 to the separate derivatives. Thus, using Rule 1,

$$\begin{aligned} d(Y^2)/db_0 &= d(b_1^2X^2)/db_0 = d(-2b_1XY)/db_0 \\ &= d(Y^2)/db_1 = d(b_0^2)/db_1 = d(-2b_0Y)/db_1 = 0. \end{aligned}$$

Note that, for instance, $b_1^2X^2$, even though it involves the unknown value b_1, is just a constant when differentiating with respect to b_0, since changes in b_0 have no effect whatever on $b_1^2X^2$.

Using Rule 2 tells us that

$$\begin{aligned} d(-2b_0Y)/db_0 &= -2Y; \quad d(2b_0b_1X)/db_0 = 2b_1X; \\ d(-2b_1XY)/db_1 &= -2XY; \quad d(2b_0b_1X)/db_1 = 2b_0X. \end{aligned}$$

Using Rule 3 tells us that

$$d(b_0^2)/db_0 = 2b_0 \quad \text{and} \quad d(b_1^2X^2)/db_1 = 2b_1X^2.$$

The end result is that

$$dE/db_0 = \Sigma(2b_0 - 2Y + 2b_1X) = 2N\,b_0 - 2\Sigma Y + 2b_1\Sigma X;$$

and

$$dE/db_1 = \sum(2b_1X^2 - 2XY + 2b_0X) = 2b_1\sum X^2 - 2\sum XY + 2b_0\sum X.$$

Finally, to find the values of b_0 and b_1 that minimize E (or maximize it—but we're confident that our procedure will yield a minimum, since there is no bound on how badly we can predict Y), we must ask what pair of values will cause these two derivatives to simultaneously equal zero. Thus we must solve the system of two simultaneous linear equations,

$$Nb_0 + \left(\sum X\right)b_1 - \sum Y = 0; \quad \text{and} \quad \left(\sum X\right)b_0 + \left(\sum X^2\right)b_1 - \sum XY = 0.$$

The solution that results (see Derivation 2.1, page 507 for details) is

$$b_1 = \sum xy \Big/ \sum x^2 \quad \text{and} \quad b_0 = \overline{Y} - b_1\overline{X}.$$

D1.2 OPTIMIZATION SUBJECT TO CONSTRAINTS

There are many situations where we wish to find optimum values of some parameter or set of parameters, but where the criterion we are interested in is not strong enough to define completely the "optimal" set of weights. For instance, we might be interested in finding a pair of weights, b_1 and b_2, that make the correlation between Y and $W = b_1X_1 + b_2X_2$ as large as possible. Clearly there is no single pair that will meet this criterion, since the correlation coefficient is unaffected by linear transformations of either or both sets of variables, so that, for instance, $b_1 = 2$ and $b_2 = 1$ will always yield precisely the same correlation of W with Y as $b_1 = 10$ and $b_2 = 5$. Thus we must limit the set of weights our optimization procedure is allowed to choose among by putting a side condition (or conditions) on the search "request." This is easily incorporated into our use of derivatives to find maxima or minima through the use of *Lagrangian multipliers*.

In order to use Lagrangian multipliers, we must first express our constraint(s) in the form $g_i(b_1, b_2, \ldots, b_m) = 0$, where g_i is some function of the unknowns whose "optimal" values we wish to identify. We then define

$$L = f(b_1, b_2, \ldots, b_m) - \lambda_1 g_1(b_1, b_2, \ldots, b_m) - \lambda_2 g_2(b_1, \ldots, b_m) - \cdots - \lambda_k g_k(b_1, \ldots, b_m),$$

where we have k different side conditions to be satisfied by the optimization process. We then take the derivatives of L with respect to $b_1, b_2, \ldots, b_m$ and set these to zero. These equations, together with the equations provided by the side conditions ($g_1 = 0$; $g_2 = 0$; and so on) enable us to solve both for the λs and for the bs. As an example, suppose we wished to choose values for b_1 and b_2 that would make $(b_1x + b_2y)^2$ as large as possible, subject to the constraint that $b_1^2 + b_2^2 = 1$. Here

$$L = b_1^2x^2 + 2b_1b_2xy + b_2^2y^2 - \lambda(b_1^2 + b_2^2 - 1);$$
$$dL/db_1 = 2b_1(x^2 - \lambda) + 2b_2xy; \quad dL/db_2 = 2b_1xy + 2b_2(y^2 - \lambda).$$

Setting derivatives equal to zero and adding the side condition gives

$$b_1(x^2 - \lambda) + b_2xy = 0; \quad b_1xy + b_2(y^2 - \lambda) = 0; \quad b_1^2 + b_2^2 = 1.$$

The first two equations each yield a value of b_1/b_2, namely,

$$b_1/b_2 = -xy/(x^2 - \lambda) = -(y^2 - \lambda)/xy;$$

whence $(x^2 - \lambda)(y^2 - \lambda) = (xy)^2$; whence λ must either equal 0 or $x^2 + y^2$. Using the value $\lambda = 0$ yields $b_1 = \pm(y/x)\, b_2$, which, together with the side condition, tells us that

$$b_1 = \pm y/\sqrt{x^2 + y^2} \quad \text{and} \quad b_2 = x/\sqrt{x^2 + y^2}.$$

Whence $(b_1x + b_2y)^2 = 0$, clearly its minimum value. Using the other value of λ gives the same expressions for the bs, except that they now both have the same sign, and

$$(b_1x + b_2y)^2 = (2xy)^2/(x^2 + y^2) = 4/(1/x^2 + 1/y^2),$$

the maximum we sought.

Digression 2

Matrix Algebra

As pointed out in Chapter 2, matrix algebra proves extremely useful in multivariate statistics for at least three reasons:

(1) Matrix notation provides a highly compact summary of expressions and equations that, if a separate symbol for each variable and each constant involved were employed, would be enormously bulky.

(2) The compactness and simplicity of matrix notation greatly facilitates memorization of expressions and equations in matrix form.

(3) The underlying similarity of the operations involved in applying a multivariate technique to a set of 2, 15, or 192 variables is much easier to see when these operations are expressed in matrix notation. Further, the similarity between univariate and multivariate techniques is much more easily seen from matrix expressions than from single-symbol formulas. These two functions combine to lessen considerably the difficulty of deriving solutions to multivariate problems.

D2.1 BASIC NOTATION

In order to reap the first two benefits it is necessary to be sufficiently familiar with the conventions of matrix notation to be able to translate expressions from matrix form to single-symbol form, and vice versa. Admittedly an adequate substitute for this skill in most situations is the ability to match matrix operations with corresponding computer algorithms as described in writeups of available library ("canned") programs, but unless the user of such computer programs is able to convert matrix expressions to single-symbol expressions in at least simple, small-matrix cases, he or she will be forever at the mercy of "bugs" in the programs being used. In order to reap the third benefit, it is necessary as well to learn the conventions for manipulating matrix expressions and equations. These conventions are highly similar to the rules of single-symbol algebra in many respects but crucially different in others. The following sections are designed to familiarize the reader with those aspects of matrix notation and matrix manipulation that are necessary in order to follow the derivations outlined in the *Primer*. The coverage of matrix algebra is by no means comprehensive, there being many techniques other than those covered here, and many alternative interpretations and applications of the techniques that are presented. The reader who wishes to delve further into matrix algebra will find study of the texts by Horst (1963) or Searle (1983) rewarding.

$$\mathbf{a} = \begin{bmatrix} a_1 \\ a_2 \\ a_3 \\ \vdots \\ a_m \end{bmatrix}$$

is a *column vector*, that is, a vertical arrangement of *m elements*. A single number or symbol, that is, a *scalar*, can be considered to be a column vector—or a row vector—containing a single element, that is, for which $m = 1$.

$$\mathbf{a}' = [a_1 \quad a_2 \quad a_3 \quad a_4 \cdots a_m]$$

is a *row vector*, a horizontal arrangement of *m* elements.

$$\mathbf{A} = [a_{ij}] = [\mathbf{a}_1 \quad \mathbf{a}_2 \quad \mathbf{a}_3 \cdots \mathbf{a}_p] = \begin{bmatrix} \mathbf{a}_1' \\ \mathbf{a}_2' \\ \mathbf{a}_3' \\ \vdots \\ \mathbf{a}_m' \end{bmatrix} = \begin{bmatrix} a_{11} & a_{12} & a_{13} & \cdots & a_{1p} \\ a_{21} & a_{22} & a_{23} & \cdots & a_{2p} \\ a_{31} & a_{32} & a_{33} & \cdots & a_{3p} \\ \vdots & \vdots & \vdots & \cdots & \vdots \\ a_{m1} & a_{m2} & a_{m3} & \cdots & a_{mp} \end{bmatrix}$$

is an $m \times p$ matrix consisting of *m* rows and *p* columns of elements, a_{ij} being the element located at the intersection of the *i*th row and the *j*th column of the matrix. The quantities *m* and *p* are called the *dimensions* of **A**.

$$\mathbf{A}' = [a_{ji}] = \begin{bmatrix} a_{11} & a_{21} & a_{31} & \cdots & a_{m1} \\ a_{12} & a_{22} & a_{32} & \cdots & a_{m2} \\ a_{13} & a_{23} & a_{33} & \cdots & a_{m3} \\ \vdots & \vdots & \vdots & \cdots & \vdots \\ a_{1p} & a_{2p} & a_{3p} & \cdots & a_{mp} \end{bmatrix}$$

is the transpose of **A**, with the rows of **A**′ being the columns of **A**. If $\mathbf{A}' = \mathbf{A}$, the matrix is said to be *symmetric*. Thus a matrix is symmetric if and only if $a_{ij} = a_{ji}$ for every *i*, *j* between 1 and *m*, *p*. As implied by the above definitions, boldface capital letters refer to matrices, while boldface lowercase letters refer to vectors (a vector being, of course, a matrix one of whose dimensions is unity). The principal exception to this general rule occurs in discussing sets of scores on predictor or outcome variables. Thus **X** will refer to a data matrix whose entries are raw scores (each row being the data for a single subject, the columns representing the different variables), X_{ij} being subject *i*'s score on predictor variable *j*, while **x** will refer to a data matrix whose entries are *deviation* scores, with x_{ij} being subject *i*'s deviation score, $(X_{ij} - \overline{X}_j)$, on predictor variable *j*. A similar distinction holds between **Y** and **y**.

D2.2 LINEAR COMBINATIONS OF MATRICES

The sum of two matrices, **A** + **B**, is defined if and only if (iff) the two matrices have identical dimensions, that is, the same number of rows and columns. In that case,

$$\mathbf{A} + \mathbf{B} = [a_{ij}] + [b_{ij}] = [a_{ij} + b_{ij}].$$

In words, the sum of the two matrices is a new matrix **C** each of whose elements c_{ij} is the sum of the corresponding elements a_{ij} and b_{ij} of the matrices being added together.

The product of a scalar times a matrix is simply a new matrix each of whose entries is equal to the corresponding entry of the matrix being multiplied times the scalar. That is,

$$c\mathbf{A} = \mathbf{A}c = c[a_{ij}] = [ca_{ij}].$$

These two definitions can be combined to define a *linear combination* of two matrices, $c_1\mathbf{A} + c_2\mathbf{B}$, as follows:

$$c_1\mathbf{A} + c_2\mathbf{B} = c_1[a_{ij}] + c_2[b_{ij}] = [c_1a_{ij} + c_2b_{ij}].$$

Of course, **A** and **B** must have identical dimensions. This notation can be useful in summarizing the process of constructing new variables that are linear combinations of original variables. For instance, if $\mathbf{X}_0$ represents an $N \times p$ matrix that records the scores of N subjects on p baseline measures of performance, and $\mathbf{X}_1$ represents the scores of the same N subjects on the same p performance measures based on an observation period twice as long as the baseline period, $\mathbf{X}_1 - 2\mathbf{X}_0$ represents a matrix of change scores that takes into account the difference in the length of the two observation periods.

D2.3 MULTIPLICATION OF MATRICES

a'b, the scalar product of two vectors, is defined only if the two vectors have the same number of elements, and is in that case equal to the *scalar* (single element, or one-dimensional vector, or 1×1 matrix) $\Sigma\, a_j b_j$. Note that if **a** and **b** are both column vectors, **ab**′ and **ba**′ are *not* scalars, but $m \times m$ matrices, with $\mathbf{ab}' = [a_i b_j]$, a square (generally nonsymmetric) matrix whose (i, j)th element is the product of the ith element of **a** with the jth element of **b** and $\mathbf{ba}' = [b_i a_j] = (\mathbf{ab}')'$. This will become clearer when matrix products are defined below. This is our first example of a difference between matrix algebra and single-symbol algebra. In single-symbol algebra, the order of the symbols in a chain of multiplications is unimportant, that is, $x \cdot y \cdot z \cdots = x \cdot z \cdot y \cdots = y \cdot x \cdot z \cdots$, and so on. In matrix algebra, however, very careful attention must be paid to the order of multiplication of matrices, since **AB** is not in general equal to **BA**—frequently only one of these two products is even defined. In more formal terms, single-symbol multiplication is commutative; matrix multiplication is not.

$$\mathbf{Ab} = \begin{bmatrix} \mathbf{a}_1' \\ \mathbf{a}_2' \\ \mathbf{a}_3' \\ \vdots \\ \mathbf{a}_p' \end{bmatrix} \times \begin{bmatrix} b_1 \\ b_2 \\ \vdots \\ b_m \end{bmatrix} = \begin{bmatrix} \Sigma\, a_{1j} b_j \\ \Sigma\, a_{2j} b_j \\ \Sigma\, a_{3j} b_j \\ \vdots \\ \Sigma\, a_{pj} b_j \end{bmatrix}.$$

Note that for the product **Ab** to be defined, the number of elements in **b** must equal the number of elements in each row of **A**. Similarly,

$$\mathbf{b}'\mathbf{A} = [b_1 b_2 b_3 \cdots b_m] \times [\mathbf{a}_1 \mathbf{a}_2 \cdots \mathbf{a}_p]$$

$$= \left[\sum_{i=1}^{m} a_{i1} b_i \sum a_{i2} b_i \sum a_{i3} b_i \cdots \sum a_{ip} b_i \right] = (\mathbf{A}'\mathbf{b})'.$$

Finally, the product of two matrices **AB** is defined if and only if the two matrices are *conformable,* that is, if and only if the number of columns of **A** (the premultiplying matrix) is equal to the number of rows of **B** (the postmultiplying matrix). If this condition is satisfied, then

$$\mathbf{AB} = \begin{bmatrix} \mathbf{a}_1' \\ \mathbf{a}_2' \\ \vdots \\ \mathbf{a}_m' \end{bmatrix} [\mathbf{b}_1 \;\; \mathbf{b}_2 \;\; \cdots \;\; \mathbf{b}_p]$$

$$= \begin{bmatrix} \mathbf{a}_1'\mathbf{b}_1 & \mathbf{a}_1'\mathbf{b}_2 & \cdots & \mathbf{a}_1'\mathbf{b}_p \\ \mathbf{a}_2'\mathbf{b}_1 & \mathbf{a}_2'\mathbf{b}_2 & \cdots & \mathbf{a}_2'\mathbf{b}_p \\ \vdots & \vdots & \cdots & \vdots \\ \mathbf{a}_m'\mathbf{b}_1 & \mathbf{a}_m'\mathbf{b}_2 & \cdots & \mathbf{a}_m'\mathbf{b}_p \end{bmatrix} = \begin{bmatrix} \sum_{k=1}^{r} a_{1k} b_{k1} & \sum a_{1k} b_{k2} & \cdots & \sum a_{1k} b_{kp} \\ \sum a_{2k} b_{k1} & \sum a_{2k} b_{k2} & \cdots & \sum a_{2k} b_{kp} \\ \vdots & \vdots & \cdots & \vdots \\ \sum a_{mk} b_{k1} & \sum a_{mk} b_{k2} & \cdots & \sum a_{mk} b_{kp} \end{bmatrix}.$$

(r: number of columns in **A** and number of rows in **B**)

In other words, the entry in the ith row, jth column of the product matrix is equal to the sum of the cross-products of the elements in row i of **A** with the elements in column j of **B**.

Until you become familiar with matrix multiplication, you may find it helpful to write the postmultiplying matrix directly above and to the right of the premultiplying matrix and draw lines through the rows of **A** and the columns of **B** as reminders of which row and column contribute to each entry in the product matrix in which the row and column lines just mentioned meet. For instance,

$$\begin{array}{cc} & \begin{bmatrix} 5 & 3 & 2 \\ 6 & 1 & 1 \\ 1 & 4 & 7 \\ 2 & 0 & 6 \end{bmatrix} \\ \begin{bmatrix} 1 & -3 & 5 & -1 \\ 7 & -1 & 6 & -4 \end{bmatrix} & \begin{bmatrix} -30 & -26 & -46 \\ -55 & -46 & -81 \end{bmatrix} \end{array}$$

Note that in general $\mathbf{AB} \neq \mathbf{BA}$. Often (as in the above example), only one of the two orders of multiplication will even be defined.

An alternative approach to being sure that a matrix product is "legal" is (as Kevin Bird of the University of New South Wales pointed out to me) to append the row and column dimensions of each matrix as left-hand and right-hand subscripts, respectively, of the letter defining it, namely, ${}_m\mathbf{A}_n$. A matrix product ${}_m\mathbf{A}_{n\,p}\mathbf{B}_q$ is permissible if and only if the two subscripts in the middle are equal ($n = p$), so that the product can be rewritten unambiguously as ${}_m\mathbf{A}_n\mathbf{B}_q$. This approach also readily gives us the dimensions of the result of a string of matrix multiplications as the initial and final subscripts; thus we know, for instance, that ${}_m\mathbf{A}_n\mathbf{B}_q\mathbf{C}_r\mathbf{D}_s$ will yield an $m \times s$ matrix.

Various products of matrices arise repeatedly in statistics. For instance, the sum of the squares of a set of numbers or symbols can be conveniently represented as

$$\mathbf{a}'\mathbf{a} = \sum a_i^2.$$

In particular, if $\mathbf{x}_j$ is a column vector containing N subjects' deviation scores on variable X_j, that is, if

$$\mathbf{x}_j' = [(X_{1,j} - \overline{X}_j), (X_{2,j} - \overline{X}_j), \cdots (X_{N,j} - \overline{X}_j)] = [x_{1,j} x_{2,j} \cdots x_{N,j}],$$

then

$$\mathbf{x}_j'\mathbf{x}_j = \sum_{i=1}^{N} x_{i,j}^2 \quad \text{and} \quad \mathbf{x}_j'\mathbf{x}_j/(N-1) = s_j^2.$$

the sample variance of variable X_j and an unbiased estimate of the variance σ_j^2 of the population from which these N observations were randomly sampled. In the case of a single variable, the matrix notation for the computation of a sample variance is not notably more compact than the corresponding single-symbol (scalar) notation. However, recall that the product of two matrices **AB** records the results of postmultiplying each row of **A** by each column of **B.** Thus if we let **x** be a data *matrix* each of whose N rows reports one of our subject's deviation scores on each of **p** variables, then **x'x** is a $p \times p$ matrix whose (i, j)th element is $\Sigma_{u=1}^{N} x_{u,i}x_{u,j}$, the sum of the cross products of the N subjects' scores on variables i and j. Further, $\mathbf{x'x}/(N - 1) = \mathbf{S}$ is the $p \times p$ *covariance* matrix whose (i, j)th element is the sample covariance, $s_{ij} = \Sigma\, x_i x_j/(N - 1)$, of variables i and j, the jth main diagonal element being $s_{jj} = s_j^2$, the sample variance of variable j. We can also record the process of computing a covariance matrix very compactly in terms of matrix manipulations as

$$\mathbf{S} = (\mathbf{X'X} - \overline{\mathbf{X}}\overline{\mathbf{X}}'\, N)/(N - 1),$$

where **X** is the $N \times p$ raw-score data matrix and $\overline{\mathbf{X}}$ is a p-element column vector whose ith entry is X_i, the mean of variable i.

Another very useful application of matrix products is to the manipulation of linear combinations (linear compounds) of the original variables in a set of data. If we define subject i's score on the new variable W as

$$W_i = w_1X_{i,1} + w_2X_{i,2} + \cdots + w_pX_{i,p},$$

we can represent the set of N scores on the new variable W as the column vector **Xw** that results when the $N \times p$ data matrix of scores on the original variables is postmultiplied by the p-element column vector **w** whose entries are the weights $w_1, w_2, \ldots, w_p$, which define the linear compound. More importantly, and as the reader will wish to verify, the variance of this linear compound s_w^2 can be expressed as a function of the variances and covariances of the original variables by the relationship

$$\begin{aligned} s_w^2 = \mathbf{w'S}_x\mathbf{w} &= w_1^2s_1^2 + \cdots + w_p^2s_p^2 + 2w_1w_2s_{12} + 2w_1w_2s_{13} + \cdots + 2w_{p-1}w_ps_{p-1,p} \\ &= [1/(N - 1)]\sum(w_1x_{i,1} + w_2x_{i,2} + \cdots + w_px_{i,p})^2 \\ &= (\mathbf{w'x'xw})/(N - 1). \end{aligned}$$

Further, the covariance of the linear compound **Xw** with the linear compound **Yb** is given by

$$\begin{aligned} s_{W,B} &= \mathbf{w'S}_{xy}\mathbf{b} = \mathbf{b'S}'_{xy}\mathbf{w} \\ &= \sum(w_1x_1 + w_2x_2 + \cdots + w_px_p)(b_1y_1 + b_2y_2 + \cdots + b_my_m)/(N - 1), \end{aligned}$$

where

$$\mathbf{S}_{xy} = [\mathbf{s}_{xy_1}\mathbf{s}_{xy_2} \cdots \mathbf{s}_{xy_m}]$$

is a $p \times m$ matrix whose (i, j)th entry is the covariance of the ith X variable with the jth Y variable.

D2.4 PERMISSIBLE MANIPULATIONS

Expressions and equations involving matrices can be manipulated much as can scalar (single-symbol) expressions, *except* that no matrix operation corresponding simply to division by a scalar exists (we must instead multiply by the inverse of the matrix, where that exists), and care must be taken to distinguish between different orders of multiplication. The following rules may be helpful in solving equations involving matrices:

Rule 1 $\mathbf{A} = \mathbf{B}$ if and only if $a_{ij} = b_{ij}$ for all i, j. In other words, two matrices are equal if and only if their elements are identical.

Rule 2 The order of additions and subtractions is immaterial. Thus

$$\mathbf{A} + \mathbf{B} - \mathbf{C} = \mathbf{A} - \mathbf{C} + \mathbf{B} = \mathbf{B} + \mathbf{A} - \mathbf{C} = \mathbf{B} - \mathbf{C} + \mathbf{A}, \text{ and so on.}$$

Note, however, that the sum or difference of two matrices exists if and only if they are of identical dimensions, that is, if they have the same number of rows and the same number of columns. The sum or difference of two matrices is simply a matrix each of whose elements equals the sum or difference of the corresponding elements of the original matrices, that is, $\mathbf{A} + \mathbf{B} = \mathbf{C}$ if and only if $c_{ij} = a_{ij} + b_{ij}$ for all i, j.

Rule 3

$$\mathbf{A}(\mathbf{B} + \mathbf{C}) = \mathbf{AB} + \mathbf{AC}; \quad \mathbf{A}(\mathbf{BC}) = (\mathbf{AB})(\mathbf{C}); \quad (\mathbf{AB})' = \mathbf{B}'\mathbf{A}';$$

but $\mathbf{AB} \neq \mathbf{BA}$ in general though there are specific choices of pairs of matrices for which the order of multiplication is immaterial (for example, a matrix and its inverse).

Rule 4 The validity of an equation is unaffected if both sides are pre- or post-multiplied by the same matrix. In other words, if $\mathbf{A} = \mathbf{B}$, then $\mathbf{AC} = \mathbf{BC}$ and $\mathbf{DA} = \mathbf{DB}$.

D2.5 INVERSES

One of the reasons for expressing the normal equations for multiple regression analysis in the form $\mathbf{R}_x\mathbf{b}_z = \mathbf{r}_{xy}$, rather than simply using a single matrix to represent the left-hand side of our set of simultaneous equations, is that we hope to be able to isolate $\mathbf{b}_z$ on the left-hand side by itself, thereby solving the set of m simultaneous equations for the previously unknown values of the b_z's. We could accomplish this isolation if we found a matrix $\mathbf{C}$ such that premultiplication of $\mathbf{R}_x$ by $\mathbf{C}$ yielded an identity matrix, for then we would have $\mathbf{CR}_x\mathbf{b}_z = \mathbf{Cr}_{xy}$

and thus $\mathbf{Ib}_z = \mathbf{b}_z = \mathbf{Cr}_{xy}$. Such a matrix is called the inverse of $\mathbf{R}_x$, $\mathbf{R}_x^{-1}$. More generally, $\mathbf{A}^{-1}$ (the inverse of **A**) is defined to be a second matrix, **C**, such that $\mathbf{CA} = \mathbf{AC} = \mathbf{I}$.

Since the major function of inverses is to permit us to solve systems of simultaneous equations of the form $\mathbf{Ax} = \mathbf{c}$, we ought to be able to deduce the properties of inverses from our knowledge (perhaps as a residue from high school algebra) of how one deals with simultaneous equations. For instance, we know that if we have fewer equations than we have unknowns, we will not be able to find solutions for all of the latter, except in the unsatisfying form of expressions for the first k unknowns that involve the last $m - k$ unknowns as well as the constants **c**. Second, in counting equations we cannot include any equation that was derived by adding together other equations in the system, since it would represent a "rehash" of information we already have. Third, if the number of equations is greater than the number of unknowns, the "extras" must either be redundant with (obtainable as a linear combination of) or inconsistent with the first m equations. Thus we can focus on systems of equations in which the number of nonredundant equations is equal to the number of unknowns. The inverse of **A** will thus be defined only for square matrices.

Detailed examination of the steps involved in solving a set of three simultaneous equations may help us derive a procedure for computing $\mathbf{A}^{-1}$. Let us consider the case (discussed briefly in Section 2.3.4) of Dick Linn's (1981) master's thesis, one aspect of which examined whether scores on the CPQ could be predicted from scores on LOC, CPT-C, and CPT-E. This leads to the following **S**-based normal equations, based on the hyperactive children only:

$$
\begin{aligned}
&[1]: & 8.667b_1 - 25.067b_2 + 13b_3 &= s_{1y} = 27{,}200;\\
&[2]: & -25.067b_1 + 678.577b_2 - 394.833b_3 &= s_{2y} = -232.150;\\
&[3]: & 13b_1 - 394.833b_2 + 521.133b_3 &= s_{3y} = 84.233.
\end{aligned}
$$

We are now going to solve these equations for b_1, b_2, and b_3 by making use of the facts that (a) multiplication of an equation by a nonzero constant does not change its validity and (b) addition or subtraction of two equations results in an equally valid third equation. This then allows us to eliminate any equation by subtracting from it a multiple of another equation in which that same unknown appears. To keep track of the process, we will label successive versions of our original equations with primes, and state our operations on equations as linear combinations of their labels. Thus, for instance, the steps needed to eliminate b_1 from equations [2] and [3] are as follows:

$$
\begin{aligned}
&[2']: [2] + (25.067/8.667)[1] \rightarrow\\
&\qquad 0 \cdot b_1 + 606.017b_2 - 357.234b_3 = 2.892s_{1y} + s_{2y},\\
&[3']: [3] - (13/8.667)[1] \rightarrow\\
&\qquad 0 \cdot b_1 - 357.234b_2 + 501.634b_3 = -1.5s_{1y} + s_{3y}.
\end{aligned}
$$

In words, we multiplied [1] by a constant, which left its coefficient for b_1 equal to b_1's multiplier in [2], and then subtracted the resulting equation from [2]; we then used the same trick to bring b_1's coefficient in [3] to zero. The rest of the process follows:

$$[3'']: [3'] - (-357.234/606.017)[2'] \rightarrow$$
$$0b_1 + 0b_2 + 291.052b_3 = .2048s_{1y} + .5895s_{2y} + s_{3y}.$$

Note that it was crucial to get a version of [2] (namely [2']) that had a zero coefficient for b_1 before attempting to eliminate b_2 from [3], since otherwise our attempt would only have reintroduced a nonzero coefficient for b_1.

$$[2'']: [2'] + (357.234/291.052)[3''] \rightarrow$$
$$0b_1 + 606.017b_2 + 0b_3 = 3.143s_{1y} + 1.7235s_{2y} + 1.227s_{3y};$$
$$[1']: [1] - (13/291.052)[3''] \rightarrow$$
$$8.667b_1 - 25.067b_2 + 0b_3 = .991s_{1y} - .0263s_{2y} - .0447s_{3y};$$
$$[1'']: [1'] - (-25.067/606.017)[2''] \rightarrow$$
$$8.667b_1 + 0b_2 + 0b_3 = 1.121s_{1y} + .0450s_{2y} + .00605s_{3y}.$$

We now have each of our unknown values of b_j isolated on the left-hand side of an equation ([1''], [2''], and [3'']), and we thus need only divide each of these equations by its only nonzero coefficient to get our final solution, namely,

$$[1''']: [1'']/8.667 \rightarrow$$
$$b_1 = .1293s_{1y} + .0052s_{2y} + .0007s_{3y} = 2.373;$$
$$[2''']: [2'']/606.017 \rightarrow$$
$$b_2 = .0052s_{1y} + .0028s_{2y} + .0020s_{3y} = -.348;$$
$$[3''']: [3'']/291.052 \rightarrow$$
$$b_3 = .0007s_{1y} + .0020s_{2y} + .0034s_{3y} = -.162.$$

The purpose of keeping s_{1y}, s_{2y}, and s_{3y} as unknowns until the very last step should now be clear: Their multipliers in our final three equations "record" the net effect of all the operations we had to go through to isolate a particular unknown. Thus, for instance, had we been sufficiently insightful at the very beginning to compute .1293 [1] + .0052 [2] + .0007 [3], we would have obtained $1b_1 + 0b_2 + 0b_3 = 2.373$ in a single operation. In other words, the multipliers of s_{1y}, s_{2y}, and s_{3y} for a given row of the [1'''], [2'''], [3'''] system of equations let us know what linear combination of the three original equations eliminates all the unknowns except b_1, b_2, or b_3, respectively. But premultiplication of **A** by a matrix **B** is equivalent to applying the linear combinations defined by the rows of **B** to the rows of **A**. Thus if we now drop the "excess baggage" of the b_j and the s_{jy} (which simply trailed along as placeholders throughout the process), it is clear that we started with

$$[\mathbf{S}_x \mid \mathbf{I}]$$

(a covariance matrix with an identity matrix of the same size concatenated to it) and then used row operations on that 3×6 matrix to get to

$$[\mathbf{I} \mid \mathbf{S}_x^{-1}].$$

Each row of $\mathbf{S}_x^{-1}$ provides the linear combinations of the original equations that will eliminate all but one of the unknowns (the one whose variance appears as the main diagonal entry of that row); that same linear combination of the constants on the right-hand side (the $\mathbf{c}$ or $\mathbf{s}_{xy}$ vector) provides the numerical value of that unknown.

This, then, provides the most meaningful procedure for computing $\mathbf{A}^{-1}$, namely,

Procedure 1 (Row operations) If row operations are applied to the combined, $m \times 2m$ matrix $\mathbf{K} = [\mathbf{A} \mid \mathbf{I}]$ so as to produce an identity matrix in the first m columns of $\mathbf{K}$, the last m columns will contain the inverse of $\mathbf{A}$.

From this immediately flows the solution to a special case, namely,

Procedure 2 (Diagonal matrix) The inverse of a diagonal matrix is a diagonal matrix whose elements are the reciprocals of the corresponding elements of the original matrix. That is,

$$\begin{bmatrix} a_{11} & & & O \\ & a_{22} & & \\ & & \ddots & \\ O & & & a_{mm} \end{bmatrix}^{-1} = \begin{bmatrix} 1/a_{11} & & & O \\ & 1/a_{22} & & \\ & & \ddots & \\ O & & & 1/a_{mm} \end{bmatrix}.$$

This is appropriate, since we begin with a set of equations of the form $a_{ii}x_i = c_i$, whence of course $x_i = c_i/a_{ii}$.

A third, often more convenient computational procedure (the cofactor method) will be presented after we discuss the computation of determinants.

D2.6 DETERMINANTS

The only "hitch" in the row-operations approach to solving equations [1], [2], and [3] above would have occurred if all of the multipliers in one of our three doubly primed equations had been zero, since division by zero produces mathematically indeterminate results. In fact, the only way in which this could occur would be if the coefficients in one of our original equations were a perfect multiple of those in another equation, so that the right-hand side of our resultant equation must *also* be zero (else the equations would be contradictory), leading to a value of 0/0 for one of our unknowns.

One indication that this has happened to us will be that the product of the main diagonal entries (m.d.e.s) of the left-hand side of [**A** | **I**] when we have reduced it to diagonal form will be zero. More generally, the magnitude of this product will tell us how close the system of equations comes to embodying a perfect linear dependence among two or more equations. We shall thus *define* the determinant of a matrix **A** (represented symbolically as $|\mathbf{A}|$) in just that way—except that we could really take the product of the m.d.e.s at the point at which we had zeroes below the main diagonal (equations [1], [2′], and [3″]), since none of our row operations after that affect the m.d.e.s. Thus we have the following:

Definition 1 (Row operations) $|\mathbf{A}|$ can be computed by using row operations (adding a constant times one row to another row) to reduce **A** to triangular form (a form in which only above-diagonal entries are nonzero) and then taking the product of the m.d.e.s. [Note that division of all of the entries in a given row by a constant is explicitly *excluded* from the row operations used to compute the determinant. Can you see why?]

From our acquaintance with inverses and the above definition, a number of properties follow, namely,

(a) $|\mathbf{A}| = 0$ if and only if (iff) there is a perfect linear dependence among the equations of the corresponding system of equations. If **A** is a covariance matrix, **S**, $|\mathbf{S}| = 0$ iff one or more of the variables are perfect linear combinations of the other variables.

(b) The value of $|\mathbf{A}|$ is unchanged by performing row operations on **A**. (But see the note directly below Definition 1.)

(c) Multiplication of any one row of **A** by a nonzero constant multiplies $|\mathbf{A}|$ by that same constant.

(d) Multiplication of the entire matrix **A** by a nonzero constant multiplies $|\mathbf{A}|$ by that constant raised to a power equal to the dimensionality of **A**; that is, $|c_m\mathbf{A}_m| = c^m|\mathbf{A}|$.

(e) If any row of **A** consists entirely of zeroes, $|\mathbf{A}| = 0$.

(f) If two rows or two columns of **A** are identical, $|\mathbf{A}| = 0$. This is certainly a desirable property, since if two of the m equations in our system are identical, they are perfectly redundant and we thus have an insufficient amount of information to solve for all m unknowns. Similarly, if the coefficients multiplying a particular pair of variables are identical in every equation, then x_i and x_j always enter into each equation as a unit, and there is therefore no way to discriminate their *relative* contributions to the values of **c**. [Note that all of our other properties phrased in terms of row operations could be phrased instead in terms of column operations. However, these have less obvious relationships to solving equations. More importantly, it is *not* possible to compute determinants and inverses via a mixture of row and column operations. You must use solely row operations or solely column operations.]

(g) The determinant of a triangular matrix is equal to the product of its main diagonal entries.

(h) $|{}_1\mathbf{A}_1| = |[a_{11}]| = a_{11}$.

(i) $$|{}_2\mathbf{A}_2| = \left|\begin{bmatrix} a_{11} & a_{12} \\ a_{21} & a_{22} \end{bmatrix}\right| = \begin{vmatrix} a_{11} & a_{12} \\ a_{21} & a_{22} \end{vmatrix} = a_{11}a_{22} - a_{12}a_{21}.$$

In words, the determinant of a 2 × 2 matrix is equal to the product of its main diagonal entries minus the product of its off-diagonal entries.

There is an additional approach to defining determinants and inverses via determinants of submatrices of the original matrix. While I have not yet succeeded in deriving this *cofactor approach* from the row-operation definitions, its computational efficiency for 3 × 3 and perhaps 4 × 4 matrices requires that we consider it.

Definition 2 (Expansion by cofactors)

$$|\mathbf{A}| = \sum_{j=1}^{m} a_{ij}(-1)^{i+j}M_{ij} \quad \text{for any choice of } i$$
$$= \sum_{i=1}^{m} a_{ij}(-1)^{i+j}M_{ij} \quad \text{for any choice of } j,$$

where the *minor* of a_{ij}, M_{ij}, equals $|\mathbf{A}_{(ij)}|$, and $\mathbf{A}_{(ij)}$ is the $(m - 1) \times (m - 1)$ submatrix that results when row i and column j are deleted from $\mathbf{A}$. The product $(-1)^{i+j}M_{ij}$ is known as the *cofactor* of $a_{ij} = C_{ij}$, so that we can also write this definition of $|\mathbf{A}|$ as

$$|\mathbf{A}| = \sum_j a_{ij}C_{ij} = \sum_i a_{ij}C_{ij}$$

for any choice of a fixed i or a fixed j. For a large original matrix, this formula must be applied to each $(m - 1) \times (m - 1)$ minor to reduce it to an expresson involving $(m - 2) \times (m - 2)$ determinants, each of these in turn being reduced by reapplication of this "cofactor definition" to an expression involving determinants of $(m - 3) \times (m - 3)$ submatrices, and so on, until we reach 2 × 2 submatrices, each of whose determinants is readily calculated as the product of the m.d.e.s minus the product of the off-diagonal elements.

Correspondingly, we have a cofactor-based definition of inverses, namely,

Definition 3 (Cofactor approach) If $|\mathbf{A}| \neq 0$, then

$$\mathbf{A}^{-1} = \frac{1}{|\mathbf{A}|}\begin{bmatrix} C_{11} & C_{21} & \cdots & C_{m1} \\ C_{12} & C_{22} & \cdots & C_{m2} \\ \vdots & \vdots & \cdots & \vdots \\ C_{1m} & C_{2m} & \cdots & C_{mm} \end{bmatrix}.$$

Note that $\mathbf{A}^{-1}$ is proportional to the *transpose* of the matrix of cofactors of $\mathbf{A}$; that is, $\mathbf{A}^{-1} = \dfrac{\mathbf{C}'}{|\mathbf{A}|}$, where $\mathbf{C} = [C_{ij}]$ is a matrix whose (i, j)th element is the cofactor of a_{ij}. Thus, for instance, via row operations we can quickly show that

$$\begin{bmatrix} a_{11} & a_{12} \\ a_{21} & a_{22} \end{bmatrix}^{-1} = \begin{bmatrix} a_{11} & a_{12} \\ a_{21} & a_{22} \end{bmatrix} \Big/ (a_{11}a_{22} - a_{12}a_{21}).$$

But the (1, 2) entry of $\mathbf{A}^{-1}$ is $-a_{12}$. which is the cofactor of a_{21}, rather than of a_{12}. This distinction is usually unimportant, since we are usually dealing with symmetric matrices and thus $\mathbf{C}' = \mathbf{C}$.

That this procedure will work can be seen by considering the nature of the entries of the product matrix, $\mathbf{P} = \mathbf{CA}/D$, where $\mathbf{C}$ is the matrix of cofactors and D is the determinant of $\mathbf{A}$. Clearly each main diagonal entry will have the form

$$P_{jj} = (1/D)\sum_k a_{kj}C_{kj} = (1/D)D = 1,$$

while each off-diagonal entry will be of the form

$$P_{kl} = (1/D)\sum_j a_{ij}C_{kj} = 0.$$

That $P_{kl} = 0$ follows from the fact that the summation $\Sigma\, a_{ij}C_{kj}$ can be seen, by the cofactor definition of the determinant, to be equal to the determinant of a matrix identical to $\mathbf{A}$ except that row k is replaced by row l. Such a matrix will have two identical rows and will thus have a determinant of zero by property (f) above.

We can combine these two definitions to provide a relatively efficient computational procedure for 3 × 3 and possibly 4 × 4 matrices. Specifically, we first compute the matrix of cofactors, $\mathbf{C}$. (A check on our work as we proceed is that $\mathbf{C}$ should be symmetric if $\mathbf{A}$ is symmetric.) We then compute $|\mathbf{A}|$ m different times by taking the sum of the cross products of the entries in each row of $\mathbf{A}$ with the corresponding entries in $\mathbf{C}$. If these m values do not agree to within round-off error, we know we have made a mistake. If they do agree "closely enough," we take the mean of the m values as our best estimate of $|\mathbf{A}|$ and divide it into the transpose of $\mathbf{C}$ to get $\mathbf{A}^{-1}$. (Do not assume that because the first two rows yield the same value for $|\mathbf{A}|$, there is no error. With symmetric matrices a mistake in computation of, say, C_{21} is apt to lead to a similar mistake in computing C_{12}, and the result will be identical mistaken values for $\Sigma_j\, a_{1j}C_{1j}$ and $\Sigma_j\, a_{2j}C_{2j}$.)

As applied to the example we used in Section D2.5, we have $\mathbf{C} =$

$$\begin{bmatrix} \begin{vmatrix} 687.517 & -394.833 \\ -394.833 & 521.133 \end{vmatrix} & -\begin{vmatrix} -25.067 & -394.833 \\ 13 & 521.133 \end{vmatrix} & \begin{vmatrix} -25.067 & 678.517 \\ 13 & -394.833 \end{vmatrix} \\ -\begin{vmatrix} -25.067 & 13 \\ -394.833 & 521.133 \end{vmatrix} & \begin{vmatrix} 8.667 & 13 \\ 13 & 521.133 \end{vmatrix} & -\begin{vmatrix} 8.667 & -25.067 \\ 13 & -394.833 \end{vmatrix} \\ \begin{vmatrix} -25.067 & 13 \\ 678.517 & -394.833 \end{vmatrix} & -\begin{vmatrix} 8.667 & 13 \\ -25.067 & -394.833 \end{vmatrix} & \begin{vmatrix} 8.667 & -25.067 \\ -25.067 & 678.517 \end{vmatrix} \end{bmatrix}$$

$$= \begin{bmatrix} 197{,}704.5 & 7930.4 & 1076.6 \\ 7930.4 & 4347.7 & 3096.1 \\ 1076.6 & 3096.1 & 5252.4 \end{bmatrix}$$

$$\text{and } \mathbf{A}^{-1} = \mathbf{C} \Big/ \begin{pmatrix} 1.5288 \times 10^6 \\ 1.5288 \times 10^6 \\ 1.5288 \times 10^6 \end{pmatrix} = \begin{bmatrix} .12932 & .00519 & .00074 \\ .00519 & .00284 & .00203 \\ .00074 & .00203 & .00344 \end{bmatrix},$$

in good agreement with our row-operations result.

I would very much appreciate hearing from any reader who has seen or has constructed a proof that the cofactor approach is derivable from the row-operations approach. This need only be shown for the definition of determinants in terms of cofactors, since we have shown above how the cofactor procedure for obtaining inverses follows from the cofactor definition of determinants.

D2.7 SOME HANDY FORMULAS FOR INVERSES AND DETERMINANTS IN SIMPLE CASES

The following formulas are useful in solving multivariate problems involving only a few variables, either for didactic purposes or in the course of "checking out" a computer program.

I. 2 × 2 Matrices

$$\begin{bmatrix} a & b \\ c & d \end{bmatrix}^{-1} = \frac{1}{D}\begin{bmatrix} d & -b \\ -c & a \end{bmatrix}, \quad \text{where } D = ad - bc;$$

$$\begin{bmatrix} w & x \\ x & y \end{bmatrix}^{-1} = \frac{1}{wy - x^2}\begin{bmatrix} y & -x \\ -x & w \end{bmatrix}.$$

II. 3 × 3 Matrices

$$\begin{bmatrix} a & b & c \\ d & e & f \\ g & h & i \end{bmatrix}^{-1} = \frac{1}{D}\begin{bmatrix} ei - fh & ch - bi & bf - ce \\ fg - di & ai - cg & cd - af \\ dh - eg & bg - ah & ae - bd \end{bmatrix},$$

where $D = a(ei - fh) + b(fg - di) + c(dh - eg)$.

$$\begin{bmatrix} u & v & w \\ v & x & y \\ w & y & z \end{bmatrix}^{-1} = \frac{1}{(uxz + 2vwy - uy^2 - xw^2 - zv^2)} \begin{bmatrix} xz - y^2 & wy - vz & vy - wx \\ wy - vz & uz - w^2 & vw - uy \\ vy - wx & vw - uy & ux - v^2 \end{bmatrix};$$

$$\begin{bmatrix} a & b & b \\ b & a & b \\ b & b & a \end{bmatrix}^{-1} = \frac{1}{a(a + b) - 2b^2} \begin{bmatrix} a + b & -b & -b \\ -b & a + b & -b \\ -b & -b & a + b \end{bmatrix};$$

$$D = (a - b)[a(a + b) - 2b^2].$$

III. Diagonal Matrices

$$\begin{bmatrix} a & 0 & 0 & \cdots & 0 \\ 0 & b & 0 & \cdots & 0 \\ 0 & 0 & c & \cdots & 0 \\ \vdots & \vdots & \vdots & \vdots & \vdots \\ 0 & 0 & 0 & \cdots & z \end{bmatrix}^{-1} = \begin{bmatrix} 1/a & 0 & 0 & \cdots & 0 \\ 0 & 1/b & 0 & \cdots & 0 \\ 0 & 0 & 1/c & \cdots & 0 \\ \vdots & \vdots & \vdots & \vdots & \vdots \\ 0 & 0 & 0 & \cdots & 1/z \end{bmatrix};$$

$$D = a \cdot b \cdot c \cdots z.$$

D2.8 RANK

We have shown that if $D = |\mathbf{A}| = 0$, then one or more rows or columns of the matrix **A** are linearly dependent on (equal to a weighted sum of) the remaining rows or columns of the matrix. It is often helpful to know how many independent rows or columns there are in a given matrix. This number is called the *rank* of the matrix and can be determined by examining *submatrices* of **A**. If an $r \times r$ submatrix of **A** (called a "minor" and obtained from **A** by deleting the row and column in which one or more elements appear) can be found whose determinant equals zero, then the rank of **A** is at most $r - 1$. The rank of a matrix can, of course, also be determined (sometimes quite conveniently) by adding or subtracting multiples of various rows or columns to other rows or columns in order to create one or more rows (columns) consisting entirely of zeros.

If **A** is symmetric, then the rank of the matrix can be determined by concentrating on the *leading principal minor determinants*, that is, the determinants of the 1×1, 2×2, 3×3, ..., $m \times m$ submatrices each of which consists of the first k rows and columns of the matrix. If the first r of these all have positive determinants while the $(r + 1) \times (r + 1)$ and larger leading principal minors all yield zero determinants, then **A** is of rank r. Moreover, the quadratic form $\mathbf{b}'\mathbf{A}\mathbf{b}$ corresponding to **A** is said to be "positive semidefinite" and can be shown to be greater than or equal to zero for any choice of coefficients for **b**. This guarantees, for instance, that the variance of any linear combination of the variables involved in a covariance matrix **a**′**Sa** will always be positive regardless of our choice of combining rule.

D2.9 MATRIX CALCULUS

It is extremely common in matrix algebra that we have to find the maximum or minimum of an expression $f(b_1, b_2, ..., b_m)$ that involves several different coefficients. We seek this extremum by differentiating $f(b_1, b_2, ..., b_m)$ with respect to each of the b_j and then setting these derivatives equal to zero. Thus we have the set of equations,

$$df(b_1, b_2, ..., b_m)/db_1 = 0;$$
$$df(b_1, b_2, ..., b_m)/db_2 = 0;$$
$$\vdots$$
$$df(b_1, b_2, ..., b_m)/db_m = 0.$$

A more compact notation would be useful in such a situation. It seems natural to rewrite the above set of equations as

$$df(\mathbf{b})/d\mathbf{b} = 0,$$

and this convention has been adopted. It must be kept in mind that $df(\mathbf{b})/d\mathbf{b}$ is nothing more than a column vector of the separate derivatives of $f(\mathbf{b})$, that is,

$$df(\mathbf{b})/d\mathbf{b} = \begin{bmatrix} df(\mathbf{b})/db_1 \\ df(\mathbf{b})/db_2 \\ \vdots \\ df(\mathbf{b})/db_m \end{bmatrix} = \begin{bmatrix} df(b_1, b_2, ..., b_m)/db_1 \\ df(b_1, b_2, ..., b_m)/db_2 \\ \vdots \\ df(b_1, b_2, ..., b_m)/db_m \end{bmatrix}.$$

Having adopted this notation, we find that it produces some very considerable economies of derivation in that we can develop some simple rules for working within that notation instead of translating $df(\mathbf{b})/d\mathbf{b}$ back into the more cumbersome notation. Dwyer and MacPhail (1948) provide a systematic listing of many of these "matrix derivatives." We shall present and justify a few of the ones we shall use most often.

(1) If $f(\mathbf{b})$ is a constant for all choices of b_1, b_2, b_3, and so forth, then

$$df(\mathbf{b})/d\mathbf{b} = d(c)/d\mathbf{b} = 0.$$

(2) If $f(\mathbf{b}) = \mathbf{a}'\mathbf{b} = a_1b_1 + a_2b_2 + \cdots + a_mb_m = \mathbf{b}'\mathbf{a}$, whence $df/db_i = a_i$; then

$$df(\mathbf{b})/d\mathbf{b} = d(\mathbf{a}'\mathbf{b})/d\mathbf{b} = \mathbf{a} = d(\mathbf{b}'\mathbf{a})/d\mathbf{b}.$$

(3) If $f(\mathbf{b}) = \mathbf{b}'\mathbf{A}\mathbf{b}$, where $\mathbf{A}$ is symmetric, then

$$df(\mathbf{b})/d\mathbf{b} = d(\mathbf{b}'\mathbf{A}\mathbf{b})/d\mathbf{b} = d\left(\sum\sum a_{ij}b_ib_j\right)\Big/ d\mathbf{b}$$

$$= d\left(\sum a_{ii}b_i^2 + 2\sum_{i>j}\sum a_{ij}b_ib_j\right)\Big/ d\mathbf{b}.$$

Clearly

$$d(\mathbf{b}'\mathbf{A}\mathbf{b})/db_h = 2a_{hh}b_h + 2\sum_{\substack{j \\ j\neq h}} a_{hj}b_j = 2\mathbf{a}_h'\mathbf{b} = 2\mathbf{b}'\mathbf{a}_h,$$

since only those terms in $\Sigma\,\Sigma\, a_{ij}b_ib_j$ that involve b_h and thus a_{hj} or a_{jh} (which are identical, since this is a symmetric matrix) will have nonzero derivatives with respect to b_h. Thus

$$d(\mathbf{b}'\mathbf{A}\mathbf{b})/d\mathbf{b} = 2\begin{bmatrix} \mathbf{a}_1'\mathbf{b} \\ \mathbf{a}_2'\mathbf{b} \\ \vdots \\ \mathbf{a}_m\mathbf{b} \end{bmatrix} = 2\mathbf{A}\mathbf{b} = 2\begin{bmatrix} \mathbf{b}'\mathbf{a}_1 \\ \mathbf{b}'\mathbf{a}_2 \\ \vdots \\ \mathbf{b}'\mathbf{a}_m \end{bmatrix}.$$

D2.10 PARTITIONED MATRICES

One common need is computation of the determinant and/or inverse of a matrix that involves two distinct sets of variables

$$\mathbf{A} = \begin{bmatrix} \mathbf{A}_{11} & \mathbf{A}_{12} \\ \mathbf{A}_{21} & \mathbf{A}_{22} \end{bmatrix}.$$

It would be convenient to have an expression for $\mathbf{A}^{-1}$ in terms of $\mathbf{A}_{11}$, $\mathbf{A}_{12}$, $\mathbf{A}_{21}$, and $\mathbf{A}_{22}$. Let us ask for a matrix

$$\mathbf{B} = \begin{bmatrix} \mathbf{B}_{11} & \mathbf{B}_{12} \\ \mathbf{B}_{21} & \mathbf{B}_{22} \end{bmatrix}$$

such that $\mathbf{BA} = \mathbf{I} = \mathbf{AB}$. We can satisfy these conditions only if, considering the entries in the first row of $\mathbf{BA}$,

$$\mathbf{B}_{11}\mathbf{A}_{11} + \mathbf{B}_{12}\mathbf{A}_{21} = \mathbf{I}, \quad \text{and} \quad \mathbf{B}_{11}\mathbf{A}_{12} + \mathbf{B}_{12}\mathbf{A}_{22} = \mathbf{0};$$

whence

$$\mathbf{B}_{12} = -\mathbf{B}_{11}\mathbf{A}_{12}\mathbf{A}_{22}^{-1};$$

and

$$\mathbf{B}_{11}\mathbf{A}_{11} - \mathbf{B}_{11}\mathbf{A}_{12}\mathbf{A}_{22}^{-1}\mathbf{A}_{21} = \mathbf{I}, \quad \text{or} \quad \mathbf{B}_{11}(\mathbf{A}_{11} - \mathbf{A}_{12}\mathbf{A}_{22}^{-1}\mathbf{A}_{21}) = \mathbf{I};$$

whence

$$\mathbf{B}_{11} = (\mathbf{A}_{11} - \mathbf{A}_{12}\mathbf{A}_{22}^{-1}\mathbf{A}_{21})^{-1}.$$

Now, substituting the expression for $\mathbf{B}_{11}$ into that for $\mathbf{B}_{12}$, we obtain

$$\mathbf{B}_{12} = -\mathbf{B}_{11}\mathbf{A}_{12}\mathbf{A}_{22}^{-1} = -(\mathbf{A}_{11} - \mathbf{A}_{12}\mathbf{A}_{22}^{-1}\mathbf{A}_{21})^{-1}\mathbf{A}_{12}\mathbf{A}_{22}^{-1}.$$

Now, using the second row of $\mathbf{BA}$, we find that

$$\mathbf{B}_{21}\mathbf{A}_{11} + \mathbf{B}_{22}\mathbf{A}_{21} = \mathbf{0}, \quad \text{and} \quad \mathbf{B}_{21}\mathbf{A}_{12} + \mathbf{B}_{22}\mathbf{A}_{22} = \mathbf{I};$$

whence

$$\mathbf{B}_{22}\mathbf{A}_{22} = \mathbf{I} - \mathbf{B}_{21}\mathbf{A}_{12}.$$

However,

$$\mathbf{B}_{21} = \mathbf{B}'_{12} = -\mathbf{A}_{22}^{-1}\mathbf{A}'_{12}(\mathbf{A}_{11} - \mathbf{A}_{12}\mathbf{A}_{22}^{-1}\mathbf{A}_{21})^{-1},$$

so that

$$\mathbf{B}_{22}\mathbf{A}_{22} = \mathbf{I} + \mathbf{A}_{22}^{-1}\mathbf{A}'_{12}(\mathbf{A}_{11} - \mathbf{A}_{12}\mathbf{A}_{22}^{-1}\mathbf{A}_{21})^{-1}\mathbf{A}_{12}.$$

Postmultiplication by $\mathbf{A}_{22}^{-1}$ thus gives

$$\mathbf{B}_{22} = \mathbf{A}_{22}^{-1} + \mathbf{A}_{22}^{-1}\mathbf{A}_{21}(\mathbf{A}_{11} - \mathbf{A}_{12}\mathbf{A}_{22}^{-1}\mathbf{A}_{21})^{-1}\mathbf{A}_{12}\mathbf{A}_{22}^{-1}.$$

Alternately, we can seek an expression in which the roles of $\mathbf{A}_{22}$ and $\mathbf{A}_{11}$ are interchanged. Beginning our analysis with the second row of the product matrix this time, we obtain

$$\mathbf{B}_{21} = -\mathbf{B}_{22}\mathbf{A}_{21}\mathbf{A}_{11}^{-1}; \qquad \text{whence} \quad \mathbf{B}_{22} = (\mathbf{A}_{22} - \mathbf{A}_{21}\mathbf{A}_{11}^{-1}\mathbf{A}_{12})^{-1}; \text{ and so on.}$$

Summarizing all these manipulations, our final expressions for $\mathbf{A}^{-1}$ are $\mathbf{A}^{-1}$

$$= \begin{bmatrix} (\mathbf{A}_{11} - \mathbf{A}_{12}\mathbf{A}_{22}^{-1}\mathbf{A}_{21})^{-1} & -(\mathbf{A}_{11} - \mathbf{A}_{12}\mathbf{A}_{22}^{-1}\mathbf{A}_{21})^{-1}\mathbf{A}_{12}\mathbf{A}_{22}^{-1} \\ -\mathbf{A}_{22}^{-1}\mathbf{A}_{21}(\mathbf{A}_{11} - \mathbf{A}_{12}\mathbf{A}_{22}^{-1}\mathbf{A}_{21})^{-1} & \mathbf{A}_{22}^{-1} + \mathbf{A}_{22}^{-1}\mathbf{A}_{21}(\mathbf{A}_{11} - \mathbf{A}_{12}\mathbf{A}_{22}^{-1}\mathbf{A}_{21})^{-1}\mathbf{A}_{12}\mathbf{A}_{22}^{-1} \end{bmatrix}$$

$$= \begin{bmatrix} \mathbf{A}_{11}^{-1} + \mathbf{A}_{11}^{-1}\mathbf{A}_{12}(\mathbf{A}_{22} - \mathbf{A}_{21}\mathbf{A}_{11}^{-1}\mathbf{A}_{12})^{-1}\mathbf{A}_{21}\mathbf{A}_{11}^{-1} & -\mathbf{A}_{11}^{-1}\mathbf{A}_{12}(\mathbf{A}_{22} - \mathbf{A}_{21}\mathbf{A}_{11}^{-1}\mathbf{A}_{12})^{-1} \\ -(\mathbf{A}_{22} - \mathbf{A}_{21}\mathbf{A}_{11}^{-1}\mathbf{A}_{12})^{-1}\mathbf{A}_{21}\mathbf{A}_{11}^{-1} & (\mathbf{A}_{22} - \mathbf{A}_{21}\mathbf{A}_{11}^{-1}\mathbf{A}_{12})^{-1} \end{bmatrix}.$$

We could now proceed to expand one or both of the above expressions by cofactors to compute the determinant of **A**, but it is much easier to reduce **A** to triangular form, as follows:

$$|\mathbf{A}| = \begin{vmatrix} \mathbf{A}_{11} & \mathbf{A}_{12} \\ \mathbf{0} & \mathbf{A}_{22} - \mathbf{D}\mathbf{A}_{12} \end{vmatrix},$$

where $\mathbf{D}\mathbf{A}_{11} = \mathbf{A}_{21}$, whence $\mathbf{D} = \mathbf{A}_{21}\mathbf{A}_{11}^{-1}$, whence

$$|\mathbf{A}| = |\mathbf{A}_{11}| \cdot |\mathbf{A}_{22} - \mathbf{A}_{21}\mathbf{A}_{11}^{-1}\mathbf{A}_{12}|.$$

Alternately, using the same procedure but getting the zero in the first row,

$$|\mathbf{A}| = |\mathbf{A}_{22}| \cdot |\mathbf{A}_{11} - \mathbf{A}_{12}\mathbf{A}_{22}^{-1}\mathbf{A}_{21}|.$$

As an example, let **S** be the $(m + 1) \times (m + 1)$ covariance matrix involving $X_1, X_2, ..., X_m$ and Y. Now $\mathbf{A}_{22} = s_y^2$; $\mathbf{A}_{12} = \mathbf{s}_{xy}$; and $\mathbf{A}_{11} = \mathbf{S}_x$, the covariance matrix of the Xs alone. Using the expressions developed above, and letting $\mathbf{C} = \mathbf{A}^{-1}$ be partitioned in the same way as **A**, we find that

$$\begin{aligned} \mathbf{C}_{22} \text{ (which is really a scalar)} &= (\mathbf{A}_{22} - \mathbf{A}_{21}\mathbf{A}_{11}^{-1}\mathbf{A}_{12})^{-1} = (s_y^2 - \mathbf{s}_{xy}'\mathbf{S}_x^{-1}\mathbf{s}_{xy})^{-1} \\ &= 1/[s_y^2(1 - R_{y\cdot x}^2)]; \end{aligned}$$

and

$$\mathbf{C}_{12} = -\mathbf{S}_x^{-1}\mathbf{s}_{xy}[s_y^2(1 - R_{y\cdot x}^2)]^{-1} = -\mathbf{b}[s_y^2(1 - R_{y\cdot x}^2)]^{-1};$$

whence

$$\mathbf{b} = -\mathbf{C}_{12}s_y^2(1 - R_{y\cdot x}^2) = -\mathbf{C}_{12}/\mathbf{C}_{22}; \quad \text{and} \quad R_{y\cdot x}^2 = 1 - 1/(s_y^2 \cdot \mathbf{C}_{22}).$$

(We have, of course, made use of the formulas for **b**, the column vector of regression coefficients for predicting Y from the Xs, and $R_{y\cdot x}^2$, the squared coefficient of multiple correlation, that were developed in Chapter 2.)

In other words, the column vector of regression coefficients is given by the last column of **C** (or the last row, since **C**, like **A**, is symmetric), where each entry has been normalized by dividing by the lower right-hand entry in **C**. However, **A** is just a covariance matrix with an "extra" variable Y added on, so that we could just as well consider it to be an $\mathbf{S}_x$ matrix that involves $m + 1$ variables, rather than m. We could then let Y be relabeled as X_{m+1}. Moreover, we could, through relabeling, bring any of the X variables into the last row and column of $\mathbf{S}_x$ and then apply the procedures outlined above. Thus the only role that positioning Y in the last row and column of **A** played was to simplify the derivation somewhat, and it should be clear that *any* given row i of the inverse of a covariance matrix is closely related to the regression of variable i on all the other variables involved in that covariance matrix, with

$$R_{i\cdot\mathbf{c}}^2 = 1 - 1/(c_{ii}s_i^2),$$

where $\mathbf{c}$ represents the set of all variables other than i, and c_{ij} is the ijth entry in $\mathbf{S}^{-1}$; and the coefficients in the regression equation that best predicts i from the other variables are simply the negatives of the entries in the ith row of $\mathbf{S}^{-1}$, divided by c_{ii}.

D2.11 CHARACTERISTIC ROOTS AND VECTORS

In the course of finding "optimal" linear combinations of variables, we shall often have recourse to the method of Lagrangian multipliers (cf. Digression 1). When applied to matrix expressions—for example, the variance or covariance of various linear combinations—this will often result in equations of the form

$$\mathbf{Cb} - \lambda\mathbf{Db} = \mathbf{0};$$

whence

$$\mathbf{D}^{-1}\mathbf{Cb} - \lambda\mathbf{Ib} = \mathbf{0},$$

whence

$$[\mathbf{D}^{-1}\mathbf{C} - \lambda\mathbf{I}]\mathbf{b} = \mathbf{0}.$$

It would therefore be extremely handy to have a systematic way of finding solutions to a matrix equation of the form $[\mathbf{A} - \lambda\mathbf{I}]\mathbf{b} = \mathbf{0}$. We know from our earlier discussions of systems of linear equations that this system has a nontrivial solution if and only if $|\mathbf{A} - \lambda\mathbf{I}| = 0$. Expanding this determinant with λ left in as an unknown produces a pth-order polynomial in λ (the *characteristic equation*) that can be solved to yield, in general, p roots (assuming $\mathbf{A}$ to be a $p \times p$ matrix) λ_i, $i = 1, 2, \ldots, p$. The ith *characteristic root* can then be "plugged into" the original matrix equation and the resulting homogeneous system of p equations in p unknowns solved for $\mathbf{b}_i$, the *characteristic vector* associated with that characteristic root λ_i. (An alternative and quite common terminology labels λ_i and b_i as the ith *eigenvalue* and *eigenvector*, respectively.) By way of illustration, the characteristic roots and vectors of the correlation matrix of the predictor variables of Data Set 1 will be obtained. The determinantal equation is

$$\begin{vmatrix} 1.0 - \lambda & .134 & .748 \\ .134 & 1.0 - \lambda & .650 \\ .748 & .650 & 1.0 - \lambda \end{vmatrix} = 0;$$

whence

$$\begin{aligned}0 &= (1.0 - \lambda)[(1.0 - \lambda)^2 - (.650)^2] - .134[.134(1.0 - \lambda) - .650(.748)] \\ &\quad + .748[(.134)(.650) - (1.0 - \lambda)(.748)] \\ &= (1.0 - \lambda)^3 - (.650^2 + .134^2 + .748^2)(1.0 - \lambda) + 2.0(.134)(.650)(.748) \\ &= (1.0 - \lambda)^3 - (.9999)(1.0 - \lambda) + .1303.\end{aligned}$$

Referring to Digression 3 (which summarizes procedures for solving cubic equations), we see that this equation is already of the form $x^3 + ax + b = 0$, whence

$$\frac{b^2}{4} + \frac{a^3}{27} = \frac{.1303^2}{4} - \frac{.9999^3}{27} = -.033,$$

whence a trigonometric solution is recommended:

$$\cos\phi = -2.59808(.1303) = -.3335 \Rightarrow \phi = 109.5°$$

and

$$1 - \lambda = \frac{1}{\sqrt{3}}(.8039, \quad -.9171, \quad .1132) = (.927, \quad -1.058, \quad .131),$$

whence

$$\lambda = (2.058, \quad .869, \quad .073).$$

Alternately, we could have gotten to this point via the *Laplace expansion,* namely,

$$|\mathbf{A} - \lambda\mathbf{I}| = (-\lambda)^p + S_1(-\lambda)^{p-1} + S_2(-\lambda)^{p-2} + \cdots + S_{p-1}\lambda + |\mathbf{A}|,$$

where $S_i = \Sigma$(principal minor determinants of size $i \times i$), and a "principal minor determinant" is the determinant of a square submatrix of **A** each of whose main diagonal elements is a main diagonal element of **A**.

Applying the Laplace expansion to our example gives

$$\begin{aligned}S_1 &= 1.0 + 1.0 + 1.0 = 3.0; \\ S_2 &= \begin{vmatrix} 1 & .134 \\ .134 & 1 \end{vmatrix} + \begin{vmatrix} 1 & .650 \\ .650 & 1 \end{vmatrix} + \begin{vmatrix} 1 & .748 \\ .748 & 1 \end{vmatrix} \\ &= 3 - .650^2 - .134^2 - .748^2 \\ S_3 &= |\mathbf{R}| = 1 + 2(.134)(.650)(.748) - (.650^2 + .134^2 + .748^2) \\ &= 1 + .1303 - .9999 = .1304.\end{aligned}$$

At this point we turn to Digression 3, finding there that we have a general cubic equation with $y = -\lambda$, $p = 3$, $q = 2.001$, and $r = .1304$. Thus the substitution $-\lambda = x - 3/3$ or $\lambda = 1 - x$ yields

$$|\mathbf{R} - \lambda\mathbf{I}| = (-\lambda)^3 + 3\lambda^2 - 2.001\lambda + .1304.$$

Letting $\lambda = 1 - x$ yields $x^3 - .9999x + .1343 = 0$, which is exactly the equation we obtained via the direct route.

Note that the product and sum of the characteristic roots are equal, respectively, to the determinant and trace (sum of the main diagonal entries) of the matrix: $(.073)(.869) \times (2.058) = .130$ and $.073 + .869 + 2.058 = 3.000$. These two properties are true of the set of characteristic roots derived from any matrix and are useful as checks on your computations. Characteristic roots have a number of other interesting and useful properties, including:

(1) The charactristic roots of a positive definite matrix (such as the covariance matrix of a set of variables none of which is linearly dependent on the others) are all positive.

(2) If an $n \times n$ symmetric matrix is positive semidefinite of rank r, it contains exactly r positive roots and $n - r$ zero roots.

(3) The characteristic roots of a diagonal matrix are the diagonal elements themselves.

With the values of the characteristic roots in hand, we now proceed to compute the characteristic vector associated with each root. For instance, to obtain the vector corresponding to our largest root 2.058, we must solve the following set of equations:

$$\begin{aligned} -1.058b_1 + .134b_2 + .748b_3 &= 0, \\ .134b_1 - 1.058b_2 + .650b_3 &= 0, \\ .748b_1 + .650b_2 - 1.058b_3 &= 0. \end{aligned}$$

Subtracting the second equation from the first gives $1.192(b_2 - b_1) + .098b_3 = 0$, whence $b_1 = .082b_3 + b_2$. Plugging this into the second equation gives $b_2 = (.661\ .924)b_3 = .715b_3$, whence $b_1 = .797b_3$. Since this is a homogeneous set of equations, the bs are specified only up to a scale factor. Using b_3 as our unit of measurement gives $\mathbf{b}_1' = (.797, .715, 1)$ as a solution to these equations and $k(.797, .715, 1)$ as the general solution, where k is any constant. It is quite common to require that $\mathbf{b}_i'\mathbf{b}_i = \Sigma\, b_i^2$ be equal to unity, and the particular vector that satisfies that side condition is referred to as the *normalized* characteristic vector. In order to "normalize" our vector, we seek a value of k such that

$$(k\mathbf{b}_1')(k\mathbf{b}_1) = k^2\mathbf{b}_1'\mathbf{b}_1 = 1;$$

whence

$$k = 1\Big/\sqrt{\sum b_i^2} = 1/\sqrt{.797^2 + .715^2 + 1} = 1/1.465 = .683.$$

In other words, the *normalized* characteristic vector corresponding to λ_i is obtained by dividing any vector satisfying the defining equations by the square root of the sum of the squares of the coefficients in that vector.

Characteristic roots of *symmetric* matrices have three very important properties (cf. Morrison, 1967, p. 63):

(1) The characteristic vectors associated with any two unequal characteristic roots are orthogonal, that is, $\mathbf{b}_i'\mathbf{b}_j = 0$ if $\lambda_i = \lambda_j$.

(2) If we define a matrix **P** whose columns are the *normalized* characteristic vectors of **A**, then $\mathbf{P'AP} = \mathbf{D}$, where **D** is a diagonal matrix (all entries zero except on the main diagonal) whose diagonal entries are the characteristic roots of **A**. Thus the characteristic vectors can be used to "diagonalize" **A**.

(3) If we transform the variables x_i involved in the quadratic form $\mathbf{x'Ax}$ to a new set of variables y_i in which each y variable is a linear combination of the x variables, the coefficients of that transformation being equal to the coefficients of the ith characteristic vector of **A** (that is, $y_i = \mathbf{x}'\mathbf{b}_i$, where $\mathbf{b}_i$ is the ith characteristic vector of **A**), then

$$\mathbf{x'Ax} = \lambda_1 y_1^2 + \lambda_2 y_2^2 + \cdots + \lambda_p y_p^2.$$

Most eigenvalue problems of practical size will require use of computer programs such as those described in Digression 4. Formulas for $p = 2$, $p = 3$, and uncorrelated cases are provided in Section 6.3.4. However, the reader could conceivably have occasion for conducting a "hand" analysis of a problem for which $p \geq 4$. If the straightforward approach to finding characteristic roots and then their vectors is adopted, the researcher will find himself or herself faced with the horrendous task of finding the roots of a fourth or higher degree polynomial. An alternate procedure (which is actually the one employed by most computer subroutines) is available in which the characteristic vectors are computed first, and the corresponding characteristic roots are inferred from them.

This approach makes use of the fact that characteristic vectors are defined only up to a scale unit, so that any p-element column vector can be represented, given appropriate choices of scale for each of the characteristic vectors, as the sum of the p characteristic vectors of **A**. Symbolically,

$$\mathbf{x}_0 = \mathbf{a}_1 + \mathbf{a}_2 + \cdots + \mathbf{a}_p,$$

where $\mathbf{a}_i$ is the ith characteristic vector of **A**. Premultiplication by **A** gives

$$\mathbf{Ax}_0 = \mathbf{Aa}_1 + \mathbf{Aa}_2 + \cdots + \mathbf{Aa}_p.$$

However, by definition of a characteristic vector,

$$\mathbf{A}\mathbf{a}_i = \lambda_i \mathbf{a}_i,$$

so that we have

$$\mathbf{x}_1 = \mathbf{A}\mathbf{x}_0 = \lambda_1\mathbf{a}_1 + \lambda_2\mathbf{a}_2 + \cdots + \lambda_p\mathbf{a}_p;$$

and, by repeated application of $\mathbf{x}_{i+1} = \mathbf{A}\mathbf{x}_i$,

$$\mathbf{x}_n = \mathbf{A}^n\mathbf{x}_0 = \lambda_1^n\mathbf{a}_1 + \lambda_2^n\mathbf{a}_2 + \cdots + \lambda_p^n\mathbf{a}_p;$$

and

$$\lambda_1^{-n}\mathbf{x}_n = \mathbf{a}_1 + \left(\frac{\lambda_2}{\lambda_1}\right)^n \mathbf{a}_2 + \cdots + \left(\frac{\lambda_p}{\lambda_1}\right)^n \mathbf{a}_p.$$

However, λ_1 is the largest root, so for large values of n, *any* trial vector premultiplied by the nth power of **A** will be very nearly equal to a constant times the first characteristic vector of **A**. In practice, $\mathbf{A}$, $\mathbf{A}^2$, $\mathbf{A}^4$, $\mathbf{A}^8$, . . . are computed by successive matrix squarings to obtain a high power of **A** before performing the premultiplication of $\mathbf{x}_0$ (which can be any nontrivial, p-element vector). $\mathbf{A}^n$ is then squared and the resulting $\mathbf{x}_{2n}$ is compared to $\mathbf{x}_n$ (after dividing each by the sum of its elements to improve comparability), the process being stopped when $\mathbf{x}_n$ and $\mathbf{x}_{2n}$ agree to within the desired limits of accuracy; at which point either of the normalized $\mathbf{x}_n$ and $\mathbf{x}_{2n}$ vectors can be taken as equal to $\mathbf{a}_1$, or the ratios of paired elements within $\mathbf{x}_n$ can be compared to the corresponding ratios in $\mathbf{x}_{2n}$ to see if agreement is reached. The ratio between $\mathbf{x}_{2n}$ and $\mathbf{x}_n$ *before* dividing each by the sum of its elements gives λ^n, or we can compute λ from **A** and **a** by using the relationship $\mathbf{A}\mathbf{a} = \lambda\mathbf{a}$.

If characteristic vectors and roots beyond the first are desired, a *residual* matrix

$$\mathbf{A} - \lambda_1\mathbf{a}_1\mathbf{a}_1' - \lambda_2\mathbf{a}_2\mathbf{a}_2' - \cdots - \lambda_{i-1}\mathbf{a}_{i-1}\mathbf{a}_{i-1}'$$

is computed based on the first $i - 1$ (normalized) characteristic roots and vectors. The iterative process is then applied to this residual matrix to obtain $\mathbf{a}_i$ and thus λ_i.

A computational example may be helpful. To keep it simple, but also to illustrate that there is nothing about the iterative method that requires a symmetric matrix, let us take the matrix product that arises in testing the main effect of visual contact for the (*CD, DC*) outcome vector in the game study used to illustrate higher order Manova in Section 4.8, namely,

$$\begin{bmatrix} .211 & .174 \\ .174 & .143 \end{bmatrix} \times 10^{-2} \qquad \mathbf{E}_{p(vc)}^{-1}\mathbf{H}_v = \begin{bmatrix} .947 & -.146 \\ -.146 & .625 \end{bmatrix}^{-1}$$

$$= \begin{bmatrix} .27566 & .22720 \\ .34279 & .28187 \end{bmatrix} \times 10^{-2} = \mathbf{A}$$

From the special formula provided in Table 4.1 (p. 150) for Manova when $p = 2$, we "predict" in advance of beginning our iterative procedure that

$$\lambda = \frac{\left\{ \begin{matrix} (.211)(.625) + (.143)(.947) - 2(.174)(-.146) \\ \pm\sqrt{[(.211)(.625) - (.143)(.947)]^2 + 4[(.174)(.947) - (.211)(-.146)] \times [(.174)(.625) - (.143)(-.146)]} \end{matrix} \right\} \times 10^{-2}}{2[(.947)(.625) - (.146)^2]}$$

$$= \frac{.318104 \pm \sqrt{.101425}}{2(.570559)} = (.558, -.0001) \times 10^{-2}.$$

(From the fact that there were only two levels of the visual contact factor, we know that $s = 1$ and there should be only one nonzero root, so the second value is attributable to roundoff errors.)

Trying the iterative approach, we make use of the fact that

$$\begin{bmatrix} a & b \\ c & d \end{bmatrix}^2 = \begin{bmatrix} a^2 + bc & b(a + d) \\ c(a + d) & d^2 + bc \end{bmatrix}$$

to help us compute

$$\mathbf{A}^2 = \begin{bmatrix} .15382 & .12667 \\ .19112 & .15733 \end{bmatrix}; \quad \mathbf{A}^4 = \begin{bmatrix} .04787 & .03941 \\ .05947 & .04909 \end{bmatrix}; \ldots;$$

$$\mathbf{A}^{16} = \begin{bmatrix} .43515 & .35875 \\ .54137 & .44632 \end{bmatrix}; \quad \mathbf{A}^{32} = \begin{bmatrix} .38357 & .31623 \\ .47720 & .39342 \end{bmatrix}.$$

Note that, until we have found **a**, it is only the relative magnitudes of the entries in these matrices that need concern us, so that constant multipliers of the form "$\times\ 10^{-36}$" have been omitted from the listed matrices.

Now, a rather simple trial vector would be $\mathbf{x}_0' = (1, 1)$, since postmultiplication by this vector simply gives us the row sums of the matrix it multiplies. Thus

$$\mathbf{x}_{16} = \mathbf{A}^{16}\mathbf{x}_0 = \begin{bmatrix} .79390 \\ .98769 \end{bmatrix} \quad \text{and} \quad \mathbf{x}_{32} = \mathbf{A}^{32}\mathbf{x}_0 = \begin{bmatrix} .69980 \\ .87062 \end{bmatrix}.$$

However, the two ratios of a_1 to a_2 are each .80379, so we can stop our iteration here, concluding that $\mathbf{a}_1'$ (the first discriminant function) = (.80379, 1) in non-normalized form. Now, returning to the basic definition of characteristic roots and vectors, we use the fact that $\mathbf{Aa} = \lambda\mathbf{a}$ to compute λ_1, as

$$\mathbf{Aa}_1 = \begin{bmatrix} .44877 \\ .55740 \end{bmatrix} \times 10^{-2} = \lambda_1 \begin{bmatrix} .80379 \\ 1.0 \end{bmatrix},$$

whence our two estimates of λ_1 are .5574 and .5583 (each $\times\ 10^{-2}$), which average to .558, in good agreement with the value computed from Table 4.3 (p. 159). Since the original matrices were rounded off to three significant digits, the slight failure of proportionality between $\mathbf{Aa}_1$ and $\mathbf{a}_1$ should not bother us.

D2.12 SOLUTION OF HOMOGENEOUS SYSTEMS OF EQUATIONS

In the preceding section, we discovered that, once a characteristic root has been computed, obtaining its corresponding characteristic vector requires solving a *homogeneous* set of equations, that is, a set of simultaneous equations of the form

$$\mathbf{Ax} = \mathbf{0}.$$

This differs from the sets of simultaneous equations we discussed in Sections D2.5 and D2.6 in that the right-hand sides of our p equations are all zero ($\mathbf{c} = \mathbf{0}$).

Clearly a homogeneous set of equations always has the trivial solution, $\mathbf{x} = \mathbf{0}$. If it is to have a nontrivial solution, $|\mathbf{A}|$ must equal zero (a fact we used in developing equations for eigenvalues), so we cannot compute $\mathbf{A}^{-1}$. (Even if we could, premultiplying both sides by $\mathbf{A}^{-1}$ would still leave us with the trivial solution.)

However, we can reduce any homogeneous set of p equations to a nonhomogeneous set of $(p - 1)$ equations by simply leaving our last variable, x_p, as an unknown and solving for the remaining variables as multiples of x_p. This is in fact as close to a solution of a homogeneous set of equations as we can get, since if a particular choice of $\mathbf{x}$ (say, $\mathbf{x}^*$) satisfies $\mathbf{Ax}^* = \mathbf{0}$, so will $c\mathbf{x}^*$, since $\mathbf{A}(c\mathbf{x}^*) = c(\mathbf{Ax}^*) = c\mathbf{0} = \mathbf{0}$. In other words, as is true of characteristic vectors, the vector of solutions to *any* homogeneous set of equations is defined only up to a scale factor (multiplicative constant).

However, defining the first $p - 1$ x_is in terms of x_p simply involves subtracting $a_{ip}x_p$ from both sides of equation i ($1 \leq i < p$), leaving us with the nonhomogeneous system of equations,

$$\mathbf{A}^*\mathbf{x}^* = -\mathbf{a}_p x_p$$

and the solution

$$\mathbf{x}^* = \mathbf{A}^{*-1}(-\mathbf{a}_p)x_p,$$

where $\mathbf{A}^*$ consists of the first $p - 1$ rows and columns of $\mathbf{A}$ and $\mathbf{a}_p$ is the first $p - 1$ elements of the pth column of $\mathbf{A}$. Our complete solution for $\mathbf{x}$ is then $\mathbf{x}' = [\mathbf{x}^{*\prime}\ 1]x_p$, or simply $\mathbf{x}' = c[\mathbf{x}^{*\prime}\ 1]$ for any choice of c.

As applied to the example of section D2.11, we have

$$\begin{bmatrix} -1.058 & .134 & .748 \\ .134 & -1.058 & .650 \\ .748 & .650 & -1.058 \end{bmatrix} \begin{bmatrix} b_1 \\ b_2 \\ b_3 \end{bmatrix} = \begin{bmatrix} 0 \\ 0 \\ 0 \end{bmatrix},$$

whence

$$\mathbf{b}^* = \begin{bmatrix} -1.058 & .134 \\ .134 & -1.058 \end{bmatrix}^{-1} \begin{bmatrix} -.748 \\ -.650 \end{bmatrix} b_3 = \frac{1}{1.1014} \begin{bmatrix} -1.058 & -.134 \\ -.134 & -1.058 \end{bmatrix} \begin{bmatrix} -.748 \\ -.650 \end{bmatrix} b_3$$

$$= \begin{bmatrix} .797 \\ .715 \end{bmatrix} b_3,$$

so that our complete solution is $\mathbf{b}' = (.797, .715, 1)c$, as before.

The pth row provides a check on our work, because the values of $\mathbf{b}$ should satisfy the last equation. In fact, .748(.797) + .650(.715) - 1.058(1) = .003, which is acceptably close to zero, given that we have been retaining only three significant digits in our calculations.

Digression 3

Solution of Cubic Equations

It frequently happens that the characteristic equation of a 3×3 matrix whose characteristic roots we seek (cf. Digression 2) will involve cubes of λ. It is therefore desirable to have a procedure for solving cubic equations so that we can obtain analytic solutions for the characteristic roots in at least the 3×3 case.

Given a cubic equation of the form $y^3 + py^2 + qy + r = 0$, first, eliminate the y^2 term through the substitution, $y = x - p/3$, whence we have

$$x^3 + ax + b = 0,$$

where

$$a = q - p^2/3 \quad \text{and} \quad b = (2p^3 - 9pq + 27r)/27.$$

Second, calculate the "discriminant," $d = b^2/4 + a^3/27$.

(i) If d is positive, there will be one real root and two imaginary roots, namely,

$$x = A + B,\ -\frac{A+B}{2} + \frac{A-B}{2}\sqrt{-3}, \quad \text{and} \quad -\frac{A+B}{2} - \frac{A-B}{2}\sqrt{-3},$$

where

$$A = (-b/2 + d)^{1/3}, \quad B = (-b/2 - d)^{1/3}.$$

(ii) If $d = 0$, there will be three real roots, of which at least two are equal, namely,

$$x = 2(-b/2)^{1/3}, \quad (b/2)^{1/3}, \quad (b/2)^{1/3}.$$

(iii) If d is negative (by far the most common case where characteristic equations are involved), there will be three real and unequal roots. In this case, use a trigonometric substitution, namely,

$$\cos\phi = -b(27/4)^{1/2}(-a^3)^{1/2} = -2.59808b/(-a)^{3/2},$$

whence, having found ϕ from a table of cosines,

$$\begin{aligned} x &= 2(-a/3)^{1/2}\cos(\phi/3),\quad 2(-a/3)^{1/2}\cos(\phi/3+120°),\\ &\quad 2(-a/3)^{1/2}\cos(\phi/3+240°)\\ &= [(-a)^{1/2}\,.86603][\cos(\phi/3),\ -\cos(60°-\phi/3),\ \sin(\phi/3-30°)]. \end{aligned}$$

The first two editions of this book included a table of the trigonometric functions involved in the above equations. However, pocket calculators with single-key trigonometric functions are now readily available and yield more accurate results than interpolation within tables, so Table D3.1 has been dropped from this edition.

Appendix A: Statistical Tables

Tables A.1 - A.4 Normal, *t*, Chi-square, and *F* Distributions

This edition of the *Multivariate Primer* omits these four tables so as to conserve space and because critical values for (extreme percentiles of) these distributions are readily available in introductory statistics textbooks and via the World Wide Web, and statistical programs that employ the corresponding significance tests almost always report the *p*-values associated with them. The online sources are especially useful when "odd" alphas that are unlikely to appear in standard tables are needed -- usually because Bonferroni adjustment (cf. Section 1.1.1) is being employed. Two such websites that are available as of this writing are the Southeast Louisiana University Critical-Value Applets (http://www.selu.edu/Academics/Faculty/vbissonnette/tables.htm) and the SurfStat Critical Value Program (http://surfstat.newcastle.edu.au/surfstat/main/tables.html). You may also request a copy of a critical-value program written in FORTRAN (a slight modification of O'Grady's, 1981 program) by emailing rharris@unm.edu.

However, tables of the null distribution of the greatest characteristic root (g.c.r.) are still *not* widely available, and programs that report g.c.r. tests (usually labeled as "ROYS" or "ROYS Criterion:") usually do not report their associated *p*-values, so tables of g.c.r. critical values *are* reported below. (Recent versions of SPSS's MANOVA program provide an option for printing g.c.r.-based confidence intervals, from which the corresponding critical values can be computed. However, I have found that the program frequently prints blank spaces or grossly incorrect bounds, and I've not yet been able to decipher the conditions under which this occurs.) The FORTRAN program used to generate these critical values (based on algorithms and computer subroutines developed by Pillai, 1965, and Venables, 1974) is available via email to rharris@unm.edu, but has not as yet been made available on a website.

The first twelve tables each provide gcr critical values for a single value of *s* and a wide range of values of the *m* and *n* parameters. The last table focuses on $n = 1000$, for two related reasons: First, the three-decimal-place entries of the preceding tables provide only one or two significant digits; the $n = 1000$ table reports its gcr critical values to five decimal places and thus three or four significant digits. Second, as I was preparing the gcr tables for the second edition of this book I noticed that the $n = 1000$ row could, for all values of *s*, be quite well approximaed by multipling the $n = 500$ row by 1/2. This led me to suspect that θ_{crit} is very nearly linear in $1/n$ beyond $n = 500$, which conjecture I verified by comparing GCRCMP results for $n = 1000$ to those for $n = 2000$. Having discovered no discrepancy greater than .001, I consider the relationship sufficiently sound to recommend linear interpolation on the reciprocal of *n* in estimating gcr critical values for $n > 500$; the entries in the last gcr table provide a more precise starting point for such extrapolation.

Table A.5 GREATEST CHARACTERISTIC ROOT DISTRIBUTION

The entries in this table are the α-level critical values of the greatest characteristic root (gcr) distribution, that is, $\theta_\alpha(s,m,n) = c$ such that $\Pr(\theta_{max} > c) = \alpha$, where θ_{max} is the largest of the s characteristic roots of $(\mathbf{H} + \mathbf{E})^{-1}\mathbf{H}$ or of $\mathbf{S}_{11}^{-1}\mathbf{S}_{12}\mathbf{S}_{22}^{-1}\mathbf{S}_{21}$; s is the rank of the matrix and equals $\min(r_1,r_2)$, where r_1 and r_2 are the ranks of **H** and **E** or of $\mathbf{S}_{11}$ and $\mathbf{S}_{22}$;

Critical Values of Distribution of gcr

$s = 1$, $\alpha = .05$ (.01)

n\m	-.5	0	1	2	3	4	5	6	8	10	15
1	.658	.776	.865	.902	.924	.937	.947	.954	.963	.970	.979
	(.841)	(.900)	(.941)	(.958)	(.967)	(.973)	(.977)	(.980)	(.985)	(.987)	(.991)
2	.500	.632	.751	.811	.847	.871	.889	.902	.921	.934	.953
	(.696)	(.785)	(.859)	(.894)	(.915)	(.929)	(.939)	(.947)	(.957)	(.964)	(.975)
3	.399	.527	.657	.729	.775	.807	.831	.850	.877	.896	.925
	(.585)	(.684)	(.778)	(.827)	(.858)	(.879)	(.895)	(.907)	(.924)	(.936)	(.954)
5	.283	.393	.521	.600	.655	.697	.729	.755	.794	.822	.868
	(.438)	(.536)	(.644)	(.707)	(.750)	(.782)	(.806)	(.825)	(.854)	(.875)	(.907)
10	.164	.239	.339	.410	.466	.511	.548	.580	.632	.672	.742
	(.265)	(.342)	(.439)	(.506)	(.557)	(.597)	(.630)	(.658)	(.702)	(.736)	(.794)
15	.115	.171	.250	.310	.360	.401	.437	.469	.521	.564	.644
	(.190)	(.250)	(.332)	(.391)	(.439)	(.478)	(.512)	(.541)	(.590)	(.628)	(.699)
20	.088	.133	.198	.249	.292	.330	.363	.392	.443	.486	.567
	(.148)	(.197)	(.266)	(.318)	(.361)	(.398)	(.430)	(.458)	(.507)	(.546)	(.622)
25	.072	.109	.164	.208	.246	.280	.310	.337	.385		
	(.121)	(.162)	(.222)	(.268)	(.307)	(.340)	(.370)	(.397)	(.443)	(.483)	(.559)
30	.061	.092	.140	.179	.213	.243	.270	.295	.340	.379	.457
	(.102)	(.138)	(.190)	(.231)	(.266)	(.297)	(.325)	(.350)	(.394)	(.432)	(.507)
40	.046	.071	.108	.139	.167	.192	.215	.237	.275	.310	.382
	(.078)	(.106)	(.148)	(.181)	(.211)	(.237)	(.261)	(.283)	(.322)	(.356)	(.428)
50	.037	.057	.088	.114	.137	.159	.179	.197	.231	.262	.328
	(.063)	(.086)	(.121)	(.149)	(.174)	(.197)	(.218)	(.237)	(.272)	(.303)	(.369)
100	.019	.029	.046	.060	.073	.085	.097	.108	.129	.148	.192
	(.032)	(.045)	(.063)	(.079)	(.094)	(.107)	(.119)	(.131)	(.153)	(.173)	(.219)
500	.004	.006	.010	.013	.015	.018	.021	.023	.028	.033	.044
	(.007)	(.009)	(.013)	(.017)	(.020)	(.023)	(.026)	(.029)	(.034)	(.039)	(.051)
1000	.002	.003	.005	.006	.008	.009	.010	.012	.014	.017	.023
	(.003)	(.005)	(.007)	(.009)	(.010)	(.011)	(.013)	(.014)	(.017)	(.020)	(.026)

$m = (|r_1 - r_2| - 1)/2$; and $n = (df_e - r_2 - 1)/2$, where df_e is the rank of the data matrix, that is, the number of degrees of freedom going into each variance or covariance estimate in the covariance matrix.

For values of *s*, *m*, and *n* not included in these tables, the user may either interpolate within the tables or use the GCRCMP program available (as FORTRAN source code) from the author to compute the appropriate critical value.

$s = 2$, $\alpha = .05$ (.01)

n\m	-.5	0	1	2	3	4	5	6	8	10	15
1	.858	.894	.930	.947	.958	.965	.970	.973	.979	.982	.987
	(.938)	(.954)	(.970)	(.978)	(.982)	(.985)	(.987)	(.989)	(.991)	(.993)	(.995)
2	.737	.792	.851	.884	.905	.919	.929	.938	.949	.957	.964
	(.850)	(.883)	(.917)	(.936)	(.948)	(.956)	(.961)	(.966)	(.972)	(.977)	(.981)
3	.638	.702	.776	.820	.849	.870	.885	.898	.916	.928	.948
	(.764)	(.808)	(.857)	(.886)	(.905)	(.918)	(.929)	(.937)	(.948)	(.956)	(.968)
5	.498	.565	.651	.706	.746	.776	.799	.818	.847	.868	.901
	(.623)	(.677)	(.745)	(.787)	(.817)	(.839)	(.857)	(.871)	(.892)	(.907)	(.931)
10	.318	.374	.455	.514	.561	.598	.629	.656	.698	.732	.789
	(.418)	(.470)	(.544)	(.597)	(.638)	(.670)	(.697)	(.720)	(.756)	(.783)	(.831)
15	.232	.278	.348	.402	.446	.483	.515	.542	.589	.627	.696
	(.312)	(.357)	(.425)	(.476)	(.517)	(.551)	(.580)	(.606)	(.648)	(.681)	(.742)
20	.183	.221	.281	.329	.369	.404	.434	.461	.508	.546	.620
	(.249)	(.288)	(.347)	(.394)	(.433)	(.466)	(.495)	(.521)	(.564)	(.600)	(.667)
25	.151	.184	.236	.278	.314	.346	.375	.401	.446	.484	.558
	(.207)	(.240)	(.293)	(.336)	(.372)	(.403)	(.431)	(.456)	(.499)	(.535)	(.604)
30	.129	.157	.203	.241	.274	.303	.330	.354	.396	.433	.507
	(.177)	(.207)	(.254)	(.293)	(.326)	(.355)	(.381)	(.405)	(.447)	(.482)	(.552)
40	.099	.122	.159	.190	.218	.243	.266	.287	.325	.359	.428
	(.137)	(.161)	(.200)	(.232)	(.261)	(.286)	(.309)	(.331)	(.369)	(.402)	(.470)
50	.081	.099	.130	.157	.180	.202	.222	.241	.275	.306	.370
	(.112)	(.132)	(.165)	(.193)	(.217)	(.240)	(.260)	(.279)	(.313)	(.344)	(.408)
100	.042	.052	.069	.084	.097	.110	.122	.134	.155	.176	.220
	(.058)	(.069)	(.088)	(.104)	(.118)	(.132)	(.145)	(.157)	(.179)	(.200)	(.246)
500	.009	.011	.014	.018	.021	.024	.027	.029	.035	.040	.052
	(.012)	(.015)	(.019)	(.022)	(.026)	(.029)	(.032)	(.035)	(.041)	(.046)	(.059)
1000	.004	.005	.007	.009	.010	.012	.013	.015	.018	.020	.027
	(.006)	(.007)	(.010)	(.011)	(.013)	(.015)	(.016)	(.018)	(.021)	(.023)	(.030)

Critical Values of Distribution of gcr (continued)

$s = 3, \ \alpha = .05 \ (.01)$

$n \backslash m$	-.5	0	1	2	3	4	5	6	8	10	15
1	.922	.938	.956	.966	.972	.976	.979	.982	.985	.988	.991
	(.967)	(.973)	(.981)	(.985)	(.988)	(.990)	(.991)	(.992)	(.994)	(.995)	(.996)
2	.837	.865	.899	.919	.932	.942	.949	.955	.963	.969	.977
	(.909)	(.925)	(.944)	(.955)	(.963)	(.968)	(.972)	(.975)	(.980)	(.983)	(.988)
3	.756	.792	.839	.868	.888	.902	.914	.922	.936	.945	.960
	(.844)	(.868)	(.898)	(.917)	(.930)	(.939)	(.946)	(.952)	(.960)	(.966)	(.975)
5	.625	.669	.729	.770	.800	.822	.840	.855	.877	.894	.920
	(.725)	(.758)	(.804)	(.834)	(.856)	(.873)	(.886)	(.897)	(.913)	(.925)	(.944)
10	.429	.472	.537	.586	.625	.656	.683	.705	.741	.770	.819
	(.520)	(.559)	(.616)	(.659)	(.692)	(.719)	(.742)	(.760)	(.791)	(.814)	(.854)
15	.324	.362	.422	.469	.508	.541	.569	.594	.635	.669	.730
	(.402)	(.438)	(.494)	(.537)	(.573)	(.603)	(.628)	(.651)	(.688)	(.717)	(.771)
20	.260	.293	.346	.390	.427	.458	.486	.511	.554	.589	.656
	(.327)	(.359)	(.410)	(.452)	(.487)	(.517)	(.543)	(.566)	(.605)	(.638)	(.698)
25	.218	.246	.294	.333	.367	.397	.424	.448	.490	.525	.594
	(.275)	(.303)	(.351)	(.389)	(.422)	(.451)	(.477)	(.500)	(.539)	(.573)	(.637)
30	.187	.212	.255	.291	.322	.350	.375	.398	.439	.473	.543
	(.237)	(.263)	(.306)	(.342)	(.373)	(.400)	(.425)	(.447)	(.486)	(.519)	(.585)
35	.164	.186	.225	.258	.287	.313	.337	.358	.397	.431	.499
	(.208)	(.232)	(.271)	(.304)	(.333)	(.359)	(.382)	(.404)	(.441)	(.474)	(.540)
40	.146	.166	.201	.232	.259	.283	.305	.326	.363	.395	.462
	(.186)	(.207)	(.243)	(.274)	(.301)	(.326)	(.348)	(.368)	(.405)	(.436)	(.501)
45	.131	.150	.182	.210	.235	.258	.279	.298	.333	.365	.430
	(.168)	(.188)	(.221)	(.250)	(.275)	(.298)	(.319)	(.338)	(.373)	(.404)	(.467)
50	.119	.136	.167	.192	.216	.237	.257	.275	.309	.339	.402
	(.153)	(.171)	(.202)	(.229)	(.253)	(.274)	(.294)	(.313)	(.346)	(.376)	(.438)
100	.062	.072	.089	.104	.118	.131	.143	.155	.177	.197	.242
	(.081)	(.091)	(.109)	(.125)	(.140)	(.153)	(.166)	(.178)	(.200)	(.221)	(.267)
500	.013	.015	.019	.022	.026	.029	.031	.034	.040	.045	.058
	(.017)	(.019)	(.023)	(.027)	(.031)	(.034)	(.037)	(.040)	(.046)	(.052)	(.064)
1000	.007	.008	.010	.011	.013	.015	.016	.018	.020	.023	.030
	(.009)	(.010)	(.012)	(.014)	(.015)	(.017)	(.019)	(.020)	(.023)	(.026)	(.033)

$s = 4$, $\alpha = .05$ (.01)

$n\backslash m$	-.5	0	1	2	3	4	5	6	8	10	15
1	.951	.959	.969	.976	.979	.982	.985	.986	.989	.991	.993
	(.979)	(.983)	(.987)	(.990)	(.991)	(.993)	(.994)	(.994)	(.995)	(.996)	(.997)
2	.888	.905	.926	.939	.949	.956	.961	.965	.971	.975	.982
	(.938)	(.947)	(.959)	(.967)	(.972)	(.976)	(.979)	(.981)	(.984)	(.987)	(.990)
3	.824	.846	.877	.898	.912	.923	.931	.938	.948	.956	.967
	(.888)	(.903)	(.923)	(.936)	(.945)	(.952)	(.958)	(.962)	(.968)	(.973)	(.980)
5	.708	.739	.782	.813	.836	.854	.868	.880	.898	.911	.933
	(.788)	(.811)	(.844)	(.866)	(.883)	(.896)	(.906)	(.915)	(.927)	(.937)	(.953)
10	.513	.547	.601	.641	.674	.700	.723	.742	.773	.798	.840
	(.595)	(.625)	(.671)	(.706)	(.733)	(.756)	(.775)	(.791)	(.817)	(.837)	(.872)
15	.399	.431	.482	.523	.558	.587	.612	.634	.671	.701	.756
	(.472)	(.501)	(.549)	(.587)	(.618)	(.644)	(.667)	(.686)	(.719)	(.745)	(.793)
20	.326	.354	.402	.441	.474	.503	.529	.552	.591	.623	.684
	(.390)	(.417)	(.463)	(.500)	(.531)	(.558)	(.582)	(.603)	(.638)	(.668)	(.723)
25	.275	.301	.344	.380	.412	.440	.464	.487	.526	.559	.624
	(.332)	(.357)	(.399)	(.434)	(.465)	(.491)	(.515)	(.536)	(.573)	(.603)	(.663)
30	.238	.261	.301	.334	.364	.390	.414	.435	.474	.507	.572
	(.289)	(.312)	(.351)	(.384)	(.412)	(.438)	(.461)	(.482)	(.518)	(.549)	(.611)
35	.210	.230	.267	.298	.325	.350	.373	.394	.431	.463	.528
	(.256)	(.276)	(.312)	(.344)	(.371)	(.395)	(.417)	(.437)	(.473)	(.504)	(.566)
40	.188	.207	.240	.269	.294	.318	.339	.359	.395	.426	.490
	(.229)	(.248)	(.282)	(.311)	(.336)	(.360)	(.381)	(.400)	(.435)	(.465)	(.527)
45	.169	.187	.218	.245	.269	.291	.311	.330	.364	.394	.457
	(.208)	(.225)	(.257)	(.284)	(.308)	(.330)	(.350)	(.369)	(.402)	(.432)	(.493)
50	.155	.171	.199	.224	.247	.268	.287	.305	.338	.367	.428
	(.190)	(.206)	(.236)	(.261)	(.284)	(.305)	(.324)	(.342)	(.374)	(.403)	(.463)
100	.082	.091	.108	.123	.137	.150	.162	.174	.196	.216	.261
	(.102)	(.112)	(.129)	(.145)	(.159)	(.172)	(.185)	(.197)	(.219)	(.240)	(.285)
500	.017	.020	.023	.027	.030	.033	.036	.039	.045	.050	.063
	(.022)	(.024)	(.028)	(.032)	(.035)	(.039)	(.042)	(.045)	(.051)	(.056)	(.070)
1000	.009	.010	.012	.014	.015	.017	.018	.020	.023	.026	.032
	(.011)	(.012)	(.014)	(.016)	(.018)	(.020)	(.021)	(.023)	(.026)	(.029)	(.036)

Critical Values of Distribution of gcr (continued)

$s =$ 5, $\alpha = .05$ (.01)

$n \backslash m$	-.5	0	1	2	3	4	5	6	8	10	15
1	.966	.971	.977	.981	.984	.987	.988	.989	.991	.992	.995
	(.986)	(.988)	(.990)	(.992)	(.993)	(.994)	(.995)	(.996)	(.996)	(.997)	(.998)
2	.919	.929	.943	.953	.959	.965	.969	.972	.976	.980	.985
	(.955)	(.961)	(.969)	(.974)	(.978)	(.981)	(.983)	(.985)	(.987)	(.989)	(.992)
3	.866	.882	.903	.918	.929	.937	.944	.949	.957	.963	.972
	(.916)	(.926)	(.940)	(.949)	(.956)	(.961)	(.965)	(.969)	(.974)	(.977)	(.983)
5	.766	.788	.821	.845	.863	.877	.888	.898	.913	.924	.942
	(.832)	(.848)	(.872)	(.889)	(.902)	(.913)	(.921)	(.928)	(.938)	(.946)	(.959)
10	.580	.607	.651	.685	.713	.735	.755	.771	.799	.820	.857
	(.653)	(.676)	(.714)	(.743)	(.766)	(.785)	(.801)	(.815)	(.837)	(.855)	(.885)
15	.462	.488	.533	.569	.599	.625	.648	.667	.701	.728	.777
	(.530)	(.554)	(.595)	(.627)	(.655)	(.678)	(.698)	(.715)	(.744)	(.768)	(.811)
20	.383	.407	.449	.485	.515	.542	.565	.586	.621	.651	.708
	(.444)	(.468)	(.507)	(.540)	(.568)	(.593)	(.614)	(.633)	(.666)	(.693)	(.744)
25	.326	.349	.388	.422	.451	.477	.500	.521	.557	.588	.648
	(.382)	(.404)	(.442)	(.473)	(.501)	(.526)	(.547)	(.567)	(.601)	(.630)	(.685)
30	.284	.305	.341	.373	.400	.425	.448	.468	.504	.535	.597
	(.334)	(.355)	(.390)	(.421)	(.448)	(.471)	(.493)	(.512)	(.546)	(.576)	(.634)
35	.252	.271	.304	.334	.360	.384	.405	.425	.460	.490	.552
	(.298)	(.316)	(.350)	(.378)	(.404)	(.427)	(.448)	(.467)	(.501)	(.530)	(.589)
40	.226	.243	.275	.302	.327	.349	.370	.389	.423	.453	.514
	(.268)	(.285)	(.317)	(.344)	(.368)	(.390)	(.410)	(.429)	(.462)	(.491)	(.550)
45	.205	.221	.250	.276	.299	.320	.340	.358	.391	.420	.481
	(.243)	(.260)	(.289)	(.315)	(.338)	(.359)	(.378)	(.396)	(.428)	(.457)	(.515)
50	.187	.202	.230	.254	.276	.296	.315	.332	.364	.392	.451
	(.223)	(.239)	(.266)	(.290)	(.312)	(.332)	(.351)	(.368)	(.399)	(.427)	(.484)
100	.101	.110	.126	.141	.155	.168	.180	.191	.213	.233	.278
	(.122)	(.131)	(.148)	(.163)	(.177)	(.190)	(.203)	(.215)	(.237)	(.257)	(.302)
500	.022	.024	.027	.031	.034	.038	.041	.044	.049	.055	.068
	(.026)	(.028)	(.032)	(.036)	(.040)	(.043)	(.046)	(.049)	(.055)	(.061)	(.075)
1000	.011	.012	.014	.016	.017	.019	.021	.022	.025	.028	.035
	(.013)	(.014)	(.016)	(.018)	(.020)	(.022)	(.024)	(.025)	(.028)	(.031)	(.039)

$s =$ 6, $\alpha = .05\ (.01)$											
$n\backslash m$	-.5	0	1	2	3	4	5	6	8	10	15
1	.975	.978	.983	.986	.988	.989	.990	.991	.993	.994	.996
	(.990)	(.991)	(.993)	(.994)	(.995)	(.996)	(.996)	(.996)	(.997)	(.998)	(.998)
2	.938	.945	.955	.962	.967	.971	.974	.977	.980	.983	.987
	(.966)	(.970)	(.976)	(.979)	(.982)	(.984)	(.986)	(.987)	(.989)	(.991)	(.993)
3	.895	.906	.922	.933	.941	.948	.953	.957	.964	.969	.976
	(.935)	(.941)	(.951)	(.958)	(.964)	(.968)	(.971)	(.974)	(.978)	(.981)	(.986)
5	.808	.825	.850	.869	.883	.895	.904	.912	.924	.934	.949
	(.863)	(.875)	(.893)	(.906)	(.917)	(.925)	(.932)	(.938)	(.947)	(.953)	(.964)
10	.633	.655	.692	.721	.744	.764	.781	.795	.819	.838	.871
	(.698)	(.717)	(.748)	(.772)	(.792)	(.809)	(.823)	(.834)	(.854)	(.869)	(.896)
15	.514	.537	.576	.608	.635	.658	.678	.696	.726	.750	.795
	(.578)	(.599)	(.633)	(.662)	(.686)	(.706)	(.724)	(.740)	(.766)	(.787)	(.826)
20	.432	.454	.491	.523	.551	.575	.596	.615	.648	.676	.728
	(.491)	(.511)	(.546)	(.576)	(.601)	(.623)	(.643)	(.660)	(.690)	(.715)	(.762)
25	.372	.392	.428	.458	.485	.509	.531	.550	.584	.613	.669
	(.426)	(.445)	(.479)	(.508)	(.533)	(.556)	(.576)	(.594)	(.626)	(.652)	(.704)
30	.326	.345	.378	.407	.433	.457	.478	.497	.531	.560	.618
	(.375)	(.394)	(.426)	(.454)	(.479)	(.501)	(.521)	(.539)	(.572)	(.599)	(.654)
35	.290	.308	.339	.366	.391	.414	.434	.453	.486	.515	.574
	(.335)	(.353)	(.384)	(.410)	(.434)	(.456)	(.475)	(.493)	(.525)	(.553)	(.609)
40	.261	.278	.307	.333	.356	.378	.397	.416	.448	.477	.536
	(.303)	(.319)	(.348)	(.374)	(.397)	(.418)	(.437)	(.454)	(.486)	(.513)	(.570)
45	.238	.253	.281	.305	.327	.348	.366	.384	.416	.443	.502
	(.277)	(.292)	(.319)	(.344)	(.365)	(.385)	(.404)	(.421)	(.452)	(.479)	(.535)
50	.218	.232	.258	.281	.302	.322	.340	.357	.387	.414	.472
	(.254)	(.269)	(.294)	(.318)	(.338)	(.358)	(.375)	(.392)	(.422)	(.448)	(.504)
100	.119	.128	.144	.158	.172	.185	.197	.208	.230	.250	.294
	(.140)	(.149)	(.166)	(.180)	(.194)	(.207)	(.220)	(.231)	(.253)	(.273)	(.318)
500	.026	.028	.031	.035	.039	.042	.045	.048	.054	.059	.073
	(.031)	(.033)	(.037)	(.041)	(.044)	(.047)	(.051)	(.054)	(.060)	(.066)	(.080)
1000	.013	.014	.016	.018	.020	.021	.023	.024	.028	.031	.037
	(.016)	(.017)	(.019)	(.021)	(.022)	(.024)	(.026)	(.028)	(.031)	(.034)	(.041)

Critical Values of Distribution of gcr (continued)

$s = 7, \alpha = .05\ (.01)$

$n \backslash m$	-.5	0	1	2	3	4	5	6	8	10	15
1	.981	.983	.986	.988	.990	.991	.992	.993	.994	.995	.996
	(.992)	(.993)	(.994)	(.995)	(.996)	(.996)	(.997)	(.997)	(.998)	(.998)	(.999)
2	.951	.956	.964	.969	.973	.976	.978	.980	.983	.986	.989
	(.973)	(.976)	(.980)	(.983)	(.985)	(.987)	(.988)	(.989)	(.991)	(.992)	(.994)
3	.916	.923	.935	.944	.950	.956	.960	.963	.969	.973	.979
	(.947)	(.952)	(.960)	(.965)	(.969)	(.973)	(.975)	(.977)	(.981)	(.983)	(.987)
5	.840	.852	.872	.887	.899	.908	.917	.923	.934	.941	.955
	(.885)	(.895)	(.909)	(.920)	(.928)	(.935)	(.941)	(.946)	(.953)	(.959)	(.968)
10	.677	.695	.726	.750	.771	.788	.802	.815	.836	.853	.882
	(.735)	(.751)	(.777)	(.797)	(.814)	(.828)	(.840)	(.851)	(.868)	(.882)	(.906)
15	.560	.579	.613	.641	.665	.686	.704	.720	.747	.769	.810
	(.619)	(.636)	(.667)	(.691)	(.713)	(.731)	(.747)	(.761)	(.784)	(.804)	(.839)
20	.475	.494	.528	.557	.582	.604	.624	.641	.671	.697	.745
	(.531)	(.549)	(.580)	(.607)	(.630)	(.650)	(.667)	(.684)	(.711)	(.734)	(.777)
25	.412	.431	.463	.491	.516	.538	.558	.576	.608	.635	.688
	(.464)	(.482)	(.512)	(.539)	(.562)	(.583)	(.602)	(.618)	(.648)	(.673)	(.721)
30	.364	.381	.412	.439	.463	.485	.505	.523	.555	.583	.638
	(.412)	(.429)	(.458)	(.484)	(.507)	(.528)	(.547)	(.564)	(.594)	(.620)	(.671)
35	.325	.342	.371	.397	.420	.441	.460	.478	.510	.538	.594
	(.370)	(.386)	(.414)	(.439)	(.462)	(.482)	(.500)	(.518)	(.548)	(.574)	(.627)
40	.294	.309	.337	.362	.384	.404	.423	.440	.471	.499	.555
	(.336)	(.351)	(.378)	(.402)	(.423)	(.443)	(.461)	(.478)	(.508)	(.534)	(.588)
45	.268	.283	.309	.332	.353	.373	.391	.407	.438	.465	.521
	(.307)	(.321)	(.347)	(.370)	(.391)	(.410)	(.427)	(.444)	(.473)	(.499)	(.553)
50	.247	.260	.285	.307	.327	.346	.363	.379	.409	.435	.491
	(.283)	(.296)	(.321)	(.343)	(.363)	(.381)	(.398)	(.414)	(.443)	(.469)	(.522)
100	.136	.145	.160	.175	.188	.200	.212	.224	.245	.265	.309
	(.158)	(.167)	(.182)	(.197)	(.211)	(.223)	(.236)	(.247)	(.269)	(.288)	(.332)
500	.030	.032	.036	.039	.042	.046	.049	.052	.058	.064	.077
	(.035)	(.037)	(.041)	(.045)	(.048)	(.052)	(.055)	(.058)	(.064)	(.070)	(.084)
1000	.015	.016	.018	.020	.022	.023	.025	.027	.030	.033	.040
	(.018)	(.019)	(.021)	(.023)	(.025)	(.026)	(.028)	(.030)	(.033)	(.036)	(.043)

$s = 8, \alpha = .05\ (.01)$

n\m	-.5	0	1	2	3	4	5	6	8	10	15
1	.985	.987	.989	.990	.992	.993	.993	.994	.995	.996	.997
	(.994)	(.994)	(.995)	(.996)	(.997)	(.997)	(.997)	(.998)	(.998)	(.998)	(.999)
2	.961	.964	.970	.974	.977	.979	.981	.983	.986	.987	.990
	(.979)	(.980)	(.984)	(.986)	(.988)	(.989)	(.990)	(.991)	(.992)	(.993)	(.995)
3	.930	.936	.946	.952	.958	.962	.965	.968	.973	.976	.982
	(.957)	(.960)	(.966)	(.970)	(.974)	(.977)	(.979)	(.980)	(.983)	(.985)	(.989)
5	.864	.874	.890	.902	.912	.920	.927	.932	.941	.948	.959
	(.903)	(.910)	(.922)	(.931)	(.938)	(.943)	(.948)	(.952)	(.958)	(.963)	(.972)
10	.713	.728	.754	.775	.793	.808	.821	.832	.851	.865	.892
	(.766)	(.779)	(.800)	(.818)	(.833)	(.845)	(.855)	(.865)	(.880)	(.892)	(.914)
15	.598	.615	.645	.670	.692	.710	.727	.741	.766	.786	.824
	(.653)	(.669)	(.695)	(.717)	(.736)	(.752)	(.767)	(.779)	(.801)	(.818)	(.851)
20	.513	.531	.561	.587	.610	.630	.648	.664	.692	.716	.761
	(.567)	(.583)	(.610)	(.634)	(.655)	(.673)	(.690)	(.704)	(.729)	(.750)	(.791)
25	.449	.466	.495	.521	.544	.565	.583	.600	.630	.655	.705
	(.499)	(.515)	(.543)	(.567)	(.588)	(.607)	(.625)	(.640)	(.668)	(.691)	(.736)
30	.398	.414	.443	.468	.491	.511	.530	.547	.577	.603	.655
	(.445)	(.460)	(.488)	(.512)	(.533)	(.552)	(.570)	(.586)	(.614)	(.639)	(.687)
35	.358	.373	.400	.425	.446	.466	.485	.501	.532	.558	.612
	(.402)	(.416)	(.443)	(.466)	(.487)	(.506)	(.523)	(.540)	(.568)	(.593)	(.644)
40	.325	.339	.365	.388	.409	.428	.446	.463	.493	.519	.573
	(.365)	(.379)	(.405)	(.427)	(.448)	(.467)	(.484)	(.500)	(.528)	(.553)	(.605)
45	.297	.311	.335	.357	.378	.396	.414	.430	.459	.485	.539
	(.335)	(.349)	(.373)	(.395)	(.415)	(.433)	(.449)	(.465)	(.493)	(.518)	(.570)
50	.274	.286	.310	.331	.351	.368	.385	.401	.429	.455	.508
	(.310)	(.323)	(.345)	(.366)	(.385)	(.403)	(.419)	(.435)	(.462)	(.487)	(.539)
100	.153	.161	.176	.190	.203	.216	.228	.239	.260	.279	.323
	(.175)	(.184)	(.199)	(.213)	(.226)	(.239)	(.251)	(.262)	(.283)	(.302)	(.346)
500	.034	.036	.040	.043	.047	.050	.053	.056	.062	.068	.081
	(.039)	(.041)	(.045)	(.049)	(.052)	(.056)	(.059)	(.062)	(.069)	(.074)	(.088)
1000	.017	.018	.020	.022	.024	.025	.027	.029	.032	.035	.042
	(.020)	(.021)	(.023)	(.025)	(.027)	(.029)	(.030)	(.032)	(.035)	(.038)	(.046)

Critical Values of Distribution of gcr (continued)

$s =$ 9, $\alpha = .05\ (.01)$

$n\backslash m$	-.5	0	1	2	3	4	5	6	8	10	15
1	.988	.989	.991	.992	.993	.994	.994	.995	.996	.996	.997
	(.995)	(.995)	(.996)	(.997)	(.997)	(.997)	(.998)	(.998)	(.998)	(.999)	(.999)
2	.968	.970	.975	.978	.980	.982	.984	.985	.987	.989	.992
	(.982)	(.984)	(.986)	(.988)	(.989)	(.990)	(.991)	(.992)	(.993)	(.994)	(.996)
3	.942	.946	.953	.959	.963	.967	.970	.972	.976	.979	.984
	(.964)	(.967)	(.971)	(.975)	(.977)	(.979)	(.981)	(.983)	(.985)	(.987)	(.990)
5	.883	.891	.904	.914	.922	.929	.935	.939	.947	.953	.963
	(.917)	(.923)	(.932)	(.939)	(.945)	(.950)	(.954)	(.957)	(.963)	(.967)	(.974)
10	.743	.756	.778	.797	.812	.825	.837	.847	.863	.876	.901
	(.791)	(.802)	(.820)	(.835)	(.848)	(.859)	(.868)	(.876)	(.890)	(.901)	(.920)
15	.632	.647	.674	.696	.715	.732	.747	.760	.782	.801	.835
	(.683)	(.697)	(.720)	(.740)	(.756)	(.771)	(.784)	(.795)	(.815)	(.831)	(.861)
20	.548	.563	.591	.614	.635	.654	.670	.685	.711	.733	.775
	(.598)	(.612)	(.637)	(.659)	(.678)	(.695)	(.709)	(.723)	(.746)	(.766)	(.803)
25	.482	.497	.525	.549	.570	.589	.606	.622	.650	.673	.720
	(.530)	(.544)	(.570)	(.592)	(.612)	(.630)	(.646)	(.660)	(.686)	(.707)	(.750)
30	.430	.445	.471	.495	.516	.535	.552	.569	.597	.622	.671
	(.475)	(.490)	(.515)	(.537)	(.557)	(.575)	(.591)	(.606)	(.633)	(.656)	(.702)
35	.388	.402	.427	.450	.471	.490	.507	.523	.552	.577	.628
	(.431)	(.444)	(.469)	(.490)	(.510)	(.528)	(.545)	(.560)	(.587)	(.611)	(.659)
40	.353	.366	.391	.413	.433	.451	.468	.484	.512	.538	.590
	(.393)	(.406)	(.430)	(.451)	(.471)	(.488)	(.505)	(.520)	(.547)	(.571)	(.620)
45	.324	.337	.360	.381	.400	.418	.435	.450	.478	.503	.555
	(.362)	(.374)	(.397)	(.418)	(.437)	(.454)	(.470)	(.485)	(.512)	(.536)	(.585)
50	.299	.311	.333	.354	.373	.390	.406	.421	.448	.473	.524
	(.335)	(.347)	(.369)	(.389)	(.407)	(.424)	(.439)	(.454)	(.481)	(.504)	(.554)
100	.169	.177	.192	.206	.219	.231	.242	.253	.274	.293	.336
	(.192)	(.200)	(.215)	(.229)	(.241)	(.254)	(.265)	(.276)	(.297)	(.316)	(.359)
500	.038	.040	.043	.047	.051	.054	.057	.060	.066	.072	.086
	(.043)	(.045)	(.049)	(.053)	(.056)	(.060)	(.063)	(.066)	(.073)	(.079)	(.093)
1000	.019	.020	.022	.024	.026	.028	.029	.031	.034	.037	.044
	(.022)	(.023)	(.025)	(.027)	(.029)	(.031)	(.032)	(.034)	(.037)	(.041)	(.048)

$s =$ 10, α= .05 (.01)

$n \backslash m$	-.5	0	1	2	3	4	5	6	8	10	15
1	.990	.991	.992	.993	.994	.995	.995	.996	.996	.997	.998
	(.996)	(.996)	(.997)	(.997)	(.998)	(.998)	(.998)	(.998)	(.999)	(.999)	(.999)
2	.973	.975	.978	.981	.983	.985	.986	.987	.989	.990	.992
	(.985)	(.986)	(.988)	(.990)	(.991)	(.992)	(.992)	(.993)	(.994)	(.995)	(.996)
3	.950	.954	.960	.964	.968	.971	.973	.975	.979	.981	.985
	(.969)	(.971)	(.975)	(.978)	(.980)	(.982)	(.983)	(.985)	(.987)	(.988)	(.991)
5	.898	.905	.916	.924	.931	.937	.941	.946	.952	.958	.967
	(.928)	(.933)	(.940)	(.946)	(.951)	(.955)	(.959)	(.962)	(.967)	(.970)	(.977)
10	.769	.780	.799	.815	.829	.840	.851	.859	.874	.886	.908
	(.812)	(.822)	(.837)	(.851)	(.862)	(.871)	(.880)	(.887)	(.899)	(.909)	(.926)
15	.662	.675	.699	.719	.736	.751	.764	.776	.797	.814	.846
	(.710)	(.721)	(.742)	(.759)	(.774)	(.788)	(.799)	(.810)	(.827)	(.842)	(.869)
20	.578	.592	.617	.639	.658	.675	.690	.704	.728	.748	.787
	(.626)	(.639)	(.661)	(.681)	(.698)	(.713)	(.727)	(.740)	(.761)	(.779)	(.814)
25	.512	.526	.551	.573	.593	.611	.627	.642	.667	.690	.734
	(.558)	(.571)	(.595)	(.615)	(.634)	(.650)	(.665)	(.678)	(.702)	(.722)	(.762)
30	.459	.473	.497	.519	.539	.557	.573	.589	.615	.639	.686
	(.503)	(.516)	(.539)	(.560)	(.579)	(.595)	(.611)	(.625)	(.650)	(.672)	(.715)
35	.416	.429	.453	.474	.494	.511	.528	.543	.570	.594	.643
	(.458)	(.470)	(.493)	(.513)	(.532)	(.549)	(.564)	(.579)	(.604)	(.627)	(.673)
40	.380	.392	.415	.436	.455	.473	.489	.504	.531	.555	.605
	(.419)	(.431)	(.454)	(.474)	(.492)	(.509)	(.524)	(.539)	(.564)	(.587)	(.635)
45	.349	.361	.383	.404	.422	.439	.455	.469	.497	.521	.571
	(.387)	(.398)	(.420)	(.439)	(.457)	(.474)	(.489)	(.503)	(.529)	(.552)	(.600)
50	.323	.335	.356	.375	.393	.410	.425	.440	.466	.490	.540
	(.359)	(.370)	(.391)	(.410)	(.427)	(.443)	(.458)	(.472)	(.498)	(.521)	(.569)
100	.185	.193	.207	.220	.233	.245	.256	.267	.287	.306	.348
	(.208)	(.215)	(.230)	(.243)	(.256)	(.268)	(.279)	(.290)	(.311)	(.329)	(.371)
500	.042	.044	.047	.051	.054	.058	.061	.064	.070	.076	.090
	(.047)	(.049)	(.053)	(.057)	(.061)	(.064)	(.067)	(.071)	(.077)	(.083)	(.097)
1000	.021	.022	.024	.026	.028	.030	.031	.033	.036	.039	.047
	(.024)	(.025)	(.027)	(.029)	(.031)	(.033)	(.034)	(.036)	(.040)	(.043)	(.050)

Critical Values of Distribution of gcr (continued)

$s = 11, \alpha = .05 (.01)$

$n \backslash m$	-.5	0	1	2	3	4	5	6	8	10	15
1	.992	.992	.993	.994	.995	.995	.996	.996	.997	.997	.998
	(.997)	(.997)	(.997)	(.998)	(.998)	(.998)	(.998)	(.998)	(.999)	(.999)	(.999)
2	.977	.979	.981	.983	.985	.986	.988	.989	.990	.991	.993
	(.988)	(.988)	(.990)	(.991)	(.992)	(.993)	(.993)	(.994)	(.995)	(.995)	(.996)
3	.957	.960	.965	.969	.972	.974	.976	.978	.981	.983	.987
	(.974)	(.975)	(.978)	(.981)	(.982)	(.984)	(.985)	(.986)	(.988)	(.990)	(.992)
5	.911	.916	.925	.932	.938	.943	.947	.951	.957	.961	.970
	(.937)	(.941)	(.947)	(.952)	(.956)	(.960)	(.963)	(.966)	(.970)	(.973)	(.979)
10	.791	.800	.817	.831	.843	.854	.863	.870	.884	.895	.915
	(.830)	(.838)	(.852)	(.864)	(.874)	(.882)	(.889)	(.896)	(.907)	(.916)	(.931)
15	.688	.700	.721	.739	.754	.768	.780	.791	.810	.825	.855
	(.733)	(.743)	(.761)	(.777)	(.790)	(.802)	(.813)	(.822)	(.839)	(.852)	(.877)
20	.606	.618	.641	.661	.678	.694	.708	.721	.743	.762	.798
	(.651)	(.663)	(.683)	(.701)	(.717)	(.731)	(.743)	(.755)	(.775)	(.791)	(.824)
25	.540	.552	.576	.596	.615	.631	.646	.660	.684	.705	.746
	(.584)	(.596)	(.617)	(.636)	(.653)	(.668)	(.682)	(.695)	(.717)	(.736)	(.774)
30	.486	.499	.521	.542	.561	.577	.593	.607	.633	.655	.699
	(.529)	(.541)	(.562)	(.581)	(.599)	(.615)	(.629)	(.642)	(.666)	(.686)	(.728)
35	.442	.454	.476	.497	.515	.532	.547	.562	.588	.610	.657
	(.482)	(.494)	(.515)	(.535)	(.552)	(.568)	(.583)	(.596)	(.621)	(.642)	(.686)
40	.405	.417	.438	.458	.476	.492	.508	.522	.548	.571	.619
	(.443)	(.455)	(.476)	(.495)	(.512)	(.528)	(.542)	(.556)	(.581)	(.603)	(.648)
45	.373	.385	.406	.425	.442	.458	.474	.488	.514	.537	.585
	(.410)	(.421)	(.441)	(.460)	(.477)	(.492)	(.507)	(.521)	(.545)	(.567)	(.613)
50	.346	.357	.377	.396	.413	.429	.444	.458	.483	.506	.554
	(.381)	(.392)	(.411)	(.429)	(.446)	(.461)	(.476)	(.489)	(.514)	(.536)	(.582)
100	.200	.208	.222	.235	.247	.259	.270	.281	.301	.319	.360
	(.223)	(.230)	(.244)	(.258)	(.270)	(.282)	(.293)	(.303)	(.323)	(.342)	(.383)
500	.046	.048	.051	.055	.058	.062	.065	.068	.074	.080	.094
	(.052)	(.053)	(.057)	(.061)	(.064)	(.068)	(.071)	(.074)	(.081)	(.087)	(.101)
1000	.023	.024	.026	.028	.030	.032	.033	.035	.038	.041	.049
	(.026)	(.027)	(.029)	(.031)	(.033)	(.035)	(.037)	(.038)	(.042)	(.045)	(.052)

$s = 12,\ \alpha = .05\ (.01)$

$n \backslash m$	-.5	0	1	2	3	4	5	6	8	10	15
1	.993	.993	.994	.995	.995	.996	.996	.997	.997	.997	.998
	(.997)	(.997)	(.998)	(.998)	(.998)	(.998)	(.999)	(.999)	(.999)	(.999)	(.999)
2	.980	.981	.984	.985	.987	.988	.989	.990	.991	.992	.994
	(.989)	(.990)	(.991)	(.992)	(.993)	(.993)	(.994)	(.994)	(.995)	(.996)	(.997)
3	.963	.965	.969	.972	.975	.977	.979	.980	.983	.985	.988
	(.977)	(.979)	(.981)	(.983)	(.984)	(.986)	(.987)	(.988)	(.989)	(.990)	(.993)
5	.921	.926	.933	.939	.944	.949	.952	.956	.961	.965	.972
	(.944)	(.948)	(.953)	(.957)	(.961)	(.964)	(.966)	(.969)	(.972)	(.975)	(.980)
10	.810	.818	.833	.845	.856	.865	.873	.880	.892	.902	.920
	(.846)	(.853)	(.865)	(.875)	(.884)	(.891)	(.898)	(.904)	(.914)	(.921)	(.936)
15	.711	.722	.740	.757	.771	.783	.794	.804	.822	.836	.864
	(.753)	(.762)	(.778)	(.792)	(.805)	(.815)	(.825)	(.834)	(.849)	(.861)	(.885)
20	.631	.642	.663	.681	.697	.711	.724	.736	.757	.774	.809
	(.674)	(.684)	(.703)	(.719)	(.733)	(.746)	(.758)	(.769)	(.787)	(.802)	(.833)
25	.565	.577	.598	.617	.634	.649	.664	.677	.699	.719	.758
	(.607)	(.618)	(.638)	(.656)	(.671)	(.685)	(.698)	(.710)	(.731)	(.749)	(.784)
30	.511	.523	.544	.563	.581	.596	.611	.625	.648	.669	.712
	(.552)	(.563)	(.583)	(.601)	(.617)	(.632)	(.646)	(.658)	(.680)	(.700)	(.739)
35	.466	.478	.499	.518	.535	.551	.566	.579	.604	.625	.670
	(.506)	(.516)	(.536)	(.554)	(.571)	(.586)	(.600)	(.613)	(.636)	(.656)	(.698)
40	.428	.439	.460	.478	.496	.511	.526	.540	.564	.587	.632
	(.466)	(.477)	(.496)	(.514)	(.531)	(.546)	(.560)	(.573)	(.596)	(.617)	(.660)
45	.396	.406	.426	.445	.461	.477	.491	.505	.530	.552	.598
	(.432)	(.442)	(.461)	(.479)	(.495)	(.510)	(.524)	(.537)	(.561)	(.582)	(.626)
50	.368	.378	.397	.415	.431	.447	.461	.474	.499	.521	.567
	(.402)	(.412)	(.431)	(.448)	(.464)	(.479)	(.493)	(.506)	(.529)	(.550)	(.595)
100	.215	.222	.236	.249	.261	.272	.283	.293	.313	.331	.372
	(.238)	(.245)	(.259)	(.271)	(.283)	(.295)	(.306)	(.316)	(.336)	(.354)	(.394)
500	.050	.052	.055	.059	.062	.065	.069	.072	.078	.084	.098
	(.055)	(.057)	(.061)	(.065)	(.068)	(.072)	(.075)	(.078)	(.085)	(.091)	(.105)
1000	.025	.026	.028	.030	.032	.034	.035	.037	.040	.043	.051
	(.028)	(.029)	(.031)	(.033)	(.035)	(.037)	(.039)	(.041)	(.044)	(.047)	(.055)

Critical Values of Distribution of gcr (continued)

α\m	.5	0	1	2	4	6	8	10	15
					n = 1000				
					s = 1				
.05	.00195	.00317	.00488	.00635	.00928	.01172	.01440	.01685	.02271
.01	.00342	.00464	.00659	.00854	.01147	.01440	.01709	.01978	.02612
					s = 2				
.05	.00439	.00537	.00732	.00903	.01196	.01489	.01758	.02026	.02661
.01	.00610	.00732	.00952	.01123	.01465	.01758	.02051	.02344	.03003
					s = 3				
.05	.00659	.00781	.00952	.01123	.01465	.01758	.02026	.02319	.02954
.01	.00854	.00977	.01196	.01367	.01709	.02026	.02344	.02637	.03320
					s = 4				
.05	.00879	.00977	.01172	.01367	.01685	.02002	.02295	.02563	.03247
.01	.01099	.01221	.01416	.01611	.01953	.02295	.02588	.02881	.03589
					s = 5				
.05	.01099	.01196	.01392	.01562	.01904	.02222	.02515	.02808	.03491
.01	.01318	.01440	.01636	.01831	.02197	.02515	.02832	.03149	.03857
					s = 6				
.05	.01318	.01416	.01611	.01782	.02124	.02441	.02759	.03052	.03735
.01	.01562	.01660	.01855	.02051	.02417	.02759	.03076	.03369	.04102
					s = 7				
.05	.01514	.01611	.01807	.02002	.02344	.02661	.02979	.03271	.03979
.01	.01782	.01880	.02075	.02271	.02637	.02979	.03296	.03613	.04346
					s = 8				
.05	.01733	.01831	.02026	.02197	.02539	.02881	.03198	.03491	.04199
.01	.02002	.02100	.02295	.02490	.02856	.03198	.03516	.03833	.04565
					s = 9				
.05	.01929	.02026	.02222	.02393	.02759	.03076	.03394	.03711	.04443
.01	.02197	.02319	.02515	.02710	.03076	.03418	.03735	.04053	.04810
					s = 10				
.05	.02124	.02222	.02417	.02612	.02954	.03296	.03613	.03931	.04663
.01	.02417	.02515	.02710	.02905	.03271	.03613	.03955	.04272	.05029
					s = 11				
.05	.02344	.02441	.02612	.02808	.03174	.03491	.03809	.04126	.04858
.01	.02637	.02734	.02930	.03125	.03491	.03833	.04175	.04492	.05249
					s = 12				
.05	.02539	.02637	.02832	.03003	.03369	.03711	.04028	.04346	.05078
.01	.02832	.02930	.03125	.03320	.03687	.04053	.04370	.04688	.05469

				$n = 1000$					
$\alpha\backslash m$	.5	0	1	2	4	6	8	10	15
				$s = 13$					
.05	.02734	.02832	.03027	.03198	.03564	.03906	.04224	.04541	.05298
.01	.03052	.03149	.03345	.03540	.03906	.04248	.04590	.04907	.05664
				$s = 14$					
.05	.02930	.03027	.03223	.03418	.03760	.04102	.04419	.04736	.05493
.01	.03247	.03345	.03540	.03735	.04102	.04443	.04785	.05103	.05884
				$s = 15$					
.05	.03125	.03223	.03418	.03613	.03955	.04297	.04614	.04932	.05688
.01	.03442	.03564	.03760	.03931	.04297	.04663	.04980	.05322	.06079
				$s = 16$					
.05	.03320	.03418	.03613	.03809	.04150	.04492	.04834	.05151	.05908
.01	.03662	.03760	.03955	.04150	.04517	.04858	.05200	.05518	.06299
				$s = 17$					
.05	.03516	.03613	.03809	.04004	.04346	.04688	.05029	.05347	.06104
.01	.03857	.03955	.04150	.04346	.04712	.05054	.05396	.05713	.06494
				$s = 18$					
.05	.03711	.03809	.04004	.04199	.04541	.04883	.05225	.05542	.06299
.01	.04053	.04150	.04346	.04541	.04907	.05249	.05591	.05908	.06689
				$s = 19$					
.05	.03906	.04004	.04199	.04370	.04736	.05078	.05420	.05737	.06494
.01	.04248	.04346	.04541	.04736	.05103	.05444	.05786	.06104	.06885
				$s = 20$					
.05	.04102	.04199	.04395	.04565	.04932	.05273	.05591	.05933	.06689
.01	.04443	.04565	.04736	.04932	.05298	.05640	.05981	.06299	.07104

Appendix B
Computer Programs Available From Author

The first two editions of this book included listings of BASIC and FORTRAN programs that I considered useful supplements to the then available statistical packages and matrix-manipulation programs. However, no one in their right mind would, at the turn of the millenium, go through the time-consuming and error-prone process of typing hundreds of lines of commands into a file from a printed listing. Rather, if you are interested in either of the programs described below, send an email note to rharris@unm.edu, and I will be glad to email back to you an ASCII file of the program(s) you request—or, if I've managed to get the programs converted to a web-friendly language by the time you "write," I may direct you to a website from which the program(s) can be downloaded.

B.1 *cvinter: p* Values and Critical Values for Univariate Statistics

This program, the "guts" of which was written by Kevin O'Grady (1981), provides *p* values and/or critical values for z, r, χ^2, t, and F distributions. These are now readily available to anyone on various websites, but you may nevertheless find it useful to have a program running on your PC or on your local computing system's mainframe computer. An additional convenience of this program is that it allows the user to specify the number of comparisons among which the total alpha for a set of tests is to be divided (evenly), so that one more source of error (slipping a decimal or simply blowing an in-your-head calculation of, say, .05/6) is eliminated.

B.2 *gcrinter: p* values and Critical Values for the Greatest Characteristic Root (gcr) Statistic

This program, based on subroutines written by Pillai (1967) and Venables (1974) provides critical values for the g.c.r. test that I argue (sections 4.5 and 5.1) is usually the preferred overall test (and base for fully post hoc followup tests) in Manova and Canona. Critical values for a wide range of values of the degree-of-freedom parameters s, m, and n are provided in Table A.5, but having a locally resident computer program avoids the need to interpolate within those tables.

Appendix C: Derivations

Where notation is in doubt, consult the chapter to which the derivation is most relevant, as indicated by the first digit of its number. For example, Derivation 2.3 uses the notation employed in chap. 2)

DERIVATION 1.1: PER-EXPERIMENT AND EXPERIMENTWISE ERROR RATES FOR BONFERRONI-ADJUSTED TESTS

Let $X_i = 1$ if the *i*th test in a set of n_t tests yields rejection of its corresponding null hypothesis. The Type I error rate for each of the n_t tests is defined as

$$\alpha_i = \Pr(X_i = 1 \mid H_0 \text{ is true}),$$

where this probability is computed relative to the sample space of all possible replications of the sampling experiment consisting of drawing a sample of observations and then conducting the n_t tests on these observations.

Clearly,

$$T = \sum_{i=1}^{n_t} X_i$$

yields a count of the total number of true null hypotheses falsely rejected in any given replication of the sampling experiment. The *per-experiment* Type I error rate (the average number of Type I errors made per replication of the sampling experiment) is defined as

$$\alpha_{pe} = E(T),$$

which equals the expected value of *T*, which in turn equals the mean of the sampling distribution of *T* as derived from the sample space described above.

However, it is well known that $E(X + Y + Z + \cdots) = E(X) + E(Y) + E(Z) + \cdots$. [This follows from the definition of $E(Y)$ as $\Sigma Y_i \cdot \Pr[Y = Y_i]$ or as $\int f(y)dy$ and the facts that $\Sigma(A + B) = \Sigma A + \Sigma B$ and $\int [f(y) + g(y)]dy = \int f(y)dy + \int g(y)dy$.] Thus

$$\alpha_{pe} = E(T) = \sum_{i=1}^{n_t} E(X_i).$$

But each X_i is a binomial variable having a mean of α_i.

$$\left[E(X_i) = \sum_{i=0}^{1} i \cdot \Pr[X_i = i] = 0 \cdot (1 - \alpha_i) + 1 \cdot (\alpha_i) = \alpha_i \cdot\right]$$

Thus $\alpha_{pe} = \Sigma\alpha_i$, so that all we need to do to set our per-experiment Type I error rate to any desired value is to lower our individual α_i so that they sum to the target value. Note that this derivation is based only on the algebraic properties of summations and integrals and does *not* require that the various tests be uncorrelated with each other.

The experimentwise Type I error rate (the proportion of replications of the sampling experiment in which one or more null hypotheses are falsely rejected) is defined as

$$\begin{aligned}\alpha_{\text{exp}} &= \Pr(X_1 = 1 \text{ or } X_2 = 1 \text{ or} \ldots X_{n_t} = 1 \mid H_0 \text{ true})\\ &= \Pr(\text{one or more individual tests falsely reject } H_0)\\ &= 1 - \Pr(\text{no test falsely rejects } H_0)\\ &= 1 - \Pr(X_1 = 0)\cdot \Pr(X_2 = 0 \mid X_1 = 0)\cdot \Pr(X_3 = 0 \mid X_1, X_2 = 0)\\ &\quad \ldots \cdot \Pr(X_{n_t} = 0 \mid X_1, X_2, \ldots, X_{n_t-1} \text{ each} = 0).\end{aligned}$$

If the tests are independent of each other, then

$$\Pr(X_i = 0 \mid X_1, X_2, \ldots, X_{i-1} \text{ each} = 0)$$

is simply equal to the unconditional (marginal) probability that test i doesn't reach significance, that is, to $(1 - \alpha_i)$. Thus

$$\alpha_{\text{exp}} = \left(1 - \prod_{i=1}^{n_t}(1 - \alpha_i)\right)$$

$$[\quad = 1 - (1 - \alpha_c)^{n_t} \quad \text{if} \quad \alpha_1 = \alpha_2 = \cdots = \alpha_{n_t} = \alpha_c]$$

when the tests are all mutually independent. For values of α_i close to zero, $1 - \Pi_{i=1}^{n_t}(1 - \alpha_i)$ is close to $\Sigma\alpha_i$, but it is always less than that sum. For instance, $1 - (1 - .05)(1 - .03)(1 - .01)(1 - .01) = 1 - .90316 = .09684 < .05 + .03 + .01 + .01 = .10$. More formally, as Myers (1979, p. 298) points out,

$$1 - (1 - a)^k = 1 - \left[1 - ka + \binom{k}{2}a^2 - \binom{k}{3}a^3 \cdots - a^k\right],$$

by the binomial expansion.

For a close to zero, a^2, a^3, and so on, are close to zero, so

$$1 - (1 - a)^k \approx 1 - (1 - ka) = ka.$$

Moreover, the terms after ka consist of pairs of terms of the form

$$\binom{k}{n}a^n - \binom{k}{n+1}a^{n+1} = \binom{k}{n}a\left[1 - \frac{k-n}{n+1}a\right],$$

which is always $\geqslant 0$, so

$$1 - (1 - a)^k \leqslant 1 - (1 - ka) = ka.$$

Unfortunately, I have not been able to uncover (or generate myself) a generalization of this proof to the cases of nonindependent tests and unequal α_i. That $\Sigma\alpha_i$ should overestimate α_{exp} is, however, intuitively reasonable, since, for instance, carrying out 30 tests at the .05 level clearly does not give us a 150% chance of rejecting at least one null hypothesis. (It *does,* however, yield an average of 1.5 false rejections of H_0 when the null hypothesis is true. This is the per-experiment rate we derived above.)

DERIVATION 2.1: SCALAR FORMULAE FOR MRA WITH ONE, TWO, AND THREE PREDICTORS

Single Predictor (Bivariate Regression)

We wish to minimize

$$\begin{aligned} E &= \sum(Y_i - \hat{Y}_i)^2 \\ &= \sum(Y_i - b_0 - b_1X_{i,1})^2 = \sum(Y^2 + b_0^2 + b_1^2X_1^2 - 2b_0Y - 2b_1X_1Y + 2b_0b_1X_1) \\ &= \sum Y^2 + Nb_0^2 + b_1^2\sum X_1^2 - 2b_0\sum Y - 2b_1\sum X_1Y + 2b_0b_1\sum X_1, \end{aligned}$$

whence

$$dE/db_0 = 2b_0N - 2\sum Y + 2b_1\sum X_1 = 0$$

if and only if (iff)

$$b_0 = \left(\sum Y - b_1\sum X_1\right)\Big/N = \bar{Y} - b_1\bar{X}_1;$$

and

$$\begin{aligned} dE/db_1 &= 2b_1\sum X_1^2 - 2\sum X_1Y + 2b_0\sum X_1 \\ &= 2b_1\left(\sum X_1^2 - \bar{X}_1\sum X_1\right) - 2\sum X_1Y + 2\bar{Y}\sum X_1 \\ &= 2b_1\sum x_1^2 - 2\sum x_1y = 0 \end{aligned}$$

iff

$$b_1 = \sum x_1y\Big/\sum x_1^2 = \sum x_1y\sqrt{\sum y^2/\sum x_1^2}\Big/\sqrt{\sum x_1^2\sum y^2} = r_{1y}(s_y/s_1).$$

Two Predictors

If we now use both X_1 *and* X_2 as predictors, we obtain

$$\begin{aligned} &\sum(Y_i - b_0 - b_1X_{i,1} - b_2X_{i,2})^2 \\ &\quad = \sum(Y^2 + b_0^2 + b_1^2X_1^2 + b_2^2X_2^2 - 2b_0Y - 2b_1X_1Y \\ &\quad\quad - 2b_2X_2Y + 2b_0b_1X_1 + 2b_0b_2X_2 + 2b_1b_2X_1X_2); \end{aligned}$$

$$dE/db_0 = 2Nb_0 - 2\sum Y + 2b_1\sum X_1 + 2b_2\sum X_2 = 0$$

iff

$$\begin{aligned} b_0 &= \overline{Y} - b_1\overline{X}_1 - b_2\overline{X}_2; \\ dE/db_1 &= 2b_1 \sum X_1^2 - 2\sum X_1Y + 2b_0 \sum X_1 + 2b_2 \sum X_1X_2 \\ &= 2b_1 \sum X_1^2 - 2\sum X_1Y + 2(\overline{Y} - b_1\overline{X}_1 - b_2\overline{X}_2)\sum X_1 + 2b_2 \sum X_1X_2 \\ &= 2b_1 \sum x_1^2 - 2\sum x_1y + 2b_2 \sum x_1x_2; \end{aligned}$$

and

$$dE/db_2 = 2b_2 \sum x_2^2 - 2\sum x_2y + 2b_1 \sum x_1x_2.$$

[Note that the present text follows the convention of letting capital letters stand for "raw" observations and lowercase letters for deviation scores. Thus, $\Sigma x_1x_2 = \Sigma(X_1 - \overline{X}_1)(X_2 - \overline{X}_2)$.] Solving for b_1 and b_2 requires solving the pair of simultaneous equations,

$$\left(\sum x_1^2\right)b_1 + \left(\sum x_1x_2\right)b_2 = \sum x_1y$$

and

$$\left(\sum x_1x_2\right)b_1 + \left(\sum x_2^2\right)b_2 = \sum x_2y;$$

whence, after some laborious algebra (made less laborious if matrix techniques to be discussed shortly are used),

$$\begin{aligned} b_1 &= \left(\sum x_1y \sum x_2^2 - \sum x_2y \sum x_1x_2\right)\Big/\left[\sum x_1^2 \sum x_2^2 - \left(\sum x_1x_2\right)^2\right] \\ &= (s_2^2 s_{1y} - s_{12}s_{2y})/(s_1^2 s_2^2 - s_{12}^2) \end{aligned}$$

and

$$b_2 = (s_1^2 s_{2y} - s_{12}s_{1y})/(s_1^2 s_2^2 - s_{12}^2),$$

where

$$s_{ij} = \left[\sum(X_i - \overline{X}_i)(X_j - \overline{X}_j)\right]\Big/(N-1) = r_{ij}s_is_j$$

and

$$s_{iy} = \left[\sum(X_i - \overline{X}_i)(Y - \overline{Y})\right]\Big/(N-1) = r_{iy}s_is_y.$$

Three Predictors

If we now add X_3 to the set of predictors, we find (as the reader will naturally wish to verify) that

$$b_0 = \overline{Y} - b_1\overline{X}_1 - b_2\overline{X}_2 - b_3\overline{X}_3;$$

and

$$\left(\sum x_1^2\right)b_1 + \left(\sum x_1x_2\right)b_2 + \left(\sum x_1x_3\right)b_3 = \sum x_1y;$$

$$\left(\sum x_1x_2\right)b_1 + \left(\sum x_2^2\right)b_2 + \left(\sum x_2x_3\right)b_3 = \sum x_2y;$$

$$\left(\sum x_1x_3\right)b_1 + \left(\sum x_2x_3\right)b_2 + \left(\sum x_3^2\right)b_3 = \sum x_3y;$$

whence

$$b_1 = [s_{1y}(s_2^2s_3^2 - s_{23}^2) + s_{2y}(s_{13}s_{23} - s_{12}s_3^2) + s_{3y}(s_{12}s_{23} - s_{13}s_2^2)]/D;$$
$$b_2 = [s_{1y}(s_{13}s_{23} - s_{12}s_3^2) + s_{2y}(s_1^2s_3^2 - s_{13}^2) + s_{3y}(s_{12}s_{13} - s_{23}s_1^2)]/D;$$
$$b_3 = [s_{1y}(s_{12}s_{23} - s_{13}s_2^2) + s_{2y}(s_{12}s_{13} - s_{23}s_1^2) + s_{3y}(s_1^2s_2^2 - s_{12}^2)]/D;$$

where

$$D = s_1^2(s_2^2s_3^2 - s_{23}^2) + s_{12}(s_{13}s_{23} - s_{12}s_3^2) + s_{13}(s_{12}s_{23} - s_{13}s_2^2)$$
$$= (s_1s_2s_3)^2 - (s_1s_{23})^2 - (s_2s_{13})^2 - (s_3s_{12})^2 + 2s_{12}s_{23}s_{13}.$$

DERIVATION 2.3: MAXIMIZING R VIA MATRIX ALGEBRA

We wish to maximize

$$R = \mathbf{b}'\mathbf{s}_{xy}/\sqrt{\mathbf{b}'\mathbf{S}_x\mathbf{b}s_y^2}.$$

Noting that when $m = 1$,

$$\mathbf{b}'\mathbf{S}_x\mathbf{b} = b_1^2 s_x^2 = \frac{\left(\sum xy\right)^2 / \sum x^2}{N-1} = \frac{\left(\sum xy / \sum x^2\right)\sum xy}{N-1} = \frac{b_1 \sum xy}{N-1} = \mathbf{b}'\mathbf{s}_{xy},$$

and wishing to maintain compatibility with the bs derived from the prediction accuracy criterion, we take as our side condition $\mathbf{b}'\mathbf{S}_x\mathbf{b} - \mathbf{b}'\mathbf{s}_{xy} = 0$. Thus, we wish to maximize $R = \sqrt{\mathbf{b}'\mathbf{s}_{xy}}$, subject to the specified constraint. (Remember that for purposes of the derivation we can ignore s_y, since it is unaffected by our choice of bs, and we can at any rate apply a standard-score transformation to the Ys to obtain $s_y = 1$.) However, we might as well maximize R^2 instead, and thereby simplify our math. Using the method of Lagrangian multipliers (cf. Digression 1) gives

$$L = \mathbf{b}'\mathbf{s}_{xy} - \lambda(\mathbf{b}'\mathbf{S}_x\mathbf{b} - \mathbf{b}'\mathbf{s}_{xy}) = \mathbf{b}'\mathbf{s}_{xy}(1 + \lambda) - \lambda(\mathbf{b}'\mathbf{S}_x\mathbf{b});$$

whence

$$dL/d\mathbf{b} = \mathbf{s}_{xy}(1 + \lambda) - 2\lambda\mathbf{S}_x\mathbf{b} = 0 \quad \text{iff} \quad \mathbf{S}_x\mathbf{b} = (1 + \lambda)\mathbf{s}_{xy}/2\lambda.$$

This last equation seems quite reminiscent of our side condition. In fact, premultiplying both sides of the equation by $\mathbf{b}'$ gives us

$$\mathbf{b}'\mathbf{S}_x\mathbf{b} = [(1 + \lambda)/2\lambda]\mathbf{b}'\mathbf{s}_{xy},$$

which is consistent with the side condition if and only if $\lambda = 1$. Plugging this value of lambda back into the equation we had before the postmultiplication by $\mathbf{b}'$ gives

$$\mathbf{S}_x\mathbf{b} = \mathbf{s}_{xy}, \qquad \text{whence} \quad \mathbf{b} = \mathbf{S}_x^{-1}\mathbf{s}_{xy},$$

the same values derived from the accuracy criterion.

whence

$$dL/db_1 = \sum x_1y - \lambda\left[2b_1\left(\sum x_1^2\right) + 2b_2 \sum x_1x_2\right]$$

and

$$dL/db_2 = \sum x_2y - \lambda\left[2b_2\left(\sum x_2^2\right) + 2b_1 \sum x_1x_2\right].$$

Setting these two derivatives equal to zero produces the two simultaneous equations,

$$2\lambda b_1 \sum x_1^2 + 2\lambda b_2 \sum x_1x_2 = \sum x_1y;$$

and

$$2\lambda b_1 \sum x_1x_2 + 2\lambda b_2 \sum x_2^2 = \sum x_2y.$$

Comparison of these two equations with the equations that resulted when we derived b_1 and b_2 from the best-prediction criterion shows that the two pairs of equations are identical if $\lambda = \frac{1}{2}$, and indeed substituting in that value for λ and the previously computed expressions for b_1 and b_2 can be readily shown to satisfy the above two equations.

We have thus shown, for the one- and two-predictor cases, that the same weights that minimize the sum of the squared errors of prediction also yield a linear combination of the original variables that correlates more highly with Y than any other linear combination of the predictor variables. We could proceed to the case in which three predictors are involved, but it will prove considerably easier if instead we pause to show how matrix algebra can be used to simplify our task.

DERIVATION 2.3: MAXIMIZING R VIA MATRIX ALGEBRA

We wish to maximize

$$R = \mathbf{b}'\mathbf{s}_{xy}/\sqrt{\mathbf{b}'\mathbf{S}_x\mathbf{b}s_y^2}.$$

Noting that when $m = 1$,

$$\mathbf{b}'\mathbf{S}_x\mathbf{b} = b_1^2 s_x^2 = \frac{\left(\sum xy\right)^2 / \sum x^2}{N-1} = \frac{\left(\sum xy / \sum x^2\right) \sum xy}{N-1} = \frac{b_1 \sum xy}{N-1} = \mathbf{b}'\mathbf{s}_{xy},$$

and wishing to maintain compatibility with the bs derived from the prediction accuracy criterion, we take as our side condition $\mathbf{b}'\mathbf{S}_x\mathbf{b} - \mathbf{b}'\mathbf{s}_{xy} = 0$. Thus, we wish to maximize $R = \sqrt{\mathbf{b}'\mathbf{s}_{xy}}$, subject to the specified constraint. (Remember that for purposes of the derivation we can ignore s_y, since it is unaffected by our choice of bs, and we can at any rate apply a standard-score transformation to the Ys to obtain $s_y = 1$.) However, we might as well maximize R^2 instead, and thereby simplify our math. Using the method of Lagrangian multipliers (cf. Digression 1) gives

$$L = \mathbf{b}'\mathbf{s}_{xy} - \lambda(\mathbf{b}'\mathbf{S}_x\mathbf{b} - \mathbf{b}'\mathbf{s}_{xy}) = \mathbf{b}'\mathbf{s}_{xy}(1 + \lambda) - \lambda(\mathbf{b}'\mathbf{S}_x\mathbf{b});$$

whence

$$dL/d\mathbf{b} = \mathbf{s}_{xy}(1 + \lambda) - 2\lambda\mathbf{S}_x\mathbf{b} = 0 \quad \text{iff} \quad \mathbf{S}_x\mathbf{b} = (1 + \lambda)\mathbf{s}_{xy}/2\lambda.$$

This last equation seems quite reminiscent of our side condition. In fact, premultiplying both sides of the equation by $\mathbf{b}'$ gives us

$$\mathbf{b}'\mathbf{S}_x\mathbf{b} = [(1 + \lambda)/2\lambda]\mathbf{b}'\mathbf{s}_{xy},$$

which is consistent with the side condition if and only if $\lambda = 1$. Plugging this value of lambda back into the equation we had before the postmultiplication by $\mathbf{b}'$ gives

$$\mathbf{S}_x\mathbf{b} = \mathbf{s}_{xy}, \qquad \text{whence} \quad \mathbf{b} = \mathbf{S}_x^{-1}\mathbf{s}_{xy},$$

the same values derived from the accuracy criterion.

DERIVATION 2.4: INDEPENDENCE OF IRRELEVANT PARAMETERS

As pointed out in Section 2.2.4, the "independence of irrelevant parameters" criterion will be satisfied if K (the matrix of coefficients relating b to y) satisfies the matrix equation Kx = I. However, the expression for b is

$$\mathbf{b} = \mathbf{S}_x^{-1}\mathbf{s}_{xy} = (\mathbf{x}'\mathbf{x})^{-1}(\mathbf{x}'\mathbf{y}) = [(\mathbf{x}'\mathbf{x})^{-1}\mathbf{x}']\mathbf{y} = \mathbf{K}\mathbf{y};$$

whence

$$\mathbf{K} = (\mathbf{x}'\mathbf{x})^{-1}\mathbf{x}' \quad \text{and} \quad \mathbf{K}\mathbf{x} = (\mathbf{x}'\mathbf{x})^{-1}(\mathbf{x}'\mathbf{x}) = \mathbf{I},$$

as asserted.

DERIVATION 2.5: VARIANCES OF b_jS AND OF LINEAR COMBINATIONS THEREOF

Now, b_j = jth element of $(\mathbf{x}'\mathbf{x})^{-1}\mathbf{x}'\mathbf{y}$ = a linear function of **y**, where the weighting coefficients are given by the jth row of $\mathbf{K} = (\mathbf{x}'\mathbf{x})^{-1}\mathbf{x}'$. However, each y_i is an independently sampled observation from the population of y scores, so the variance of b_j should be given by

$$\left(\sum_{i-1}^{N} k_{j,i}^2\right)\sigma_y^2,$$

which we can of course estimate by

$$\left(\sum_{i-1}^{N} k_{j,i}^2\right) MS_{res} = (\mathbf{k}_j{}'\mathbf{k}_j) MS_{res},$$

where $\mathbf{k}_j{}'$ is the jth row of $\mathbf{K} = \mathbf{e}'\mathbf{K}$, where **e** is a vector whose only nonzero entry occurs in the jth position. Thus the variance of b_j is

$$(\mathbf{e}'\mathbf{K}\mathbf{K}'\mathbf{e})MS_{res} = \mathbf{e}'(\mathbf{x}'\mathbf{x})^{-1}\mathbf{x}'\mathbf{x}(\mathbf{x}'\mathbf{x})^{-1}\mathbf{e}\cdot MS_{res} = \mathbf{e}'(\mathbf{x}'\mathbf{x})^{-1}\mathbf{e}\cdot MS_{res}$$
$$= d_{jj}\cdot MS_{res}$$

where d_{jj} is the jth main diagonal entry of $(\mathbf{x}'\mathbf{x})^{-1}$. Thus $(b_j - \alpha)^2/(d_{jj}\cdot MS_{res})$ is directly analogous to the familiar expression $[(x - \mu_x)/\sigma_x]^2 = = F$. A highly similar argument can be made for the reasonableness of the test for a contrast among the βs, since a linear combination (for example, a contrast) of the bs is in turn a linear combination of the ys, so that the variance of **a'b** is estimated by

$$\text{var}(\mathbf{a}'\mathbf{b}) = \mathbf{a}'(\mathbf{x}'\mathbf{x})^{-1}\mathbf{x}'\mathbf{x}(\mathbf{x}'\mathbf{x})^{-1}\mathbf{a}\cdot MS_{res} = \mathbf{a}'(\mathbf{x}'\mathbf{x})^{-1}\cdot MS_{res}.$$

DERIVATION 2.6: DROP IN $R^2 = b_{z_i}^2(1-R_{i\bullet oth}^2)$

From the second of the two formulae for the inverse of a partitioned matrix (Section D2.8), we have

$$\begin{aligned}\mathbf{A}_{11} &= \mathbf{R}_{\text{oth}}; \quad \mathbf{A}_{21} = \mathbf{r}'_{\text{oth}\bullet m};\\ \mathbf{A}_{12} &= \mathbf{r}_{\text{oth}\bullet m}; \quad \mathbf{A}_{22} = \mathbf{I};\\ \mathbf{A}_{21}\mathbf{A}_{11}^{-1}\mathbf{A}_{12} &= \mathbf{r}'_{\text{oth}\bullet m}\mathbf{R}^{-1}_{\text{oth}}\mathbf{r}_{\text{oth}\bullet m}\\ &= R^2_{\text{oth}\bullet m}; \quad \text{and}\end{aligned}$$

$$\mathbf{A}_{11}^{-1}\mathbf{A}_{12} = \mathbf{R}^{-1}_{\text{oth}}\mathbf{r}_{\text{oth}\bullet m} \bullet \mathbf{b}_{\text{oth}\bullet m}, \text{ whence}$$

$$\mathbf{R}_x^{-1} = \begin{bmatrix} \mathbf{R}^{-1}_{\text{oth}} + \mathbf{b}_{\text{oth}\bullet m}\mathbf{b}'_{\text{oth}\bullet m} & -\mathbf{b}_{\text{oth}\bullet m}/(1-R^2_{\text{oth}\bullet m}) \\ -\mathbf{b}_{\text{oth}\bullet m}/(1-R^2_{\text{oth}\bullet m}) & 1/(1-R^2_{\text{oth}\bullet m}) \end{bmatrix}.$$

Thus

$$R^2_{\text{with}} = \begin{bmatrix}\mathbf{r}'_{\text{oth}\bullet y} & r_{m\bullet y}\end{bmatrix}\mathbf{R}_x^{-1}\begin{bmatrix}\mathbf{r}_{\text{oth}\bullet y}\\ r_{m\bullet y}\end{bmatrix}$$

$$= \begin{bmatrix} \mathbf{r}'_{\text{oth}\bullet y}\mathbf{R}^{-1}_{\text{oth}} + \mathbf{r}'_{\text{oth}\bullet y}\mathbf{b}_{\text{oth}\bullet m}\mathbf{b}'_{\text{oth}\bullet m}/(1-R^2_{\text{oth}\bullet m}) & -\mathbf{r}'_{\text{oth}\bullet y}\mathbf{b}_{\text{oth}\bullet m}/(1-R^2_{\text{oth}\bullet m}) \\ -r_{m\bullet y}\mathbf{b}'_{\text{oth}\bullet m}/(1-R^2_{\text{oth}\bullet m}) & r_{m\bullet y}\mathbf{b}'_{\text{oth}\bullet m}/(1-R^2_{\text{oth}\bullet m}) \end{bmatrix}\begin{bmatrix}\mathbf{r}_{\text{oth}\bullet y}\\ r_{my}\end{bmatrix}$$

$$= \mathbf{r}'_{\text{oth}\bullet y}\mathbf{R}^{-1}_{\text{oth}}\mathbf{r}_{\text{oth}\bullet y} + 1/(1-R^2_{\text{oth}\bullet m})\left\{\begin{matrix}\mathbf{r}'_{\text{oth}\bullet y}\mathbf{b}_{\text{oth}\bullet m}\mathbf{b}'_{\text{oth}\bullet m}\mathbf{r}_{\text{oth}\bullet y} - r_{my}\mathbf{b}'_{\text{oth}\bullet m}\\ -\mathbf{r}'_{\text{oth}\bullet y}\mathbf{b}_{\text{oth}\bullet m}r_{my} + r^2_{my}\end{matrix}\right\}$$

$$= R^2_{\text{without}} + \left(-\mathbf{r}'_{\text{oth}\bullet y}\mathbf{b}_{\text{oth}\bullet m} + r_{my}\right)^2/(1-R^2_{\text{oth}\bullet m}).$$

However, we know that the last (mth) row of R_x^{-1}, post-multiplied by r_{xy}, gives b_{z_m}, so

$$b_{z_m} = [1/(1-R^2_{\text{oth}\bullet m})](-\mathbf{b}'_{\text{oth}\bullet m}\mathbf{r}_{\text{oth}\bullet y} + r_{my}), \quad \text{whence}$$

$$(r_{my} - \mathbf{r}'_{\text{oth}\bullet y}\mathbf{b}_{\text{oth}\bullet m}) = b_{z_m}(1-R^2_{\text{oth}\bullet m}), \text{ and}$$

$$\begin{aligned}\Delta R^2 = R^2_{\text{with}} - R^2_{\text{without}} &= \frac{\left[b_{z_m}(1-R^2_{\text{oth}\bullet m})\right]^2}{(1-R^2_{\text{oth}\bullet m})}\\ &= b^2_{z_m}(1-R^2_{\text{oth}\bullet m}).\end{aligned}$$

DERIVATION 2.7: MRA ON GROUP-MEMBERSHIP VARIABLES YIELDS SAME F AS ANOVA

The most straightforward way of showing this equivalence is to show that the models (and thus optimization tasks) are identical. This was in fact done in Section 2.7. It should be pointed out, however, that the equivalence extends to the distribution of the ε_i in MRA and in Anova—both models assume them to be independently and normally distributed, with the same variance for all possible combinations of the fixed x_is. Thus the significance test procedures must be equivalent, too. We shall, however, show the equivalence at the detailed, algebraic level for the equal-n case.

Referring to the definition of $(1,0,-1)$ group-membership variables for one-way Anova in Section 2.7, we see that the jth main diagonal entry (m.d.e.) of $\mathbf{X}'\mathbf{X} = n_j + n_k$, while the (i, j)th off-diagonal entry (o.d.e.) $= n_k$. Thus the jth m.d.e. of $\mathbf{x}'\mathbf{x} = (n_j + j_k) - (n_j - n_k)/N$, and the (i, j)th o.d.e. of $\mathbf{x}'\mathbf{x} = n_k - (n_i - n_k)(n_j - n_k)/N$. In the equal-$n$ case, these expressions simplify greatly to yield

$$\mathbf{x}'\mathbf{x} = \begin{bmatrix} 2n & n & \cdots & n \\ n & 2n & \cdots & n \\ \cdot & \cdot & \cdots & \cdot \\ \cdot & \cdot & \cdots & \cdot \\ n & n & \cdots & 2n \end{bmatrix}$$

The inverse of this highly patterned matrix is easily found (from the symmetry that will require that the inverse also have identical m.d.e.s and identical o.d.e.s) to have an inverse each of whose m.d.e.s $= (k - 1)/(nk)$ and each of whose o.d.e.s $= -1/(nk)$. Further, the jth term of $\mathbf{X}'\mathbf{Y}$ equals, in both the general and the equal-n cases, $T_j - T_k$, so $\mathbf{x}'\mathbf{y} = (n_j\overline{X}_j - n_k\overline{X}_k) - \overline{X}(n_j - n_k)$ in the general case and $T_j - T_k$ in the equal-n case. Thus $b_j = (1/n_k)[(k - 1)(T_j - T_k) - \Sigma(T_i - T_k)]$.

But

$$(k - 1)(T_j - T_k) - \sum_{i \neq j}(T_i - T_k) = kT_j - (k - 1)T_k - \sum_{i=1}^{k} T_k + (k - 1)T_k$$
$$= n_k(\overline{X}_j - \overline{X}),$$

so that

$$(\mathbf{y}'\mathbf{x})(\mathbf{x}'\mathbf{x})^{-1}(\mathbf{xy}) = \sum(T_j - T_k)(\overline{X}_j - \overline{X}) = n\sum(\overline{X}_j - \overline{X})^2,$$

which is exactly the formula for SS_{among} in Anova. But s_y^2 in MRA is clearly the same as SS_{total} in Anova, so SS_w in Anova $= SS_{total} - SS_{among} = \Sigma y^2 - (\mathbf{y}'\mathbf{x})(\mathbf{x}'\mathbf{x})^{-1}(\mathbf{x}'\mathbf{y})$ = the sum of squares for departure from regression in MRA. The Anova and MRA summary tables are thus numerically and algebraically equivalent, as must be the significance test based on this summary table.

DERIVATION 2.8: UNWEIGHTED MEANS AND LEAST SQUARES ANOVA ARE IDENTICAL IN THE 2^n DESIGN

We shall lean heavily on Searle's (1971) proof that testing H_0: $\beta_j = 0$ in an MRA model with contrast-coded group-membership variables ($X_i = c_j$ for all subjects in group j, with the c_j defined such that $\Sigma c_j = 0$) is equivalent to testing the null hypothesis that $\Sigma c_j \mu_j = 0$ and thus to comparing $F_{\text{contr}} = (\Sigma c_j \bar{X}_j)^2/[\Sigma(c_j^2/n_j)\, MS_{\text{error}}]$ to $F_\alpha(1, N - k)$.

Thus all we need to do to establish the equivalence of UWM and least-squares approaches is to show that the UWM test of factor i^* is equivalent to the least-squares test of the contrast corresponding to factor i^*.

Each test of an effect corresponds to a single-*df* contrast among the 2^n groups. Testing H_0: $\Sigma c_j \mu_j = 0$ under the least-squares approach involves the use of

$$F_{\text{contr}} = \left(\sum c_j \bar{X}_j\right)^2 \Big/ \left[MS_w\left(\sum c_j^2/n_j\right)\right] = \left(\sum c_j \bar{X}_j\right)^2 \Big/ \left[MS_w\left(\sum 1/n_j\right)\right],$$

since all c_j are $\pm$ 1. Whereas under the unweighted mean approach, we compute

$$F_{\text{contr}} = \tilde{n}\left(\sum c_j \bar{X}_j\right)^2 \Big/ \left[MS_w\left(\sum c_j^2\right)\right] = \tilde{n}\left(\sum c_j \bar{X}_j\right)^2 /[k\, MS_w],$$

where $k = 2^n$ = total number of groups. But

$$\tilde{n} = k \Big/ \sum(1/n_j) F_{\text{contr}}^{(\text{UWM})} = \left(\sum c_j \bar{X}_j\right)^2 \Big/ \left[MS_w\left(\sum 1/n_j\right)\right] = F_{\text{contr}}^{(\text{LSQ})}.$$

Note that the above equivalence depends only on two assumptions: (1) All c_j for the contrast corresponding to each effect tested are ± 1. It may thus appear that we could generalize the equivalence to other sets of orthogonal contrasts such that all $c_j = \pm 1$. However, I strongly suspect that any set of $2^n - 1$ contrasts meeting this condition must in fact involve a 2^n factorial breakdown of SS_{among}. (2) The hypotheses to be tested are contrasts among the population means. In certain predictive situations, we might instead be interested in testing hypotheses of the form H_0: $\Sigma\, c_j n_j \mu_j = 0$; for example, the mean of all subjects in the population at level 1 of A = the corresponding mean for A_2, irrespective

DERIVATION 3.1: T^2 AND ASSOCIATED DISCRIMINANT FUNCTION

SINGLE-SAMPLE T^2

The univariate t-ratio computed on the combined variable W is given by

$$t(\mathbf{a}) = (\mathbf{a}'\overline{\mathbf{X}} - \mathbf{a}'\boldsymbol{\mu}_0)/\sqrt{\mathbf{a}'\mathbf{S}\mathbf{a}/N} = \mathbf{a}'(\overline{\mathbf{X}} - \mathbf{a}'\boldsymbol{\mu}_0)/\sqrt{\mathbf{a}'\mathbf{S}\mathbf{a}/N}. \quad (3.2)$$

Since absolute values are a tremendous inconvenience mathematically, we shall seek instead to maximize the square of $t(\mathbf{a})$, which is of course equal to

$$t^2(\mathbf{a}) = [(\mathbf{a}'\overline{\mathbf{X}} - \mathbf{a}'\boldsymbol{\mu}_0)]^2/(\mathbf{a}'\mathbf{S}\mathbf{a}/N) = N\mathbf{a}'(\overline{\mathbf{X}} - \boldsymbol{\mu}_0)(\overline{\mathbf{X}} - \boldsymbol{\mu}_0)'\mathbf{a}/(\mathbf{a}'\mathbf{S}\mathbf{a}/N). \quad (3.3)$$

Our side condition is expressed as $\mathbf{a}'\mathbf{S}\mathbf{a} - 1 = 0$, whence we need to maximize

$$h(\mathbf{a}) = N\mathbf{a}'(\overline{\mathbf{X}} - \boldsymbol{\mu}_0)(\overline{\mathbf{X}} - \boldsymbol{\mu}_0)'\mathbf{a} - \lambda(\mathbf{a}'\mathbf{S}\mathbf{a} - 1).$$

Differentiating $h(\mathbf{a})$ with respect to **a**—which is equivalent to differentiating it with respect to each a_j separately and then putting the p different derivatives into a single column vector—we obtain

$$h(\mathbf{a})/d\mathbf{a} = 2N(\overline{\mathbf{X}} - \boldsymbol{\mu}_0)(\overline{\mathbf{X}} - \boldsymbol{\mu}_0)'\mathbf{a} - 2\lambda\mathbf{S}\mathbf{a},$$

which equals the zero vector if and only if

$$[N(\overline{\mathbf{X}} - \boldsymbol{\mu}_0)(\overline{\mathbf{X}} - \boldsymbol{\mu}_0)' - \lambda(\mathbf{S}]\mathbf{a} = \mathbf{0}),$$

whence

$$[N\mathbf{S}^{-1}(\overline{\mathbf{X}} - \boldsymbol{\mu}_0)(\overline{\mathbf{X}} - \boldsymbol{\mu}_0)' - \lambda(\mathbf{I}]\mathbf{a} = \mathbf{0}).$$

Thus our problem is reduced to that of finding the characteristic roots and vectors (see Digression 2) of the matrix $N\mathbf{S}^{-1}(\overline{\mathbf{X}} - \boldsymbol{\mu}_0)(\overline{\mathbf{X}} - \boldsymbol{\mu}_0)'$. Note that this matrix will generally *not* be symmetric.

Section D2.12 (in Digression 2) discusses procedures for solving eigenvalue problems. However, a pair of quite simple expressions for T^2 and its associated **a** vector are available that bypass the need for the full panoply of eigenvalue-eigenvector techniques.

First, note that premultiplication of Equation (3.4a) by $\mathbf{a}'$ yields $N\,\mathbf{a}'\Delta\Delta'\mathbf{a} - \lambda\,\mathbf{a}'\mathbf{S}\mathbf{a} = 0$, where $\Delta = (\overline{\mathbf{X}} - \boldsymbol{\mu}_0)$, whence $\lambda = N\,\mathbf{a}'\Delta\Delta'\mathbf{a}\,/\mathbf{a}'\mathbf{S}\mathbf{a} = t^2(\mathbf{a})$. That is, λ is equal to the squared t ratio computed on **a**. But this is exactly what an **a** vector that satisfies (3.4a) maximizes, so clearly $\lambda = T^2$. This leaves us, then, with the matrix equation $N\Delta\Delta'\mathbf{a} = T^2\mathbf{S}\mathbf{a}$. Given the dimensionality of the matrices and vectors involved, it seems clear that we must have either

$$N\Delta\Delta' = T^2\mathbf{S}$$

or

$$N\Delta'\mathbf{a} = T^2 \quad \text{and } \mathbf{a} = \mathbf{S}^{-1}\Delta.$$

Taking the first proposed solution would give us $T^2 = N\mathbf{S}^{-1}\Delta\Delta'$, which would equate a scalar to a $p \times p$ matrix and thus cannot be correct. Taking the second approach gives us (by premultiplying the second equation by $\mathbf{S}^{-1}$ and then substituting the resulting expression for $\mathbf{a}$ into the first equation) $t^2(\mathbf{a})$ maximized by $\mathbf{a} = \mathbf{S}^{-1}(\overline{\mathbf{X}} - \boldsymbol{\mu}_0)$, which yields

$$T^2 = [N(\overline{\mathbf{X}} - \boldsymbol{\mu}_0)'\mathbf{S}^{-1}(\overline{\mathbf{X}} - \boldsymbol{\mu}_0). \tag{3.5}$$

As a final check of equation (3.5), substituting these two expressions back into Equation (3.4) yields

$$N\Delta\Delta'\mathbf{a} - T^2\mathbf{S}\mathbf{a} = N\Delta\Delta'\,\mathbf{S}^{-1}\Delta - N\Delta'\,\mathbf{S}^{-1}\Delta\Delta = \Delta\, T^2 - T^2\Delta = \mathbf{0},$$

thus establishing that Equation (3.5) indeed satisfies (3.4a).

Equation (3.5) should be relatively easy to remember, since it is such a direct generalization of the formula for the univariate single-sample t. Specifically, the univariate t formula can be written as

$$t = (\overline{X} - \mu_0)/s_{\overline{X}} = (\overline{X} - \mu_0)/(s/\sqrt{N});$$

whence

$$\begin{aligned} t^2 &= (\overline{X} - \mu_0)^2/(s^2/N) = N(\overline{X} - \mu_0)(\overline{X} - \mu_0)/s^2 \\ &= N(\overline{X} - \mu_0)(s^2)^{-1}(\overline{X} - \mu_0); \end{aligned}$$

so that Equation (3.5) is obtained from the formula for t^2 by replacing s^2 with $\mathbf{S}$ and $(\overline{X} - \mu_0)$ with $(\overline{\mathbf{X}} - \boldsymbol{\mu}_0)$, keeping in mind the matching up of dimensions of various matrices needed to yield a single number as the end result of applying Equation (3.5).

Yet another procedure for computing T^2 is available. Using the results in Digression 2 with respect to determinants of partitioned matrices, it can be shown that

$$\begin{aligned} T^2 &= (N-1)\left(\frac{|\mathbf{A} + N(\overline{\mathbf{X}} - \boldsymbol{\mu}_0)(\overline{\mathbf{X}} - \boldsymbol{\mu}_0)'|}{|\mathbf{A}|} - 1\right) \\ &= \frac{|\mathbf{S} + N(\overline{\mathbf{X}} - \boldsymbol{\mu}_0)(\overline{\mathbf{X}} - \boldsymbol{\mu}_0)'|}{|\mathbf{S}|} - 1, \end{aligned} \tag{3.6}$$

where $\mathbf{A} = \mathbf{x}'\mathbf{x} = (N-1)\mathbf{S}$ is the cross-product matrix of deviation scores on the p outcome measures. Either of these two determinantal formulas avoids the need to invert $\mathbf{S}$.

Two-Sample T^2

The development of the appropriate multivariate test follows that of the single-sample T^2 point for point. The matrix analog of $\overline{X}_1 - \overline{X}_2$ is $\overline{\mathbf{X}}_1 - \overline{\mathbf{X}}_2$, while the matrix analog of Σx_i^2 is $\mathbf{A}_i$, so that s_c^2 becomes

$$\mathbf{S}_c = (\mathbf{A}_1 + \mathbf{A}_2)/(N_1 + N_2 - 2).$$

Thus the two-sample T^2 can be computed from any of the following formulas:

$$T^2 = [N_1N_2/(N_1 + N_2)](\overline{\mathbf{X}}_1 - \overline{\mathbf{X}}_2)'\mathbf{S}_c^{-1}(\overline{\mathbf{X}}_1 - \overline{\mathbf{X}}_2) \tag{3.9}$$

$$= \text{single nonzero root of } [N_1N_2/(N_1 + N_2)](\overline{\mathbf{X}}_1 - \overline{\mathbf{X}}_2)(\overline{\mathbf{X}}_1 - \overline{\mathbf{X}}_2)'\mathbf{S}_c^{-1} \tag{3.10}$$

$$= \frac{|\mathbf{S}_c + N_1N_2(\overline{\mathbf{X}}_1 - \overline{\mathbf{X}}_2)(\overline{\mathbf{X}}_1 - \overline{\mathbf{X}}_2)'/(N_1 + N_2)|}{|\mathbf{S}_c|} - 1. \tag{3.11}$$

DERIVATION 3.2: RELATIONSHIP BETWEEN T^2 AND MRA

We shall demonstrate the equivalence between Student's t and r_{xy} and then generalize this relationship to linear combinations of the predictors in MRA and of the dependent variables in T^2.

Two-Sample t versus Pearson r with Group-Membership Variable

Let $X = 1$ if the subject is in group 1; 0 if he or she is in group 2. Thus $\Sigma x^2 = \Sigma X = n_1$; $\Sigma XY = \Sigma Y_1$, the sum of the scores on the dependent variable of all subjects in group 1. Thus

$$r_{xy}^2 = \frac{\left[\sum Y_1 - n_1\left(\sum Y_1 + \sum Y_2\right)\right]^2}{(n_1 - .n_1^2/N)\sum y^2}$$

$$= \frac{\left[\left(n_2\sum Y_1 - n_1\sum Y_2\right)\Big/N\right]^2}{(n_1 n_2/N)\sum y^2}$$

$$= \frac{(n_1 n_2/N)(\overline{Y}_1 - \overline{Y}_2)^2}{(n_1 n_2/N)\sum y^2}$$

$$= [n_1 n_2/(n_1 + n_2)](\overline{Y}_1 - \overline{Y}_2)^2 \Big/ \sum y^2.$$

But

$$\overline{Y}_1 - \overline{\overline{Y}} = \overline{Y}_1 - (n_1\overline{Y}_1 + n_2\overline{Y}_2)/N = n_2(\overline{Y}_1 - \overline{Y}_2)/N$$

and

$$\overline{Y}_2 - \overline{\overline{Y}} = n_1(\overline{Y}_2 - \overline{Y}_1)/N.$$

Thus

$$\begin{aligned}
\sum y^2 &= \sum[(Y_1 - \overline{Y}_1) + (\overline{Y}_1 - \overline{\overline{Y}})]^2 + \sum[(Y_2 - \overline{Y}_2) + (\overline{Y}_2 - \overline{\overline{Y}})]^2 \\
&= \sum y_1^2 + n_1(\overline{Y}_1 - \overline{\overline{Y}})^2 + \sum y_2^2 + n_2(\overline{Y}_2 - \overline{\overline{Y}})^2 \\
&= \sum y_1^2 + \sum y_2^2 + (\overline{Y}_1 - \overline{Y}_2)^2[(n_1 n_2^2 + n_2 n_1^2)/N^2] \\
&= \sum y_1^2 + \sum y_2^2 + (n_1 n_2/N)(\overline{Y}_1 - \overline{Y}_2)^2.
\end{aligned}$$

Thus

$$r_{xy}^2 = \frac{(n_1n_2/N)(\overline{Y}_1 - \overline{Y}_2)^2}{\sum y_1^2 + \sum y_2^2 + (n_1n_2/N)(\overline{Y}_1 - \overline{Y}_2)^2},$$

and our test for the statistical significance of this r_{xy}^2 is

$$\begin{aligned} t^2 &= (N-2)r^2/(1-r^2) \\ &= (N-2)(n_1n_2/N)(\overline{Y}_1 - \overline{Y}_2)^2 \Big/ \left(\sum y_1^2 + \sum y_2^2\right) \\ &= \frac{(\overline{Y}_1 - \overline{Y}_2)^2}{\left[\left(\sum y_1^2 + \sum y_2^2\right) \Big/ (N-2)\right](1/n_1 + 1/n_2)}; \end{aligned}$$

but this is exactly equal to the formula for an independent-groups t test of the difference between $\overline{Y}_1$ and $\overline{Y}_2$, so that the two tests are equivalent.

It also follows that $r^2 = t^2/[t^2 + (N - 2)]$, so that we can readily obtain a measure of percent of variance accounted for from any two-sample t test.

Example

Group 1	Group 2
2	6
3	7
4	8
$\overline{Y}_1 = 3$	9
	10
	$\overline{Y}_2 = 8$

$$t^2 = \frac{(3-8)^2}{[(2+10)/6](1/3+1/5)} = 25/[2(8/15)] = 23.4375$$

X	Y
0	2
0	3
0	4
1	6
1	7
1	8
1	9
1	10

$\Sigma X = \Sigma X^2 = 5$

$\Sigma y^2 = 359 - 49^2/8 = 58.875$

$\Sigma x^2 = 5 - 25/8 = 1.875$

$\Sigma xy = 40 - 5(49)/8 = 9.375$

$$r_{xy}^2 = 9.375^2/1.875(58.875) = .79618$$

$$t^2 = 6(.79618/.20382) = 23.4377$$

Single-Sample t Test versus "Raw-Score" r_{xy}

Let $X = 1$ for all subjects, whence $\Sigma X = \Sigma X^2 = N$ and $\Sigma XY = \Sigma y$. Then the zero-intercept r^2 (using raw scores, instead of deviation scores) $= r_0^2 = (\Sigma XY)^2/[(\Sigma X^2)(\Sigma Y^2)] = (\Sigma Y)^2/(N\Sigma Y^2)$, and t^2 for the hypothesis that $\rho_0^2 = 0 = (N-1)r^2/1 - r^2 = (N-1)(\Sigma Y^2)/[N\Sigma Y^2 - (\Sigma Y)^2] =$

$$\frac{\left(\sum Y\right)^2 \Big/ N}{\left[\sum Y^2 - \left(\sum Y\right)^2 \Big/ N\right] \Big/ (N-1)} = N\overline{Y}^2/s_y^2$$

but this is precisely equal to the single-sample test of H_0 that $\mu_Y = 0$. [Note that we use $N - 1$, rather than $N - 2$, as the *df* for tests of zero-intercept correlations, since only *one* parameter has to be estimated from the data.]

Example H_0 that the population mean for Group 2 above is zero, is tested via

$$t^2 = N\bar{Y}^2/s_y^2 = 5(8)^2/(10/4) = 320/2.5 = 128.$$

On the other hand, $r_0^2 = 40^2/[5(330)] = .969697$, and t^2 for H_0 that the corresponding population "correlation" is zero is given by $t^2 = 4(.969697/.030303) = 128.000$.

T^2 versus MRA

MRA consists of finding that linear combination of the predictors which has the maximum squared univariate correlation with the outcome measure. But if Y is one of the special cases mentioned above—a 0–1 group membership variable or a constant for all subjects—then each of these squared correlations for a particular linear combination is related monotonically to the squared univariate two- or single-sample t^2. Thus maximizing r_{xy}^2 in these cases is equivalent to maximizing the corresponding t^2, which is the task of T^2 analysis, so that the combining weights employed in T^2 must be functionally equivalent to (i.e., a constant times) the MRA combining weights. All that remains, then, is to show that the F tests for the statistical significance of T^2 versus its MRA counterpart, R^2, are numerically identical and have the same *df*.

A. Single-Sample Case

For MRA, $F = (R^2/m)/[(1 - R^2)/(N - m)]$ with m and $N - m$ *df*. For T^2, $F = [(N - p)/p(N - 1)]T^2$ with p and $N - p$ *df*. But, since R^2 and T^2 are based on the same "emergent" variable, $(N - 1)R^2/(1 - R^2) = T^2$, so that $F_{T^2} = [(N - p)/p]R^2/(1 - R^2) = (R^2/p)/[(1 - R^2)/(N - p)] = F_{R^2}$.

B. Two-Sample Case

By the same reasoning, $T^2 = (N - 2)R^2/(1 - R^2)$, so that $F_{T^2} = [(N - p - 1)/p(N - 2)]T^2 = (R^2/p)/[(1 - R^2)/(N - p - 1)] = F_{R^2}$ with p substituted for m.

DERIVATION 4.1: MAXIMIZING $F(\mathbf{a})$ IN MANOVA

The task is to choose an **a** such that $F(\mathbf{a})$ is maximized. However maximizing $F(\mathbf{a})$ is equivalent to maximizing $\mathbf{a'Ha}/\mathbf{a'Ea}$, which is just $F(\mathbf{a})$ with the bothersome constant "stripped off." Further, we know that the univariate F ratio is unaffected by linear transformations of the original data, so that we are free to put a side condition on **a** so that it may be uniquely defined. The most convenient restriction is $\mathbf{a'Ea} = 1$, whence, applying the method of Lagrangian multipliers (Digression 1) and the matrix differentiation procedures discussed in Digression 2, we obtain

$$L = \mathbf{a'Ha} - \lambda\mathbf{a'Ea};$$

whence

$$\begin{aligned} dL/d\mathbf{a} &= 2\mathbf{Ha} - 2\lambda\mathbf{Ea} \\ &= \mathbf{0} \text{ iff } [\mathbf{H} - \lambda\mathbf{E}]\mathbf{a} = \mathbf{0} \qquad (4.9) \\ &\quad\;\; \text{iff } [\mathbf{E}^{-1}\mathbf{H} - \lambda\mathbf{I}]\mathbf{a} = \mathbf{0}. \qquad (4.10) \end{aligned}$$

However, Equation (4.9) is precisely the kind of matrix equation characteristic roots and vectors (cf. Digression 2) were designed to solve. We need only solve the determinantal equation, $|\mathbf{E}^{-1}\mathbf{H} - \lambda\mathbf{I}| = 0$, for λ, use this value to solve Equation (4.10) for **a**, and then plug into Equation (4.8) to obtain the desired maximum possible F. However, $\mathbf{E}^{-1}\mathbf{H}$ will in general have a rank greater than or equal to 2 (usually it will equal either $k - 1$ or p, whichever is smaller), so that $\mathbf{E}^{-1}\mathbf{H}$ will have more than one characteristic root and associated vector. Which shall we take as our desired solution, and what meaning do the other roots and vectors have?

Returning to Equation (4.9) and premultiplying both sides of that equation by $\mathbf{a}'$, we find that each λ satisfying Equations (4.9) and (4.10) also satisfies the relationship

$$\mathbf{a'Ha} - \lambda\mathbf{a'Ea} = 0,$$

whence

$$\lambda = \mathbf{a'Ha}/\mathbf{a'Ea} = (k - 1)F(\mathbf{a})/(N - k); \qquad (4.11)$$

that is, each characteristic root is equal (except for a constant multiplier) to the univariate F ratio resulting from using the coefficients of its associated characteristic vector to obtain a single combined score for each S. Since this is what we are trying to maximize, the *largest* characteristic root is clearly the solution to our maximization problem, with the characteristic vector associated with this largest root giving the linear combining rule used to obtain this maximum possible $F(\mathbf{a})$.

The *i*th largest characteristic root is the maximum possible univariate *F* ratio obtainable from any linear combination of the *p* original variables that is uncorrelated (in all three senses—see the next paragraph) with the first $i - 1$ "discriminant functions," the characteristic vector associated with the *i*th largest root giving the coefficients used to obtain this maximum possible *F*.

What was meant by the phrase, "in all three senses," in the last paragraph? Simply that any two discriminant functions, $\mathbf{Xa}_i$ and $\mathbf{Xa}_j$, are uncorrelated with each other whether we base this correlation on the total covariance matrix $[(\mathbf{H} + \mathbf{E})/(N - 1)]$ or on the variances and covariances of the means $[\mathbf{H}/(k - 1)]$ or on the within-cells variance–covariance matrix $[\mathbf{E}/(N - k)]$.

To see this, consider the fundamental side condition that the correlation between $\mathbf{Xa}_i$ and $\mathbf{Xa}_j$ be zero when computed in terms of the total covariance matrix. This will be true if the numerator of that correlation, $\mathbf{a}_i' (\mathbf{H} + \mathbf{E}) \mathbf{a}_j$, $= \mathbf{0}$. But if Equation (4.9) is true, then so is

$$[(1 + \lambda)\mathbf{H} - \lambda(\mathbf{H} + \mathbf{E})]\mathbf{a}_j = \mathbf{0},$$

and so is

$$[\mathbf{H} - \theta(\mathbf{H} + \mathbf{E})]\mathbf{a}_j = \mathbf{0},$$

where $\theta = \lambda/(1 + \lambda)$. (Do you recognize the transformation we perform before looking up gcr critical values for stripped *F* ratios?) But premultiplication of this last equation by $\mathbf{a}_i'$ leaves us with $\mathbf{a}_i'\mathbf{H}\mathbf{a}_j = \mathbf{a}_i'(\mathbf{H} + \mathbf{E})\mathbf{a}_j = \mathbf{0}$. The same premultiplication of Equation (4.9) tells us that

$$\mathbf{a}_i'\mathbf{H}\mathbf{a}_j = \lambda\mathbf{a}_i'\mathbf{E}\mathbf{a}_j,$$

so $\mathbf{a}_i'\mathbf{E}\mathbf{a}_j$ must also be zero.

DERIVATION 5.1: CANONICAL CORRELATION AND CANONICAL VARIATES

The correlation between u and v is given by

$$r(\mathbf{a}, \mathbf{b}) = (\mathbf{a}'\mathbf{S}_{xy}\mathbf{b})/\sqrt{(\mathbf{a}'\mathbf{S}_x\mathbf{a})(\mathbf{b}'\mathbf{S}_y\mathbf{b})}. \tag{5.1}$$

To eliminate the mathematically inconvenient square root we seek a method of maximizing r^2, knowing that this will maximize the absolute value of r as well.

We establish the side conditions that $\mathbf{a}'\mathbf{S}_x\mathbf{a}$ (the variance of our linear combination of predictor variables) = $\mathbf{b}'\mathbf{S}_y\mathbf{b}$ (the variance of the linear combination of outcome variables) = 1. Applying the methods of Lagrangian multipliers and matrix differentiation (cf. Digressions 1 and 2), we find that

$$L = (\mathbf{a}'\mathbf{S}_{xy}\mathbf{b})^2 - \lambda(\mathbf{a}'\mathbf{S}_x\mathbf{a} - 1) - \theta(\mathbf{b}'\mathbf{S}_y\mathbf{b} - 1);$$

whence

$$dL/d\mathbf{a} = 2(\mathbf{a}'\mathbf{S}_{xy}\mathbf{b})\mathbf{S}_{xy}\mathbf{b} - 2\lambda\mathbf{S}_x\mathbf{a}.$$

and

$$dL/d\mathbf{b} = 2(\mathbf{a}'\mathbf{S}_{xy}\mathbf{b})\mathbf{S}_{xy}'\mathbf{a} - 2\theta\mathbf{S}_y\mathbf{b}. \tag{5.2}$$

Setting the Equations (5.2) equal to zero leads to a homogenous system of two simultaneous matrix equations, namely,

$$\begin{aligned} -\lambda\mathbf{S}_x\mathbf{a} + (\mathbf{a}'\mathbf{S}_{xy}\mathbf{b})\mathbf{S}_{xy}\mathbf{b} &= \mathbf{0}, \\ (\mathbf{a}'\mathbf{S}_{xy}\mathbf{b})\mathbf{S}_{xy}'\mathbf{a} - \theta\mathbf{S}_y\mathbf{b} &= \mathbf{0}. \end{aligned} \tag{5.3}$$

If we premultiply the first equation by $\mathbf{a}'$ and the second equation by $\mathbf{b}'$ we obtain (keeping in mind that $\mathbf{a}'\mathbf{S}_{xy}\mathbf{b}$ is a scalar)

$$-\lambda\mathbf{a}'\mathbf{S}_x\mathbf{a} + (\mathbf{a}'\mathbf{S}_{xy}\mathbf{b})^2 = 0,$$

and

$$(\mathbf{a}'\mathbf{S}_{xy}\mathbf{b})^2 - \theta\mathbf{b}'\mathbf{S}_y\mathbf{b} = 0;$$

whence, keeping in mind our side conditions,

$$\lambda = \theta = r^2(\mathbf{a}, \mathbf{b}).$$

In other words, each of our Lagrangian multipliers is equal to the desired maximum value of the squared correlation between u and v. For purposes of simplifying

Equations (5.3), we note that $(\mathbf{a}'\mathbf{S}_{xy}\mathbf{b}) = \sqrt{\lambda}$ since by our side condition $\mathbf{b}'\mathbf{S}_y\mathbf{b} = \mathbf{a}'\mathbf{S}_x\mathbf{a} = 1$; whence Equations (5.3) become

$$-\sqrt{\lambda}\mathbf{S}_x\mathbf{a} + \mathbf{S}_{xy}\mathbf{b} = 0,$$
$$\mathbf{S}'_{xy}\mathbf{a} - \sqrt{\lambda}\mathbf{S}_y\mathbf{b} = 0. \tag{5.4}$$

However, a nontrivial solution of Equations (5.4) exists only if the determinant of the coefficient matrix vanishes. Using the results of Digression 2 on partitioned matrices, we can express this condition in one of two alternative ways:

$$|-\lambda\mathbf{S}_x + \mathbf{S}_{xy}\mathbf{S}_y^{-1}\mathbf{S}'_{xy}| = 0; \quad \text{or} \quad |-\lambda\mathbf{S}_y + \mathbf{S}'_{xy}\mathbf{S}_x^{-1}\mathbf{S}_{xy}| = 0. \tag{5.5}$$

The roots of Equations (5.5) are the characteristic roots of the matrices $\mathbf{S}_x^{-1}\mathbf{S}_{xy}\mathbf{S}_y^{-1}\mathbf{S}'_{xy}$ and $\mathbf{S}_y^{-1}\mathbf{S}'_{xy}\mathbf{S}_x^{-1}\mathbf{S}_{xy}$. The characteristic vector associated with the gcr of the first matrix gives the canonical coefficients for the left-hand set of variables, while the characteristic vector associated with the gcr of the second matrix gives the canonical coefficients for the right-hand set of variables, and the gcr of either matrix gives the squared canonical correlation between the two sets of variables.

DERIVATION 5.2: CANONICAL CORRELATION AS "MUTUAL REGRESSION ANALYSIS"

Let us get a bit more explicit about the relationship between the matrix equation defining the canonical variate for the *Y*s (namely, the second of Equations 5.2, since this equation leaves us with the correct number of entries in our characteristic vector) and the problem of predicting scores on the *Y* variables' canonical variate from scores on the *p* *X*s.

Consider the task of predicting a linear combination of the *Y*s, $\mathbf{b}'\mathbf{Y}$, on the basis of knowledge of the *X* scores. From Chapter 2, we know that

$$s_W^2 R_{W\cdot X}^2 = \mathbf{s}_{xw}'\mathbf{S}_x^{-1}\mathbf{s}_{xw},$$

where $W = \mathbf{b}'\mathbf{Y}$. From by now familiar formulas for the variances and covariances of linear combinations of variables, we know that

$$s_W^2 = \mathbf{b}'\mathbf{S}_y\mathbf{b} \quad \text{and} \quad \mathbf{s}_{xw} = \mathbf{S}_{xy}\mathbf{b},$$

whence

$$(\mathbf{b}'\mathbf{S}_y\mathbf{b})R_{\mathbf{b}'\mathbf{Y}\cdot X}^2 = \mathbf{b}'\mathbf{S}_{xy}'\mathbf{S}_x^{-1}\mathbf{S}_{xy}\mathbf{b},$$

whence

$$\mathbf{b}'[\mathbf{S}_y R_{\mathbf{b}'\mathbf{Y}\cdot X}^2 - \mathbf{S}_{xy}'\mathbf{S}_x^{-1}\mathbf{S}_{xy}]\mathbf{b} = 0,$$

which would certainly be satisfied if

$$[\mathbf{S}_y R_{\mathbf{b}'\mathbf{Y}\cdot X}^2 - \mathbf{S}_{xy}'\mathbf{S}_x^{-1}\mathbf{S}_{xy}]\mathbf{b} = \mathbf{0},$$

that is, if

$$[\mathbf{S}_y^{-1}\mathbf{S}_{xy}'\mathbf{S}_x^{-1}\mathbf{S}_{xy} - R_{\mathbf{b}'\mathbf{Y}\cdot X}^2\mathbf{I}]\mathbf{b} = \mathbf{0},$$

which is precisely the relationship that holds when we have found a solution to the matrix equation that defines R_c^2 and the corresponding canonical variate. A perfectly symmetric argument establishes that the matrix equation defining the canonical variates for the set of *X* variables implies an equality between canonical *R* and the multiple *R* resulting from predicting a particular linear combination of the *X* variables (one of the left-hand canonical variates) from knowledge of scores on the *Y* variables.

DERIVATION 5.3: RELATIONSHIP BETWEEN CANONICAL ANALYSIS AND MANOVA

One-Way Manova

Derivation 2.7 showed the equivalence of one-way Anova to MRA in which the predictors are (1,0, −1) group-membership variables. Thus, having constructed $k - 1$ group-membership variables (or contrast scores) to represent the differences among the k levels of our independent variable, the SS_b for any one dependent variable, $SS_{b,j}$, is given by

$$(\mathbf{y}_j'\mathbf{x})(\mathbf{x}'\mathbf{x})^{-1}(\mathbf{x}'\mathbf{y}_j) = (N - 1)\, \mathbf{s}_{xy_j}'\mathbf{S}_x^{-1}\mathbf{s}_{xy_j},$$

and $SS_{w,j}$ is

$$(N - 1)[\mathbf{s}_{y_j}^2 - \mathbf{s}_{xy_j}'\mathbf{S}_x^{-1}\mathbf{s}_{xy_j}].$$

These are of course the main diagonal entries of the **H** and **E** matrices, respectively, of Manova. From this insight it is trivial to show that

$$(\mathbf{y}'\mathbf{x})(\mathbf{x}'\mathbf{x})^{-1}(\mathbf{x}'\mathbf{y}) = \mathbf{H} = (N - 1)\mathbf{S}_{xy}'\mathbf{S}_x^{-1}\mathbf{S}_{xy},$$

and

$$\mathbf{E} = (N - 1)[\mathbf{S}_y - \mathbf{S}_{xy}'\mathbf{S}_x^{-1}\mathbf{S}_{xy}],$$

so that finding the linear combination of the Ys that maximizes $F(\mathbf{a})$ is a matter (Equation 4.9) of solving the equations

$$[\mathbf{H} - \lambda\,\mathbf{E}]\,\mathbf{a} = \mathbf{0}.$$

That is, in the present application,

$$(N - 1)[\mathbf{S}_{xy}'\mathbf{S}_x^{-1}\mathbf{S}_{xy} - \lambda(\mathbf{S}_y - \mathbf{S}_{xy}'\mathbf{S}_x^{-1}\mathbf{S}_{xy})]\,\mathbf{a} = \mathbf{0},$$

which matrix equation is true if and only if ($\Leftrightarrow$)

$$[(1 + \lambda)\mathbf{S}_{xy}'\mathbf{S}_x^{-1}\mathbf{S}_y - \lambda\mathbf{S}_y]\,\mathbf{a} = \mathbf{0}$$

$$\Leftrightarrow \left[\mathbf{S}_{xy}'\mathbf{S}_x^{-1}\mathbf{S}_y - \left(\frac{\lambda}{1 + \lambda}\right)\mathbf{S}_y\right]\mathbf{a} = \mathbf{0}$$

$$\Leftrightarrow [\mathbf{S}_{xy}'\mathbf{S}_x^{-1}\mathbf{S}_y - \theta\mathbf{S}_y]\,\mathbf{a} = \mathbf{0},$$

which is precisely the second of Equations (5.2), except for the label on our characteristic vector.

Factorial Manova

We learned in Section 2.7 that the null hypothesis that a particular effect in an unequal-n factorial design could be tested in terms of the significance of the drop in R^2 that resulted when the set of group-membership variables representing that effect was dropped from the full model. This is equivalent to comparing

$$F = \frac{(SS_b - SS_{red})/df_{effect}}{SS_{error}/df_{error}}$$

to $F_\alpha(df_{effect}, df_{error})$, where SS_b is the total between-groups sum of squares and would be, in a balanced design, the sum of the SS_{effect}s for all effects in our design. In the unequal-n design it is most readily computed as $\Sigma y^2 (1 - R^2)$, where R^2 is computed on the basis of all $k - 1$ group-membership variables in the analysis. SS_{red} is computed similarly, but on the basis of the squared multiple correlation between Y and the group-membership variables *other* than those for the effect we are testing. $SS_b - SS_{red}$ will be listed as the sum of squares for the effect in an $SPSS^\times$ ANOVA run, provided that "METHOD = SSTYPE (UNIQUE)" is specified, and may thus be labelled as SS_{effect}, even though its computation is not quite as straightforward as for the two SSs between which it is the difference. (In SAS's PROC GLM, SS_{effect} would be listed as the Type III SS for the effect.)

The equivalent Manova problem would thus involve finding

$$\max_{\mathbf{a}} F(\mathbf{a}) = \frac{df_{error}}{df_{effect}} \lambda_{max},$$

where λ_{max} is the greatest characteristic root (gcr) of $\mathbf{E}^{-1}[\mathbf{H}_b - \mathbf{H}_{red}]$. But from our arguments for the one-way case, it follows that

$$\mathbf{H}_b = (N - 1)\mathbf{S}_{xy}'\mathbf{S}_x^{-1}\mathbf{S}_{xy} = \mathbf{P}_{full},$$

where $\mathbf{X}$ is the $N \times k - 1$ matrix of scores on all $k - 1$ group-membership variables, including the df_{effect} group-membership variables representing the effect that we are testing, and where our discussion in Derivation 5.2 shows that $\mathbf{P}_{full}$ (for the covariance matrix of the *predicted* scores on the Ys) is an appropriate abbreviation for $\mathbf{S}_{xy}'\mathbf{S}_x^{-1}\mathbf{S}_{xy}$. $\mathbf{H}_{red} = (N - 1)\mathbf{P}_{red}$ is defined similarly but employing $\mathbf{X}_{red}$, the $N \times (k - 1 - df_{effect})$ group-membership variables *other than* those representing the effect we are testing.

Thus our Manova test of the effect consists of finding (and testing for statistical significance) the gcr (λ) of

$$\begin{aligned} &\mathbf{E}^{-1}[\mathbf{H}_b - \mathbf{H}_{red}] \\ &= \text{the gcr of } [\mathbf{S}_y - \mathbf{P}_{full}]^{-1}[\mathbf{P}_{full} - \mathbf{P}_{red}]. \end{aligned}$$

$\mathbf{H}_{effect} = \mathbf{H}_b - \mathbf{H}_{red}$ will be listed in $SPSS^\times$ MANOVA and in SAS's PROC

GLM as the hypothesis matrix for that effect, provided that printing of hypothesis matrices is requested and METHOD = SSTYPE(UNIQUE) IS SPECIFIED.

Note that the θ we compute in testing this effect (i.e., the squared canonical correlation between our dependent variables and the group-membership variables for this effect, corrected for all other effects) equals

$$\frac{SS_{\text{effect}}}{SS_{\text{effect}} + SS_w},$$

computed for scores on the discriminant functions. This will *not* be the same as $\Delta R_c^2 = R_{\text{c,full}}^2 - R_{\text{c,red}}^2$, since

$$\Delta R_c^2 = \frac{SS_b}{SS_{\text{total}}} - \frac{SS_{\text{red}}}{SS_{\text{total}}} = \frac{SS_b - SS_{\text{red}}}{SS_{\text{total}}} = \frac{SS_b - SS_{\text{red}}}{SS_w + SS_b} = \frac{SS_{\text{effect}}}{SS_w + SS_{\text{effect}} + SS_{\text{red}}}.$$

In the Second Edition (pp. 531-539) an extremely unbalanced 2 x 2 factorial design was analyzed via SPSS^X (sic) MANOVA to demonstrate these relationships. On the basis of this demonstration the reader was urged to

Note that **E** (the "WITHIN CELLS SUM-OF-SQUARES AND CROSS-PRODUCTS," or SSCP, matrix) for each of the Manova tests (those using our "class variables," *A* and *B*) is equal to the corresponding matrix from the Canona employing all three *X*s as predictors, but *not* to the within-cells (residual) matrix for any of the two-predictor Canonas, each of which is, in Manova terminology, $\mathbf{E} + \mathbf{H}_{\text{effect}}$ for the effect being tested. In other words, the Manova analysis uses $\mathbf{E} = (N-1)\mathbf{S}_y - \mathbf{H}_b$ as the error matrix for all tests, while each Canona uses $(N-1)\mathbf{S}_y - \Sigma\mathbf{H}_{\text{effect}}$ for only those effects whose group-membership variables are included in the analysis. The θs ("ROYS LARGEST ROOT CRITERION" or the square of "CANON. COR.") for the Manova analysis are thus in each case the proportion of "error plus effect" variance that effect accounts for, while the R_c^2 for the corresponding Canona is the proportion of total variance (error plus all effects) that particular effect accounts for.

Finally, to demonstrate construction of the Manova test from the Canona matrices, note that the test of the *AB* interaction in the Manova analysis is equivalent to testing the gcr (θ) of

$$[\mathbf{S}_y - \mathbf{P}_{x_1,x_2}]^{-1}[\mathbf{P}_{x_1,x_2} - \mathbf{P}_{x_1,x_2,x_3}].$$

$\mathbf{S}_y$ is not listed directly in the MANOVA output for the canonical analysis, but it is obtainable as $(N-1)$("HYPOTHESIS SSCP" matrix + "WITHIN CELLS SSCP" matrix) for any of the Canona analyses, while each **P** matrix is simply $N - 1$ times the "HYPOTHESIS SSCP" matrix for the appropriate Canona.

DERIVATION 6.1: PRINCIPAL COMPONENTS

$$s^2_{PC_1} = [1(N - 1)]\sum(b_{i,1}x_1 + b_{1,2}x_2 + \cdots + b_{i,m}x_m)^2$$
$$= \mathbf{b}_i'\mathbf{S}_x\mathbf{b}_i, \qquad (6.2)$$

where $\mathbf{S}_x$ is the variance–covariance matrix of the *X*s. Using the method of Lagrangian multipliers (Digression 1), we find that, for $\mathbf{b}_1$,

$$L = \mathbf{b}_1'\mathbf{S}_x\mathbf{b}_1 - \lambda(\mathbf{b}_1'\mathbf{b}_1 - 1)$$

and, referring to Section D2.10,

$$dL/d\mathbf{b}_1 = 2\mathbf{S}_x\mathbf{b}_1 - 2\lambda\mathbf{b}_1 = \mathbf{0} \quad \text{iff} \quad [\mathbf{S}_x - \lambda\mathbf{I}]\mathbf{b}_1 = \mathbf{0},$$

where $\mathbf{0}$ is a column vector each of whose *m* entries is zero.

Thus we have a set of *m* homogenous equations in *m* unknowns. In order for this set of equations to have a nontrivial solution, the determinant $|\mathbf{S}_x - \lambda\mathbf{I}|$ must equal 0. As pointed out in Section D2.12, computing the determinant with λ left in as an unknown produces an *m*th degree polynomial in λ that, when set equal to zero, constitutes the characteristic equation. Solving this equation for λ produces *m* roots, some of which may be zero if there is a linear dependence among the original variables. Any one of the nonzero roots can be "plugged into" the matrix equation, and the resulting set of equations solved for the coefficients of $\mathbf{b}$. Which of these *m* sets of coefficients is to be taken as PC_1?

Recall that each root of the characteristic equation λ_i makes the matrix equation $[\mathbf{S}_x - \lambda_i\mathbf{I}]\mathbf{b}_i = \mathbf{0}$ true, whence $\mathbf{b}_i'\mathbf{S}_x\mathbf{b}_i = \lambda_i\mathbf{b}_i'\mathbf{b}_i$. However, we require that $\mathbf{b}_i'\mathbf{b}_i = 1$, whence $\lambda_i = \mathbf{b}_i'\mathbf{S}_x\mathbf{b}_i$. In other words, each root of the characteristic equation is equal to the variance of the combined variable generated by the coefficients of the characteristic vector associated with that root. Since it is this variance that we set out to maximize, the coefficients of the first principal component will be the characteristic vector associated with the *largest* characteristic root. As it turns out, PC_2 is computed via the characteristic vector corresponding to the second largest characteristic root, and in general PC_i has the same coefficients as the characteristic vector associated with the *i*th largest characteristic root λ_i. To show this for the case of PC_2, consider the problem of maximizing $\mathbf{b}_2'\mathbf{S}_x\mathbf{b}_2$ subject to the *two* side conditions that $\mathbf{b}_2'\mathbf{b}_2 = 1$ and that $\mathbf{b}_2'\mathbf{b}_1 = 0$. This leads to

$$L = \mathbf{b}_2'\mathbf{S}_x\mathbf{b}_2 - \lambda(\mathbf{b}_2'\mathbf{b}_2 - 1) - \theta(\mathbf{b}_2'\mathbf{b}_1),$$

whence

$$dL/d\mathbf{b}_2 = 2\mathbf{S}_x\mathbf{b}_2 - 2\lambda\mathbf{b}_2 - \theta\mathbf{b}_1 = \mathbf{0} \quad \text{iff} \quad 2[\mathbf{S}_x - \lambda\mathbf{I}]\mathbf{b}_2 = \theta\mathbf{b}_1;$$

whence

$$2\mathbf{b}_1'[\mathbf{S}_x - \lambda\mathbf{I}]\mathbf{b}_2 = \theta\mathbf{b}_1'\mathbf{b}_1;$$

whence

$$\theta = 2\mathbf{b}_1'\mathbf{S}_x\mathbf{b}_2 .$$

However, from the definition of $\mathbf{b}_1$, namely, $\mathbf{S}_x\mathbf{b}_1 = \lambda_1\mathbf{b}_1$, we have

$$\mathbf{b}_2'\mathbf{S}_x\mathbf{b}_1 = \lambda_1\mathbf{b}_2'\mathbf{b}_1 = 0;$$

whence we know that $\theta = 0$, and $\mathbf{b}_2$ must satisfy the equations

$$[\mathbf{S}_x - \lambda\mathbf{I}]\mathbf{b}_2 = \mathbf{0}.$$

Thus $\mathbf{b}_2$ will be one of the characteristic vectors of $\mathbf{S}_x$, and in particular, given our goals, will be the characteristic vector associated with the second largest characteristic root.

As one step in the above derivations, we showed that $\mathbf{b}_2'\mathbf{S}_x\mathbf{b}_1$, which is of course the covariance of PC_1 and PC_2, is equal to zero. Using the same approach, it can be proved that in general the orthogonality of (lack of correlation among) the characteristic vectors of $\mathbf{S}_x$ is paralleled by the absence of any correlation among the *PC*s computed from those sets of coefficients.

DERIVATION 6.2: *PC* COEFFICIENTS DEFINE BOTH COMPONENTS IN TERMS OF *X*s AND *X*s IN TERMS OF *PC*s

The coefficients used in the linear combination of *PC*s that "generates" or "reproduces" a particular original variable bear a close relationship to (and are in fact identical to) the coefficients in that row of the *factor pattern.* To show that this is correct, consider the equation defining PC_j, namely,

$$\mathbf{X}\mathbf{b}_j = b_{j,1}\mathbf{X}_1 + b_{j,2}X_2 + \cdots + b_{j,m}X_m = PC_j. \tag{6.5}$$

Our assertion is that

$$X_i = b_{1,i}PC_1 + b_{2,i}PC_2 + \cdots + b_{m,i}PC_m. \tag{6.6}$$

If this is correct, then substitution of expressions like Equation (6.6) for each occurrence of an X_i in Equation (6.5) should "satisfy" the equation; that is, it should produce an expression that equals PC_j. Carrying out these substitutions gives

$$\begin{aligned}
&b_{j,1}\,(b_{1,1}PC_1 + b_{2,1}PC_2 + \cdots + b_{j,1}PC_j + \cdots + b_{m,1}PC_m)\\
&+ b_{j,2}\,(b_{1,2}PC_1 + b_{2,2}PC_2 + \cdots + b_{j,2}PC_j + \cdots + b_{m,2}PC_m)\\
&+ b_{j,m}\,(b_{1,m}PC_1 + b_{2,m}PC_2 + \cdots + b_{j,m}PC_m + \cdots + b_{m,m}PC_m)\\
&\qquad\qquad\qquad \|\\
&PC_1\,(b_{j,1}b_{1,1} + b_{j,2}b_{1,2} + \cdots + b_{j,m}b_{1,m})\\
&+ PC_2\,(b_{j,1}b_{2,1} + b_{j,2}b_{2,2} + \cdots + b_{j,m}b_{2,m})\\
&+ PC_j\,(b_{j,1}^2 + b_{j,2}^2 + \cdots + b_{j,m}^2)\\
&+ PC_m\,(b_{j,1}b_{m,1} + b_{j,2}b_{m,2} + \cdots + b_{j,m}b_{m,m})\\
&\qquad\qquad\qquad \|\\
&\sum_{i=1}^{m} \mathbf{b}_j'\mathbf{b}_i PC_i = PC_j.
\end{aligned}$$

However, we know that the *PC*s are orthogonal, that is, that $\mathbf{b}_j\mathbf{b}_i = 1$ if $i = j$ but zero otherwise, so that $\Sigma\ \mathbf{b}_j'\mathbf{b}_i PC_i = PC_j$, and the assertion is proved.

Rather than working with the symbols for the *PC*s and the *X*s, we could have worked with each subject's score on each of these variables and conducted a multiple regression analysis on the relationship between scores on the *PC*s (treated as predictor variables) and scores on the original variables (each X_i being treated in turn as the outcome variable). Let $\mathbf{x} = [\mathbf{x}_1, \mathbf{x}_2, \ldots, \mathbf{x}_m]$ be the $N \times m$ matrix

of deviation scores on the X variables; $\mathbf{pc} = [\mathbf{pc}_1, \mathbf{pc}_2, \ldots, \mathbf{pc}_m]$ be the $N \times m$ matrix of deviation scores on the principal components; and $\mathbf{B} = \mathbf{b}_1, \mathbf{b}_2, \ldots, \mathbf{b}_m$ be the factor pattern, that is, the $m \times m$ matrix of the characteristic vectors of $\mathbf{S}_x$. The multiple regression analysis seeks, for each X_i, a vector $\mathbf{c}_i$ such that $\mathbf{pc}\ \mathbf{c}_i = \mathbf{x}_i$. More generally, the results of all m of these regression analyses can be "stored" in the matrix $\mathbf{C} = [\mathbf{c}_1, \mathbf{c}_2, \ldots, \mathbf{c}_m]$. The analysis of the relationship between $\mathbf{c}_i$ and $\mathbf{pc}$ takes the form

$$\mathbf{c}_i = (\mathbf{pc}'\mathbf{pc})^{-1}(\mathbf{pc}'\mathbf{x}_i);$$

or, more generally,

$$\mathbf{C} = (\mathbf{pc}'\mathbf{pc})^{-1}(\mathbf{pc}'\mathbf{x}).$$

Thanks to the orthogonality of the *PC*s, $\mathbf{pc}'\mathbf{pc} = \mathbf{D}$, a diagonal matrix whose main diagonal entries are the characteristic roots of $\mathbf{S}_x$, so that $(\mathbf{pc}'\mathbf{pc})^{-1} = \mathbf{D}^{-1}$ is a diagonal matrix whose main diagonal entries are the reciprocals of the λ_i. However, by the definition of the *PC*s, $\mathbf{pc} = \mathbf{xB}$, whence $\mathbf{pc}' = \mathbf{B}'\mathbf{x}'$, whence

$$\mathbf{pc}'\mathbf{x} = \mathbf{B}'(\mathbf{x}'\mathbf{x}) = (\mathbf{x}'\mathbf{xB})',$$

so that

$$\mathbf{C} = \mathbf{D}^{-1}\mathbf{B}'(\mathbf{x}'\mathbf{x}) = \mathbf{D}^{-1}(\mathbf{x}'\mathbf{xB})'.$$

However, the characteristic vectors were obtained as solutions to the equations $[\mathbf{x}'\mathbf{x} - \lambda_i\mathbf{I}]\ \mathbf{b}_i = 0$, whence we know that $\mathbf{x}'\mathbf{x}\mathbf{b}_i = \lambda_i\mathbf{b}_i$ and therefore $\mathbf{x}'\mathbf{xB} = \mathbf{BD}$, whence

$$\mathbf{C} = \mathbf{D}^{-1}\mathbf{D}'\mathbf{B}' = (\mathbf{D}^{-1}\mathbf{D})\ \mathbf{B}' = \mathbf{B}',$$

as was initially asserted.

DERIVATION 6.3: WHAT DOES ROTATION OF LOADINGS DO TO COEFFICIENTS?

We shall prove Equation (6.12) by showing that its application

(a) yields z scores on the rotated *PC*s that are uncorrelated, and

(b) yields correlations between the original variables and the *PC*s that are identical to the correlations contained in the rotated factor structure.

The proof shall actually be given only for a single pair of *PC*s (indeed, only the first two, since the only property of the *PC*s essential to the proof is their zero intercorrelations). Since part of the proof establishes that the scores on the rotated *PC*s are uncorrelated, too, the proof can easily be extended to the scores obtained after rotating a second pair of axes—including possibly one of the new axes obtained from the just-rotated pair of *PC*s—and so on. First, consider the correlation between h_1 and h_2, the rotated axes corresponding to the first two *PC*s. Applying Equation (6.14), we have

$$z_{h_1} = \cos\theta\, z_{\text{pc}_1} + \sin\theta\, z_{\text{pc}_2};$$

and

$$z_{h_2} = -\sin\theta\, z_{\text{pc}_1} + \cos\theta\, z_{\text{pc}_2}.$$

Thus

$$\begin{aligned} r_{h_1,h_2} &= \sum z_{h_1 h_2}/(N-1) \\ &= [1/(N-1)]\sum[(\cos\theta\, z_{\text{pc}_1} + \sin\theta\, z_{\text{pc}_2})(-\sin\theta\, z_{\text{pc}_1} + \cos\theta\, z_{\text{pc}_2})] \\ &= [1/(N-1)]\sum[-\sin\theta\cos\theta\, z^2_{\text{pc}_1} + (\cos^2\theta - \sin^2\theta)\, z_{\text{pc}_1} z_{\text{pc}_2} \\ &\quad + \sin\theta\cos\theta\, z^2_{\text{pc}_2}]. \end{aligned}$$

However,

$$\sum z^2_{\text{pc}_1} = \sum z^2_{\text{pc}_2} = N-1 \quad \text{and} \quad \sum z_{\text{pc}_1} z_{\text{pc}_2} = (N-1)\, r_{\text{pc}_1,\text{pc}_2},$$

so that we have

$$r_{h_1,h_2} = (\cos^2\theta - \sin^2\theta)\, r_{\text{pc}_1,\text{pc}_2} = 0,$$

since we know that any two *PC*s are uncorrelated.

Next, consider the correlation between, say, X_1 and h_1:

$$r_{x1,h1} = \sum z_1 z_{h1}/(N-1) = [1/(N-1)]\sum z_1(\cos\theta\, z_{\text{pc}_1} + \sin\theta\, z_{\text{pc}_2})$$
$$= [1/(N-1)]\left[\cos\theta \sum z_1 z_{\text{pc}_1} + \sin\theta \sum z_1 z_{\text{pc}_2}\right]$$
$$= \cos\theta\, r_{x1.\text{pc}_1} + \sin\theta\, r_{x1.\text{pc}_2}.$$

This final expression is exactly the expression for the loading of X_1 on the first rotated *PC*, that is, the entry in the first row, first column of the rotated factor structure. The extension to other *X*s and the second rotated *PC* should be obvious. It should also be clear from the above why it is *not* true that rotation through an angle θ leads to the same transformation being applied to the *raw score* coefficient matrix. For example, $\Sigma\, \text{pc}_1^2 \neq \Sigma\, \text{pc}_2^2$, so that we would have had a nonzero term, $\sin\theta \cos\theta(\Sigma\, \text{pc}_2^2 - \Sigma\, \text{pc}_1^2)$, in the numerator of the expression for $r_{h1,h2}$.

DERIVATION 7.1: NEAR EQUIVALENCE OF PCA AND EQUAL-COMMUNALITIES PFA

To see this, consider the general problem of performing an equal-communalities PFA. We seek the eigenvalues and eigenvectors of

$$\mathbf{R}_c = \mathbf{R} - \mathbf{D}(1 - h_j^2) = \mathbf{R} - (1 - c)\mathbf{I}$$

in the equal-communalities case. Thus we seek solutions to the matrix equation

$$[\mathbf{R}_c - \lambda\mathbf{I}]\mathbf{a} = \mathbf{0};$$

that is, to

$$[\mathbf{R} - (1 - c)\mathbf{I} - \lambda\mathbf{I}]\mathbf{a} = \mathbf{0};$$

whence to

$$[\mathbf{R} - (\lambda + (1 - c))\mathbf{I}]\mathbf{a} = \mathbf{0};$$

that is, to

$$[\mathbf{R} - \lambda^*\mathbf{I}]\mathbf{a} = \mathbf{0}, \quad \text{where} \quad \lambda^* = \lambda + (1 - c).$$

However, λ_j^* is the jth characteristic root (eigenvalue) of $\mathbf{R}$. Clearly, then, the characteristic vectors of $\mathbf{R}_c$ are the same as those of $\mathbf{R}$. However, when we multiply by $\sqrt{\lambda_j}$ to obtain the loadings on factor j, we multiply by

$$\sqrt{\lambda^* - (1 - c)} = \sqrt{\lambda^*} \cdot \sqrt{1 - (1 - c)/\lambda^*}.$$

QED

References

Adams, J. S. (1965). Inequity in social exchange. In L. Berkowitz (Ed.), *Advances in experimental social psychology* (Vol. 2). New York: Academic Press.

Anderson, N. H. (1981). Scales and statistics: Parametric and nonparametric. *Psychological Bulletin, 58,* 305–316.

Anderson, T. W. (1958). *Introduction to multivariate statistical analysis.* New York: Wiley. (2nd ed. 1984.)

Anderson, T. W. (1963). Asymptotic theory for principal component analysis. *Annals of Mathematical Statistics, 34,* 122–148.

Anscombe, F. J. (1973). Graphs in statistical aaalysis. *American Statistician, 27,* 17–21.

Arbuckle, J.L. (1995). *Amos for Windows: Analysis of moment structures. Version 3.5.* Chicago: SmallWaters Corp.

Atkinson, R. C., Bower, G., & Crothers, E. J. (1965). *Introduction to mathematical learning theory.* New York: Wiley.

Bartlett, M. S. (1939). A note on tests of significance in multivariate analysis. *Proceedings, Cambridge Philosophical Society, 35,* 180–185.

Bartlett. M. S. (1954). A note on the multiplying factors for various chi-square approximations. *Journal of the Royal Statistical Society, Series B*, 116, 296–298.

Bednarz, J. C. (1981). *I. Productivity, nest sites, and habitat of red-shouldered and red-tailed hawks in Iowa. II. Status, habitat utilization, and management of red-shouldered hawks in Iowa.* MS thesis, Iowa State University, Ames.

Bednarz, J. C., & Dinsmore, J. J. (1981). Status, habitat use, and management of red-shouldered hawks in Iowa. *Journal of Wildlife Management, 45,* 236–241.

Bentler, P. M. (1995). *EQS structural equations program manual.* Encino, CA: Multivariate Software.

Bentler, P. M., & Bonnett, D. G. (1980). Significance tests and goodness of fit in the analysis of covariance structures. *Psychological Bulletin, 88,* 588–606.

Bentler, P. M., & Wu, E. J. C. (1995). *EQS for Windows user's guide.* Encino, CA: Multivariate Software.

Bird, K. D. (1975). Simultaneous contrast testing procedures for multivariate experiments. *Multivariate Behavioral Research, 10,* 343–351.

Bird, K. D., & Hadzi-Pavlovic, D. (1982). *New tools for the analysis of interactions.* Unpublished manuscript. University of New South Wales.

Bird, K. D., & Hadzi-Pavlovic, D. (1983). Simultaneous test procedures and the choice of a test statistic in MANOVA. *Psychological Bulletin, 93,* 167–178.

Bishop, Y. M. M., Fienberg, S. E., & Holland, P. (1978). *Discrete multivariate analysis : theory and practice* Cambridge, MA: MIT Press.

Bochner, S., & Harris, R. J. (1984). Sex differences in the attribution of socially desirable and undesirable traits to out-group members. *International Journal of Psychology, 19,* 207–215.

Bock, R. D. (1975). *Multivariate statistical methods in behavioral research.* New York: McGraw-Hill.

Bock, R. D. (1991). Prediction of growth. In L. M. Collins & J. L. Horn (Eds.), *Best Methods for the Analysis of Change* (pp. 126–136). Washington, D.C: American Psychological Association.

Boik, R. J. (1981). A priori tests in repeated measures designs: Effects of non-sphericity. *Psychometrika, 46,* 241–255.

Boik, R. J. (1986). Testing the rank of a matrix with applications to the analysis of interaction in ANOVA. *Journal of the American Statistical Association, 81*, 243–248.

Boik, R. J. (1993). The analysis of 2-factor interactions in fixed effects linear models. *Journal of Educational Statistics, 18*, 1–40

Boneau, C. A. (1960). The effects of violations of assumptions underlying the *t* test. *Psychological Bulletin, 57*, 49–64.

Bonferroni, C. E. (1936a). Teoria statistica delle classi e calcolodelle probabilità. *Pubblicazioni lnstitutio Superiore di Scienze Economiche e Commerciali Firenze, 8,* 1–62. (Michael Dewey, Nottingham University, lists this as a 1935 publication.)

Bonferroni, C. E. (1936b). Il calcolo delle assicurazioni *su* gruppi di teste. In *Studii in onore de1 prof. S. 0. Carboni.* Roma.

Box, G. E. P. (1949). A general distribution theory for a class of likelihood criteria. *Biometrika, 36,* 317– 346.

Box, G. E. P. (1950). Problems in the analysis of growth and wear curves. *Biometrics, 6,* 362–389.

Bradu, D., & Gabriel, K. R. (1974). Simultaneous statistical inference on interactions in 2-way Anova. *Journal of the American Statistical Association, 69,* 428–436.

Braver, S. L. (1975). On splitting the tails unequally: A new perspective on one versus two-tailed tests. *Educational & Psychological Measurement*, *35*, 283–301.

Brien, A. J. & Simon, T. L. (1987). The effects of red blood cell infusion on 10-km race time. *Journal of the American Medical Association, 257*, 2761–27655

Brown, R. (1965). *Social psychology.* Glencoe, IL: Free Press.

Browne, M. W. (1979). The maximum-likelihood solution in inter-battery factor analysis. *British Journal of Mathematical and Statistical Psychology, 32,* 75–86.

Byrne, B. M. (1994). *Structural equation modeling with EQS & EQS–Windows: Basic concepts, applications & programming.* Thousand Oaks, CA: Sage.

Byrne, B. M. (2001). *Structural equation modeling with Amos: Basic concepts, applications & programming*. Mahwah, NJ : Lawrence Erlbaum Associates.

Carroll, J. B. (1953). An analytical solution for approximating simple structure in factor analysis. *Psychometrika, 18,* 23–38.

Carroll, J. D. (1968). Generalization of canonical correlation analysis to three or more sets of variables. *Proceedings of the 76th Annual Convention, American Psychological Association*, *3*, 227–228.

Cattell, R. B., & Vogelman, S. (1977). A comprehensive trial of the scree and KG criteria for determining the number of factors. *Multivariate Behavioral Research, 12,* 209–325.

Cattin, P. (1980). Note on estimation of the squared cross-validated multiple correlation of a regression model. *Psychological Bulletin, 87,* 63–65.

Cliff, A., & Krus, D. J. (1976). Interpretation of canonical analysis: Rotated versus unrotated solutions. *Psychometrika, 41,* 35–42.

Cliff, N. (1996). Answering ordinal questions with ordinal data using ordinal statistics. *Multivariate Behavioral Research, 31,* 331–350.

Cohen, B. P. (1963). *Conflict and conformity.* Cambridge, MA: MIT Press.

Cohen, B. P. (1977). *Statistical power analysis for the behavioral sciences* (rev. ed.). New York: Academic Press.

Cohen, J., & Cohen, P. (1975). *Applied multiple regression for the behavioral sciences.* New York: Wiley.

Collins, L. M. & Horn, J. L. (Eds.). (1991). *Best methods for the analysis of change* . Washington, DC: American Psychological Association.

Comrey, A. L. (1973). *A first course in factor analysis.* New York: Academic Press. (2nd ed. 1992, Lawrence Erlbaum Associates, Mahwah, NJ.)

Cooley, W. W., & Lohnes, P. R. (1971). *Multivariate data analysis.* New York: Wiley.

Cotton, J. W. (1967). *Elementary statistical theory for behavior scientists.* Reading, MA: Addison-Wesley.

Cox, C. M., Krishnaiah, P. R., Lee. J. C., Reising, J., & Schuurman, F. J. (1980). A study on finite intersection tests for multiple comparison of means. In P. R. Krishnaiah (Ed.), *Multivariate analysis V* (pp. 435–466). Amsterdam: North-Holland.

Cramer, E. M., & Nicewander, W. A. (1979). Some symmetric, invariant measures of multivariate association. *Psychometrika, 44,* 43–54.

Cramer, H. (1946). *Mathematical methods of statistics.* Princeton, NJ: Princeton University Press.

Creager, J. A. (1969). The interpretation of multiple regresssion via overlapping rings. *American Educational Research Journal, 6,* 706–709.

Cronbach, L. J., & Furby, L. (1970). How should we measure change -- or should we? *Psychological Bulletin, 74,* 68–80.

Cureton, E. E. (1976). Studies of the Promax and Optres rotations. Multivariate Behavioral Research 11, 449–460.

Daniel, C., & Wood, F. S. (1971). *Fitting equations to data.* New York: Wiley.

Darlington, R. B. (1968). Multiple regression in psychological research and practice. *Psychological Bulletin, 79,* 161–182.

Darlington, R. B. (1978). Reduced-variance regression. *Psychological Bulletin, 85,* 1238–1255.

Dawes, R. M., & Corrigan, B. (1974). Linear models in decision making. *Psychological Bulletin, 81,* 95–106.

Degroot, M. H., & Li, C. C. (1966). Correlations between similar sets of measurements. *Biometrics, 22,* 781–790.

Delaney, H. D., & Maxwell, S. E. (1981). On using analysis of covariance in repeated-measures designs. *Multivariate Behavioral Research, 16,* 105–124.

Dempster, A. P. (1969). *Elements of continuous multivariate analysis.* Menlo Park, CA: Addison-Wesley.

DeSarbo, W. S. (1981). Canonical-redundancy factoring analysis. *Psychometrika, 46,* 307–329.

Deutsch, M. (1960). The effect of motivational orientation upon threat and suspicion. *Human Relations, 13,* 122–139.

Donaldson, T. S. (1968). Robustness of the *F*-test to errors of both kinds and the correlation between the numerator and denominator of the *F*-ratio. *American Statistical Association Journal, 63,* 660–676.

Doob, A. N., & MacDonald, G. E. (1979). Television viewing and fear of victimization: Is the relationship causal? *Journal of Personality and Social Psychology, 37,* 170–179.

Draper, N. R., & Smith, H. (1981). *Applied regression analysis* (2nd ed). New York: Wiley. (1st ed., 1966.)

Drasgow, F., & Dorans, N. J. (1982). Robustness of estimators of the squared multiple correlation and squared cross-validity coefficient to violations of multivariate normality. *Applied Psychological Measurement, 6,* 185–200.

Dwyer, P. S. (1939). The contribution of an orthogonal multiple factor solution to multiple correlation. *Psychometrika, 4,* 163–171.

Dwyer, P. S., & MacPhail. M. S. (1948). Symbolic matrix derivatives. *Annals of Mathematical Statistics, 19,* 517–534.

Estes, W. K. (1957). Of models and men. *American Psychologist, 12,* 609–617.

Eye, A. (1988). Some multivariate developments in nonparametric statistics. In J. R. Nesselroade, R. B. Cattell, et al (Eds.), *Handbook of multivariate experimental psychology, 2nd. ed.* (pp. 267–398). NY: Plenum Press.

Ferguson, G. A. (1954). The concept of parsimony in factor analysis. *Psychometrika, 19,* 281–290.

Ferraro, D. P., & Billings, D. K. (1974a). Illicit drug use by college students. Three year trends. 1970–1972. *Journal of the Addictions, 9,* 879–883.

Ferraro, D. P., & Billings, D. K. (1974b). Marihuana use by college students. Three year trends. 1970–1972. *Journal of the Addictions, 9,* 321–327.

Finn, J. D. (1974). *A general model for multivariate analysis.* New York: Holt-Rinehart-Winston.

Flint, R. A. (1970). *The relative importance of structure and individual differences in determining behavior in two person games.* Unpublished doctoral_dissertation, University of New Mexico, Albuquerque.

Frankel, M. R. (1971). *Inference from survey samples: An empirical investigation.* Ann Arbor: University of Michigan Press.

Gabriel, K. R. (1969). A comparison of some methods of simultaneous inference in Manova. In P. R. Krishnaiah (Ed.), *Multivariate analysis-II* (pp. 67–86). New York: Academic Press.

Gabriel, K. R. (1970). On the relation between union intersection and likelihood ratio tests. In R. C. Bose, I. M. Chakravarti, P. C. Mahalanobis, C. R. Rao, & K. J. C. Smith,

(Eds.), *Essays in probability and statistics,* (pp. 251–266). Chapel Hill: University of N. Carolina Press.

Gaito, J. (1980). Measurement scales and statistics: Resurgence of an old misconception. *Psychological Bulletin, 87,* 564–567.

Games, P. (1971). Multiple comparisons of means. *American Educational Research Journal, 8,* 551–565.

Glass, G. V., & Stanley, J. C. (1970). *Statistical methods in education and psychology.* Englewood Cliffs, NJ: Prentice Hall.

Gleason, T. C. (1976). On redundancy in canonical analysis. *Psychological Bulletin, 83,* 1004–1006.

Gollob, H. F. (1968). A statistical model which combines features of factor analytic and analysis of variance techniques. *Psychometrika, 33,* 73–115.

Gonzales, R., Carabajal, J., & Roll, S. (1980). *Early memories of Anglos students versus traditional and acculturated Chicanos.* Unpublished manuscript, University of New Mexico.

Goodman, L. A., & Kruskal, W. H. (1954). Measures of association for cross-classifications. *Journal of the American Statistical Association, 49,* 732–764.

Gorsuch, R. L. (1970). A comparison of biquartimin, maxplane, promax, and varimax. *Educational & Psychological Measurement 30*, 1970, 861–872.

Gorsuch, R. L. (1983). *Factor analysis.* (2nd ed.). Hillsdale, NJ: Lawrence Erlbaum Associates..

Green, B. F. (1976). On the factor score controversy. *Psychometrika. 41*, 263–266.

Green, S. B. (1991). How many subjects does it take to do a regression analysis? *Multivariate Behavioral Research, 26,* 499-510,

Green, S. B., Salkind, N. J., & Akey, T. M. (2000). *Using SPSS for Windows: Analyzing and understanding data* (2nd ed.). Englewood Cliffs, NJ: Prentice-Hall.

Greenhouse, S. W., & Geisser, S. (1959). On methods in the analysis of profile *data. Psychometrika, 24,* 95–112.

Greenwald, A. G., Gonzalez, R., Harris, R. J., & Guthrie, D. (1996). Effect sizes and *p*-values: What should be reported and what should be replicated? *Psychophysiology, 33*, 175–183.

Grice, G. R. (1966). Dependence of empirical laws upon the source of experimental variation. *Psychological Bulletin, 66,* 488–498.

Grice, J. W. (1995). *The relative utility of loading and regression weights for computing factor scores.* Unpublished doctoral dissertation, University of New Mexico.

Grice, J. W., & Harris, R. J. (1998). A comparison of regression and loading weights for the computation of factor scores. *Multivariate Behavioral Research, 33,* 221–247.

Guttman, L. (1953). Image theory for the structure of quantitative variates. *Psychometrika, 18,* 277–296.

Guttman, L. (1955). The determining of factor score matrices with implications for five other basic problems of common-factor analysis. *British Journal of Statistical Psychology, 8,* 65–81.

Guttman, L. (1956). "Best possible" systematic estimates of communalities. *Psychometrika, 21,* 273–285.

Hakstian, A. R. (1974). A further comparison of oblique factor transformation methods. *Psychometrika, 39*, 429–444.

Harlow, L., Mulaik, S., & Steiger, J. (Eds.). (1997). *What if there were no significance tests?* Mahwah, NJ: Lawrence Erlbaum Associates.

Harman, H. H. (1976). *Modern factor analysis* (3rd ed.). Chicago: University of Chicago Press.

Harris, C. W. (1963). *Problems in measuring change.* Madison: University of Wisconsin Press.

Harris, C. W., & Knoell, D. M. (1948). The oblique solution in factor analysis. *Journal of Educational Psychology, 39,* 385–403.

Harris, M. B. (1998). *Basic statistics for behavioral science research* (2nd Ed.). Needham Heights, MA: Allyn & Bacon.

Harris, M. B., Harris, R. J., & Bochner, S. (1982). Fat, four-eyed, and female: Stereotypes of obesity, glasses, and gender. *Journal of Applied Social Psychology, 12,* 303–516.

Harris, R. J. (1969). Deterministic nature of probabilistic choices among identifiable stimuli. *Journal of Experimental Psychology, 79,* 552–560.

Harris, R. J. (1973a). *Repeated-battery canonical correlation analysis*. Unpublished manuscript, University of New Mexico, Albuquerque.

Harris, R. J. (1973b). *The useless power* of *likelihood-ratio tests in multivariate analysis* of *variance and canonical correlation.* Unpublished manuscript, University of New Mexico, Albuquerque.

Harris, R. J. (1974, April). *Mutivariate analysis of variance.* Paper presented at meetings of American Educational Research Association, Chicago.

Harris, R. J. (1976a). Handling negative inputs: On the plausible equity formulae. *Journal* of *Experimental Social Psychology, 12,* 194–209.

Harris, R. J. (1976b). The invalidity of partitioned-*U* tests in canonical correlation and multivariate analysis of variance. *Multivariate Behavioral Research, 11,* 353–365.

Harris, R. J. (1976c, September). *Optimal decomposition of interactions into contrasts* of *contrasts.* Presented at 84th annual meetings of the American Psychological Association, Washington, DC.

Harris, R. J. (1985a). Multivariate statistics: When will experimental psychology catch up? In S. Koch & T. Leary (Eds.), *A century* of *psychology as a science* (pp. 678–697). NY: McGraw-Hill.

Harris, R. J. (1985b). *A primer of multivariate statistics* (2nd ed.). Orlando, FL: Academic Press. (1st ed., 1975.)

Harris, R. J. (1989). A canonical cautionary. *Multivariate Behavioral Research 24,* 17–39.

Harris, R. J. (1993). Multivariate analysis of variance. In L. K. Edwards (Ed.), *Applied analysis of variance in behavioral science*. New York: Marcel Dekker.

Harris, R. J. (1994). *An analysis of variance primer.* Itasca, IL: F. E. Peacock.

Harris, R. J. (1997a). Reforming significance testing via three-valued logic. In L. Harlow, S. Mulaik, & J. Steiger (Eds.), *What if there were no significance tests?* (pp. 145–174). Mahwah, NJ: Lawrence Erlbaum Associates.

Harris, R. J. (1997b). Significance tests have their place. *Psychological Science, 8,* 8–11.

Harris, R. J. (1999a, April). Apparent confirmation of incorrect factor interpretations. In B. Bolton (Chair), *Multivariate Measurement, Annual Workshop of the Society for Applied Multivariate Research,* Albuquerque.

Harris, R. J. (1999b). Beware Canona via SPSS Manova. *Tie Line* (Newsletter of the Society for Applied Multivariate Research), November 11.

Harris, R. J., Flint, R. A., & Everitt. G. (1970, May). *Directive versus nondirective instructions in the prisoner's dilemma.* Paper presented at Rocky Mountain Psychological Association, Denver, CO.

Harris, R. J., & Joyce, M. A. (1980). What's fair? It depends on how you phrase the question. *Journal* of *Personality and Social Psychology, 38,* 165–179.

Harris, R. J., & Millsap, R. E. (1993). Leakage of normalized variance under rotation of principal components: Once more, slowly. Presented at Society for Multivariate Experimental Psychology, San Pedro, CA, 10/28/93.

Harris, R. J., & Quade, D. (1992). The *M*inimally *I*mportant *D*ifference *S*ignificant criterion for sample size. *Journal of Educational Statistics, 17,* 27–49.

Harris, R. J., Tuttle, W. A., Bochner. S., & Van Zyl, T. (1982). *Is there an equality norm? Explorations* of *explanations of the phrasing effect.* Unpublished manuscript, University of New South Wales.

Harris, R. J. & Vigil, B. V. (1998, October). *Peekies: Prior information confidence intervals.* Presented at meetings of Society for Multivariate Experimental Psychology, Woodside Lake, NJ.

Harris, R. J., Witttner, W., Koppell, B., & Hilf, F. D. (1970). MMPI scales versus interviewer ratings of paranoia. *Psychological Reports, 27,* 447–450.

Harwell, M. R. & Serlin, R. C. (1989). A nonparametric test statistic for the general linear model. *Journal of Educational Statistics, 14,* 351–371.

Hays, W. L., & Winkler, R. L. (1971). *Statistics: Probability, inference, decision.* San Francisco: Holt, Rinehart and Winston.

Heck, D. L. (1960). Charts of some upper percentage points of the distribution of the largest characteristic root. *Annals of Mathematical Statistics, 31,* 625–642.

Hendrickson, A. E., & White, P.O. (1964). Promax: A quick method for rotation to oblique simple structure. *British Journal of Statistical Psychology 17*, 65–70.

Herzberg, P. A. (1969). The parameters of cross-validation. *Psychometrika, Monograph Suppiement, 34*(2, Pt. 5, No. 16).

Hinman, S., & Bolton, B. (1979). *Factor analytic studies, 1971-1975.* Troy, NY: Witson.

Holland, T., Levi, M., & Watson, C. (1980). Canonical correlation in the analysis of a contingency table. *Psychological Bulletin, 87,* 334–336.

Horst, P. (196la). Generalized canonical correlations and their applications to experimental data. *Journal* of *Clinical Psychology Monograph Supplement, 14,* 331–347.

Horst, P. (196lb). Relations among *m* sets of measures. *Psychometrika, 26,* 129–149.

Horst, P. (1963). *Matrix algebra* for *social scientists.* New York: Holt. (Reprinted 1983 by Irvington Publishers.)

Horst, P., & Edwards, A. L. (1981). A mathematical proof of the equivalence of an unweighted means analysis of variance and a Method I regression analysis for the 2^k unbalanced factorial experiment. *JSAS Catalog* of *Selected Documennts in Psychology, 3,* 76 (Manuscript No. 2354).

Horst, P., & Edwards, A. L. (1982). Analysis of non-orthogonal designs: The 2^k factorial experiment. *Psychological Bulletin, 91,* 190–192.

Hotelling, H. (1931). The generalization of Student's ratio. *Annals of Mathematical Statistics, 2,* 360–378.

Huberty, C. J. (1975). Discriminant analysis. *Review of Educational Research, 45,* 543–598.

Huberty, C. J. (1984). Issues in the use and interpretation of discriminant analysis. *Psychological Bulletin, 95*, 156–171.

Huynh, H. (1978). Some approximate tests in repeated-measures designs. *Psychometrika, 43,* 1582–1589.

Huynh, H. (1982). A comparison of four approaches to robust regression. *Psychological Bulletin, 92,* 505–512.

Huynh, H., & Feldt, L. S. (1970). Conditions under which mean square ratios in repeated measurements designs have exact *F* distributions. *Journal* of *the American Statistical Association, 65,* 1582–1589.

Isaac, P. D., & Milligan, G. W. (1983). A comment on the use of canonical correlation in the analysis of contingency tables. *Psychological Bulletin, 93*, 378–381.

Ito, K. (1969). On the effect of heteroscedasticity and nonnormality upon some multivariate test procedures. In R. R. Krishnaiah (Ed.), *Multivariate analysis II* (pp. 87–120). New York: Academic Press.

Ito, K., & Schull, W. J. (1964). On the robustness of the T_0^2 test in multivariate analysis of variance when variance-covariance matrices arc not equal. *Biometrika, 51*, 71–82.

James, L. R., Mulaik, S. A., & Brett, J. M. (1982). *Causal Analysis: Assumptions, Models, and Data.* Beverly Hills, CA: Sage.

James, L. R. & Singh, K. (1978). An intro to the logic, assumptions, and basic analytic procedures of 2-stage least-squares. *Psychological Bulletin, 85*, 1104–1122.

Jennrich, R. I., & Sampson, P. F. (1966). Rotation for simple loadings. *Psychometrika, 31,* 313–323.

Johnson, D. E. (1976). Some new multiple comparison procedures for the 2-way analysis of variance model with interaction. *Biometrics, 32,* 929–934.

Joliffe, I. T. (1986). *Principal component analysis.* New York: Springer-Verlag.

Jones, L. V. (1966). Analysis of variance in its multivariate developments. In R. B. Cattell (Ed.), *Handbook of multivariate experimental psychology* (pp. 242-266). Chicago: Rand McNally.

Jöreskog, K. G. (1962). On the statistical treatment of residuals in factor analysis. *Psychometrika. 27,* 335–354.

Jöreskog, K. G. (1967). Some contributions to maximum likelihood factor analysis. *Psychometrika, 32,* 443–482.

Jöreskog, K. G. (1969). A general approach to confirmatory maximum likelihood factor analysis. *Psychometrika, 34,* 183–202.

Jöreskog, K. G. (1970). A general method for analysis of covariance structures. *Biometrika, 57,* 239–251.

Jöreskog, K. G. (1973). Analysis of covariance structures. In P. R. Krishnaiah (Ed.), *Multivariate anatysis III* (pp. 263–285). New York: Academic Press.

Jöreskog, K. G. (1978). Structural analysis of covariance and correlation matrices. *Psychometrika, 43,* 443–477.

Jöreskog, K. G., & Sörbom, D. (1982). *LISREL V Estimation of linear structural equations by maximum-likelihood methods.* Chicago: National Educational Resources.

Kaiser, H. F. (1956). *The varimax method of factor analysis.* Unpublished PhD thesis, University of California at Berkeley.

Kaiser, H. F. (1958). The varimax criterion for analytic rotation in factor analysis. *Psychometrika, 23,* 187–200.

Kaiser. H. F. (1960a). The application of electronic computers to factor analysis. *Educational and Psychological Measurement, 20,* 141–151.

Kaiser. H. F. (1960b). Directional statistical decisions. *Psychological Review 67,* 160–167.

Kaiser, H. F. (1963). Image analysis. In C. W. Harris (Ed.), *Problems in measuring change* (pp. 156–166). Madison: University of Wisconsin Press.

Kaiser, H. F., & Horst, P. (1975). A score matrix for Thurstone's box problem. *Multivariate Behavioral Research, 10,* 17–25.

Kendall, M. G. (1968). On the future of statistics: A second look. *Journal of the Royal Statistical* Society, *Series A, 131,* 182–192.

Kendall, M. G.. & Stuart, A. (1961). *The advanced theory of statistics, Vol. 2. Inference and relationship.* London: Hafner.

Kendall, M. G., & Stuart, A. (1968). *The advanced theory of statistics, Vol. 2. Inference and relationship* (rev. ed.). London: Hafner.

Kenny, D. A. (1974). A test for a vanishing tetrad: The second canonical correlation equals zero. *Social Science Research, 3,* 83–87.

Kenny, D. A. (1979). *Correlation and causality.* Toronto: Wiley.

Keren, G., & Lewis, C. (1977). A comment on coding in nonorthogonal designs. *Psychological Bulletin, 84,* 346–348.

Keselman, H. J. (1982). Multiple comparisons for repeated measures means. *Multivariate Behavioral Research, 17,* 87–92.

Keselman, H. J., Rogan, J. C., & Games. P. A. (1981). Robust tests of repeated-measures means. *Educational and Psychological Measurement, 41,* 163–173.

Kirk, R. E. (Ed.). (1972). *Statistical issues: A reader for the behavioral* sciences. Belmont, CA: Wadsworth.

Kirk, R. E. (1982). *Experimental design: Procedures for the behavioral* sciences. (2nd ed.) Monterey, CA: Brooks/Cole.

Kirk, R. E. (1995). *Experimental design: Procedures for the behavioral* sciences. (3rd ed.) Monterey, CA: Brooks/Cole.

Kirk, R. E. (1998). *Statistics: An Introduction* (2nd ed.). Fort Worth, TX: Harcourt.

Klockars, A. J., & Hancock, G. R. (1998) A more powerful post hoc multiple comparison procedure in analysis of variance. *Journal of Educational & Behavioral Statistics, 23,* 279-289.

Knapp, T. R. (1978). Canonical correlation analysis: A general parametric system. *Psychological Bulletin*, 85, 410–416.

Korin, B. P. (1972). Some comments on the homoscedasticity criterion, *M*, and the multivariate analysis of variance tests T^2, *W* and *R. Biometrika, 59,* 215–217.

Krishnaiah, P. R. (1965). On the simultaneous ANOVA and MANOVA tests. *Annals of the institute of Statistical Mathematics, 17,* 35–53.

Krishnaiah, P. R. *(Ed.)* (1969). *Multivariate analysis-II.* New York: Academic Press.

Krishnaiah, P. R. (1979). Some developments on simultaneous test procedures. In P. R. Krishnaiah (Ed.), *Developments in statistics* (Vol. 2, pp. 157 ff.). San Francisco: Academic Press.

Krishnaiah, P. R. (Ed.) (1980). *Analysis of Variance* (Vol. 1 of *Handbook of statistics*). NewYork: Elsevier North-Holland.

Lancaster, H. O. (1963). Canonical correlations and partitions. *Quarterly Journal of Mathematics, 14,* 220–224.

Lance, C. E., Cornwall, J. M., & Mulaik, S. A. (1988). Limited info parameter estimates for latent or mixed manifest and latent variable models. *Multivariate Behavioral Research, 23*, 171–187.

Laughlin, J. E. (1978). Comment on "Estimating coefficients in linear models: It don't make no never-mind." *Psychological Bulletin, 85,* 247–253.

Laughlin, J. E. (1979). A Bayesian alternative to least squares and equal weighting coefficients in regression. *Psychometrika*, *44*, 1979, 271–288.

Lawley, D. N. (1940). The estimation of factor loadings by the method of maximum likelihood. *Proceedings of the Royal Society of Edinburgh, 60,* 64–82.

Lawley, D. N. (1963). On testing a set of correlation coefficients for equality. *Annals of Mathematical Statistics, 34*, 149–151.

Lawley, D. N., & Maxwell, A. E. (1963). *Factor analysis as a statistical method.* London: Butterworth.

Leventhal, L. (1999a). Answering two criticisms of hypothesis testing. *Psychological Reports*, *85*, 1999, 3–18.

Leventhal, L. (1999b). Updating the debate on one- versus two-tailed tests with the directional two-tailed test. *Psychological Reports, 84,* 707–718.

Leventhal, L., & Huynh, C-L. (1996). Directional decisions for two-tailed tests: Power, error rates, and sample size. *Psychological Methods 1*, 278-292.

Lindquist, E. F. (1953). *Design and analysis of* experiments *in psychology and education.* Boston: Houghton-Mifflin.

Linn, R. T. (1980, May). *Electromyographic biofeedback and hyperkinesis.* Unpublished master's thesis, University of New Mexico.

Linn, R. T., & Hodge, G. K. (1981). *Attention, Conners' Parent Questionnaire, and locus of control in hyperactive children.* Unpublished manuscript, University of New Mexico.

Little, T. D., Schnabel, K. U., & Baumert, J. (2000). *Modeling longitudinal and multilevel data: Practical issues, applied approaches, and specific examples.* Mahwah, NJ: Lawrence Erlbaum Associates.

Loehlin, J. C. (1998). *Latent variable models: An Introduction to factor, path, & structural analysis* (3rd ed.). Mahwah, NJ: Lawrence Erlbaum Associates.

Lord, F. M. (1953). On the statistical treatment of football numbers. *American Psychologist, 8,* 750–751.

Maki, J. E., Thorngate, W. B., & McClintock, C. G. (1979). Prediction and perception of social motives. *Journal of Personality and Social Psychology, 37*, 203–220.

Marks, M. R. (1966, September). *Two kinds of regression weights that are better than betas in crossed samples.* Paper presented at convention of the American Psychological Association, New York.

Martin, J. K., & McDonald, R. P. (1975). Bayesian estimates in unrestricted factor analysis: A treatment of Heywood cases. *Psychometrika, 40,* 505–517.

Mathworks, Inc. (1998a). *The student edition of MATLAB : The language of technical computing, Version 5*. Englewood Cliffs, NJ: Prentice Hall.

Mathworks, Inc. (1998b). *Getting started with Matlab, Version 5.* Available online: http://www.mathworks.com/access/helpdesk/help/pdf_doc/matlab/getstart.pdf

Maxwell, S. E.. & Avery, R. D. (1982). Small sample profile analysis with many variables. *Psychological Bulletin, 92,* 778–785.

Maxwell, S. E., Delaney, H. D. (2000). *Designing experiments and analyzing data: A model comparison perspective*. Mahwah, N.J.: Lawrence Erlbaum Associates. (Originally published 1990, Belmont, CA: Wadsworth.)

Mawell, S. E., Delaney, H. D., & Dill, C. A. (1983). Another look at ANCOVA *vs.* blocking. *Psychological Bulletin, 95,* 136–147.

Maxwell, S. E., Delaney, H. D., & Manheimer, J. M. (1985). ANOVA of residuals and ANCOVA: Correcting an illusion by using model comparisons and graphs. *Journal of Educational Statistics. 10,* 197–209.

McArdle, J. J. (1996). Current directions in structural factor analysis. *Current Directions, 5,* 11–18.

McArdle, J. J., & Hamagami, F. (1996). Multilevel models from a multiple group structural equation perspective. In G.A. Marcoulides & R.E. Schumacker (eds.), *Advanced structural equation modeling: Issues and techniques* (pp. 89–124). Mahway, NJ: Lawrence Erlbaum Associates.

McCall, R. B.. & Appelbaum, M. I. (1973). Bias in the analysis of repeated-measures designs: Some alternative approaches. *Child Development, 44,* 401–415.

McDonald, R. P. (1974). The measurement of factor indeterminacy. *Psychometrika, 39,* 203–222.

McDonald, R. P. (1977). The indeterminacy of components and the definition of common factors. *British Journal of Mathematical and Statistical Psychology,* 30, 165–176.

McDonald, R. P., & Mulaik, S. A. (1979). Determinacy of common factors—A nontechnical review. *Psychological Bulletin, 86,* 297–306.

McFatter, R. M. (1979). The use of structural equation models in interpreting regression equations including suppressor and enhancer variables. *Applied Psychological Measurement, 3,* 123–135.

McGuire, W. J. (1973). The Yin and Yang of progress in social psychology: Seven koan. *Journal of Personality and Social Psychology, 26,* 446–456.

McGuire, W. J., & Papageorgis, D. (1961). The relative efficacy of various types of prior belief-defense in producing immunity against persuasion. *Journal of Abnormal and Social Psychology, 62,* 327–337.

Melgoza, B., Harris, R. J., Baker, R., & Roll, S. (1980). Academic preference and performance of Chicano and Anglo college students. *Hispanic Journal of Behavioral Sciences, 2,* 147–159.

Mendoza, J. L., Markos, V. H., & Gonter, R. (1978). A new perspective on sequential testing procedures in canonical analysis: A Monte Carlo evaluation. *Multivariate Behavioral Research, 13,* 371–382.

Mendoza, J. L., Toothaker, L. E., & Nicewander, W. R. (1974). A Monte Carlo comparison of the univariate and multivariate methods for the group by trials repeated-measures design. *Multivariate Behavioral Research, 9,* 165–177.

Meredith, W. (1964). Canonical correlations with fallible data. *Psychometrika, 29,* 55–66.

Meredith, W., & Tisak, J. (1982). Canonical analysis of longitudinal and repeated-measures data with stationary weights. *Psychometrika, 47,* 47–67.

Meredith, W, & Tisak, J. (1990). Latent curve analysis. *Psychometrika, 55,* 107–122.

Messick. D. M., & Van de Geer, J. P. (1981). A reversal paradox. *Psychological Bulletin, 90,* 582–593.

Messick, S. (1981). Denoting the base-free measure of change. *Psychometrika, 46,* 215–217.

Moler, C. (1982). *MATLAB user's guide* (Tech. Rept. No. CS8l-1, Rev). Department of Computer Science, University of New Mexico, Albuquerque.

Mood, A. M., & Graybill, F. A. (1963). *Introduction to the theory of statistics* (2nd ed.). Toronto: McGraw-Hill.

Morris, J. D. (1982). Ridge regression and some alternative weighting techniques: A comment on Darlington. *American Statistician, 33,* 15–18.

Morrison, D. F. (1976). *Multivariate statistical methods* (2nd ed.). San Francisco: McGraw-Hill. (3rd ed., 1990.)

Mosteller. F., & Tukey, J. W. (1968). Data analysis. including statistics. In G. Lindzey & E. Aronson (Eds.). *Handbook of social psychology* (2nd ed., Vol. 2, pp. 80–203). Reading, MA: Addison-Wesley.

Mosteller, F. & Tukey, J. W. (1977). *Data onalysis and regression.* Reading, MA: Addison-Wesley.

Mudholkar, G. S., & Subbaiah, P. (1980). MANOVA multiple comparisons associated with finite intersection tests. In P. R. Krishnaiah (Ed.), *Multivariate analysis V* (pp. 467–482). Amsterdam: North-Holland.

Mulaik, S. A. (1972). *The foundations of factor analysis.* San Francisco: McGraw-Hill.

Mulaik, S. A. (1976). Comments on "The measurement of factor indeterminacy." *Psychometrika, 41,* 249–262.

Muller, K. E. (1981). Relationships between redundancy analysis, canonical correlation, and multivariate regression. *Psychometrika, 46,* 139–142.

Myers, J. L. (1979). *Fundamentals of experimental design.* (3rd ed.) Sydney: Allyn & Bacon.

Newhaus, J., & Wrigley, C. (1954). The quartimax method: An analytical approach to orthogonal simple structure. *British Journal of Statistical Psychology, 7,* 81–91.

Nicewander, W. A., & Price, J. M. (1978). Dependent variable reliability and the power of significance tests. *Psychological Bulletin, 83,* 405–409.

Nicewander, W. A., & Price, J. M. (1983). Reliability of means and the power of statistical tests: Some new results. *Psychological Bulletin, 95,* 524–533.

Norris, R. C., & Hjelm. H. F. (1961). Non-normality and product moment correlation. *Journal of Experimental Education, 29,* 261–270.

Norton, D. W. (1953). *An empirical investigation of some effects of non-normality and heterogeneity on the F-distribution.* Unpublished doctoral dissertation, State University of Iowa, Iowa City.

O'Grady, K. E. (1981). Probabilities and critical values for *z,* chi-square, *r, t,* and *F.* *Behavior Research Methods & Instrumentation, 13,* 55–56.

Olkin, I., & Pratt, J. W. (1958). Unbiased estimation of certain correlation coefficients. *Annals of Mathematical Statistics, 29,* 201–211.

Olson, C. L. (1974). Comparative robustness of six tests in multivariate analysis of variance. *Journal of the American Statistical Association, 69,* 894–908.

Olson, C. L. (1976). On choosing a test statistic in multivariate analysis of variance. *Psychological Bulletin, 83,* 579–586.

Ottaway, S. A. & Harris, R. J. (1995, April). A simple finite-intersection procedure for conducting posthoc pairwise and complex contrasts. Presented at meetings of Rocky Mountain Psychological Association, Boulder, CO.

Otto, M. W., & Dougher, M. J. (1985). Induced stress, naloxone, and pain. Unpublished manuscript, University of New Mexico, Albuquerque.

Overall, J., & Klett, C. J. (1972). *Applied multivariate analysis.* San Francisco: McGraw-Hill. (2nd ed. published 1983, R. E. Krieger, Huntington, NY.)

Overall, J. E., & Spiegel, D. K. (1969). Concerning least squares analysis of experimental data. *Psychological Bulletin, 72,* 311–322.

Overall, J. E., & Spiegel, D. K. (1973a). Comment on Rawlings' nonorthogonal analysis of variance. *Psychological Bulletin, 79,* 164–167.

Overall, J. E., & Spiegel, D. K. (1973b). Comment on "Regression analysis of proportional cell data." *Psychological Bulletin, 80*, 28–30.

Parkinson, A. (1981). *Structured characteristics of territories of 3 species of damselfish (Eupomacentrus).* Unpublished manuscript, Department of Biology, University of New Mexico, Albuquerque.

Pillai, K. C. S. (1965). On the distribution of the largest characteristic root of a matrix in multivariate analysis. *Biometrika, 52,* 405–414.

Pillai, K. C. S. (1967). Upper percentage points of the largest root of a matrix in multivariate analysis. *Biometrika, 54,* 189–194.

Pothoff, R. F., & Roy, S. N. (1964). A generalized multivariate analysis of variance model useful especially for growth curve problems. *Biometrika, 51,* 313–326.

Prentice-Dunn, S., & Rogers, R. W. (1982). Effects of public and private self-awareness on deindividuation and aggression. *Journal of Personality and Social Psychology, 43,* 503–513.

Press, S. J. (1972). *Applied multivariate analysis.* San Francisco: Holt.

Press, S. J. (1982). *Applied multivariate analysis: Using Bayesian and frequentist methods of inference.* (2nd ed.). Huntington, N.Y.: R.E. Krieger.

Price, R. B. (1977). Ridge regression: Application to nonexperimental data. *Psychological Bulletin, 84,* 759–766.

Pruzek, R. M. (1971). Methods and problems in the analysis of multivariate data. *Review of Educational Research, 41,* 163–190.

Pruzek, R. M. (1973, February). *Small sample generalizability of factor analytic results.* Presented at Annual Meetings of American Educational Research Association. New Orleans.

Pruzek, R. M., & Frederick, B. C. (1978). Weighting predictors in linear models: Alternatives to least squares and limitations of equal weights. *Psychological Bulletin, 85,* 254–266.

Pruzek, R. M., & Rabinowitz, S. N. (1981). A class of simple methods for exploratory structure analysis. *American Educational Research Journal, 18,* 173–189.

Pugh, R. C. (1968). The partitioning of criterion score variance accounted for in multiple correlation. *American Educational Research Journal, 5,* 639–646.

Rao, C. R. (1955). Estimation and tests of significance in factor analysis. *Psychometrika, 20,* 93–155.

Rao, C. R. (1959). Some problems involving linear hypotheses in multivariate analysis. *Biometrika. 46,* 49–58.

Rawlings, R. R., Jr. (1972). Note on nonorthogonal analysis of variance. *Psychological Bulletin, 77,* 373–374.

Rawlings, R. R., Jr. (1973). Comments on the Overall & Spiegel paper. *Psychological Bulletin, 79,* 168–169.

Rogan, J . C ., Keselman, H . J ., & Mendoza, J. L. (1979). Analysis of repeated measurements, *British Journal* of *Mathematical* and *Statistical Psychology, 32,* 269–286.

Rosenberg, S. (1968). Mathematical models of social behavior. In G. Lindzey & E. Aronson (Eds.), *Handbook of social psychology* (2nd ed., Vol. I, pp. 245–319). Menlo Park, CA: Addison-Wesley.

Roy, S. N. (1957). *Some aspects of multivariate analysis.* New York: Wiley.

Roy, S. N., & Bose, R. C. (1953a). On a heuristic method of test construction and its use in multivariate analysis. *Annals of Mathematical Statistics, 24,* 220–238.

Roy, S. N., & Bose, R. C. (1953b). Simultaneous confidence interval estimation. *Annals of Mathematical Statistics, 24,* 513–536.

Rozeboom, W. W. (1966). *Foundations of the theory of prediction.* Homewood, IL: Dorsey.

Rozeboom, W. W. (1979). Ridge regression: Bonanza or beguilement? *Psychological Bulletin, 86,* 242–249.

Ryan, T. A. (1980). Comment on "Protecting the overall rate of Type 1 errors for pairwise comparisons with an omnibus test statistic," or Fisher's two-stage strategy still does not work. *Psychological Bulletin, 88,* 354–355.

SAS Institute, Inc. (1986). *Guide to the SAS usage notes and sample library: Version 6.* Cary, NC: Author.

SAS Institute, Inc. (1999). *SAS/STAT user's guide. SAS OnlineDoc, Version Eight.* Cary, NC: Author. (Online access by license to your institution; url will be site specific.)

Saunders, D. R. (1953). *An analytic method for rotation to orthogonal simple structure.* Research Bulletin RB 53-10. Princeton, NJ: Educational Testing Service.

Scarr, S., & Weinberg, R.A. (1986). The early childhood enterprise: Care and education of the young. *American Psychologist, 41,* 1140–1146.

Schmidt, F. L. (1971). The relative efficiency of regression in simple unit predictor weights in applied differential psychology. *Educational and Psychological Measurement, 31,* 699–714.

Searle, S. R. (1966). *Matrix algebra for the biological* sciences. New York: Wiley.

Searle, S. R. (1971). *Linear models.* Toronto: Wiley.

Searle, S. R. (1982). *Matrix algebra useful for statistics*. New York: Wiley.

Selltiz, C., Jahoda, M., Deutsch, M., & Cook, S. W. (1959). *Research methods in social relations* (rev. ed.). New York: Holt.

Senders, V. L. (1958). *Measurement and statistics: A basic text emphasizing behavioral science applications.* New York : Oxford University Press.

Sheu, C-F. & O'Curry, S. (1996). Implementation of nonparametric multivariate statistics with S. *Behavior Research Methods, Instruments, & Computers, 28,* 315–318.

Siegel, S, (1956). *Nonparametric statistics for the behavioral sciences*. New York: McGraw-Hill.

Singer, J. D. (1998). Using SAS PROC MIXED to fit multilevel models, hierarchical models, and individual growth models. *Journal of Educational & Behavioral Statistics, 23*, 323–355.

Skellum, J. G. (1969). Models, inference, and strategy. *Biometrics, 25,* 457–475.

Smith, E. R. (1978). Specification and estimation of causal models in social psych: Comment on Tesser & Paulhus. *Journal of Personality and Social Psychology, 36*, 34–38.

Späth, H. (1980). *Cluster analysis algorithms.* NY: Wiley.

Spearman, C., & Holzinger, K. J. (1925). Note on the sampling error of tetrad differences. *British Journal of Psychology, 16,* 86–89.

SPSS, Inc. (1983). *SPSS user's guide*. St, Louis, MO: McGraw-Hill.

Steele, M. W., & Tedeschi, J. (1967). Matrix indices and strategy choices in mixed-motive games. *Journal of Conflict Resolution, 11*, 198–205.

Steiger, J. A. (Ed.). (1971). *Readings in statistics for the behavioral scientist.* Dallas, TX: Holt.

Steiger, J. H. (1979a). Factor indeterminacy in the 1930s and 1970s: Some interesting parallels. *Psychometrika, 44,* 157–167.

Steiger, J. H. (1979b). Relationship between external variables and common factors. *Psychometrika, 44,* 93–97.

Steiger, J. H. (1980). Tests for comparing elements of a correlation matrix. *Psychological Bulletin, 87,* 245–251.

Stevens, S. S. (1951). Mathematics, measurement, and psychophysics. In S. S. Stevens (Ed.). *Handbook of experimental psychology* (pp. 1-49). New York: Wiley.

Stevens, S. S. (1968). Measurement, statistics and the schemapiric view. *Science, 161,* 849–856.

Stewart, D., & Love, W. (1968). A general canonical correlation index. *Psychological Bulletin, 70,* 160–163.

Tabachnick, B. G. & Fidell, L. S. (1997). *Using Multivariate Statistics* (3rd ed.). Reading, MA: Addison-Wesley. (4th ed., 2001, Allyn and Bacon, Boston.)

Tatsuoka, M. M. (1971). *Multivariate analysis: Techniques for educational and psychological research.* New York: Wiley.

Tatsuoka, M. M. (1972). *Significance tests: Univariate and multivariate* (Booklet LB 332). Champaign, IL: Institute for Personality and Ability Testing.

Tatsuoka, M. M., & Lohnes, P. R. (1988). *Multivariate analysis: Techniques for educational and psychological research* (2nd ed.). New York: Macmillan.

Tesser, A. & Paulhus, D. (1978). On models and assumptions: A reply to Smith. *Journal of Personality and Social Psychology, 36* , 40–42

Thompson, B. (1990). Finding a correction for the sampling error in multivariate measures of relationship: A Monte Carlo study. *Educational and Psychological Measurement, 50,* 15–31.

Thompson, B. (1991). A primer on the logic and use of canonical correlation analysis. *Measurement and Evalutation in Counseling and Development, 24,* 80–93.

Thurstone, L. L. (1947). *Multiple factor analysis.* Chicago: Univ. of Chicago Press.

Timm, N. H. (1975). *Multivariate analysis, with applications in education and psychology.* Monterey, CA: Brooks/Cole.

Timm, N. H., & Carlson, J. E. (1975). Analysis of variance through full rank models. *Multivariate Behavioral Research Monographs, No.* 75-l.

Trendafilov, N. T. (1994). A simple method for Procrustean rotation in factor analysis using majorization theory. *Multivariate Behavioral Research 29*, 385–408.

Tryon, R. C., & Bailey, D. E. (1970). *Cluster analysis.* New York: McGraw-Hill.

Tukey, J. W. (1953). The problem of multiple comparisons. Unpublished manuscript (mimeo), Princeton University. Tukey, J. W. (1977). *Exploratory data analysis.* Menlo Park, CA: Addison-Wesley.

Van de Geer, J. P. (1971). *Introduction to multivatiate analysis for the social sciences.* San Francisco: Freeman.

Van den Wollenberg, A. L. (1977). Redundancy analysis: An alternative for canonical correlation analysis. *Psychometrika, 42,* 207–219.

Velicer, W. F., & Jackson, D. N. (1990). Component analysis versus common factor analysis: Some issues in selecting an appropriate procedure. *Multivariate Behavioral Research 25*, 1–28.

Venables, W. N. (1973). Computation of the null distribution of the largest or smallest latent roots of a *beta* matrix. *Journal of Multivariate* Analysis, 3, 125–131.

Venables, W. N. (1974). Null distribution of the largest root statistic. *Journal of the Royal Statistical Society, Series C: Applied Statistics, 23, Algorithm AS77*, 458–465.

Wackwitz, J. H., & Horn, J. L. (1971). On obtaining the best estimates of factor scores within an ideal simple structure. *Multivariate Behavioral Research, 6*, 389–408.

Wainer, H. (1976). Estimating coefficients in linear models: It don't make no nevermind. *Psychological Bulletin, 83,* 213–217.

Wainer, H. (1978). On the sensitivity of regression and regressors. *Psychological Bulletin, 85,* 267–273.

Ward, J. H. (1969). Partitioning of variance and contribution or importance of a variable: A visit to a graduate seminar. *American Educational Research Journal*, *6*, 467–474.

Wechsler, D. (1974). *Manual for the Wechsler Intelligence Scale for Children–Revised.* New York: Psychological Corporation.

Wherry, R. J. (1931). A new formula for predicting the shrinkage of the coefficient of multiple correlation. *Annals of Mathematical Statistics, 2,* 440-451.

Whittle, P. (1952). A principal component and least squares method of factor analysis. *Skand Aktuarietidskift, 35,* 223–239.

Wilkinson, L. (1977). Confirmatory rotation of MANOVA canonical variates. *Multivariate Behavioral Research, 12*, 487–494.

Wilkinson, L. & APA Task Force on Statistical Inference (1999). Statistical methods in psychology journals: Guidelines and explanations. *American Psychologist, 54*, 594–604.

Wilks, S. S. (1932). Certain generalizations in the analysis of variance. *Biometrika. 24,* 471–494.

Wilks, S. S. (1935). On the independence of *k* sets of normally distributed statistical variables. *Econometrica,* 3, 309–326.

Winer, B. J. (1971). *Statistical principles in experimental design* (2nd ed.). San Francisco, CA: McGraw-Hill. (1st ed., 1962.)

Wishart, J. (1931). The mean and second moment coefficient of the multiple correlation coefficient in samples from a normal population. *Biometrika, 22,* 353–361.

Wishner, J. (1960). Reanalysis of "Impressions of personality." *Psychological Review, 67,* 96–l 12.

Wolf, G., & Cartwright, B. (1974). Rules for coding dummy variables. *Psychological Bulletin, 81.* 173–179.

Wrigley, C. (1958). Objectivity in factor analysis. *Educational and Psychological Measurement, 18,* 463–476.

Index

NOTE: Figure pages are indicated by ***italics***, tables by the letter "t" after the page number, and footnotes by the letter "n" after the page number.

A

B

C

D

E

F

G

M

N

Q

R

U

V

W

Z